FARM IMPLEMENTS AND CONSTRUCTION

THE FARMER'S WORTH is measured by his results. His results depend largely upon the ability with which he judges, buys, uses, and takes care of his equipment. In the marvellous development of agriculture throughout the ages, no factor has been more important than the invention and perfection of improved tools, implements, buildings, and machines.

FARM IMPLEMENTS AND CONSTRUCTION

A Complete Manual of Successful Farming Written by Recognized Authorities in All Parts of the Country; Based on Sound Principles and the Actual Experience of Real Farmers—"The Farmer's Own Cyclopedia"

EDITED BY

E. L. D. SEYMOUR, B. S. A.

The Selection and Use of Vehicles, Implements and Farm Machinery; Farm Power and Its Applications; The Principles and Practices of Farm Engineering; The Arrangement, Design, Construction and Equipment of Farm Buildings

Fredonia Books
Amsterdam, The Netherlands

Farm Implements and Construction

Edited by
E. L. D. Seymour

ISBN: 1-4101-0701-9

Reprinted from the 1919 edition

Fredonia Books
Amsterdam, The Netherlands
http://www.fredoniabooks.com

FARM IMPLEMENTS
AND CONSTRUCTION

CONTENTS

FULL PAGE ILLUSTRATIONS

(See page 482 for index of text illustrations)

The Farmer's Worth is measured by his results. His results depend largely upon the ability with which he judges, buys, uses and takes care of his equipment. In the marvellous development of agriculture throughout the ages, no factor has been more important than the invention and perfection of improved tools, implements, buildings and machines.

Frontispiece (in color)

AUTHORS

A List of the Men and Women Who Have Written This Volume, Together With the Subjects on Which They Have Written, and the Pages on Which Their Contributions Appear.

PAGE

H. COLIN CAMPBELL, C.E. and Min. E. (Columbia) M.D. (College of Physicians and Surgeons), Consulting Concrete and Sanitary Engineer; Director, Editorial Bureau, Portland Cement Association. Author of "Concrete on the Farm and in the Shop" and other text books, numerous pamphlets, and articles in farm and technical papers. Member Editorial Staff, *Farm Engineering*.

R. P. CLARKSON, Professor of Engineering, Acadia University, Nova Scotia; Vice-President and Managing Director, Cape Split Development Co., Ltd.; Director, Bay of Fundy, Tide-Power, Ltd. Formerly Assistant Manager, Worcester Boiler Works; Designer of Special Machinery, United Shoe Machinery Co.; Instructor in Mechanical Engineering, University of Vermont; Examiner of Patents, Electrical Engineering Expert, U. S. Government. Inventor Clarkson Lowhead Turbine. Author, "Practical Talks on Farm Engineering," "Essential Elementary Electrical Engineering," and numerous articles for scientific, technical, and agricultural publications.

Frank H. DEMAREE, B.S.A., Farm Adviser of Grundy County, Ill., Graduate of Purdue University. Formerly Assistant Professor in Crop Production, University of Missouri; Agronomist, Organizer and Manager of Demonstration Farm of the J. I. Case Plow Works, Racine, Wis.; Member Crop Improvement Committee of the National Council of Grain Exchanges.

AUTHORS
(Continued)

J. L. EDMONDS, B. S., Assistant Professor and Assistant Chief of Horse Husbandry, College of Agriculture of the University of Illinois; in charge, since 1910, of the horse breeding and experimental horse feeding work, and one of the Animal Husbandry Department farms. Formerly in charge of horse work at the Minnesota Agricultural College; previous to that engaged in practical work on a number of general and stock farms.

K. J. T. EKBLAW, M.S., M.E., Professor of Agricultural Engineering, Kansas State Agricultural College. Formerly Associate Professor of Farm Mechanics, University of Illinois; Lecturer before Farmers' Institutes on agricultural engineering subjects. Author of "Farm Structures," "Farm Concrete," and numerous building bulletins and contributions to agricultural periodicals.

JOHN MARCUS EVVARD, M.S., Assistant Chief in Animal Husbandry, Associate Professor in Animal Husbandry, Iowa State College; Chief in Swine Production, Iowa Experiment Station; Member U. S. Livestock Industry Committee; Chairman Swine Commission, U. S. Food Administration; Member, Advisory Committee of Food Administration in charge of Meat and Packing Regulations. Formerly Assistant Dean and Director, University of Missouri; Contributing Editor to *National Swine Magazine.* Fellow in the Iowa Academy of Science; Member, American Association for the Advancement of Science; American Genetic Association; U. S. Livestock Sanitary Association; American Chemical Society; Vice-President, American Society of Animal Production. Author of numerous bulletins and articles in the farm press.

RUBY WESTLAKE (Mrs. W. K.) FREUDENBERGER, B.S., M.S., Member of the Nevada Non-militant Equal Suffrage Society. Practical farm housewife and student of farm home problems, formerly member of the Nevada Commission for the promotion of Uniform Legislation in the United States; and Teaching Fellow in the University of Missouri. Member of the Daughters of the American Revolution. Author of "Home Names," and numerous articles on social and home problems of country life.

W. K. FREUDENBERGER, B.S. in Electrical Engineering. With State Public Service Commission of Missouri. Has followed profession of Electrical Engineering since graduation, making a specialty of electrical power work. Formerly connected with U. S. Steel Corporation, American Smelting and Refining Company, and Colorado Fuel and Iron Co., in charge of electric light and power; Chief Engineer of State Public Service Commission of Nevada. Member, American Society Agricultural Engineers; American Institute Electrical Engineers; and American Society of Civil Engineers.

C. F. GOBBLE, B.S., Assistant Professor of Animal Husbandry, Purdue University School of Agriculture. Has had an extensive practical farm experience.

CURTIS HILL, City Engineer, Kansas City, Mo., Formerly in Municipal Engineering Department of St. Louis, Mo.; State Highway Engineer of Missouri. Member of the American Society of Civil Engineers.

C. V. HULL, Charles City, Ia., student of and practical authority on automobiles, gas engines, and tractors. Graduate in science of De Pauw University; formerly with the Hart-Parr Gas Tractor Company, doing engine erecting, electrical work and power testing; carried on traction-engineering school and traveled among farmers using tractors. Contributor to trade and farm papers.

M. A. R. KELLEY, B.S., in Agricultural Engineering and in Mechanical Engineering. In charge of the Service Department of the Louden Machinery Company, Fairfield, Iowa. Formerly in charge of Agricultural Engineering work at University of Missouri. Member of American Society of Agricultural Engineers, and Chairman of its Farm Buildings Equipment Committee.

H. L. KEMPSTER, B.S.A., Professor of Poultry Husbandry, Missouri College of Agriculture. Formerly Assistant Professor of Poultry Husbandry, Michigan Agricultural College from which he graduated after specializing in poultry. Author of numerous bulletins on different phases of poultry keeping, always from the farmer's viewpoint.

AUTHORS

(Continued)

E. W. LEHMANN, E.E., B.S., in Agricultural Engineering, Professor of Agricultural Engineering, University of Missouri. Formerly Instructor in Physics at the Texas A. & M. College; and Professor of Agricultural Engineering at Iowa State College. Has taught surveying, drainage, farm sanitation, machinery, and concrete construction, and supervised practical drainage prospects in Iowa and Missouri.

H. H. NIEMANN, Manager of the Architectural Department of the Louden Machinery Co., Fairfield, Iowa; Chairman, Committee of Farm Structures of the American Society of Agricultural Engineers. Contributor to the *American Carpenter and Builder* and other technical magazines on farm building construction.

RAYMOND OLNEY, Editor of *Power Farming*. Formerly tractor operator on tractor experimental farm; Assistant in Agricultural Engineering, Iowa Experiment Station; Power-farming expert for large tractor concern. Author of numerous articles on the application of mechanical power to farming operations.

F. F. ROCKWELL, Manager of the Nurserymen's National Service Bureau. Formerly consulting agriculturist with W. Atlee Burpee and Co. and manager of Burpee's Castle Valley Farm and orchards; also in market garden and greenhouse business at Woodstock, Conn. Contributor to agricultural and horticultural press. Author of "Home Vegetable Gardening," "Gardening Indoors and Under Glass," "Making a Fruit Garden," "The Pocket Garden Guide," "The Key to the Land." "Around the Year in the Garden," etc.

J. H. SQUIRES, Ph. D., Agronomist, Agricultural Division, the du Pont de Nemours Powder Company, Wilmington, Del. Formerly with U. S. Department of Agriculture; Assistant in New York State College of Agriculture; Professor of Agronomy, New Mexico Agricultural College.

W. A. STOCKING, M.S.A., Professor of Dairy Bacteriology, and Head of Dairy Department, New York State College of Agriculture. Formerly Assistant Professor of Dairy Bacteriology, Cornell University; Acting Director, New York State College of Agriculture; Farm Superintendent and Instructor in Agriculture and Professor of Dairy Bacteriology, Connecticut Agricultural College; Dairy Bacteriologist, Storrs Experiment Station. Author of "Manual of Milk Products," and of numerous bulletins on dairy subjects. Member of New York State Dairyman's Association, American Medical Milk Commission, Certified Milk Producers' Association of America, New York State Agricultural Society, Society of American Bacteriologists; president of American Dairy Science Association.

J. KELLY WRIGHT, Lecturer for the Missouri State Board of Agriculture. Organizer of the first boys' corn show held in the state. Formerly County School Commissioner. Author of numerous farmers' bulletins. Has made a special study of silos, their construction and use.

H. C. RAMSOWER, B.Sc. in Agr.; special student in Engineering. Professor of Agricultural Engineering, College of Agriculture, Ohio State University since 1913. Formerly Assistant Professor of Agronomy at that institution. Sec'y-Treasurer, American Society of Agricultural Engineers; author of "Equipment for the Farm and Farmstead"; since 1909 owner and manager of a large farm.

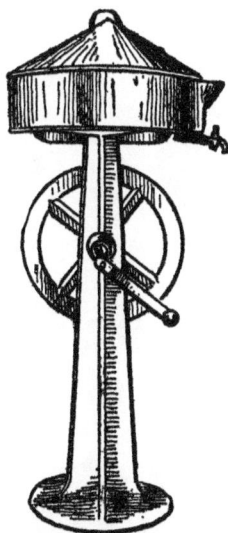

The original Babcock tester

FARM IMPLEMENTS AND CONSTRUCTION

PART I
—

Farm Vehicles

IT IS common knowledge that the materials with which the farmer carries on his work are plants, animals, and the soil. It is less generally realized, or at least appreciated, that in using them, the farmer has to practice in no small degree the applied sciences of mechanics and engineering. This is a condition of relatively recent growth. For many years farming retained its early simplicity; practically all its tasks were accomplished by hand or horse or ox power with the help of crude, cumbersome tools. The muscles of man largely measured the magnitude of his accomplishments and consequently his agricultural success.

But times have indeed changed. Analytical and inventive genius have applied themselves to the problem of creating machines to do farm work—both because man power is becoming scarce, and in order that the farmer might employ less of his energy in manual labor and more in mental achievement and planning. New standards of living have arisen to make farmers ready and anxious for more comforts and conveniences in their homes and barns—and better buildings in which to install them. Modern investigations have discovered, simplified and perfected engineering methods for improving soil conditions and reclaiming waste lands. All these developments have not only offered the farmer new opportunities, but have also demanded of him new abilities, and shouldered him with new responsibilities. To-day the most successful farmer is not he who can cradle the most grain or build the smoothest hay stack, milk the most cows in a day or work best side by side with his hired hands; rather it is the man who has the most complete and best cared for outfit of farm machinery; who can run and repair the most efficient power plant; who can design and construct the most economical barn or poultry house; who can most effectively drain a marshy meadow or irrigate a leachy, drought-ridden slope. There is no industry to-day (outside of those that are essentially and technically mechanical) that makes such varied and extensive and important use of mechanical and engineering principles. It is for this reason that the entire present volume is devoted to the different phases of this subject.

These fall into five groups as shown in the index: first, Vehicles, including the tractor (which is that and much more); second, Machinery; third, Power, where to get it and how to use it; fourth, what may be called the Civil Engineering of the farm; and fifth, the Design, Construction, and Equipment of Buildings, from the dwelling to the least of the wagon sheds and farm ice houses. Since there are many professional men whose sole interests lie along one or another

3

of these directions, it is not to be expected that the farmer—who as such is a specialist in a very complex and different industry—will become an expert in all, or any one of them. But as he becomes reasonably familiar with their principles and methods, he is becoming ever more independent, better equipped, and more assured of ultimate success and prosperity.

The wide range between a one-horse light runabout or a dump cart and a six-cylinder touring car, a three-ton truck or a 40 H. P. gas tractor—each of which is no longer a mystery to many a real farmer—is but one illustration of the diversity of interests and abilities that marks the twentieth century tiller of the soil. —EDITOR.

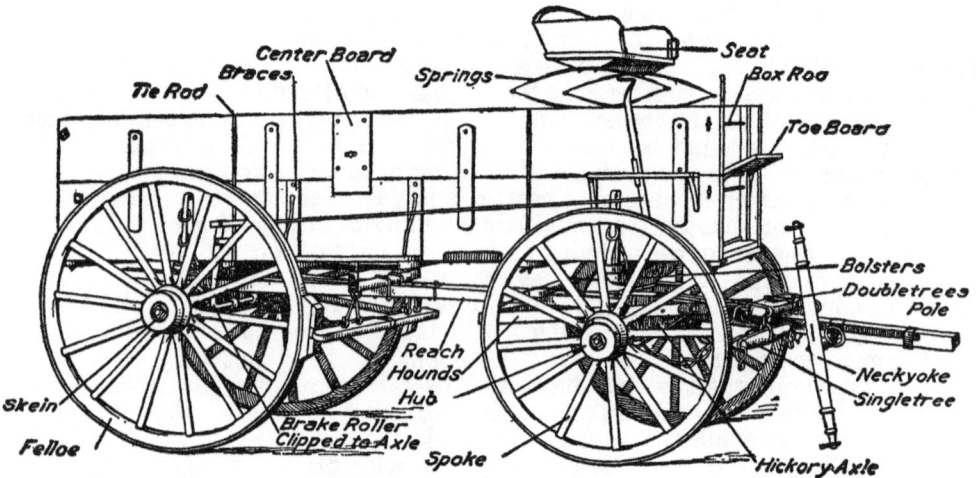

FIG. 1. Typical heavy farm wagon,

CHAPTER 1

Horse-Drawn Work Vehicles for the Farm

By E. W. LEHMANN, Professor of Agricultural Engineering, University of Missouri, who was born on a farm in southern Mississippi and lived there until he entered college. Since then he has given his summers to practical farm work and now owns a half interest in a farm adjoining the old home place. His technical experience and teaching have taken him to the agricultural colleges of Mississippi, Texas, Iowa, Wisconsin, Cornell, and Missouri, where he has become familiar with the mechanical features of the problems with which he is equally well acquainted as a practical farmer.

The farm wagon is so common an implement that the average person takes it for granted, and fails to realize the value of scientifically correct design and construction. This chapter throws a new light on the subject, and should enable the farmer to get better value for his money, both in buying and keeping his work vehicles.—EDITOR.

THOUSANDS of years ago man had few possessions, and, therefore, very little need of vehicles. As he acquired clothing, weapons, and other possessions, and as he changed his place of residence, some means of transportation became necessary. It was probably some hunter who created the first vehicle when he found how much easier it was to drag his game home on a branch of a tree than to carry it on his back. Another hunter, accidentally or otherwise, may have placed a round stick under his sled and thus started the idea of a wheel, which is the basis of all our systems of transportation. There was a gradual development from this roller device to the wooden wheel made of a sawed off section of the trunk of a tree with a hole in the centre to receive the axle. But it was the early Egyptian who perfected the spoked wheel.

Improved roads in every country have produced great changes in types of vehicles. The early Roman road especially made possible a great advance. In all cases the development of methods of transportation follows the settlement of a country. Before roads were built in America, the early settler hauled all his equipment and household effects on pack horses over bridle paths through the forests. Only on the introduction of the stage coach and heavy wagons for hauling freight were the turnpikes and state and toll roads built.

FIG. 2. Prairie schooner

As farmers acquire more modern types of vehicles the country roads are steadily improved. But there are still backward sections of the United States where instead of wheeled vehicles, sleds, called "slides," are still used. In parts of the South, negroes use on the farm crude wooden wagons, which represent an improvement over the slide, but of which the wheels, like those of the wagons of past ages, are disks cut from the trunks of large trees. The black gum being very tough wood, and not splitting easily, is well suited for this purpose. These wagons are made entirely of wood except for the linchpins (usually large nails) that hold the wheels on the spindle.

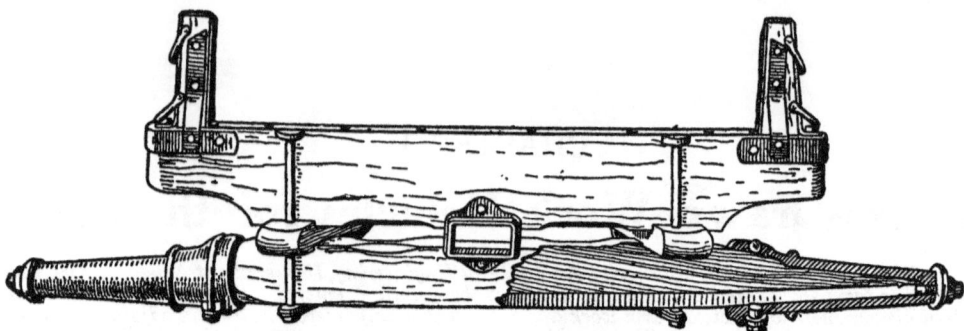

FIG. 3. Thimble skein hind gear, a sectional view of part of it showing axle reinforcement

The stone-boat originated in the eastern states, where it was used in clearing fields of stones.   For moving plows, harrows, materials for concrete or masonry work, barrels of spray mixture and other heavy or bulky articles about the farm it is more convenient than a wagon.   Its frame should be of 2 x 4s (for runners) with boards fastened across and the hitch fastened to one end.   A better boat is made by cleating heavy planks together lengthwise and bolting them at the front end to a special metal head made for the purpose with bolt holes and one for the hitch.   A wagon with a platform hung below the axle is an improvement on a stone-boat for some purposes.

The modern wagon has been developed to meet a need that exists on practically every farm.   Because it is used more throughout the year than any other farm implement (to give only one reason), every farmer should know something about the construction and essential points of a well-made wagon.

Materials for wagon construction.   The durability and strength of a wagon depend to a large degree on the materials of which it is made.   Every well-built wagon is made of selected, air-dried materials, air-drying or seasoning of lumber requiring about 3 years.   Lumber that is dried rapidly in a kiln is not uniformly cured, is usually more brash, and will not resist as great a shock or strain as air-dried material.   Black birch or white oak is usually used for hubs; white oak for spokes, felloes, bolsters, sand boards and hounds; hickory is used for axles in the best wagons, but because of the high price of the best grades, some makers use maple, oak or cheaper grades of hickory.   If the wooden parts are soaked in linseed oil before being assembled, they will not shrink and decay when exposed to the weather as readily as if untreated.   All metal should be of a good grade of mild steel and all parts of ample size to give strength.   Nearly every wagon is built to carry a certain load which is specified as its *capacity*.   The capacity ratings of a reputable firm are an aid to the farmer in selecting a wagon for a particular service.

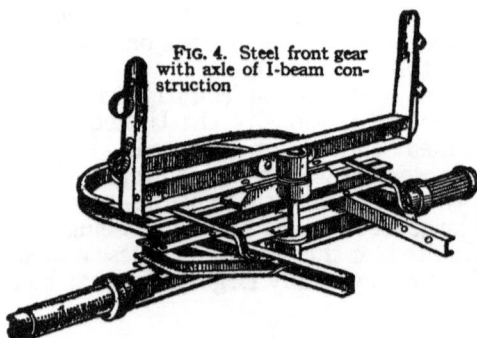

FIG. 4. Steel front gear with axle of I-beam construction

Draft of Wagons

Height of wheels.   The nature of the road bed is the biggest factor affecting the pull required to move a wagon.   Features of construction which affect its draft are the height of the wheels, the width of the tires and the efficiency of the bearings.   The following tests (made at the Missouri Agricultural Experiment Station) compare the draft of a 2,000-pound load on high, medium, and low wheels on different road surfaces.   The heights of the wheels were as follows: the high, 44 inches (front) and 56 inches (rear); the medium, 36 inches (front) and 40 inches (rear); the low, 24 inches (front) and 28 inches (rear); the wheels were of steel with 6-inch tires. The figures represent pounds of pull required per ton of load.

While the results are in favor of the high wheels there are other factors which would determine the height to select. The medium height is generally preferred as it is more convenient.

Tires. The effect of the width of the tire on draft is shown in the following table (from Bulletin 39, Missouri Experiment Station). The wheels used were of the same height and provided with 1½-inch and 6-inch tires. Again the figures are pounds of pull required per ton of load.

The summary of the results of these experiments states that the wide tire gave lighter draft except under the following conditions: (a) when the earth road was muddy, sloppy and sticky but firm underneath, (b) when the mud was deep and stuck to the wheels, (c) when the road was covered with deep, loose dust, and (d) when the road was badly rutted with the narrow tire. Another very important advantage of the wide tire is that it

ROAD SURFACE	WIDTH OF TIRE	
	1½-INCH	6-INCH
Broken stone road; hard, smooth; no dust	121	98
Gravel road; hard and smooth	182	134
Gravel road; wet, loose sand, 1 to 2½ inches deep	246	254
Earth road, loam; dry dust, 2 to 3 inches deep	149	109
Earth road, loam; dry and hard; no dust .	90	106
Earth road, loam; stiff mud, dry on top, spongy underneath	497	307
Earth road, clay; sloppy mud 3 to 4 inches deep; hard below	286	406
Earth road, clay; stiff, deep mud . . .	825	551
Plowed land harrowed smooth and compact	466	323

FIG. 5. Parts of a roller bearing

does not destroy the road surface as rapidly as the narrow tire.

Bearings. The efficiency of the bearings in relation to draft is usually a small factor.

However, all wagon spindles should be properly greased at intervals. The draft is slightly greater for large axles than for the small ones. Some manufacturers are reducing the axle friction by means of ball and roller bearings. The fact that such bearings are being used so efficiently in tractors and automobiles leads one to believe they should be equally efficient in a wagon.

ROAD SURFACE	WHEELS		
	HIGH	MEDIUM	LOW
Macadam; slightly worn clean, fair condition .	57	61	70
Dry gravel road; sand 1 inch deep; some loose stones . . .	84	90	110
Earth road; dry and hard	69	75	99
Earth road; thawing; half inch sticky mud .	101	119	139
Timothy and bluegrass sod, dry, grass cut .	132	145	179
Timothy and bluegrass; wet and spongy . .	173	203	281
Corn field; flat culture across rows, dry on top	178	201	265
Plowed ground, not harrowed; dry and cloddy	252	303	374

FIG. 6. Two types of spokes: *above*, regular spoke with square tenon; *below*, Milburn spoke with double slope shoulder tenon.

The strength of a wagon. Every well-built wagon is designed to resist hard usage. All parts should be made of ample size and strength. All irons should be attached to resist strains instead of merely for looks, and in a way that will not weaken the wooden parts. Clips are much better than bolts and should be used whenever possible. In building a wagon to suit the needs of a particular locality, the local conditions must be kept in mind. Different sorts of wagons are built for mountainous and prairie sections.

Wagon Construction

(See page 11 for standardized schedule)

The wheels. The wheel is the most important part of a wagon. It is probably the first part to break in a poorly built wagon and the most expensive to repair. The hub is the heart of the wheel. There is a great deal of discussion as to the relative merits of oak and black birch for hubs, but both are well suited to this use if properly handled. Black birch does not check as badly as oak but is claimed to be more liable to dry rot. The common type of hub is turned to proper shape and given a severe heat test. It is then mortised to receive the spokes, after which heavy iron bands are forced on. Forcing the bands on cold is the best practice, since the wood is then not charred. Next the spokes are dipped into glue and forced into the hub. Most failures in wheels are at the point where the spokes enter the hub. There are two types of spokes— that with a double slope shoulder tenon and that with a square tenon. The former is said to be much stronger and more durable. It does not matter greatly whether the felloes are bent or sawed, but they should be well doweled and the tire bolted to them. A rivet should be placed on either side of each spoke to prevent splitting. Tires should be heavy and for the average farm wagon not less than 2 or 3 inches wide; the 4-inch tire is growing in favor. All tires must be oval-edged. When the tires are shrunk on the wheel the spokes are sprung slightly; this is called dishing the wheel. To make all the wheels uniform when the spokes are forced in they are given a little dish, and when the tire is shrunk on it adds to the dish.

A type of wheel used on both wagons and buggies is known as the Sarven patent wheel. A great many more spokes are used in it than the ordinary wood hub wheel and these are driven so as to form a solid arch around the entire hub. The hub is covered with metal flanges, the two sides being bolted together with eight or ten bolts. The metal covering protects the hub and prevents its decay. This type of wheel is very strong; the schedule

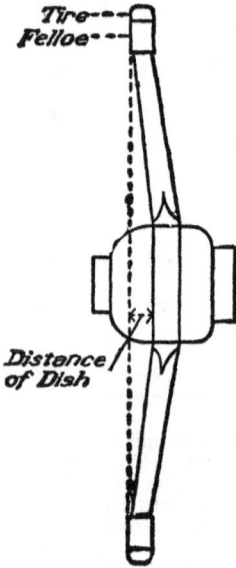

FIG. 7. The dish is the essential feature of wheel construction.

Tire
Felloe

Distance of Dish

permits its use only on steel axles such as are used for western trade.

In the carrying out of the plan for the standardizing of wagons, there were adopted recommendations made by the United States Department of Agriculture after much experimentation. These had to do with the widths of

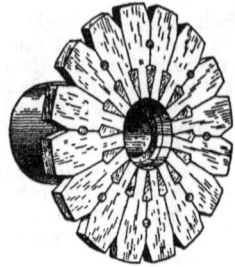

FIG. 8. Section of hub of a Sarven patent wheel

tires and the diameters of skeins for wagons of the different standardized classes and capacities. The dimensions finally decided upon are as follows:

TYPE AND CLASS OF WAGON		CAPA-CITY (Lbs.)	TIRES (IN.)		DIAM-ETER OF SKEIN (IN.)
			WIDTH	THICK-NESS	
Class A. Farm Wagons and Gears	Light	1,500	2, 3 or 4	⅜ to ⅞	2⅛
	Medium	3,000			2¾
	Standard	4,500			3⅛
	Heavy	6,000			3¼
Class B Valley Wagons and Gears	Medium	3,000	2, 3 or 4	⅜ to ⅞	2¾
	Standard	4,500			3⅛
	Heavy	6,000			3¼
Class C Mountain Wagons and Gears	Light	2,500	2, 3 or 4	⅜ to ⅞	2⅛
	Medium	4,000			3⅛
	Standard	5,500			3⅛
	Heavy	7,000			3¼
Class D One-horse Wagons and Gears	Light	1,000	1⅛ to 4	₁₆⁄ to ⅜	2⅛
	Medium	1,250			2
	Heavy	1,500			2

The standard heights for both wood and steel wheels for farm wagons are 40-44 inches and 44-48 inches. Wood wheels for trucks are to be made only 36 and 40 inches, and steel wheels for trucks only 28 and 32 inches.

The axle. The axle is another important feature in wagon construction since it is the site of many breaks. Axles are made of both steel and wood. There are several types of steel axles, the commonest three being the hollow, the solid square, and the I-beam. On account of the tubular shape of the hollow axle it will stand equally well a strain from any direction, and for this reason it is most desirable in hilly or mountainous sections. It is comparatively a light axle. The I-beam axle is of the same construction as the steel beams in structural work and is very strong. The greatest demand, however, is for wagons with wooden axles which are more

FIG. 9. Typical hub of wheel

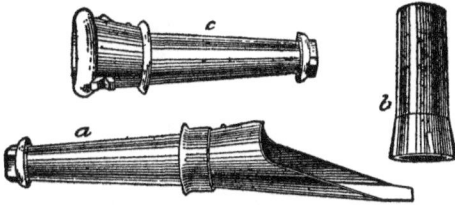

FIG. 10. Three types of skeins: *a* steel skein; *b* box; *c* cast skein

economical and which, according to manufacturers who build wagons with both steel and wooden axles, give lighter draft. Hickory is the best axle material. Wooden axles are provided with either cast-iron or steel skeins (that is, the sections which carry the wheels).

The cast-iron skein is used exclusively on wagons built for central, southern and eastern trade and the steel skein on wagons for western trade. The throat of the skein should be large so as to prevent the entrance of sand and dust into the boxing. The skein, sandboard, and bolster should be fastened to the axle with clips rather than bolts whenever possible. (Fig. 3.)

Many axles are provided with a steel truss that not only helps to carry the load but also holds the skeins in place. One of the best types of reinforcement is to have a bar of steel in a groove on the bottom of the axle. The end of the bar passes through the point of the skein and is finished to take a nut which holds the skein rigidly in place. (Fig. 3).

A proper "pitch" or set of skein is needed to overcome the effect of the dished wheels. The point of the skein is turned down until the supporting spokes are always vertical. To prevent the wheel from crowding out on the linchpin or axle nut the axle must be given some "gather." This means that the skein must be put on the axle so that the points of the skein are a little forward. The proper adjustment of the dish, set, and gather go a long way toward making a smooth-running, light-draft wagon. When the wheels are on, the set is measured by determining the difference in distance between the wheels at the top and at the bottom. This varies from $2\frac{1}{2}$ to $3\frac{1}{2}$ inches on the front gear and $3\frac{1}{2}$ to $4\frac{1}{2}$ inches on the rear gear for different sized skeins. The gather is measured

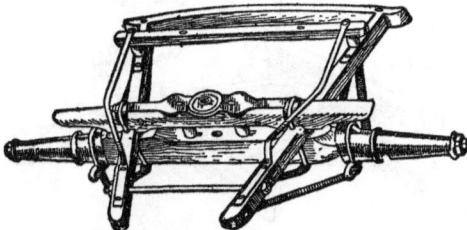

FIG. 11. Drop tongue front gear, with square hounds

as the difference in distances between the front and rear points on the wheel and is usually $\frac{1}{2}$ inch on the front gear and $\frac{5}{8}$ inch on the rear gear. In the modern wagon factory the skeins are forced on by powerful presses, and gauges are used to see that each skein is equally distant from the centre of the axle and has proper set and gather.

Kinds of gears. Only one gear with but one height of bolster is now made for each capacity of wagon. The front gear may be of drop, slip or coach-tongue type. Gears are made with either square or bent hounds. The square hounds are usually recommended because they are much easier to repair in case of an accident.

Bolsters. The bolsters should be well ironed to resist rough usage; if ironed full length on top, it is an advantage. They should be provided with suitable stakes rigidly fastened to them by irons. A suitable wearing plate should be provided between the sandboard and bolster on the front gear to

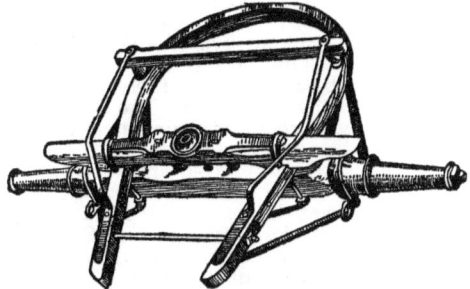

FIG. 12. Drop tongue front gear, with bent hounds, still in use in some localities

make turning easy. The new style cup and saucer bolster plate is one of the best types, and it takes the strain off the king bolt and prevents the sandboard from splitting. On some wagons a fifth wheel is provided between the bolster and sandboard. With a properly constructed fifth wheel the strain is better distributed between the front bolster and sandboard than with the old style, straight bolster plates. While there are many good features about the fifth wheel for use in wagons they have not come into general use.

The reach or coupling pole should be flexible and made of good, tough, straight-grained oak so that it will resist any strain when one wheel drops into a hole or rises over an obstruction. Reach boxes should be provided on wagons with wooden axles to prevent vibration and wear on the axle and sand bolster. The reach construction is restricted to the rectangular type only.

Tongues must also be made from tough, straight-grained timber and be both strong and elastic. The drop tongue is not as easily replaced as the stiff tongue. A special tongue and front gear is provided for working

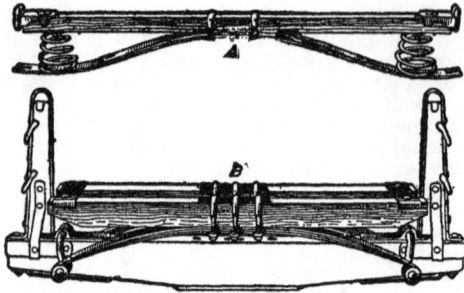

FIG. 13. Two types of spring bolsters: *a* one-horse single; *b* two-horse double

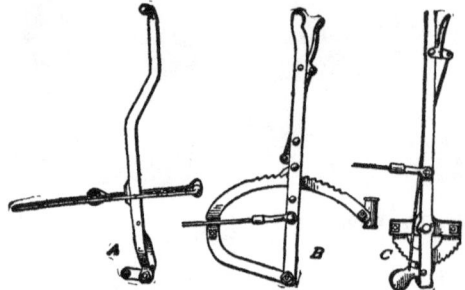

FIG. 15. Three types of brake levers: *a* Colorado foot lever; *b* Giesler handle; *c* Palla handle

oxen. When these are used the draft is from end of the tongue instead of from the front gear by traces and whiffletrees.

The wagon box. Wagon boxes are now made in one width only to fit between stakes set 38 inches apart, and are without foot boards. The sides and ends are best made of poplar, though some manufacturers use yellow cottonwood. Long-leaf pine makes a good bottom, but all bed sills should be of oak. All points on the bed where the bolster rubs should be ironed, and there should be a rub iron for the front wheel. The bed should be provided with rods and chains to prevent its spreading when heavily loaded.

The bed is often removed and the running gear used for hauling logs or lumber; or a special rack may be put on for hauling hay. It is such a task for a man or even 2 men to remove a bed and put on a heavy hay rack, that there has been developed a convertible wagon bed which is a great labor saver. (See Figs. 21, 22 and 23.) Such a bed must necessarily be well-braced and strongly built of the best material obtainable. A farmer who has a shop and is handy with tools could make such a bed during his spare time.

Bolster springs. It is a mistake that springs are so little used on farm wagons. Certainly all light wagons should be equipped with them, especially if the wagon is used in delivering garden truck, fruit, or other easily-bruised products. Springs lessen the effect of bumps, make possible faster driving and

the hauling of a larger load, and give the driver more comfort and satisfaction. The two types in use are the coil and the leaf spring, the latter seeming to give better satisfaction.

Brakes are a necessity in hilly country. There are two types. The *box brake* is used most on light one- and two-horse wagons; the objection to this type is that it tends to weaken the box. The *gear brake* is used on both light and heavy wagons. It does not weaken the box and, further, it may be used when the gear without the box is used for hauling.

Lock chains. In some sections instead of a brake, chains are used to lock one of the rear wheels. These are attached to the wagon bed and fastened to the wheel by means of a latch. The disadvantage is that this device puts undue strain on both the wheel and the bed, and is inconvenient to lock and unlock when the wagon is heavily loaded.

Width of wagon track. In the past all wagons were constructed to meet local requirements. Two widths of wagon tracks gradually developed: (1) the narrow or regular measuring 4½ feet from centre to centre of tires; and (2) the wide, measuring 5 feet. The regular width was used nearly altogether in the northern states and the wide width in the western states. As already mentioned, all wagons are now constructed with a 56-inch track, which is the standard tread for automobiles.

Painting the wagon. In careful wagon making, all gear parts when

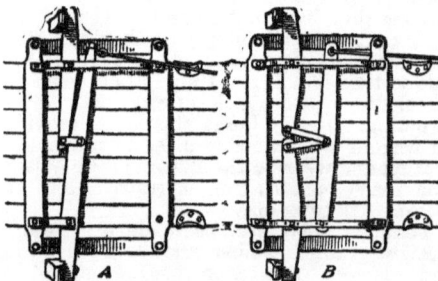

FIG. 14. Bottom views of two types of brakes: *a* joint box brake; *b* three-piece box brake

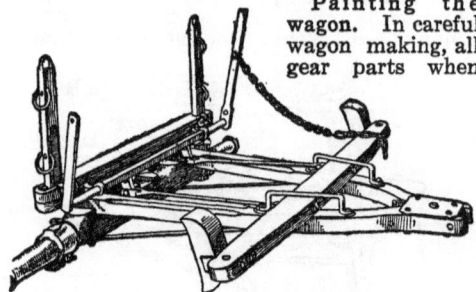

FIG. 16. Eastern gear brake

finished are soaked in linseed oil and put into a warm room for from 3 to 8 months so that the oil will penetrate into the wood before the paint is applied. One coat of paint should be applied before the ironing is done to keep the moisture from entering the wood under the irons. After ironing, at least 2 more coats of red lead should be applied and these followed by a coat of wagon varnish. Many farm implements are given a coat of paint by dipping, but this is not good practice; a wagon especially should be carefully painted and varnished all over by hand.

The Care of a Wagon

The first point to consider in the care of a wagon is proper housing. Many wagons have, no doubt, been injured more by being exposed to the weather than by use. Properly cared for wagons have been known to last for more than 25 years.

Lubrication. The spindles should be greased at regular intervals. Before the grease is applied the spindle or skein should be wiped off with a rag so that all of the grit and dirt is removed before fresh grease is applied. Do not wait to grease until the wagon screeches; it is best to use less grease at a time and apply it oftener. Too much at a time is not only wasted but

FIG. 17. Plantation or dump cart

gathers dust and in some cases does more harm than good. Use a good grade of axle grease and not one that "gums."

Care of wheels. These should be carefully watched to see that the tires stay tight. The pounding over a stony road will eventually lengthen a tire and with the shrinking of the wood in summer it may run off the wheel. A tire should be reset by a competent blacksmith as soon as it begins to work loose. In doing this he should cut enough out of the felloe to allow it to draw up snugly on the spokes and force them firmly into the hubs. If too little is cut out the wheel becomes felloe bound and the spokes will work loose. If too much is cut out, when the tire is shrunk on the wheel will be over dished.

See that all bolts are kept tightened and that any that are lost are replaced immediately; otherwise, the wagon is weakened. It will add much to the life of a wagon if it is repainted every few years with a coat of red lead and oil.

Types and Sizes of Wagons

Wagon construction standardized. The construction of wagons is now on a standardized basis as a result of the efforts of the Conservation Division of the War Industries Board and the coöperation of the wagon manufacturers. The standardized schedule went into effect January 1, 1919, with the exception of portions referring to wheel construction and the manufacture of parts for the repair of wagons still in use.

The standardized manufacture of wagons makes it possible to save material and labor through the elimination of unnecessary types and sizes. It means greater economy in cost of handling from the standpoints of both manufacturer and local distributer. A smaller

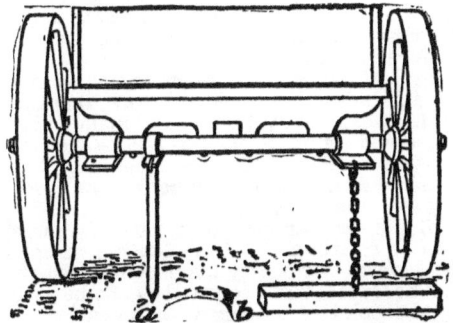

FIG. 18. Devices for holding a wagon on a steep hill

FIG. 19. Farm truck or handy wagon suitable for use with convertible body. (See Figs. 21, 22, 23.)

stock can be carried by the latter and better service given to the user. The final result is a better wagon at a saving for the farmer.

The principal features of standard built wagons are: (1) the four standard capacities of the different classes; (2) the marking of the capacity on the gear of each wagon; (3) the standard width of track of 56 inches which is the same track made by automobiles; (4) the standard width between stakes, making wagon boxes interchangeable; (5) the standard width of boxes; and (6) the standard gear for each capacity, the front gear to be made with drop slip or coach tongue type. Other minor details are also made standard.

Special Types of Farm Wagon

The handy wagon or farm truck. The handy wagon is a low-wheeled, broad-tired wagon often needed on every farm. A suitable rack fits it for hauling almost any sort of farm load. Handy wagons can be secured with either metal or wooden wheels, but the latter type are, no doubt, better suited for hauling through mud as there is less tendency for the wheel to fill with mud above the tire.

The handy under-hung wagon is a standard high-wheeled wagon that has been remodeled to bring the bed nearer the ground for convenience in loading and unloading. The wagon-bed proper can be made almost any width or length desired depending on the strength of the remodeled wagon and the size of the team. The remodeling consists of removing the rear skeins and fitting them on a 4 x 6-inch axle long enough to accommodate the width of bed desired. The beams and all other parts should be made of strong materials and well ironed.

Spring wagons are much better adapted for all kinds of light delivery work than those without springs. They can make better speed and are much used in truck- and fruit-growing sections. These wagons may be of either the one- or the two-horse type according to the capacity desired.

Dump carts or plantation carts are largely used in certain parts of the South, principally on account of the easy manner in which they can be loaded, and because they can be gotten to the point of loading with much less difficulty than can a wagon. They are still used

in many southern cities for drayage work and almost altogether for hauling sugar cane on many large Louisiana sugar plantations. One type of dump cart is made by taking an ordinary 2-wheel cart, shortening the tongue and fastening it to the forward axle of a wagon with a king bolt. It can be turned easier than a cart of the ordinary type, and is a convenient cart for hauling about the farm.

Modern sleds. A great change has been made in the sled since it was first used by our forefathers. Climatic conditions, of course, have much to do with the type in use. Throughout the North where the snow covers the ground for several months, the bed is removed from the wagon and placed on a bob sled. Such a sled with 2 knees in each bob would have a capacity of about 2 tons, depending on the materials used and its construction. After a sled is started there is very little draft so a team can pull a much greater load on one than on a wagon. To get the sled in motion a lead team is sometimes used, but a bar in the hands of a good driver is usually all that is necessary. In hilly regions some provision must be made to hold the load back when going down the slope; a chain attached to the runner and dropped beneath it is often used. To prevent a sled from sliding backward a curved spike is often bolted to the sled. In northern logging camps where sleds are much used, the heavy work calls for strong runners under a very strong frame. Northern farmers who haul out manure in winter and do not have a manure spreader will find a dump sled a great convenience. Such a sled is made by using the front bob of a double sled and raising the framework to a sufficient height to carry the box which is secured to it by means of eyebolts, staples, and pin.

Hitches

In hitching to the ordinary farm wagon, the straight 2-horse hitch is used. In using 3 horses in some localities they are hitched abreast by means of a 3-horse evener. In other localities 1 horse is hitched to the point of the tongue as a leader by means of a single tree. The latter method is also used in hitching 4 horses. The relation of the construction of the wagon to its draft has already been mentioned. It is well to remember that the line of draft, the length of the traces, and the position of the load in the wagon also affect

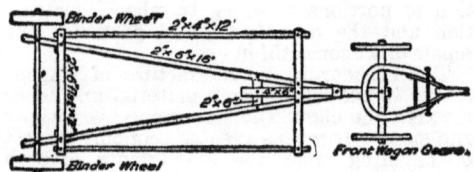

FIG. 20. Home-made fodder wagon and rack

the draft. The proper line of draft varies with the type of road surface. If this is hard and does not yield to the weight of the load, the line of draft of the trace should be parallel to the general surface of the roadway. If the wheels sink into the road surface, the line of draft should be raised. Shortening the traces will do this. However, any adjustment of the traces made to change the line of draft should not interfere with the angle they make with the hames. The traces should always be as nearly as possible at right angles to the hames so the collar will bear uniformly against the shoulder. If the point of the hitch is very low there is a tendency to carry the greater part of the load on the withers, which is objectionable.

Loading. The proper distribution of a load on a wagon is a much discussed question. Where only a part of a load is carried, it is usually placed well forward on the wagon, especially if the roads are soft. This, in effect, raises the line of draft so the team will tend to lift the load out of the mud. It makes little difference whether the load is forward or

FIG. 22. A convertible wagon-bed, with sides fully extended for hauling loose fodder, etc.

FIG. 21. A convertible wagon-bed, with end gate down

back if the road is hard and does not yield to the wheels. When the wagon is fully loaded it is better to place more weight over the rear wheels for two reasons: (1) by lightening the load forward, a certain amount of play in the front axle over rough ground is permitted; (2) the rear wheels are of larger diameter than the front wheels, and have a greater base of support and, therefore, do not cut into the roadbed as badly as the front wheels. Thus there is less draft for the larger wheels.

Double trees. To have a load uniformly divided between 2 horses where a 2-horse evener is used, it is necessary to have the 3 holes in the evener in a straight line. With an evener so constructed, it does not matter in what position the horses are, the load will be equally divided. If the centre hole is ahead or behind the line of the two end holes, the horses will not pull the same except when they are even. In one case, the horse ahead will be pulling more and in the other case less.

FIG. 23. A convertible wagon-bed, as used for transporting livestock

FIG. 24. French state chariot of 13th century

CHAPTER 2

FIG. 25. Scythian house on wheels

Horse-Drawn Pleasure Vehicles

By PROFESSOR E. W. LEHMANN, *University of Missouri (See Chapter 1). With the increased use of machinery in modern times, there is a growing tendency to think of the automobile as the only real pleasure vehicle. Certainly it has its place (See Chapter 4), but certainly also it is not going to entirely replace the horse as a means of obtaining pleasure and recreation any more than the tractor has entirely replaced or is going to entirely replace the horse as a source of power. But pleasure driving hereafter is going to be restricted largely to the open country; it is to be one of the rewards of those who live there. It is therefore those who live on farms who should know the most about the essentials and possibilities of pleasure vehicles.*—EDITOR.

THE development of pleasure vehicles. The early pleasure vehicles were very elaborate in their design and finish. Along about the sixteenth century the nobles of Europe supplied themselves with the most beautiful and extravagantly adorned coaches; often thousands of dollars were spent on a single coach. During the seventeenth, eighteenth, and nineteenth centuries the construction of all types of pleasure vehicles gradually became more simple and efficient.

The carriage industry in America has existed since early Colonial times and, up to the census of 1904, its growth about kept pace with the increase in population. However, the advent of the automobile has had a decidedly retarding influence on the industry. Many companies have been forced to close their factories, while others have turned their attention wholly or in part to the manufacture of automobiles.

There were three types of early American pleasure vehicles. One was a heavy cumbrous 2-horse outfit that had come into limited use in England about 1600; this type was not a success. The other two were lighter and better suited to American conditions at that time. One of these—an open two-wheeled outfit—was called a chair. The other, called a chaise (and later a "one-horse shay"), was simply a chair provided with a top. Few springs were used in those days. The vehicles were swung on braces of either wood or leather. Steel springs were used as early as 1670, but it was not until 1804, when Obadiah Elliott patented the elliptical spring, which is now so widely used on all types of vehicles, that they were used to any extent. The invention of this type of spring, along with the

FIG. 26. George Washington's private coach

FIG. 27. One-horse shay

FIG. 28. President Lincoln's state coach

14

increased building of improved roads, brought about the real development of the carriage industry.

Two factors affect the design of pleasure vehicles: they are utility and fashion. The usefulness of the horse-drawn vehicle is given more attention now than formerly. Practically all horse-drawn pleasure vehicles used at the present time represent forms of either the common top buggy or the road wagon. The private coach of Washington and the carriage provided for LaFayette were much more elaborate than Lincoln's but not more useful. A description of President Harrison's buggy tells very clearly of its useful features. "It is for one horse and has a large roomy place under the seat big enough to take in a minnow basket, a watermelon, a basket of lunch (even the catsup bottle may reach above the handle), and a great many other things hardly worth mentioning. Back of the seat the place is all boxed up so that the boys at the camp meeting can not steal anything from behind. There is room under the seat for his shotgun and quail bag." President Harrison considered very carefully the utility of his carriages.

Types of Pleasure Vehicles

There are many types of sporting carriages and vehicles used by the wealthy class of the cities and by men and women who show horses. Coaching, a sport indulged in, in America, only by the rich, is confined to New York and the neighboring cities. It requires a great deal of skill and judgment to handle a coach with four or more horses attached. All these show vehicles are carefully and accurately made. Mr. H. H. Salmon in the official Horse Show Blue Book makes the following statement: "In vehicles the standards are already so excellent that the provision of the first-class carriage maker may be accepted as correct." In other words, the broughams, victorias, hansoms, coaches, phaetons, and carts of all sorts are practically perfect as made by the best carriage makers.

The *brougham* or *coupé* is a closed carriage with a seat for the driver in front, ordinarily called a cab or "bus." The *victoria* is an open-hooded carriage with two seats. It is used for park driving in the cities. The *hansom* (invented in 1834 by Joseph Hansom) is a two-wheeled vehicle with the driver's seat behind and above the body. The *phaeton* is used a great deal, but the basket type is not as common as the ordinary type. The phaeton may be provided with either a folding top or a canopy. The *landau* is a large carriage for 4 passengers with a folding top. The *rockaway* resembles the *landau*, but is higher, lighter, and less costly. It is quite suitable for country use and for small towns.

Carts. There is a distinct class of 2-wheeled vehicles used for pleasure and in breaking colts and exercising horses both in the country and in the cities. An essential feature of a light-draft pleasure cart is a proper balance such that its weight is removed from the horse's back, thereby making it easier to pull. In this it differs from heavy one-horse work carts. With them weight is sometimes put on the horse's back to increase its traction.

The *gig* is a common type of cart with one seat. The Irish *jaunting car* has two seats placed back to back. Road carts of various types are useful, convenient, and easy on the horse but not very comfortable. The *trotting sulky* is a type of cart representing a purely American product. It is made light of draft by being provided with ball bearings and with either solid or pneumatic rubber tires. The improvement of the modern sulky has done much to lower trotting records.

Types found on farms. The *surreys* and *carryalls* are being displaced by automobiles in many localities. There is more general use of these types of vehicles throughout the South and East than through the North and West. Surreys are usually provided with either extension or canopy tops, but a light, open surrey is also used. A surrey should be well constructed, of first class material and provided with 2 or 3 strong elliptical springs. (For easy riding the latter type is best.)

The greatest demand for horse-drawn pleasure vehicles at this time is for top *buggies* and *runabouts* or open buggies. The latter are sometimes called driving wagons. The platform spring wagon is also used in some sections. This type of carriage is pro-

FIG. 29. Surrey

Fig. 30. Rockaway

vided with 2 seats, the body is supported on a special, half- or full-platform spring. In the half-platform spring type, the rear end of the body is supported by 4 springs arranged as a platform.

There is great variety in the types and designs of runabouts or open buggies. One of the most noticeable differences is found in the styles of bodies and seats. There are 2 general classes of bodies or boxes. The *piano box* is either narrow or wide and has the same height of panel all around; the *corning body* has low panels just back of the dash board. Top buggies also have these two types of bodies. Nearly all of the cheap runabouts are provided with piano bodies, but some of the highest-priced buggies have the same type. As to variety in the seats, some types of vehicle have solid one-piece seats; others, stick seats; and still others have combinations of these.

Essentials of Good Buggy Construction

The essentials of a good buggy for pleasure driving are: (1) lightness of construction as well as light draft; (2) neatness of design; (3) excellent and durable finish; (4) strong, well-braced construction; (5) good springs, making easy riding; (6) a reliable fifth wheel; (7) well-secured clips; (8) a strong, well-made body wide enough for comfort; (9) a neat dash and boot; and (10) a seat plenty wide and provided with good springs.

The material used in a buggy is selected and seasoned with even greater care than that put into a wagon. To secure lightness, the very best and strongest materials must be used. The hickory used in carriage construction is from the younger trees, as it is stronger and more elastic. A distinct advantage is claimed for American carriage

Fig. 31. Phaeton

builders over those of other countries on account of the superior wood available here, and much American wood has been exported for foreign trade. The foreign design in carriage building is much heavier than the American.

Wheels. The wheels are as important a part of the pleasure vehicle as they are of the wagon. The great majority of buggy wheels are of the Sarven patent type. Some hubs are made larger in the centre which gives them greater strength where the spokes are driven which is usually the weak part of a wheel. Buggy hubs are quite often made of elm, the spokes and rims being made of straight-grain hickory. In wheel construction the tiring plays an important part. The rims are liable to be charred if the tires are put on hot from a flame. On some types of buggies the tires are heated in a tank of boiling water, which gives uniformity of heat around the entire tire, and results in the production of a better wheel than the earlier, cruder methods. A buggy tire should always be bolted to the rims, the heads of the bolts being counter sunk into the tire.

An important step in the development of pleasure vehicle construction was the introduction of rubber tires. The solid tire was first employed in the 'fifties, but did not come into general use on pleasure vehicles until the invention of the grooved tire many years later. The pneumatic tire was introduced in 1890, but, except for trotting sulkies and light runabouts, it has not found popular favor. On good macadam

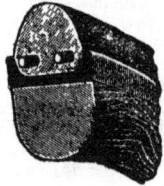

Fig. 32. Solid rubber tire showing the internal wire construction.

roads and on paved streets, rubber tires are more elastic and much less noisy than steel tires; on dirt roads they offer no special advantages. The amount of pull required for rubber-tired buggies is quite a little less than for steel tires, especially in the case of vehicles traveling at high speed. At low speeds the draft with steel and rubber tires is about the same.

The type of rubber tire shown at Fig. 32 illustrates the internal wire construction, by means of which the wires are protected from moisture and all danger of rusting. The channel is entirely filled with rubber preventing the entrance of dirt and moisture both of which shorten the life of a tire. Extreme care must be observed in cutting and putting on rubber tires of this sort. The wire wheel is used to a certain extent, principally on trotting sulkies and runabouts used for park driving in cities; but very little on the farm. The principal advantage of wire wheels is lightness. Such wheels are usually provided with either roller or ball bearings and pneumatic rubber tires, which together reduce the draft to the lowest possible point.

Axles. Buggies are usually provided with either arch or drop axles. The true arch axle is a little more expensive. Best axles are of solid steel capped with straight-grained hickory. The type of axle called the long-distance axle should be selected. It has 2 lengthwise and 2 crosswise grooves on the spindle which insure positive and thorough lubrication. It should also have a collar to protect the spindle from dirt and grit. The axle must have the proper set at all times so that the under spoke is always vertical and so the box will not wear and will hold grease. To test the proper set and gather on an axle, place the wheels on the spindle just enough to give them holding power, then pull the buggy forward and watch the wheels go in place. Proper set and gather of axle insure a light-running vehicle.

The reach. The bars connecting the front and rear running gear are called the reach. There are several types in use. The double bar or twin reach is used a great deal, but there are also the diverging double reach and the single reach. All buggies made by hand many years ago had a single reach. The double reach in addition to having all of the features of the other types is more flexible, and has a more positive recovery should the wheel strike an obstruction. The reach

FIG. 33. Dust-proof axle

must be of correct length, or troubles with the fifth wheel will result. If the reach becomes bent or sprung out of alignment, it will cause the upper and lower parts of the fifth wheel to fit together unevenly, thereby causing undue wear. Proper attachment of the reach also insures that the rear wheels will track with the front wheels making lighter draft. It is claimed by some that an elastic reach is better than a stiffer one, but the latter type is in more general use. A type which is highly recommended is the full length hickory reach with channel iron support. Every type of reach should be well braced and its connections are better made with clips than bolts.

Fifth wheel. The best fifth wheels are made of wrought iron, which is tough and strong. There should be no strain on the king bolt when in place. It should simply hold the 2 parts of the fifth wheel snugly together.

Shafts and poles. Shafts should be made of a good grade of straight-grained hickory. A shaft must be elastic so it can bend without breaking. Heavily ironed shafts and poles are not always an indication of quality. The

FIG. 34. Buggy

greatest cause of failure of shafts is the practice of backing or cramping the buggy, which brings a strain on the bend of the shaft causing it to splinter and break. A good type of shaft has the bend properly strengthened by trusses or braces. It is an advantage to have quick couplers connecting shaft to axle. This is especially true when it is desirable to change from shaft to a pole. With standard couplers there is no noise from loose bolts.

Springs. The type, strength, and elasticity of the springs determine the easy-riding qualities of a buggy; the type also affects its draft and its length of life. A stiff set of springs not only makes hard riding but also throws a strain on some other part of the buggy, eventually shortening its life. Springs must be firmly set in place on both the front and rear axles to prevent any swaying forward or backward.

Box or body. The piano type of body is in most general use on top and open buggies. The framing of buggy bodies is of oak or ash and the sides and panels of clear yellow poplar. Many buggy bodies are made by putting screws through the panels from the outside and filling the screw holes with wooden plugs. After such buggies have been in use for several years the plugs swell and eventually drop out. The bodies made by attaching the panels with glue and screws from the inside have a smooth surface for paint and are more satisfactory in the end. Every body should be well braced and supported, especially at the corners and the support for the seat, to prevent it spreading. A good dash brace is essential.

The seat. There is a variety of types of seats; that selected should be considered

FIG. 35. Runabout

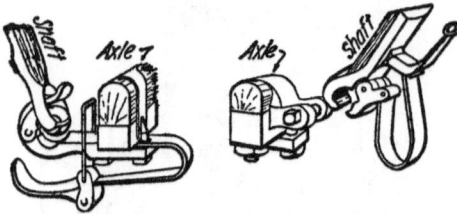

FIG. 36.   Two types of quick couplers

from a standpoint of comfort and utility as well as of beauty. There are plain driving wagon seats, solid one-piece, oval-back seats, stick seats, combination stick and panel seats, the automobile type and others. Some are made of wood with carved sunken panels, others of metal lined with hardwood. The design of the seat should not be considered an important factor in selecting a buggy, but it can add much to its looks. The style of buggy varies to a certain extent with the locality. In many places the automobile pattern of seat is most common, while in other localities a buggy with this type of seat could hardly be sold. Through the South and Southeast the seat with low panel sides and low panel back is still used. Every seat should be well ironed and braced and the cushion supplied with a sagless spring. A good cushion goes far toward making a buggy comfortable and easy riding.

Top. The top is often one of the first parts of a buggy to become shabby and go to pieces, because of the poor material used in its construction. The top sags because the bows have not been properly braced; or because some of the folding joints break on account of flimsy construction. The props should be of steel for strength. If covered with a poor grade of material, a leaky top results. The best buggies are provided with full leather tops. A cheaper top is the "quarter leather" type, which is a top with leather sides above the curtains, while the roof is made of rubber or oil cloth. The poorest grade of tops are covered entirely with rubber or oil cloth. The covering for both the seat and top should be in keeping with the style, construction, and quality of the vehicle as a whole.

Painting a buggy. The painting of a buggy is very important and should be done only by an expert. The best results can be obtained only by using the best of materials and allowing each coat to harden sufficiently before another coat is applied. In all 12 to 18 coats should be applied. Several coats of filler should be used and between coats the surface should be well sandpapered. The finishing coat should be of the best quality varnish made of pure gum. On all of the best grade of buggies the outside irons such as steps, dash braces, and toe rails are given two coats of japan and thoroughly baked. This treatment prevents the unsightly appearance of rust on these parts shortly after the vehicle has been put into service.

FIG. 37.   Cross-section showing construction of Sarven wheel as used on light vehicles.

Care of Buggies and Carriages

The first requirement in the proper care of a buggy or carriage is to have it properly sheltered from the sun and rain when not in use. The common practice on many farms of keeping the carriage in a space provided in the barn next to the stables is not a good one, for the ammonia in the manure fumes is harmful to the varnish. It is best, therefore, to provide a shed apart from the barn in which to keep the carriages.

To keep a good finish on a buggy, mud should not be allowed to dry on its surface. Care must

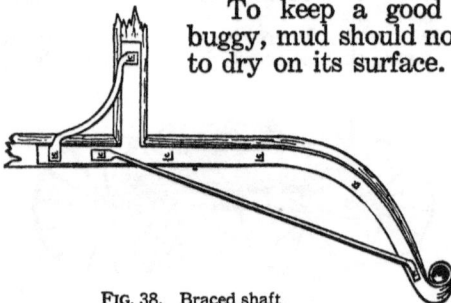

FIG. 38.   Braced shaft

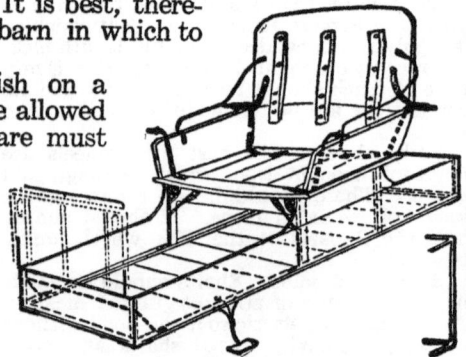

FIG. 39.   Piano-box body showing metal bracing

be observed in removing mud and grit to avoid scratching the varnish.   It is best to soak the mud loose and rinse it off by means of a hose or by dashing water on the surface with a bucket.   The use of cold, soft water is best.

To clean a varnished surface or leather dash that looks "smoky," use a mixture of 1 part linseed or olive oil and 4 parts good vinegar.   See that all dirt is removed from the surface and that it is dry, then apply the mixture liberally with a soft cloth until the surface is thoroughly wet; finally wipe clean and polish with a dry cheese cloth.   The mixture must be shaken frequently while being used.   Where the buggy has been used for a year or two the dash, top, and seat will become marred and discolored.   To restore a new black appearance a dressing, such as is made by a number of large varnish companies, should be applied.

If a buggy is used a great deal, it should be revarnished each year.   Before revarnishing see that all parts are tightened, then rub the surface with ground pumice stone and water until all the gloss is removed and a clean, smooth surface is secured.   Have the surface thoroughly dry, then carefully apply a new, even coating of varnish.

All clips and bolts should be kept tight.   The buggy that is neglected along this line becomes in a few years a veritable "rattle trap."   The king bolt and attachment of the fifth wheel to the gear should be watched.   Grease the fifth wheel occasionally, but do not apply too much oil as it will collect dust and cause undue wear.   It is well to always clean the spindle before it is greased.   A trouble sometimes experienced is that the boxing in the hub becomes loose; this occurs mostly in the cheaper grades of vehicles.   Care must be observed in driving the box back into place to prevent breaking it.   This can be done after proper wedging by using a piece of hardwood placed on the box and by driving with a heavy hammer.

Buggies that have been used a number of years can often be put in first-class condition at a comparatively small expense.   The chief thing required is a little interest on the part of the owner.   A new dash and probably a few other parts will need to be replaced.   After all parts are thoroughly tightened, go over the surface with sandpaper then apply one or two coats of prepared carriage paint which can be secured at a reasonable price.   The results will be most gratifying.

Selecting a buggy or carriage.   In selecting a buggy or carriage many factors should be considered.   Its usefulness, that is, whether or not it will meet the needs of the family, is one of the first.   A carriage should be selected that has some style.   It should be strong, well-built, and have ample room.   It should also be easy to pull and have easy riding qualities.   It should be well designed with all parts of sufficient strength.   Since it is almost impossible to recognize all of the good and poor qualities of a vehicle when examining it on a dealer's floor, it is best to select a carriage that is known to be of good quality and manufactured by a company that has been in the business a great many years, and that is likely to stay in it for some time to come.

CHAPTER 3

The Use and Care
of Farm Harness

By J. L. EDMONDS, *Assistant Professor, Horse Husbandry, College of Agriculture of the University of Illinois, where since 1910 he has had charge of the horse breeding, the experimental horse feeding, and one of the Animal Husbandry Department farms. He graduated from the Ohio College of Agriculture and then spent 2 years in charge of horse work at the Minnesota College. He has obtained additional practical experience by working on a number of general farms, and the following large stock farms where horses are an especially important feature: Irvington Farm, Sewickley, Penn., Meadow Lawn Farm, St. Cloud, Minn., and Oak Hill Farm, Aldie, Va.—EDITOR.*

S KILL in the selection, fitting, and care of farm work harness contributes much to the economical use of horse power on the land. The horse that works comfortably will do his work on less feed than will the horse that is irritated by poorly-chosen and improperly-fitted harness; furthermore, it is less likely to form bad habits. Injuries caused by ill-fitting harness increase the possibility of a horse becoming restless when being harnessed and hitched, and may lead to the more serious vices of balking and running away. Making a good appearance with a team has a tendency to improve not only the horsemanship of the driver but also the quality of his other work. Because of the direct and indirect benefits dependent upon it, this subject of harness is worthy of careful consideration.

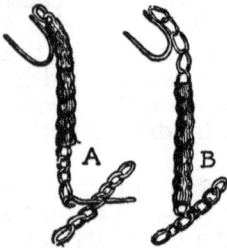

FIG. 40. Trace supporter; the sliding leather sleeve secures the hook.

Harness making. Harness values are quite largely determined by the quality of the material and workmanship employed in their making. The best harness leather is made from steer and heifer hides which are free from cuts, scars, and rough grain. Packer steer hides make the best grade of harness leather; country butcher hides make the cheaper grades, the reasons being that the best cattle usually go to the larger markets, and the big packers located there exercise more care and skill in taking off and handling the hides. After the hair has been removed, these hides are carefully tanned by being soaked in a series of bark baths of gradually increasing strengths. Most harness leather at the present time is tanned with combinations of materials blended in varying proportions; the principal ones employed are oak bark, hemlock bark, chestnut wood extract, and quebracho extract. There is very little harness leather which is either pure oak or pure hemlock tanned. It takes approximately 5 months properly to tan and finish harness leather, although the cheaper grades are rushed through faster.

The quality of the leather is largely determined by its feel, and can be much more readily judged in the side than after being made up into harness. Good leather has a firm and mellow feel. The strength of a strap depends upon the part of the hide from which it is taken and upon the way it is cut. Neck, belly, and leg pieces have not the strength of those cut from the back and side; cutting across the grain very materially weakens the strength of a strap. For reasons of economy, the harness maker sometimes uses weak pieces where they should not be used. Because these things are not easily determined after leather is made into harness, the honesty of the manufacturer counts for much.

Metal parts. Metal mountings, buckles, etc., vary greatly with respect to their design and the materials from which they are made. Commonly, they are either of brass, japanned white metal, or nickel-plated. Brass is the showiest material and is almost universally used on heavy show harness and harness of which the use has advertising value. Too much brass on a farm harness looks out of place, and its proper care requires time and labor that might be more profitably used elsewhere; brass does not rust, but tarnishes quickly. Where the teamster has but a single team to look after, as is often the case in the city, this brass may well be worth the effort needed to keep it bright because of the extra attention it attracts. On the farm harness, it is well to confine the brass to the rosettes and brow band on the bridle and, possibly, to buckle shields; the balance of the hardware may well be japanned, thus presenting a neat appearance without requiring an undue amount of time for cleaning mountings. In this connection it should be said that the use of large numbers of celluloid rings is to be discouraged as they do no good but are in the way, and also add useless weight to the harness. In general, the same considerations hold true for the farm buggy harness as for the work equipment. Genuine rubber mountings or trimmings (which are really iron parts covered with hard vulcanized rubber) are to be preferred on light harness. A harness so trimmed is not only more easily kept clean (since wiping off is all that is required) but also looks better than the one on which cheap nickel and brass mountings are used.

Hand vs. machine made harness. The best-appearing and highest-priced harness was formerly made throughout by skilled hand labor, and some high-class custom shops still make their best harness that way. A point is finally reached, however, beyond which additional hand labor adds mainly if not solely to the appearance rather than the strength of the harness. From the standpoint of utility, the use of improved machines for stitching is entirely satisfactory. They draw stitches tighter and lock them more securely than can be done by hand. In these machines the thread is worked through hot wax which thoroughly penetrates the thread. With hand-stitching the wax is put on cold, and hence gets no farther than the surface. Hand-stitching is always necessary in the case of small parts, such as around buckles, rings, etc. Plain stitching should always be chosen. The employment of scrolls and other fancy designs detracts from the appearance of the harness and adds nothing to its strength. In the end, and all along the line, it pays to pay for good material and good workmanship. The so-called cheap harness always looks the part and is short-lived.

Harness styles. All harness should be neat and appropriate in design and so constructed as to be heavy enough to withstand severe strain. Some farm harness, however, are made heavier than need be; this (especially with regard to those parts which do not bear the heavy strain of the load) involves needless expense and makes them cumbersome. The particular style of harness to be chosen depends upon the use to which it is to be put. Prevailing styles, if by chance there be enough similarity to permit this statement, vary greatly in different sections of the country, sometimes with considerable reason and sometimes without much. A mountainous country requires a different style from that which might serve well in the level plains country. Again, a harness of few parts might answer the purpose when used largely for plowing, harrowing, and similar jobs, while much pulling and backing of heavily-loaded wagons would require a much more complicated outfit. Obviously, much would be gained if a style well suited to local requirements were selected for use on a given farm and all harness purchased thereafter were chosen to conform closely to the original pattern.

Fitting Harness

An ill-fitting harness lessens both the

FIG. 41.   Plain ring bit

FIG. 42.   Stiff mouth piece, half check Dexter snaffle.

FIG. 43.   Jointed mouth piece, half check Dexter snaffle.

FIG. 44.   Rubber mouth piece

FIG. 45.   Humane bit, a very easy type, used for breaking colts.

FIG. 46.   Double bar, twisted wire bit, used for horses hard to hold.

FIG. 47.   Rockwell bit, a very useful bit for hard-mouthed horses.

FIG. 48.   J. I. C. bit; a severe bit

FIG. 49.   Bit to prevent tongue lolling

FIG. 50.   Over-check bit

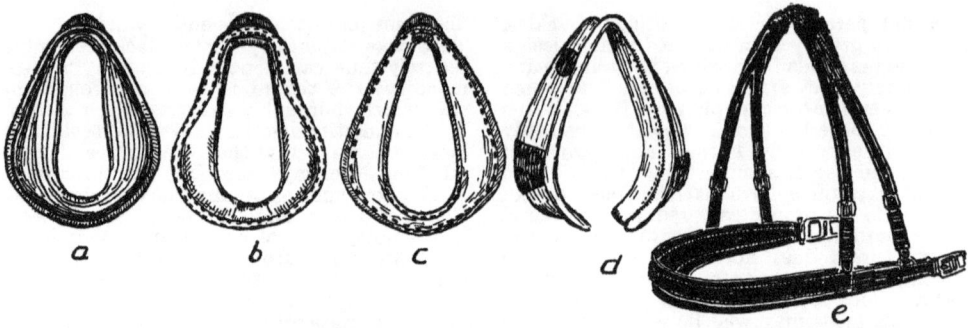

FIG. 51. Several types of collars: *a* well-made leather collar; *b* half sweeney collar; *c* narrow-topped, heavy mule collar; *d* cheap, open-throat plow collar; *e* heavy breast collar

quantity and the quality of work a horse is capable of performing; and, as has been pointed out, it may be the means by which an honest, free worker is made into an unreliable one, or even into a balker. Proper attention to fitting harness is particularly important in the case of farm horses; the working season in the spring is so short as to make it imperative that all losses of time due to poorly fitting harness be prevented.

The bridle. The factors which govern the fitting of the bridle are the shape of the horse's head and his disposition. The cheek-pieces should be so adjusted that the bit will not be so low in the horse's mouth as to bother him and permit him easily to get his tongue over it; on the other hand, it should not be so high as to raise the corners of the mouth and pinch the cheeks. It is generally advisable to adjust the bit rather high in a young horse's mouth so as to prevent his getting his tongue over it. The best plan is to fit the horses that are worked regularly with strong, jointed snaffle bits. Good horsemen, however, find it necessary, occasionally, to resort to something more severe, such as a double-twisted wire or a "bicycle" bit. Needless to say, pressure should be strongly applied to these bits only when the horse attempts to bolt. Mouths are easily ruined by severe bits used by unskilled drivers. The

brow band must not pinch the thin skin at the base of the ears. Blinkers or blinds must be kept in place and must not fit too closely in front. The propriety of training and working some horses without blinds is not questioned. However, experience with large numbers of work horses has convinced the writer that the great majority of horses work more pleasantly and are less likely to "loaf on the job" when blinds are used. The moderate use of side check-reins or plain bearing-reins is to be recommended for work horses to prevent their getting their heads down to eat grass when stopped, and also to prevent their bridles being caught on the end of the pole.

The collar. For heavy work, well-made leather collars give the longest service. No part of the harness deserves more careful fitting than the collar. With the horse holding his head in the position in which he keeps it when at work, a collar should so fit that, when pressed firmly back with the hands, it has an even contact or bearing against all parts of the shoulders, and leaves enough space at the windpipe to insert the flat of the hand. By making use of some one of the many different styles of collars, and keeping them clean, it is possible to fit properly almost all horses and keep their necks and shoulders in shape for work. Hame straps should be properly adjusted and buckled as tightly as possible at the bottom; failure to do this has spoiled many new collars. A short trial will show what adjustment of the hame tugs is necessary to bring the pressure at the proper points. If a new leather collar is wrapped over night with wet gunny sacks before using, it will shape to the horse's neck much more quickly than if not so treated. A considerable saving in collars will

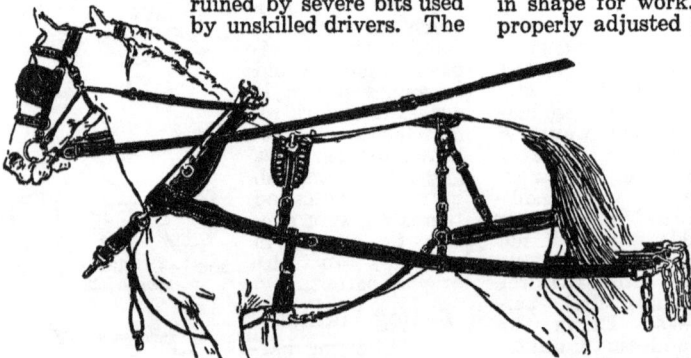

FIG. 52. A team harness with back pad and strap

result from putting them on and taking them off over the head rather than by unbuckling them at the top, as is generally practised. Such careless handling is the reason for many collars breaking at the throat.

Sweat pads are a necessary evil since horses undergo considerable shrinkage in weight during some seasons of the year, when it becomes necessary either to use a sweat pad or to change the collar. The cutting away of portions of sweat pads sometimes relieves galled spots. A still better way of handling such trouble is to use a small cork-filled pad between the collar and the sweat pad. Wabash pads are useful, in the case of sore necks, to remove the pressure of the collar from the injured surface. The use of smooth deer-skin pads or zinc pads does much to protect the neck from becoming galled.

Breast collars are useful for light work. The shoulder strap should be so adjusted as not to allow the breast collar to interfere with the windpipe, or be so low as to hinder movement. Any extra-heavy breast collar which has been lined with sheep skin is useful in place of the regular work collar where necks or shoulders are galled. Draft stallions frequently develop so much neck and shoulder that they can be best worked in a heavy breast collar. Patent horse collars, referred to as "humane collars," are useful for field work, and as a change in case of sore shoulders or necks.

The right adjustment of some of the other parts of the harness may be worth mentioning,

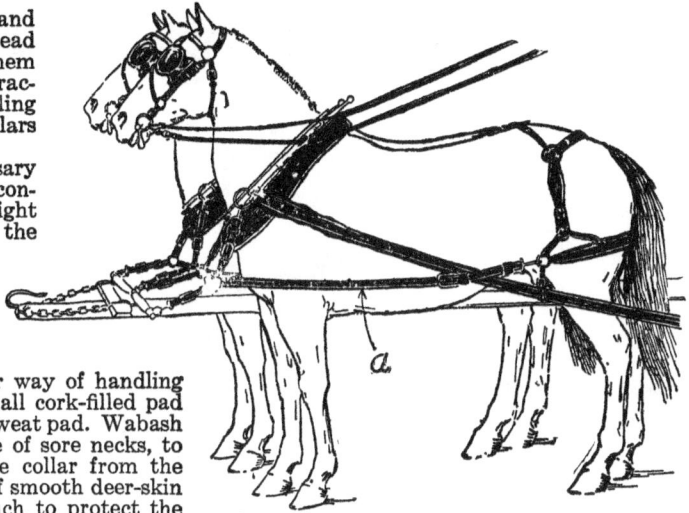

FIG. 53. An extra-heavy truck harness made with "Boston backers" (a) which are good equipment where heavy loads are to be backed or held back when going down steep grades.

although it should come easily. The saddle should fit the back, and the back strap should not be too short. The crupper, when it is present, should be of good size, smooth, and well-stuffed. See that the breeching is not too low or too tight. Careful observation of the methods employed by skilful teamsters will be of much aid to the young or inexperienced farmer. In putting a horse to a vehicle, remember that the lines should be taken down and adjusted first. The careful observing of this right order in "hitching up" has prevented many accidents.

The Care of Harness

At the outset, it should be said that a harness cannot be properly cared for unless one has a suitable place in which to hang it. In damp stables it molds quickly. The presence of mold indicates that moisture is taking the place of the oil, upon which the life of the leather depends. A harness should not be stored where the ammonia from the manure can reach it. However, in regularly cleaned, airy work-horse stables, most of us prefer to have the harness hung on a hook back of each horse; or by means of a rope and pulley, to haul it up and out of the way on the post at the rear of the stall partition. In stables where a number of teamsters are employed, too much time is wasted in going to and from a central harness room; such a room should be provided, however, for the storing of supplies and extra sets of harness. In this room should be a bench and materials for making small repairs. Valuable harness should be kept in tight cases in a room where there is some artificial heat.

At least twice a year all work harness should be entirely taken apart (particular attention being paid to the straps at the buckles) and then cleaned and oiled. At these times, all needed repairs should be made. In cleaning harness, as little water as possible should be used; warm, soft water is best, but hard water may be used if a handful or two of salsoda is added to each tubful. Some harness is so dirty that sponging alone will not remove the dirt; in this case it should be soaked for 15 minutes, then scrubbed with soap and a brush, rinsed, wiped with rag or chamois, and hung on a wooden horse to dry in a warm place but away from the stove. As soon as it is dry, apply neat's-foot oil (diluted one half with a good standard harness oil, or about one fourth with kerosene) with a rag or a piece of sponge. Several applications of oil

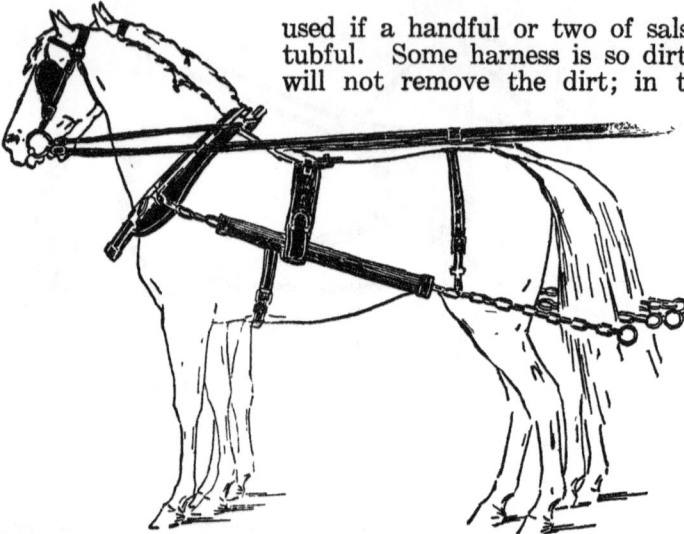

FIG. 54. A set of chain harness which is suitable for field work. They are, ordinarily, made for small horses and mules and have to be specially made when intended for use on horses approaching drafters in size.

are generally needed, and it will pay to rub it well into the leather with the hands. Neat's-foot oil, when used straight, is likely to cause the work harness to become too stretchy. It may be made black by adding a tablespoonful of lamp-black per pint. When the leather shows up very red after washing, give it a coat of edge blacking before oiling. Under no circumstances is it advisable to use a drying oil, such as linseed oil. Low grade vaseline is useful for smearing over a harness which is to be stored for a considerable length of time. After the oil has soaked in, sponge the straps with a good grade of castile soap. The frequent sponging over of a harness and the use of any of the good dressings are to be recommended.

When a brilliant black finish to the harness is desired, it becomes necessary to use some one of the standard harness "compositions" which are quite similar to the best pastes used for polishing black shoes. There is, in fact, no objection to using shoe polish, except for the extra expense entailed in purchasing it in small boxes at retail stores. The paste should be evenly applied to the harness with a dauber then polished with an ordinary blacking brush, and finally with a flannel rag.

For cleaning the metal mountings, one may use most any of the liquid or paste brands of metal polish on the market. We

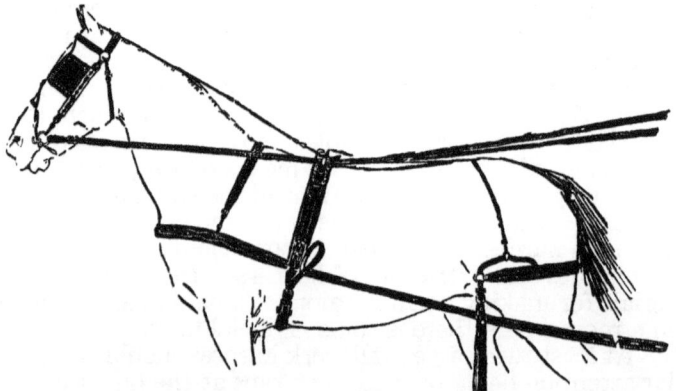

FIG. 55. A light, single-buggy harness. For use on heavy roads, it is better equipped with collar and hames instead of the breast collar as shown; a side check is more humane and on most horses just as useful as the over-check shown.

find the paste to be more economical because it does not evaporate nearly as readily as do the liquid polishes. Steel bits are cleaned by washing with soap and water, then smearing over with a cake of soap and polishing with silver sand. The soap film makes the sand stick. The fingers are of most service in rubbing the sand on the bits; a soft pine stick can be used in parts too small for the fingers. After sanding, rinse the bit, dry with a cloth, and burnish with a small steel burnisher. Forged steel bits are the strongest and also the best looking if they are kept clean. Careful drying and wiping with an oily rag after using will prevent their rusting.

It is so satisfactory to use a harness which is kept in first-class shape that one is surprised that more people do not make the effort to keep their equipment so. In the end, proper care will save both time and money. On the farm much of this work may be done at times of the year when work is slack. Minor repairs may be made at home if one has provided himself with a small repair outfit; those sold at moderate prices by the large supply houses are very satisfactory for small repairs.

Among accessories which contribute much to a horse's comfort are fly nets in summer and blankets in winter. Well-made cotton cord nets give satisfaction. Covers made of old gunny sacks are much to be preferred to none at all and should be used when economy is of first importance. Ordinarily, farm horses are healthier and better off generally, if not blanketed in the stable. A heavy blanket should, however, be used to cover the horse warmed up from work when he is stopped in the open for the taking on or putting off of a load. Heavy, waterproof, duck blankets should be worn by horses while worked in the rain in cold weather.

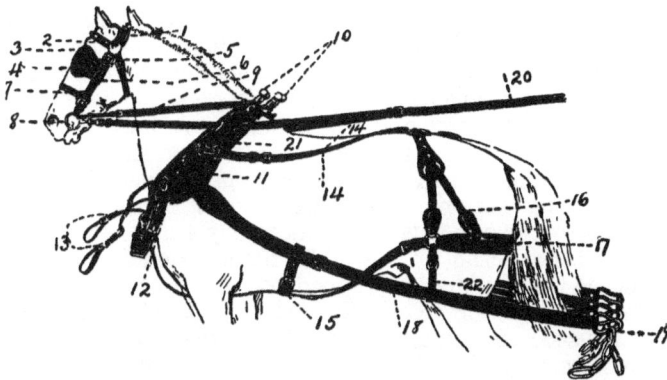

Fig. 56. Double truck harness: 1 crown-piece; 2 brow-band; 3 winker stay; 4 Concord blind; 5 throat-latch; 6 check-piece; 7 nose-band; 8 bit; 9 flat-side rein; 10 "Dandy" ball-top hames; 11 Concord bolt; 12 breast strap; 13 martingales; 14 forked back strap; 15 belly-band; 16 hip straps; 17 breeching; 18 traces; 19 heel chains; 20 lines; 21 leather collar; 22 lazy strap.

CHAPTER 4

Motorcycles and Light Automobiles
on the Farm

By C. V. HULL, *Charles City, Ia. Born on a farm in Ogle County, Illinois, he worked at truck farming while at high school, then went into automobile and gas engine work at Harvard, Ills., giving much time to farmers' outfits and problems. He spent two years in the Science Department of Depauw University, one year on the farm, and two more at automobile work and teaching school. For nearly eight years he was with the Hart-Parr Gas Tractor Co., doing engine and electrical work and, in addition, answering the inquiries received by its Service Department and solving the problems of gas engine and tractor owners. He has carried on a tractor engineering school, contributed to numerous farm and trade papers and in other ways developed the idea of making the practical study of gas engines and their use his lifework.*—EDITOR.

THE motorcycle. The motorcycle is speedy and inexpensive in operation. It requires but little storage room and can be used on poor roads. For these reasons it is adapted to the use of single men on the farm. On a motorcycle they can go to town easily and quickly after the chores are done; the farmer boy, too, can get away for an afternoon at a ball game or to go hunting. If a side car is used, two persons may go on a pleasure trip. In general the motorcycle may be used for pleasure just as a horse and single buggy are used, with the advantage that much greater distances may be covered in a given time.

As a business machine, the motorcycle is of great use when hurried trips to town are necessary. In sections where hired men insist on the use of a horse, a man who owns a motorcycle is entitled to more pay than others because he needs no horse. On the other hand, it is well to have a definite understanding with the motor cyclist, so that he may not spend too much time on the road.

The capacity of the motorcycle is limited to 2 or at most 3 passengers even when a side car is used; and unless the roads are good, the side car must be left at home and the 2 persons ride tandem. The motorcycle can be used but little for carrying produce, though it is a fine thing for the rural mail carrier. Few women ride motorcycles but the side car can be used by a woman quite comfortably if the roads are good. For its capacity, the motorcycle gives a great deal of pleasure and is very serviceable. Its principal limitation is lack of carrying capacity. The first cost of the motorcycle is reasonable and is the principal one, and if properly cared for, the machine entails but a small upkeep.

Before purchasing a motorcycle, the farmer should consider several factors. One is, of course, the first cost. Another is the amount of use he can make of it both for himself and for his family. If he is purchasing for the boys, he should consider whether they are capable of caring for the machine and using it wisely. A third factor is the real value of the machine for business purposes. In some cases it can be used for many trips and for long rides.

The motorcycle requires systematic care. The oil supply for the engine must be made up at regular intervals, otherwise the motor will have alternating feasts and famines of oil. The clutch and gears must be cared for according to the maker's instructions. The very best oil and grease must be used, because the motor is air-cooled. Besides, all parts of the mechanism are small and run at high speed. The magneto needs regular but slight oiling. Occasionally the contact points need cleaning and setting. All nuts and bolts must be watched and kept tight.

FIG. 57. The side car increases the usefulness of the motorcycle whether for work or pleasure

Speeding and fast riding cause many motorcycle accidents. The rider must watch the road closely because his machine is easily unbalanced. Curves must be taken with discretion, to avoid the skidding of the machine from under the rider. Mud and sand must be negotiated with care, otherwise a nasty spill may result.

When a side car is used, it must be so fastened that the front wheel is not pulled out of line.

The motorcycle requires careful handling if serious accidents are to be prevented. One should learn to ride by practising on little-used roads before he attempts to go where there is much traffic.

Light Automobiles

At the beginning of the present century, the automobile was an experimental machine; 5 years later it was a luxury; 5 years more and it was common, although its electrical equipment was a new feature; to-day the automobile is a practical, efficient machine. Automobile riding and touring are popular pleasures. More than that, light automobiles are much used for business purposes. This is as true of farming as it is of city work.

For real pleasure the farmer can buy nothing equal to a light automobile. With it the whole family can go to town for an entertainment or any special day; and they can go comfortably and quickly with the assurance that the horses at home are resting. The farmer gains new ideas from trips in his car. The opportunity to go and see and learn is one of the greatest benefits which come to the farmer from the ownership of a light car.

The light car is a boon to the farmer's wife. Also in many cases she learns to drive. In this there is a double advantage: the wife can often get away for a half day for some neighborhood social affair, or the wife or daughter may go to town during the busy season and get supplies. This means a saving of time and, in some instances, of crops as well. The first advantage is worth more than casual consideration because it is absolutely necessary to relieve the monotony of the lives of farm women. Until the coming of the light car, the life of women on the farm was in many cases very monotonous and unpleasant. Beyond question the pleasure which comes to farm women because of the automobile is a vital factor in farm economics.

For the farming business the light car is very valuable. It saves time and horseflesh, and often makes actual money for the owner. One has but to visit

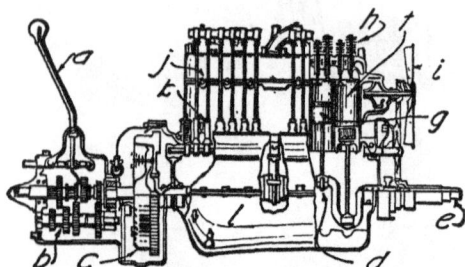

Fig. 58. Section of a typical 6-cylinder engine showing important parts: *a* transmission lever; *b* sliding gear, selective transmission; *c* flywheel, and cone clutch combined; *d, e* crankshaft; *f* cylinder with valves (*h*) in head; *g* piston; *i* fan of cooling system; *j* spark plug; *k* rod operated by secondary shaft operating valve.

the towns and cities of the agricultural sections to see the use of automobiles for farm purposes.

If the farmer is near a city, he can often use his light car to deliver seasonable produce. The market gardener or truck farmer can make quick deliveries with fresh goods. Much better time can be made with the automobile than with horses, and the supplies reach the market in much better condition. Besides, more time can be given to actual field work, and the horses are fresher for it.

The saving of horses when a light automobile is used is a very considerable item. Several years ago three young farmers in northern Illinois bought light cars. They were used for everything but the heavy hauling, groceries, light supplies, eggs, poultry, and light produce being carried in them. One of these men declared that they had more than saved the cost of the cars in the increased value of the horses and the greater amount of work which could be done with them. Cars are used so generally by some Iowa farmers that the heavy teams are seldom driven to town. The same statement is true of sections of the Dakotas, Minnesota, Wisconsin, Illinois, and Missouri.

The automobile is also employed by the farmer to attend auction and other sales. By its use the owner enlarges his market and increases his buying power. In this way he sees new breeds and strains and often comes in contact with men who have ideas differing from those of men in his own immediate neighborhood.

The value of the education which comes to the farmer from the labors of county agents and university extension men cannot be estimated. It is to be hoped that this work may be extended to many more states. It is giving the farmer a finer and better conception of his work. Imagine, if you can, a 25-mile trip with horses to the various farms in a county! But for the automobile, such trips would be out of the question.

Disadvantages. There are, however, some disadvantages which must be considered. The amount of money invested in addition to the horses used is an item of expense in interest and depreciation. One Iowa farmer who drove a light car disposed of it because he thought it less profitable than horses. He had a number of colts at all times, and used them on the road while breaking them. He contended that his horses needed the exercise and that he made more money with them than he could with a car. His conclusions are worth considering, especially where the farm can sustain a large number of horses.

The road problem is in some sections a very serious one. In the northern part of the Mississippi Valley the winter seasons are so severe that the car must stand idle, with interest and depreciation adding to the mileage costs. In justice it must be stated that some light cars can be used practically all the time that there is real need for them in the sections with bad spring roads. The fact that Iowa with notoriously bad roads has one automobile to every nine inhabitants indicates that road conditions are not so serious as they seem.

The purchaser of a light car must not attempt to race or speed with it. Such practice is unsafe even for the most skilful drivers. Light cars are very hard to control and often turn turtle or skid badly at high speed. The driver of a light car should compare his speed with that of the horses he passes rather than with the big 6-, 8-, and 12-cylinder cars which pass him. High speed means that the driver must watch more closely and move more quickly. The danger is greatly increased because the car tips and rolls more easily.

The notion that a car runs more steadily and hugs the road better at high speed is a faulty one. Naturally any object moving at high speed tends to travel in a straight line, even if it is passing over a rough surface; but if a car strikes an obstruction or turns over at high speed the theory is quite apt to leave a twisted mass of steel, iron, and wood. If any one caution should be printed and pasted on the wind shield it is that most familiar one, "Safety First."

Overloading is also to be avoided. While the car may for a time stand the hard service, parts are strained and will sooner or later break; and in case of an accident an overload of people adds greatly to the chances of serious injury to the passengers. Besides, the depreciation is greater than it should be.

Neither is it advisable to use a light car for heavy dray or tractor work. The general opinion of automobile men seems to be that a truck must be specially designed for its work, and that heavy hauling and plowing should be done with tractors. The farmer should realize that the automobile is not a general-purpose machine.

Unnecessary use means expense. The automobile on the farm should not be used any more than is necessary. It is a waste of time and money to make two trips to town when one will do. So many people, both urban and rural folk, seem to think that it makes no difference if something is forgotten because it is so easy to go again for it. This idea may soon lead to the habit of using the car much, that the time which it saves over the use of horses is more than lost in the number of trips made. Of course if one rides for pleasure he may look at the matter differently; but for the farmer who considers that his car is a part of his business, and who uses it to add to his profit, the habit of using the automobile to excess will not help to pay the dividends. It will, on the contrary, soon prove to be a decidedly expensive practice. If the automobile is really to pay the farmer, he must plan his work with it as carefully as he does that done with horses.

Another item which must be considered is that of expense per mile. Horses, whether idle or in use, must be fed and cared for. The expense per mile or per day must be figured for 300 days or perhaps for 365 days per year. With the automobile, there is an actual cost for gasoline, oil, tire wear, and general depreciation while running. The purchaser of a car should realize this fact. While it costs practically nothing to drive the horse, every mile that a car travels means as much cost which would not arise if the automobile were idle. This cost, of course, varies with different automobiles and drivers. The best of care and capable driving cannot eliminate it. Opinions and statistics as to the actual cost per mile vary so much that accurate figures cannot be given. However, it is advisable before purchasing to find out something about the cost of running a car, interest and depreciation considered, as well as the

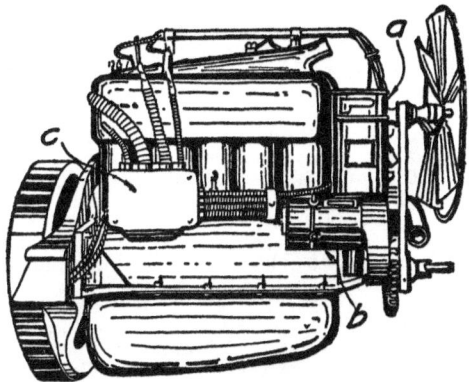

Fig. 59. Side view of a 4-cylinder engine showing the vertical generator (a) and the horizontal motor (b) of a two unit outfit, and the electrical wiring junction box (c).

actual mileage cost. This must be done if the car is to be used as a business unit.

What Car to Buy

When the various advantages and disadvantages of a light car for the farm have been considered the farmer naturally asks, "What car shall I buy?" The neighbors can offer many suggestions, and naturally each will have his favorite car. Usually there is some basis for the various opinions. The farmer who is planning to purchase a car should consider carefully the various features of one which can be used for both business and pleasure.

Weight. In the first place the automobile should be a light one; for a light car can travel over worse roads than a heavy one and will make better time with less fuel. If the roads are heavy with mud, the large automobile is very hard to steer. If it gets off the beaten course to the side of the road, it is very hard indeed to get back. The light car is not so apt to get off the beaten track on a wet and muddy road, but if it does it can usually be brought back without great trouble. A great many commercial travelers use light automobiles, because with heavy ones they are not able to get over the roads after a rain. Also, the light car is much cheaper to buy and can be run for less money than a heavy one. The difference in gasoline, oil, and tire costs is a considerable one. From a purely business standpoint, no farmer can afford to buy a large car owing to its greater upkeep costs.

Regarded purely as an investment, the farmer's car should not be an expensive one. It should be such a car as can be replaced every 3 or 4 years without great loss. It should be a car in which produce, eggs, and light supplies can be hauled without serious injury to the seats or finish. With fair handling there is no reason why a car should not be used for both business and pleasure. Generally speaking, the farmer cannot afford to own a car unless he does some hauling and carrying with it; hence the advisability of getting a reasonably cheap light car.

Simplicity. Simplicity of construction should be a prime consideration. This means that the various parts should be easily inspectable and cared for. The farmer who has not many tools will find it difficult to make even simple repairs unless the automobile is "get-at-able." The various oiling devices should be simple in operation and easy to care for. The spark coils, switches, and electrical equipment should be so located that they can be easily examined. The electrical equipment of the farm car should be as simple as possible.

Preferably the general design of the farmer's car should be simple and plain. Unnecessary tool boxes, luggage carriers, and lights should be accepted with caution. Any unusual type of body will make it harder to use the car for general work and to dispose of it when a new one is desired.

Strength and durability. It goes without saying that the car must be strong, even if it is light and simple. The farmer who plans to use his car for both business and pleasure must have a machine which will stand more abuse than one used for pleasure only. It should be strong enough to haul light loads or even to pull a trailer in an emergency. Also, the construction must be heavy enough to permit the machine being used on rough and bad roads without frequent breakdowns.

Durability is also a prime factor. The average farmer cannot replace his car every second or third year. He must buy a durable machine if he expects to get profitable service.

This factor can be easily determined by a canvass of the machines in the neighborhood. In many cases, the old cars are not really durable but have lasted well because of a heavy expense bill for repairs. A durable car is one which stands hard service without excessive repair bills.

If the car is to be used for hauling and other business work, it must be dependable. The proper course is to judge the dependability of any automobile by the average action and life of the machine and then to buy a car of a make which is giving good service in the hands of several users. However, the reliability of an automobile depends to a surprising extent upon the ability of its caretaker. No car can be expected to be reliable unless it is properly cared for.

The farmer's automobile must be serviceable unless it is bought purely for pleasure. While it may not be advisable to use the automobile as a dray, it ought to be of use in a hundred other ways. A serviceable car is one which can be used for going out to the field, down to the pasture, or up to the wood lot. It is not truly serviceable unless a few light articles can be loaded on it and taken where needed. It would be a mistake to drive a team a mile or two to repair a fence when the few tools required could be loaded in and the fence repaired in a third of the time by the use of a car. It is a serious blunder to tie up a lot of money in a car which is too nice for practical farm work. If a car is to be a paying investment, it must be serviceable.

Cost. The cost of the car is one of the first items to be considered. A National Tractor demonstration was held in 1916 at Cedar Rapids, Iowa. Thousands of farmers came in cars. These cars were of various prices, but very few were of the really ex-

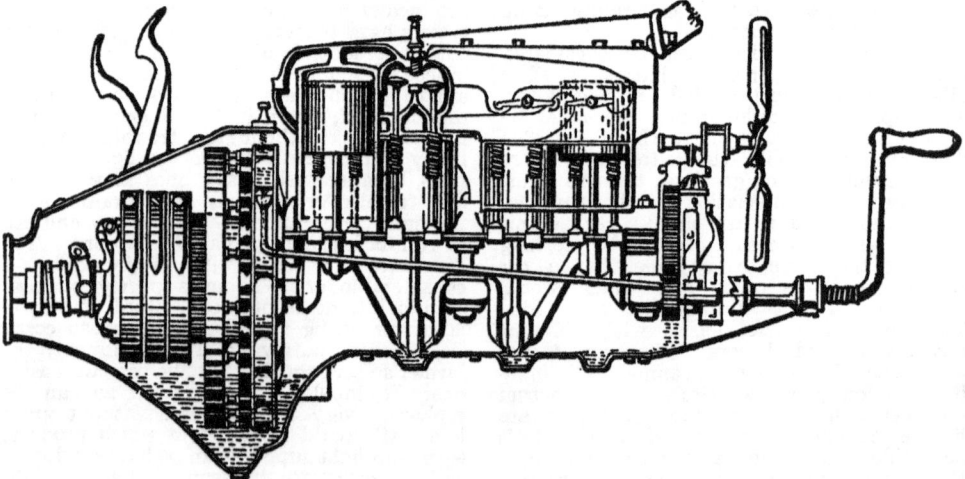

FIG. 60. Section of a popular 4-cylinder engine showing a finely designed planetary transmission using a multiple disc clutch. Compare Fig. 58.

pensive class. Even in the prosperous Mississippi Valley, there are not many farmers who drive cars listing over $2,500. The practical car for the farmer is one which costs much less.

From a practical standpoint, the automobile has no place in farm economy unless it effects a saving. That a high-priced car can be used enough for actual farm work to pay, is hardly possible, so from an economic point, the cheaper automobiles are the better for farm service. Of course the farmer who buys for pleasure and can afford to do so ought to buy a comparatively large car because it will ride easier and go faster than a lighter, cheaper one.

Fortunately several makes of light, reliable cars listed at $1,000, more or less, are now on the market. In many cases, these smaller cars have proved more durable than the larger, more expensive makes. They are strong and efficient. They have been sold long enough to give ample opportunity to judge the various features in their design and to tell whether they are suited for any particular section of the country. Most of the reasonably priced cars meet the requirements of the average farmer. They are serviceable and finished so well that they can be used for pleasure as well as for farming purposes.

Naturally the farmer who plans to purchase a car first decides how much money he can afford to invest. For the man who plans for $600 or less, the choice lies between second-hand cars or a new one without electrical starting equipment. Generally, though exceptions may be found, the second-hand car does not pay. The electrical starting device is a great convenience but not a necessity. Probably half the cars used by farmers are of a make the engine of which must be cranked by hand. In some cases, cars of this make are equipped with extra electrical parts for power starting. However, this adds to the complication and increases the weight as well as the cost of the machine.

There are a number of makes in the $750 class. These are in most cases very satisfactory machines, being fully equipped electrically for lighting and starting. Cars of this grade are neat and trim in appearance and well finished. Farmers can get excellent service from cars costing about $1,000.

Another popular class is that of cars listing at about $1,500. In most cases cars of this class have longer wheelbases and a more roomy body than those of the above-mentioned classes. They ride easier and travel faster with comparative safety. There are more refinements in the body and chassis, some of which are, however, more or less unnecessary. The expense of operation is greater, while actual service is not increased. Of course the car rides easier, the body is roomier, and the cushions are deeper. But none of these advantages adds to the efficiency of the car for farm use. However, the real question to be decided about the car of this class is whether the farmer can afford to purchase and run it.

Cars costing from $1,500 to $2,000 constitute the last class which the farmer can afford to consider, unless he buys the car solely for a pleasure vehicle with the expectation of spending considerable money for its operation. An automobile of this class is quite heavy. Muddy roads and rainy weather will often prevent its use. Tires, too, become a heavy item of expense both for upkeep and renewals. The gasoline consumption may run to 3 times that of the light car in the first class. Because road conditions may prevent the use of large cars, they are not so serviceable and dependable as the smaller ones. Cars of the $1,500 class are rather heavy, though it must be conceded that they are comfortable and "classy."

The big heavy car is out of the question for the farmer. It costs too much, is hard to handle on poor roads, and requires too much attention. The farmer must consider all these things before he buys his car. First cost, upkeep, depreciation, and interest are important. If the farmer considers them he will logically purchase a light, reasonably priced car which he can use rain or shine most of the year.

Points of a Good Car

The engine. When the decision to purchase has been made there come problems of design and construction. One salesman recommends the 4-cylinder motor. He states that the best-known machine in the world has a 4-cylinder motor, and that many of the old, experienced companies still continue to build 4- and 8-cylinder motors. He talks glibly of the simplicity and compactness of the 4, and assures the customer that it is bound to be the final type in standardized cars. All of this is more or less true, yet there are other types to consider. The 6-cylinder motor is very popular. It is theoretically a well-balanced motor and may run at high speeds without excessive vibration. The

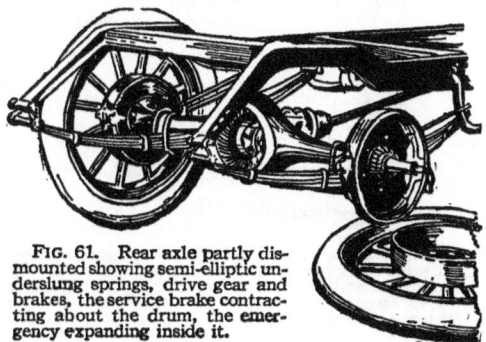

FIG. 61. Rear axle partly dismounted showing semi-elliptic underslung springs, drive gear and brakes, the service brake contracting about the drum, the emergency expanding inside it.

torque or turning force is very steady. The long-stroke, 6-cylinder engine is a good puller at low speeds because of the greater number of strokes per revolution. The cylinders are smaller than they are in a 4-cylinder of the same power and speed. The number of parts and their complication are not increased enough to be objectionable.

On the other hand, the 6-cylinder engine must be quite long. This may mean a longer wheelbase and added expense or a more compact seating arrangement. It also means complication in the ignition devices. There are more small parts to get out of order and cause misfiring.

The 8-cylinder engine has good balance, runs smoothly, and takes only the space of the 4. The torque is excellent and the engine is very flexible. It is, however, a complicated motor the cylinders of which must set at an angle. This makes an extremely efficient lubricating arrangement an absolute necessity. In the very nature of the case, the engine cannot be as simple and accessible as a 4- or 6-cylinder. The 12-cylinder engine is not a farmer's type.

After all, the important feature of the engine is its efficiency. So long as the engine is

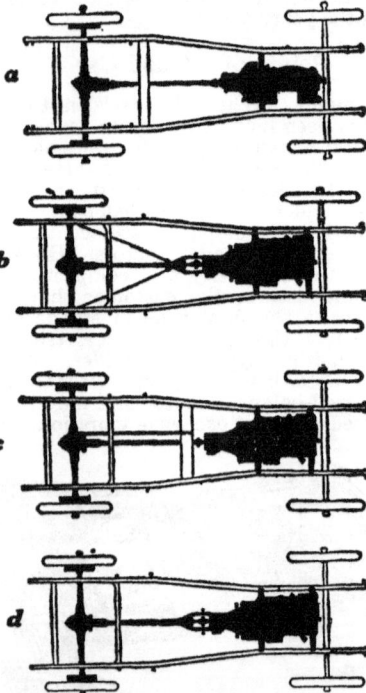

FIG. 62.   Types of drive: *a* Hotchkiss simple and light, giving drive and torque through rear springs; *b* drive by torque tube and radius rods strengthened by yoke; *c* drive by torque beam fastened to rear axle and a frame member; *d* typical torque tube and yoke drive. (Oakland Motor Co.)

dependable and durable the number of cylinders does not matter so much.

Transmission.   Two forms of transmission are generally found on farmer-owned cars: the planetary type and the selective gear type. The friction type has never become popular among farmers and the magnetic type is too expensive for most of them. This is true also of magnetic control of gear sets. In one popular car in which it is used entirely, planetary transmission is quite satisfactory. It is rather noisy, and the bands require occasional adjustment.

The selective type has the advantage of giving 3 speeds ahead instead of 2. In high-priced cars, 4 speeds are sometimes arranged for. The selective type gives a positive drive at all times and is comparatively quiet if kept in good condition. The purchaser should, however, make a careful examination of the transmission in order to be assured that it is well made and easily cared for.

Control.   The majority of light automobiles are steered from the left. Since the general custom is to pass on the right, left steering control makes it easier to gauge the distance between automobiles and rigs, though it is easier to see the roadside from a right controlled car. But with a left-hand steering wheel, the gear-shifting levers and emergency brake may be placed in the centre. The clutch is then operated by the left foot, and brake and foot throttle by the right foot. Also, the various switches and adjusting screws are more easily reached with the right hand. This arrangement is almost standard for medium-priced cars.

The well-known cheaper car with planetary transmission has a left-hand steering wheel. In this one, however, the hand-brake lever is on the left hand, while the clutch is operated by the left foot, the service brake by the right foot, and the reverse by either. The purchaser of a really good car of reasonable price must take this arrangement or get a higher-priced car.

The foot throttle should be convenient for the average man and easy in action. The preferable kind is one which moves quite a bit from light to full feed of fuel. If the car has no foot throttle, it should be equipped with one, to make it easier to handle in tight places.

Gasoline feed systems.   Gravity feed from fuel tank to carburetor is the simple way; but in some types the tank is so far from the carburetor that fuel does not flow when the machine is on a heavy grade. If the fuel tank is under the dash, and hence just above the motor, gravity feed will be good, but the fire risk will be increased.

The vacuum feed system is a device for conveying fuel to the carburetor. The arrangement is such that the vacuum in the intake manifold is used to suck fuel from the supply tank to a cup which feeds the carburetor. The vacuum feed system is really an

Whether for business trips or pleasure drives, the horse and carryall or democrat remain the preferred means of transportation for many farmers

However, the automobile, having been brought within reach of the average farmer, in both cost and simplicity of operation, is being accepted by him more and more readily

THE PLEASURE VEHICLE IS AN IMPORTANT FACTOR IN FARM LIFE. IT KEEPS COUNTRY FOLK IN TOUCH WITH EACH OTHER AND WITH THE OUTSIDE AS NOTHING ELSE CAN

The ox team is an almost extinct relic of the plodding but persistent progress of pioneer days

A twelve-mule-team load of grain in California

The sort of outfit that can do, at one trip, the hauling for an entire community

THE LIMITS OF WAGON TRANSPORTATION EFFICIENCY ARE DETERMINED ONLY BY THE EXTENT OF THE FORCES THAT A MAN CAN CONTROL. EACH OF THESE IS A ONE-MAN-DRIVEN LOAD

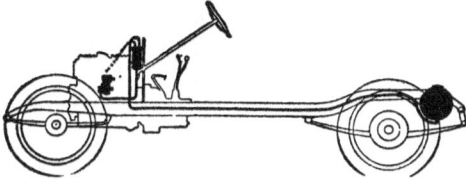

Fig. 63. Typical vacuum feed system. Fuel tank at rear and vacuum tank in front higher than carburetor. (Oakland Motor Co.)

intermittent vacuum pump whose valves are operated by the gasoline it pumps. This system permits the location of the fuel tank low down at the rear of the car. While it adds to the complication of the car it is dependable. It largely removes the danger of a destructive fire, and makes filling a simple job without much risk should gasoline be spilled.

Wheelbase. The wheelbase varies considerably in different makes. The light, comparatively cheap car has of necessity a short wheelbase, in some makes not much more than 100 inches. In medium types, the wheelbase is 120 inches more or less, while in the large, heavy cars it is as much as 144. The longer the wheelbase the easier the car rides, other conditions being equal. For general-purpose farm work, where an automobile must be driven over all sorts of roads and in tight places, one with a very long wheelbase requires too much space for turning. The farmer should get a car with a medium wheelbase. He may put shock absorbers on the short car and get a serviceable, comfortable machine.

Seating. The seating arrangement is a matter of choice. Probably a 2-seated touring car is best. It gives the most room and is a standardized construction. The back seat space can be used for carrying articles of considerable size and weight. Then, if the whole family wish to go for a pleasure trip, there is room enough. Beyond question the 5- or 7-passenger touring car is the one for farmers.

Electrical Equipment. If complete electrical equipment is desired, there are a number of reasonably priced cars which have it. In one group one motor generator is used. The arrangement of switches and gears is such that the same electrical unit is used to start the engine and to charge the storage batteries. This arrangement demands 2 driving devices for the unit and must be back-geared to the engine to get enough purchase to start it. But when used as a generator of current instead of a starting motor, it must run at high speed. Many ingenious devices are employed to get this double use of the unit. In a second group or type of cars, both motor and dynamo are used. In this case the motor is idle and not coupled to the engine except for starting. The generator is permanently

coupled up or chain-driven at all times. Both single- and double-unit systems give excellent results in service. The best plan for the man who does not understand electrical equipment is to judge it by the performance of cars which he can watch. No matter what car is bought, the electrical starter should be heavy enough to run the engine in cold weather. The storage battery should be one of a standard make. The majority of cars are equipped with 6-volt batteries, though 12-volt batteries are also used. There is less chance of arcing (burning out) with the former, but the latter is lighter for the same amount of work.

Tires. For convenience, demountable rims or quickly detachable tires are to be preferred to clincher types. However, in most cases, the lower-priced cars have but one style of tire equipment.

Springs. The cantilever-spring suspension is popular. The manufacturers who use it claim that the springs work in such a way that the wheels tend to roll rather than bounce over bumps. Some of the most comfortable cars have elliptical springs in various combinations and forms. One can best decide about springs by actual trips in cars with various types of suspension.

Steering gear. The irreversible steering gear is best. With this the front wheels can easily be turned by the steering wheel, although the steering wheel cannot be turned by the front wheels. This feature gives safety; for even if the wheels strike a rut, or the grip on steering wheel loosens they will not turn aside.

Cooling systems. Thermo-syphon cooling, in which the water moves through the system naturally, is good. As the water becomes heated, it rises to the top of the radiator, where it cools, settles to the bottom again, and passes to the cylinders. With this plan the rapidity of circulation depends upon the work the engine is doing.

With pump circulation the water is forced through the cooling system by a pump which is connected to the engine. While there is another part to care for, the water is positively circulated. The cooling plan and the type of radiator are not so important as certain cooling. The prospective purchaser should assure himself that the cooling system will maintain the proper temperature without evaporating too much water.

Lubrication. In lubrication the important thing is efficiency. The motor should be well lubricated at any reasonable speed without waste or carbonization. It is a good thing to have an oil indicator on the dash, so that one can easily tell whether oil is being circulated and fed properly. Both splash and force-feed systems are successfully used, as well as a combination of the two.

Clutch. Both cone and multiple-disc clutches are used on standard cars. The superiority of either depends upon how well it is made. Both give excellent service if properly cared for. The method of adjusting the clutch should be carefully studied. However, the kind of clutch is not so important as sure and easy action.

Rear axles. Rear axles are described as dead, semifloating, three quarter floating, and floating. In the dead axle the rear wheels are driven by gears or by chain sprockets which are mounted on a cross shaft which carries also a differential gear. This construction is much used on trucks and small chain-driven automobiles. In the semifloating type, the wheel bearings are inside the rear-axle housing, the shafts carrying part of the load. In the floating type, the rear-axle shaft bearings are outside the axle housing; the shafts may be removed by taking off the hub caps. The three-quarter floating construction is quite similar, except that the shafts are securely fastened to the wheels. About half of the well-known makes have floating axles. Because one make of light car is sold in so large

numbers, it is quite probable that more than half of all automobiles in use have the semi-floating rear-axle construction. The real question regarding rear-axle construction is, are the bearings strictly high-grade and put in a well-designed axle structure?

Brakes. These are decidedly important. In some models, both expanding and contracting brakes are used. Generally the service brake (that used for ordinary running), is of the contracting type. The emergency brake is quite apt to be of the expanding type. In one planetary type of transmission, the service brake is part of the planetary system. Pressure on the brake lever clamps the brake band about a drum which is connected to the longitudinal drive shaft. The emergency brakes of this automobile are of the expanding type. These types are all efficient if kept in good condition. However, the prospective purchaser should insist upon a trial of the brakes to make sure that they are strong enough to be dependable.

How to Drive an Automobile

One must have considerable experience to drive successfully. The new driver should write across the windshield "Safety First!" The first driving should be done away from traffic. A good plan is to block up the rear wheels, and practise starting, gear changing, etc., until one is thoroughly familiar with the operations. In any case it is advisable to practise driving, stopping, starting, and reversing until able to do them almost without thinking. Above all things, the new driver should go slowly. After one has driven a bit he will be tempted to drive at high speed. Yielding to this desire causes many accidents.

Change gears quickly. When changing gears, move the levers with a quick, strong motion. Do not hesitate an instant. Practise shifting until the gears can be engaged with a quick snap instead of a rubbing and burring. Always push the clutch pedal far enough to release the engine. In making a change of speeds, move the clutch and gear shift together. A good driver does this so quickly that the engine has no time to speed up. The clutch should be engaged easily. Slamming in the clutch is bad practice and strains the transmission. If the planetary transmission is

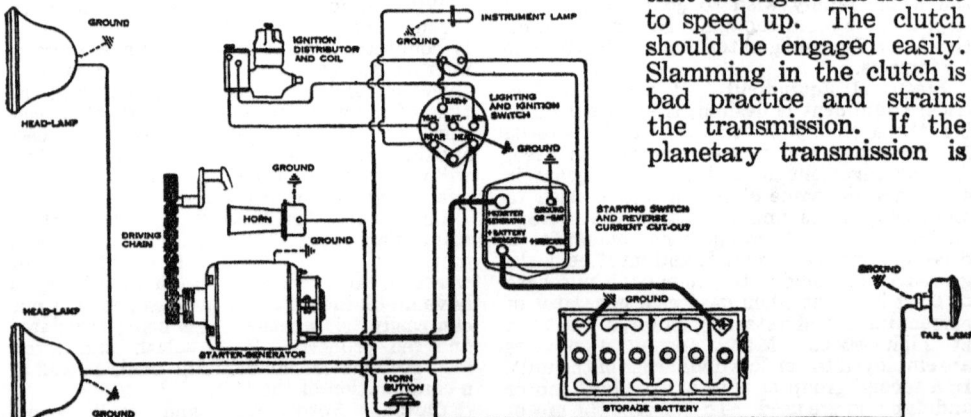

FIG. 64.   Wiring plan of electric starting and lighting system using a single unit for starting and generating. (North East Electric Co.)

used, the pedals must be learned so that they can be operated without confusion. In any case racing the engine or allowing it to run at excessive speed, is a mechanical blunder, especially if the operator lets the clutch engage quickly.

Care on the road. On the road the smoothest track should always be chosen. Experienced drivers often find the best track at one side of the main traveled track. If possible they avoid ruts or else travel slowly while in them. Naturally curves, hidden turns, and down grades will be taken cautiously, with the machine under control. Railway crossings, culverts, and viaducts are always dangerous, and the negotiation of them should never be attempted without looking carefully for possible danger.

One more suggestion for the sake of comfort: drive so that the people in the back seat are not bounced about. One is not really a good driver until he can handle an automobile so that the back seat is fairly comfortable.

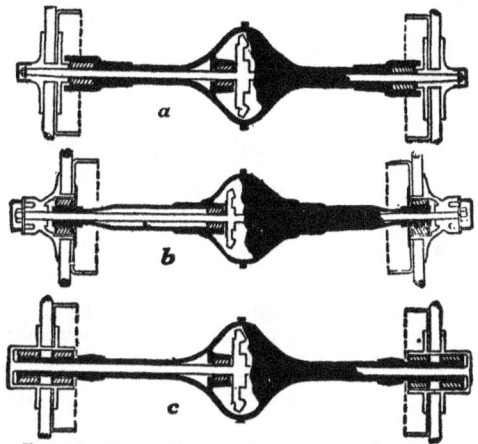

Fig. 65.   Types of rear axle: *a* Semi-floating, bearings at outer ends rest directly upon axle shafts, inside axle housing; *b*, three-quarter-floating, bearings on wheels roll on axle housing; *c* full-floating, two sets of bearings in each wheel supported on axle housing. Bearings shown by cross lines. (Oakland Motor Co.).

How to Care for the Car

Engine. Three things about the engine are important: lubrication, cooling, and adjusting. If the proper supply of oil is kept in the oil chamber, heating of bearings will seldom occur. An ample supply of water should be maintained in the cooling system at all times. Any unusual knocking or pounding should be located and the cause removed at once. If these 3 items are attended to, there will be little wear of, or damage to, the engine.

Bearings. The various bearings should be adjusted and oiled as the manufacturer directs. A bearing must be snug but not tight. If too loose, a plain bearing pounds; if too tight, it heats. Here again one must go carefully. Generally the inexperienced man will find that it pays best to have an expert adjust a loose crank pin or crank-shaft bearing.

Electrical equipment. Unless one is an electrician, it is well to let the electrical equipment alone unless dead sure that the proper adjustment is being made. Of course spark plugs must be cleaned and set, contact points must be smoothed off, and distilled water supplied to the storage battery. The average owner will make money and save time if he has a competent man to care for the magneto, battery, and accessories. This applies as well to the other accessories such as the vacuum feed system. Don't tinker!

Generally the transmission system requires little care except thorough lubrication. Looseness of parts or difficulty in gear shifting may mean trouble if the cause is not removed. The engine clutch should be just tight enough to prevent slipping and no tighter.

The steering knuckles must be kept in proper alignment to prevent excessive tire wear and hard steering. Brakes should be adjusted with care. If too loose, they will not set tight enough to hold; if too tight, the linings will wear rapidly.

Lubricating material should be placed between the leaves of the springs. This will prevent squeaking, and make the springs work more easily. On rough roads or with heavy loads, the strains on the springs are enormous. In such cases it is much better to shift to low or intermediate gear and drive slowly, than to chance breaking a spring by going on high.

Tires. While tires do not ride as easily when properly inflated, they wear longer and the automobile travels with less power. If the machine is to be idle for some time, it is a good plan to take the weight off the tires and reduce the pressure in them. The new driver should remember that both casing and inner tube will be ruined if driven when flat from a puncture or blow out. Ruts and fast driving over pavements are hard on tires, especially if they are a bit soft.

Body. The body requires care if it is to retain its finish. Mud must be washed off before it hardens and sets. Really it should be soaked loose, rather than rubbed off. A sponge is a great aid to the hose for this work. After washing, the body should be dried carefully and then polished with a chamois skin. Some one of the high-grade body polishes will add to the appearance

FIG. 66. The larger the farm, the more useful the automobile in keeping the owner in touch with its various activities.

and protect the finish as well. If the body is varnished over, use a high-grade varnish; for an inferior one will not stand the dirt, oil, and water.

Starting troubles. If the engine does not start, the cause may be one of the following: (1) No spark or dirty spark plugs; (2) throttle closed too much or too little; (3) lack of fuel (vacuum feed cup and carburetor cup empty, or water and dirt in pipes, etc.); (4) improper adjustment of carburetor; (5) engine and fuel too cold; (6) poor compression.

Spark trouble may be due to one of the following causes: (1) plugs full of carbon or warped; (2) dirty contact points on (a) high-tension magneto; (b) spark coils; (c) timer; (3) dirt on parts of high-tension distributor; (4) broken wire or poor dirty connection; (5) exhausted storage battery or run-down dry cells; (6) timing mechanism has slipped.

If the throttle is too tightly closed, the mixture may be too rich and flood the cylinders. If opened too wide the air velocity may not be sufficient to draw fuel. A few trials will determine the best point for setting the throttle to start.

Lack of fuel is a frequent but scarcely excusable cause of trouble. If the tank runs dry on a car equipped with vacuum feed, the engine must be turned over several times or the carburetor bowl filled. The various pockets and traps should be drained occasionally to remove water and dirt from the fuel lines.

Improper adjustment of the carburetor may cause starting trouble. If an inexperienced owner tries to adjust the carburetor he should go slowly and be sure that each move helps. By all means read the manufacturer's suggestions before trying to adjust the carburetor of a balky engine.

In cold weather an engine may start hard or even refuse to start. This is because the engine and fuel are so cold that the fuel will not vaporize readily. The remedy is to warm the engine or use something for starting which vaporizes more readily than common gasoline. High-test gasoline, warm water in the cooling system, or a warm garage will help greatly. A warm garage is perhaps the best way to end this trouble.

When the engine has been used for some time, the valves and rings on the piston may become defective so that they do not hold compression well. Poor compression adds to starting trouble.

Overheating. The engine may heat from one of the following causes: (1) improper cooling; (2) poor lubrication; (3) retarded ignition; (4) poor fuel mixture; (5) carbon.

If the engine overheats be sure that there is plenty of water in the cooling system. Feel the radiator and various parts to make sure that the circulation is good. Note that the pump, if one is used, is running properly and that the rubber hose connections are not flattened or cramped.

If a splash oiling system is used, be sure that there is enough oil. If a pump is used, be sure that it is running and that the pump valves are seating. In case of doubt drain out the oil and put in a fresh supply.

Late ignition will cause heating, especially when an engine has been run considerably and the compression is poor. It is best to run with the spark advanced as much as possible without pounding or clicking in the cylinder.

A poor mixture of fuel and air may cause heating. If an engine continues to heat and pound after other causes have been investigated, it is probably due to a too rich mixture from a carburetor with the auxiliary air valve set too tight.

Carbon in the cylinders may cause pounding and heating. The remedy is to remove it either by scraping or the use of the acetylene flame. A change of oil may be helpful, too.

Other troubles. If the automobile steers hard, the front wheels should be lined up. It is best to follow the manufacturer's instructions in this case.

Jerking and jumping when running at slow speed may indicate slipping of the clutch or misfiring. Tighten the clutch in the first case, and clean the spark plugs in the second. If the clutch is too tight it may be hard to shift the gears because the clutch will keep the shafts turning. The remedy is to loosen the clutch a bit. A tight clutch which slips badly must have new facings or new discs.

Grinding or racket in the transmission case indicates worn bearings, worn gears, or slipped parts. A good mechanic should be able to put the gear set in order. A decided pound in the gear set may indicate either the presence of some foreign matter between the gear teeth or a loose bearing. This pound should be distinguished from the hammering of loose engine bearings. Both sounds differ from the peculiar ringing sound of preignition or early firing of the cylinders. The man who plans to do his own repair work must learn to pick out the various knocks and to tell them from preignition pounds.

A noisy differential may be due to worn teeth or to poor adjustment. The remedy, if a proper adjustment does not help, is to buy new parts.

Brakes should be used with judgment.

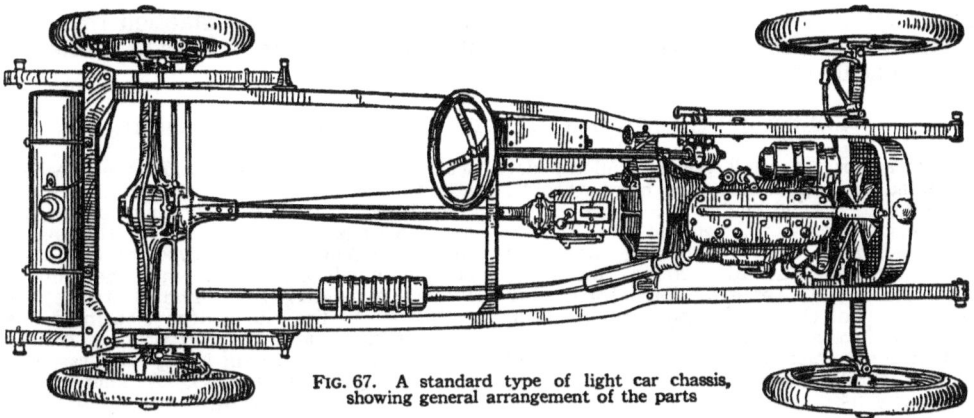

FIG. 67. A standard type of light car chassis, showing general arrangement of the parts

They ought to be set gradually except in emergencies. Never set the brakes hard on a wet pavement, for the car is almost certain to skid sideways. It is better to reduce the speed of the car gradually by shutting off the engine. The special or emergency brake should be set if the car is left where there is any chance for it to roll. The service brake is the proper one to use for ordinary running.

If the engine stops while on the road, look at the fuel tank and the carburetor. Then test the ignition, if necessary. If the motor heats, fill the radiator, look at the oil supply, and make sure that cooling water and lubricating oil are circulating. Always carry a piece of wire, well insulated, with which to splice out any broken cables. Be sure, too, to have tire repairs and replacements.

In case the electric starter fails or the hand crank is lost, jack up one rear wheel, throw in the high gear, and crank the engine with the rear wheel. Be careful in doing this.

If gears are stripped or keys sheared off it is best to get a first-class repair man at once. When the trouble comes from the electrical equipment, call a man who knows how to repair it. This should be done also in case of serious breakage or heated bearings. In general, it pays the owner to call the repair man for any job which he does not understand. It is, however, true that many repair men are not competent. For this reason one must pick his garage with care and then insist on first-class service or no pay. It not infrequently happens that the best repair man is one who is in a small shop alone and does not depend much on hired help.

How to get the most from an automobile. Insurance should be carried, for gasoline is always liable to burn. It is dangerous when handled carelessly. As a further precaution, a first-class fire extinguisher of the type in which the liquid is pumped should be carried. To reduce the chance of using either insurance or extinguisher, one must be careful in filling the tank and be dead sure there are no leaks.

The autoist should give his car constant attention. Careful and systematic lubrication, regular filling of the radiator, adjustment when needed, and decent driving, are the chief requirements for success with an automobile. The average man can take care of these requirements without difficulty. In case of doubt he has only to consult his instruction book or to refer the trouble to a man in whom he has confidence. The farmer autoist should remember that he is driving a perfected mechanical creation. He can then understand the importance of removing every needless noise or harmful pound if he expects to get the worth of his money.

If the driver is a "Safety First" man and gives reasonable attention to his car, it will prove a profitable and pleasurable investment; but the man who buys a car and runs it without a regular schedule for its care, and regardless of its mechanical condition, is sure to be disappointed. Assuredly a man need not be a mechanical genius to become a successful driver. It is equally certain that he must be careful and alert to get the best results. The farmer cannot get satisfaction from his automobile unless he is willing to give to it better care than he does to his heavy farm machinery.

Common Automobile Terms and What They Mean

Accelerator. A small foot-controlled pedal intended to supplement the hand throttle lever for increasing engine speed by increasing the gas supply to the engine.

Back-firing. (1) A popping noise in the carburetor due to mixture containing too much air; or (2) an explosion in the muffler of an unburnt charge of vapor that has been discharged from a "skipping" cylinder.

Cantilever Spring. A form of spring that is similar in form to the side spring ordinarily used on a buggy, except that, instead of carrying the load at the centre and being shackled to each axle at the ends, it is attached to the frame at the centre, and shackled to the frame at the front end and to the car axle at the rear. The centre support is a swiveling form to permit of spring deflection under load. This form is nearly always used for rear suspension.

Carburetor. A simple device attached to the gasoline engine for the purpose of producing an explosive gas for the engine by vaporizing a liquid fuel, such as gasoline, and mixing the resulting vapor with air in proportions necessary to secure extremely rapid combustion in the cylinders.

Clutch. A releasable driving connection interposed between the engine and the driving gearing in all gasoline automobiles and employed to disconnect the engine from the power-transmission system when it is desired to stop the vehicle without stopping the engine, and, after the car has been stopped to secure a gradual start by engaging the clutch gradually when it is desired to resume motion.

Chassis. Name given to all parts of the automobile mechanism when the body is removed. It includes the frame, wheels, axles, springs, power plant, change-speed gearing, power-transmission system, and control elements.

Cylinders Cast en Bloc. The old method of casting engine cylinders was to make them individual members. The new way is to cast 4 or 6 cylinders in one large casting to make a more compact design, all cylinders having a common water jacket and, in some cases, one large removable cylinder head.

Differential. An assembly of gearing in the rear axle that permits the driving wheels to turn at different speeds when rounding curves. As the outer wheel covers more ground, it must turn faster than the inner or pivot wheel, and, at the same time, the power of the engine must be delivered to each wheel in proper proportion. If no differential gear was fitted, there would be considerable wear on the tires, as the difference in wheel speed would result in the tires slipping.

Ignition. In order to produce power the explosive gas in the cylinders of the engine must be exploded at the proper time, so that the piston will be forced down in the cylinder by the expanding gas. The electric spark employed to fire the charge is termed the *ignition spark*, and the electrical appliances producing it are parts of the *ignition system*.

Knocking or Pounding. A sharp, metallic, clanking noise in the engine while running, indicating a loose bearing or other interior part of the mechanism, or serious engine overheating because of poor cooling or failure of the lubrication system to act. It is often caused by carbon deposits in the combustion chamber of the engine.

Magneto. A simple form of dynamo that produces electric current for ignition purposes and that may time it and distribute it to the cylinders so that they will fire in regular sequence. A simple device for converting mechanical energy into electricity. It can be used in place of batteries in the ignition system.

Misfiring. Irregular engine action, in which the cylinders do not fire regularly, due for the most part to faulty carburetion or ignition. It is accompanied by considerable vibration and loss of power.

Over-size Tires. Tires that are larger than standard rim sizes but which will fit rims intended for smaller tires when such are overloaded. For example, the oversize casing 33 inches by 4 inches is intended to go on a rim designed for 32 inches by $3\frac{1}{2}$ inches casings. Over-size tires give longer life, reduce chances of tire trouble, and contribute to easier riding.

Suspension, Three-point. An engine can be held in the motor car frame by 3 or 4 supports. When held by 3 hangers, one is at the front end and in the centre, the other two are at the rear, one on each side. The other method is to use 4 hangers, one on each corner of the engine base. A 3-legged milking stool will rest more steadily on uneven ground than a 4-legged one. Similarly, a 3-point-supported engine will not be strained by frame distortion as much as one held by 4 points.

Timer. A mechanically operated switch driven by the engine and employed to distribute the ignition current to the various cylinders in their proper firing order.

Torque. The twist produced in a shaft or other member transmitting power. The greater the power transmitted at a given speed, the greater the torque in the shaft. A measure of power introducing a leverage factor as the torque is greater at 1 inch from the shaft centre than it is at 6 inches from the centre. The torque is always less at the rim of a wheel than it is at the hub, assuming that the wheel is driven by the hub.

Transmission, Planetary. A form of change-speed gearing in which the gearing is so arranged that the speed-reducing gears surround the driving gear in the same way that various planets encircle the sun. The centre gear is called the "sun" gear, as it is the centre of the planetary system; the others are called "planet" gears.

Transmission, Selective. A form of change-speed mechanism in which sliding gears are meshed with each other, one set being carried on a main shaft, the other on a lay shaft. Various ratios of speed are secured by meshing together gears of different diameters. The selective system permits of engaging any desired pair of gears without first going through another ratio, as is necessary in the progressive system.

Tread. The distance between the points of contact of the front or rear tires measured along the ground and parallel with the axle corresponds to the "track" of a horse-drawn vehicle. The standard tread for automobiles is 56 inches.

Universal Joint. A form of driving coupling employed to join the engine to the rear axle driving shaft so as to allow some vertical and horizontal motion between the driving source and the driven member without straining the driving shaft or losing much power. An important part of all automobiles.

Wheelbase. The distance between the point of contact of the front tires and the ground and that of the rear tires and the ground. Moderate and long wheelbase cars are easier riding than short ones.

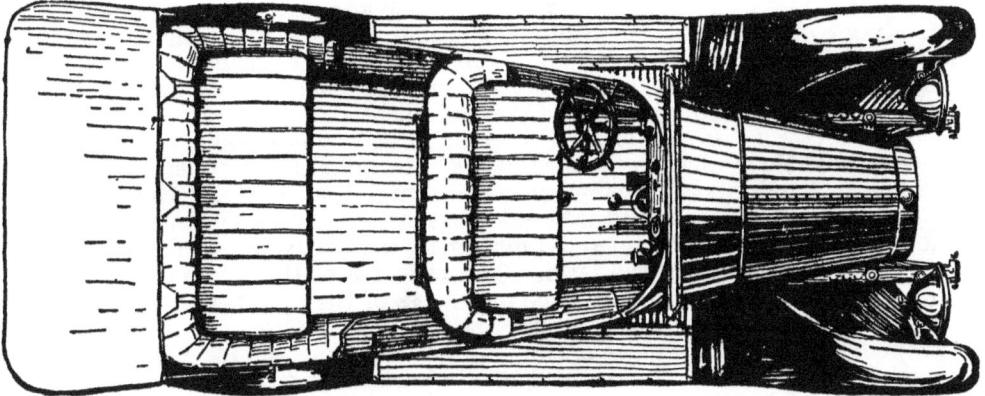

FIG. 68. Top view of a standard 5-passenger, left-hand-steering touring car body—an excellent all-round type for the farmer

FIG. 69. The motor truck enables the market gardener to make two or more trips to town where he used to make but one; and to meet special, profitable demands as he never could before

CHAPTER 5

Motor Trucks and Trailers on the Farm

AUTHORITIES on agricultural matters have computed that much more power is needed to carry on the farming operations essential in raising the produce required to feed our population than is utilized in all of the manufacturing establishments of this country. It has been many years since either man power or animal power was used in our industrial establishments; and just as the use of mechanical power has modernized manufacturing, it is equally essential that wherever possible mechanical power be used in modern farming operations to the exclusion of hand labor or animal power. Agriculture is one of the country's most important industries and it is evident that the same economic arguments that make the use of power-operated machinery imperative in the factory render it of equal importance on the farm. The wide use of the internal-combustion engine for power on the farm and the usefulness of the passenger-carrying types of automobiles are fully considered elsewhere in this treatise. In this chapter it is proposed to outline the place of the motor truck in modern farming operations and the farm work for which it is best fitted. Various types, sizes, and costs will be considered briefly, and endeavor will be made to outline some of the salient points to be considered in properly maintaining the motor truck.

Gasoline engine easy to master. Mechanical transportation has been made possible by the great development which has taken place in the last decade and a half in the internal-combustion, or, as it is familiarly called, the "gasoline" engine, and especially in its application as a propulsive agent for all types of motor vehicles. The big advantage of this type of prime mover is that it can be mastered by any person having the slightest knowledge of mechanics, and as most farmers are naturally mechanical they should experience no difficulty in obtaining from motor vehicles service more than commensurate with the amount of money invested in them.

The work such vehicles are doing on the farm is apt to be judged more by the spectacular performances or the unusual tasks than by the every-day, humdrum work that must be done regularly and as a matter of routine. Transportation is one of the daily tasks of the farmer that lacks the spectacular element, and for some time it was the one branch of farming that was neglected. Farmers who had thousands of dollars tied up in farm machinery that was used only at certain

42

seasons of the year hesitated to invest money in a motor truck that could be used at all seasons.

The early forms of trucks were far from being reliable, and it is only in recent years that a refinement in structural detail has made the heavy-capacity vehicles dependable and durable. Another factor that has retarded the adoption of the motor truck by the farmer has been the slow development of good highways in some sections of the country. In those states where improved roads are found, absolutely no difficulty has materialized that would prevent the economical use of motor trucks. Indeed, the one essential that controls the amount of service to be obtained from any truck is good roads, so that in districts where the highways are of an indifferent character the farmer will need to give the matter of purchasing a truck of conventional design very careful consideration. Still, even if road conditions are not of the best a motor truck will give good service, and special forms such as the 4-wheel-drive and types using the "caterpillar" tread or track-laying method of power transmission may be used under conditions that would stall the conventional 2-wheel-drive form of trucks. In some sections of the country it is possible to use mechanical power exclusively; in others the motor truck will prove practical for certain tasks and horses better adapted for others; and there are sections where a motor truck would not give satisfactory service and where animal power would have to be depended upon entirely.

The ideal motor truck. The tasks of the farmer are many and varied. Considerable work must be done during relatively short periods and only at certain seasons. The method of transportation used must be one that can be employed in various kinds of farm work, besides hauling produce to the market and returning with the supplies that are necessary to carry on the farm operations. The ideal motor truck for the farmer must be essentially simple in construction, easy to operate and care for, and economical to maintain. The motor truck has the advantage over animal power that it does not entail any maintenance expense except when actually in use and that it is ready for any task at short notice. When horses are used, unless some are held in reserve, it is not possible to make more than one trip to town and back if a farm is located outside of a radius of 10 miles, and generally the greater part of the day is consumed in making that one trip. With a motor truck several trips can be made, if necessary, in the same time, and if the work can be accomplished on one trip, it is done in so much less time that a longer period is available for doing other necessary tasks. All farms do not need gas tractors, but it is safe to say that almost any farm can use a motor vehicle to advantage. One feature is that on occasion motor transportation can be used by the women of the farm as well as by the men. The automobile truck brings the farmer closer to his market and greatly increases the radius of the circle of markets for the sale of his produce.

Choosing a truck. The important point to consider in selecting a motor truck for farm use is the character of the work it is expected to do. It would be inadvisable to purchase a large truck if the work could be done with a small one, and it would be a serious mistake to purchase a truck of less capacity than needed and overload it in attempting to make it do work that it was not intended for. On small farms, such as those devoted to the raising of poultry or dairy products, it is not necessary to use large trucks. A safe criterion to go by in selecting a truck is the number of horses that would normally be required to do the work expected of the motor vehicle. A 1-ton truck of good design will do the same amount of work as 3 horses in the same time, provided that conditions are favorable to the use of motor trucks. It is stated that the average work of a team of horses used for transportation purposes is about 20 miles a day. A truck can easily cover twice that mileage, and in emergencies 3 times the distance can be covered.

A 2-ton truck can do the work of 6 horses in many lines, and in a few, such as hauling, under good road conditions it can do more. As a rule, a 2-ton truck compares very favorably with 6 horses. While it is difficult to give maintenance figures that will apply to all localities for either the upkeep of horses or for

FIG. 70. The breeder who takes his stock to fairs and shows by truck avoids the expense, trouble, and danger of infection involved in railroad transportation.

that of motor trucks, it is safe to say that it will cost at least $5 per day for 1 team capable of traveling 20 miles. This is equivalent to 25 cents per mile. A consideration of the operating cost of various types of trucks shows that a 2-ton truck will travel 40 miles a day without difficulty at a cost of operation, including depreciation, maintenance, etc., of about $8 per day. If the truck travels 40 miles, the cost is about 20 cents per mile.

A 2-ton truck will easily average 40 miles a day; in fact many are making 60 miles and more. Most of the cost figures given in the past have been based on the use of gasoline as fuel; but many trucks of recent development operate satisfactorily on the cheaper products such as kerosene or distillate, the latter being a very plentiful and common fuel in the western states, while the former can be easily obtained anywhere in this country.

Why motor truck is displacing horse. The reason that the motor truck is displacing the horse wherever mechanical power can be used is that it is now unprofitable to keep horses owing to the increase in their cost—about 150 per cent during the last decade. The maintenance expenses, also, have all increased, the cost of feed, harness, shoeing, barn construction, and other incidental expenses having become greater. In spite of the increased cost, the horse of to-day is no more powerful than that of 25 or 30 years ago, when horseflesh was cheap. Thomas A. Edison says that the horse is the most inefficient power producer known and the poorest motor ever built. Its food consumption is 10 pounds for every hour of work and the average horse eats 6 tons of food per year. In fact, it consumes the entire output of 5 acres of land. Its thermal efficiency is very low, that is, the amount of power obtained for the food consumed is but 2 per cent. Compare this with the power plant of the average motor truck, which has a thermal efficiency of 25 per cent in most forms, and it will be evident that the user of mechanical power has a pronounced advantage.

Analysis of a recent government report shows that a farm horse averages less than 4 hours' work per day, and that from tiring his efficiency is greatly reduced in 5 or 6 hours. A motor truck can be operated continuously. It has been stated that if the food required by the horses and mules now being used in the United States were raised on one large farm that this would have an area as large as Indiana, Illinois, Iowa, and Ohio combined. It is evident that in periods of food shortage this area could be used to much greater

Fig. 71. One form of motor truck abuse is uneven loading which flattens the springs, distorts the tires, and often results in serious upsets.

economic advantage in feeding our population than in providing fuel for inefficient animal power.

The farm horse is a contemporary of the spade and the scythe. The motor truck is a contemporary of the tractor plowing outfit and the automatic binder and harvester. Horses that are used with great care will not work more than half of the working days of the year, and if more service than this is expected, it can be done only at the expense of the animals' endurance. If a motor truck is taken care of, it will be in working order about 85 per cent of the time, if one considers the time lost in maintenance and in making necessary repairs. It should be borne in mind that in this period the motor truck may be worked 24 hours a day, if necessary. Another consideration which shows an important difference between a living organism and a mechanical contrivance is that a horse is depreciating in value, that is, he is growing older all the time, even if most of that time is spent in a stall. The motor truck depreciates only when in use, as when it is idle there is practially no expense to be considered except the cost of storage and the interest charges on the investment.

The mistake that is often made by the farmer who contemplates using motor trucks is that he compares motor-truck service with his previous experience in using animals, and in so doing he is apt to lose sight of the many marked advantages that a motor truck has over the horse in the field of transportation.

As regards the comparative cost of maintaining horses and trucks, it has been found that, when all of the expenses of a light double team for 1 month were considered, the average miles per day numbered but 12 and the cost per mile was 38 cents. One heavy double team made 10 miles per day average and cost 64 cents per mile to operate. A 1½ -ton motor truck averaged 36 miles per day and cost 23½ cents per mile to operate. A 3-ton motor truck averaged 30 miles per day and cost 36½ cents per mile. The costs per ton mile (that is, per ton of load hauled one mile) were respectively 36, 32, and 12 cents.

Types and sizes of trucks. It is not possible to give an opinion as to the best size of truck to use on a farm of certain acreage, because the nature of the work done on two farms of identically the same size would vary widely according to the section in which they were located, the character of the soil, and the possible market for the product. Some small farms are cultivated intensively and raise garden truck for city markets in such large quantities as to make the use of a motor truck imperative. Motor transportation not only insures the prompt delivery of perishable products, but makes it possible to seek a market farther away than the original one, if it is found that an unusual quantity of the same class of produce has been brought in. A farm producing a bulky product would require a larger motor truck than one of the same size which shipped its product in a more concentrated form, such as a milk or a poultry farm. The best way to determine the most useful size of truck is to compare its prospective work with that accomplished by horses, bearing in mind the fact that on many farms it will still be necessary to keep some of the animals for work that a truck is unable to do as cultivating, pulling mowing machines, hauling over meadows and through roadless fields and forests, etc.

Types and their cost. The types of trucks available, their initial cost and maintenance expense, vary as widely as the character of work these trucks can do. On a small farm where no more than 2 horses are kept for transportation it may be possible to dispense with both animals and replace them with a small truck of about 1,500 pounds capacity. As a general rule, trucks having a capacity ranging from 500 pounds to 1,500 pounds are usually converted standard passenger-carrying chassis that have been fitted with light bodies of the express type. Small trucks are generally equipped with pneumatic tires, and most of them have about the same speed as pleasure cars of equal power and cost but little more to operate. A light vehicle of this form having a capacity of 750 pounds will cost about 7 cents per mile to operate, this including cost of tires, depreciation, and interest on the investment, as well as the cost of fuel and oil. The mistake is often made by truck salesmen of quoting figures considerably lower than those of the actual maintenance cost, only such items as are apparent, as fuel and oil consumption and tire depreciation, being considered. The truck purchaser expects to duplicate the figures given by the optimistic salesman and is disappointed when he finds that the actual operating expense is two or three times what he has been expecting.

No figure is given for repairs in the estimates considered because this is an unknown factor in which the personal element enters. The depreciation considered is only normal wear and tear, and it is considerably lower than it would be if a truck were carelessly operated, poorly inspected and lubricated, and continuously overloaded. The average of a set of maintenance figures shows that the cost of running the average light truck of 1,500 pounds capacity is between 12 cents and 14 cents per mile. Trucks having a carrying capacity of 1 ton or more are usually built along lines that are a departure from those used in passenger-carrying chassis. All of the parts are more substantial; and special features of truck construction are noticed, such as heavy frames and springs, more substantial wheels and axles, the use of hard rubber tires instead of the pneumatic types employed on the lighter trucks, special power-transmission systems such as worm and internal gear drive and, as a concession to the

Fig. 72. Modern devices for hitching an ordinary wagon to a pleasure car provide increased hauling capacity at minimum cost.

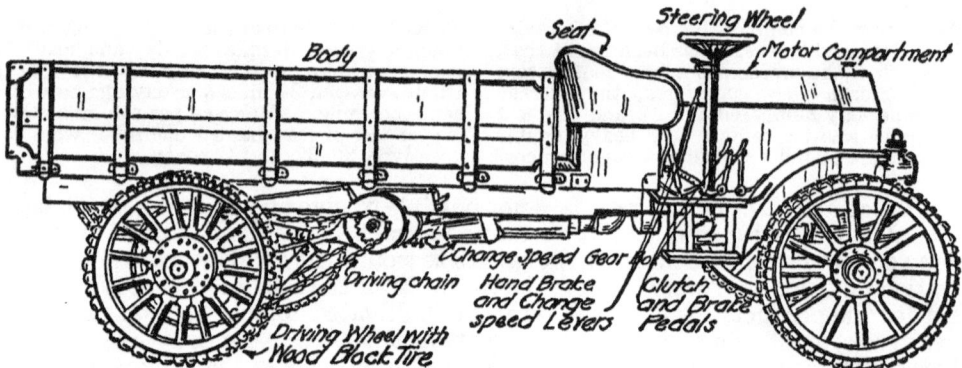

FIG. 73.   A good heavy farm truck of the chain drive type

heavier load, the less resilient springs and the solid rubber tires, a materially reduced maximum speed of operation.   When solid tires are used the factor of tire expense is reduced, but that of mechanical depreciation is increased.   The cost of operating a 1-ton truck of real truck construction will not be very much more than that of the converted pleasure-car chassis of considerably less capacity, and it is safe to figure on about 20 cents per mile as the operating cost.

As trucks augment in capacity there is not a directly proportional increase in operating expense, because there is very little difference in the size of the engine used in a 1-ton truck and in one of twice that capacity.   The increase in weight is compensated by gearing the truck lower and reducing its speed.   In this way the power plant will use but little more fuel and oil than in the lighter truck, and instead of the cost of operation being twice as much for a 2-ton truck as it would be for one of half the capacity, the increase in cost is but 33⅓ per cent., or about 27 cents per mile for the 2-ton job.   A good 2,000-pound capacity truck can be purchased for $1,500. A reliable truck of twice this capacity may be purchased for approximately $2,000.   Hence, it is economy to purchase a truck somewhat in excess of the actual requirements rather than one of just the capacity needed.   In figuring the probable cost of operation of trucks of greater capacity than 2 tons one need only add 5 cents to the basic figure of about 30 cents per mile for each 1,000-pound increase in truck size.   A 3-ton truck therefore would cost about 40 cents per mile to operate.   Truck operation figures can be kept to a minimum if the work done is intelligently planned so that return trips can be utilized instead of bringing the truck back from market empty. The cost of truck operation has been materially reduced by many farmers who have brought supplies back from town for their less fortunate neighbors at a nominal figure rather than drive the truck home empty.

Points of a Good Truck

There are certain characteristics which have become standardized on practically all trucks of modern manufacture.   Of the converted pleasure cars but little need be said here, the characteristics of the passenger-carrying automobile having been fully considered in their proper place.   Practically all trucks, regardless of capacity, are now provided with 4-cylinder 4-cycle engines of the water-cooled form.   While a truck engine is similar in principle to that used on a pleasure car, it is considerably heavier, and the parts are of more substantial proportions because the engine is of a medium-duty type rather than of the light-duty, high-speed type that can be used successfully in passenger-carrying vehicles. The bearings are of greater area because they do more work, and the cooling system has more of a margin regarding heat-absorbing capacity when used on a medium-duty power plant.   At the same time, the construction of a truck engine is not nearly as heavy as that of a tractor power plant or of an engine intended for stationary work.

Position of power plant.   There are two methods of placing the power plant relative to the operator's seat.   In all types of commercial vehicles the motor is installed at the front end of the chassis, and in some cases it is under a hood as in pleasure cars, but in other designs it is placed under the floor boards or the driver's seats which are elevated for that purpose.   Most of the modern trucks have the engine located under the hood.   The advantages of the motor-under-the-seat location may be well summed up by saying that it makes possible more loading space for a given carrying capacity and wheelbase.   The shorter wheelbase vehicle has advantages, if it is to be operated in congested city traffic, as it is more easily maneuvered when driving in narrow thoroughfares, taking corners, or backing up to a loading platform.   The main advantage advanced for the motor-in-front type of trucks is accessibility of the power plant, which may be easily reached by raising the hood. This feature is by no means lost when the motor is placed under the seat, because all average adjustments may be easily made by raising the floor boards or by opening hinged doors which are placed at the side of the motor compartment. Some makers who install the motor under the seat arrange the parts in such a way that they may be removed as units, permitting ready access to the motor and making for prompt removal in event of overhauling. In such a truck, a dash unit which includes the radiator, control levers, fuel tank and a frame for the support of the floor boards may be removed after a seat unit, which is separate and designed to fit over the dash unit when that is in place on the chassis, is taken out of the way.

Combination motor truck and tractor. For farm use, combination vehicles which incorporate some of the features of the motor truck in conjunction with those of the tractor have been favorably received by farmers. The farm truck (Fig. 73) is a very good form for general use. It is said that it will carry a load of 3 tons in the wagon bed and that it has a drawbar for pulling plows and other machines requiring draft. A pulley is located at the front end for driving stationary machinery. The weight of this truck is carried on four wheels, and, when used as a tractor, a load must be placed over the driving wheels to secure proper traction. The machine is spring-mounted and is adapted to speeds ranging from 3 to 15 miles per hour. It may, therefore, be used for a wide variety of work. The wheels may be provided with solid rubber tires for ordinary road use or with a series of wooden plugs to adapt the tractor to field work. Traction on soft ground may be secured by an extension rim having a number of

mud lugs attached, these being so fastened to the wheels that they will automatically grip the soil when the wheels sink to a certain depth. A machine of this nature will haul grain, hay, stock, milk, fruit, vegetables, or any kind of farm produce to the market, and it will bring back coal, lumber, fertilizer, or any other kinds of merchandise needed by the farmer. Provided with a 40-horse-power motor, it has sufficient tractive power to pull three 14-inch plows and a harrow in ordinary stubble plowing and will complete an acre per hour. It will pull 2 discs or 2 spike harrows. 2 seeders, 2 binders, corn planters, a road grader, a train of loaded wagons, or any other machinery that does not require a draft greater than that furnished by 10 horses. A general-purpose machine of this character is a valuable piece of equipment for both small and medium-sized farms, as it will not only haul loads of all kinds in its wagon bed but also can be used for field work to some extent and serve as a portable belt power plant when desired.

While the conventional rear-wheel-drive truck is adequate for fully 75 per cent of hauling requirements, there are conditions where the use of a 4-wheel-drive truck would be advisable. In this machine all 4 wheels combine directive as well as tractive functions, which means that they are all movable for steering purposes and are driven by the engine power for transmission purposes. Inasmuch as a separate brake is provided on each wheel and a special form of differential is employed that delivers the power to the wheel having the greatest traction, it is hardly possible for this truck to become stalled in mud or sand, and it is practically impossible for it to get out of control. The four-wheel-drive principle is especially valuable if the truck is to be used for hauling trailers or for doing other drawbar work.

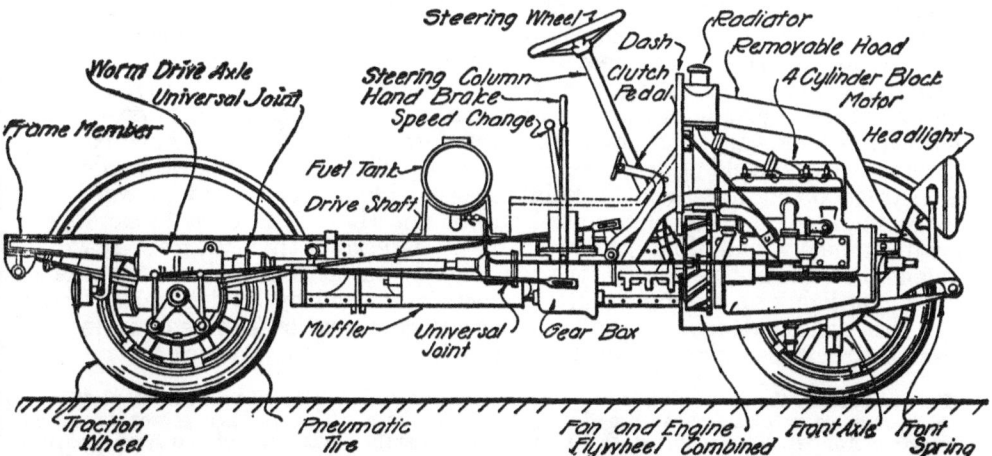

FIG. 74. Truck chassis of medium-duty, shaft drive, 1½ ton type. Pneumatic tires are a great advantage in hauling eggs, fruit, and other easily injured products

A very popular form of truck tractor has the conventional front-wheel construction, but has the rear traction members in the form of 2 endless treads or flexible steel tracks passing over wheels and therefore lays its own road as it goes along. These treads offer a much larger area of contact to support the weight, and for that reason they are much better suited for work on soft ground than the usual truck wheel would be. There is considerable mechanical friction, and this form of drive is suited only to low-speed vehicles intended for use on the field rather than on the road. Various attachments for the conversion of low-priced pleasure-car chassis into trucks of 1,500 to 2,000 pounds capacity are marketed, and when these are used it is possible to obtain a 1-ton truck by such conversion for less than $700. These attachments can be regarded only as a makeshift, however, and the converted job cannot in any sense of the word be considered equal in strength, durability, or reliability to specially designed trucks intended for commercial work only.

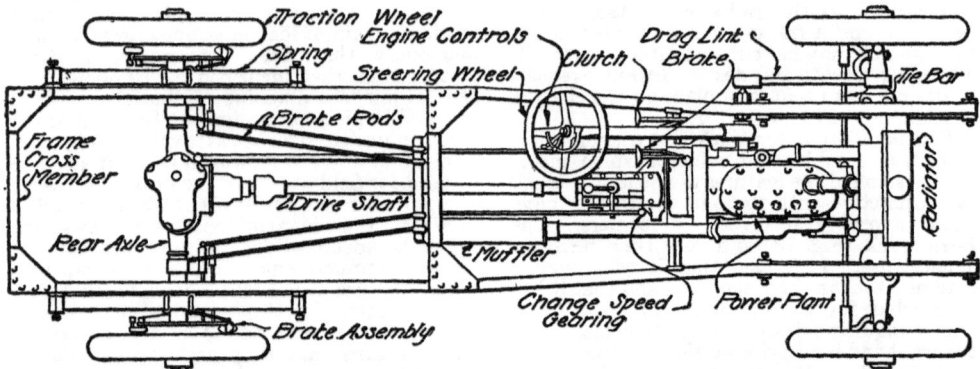

FIG. 75.   Truck chassis built on pleasure car lines, for light and relatively fast farm work.   Compare with Fig. 76

The best truck to buy. The best type of truck for the farmer to buy is one built on truck-engineering principles rather than the types that follow pleasure-car practice.   There is a wide range of body designs for the farmer to select from. For general use, the express-type body is perhaps the most useful, but for trucks of more than 2 tons capacity the writer would recommend a platform-type body to which the type of siding best adapted for the work in hand may be easily attached. For certain work, such as hauling farm machinery, stationary engines, etc., stakes would be all that would be needed.   High-rack sides or express-body panels could be easily installed in place of the stakes and in the same sockets.   The former could be used in transporting relatively light but bulky produce, while the latter could be used for the heavier materials such as coal, fertilizer, etc.   By having detachable bows, as on an army wagon, and by having sockets to receive the bows attached to the high racks at the sides, it would be possible to protect the load with a large tarpaulin cover and transform the truck into a covered body patterned after the old prairie-schooner design.   The high-rack sides are also useful in transporting livestock.   Special racks having a pronounced outward flare can be employed on the common wagon bed and the truck be made capable of carrying a heavy load of hay, straw, or other material of that nature.   It will be evident that it is not necessary for the farmer to purchase a variety of trucks, as by standardizing on a simple platform-bed design and by having various body sides that can be attached at will, almost any combination can be obtained and the same truck used for carrying materials of widely different nature.

Instead of using the lighter forms of pressed-steel frames, as in pleasure-car practice, the average motor truck employs steel I-beams or channel sections which are well braced by substantial cross members of the same form and provided with liberal gusset plates and brace bars. The average motor truck has considerable overhang of the frame, as it is desirable not to have too long a wheelbase, and yet it is not considered advisable to sacrifice on the load-carrying platform. While the unit-power-plant construction, in which the change-speed gearing and the engine are combined in 1 unit, is very popular in pleasure cars and the lighter trucks, in practically all trucks of large capacity the change-speed gearing is placed amidships or located in such a way as to make possible the use of shorter driving shafts than would be the case if the gear box were placed well forward. This construction is well shown in the view of the large-capacity chassis (Fig. 76) which also outlines the conventional disposition of the important components and which details the pronounced overhang of the rear end of the frame.

Formerly, chain drive was very popular, but practically all trucks of modern development use shaft drive, the rear axle having a single worm and worm-gear reduction, or, in the composite-axle type, the truck may have a double-gear reduction in which the primary reduction of speed is through bevel gears, while the secondary speed reduction is secured between spur pinions on the ends of the axle shafts and the large internal gears attached to the wheels to which the final delivery of power is made. Chain drive has important advantages, one of which is that it is possible to obtain different final gear reductions to suit different operating positions. This is a very difficult thing to do with any other form of final drive. If a chain-driven truck is to be used for carrying heavy loads and for drawbar work, it is possible to reduce materially the gear ratio and increase the draft by substituting smaller drive sprockets on the truck countershaft. If the working conditions are less severe, and more speed is desired, the small sprockets may be replaced by larger ones.

The usual ratio of final drive on light trucks is about 8 turns of the engine crankshaft for each revolution of the rear wheels. On heavy trucks the reduction on the direct drive may be as high as 15 turns of the engine to 1 revolution of the rear wheel. In order to prevent overspeeding, which is detrimental to the endurance of the truck mechanism and results in rapid depreciation of the power plant, automatic governors are fitted to the engine to restrict it to a certain definite speed which cannot be exceeded. On the lighter trucks 3 forward speeds and 1 reverse are considered sufficient in the gear set, but in the change-speed gearing of heavier trucks, 4 forward speeds and, in some cases, 2 reverse speeds have been provided. The spring suspension of practically all motor trucks is the same, as semielliptic springs are used on both front and rear ends. This is the simplest and strongest form of spring, and it is very easy by watching the deflection of the spring to determine whether or not the truck is overloaded.

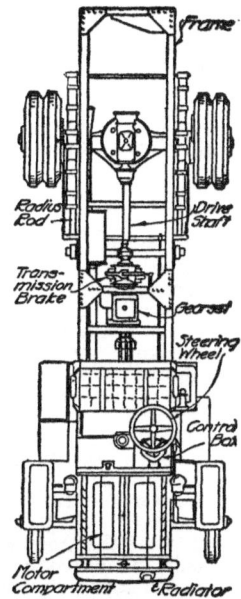

FIG. 76. Chassis of heavy, double, wheeled, shaft-driven farm or commercial truck. One of the essentials of profitable truck operation is good roads.

Care of the Truck

Avoid overloading and overspeeding. Overloading and overspeeding are the two most common causes of rapid truck depreciation. Both of these faults are often noticed in combination, producing doubly injurious results. While most trucks are designed to withstand a certain margin of overload, it is not advisable to load them beyond their capacity, except in an emergency when this is unavoidable. Overspeeding is a result of carelessness and should never be tolerated. In driving a truck, especially a loaded one, the operator should bear in mind that a very large mass is in motion and that it cannot be easily stopped if a sudden halt is to be made, and that it should not be allowed to attain high speed when descending hills, as it may get out of control. Considerably more care and judgment is needed in driving a truck than is usually exercised in operating passenger-carrying vehicles. It should be remembered that the load is carried on stiff springs which cannot have any great degree of flexibility or a large radius of action without being seriously weakened, and that the hard rubber tires with which most trucks are provided do not shield the mechanism from road shocks as much as pneumatic tires do on the lighter vehicles. This means that when a truck is operated on rough or bumpy roads it must be driven carefully and slowly, if rapid depreciation of the mechanism is to be avoided. More trucks are worn out from abuse than by normal use. Because of the severe vibratory stresses that are present in the truck mechanism it is necessary that these be inspected very frequently and any loose bolts, nuts, or other fastenings be tightened. It is also important that lubrication of the chassis parts be carried out systematically.

Lubrication. The average truck owner is not apt to overlook the lubrication of his engine because, for the most part, automatic lubricating systems are provided which call only for replenishing the oil supply from time to time on the part of the operator. There are numerous points about the truck chassis that are apt to be neglected. The manufacturer of the truck has provided compression grease cups at all of the principal points, and it is important that these be screwed down periodically to supply grease to the parts that carry considerable load even though the motion is relatively slight. For example, beginning at the front of the truck one will find grease cups on the front spring supports and on the rear shackles of the front springs. These are intended to lubricate the spring-supporting bolts which are subjected to a heavy, continuous load all the time the truck is in operation. It is seldom that these points receive adequate attention, and the result is that the bolts wear out very quickly.

The steering knuckles on the front axle and the various joints on the tiebar which joins the two steering knuckles and those on the drag link which connects the steering arm on the front axle to that of the steering gear, are also continually in action while the truck is being

driven, and should receive copious lubrication by screwing down the grease cups every 2 or 3 days if the truck is continually in service. Similarly, the rear spring shackles and shackle bolts need this attention. A number of grease cups are placed on parts of the mechanism which are easily reached without getting under the truck. These should never be neglected; they should, on the other hand, be inspected regularly and frequently to make sure that the parts are getting enough grease. Whenever a more thorough inspection is made for loose parts or fastenings, special care should be exercised to lubricate the inaccessible points as well as those that are easily reached.

Frequent inspection. It is desirable to make a thorough inspection of the mechanism at least once a weak, if a truck is in constant use. One cannot give too much care to the inspection and adjustment of such important control elements as the parts of the steering system and the brakes. Brakes must be kept properly adjusted, so that the truck can be brought to a stop without the expenditure of too great effort on the part of the operator. At the slightest suspicion of depreciation in the brake linings or in the brake-operating linkage, prompt steps should be taken to supply new parts. It will be advisable to lubricate the spring leaves from time to time, as, if this precaution is neglected and the springs get rusty, broken spring leaves will result. Special tools are procurable for spreading the spring plates apart when the spring is relieved of the truck load by jacking up under the frame, and the lubricant, which is graphite grease or graphite and oil, can be introduced without difficulty. The change-speed gearing, the differential casing, and the wheel bearings of practically all trucks are so housed that a large quantity of grease may be packed in to insure lubrication for extended periods.

Shelter. The one point in the care of their motor trucks, tractors, and other farm machinery about which nearly all farmers err is in not providing sufficient protection from the elements. Items of equipment that represent considerable money outlay are allowed to stay out in the weather, sometimes because of lack of space for their storage and at other times through sheer neglect. It is not uncommon for the traveler through our agricultural sections to see mowing machines, hay rakes, plows, harrows, and similar farm machinery lying in the fields at seasons of the year when they should be under cover. The farmer who

Fig. 77. A fifth wheel arrangement can be attached to the chassis of either truck or pleasure car so as to provide motive power for any number of semitrailers to be used one at a time. This enables one load to be hauled while another is loaded or unloaded.

The motor cycle is a splendid emergency work vehicle, but it often renders its greatest service as a source of pleasure

While the automobile is primarily a pleasure vehicle it often becomes invaluable in supplementing the farm's work-horse power

THE MODERN MOTOR VEHICLE HAS WON ITS HIGH AND PERMANENT POSITION IN THE FARMER'S ESTEEM BECAUSE OF ITS DUAL PURPOSE NATURE

The logical place for the truck is on the road, where long, heavy hauls, in order to be profitable, must be made swiftly

But here, as in the automobile, elasticity of usefulness makes it available for occasional field work as well

THE BETTER THE FARMER KNOWS HIS MACHINES AND VEHICLES, THE FARTHER HE IS ABLE TO EXTEND THEIR RANGE OF USEFULNESS

does not provide proper housing facilities for his motor trucks will have to pay maintenance and repair charges out of all proportion to the work accomplished by the machines. An automobile contains considerable machinery which may not be exposed to the weather with impunity, as serious trouble will develop in the ignition and carburetion systems if these become water-soaked. Brightly polished metal parts will rust, and dampness will cause corrosion or rusting of parts that will not function properly unless they are clean.

Truck should have special garage. Any farmer investing in a motor truck should provide a special garage in which it can be properly cared for, and should not house it in any shed or stable as he has been accustomed to do with his other rolling stock. A motor truck must be cared for. From costly experiences in the past, the farmer who neglects his horses knows what to expect. Many believe that inasmuch as a motor truck is machinery it needs no care. It has been stated that the chores incidental to keeping a horse in condition consume about half an hour per day per horse, which is equal to about 20 full, nine-hour days in a year. This same amount of time intelligently spent in a motor-truck upkeep will enable the user to obtain reliable service from a machine that may replace from 3 to 8 horses. The garage for the motor truck should be well-

Fig. 78. Convertible body attached to a light truck. This basket rack body, especially fitted for hauling fruit and truck crops has a one-ton capacity.

lighted, have a concrete floor, be provided with a workbench and a complete set of tools for proper truck upkeep, and must be dry. If the motor truck is kept in such a workshop the work of upkeep will be pleasanter. If a truck is kept in a tumbledown, leaky shed or outhouse there will be no inducement for working on it. Many odd jobs can be done on rainy days, if the truck is properly housed. All of the tools used in motor-truck upkeep are equally useful in the repair and maintenance of other farm machinery, so that the amount invested in a complete mechanical equipment is money well spent, and the returns are usually large when compared with the investment.

What Trailers Are and How to Use Them

The capacity of any automobile can be greatly increased by using a trailer, and on many small farms the indispensable pleasure car can be utilized to good advantage and made to do useful work by using it as a tractor for pulling a trailer. There are 3 forms of trailers devised for use with automobiles. The simplest of these is the 2-wheel form which can be attached to the rear end of any pleasure car and which will easily carry loads ranging from 500 to 1,200 pounds. The wheels are placed at the centre of the body and will carry practically all of the load, if it is properly distributed so that the towing tongue or fastening is not loaded to any appreciable extent. These small trailers use artillery-type wooden wheels with solid rubber tires, and good springs, so that the trailer will ride easily. A small model, which has a capacity of 800 pounds, has a solid panel body, 51 by 72 inches with 10-inch high sides, and a 6-inch flare board. It weighs about 275 pounds. A 1,200 pound model has a stake body 4 by 8 feet with removable side and end gates. This larger trailer weighs about 350 pounds. In order to facilitate attachment automatic couplers are used which can be attached or removed in a few seconds.

The semitrailer type has but 2 wheels, and cannot be used without specially built tractors to support the front end. A very practical form of fifth-wheel arrangement is marketed which may be carried on the chassis of a light pleasure car when the rear portion of the body is removed, and with this combination the ordinary forms of farm wagons may be hauled by merely removing the front wheels and axle at the king bolt and attaching the king bolt to the special table, which permits of universal action, carried on the tractor. Special automobiles of short wheelbase pattern are used and are employed purely as a tractor unit.

FIG. 79. With a two-wheeled trailer one can carry an extra load to or from town, or all the equipment needed for a ten days' camping out vacation.

For example, a truck chassis that would carry only 3 tons will haul nearly twice that weight if it be employed as a semi-trailer-tractor combination.

Another form of trailer is a 4-wheel type having all wheels carried on steering knuckles so joined by interconnected levers and drag links that the wheels will track when rounding corners. This is a very useful feature, making it possible to use several of these substantial trailers so that they will form a road train behind any motor truck or tractor providing the required draft. With this construction, the trailers carry all the load, and the towing truck is called upon to carry only that portion of the weight which is loaded into the body to secure adequate traction. In the semitrailer form it is evident that a portion of the load must be supported by the tractor rear wheels, so that what we really have in this combination is a 6-wheel automobile so jointed in the middle that the front and rear may assume different angles to make steering easier or to compensate for inequalities in the road surface.

What a truck can haul. The amount of load a truck can haul on a trailer depends entirely upon the road conditions, and as a general thing it may be stated that any truck of good design will pull over reasonably good level highways, a loaded trailer having the same capacity as the rated load of the towing truck. Of course, on sandy roads or in country where there are many hills it would not be possible to use a trailer nearly as heavy as could be employed under favorable conditions, and it is conceivable that motor trucks may be employed under such conditions that would make it impossible to use a trailer. Considerable judgment is needed in this regard, and before purchasing a trailer of the heavier types it will be well to consult somebody familiar with motor trucks to make sure that the truck mechanism would not be unduly stressed if a loaded trailer were attached to and drawn by the truck.

Some truck manufacturers have provided towing hooks and drawbar attachments and have strengthened the frame construction so that trailers may be used without injuring the towing-vehicle chassis. Certain road tractors have been contrived which are designed especially for trailer work. Some have winch attachments driven by the engine so that they can pull the loads slowly over poor roads and over grades where the truck would not have sufficient traction to haul them in the conventional manner. There is considerable difference in the road resistance of a good asphalt or hard gravel road, which will range from 25 to 50 pounds drawbar pull per ton to be moved, and an ordinary country sand road, which offers such resistance that 150 pounds pull per ton is needed, or a soft dirt road or one having sand 2 or 3 inches deep, which will necessitate a drawbar pull of 275 to 300 pounds. Chain-drive or internal-gear, double-reduction, axle-driven trucks are much better adapted for towing trailers than those provided with worm drive.

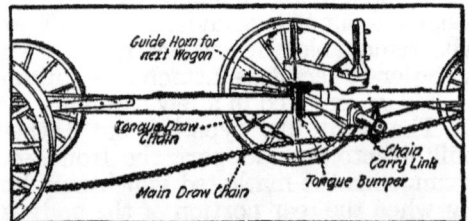

FIG. 80. A device for hitching up a number of trucks or wagons. The essential parts are shown in black.

FIG. 81. The tractor is not going to replace the horse—entirely; but when nearly as much work as can be done by two men and eight horses (as shown here) can be done by an outfit such as that shown in Fig. 82 below, and in less time, it is fitting that the farmer should realize what the machine has to offer

CHAPTER 6

The Farm Tractor

By RAYMOND OLNEY, *Editor of "Power Farming," who was born and reared on a farm in New York State and later graduated from the mechanical engineering course of Cornell University. He then spent a season on a tractor experimental farm in the Middle West as tractor operator, attempting to determine to what extent a tractor is practical and economical for general-purpose farm work. Later he took a post-graduate course in agricultural engineering at the Iowa State College, at which time he was also employed as assistant in agricultural engineering of the Iowa Agricultural Experiment Station. Still later he spent a year and a half as power-farming expert for one of the large tractor concerns, and for the past 4 years has been engaged in farm-paper editorial work, specializing on the application of mechanical power to farming operations.*—EDITOR.

REASONS for the tractor. There is a real need on a very large percentage of American farms for the farm tractor. While it is not yet a perfect machine, there are several makes that are giving satisfaction. In spite of many weaknesses (which are naturally to be expected in any new machine), the tractor has proved itself sufficiently successful so far, and it is quite generally recognized as having come to stay. Machines are now being purchased each year by the tens of thousands, and this leads to the conclusion that there is a great, actual need of them.

The world need of an unfailing supply of food has resulted in a need of more power to produce it; and, because of labor and other conditions, farmers have been forced in thousands of cases to buy tractors to maintain the ordinary volume of production on their farms, to say nothing of increasing it. The production of food in the United States has not kept pace with the growth in population. During the past 10 years our population has increased about 29 per cent, while the production of wheat and corn has remained practically stationary, and that of meat has actually declined. Thinking men are beginning to recognize that mechanical power in the shape of the tractor is needed to help farmers increase production in proportion to the increased need of food by our own population, as well as that of foreign countries.

FIG. 82. Whether in peace time or war time, profitable farming demands all possible saving of man power

FIG. 83. A thoroughly efficient, correctly designed tractor, when rightly adjusted, should carry on its work under normal conditions without the need of man's guidance.

The tractor and farm labor. Probably the strongest evidence of the need of the tractor is the solution—partial, at any rate—which it seems to offer of the farm labor problem. It is not by any means a cure-all for the labor shortage, but it has certainly been the means of bringing relief to a great many farmers. Since purchasing his tractor, a certain farmer and his son have done the farm work more easily than ever before, and without hired labor except possibly for a few days at rush times when extra hands are always needed. This experience is similar to that of thousands of tractor owners.

The tractor makes it possible to get along with less help, because with the machine one man can direct easily and efficiently much more power than he could if he were using horses. A tractor that pulls 4 plows requires but one man for plows and all. The same plows pulled by teams would call for at least 2 men, and possibly 3 or 4, depending on whether the units were of 1 or of 2 plows each. For some jobs, particularly harvesting grain, one man is needed on the tractor and one on the implement or machine that is being hauled—in other words, an extra man is required. But this extra labor is being gradually rendered unnecessary, by improved attachments which enable the operator to drive the tractor from the seat of the implement being used.

FIG. 84. The tractor may be used in harvesting a crop—

The supply of horses and mules in this country has been depleted to some extent by the World War, and this together with the greatly increased demands upon farmers for greater production, is revealing the fact that not only do we not have enough animals to supply the needed power, but also that they cannot be produced fast enough for this purpose. Aside from this, work animals the country over have increased in price to a considerable extent; also, it costs much more to keep a horse now than it did a few years ago.

On the other hand, the cost of a tractor and the expense of operating it compare very favorably with animal power. In some cases it has proved much more expensive than horses, while in others it has been cheaper. The expense varies, depending on a great many conditions. An advantage of the tractor, particularly at a time when there is an unprecedented demand for farm products, is that it can be produced quickly in large numbers; the annual output is even now reckoned in tens of thousands.

The tractor and better farming. Farmers who own tractors, with but comparatively few exceptions, are agreed that the

FIG. 85. —And then hitched up as a stationary engine to shred, thresh, or bale it

machine enables them to do better farming. There have, of course, been many miserable failures, but these were not due entirely to the tractor. It is a fact that a suitable machine in the hands of a good manager will, especially in such work as plowing and preparing the seedbed, do better work than is possible with teams. If one is careful to get a machine that has plenty of power for the number of plows used with it, he can plow deeper than he would care to with horses. In disking and fitting the ground for a crop, which is usually horse-killing work, the implements may be adjusted and operated to do the work as it should be done without having to fear that the tractor is being overworked, provided, of course, it is not overloaded.

It is not alone the concentration of power in the tractor that enables one to do better farm work with it. The fact that, as a rule, the machine permits of doing the work at the right time, when soil, weather, or crop conditions are just right, is a still more important factor. The work can be done at the right time, because the tractor usually works faster, in all kinds of weather, than horses, and, where desirable, it can be worked night and day by providing two shifts.

While the farmer may be able to do better farming with a tractor, the fact that it supplies no manure with which to keep up the fertility of his land must not be lost sight of. Furthermore, the wear and tear on implements used with a tractor is greater than with horses. These are important factors that prospective tractor buyers must consider.

Tractor enthusiasts have a great deal to say about the tractor permitting the farmer to handle a larger acreage with the same amount of labor. While this can be done successfully under good management, in

FIG. 87. Disking and seeding at one operation means maximum results just when they are most essential

not a few instances the attempt has resulted in disaster. Under no circumstances should it be undertaken if it involves lowering the quality of the work done, which too often has been the case.

The tractor and larger farms. It is true, however, that already the tractor has resulted in a tendency toward larger farms. In this connection a statement of Arnold P. Yerkes, assistant agriculturist in the United States Department of Agriculture in charge of tractor investigations, is of interest. He says: "The introduction of labor-saving machines has a tendency, other things remaining equal, to increase the size of our farms. This has always been true in the past and undoubtedly will be true in the future. Our investigations show that the tractor is having a decided influence in this direction. A large percentage of farmers, after purchasing a tractor, increase the size of their farms to a very considerable extent."

FIG. 86. A small tractor is a great help in caring for farm roads

Tractor and the Horse

Much as the tractor is needed on a very large percentage of farms, the time will probably never come when horses will cease to be needed to do farm work. In other words, the tractor has come to *supplement* animal power, not to *displace* it. There will no doubt be farms where horses can be dispensed with practically entirely; but these will be very few, and they will not be farms where a diversity of crops is grown.

There is need for both the horse and the tractor on most farms, as there is work for which one is better suited than the other. There are certain kinds of work, especially the light jobs, in which a team of horses is more flexible or otherwise better adapted than tractors as at present designed. In such work as planting and cultivating rowed crops, and in haying operations, such as mowing, tedding, and raking, horses are more suitable than most tractors; although there are a few machines on the market which are specially designed for the purpose, and which are proving very successful, particularly for planting and cultivating. The improvement in these machines has been rapid, and where any considerable amount of cultivating is to be done they will prove strong competitors of the horse.

But even with new designs of tractors that are practical and economical for

FIG. 88. The caterpillar type is especially valuable for fitting low, moist, soft land earlier than horses could perhaps work on it.

FIG. 89. This tractor and this team are hauling the same amount of machinery, but the tractor can haul it for a longer time, can drive other kinds of machines between harrowing jobs, requires far less space and care when not in use and, ordinarily, will work day after day, in any weather without needing rest.

light work, there is still a place on the farm for the horse, for even in the case of tasks to which a tractor is suited, it is not good economy to use it for light work that teams will do just as easily, cheaply, and quickly, inasmuch as the teams have to be fed and cared for whether they are worked or not, while the tractor consumes fuel and oil only when it works.

Advantage of tractor over horse. Where the tractor has a distinct advantage over the horse is in heavy work; and one of the greatest things that can be said in its favor is that it is a very practical, effective means of relieving horses of the drudgery of farm work. Frequent stops for rest are always necessary in plowing and other heavy work when teams are soft in the spring or when the weather is hot. But with a tractor these are not requisite; for the machine never gets tired, and, if the work has to be rushed, the engine can be worked 20 hours a day as well as 10. The special advantage in this is that, when putting in a crop, for example, the owner of a tractor can make the most of favorable soil and weather conditions by working his outfit double time.

If on every farm there were kept as many horses as were required to furnish sufficient power at certain rush times, on the average farm there would be several head more than would be needed most of the time. Except for a few weeks each year, these extra horses would be an expense to the owner. It is the fact that more power is needed on most farms than is available, and here is where the tractor has solved many a perplexing problem. The tractor takes the place of these extra horses, and while not working it is no expense to the owner. An Indiana farmer on being asked how many horses his tractor displaced, said: "None at all. I did not buy it with that purpose in view. I wanted it to take the place of those animals I did not have and which I seriously needed, especially for the short, heavy season of spring plowing and fitting the ground for seeding."

The following is from a summary of reports received by the United States Department of Agriculture from 200 Illinois tractor owners, and represents the general opinion on the relative usefulness of horse and tractor:

"The tractor has not displaced horses to the extent commonly expected by purchasers, but its greatest advantage lies in the fact that it does the heavy work quickly, and thus completes it within the proper season, since it places at the farmer's command a large amount of power when needed."

FIG. 90. Showing the bulk of a year's feed for 8 horses, compared with that of a year's fuel for a typical 8-16 oil tractor. (International Harvester Co.)

The tractor and the horse-raising industry. Because the tractor can relieve the horse of the drudgery of farm work, and do this work better, quicker, and cheaper, many have felt that it will have a tendency to injure the horse-breeding industry. On the contrary, there is evidence that the tractor is tending to stimulate the production of good horses. In many cases the combination of a tractor and brood mares for use on a farm has proved a practical and profitable one; the tractor relieves the mares of the heaviest farm work and makes possible the raising of better colts. Horses are money-makers for farmers; they are not a dead expense, as some tractor enthusiasts would have the public believe.

FIG. 91. How horse labor is distributed on a typical 200-acre corn belt farm. Though the use of the team may be limited by weather, etc., its upkeep continues with little or no reduction. (Farmers' Bulletin 719.)

There is no reason to suppose that the horse will not continue to be an important source of farm power.

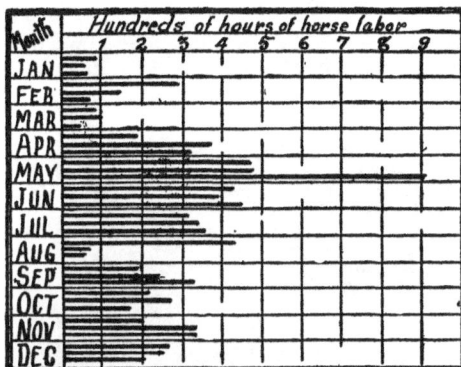

8 Horses in one year

8-16 oil tractor in 100 days

FIG. 92. 8,000 horse power hours can be delivered by 8 horses in one year; and by an 8-16 oil tractor in 100 days. On this basis alone, there is but one choice possible. (International Harvester Co.)

The tractor has its advantages over the horse, and vice versa. It is not subject to death, and if it breaks down, it can be replaced piecemeal for a time. Yet it cannot reproduce itself as does the horse, nor convert the products of the farm, cheap roughage as well as grain, into power and fertility. The tractor has not yet reached the stage where it is as reliable as the horse.

Conditions that Make the Tractor Practical and Economical

The natural question of most men considering the purchase of a tractor is: When is it practical and economical to own one? There are, of course, a number of conditions, any one of which might justify the buying of a machine. But in most cases it is the need of more power that first raises the question of whether to buy or not to buy. Whether this need and several others will warrant investment in a tractor is a matter that the farmer must decide for himself.

The over-enthusiastic tractor salesman may advise selling off some of the horses on the farm to help pay for the tractor. This may not be the wise thing to do, especially if the tractor is being bought to furnish additional power. Displacing some of the horses is not necessarily essential to make the tractor practical and economical.

On most farms there are none too many horses in the first place, and the problem of extra power may be solved by buying a tractor. However, if the

FIG. 93. A two-row tractor cultivator, capable of considerable speed variation and careful work along the crop rows.

tractor is adapted to a variety of work and will furnish cheaper power, some of the horses may be dispensed with. Conditions vary so widely on different farms that it is impossible to lay down any hard and fast rule on this point.

FIG. 94. Testing out a tractor to show its flexibility and resistance to unusual strain

Labor shortage. A shortage of labor might decide the question of the economy of buying a tractor, especially on the larger farms. If it were impossible to get sufficient help to handle the teams, one man and a tractor outfit could take the place of several men and teams, especially for such work as plowing, harrowing, and seeding; and instead of curtailing his farm operations, this arrangement might even enable the farmer to extend them. If the tractor should enable the farmer and his family to do as much work or more at less expense than when he relied upon hired help and horses entirely, it would be an economical investment for him.

Size of farm. For a tractor to be practical and economical, the farm must be of sufficient size to make it pay, and the lay of the land must be suitable for tractor use. Some farmers have found that they can make a tractor pay on 50 acres, but such cases are very few. Other tractor owners are of the opinion that a farm should be from 200 to 300 acres in extent to make power farming pay. The power requirement and many other conditions are so variable on different farms, that it would be impossible to say that one must have a certain, definite acreage under cultivation to make a tractor a profitable investment. If on any farm considerable additional power is needed, and if the tractor can be used on an average 100 days a year, or close to that, other

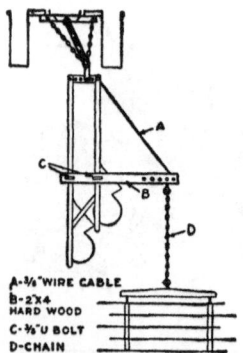

A-¾" WIRE CABLE
B-2"X4" HARD WOOD
C-¾" U BOLT
D-CHAIN

FIG. 95. Tractor hitch for 2 plows and a peg-tooth harrow

conditions being favorable, it is pretty safe to say that the employment of the machine will prove both practical and economical.

Many farmers have bought tractors more for the novelty of the thing than anything else, and the result has been failure. Buying a tractor means a big investment for the average man, and under no circumstances should he put his money into one unless he is sure that he can make it pay —and pay better than farming with horses alone.

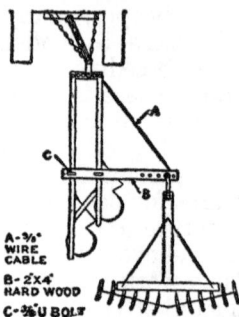

A-¾" WIRE CABLE
B-2"X4" HARD WOOD
C-⅝" U BOLT

FIG. 96. For 2 plows and a disc harrow

A tractor is neither practical nor economical on a farm that is very hilly. Just how hilly a farm has to be to make it impractical can usually be found out quickly enough by trying the tractor out on the hills before the prospective purchaser places his order.

One of the principal objections to most tractors is that they do not have enough reserve power. They will pull their stipulated number of plows very satisfactorily on level ground, but when they come to a grade, the pull required is increased considerably, and the load may become so great as to make it necessary to pull out a plow or two. The power required in a tractor to do good work on hilly ground may be so great that it would not be practical to attempt to use one. Also, the hills may be so steep that it would be impossible to use a tractor owing to the difficulty of keeping it right side up. Another problem in this connection which tractor designers have not yet entirely solved is the proper type of wheel lug that will effectively prevent tractors from slipping sideways when used on hills, although the tracklaying or caterpillar type of machine has proved very successful in this respect.

Type of farming. The type of farming followed will determine to some extent whether or not one

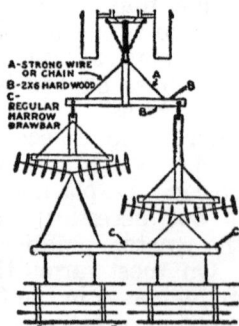

A-STRONG WIRE OR CHAIN
B-2X6 HARD WOOD
C-REGULAR HARROW DRAWBAR

FIG. 97. For 2 disc and 2 peg-tooth harrows

would be justified in purchasing a tractor. As a general rule, where there is a considerable amount of plowing and other heavy work, to which a tractor is well adapted, and provided other conditions are favorable, it is safe to say that a tractor can be made an economical investment.

When the tractor first came into use to any noticeable extent, it was used almost entirely on the large grain ranches of the West. For the most part, the tractor (and this includes also the larger machines) is better suited to grain growing than any other type of farming. This is due principally to the fact that the types of tractors developed thus far, can, with but few exceptions, be easily adapted to practically all of the implements and machines needed in grain farming. Suitable engine-gang plows have been developed, and by means of hitches designed especially for the purpose, discs, drills, binders, etc., can be conveniently and satisfactorily used with a tractor. In fact, there are scarcely

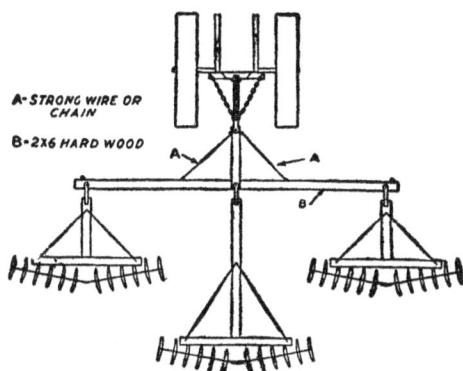

FIG. 99. Hitch for 3 disc harrows abreast

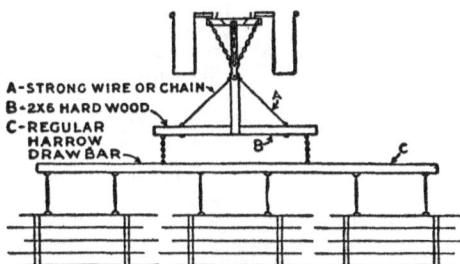

FIG. 98. Hitch for 3 peg-tooth harrows abreast

any operations connected with grain farming that the machine will not perform economically. The writer has even used it for hauling bundles from the shock to the thresher.

In the case of general farming, while there are more limitations in the use of the tractor than in grain farming, whether or not it would be economy for a farmer to own one would depend upon several conditions, principal of which, as pointed out above, is the possibility of using it for a variety of operations and keeping it busy a reasonable number of days each year.

One of the chief complaints that farmers operating small to medium-sized farms make, with respect to most designs of general-purpose tractors, is that they are not adapted to the cultivation of rowed crops. This is a serious disadvantage in a great many cases. In view of this limitation, the man who is considering the purchase of a tractor must have sufficient work aside from cultivating to make it profitable before he decides to invest. The tractor will, in many cases, under these conditions, relieve the teams of other work, so that they can be kept busy

cultivating most of the time, at least when cultivation is necessary.

Though there are some makes of small tractor that are well adapted to cultivating, as well as plowing and other general farm work, there is still considerable doubt whether a tractor that gives maximum satisfaction for general work can be equally satisfactory for cultivating.

For a tractor to be practical and economical for fruit farming, the principal consideration is suitability of type. The machine must be built low, so that it can work fairly close to the trees, and pass under the branches without doing serious damage. The fact that the tractor can be used very satisfactorily with discs and other types of harrow makes it a very practical machine for orchard cultivation. If it is properly designed, it will work among the trees to much better advantage than teams.

The owner and operators. All other conditions being favorable, if the owner is not a fairly good hand with machinery, or if he is a poor manager, the tractor is quite apt not to be an economical investment for him. With the right management, however, there is probably only a very small percentage of farms in this country, on which a tractor will not eventually be both practical and economical. While the present-day machines are profitable investments for a large number of farmers, there is needed a

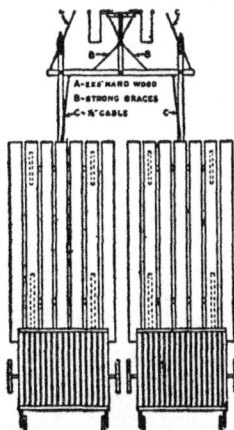

FIG. 100. Hitch for 2 hay loaders

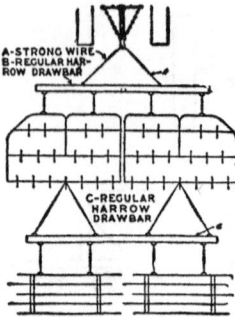

FIG. 101. Hitch for 2 spring-tooth and 2 peg-tooth harrows.

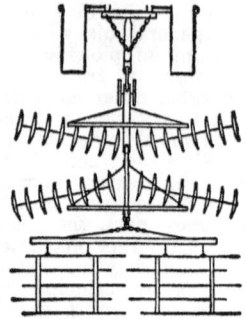

FIG. 102. A tandem hitch for disc and peg-tooth harrows.

more suitable type as well as machines of better design and construction, before the economic value of the tractor to the great body of American farmers becomes established. But the limiting factor in any case is the "man behind." There are some men who, with the crudest and most poorly-built tractors on the market, have been highly successful in their power-farming operations, while others with the best machines obtainable have failed completely.

The farmer who is on the point of buying a tractor must remember that it is very different from the ordinary run of machines, and that to get even fairly good service out of it requires more intelligent care and operation than with any other piece of farm equipment. This is due to some extent to the fact that tractors are still in the early stages of development; as they are improved they will become more reliable and require less attention. Herein lies the reason why running a tractor is not a job for either a young boy or an inexperienced man. On many farms the owner himself or a grown son will handle the tractor, and if he has taken the pains to get the instruction necessary to handle it properly, he should have very little difficulty, provided the machine is a good one.

On the other hand, a large number of farmers must depend on hired help to run their tractors, and in many cases this help will be extremely incompetent. Under such conditions, the purchase of a tractor may be of questionable advantage, for success with it is not to be had without efficient operation. An incompetent laborer may abuse a team and still get fairly good work out of it, but to abuse or neglect a tractor usually means disaster at the very start.

"Is the tractor a practical and economical investment for the tenant farmer?" is a question that is often asked. As a general thing it can be answered in pretty much the same way as for men who own their farms, as outlined in preceding paragraphs. If one is a good manager and not in that class of renters who change their location every year or so, and if other conditions, such as size of farm, type of farming, lay of land, financial standing, and so forth, are favorable, he will probably find a tractor a paying investment. But if it is necessary for the farm owner to consider very thoroughly every angle of the proposition before buying a tractor, it is doubly so for the renter.

Putting the Tractor to Work

One of the questions that the farmer will probably ask before buying his tractor is, what kind of work will it do on my farm? As already pointed out, the extent to which it can be adapted to different operations will, of course, determine whether or not it will be a profitable investment. To give one who is thinking of buying a tractor an idea of what can be done with one, the following list of the more important uses to which tractors have already been put, should be helpful:

Plowing
Disking
Hauling harrows
Soil packing
Leveling land
Seeding
Planting and listing corn
Cultivating rowed crops
Hauling corn binders and pickers and silage harvesters
Filling silos
Husking and shredding and shelling corn
Grinding feed
Cutting fodder
Planting and digging potatoes
Mowing

Raking, tedding, loading, and hauling hay
Hoisting and baling hay
Hauling grain binders and headers
Combined harvesting and threshing
Hauling sheaves to thresher
Driving grain separator
Elevating grain
Hulling clover and alfalfa
Threshing beans
Road hauling
Grading, rolling, and dragging roads
Driving lime pulverizers and concrete mixers
Hauling lime spreaders and manure and straw spreaders
Harvesting beets
Clearing land

Sawing wood
Logging
Driving sawmills
Pulling fence posts
Stretching wire fence
Digging and filling ditches
Drilling wells

Pumping water
Moving buildings
Driving the farm powerhouse
Hauling spray rigs
Excavating for basements, pits, etc.
Digging nursery stock
Drilling holes for blasting.

Tractors have actually performed all the operations in the above list, and in a satisfactory manner as far as the mechanical operation itself is concerned. This does not mean, however, that each one was necessarily a success from an economical standpoint. Very often tractor owners, and especially the more irresponsible members of the farm force, will use the tractor for work that a team would do more easily and more economically.

In going over the list of uses, one tractor owner will decide that it is best to do certain operations with teams and others with the tractor, while another owner may decide differently. Just how far the individual owner can go in adapting his machine to a variety of work varies with conditions, and this is a problem that depends largely upon his own initiative for solution. This is not so important, however, as it is to know that the machine will perform practically and economically a few of the more important operations, such as plowing, disking, harrowing, seeding, packing, harvesting, threshing, etc.

The tractor for plowing. Plowing is by far the hardest of all farm work for horses, and it is the sort of drudgery for which most farmers need a tractor.

In view of this, it is indeed fortunate that on the whole the tractor is better adapted to plowing than any other farm operation. It is also a fact that

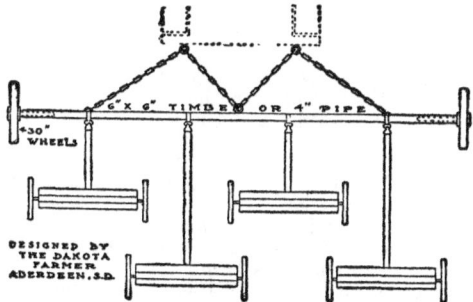

FIG. 103. A wheel and axle hitch for four seeders or other similar machines

a good tractor outfit in the hands of a capable man will, almost without exception, do a much better job of plowing, and do it in less time and at less expense, than horses. This advantage of the tractor is realized by power farmers more when plowing for winter wheat or in the spring when the season is late and the work is rushed. It is entirely conservative to say that there is a fairly large percentage of farms in this country on which a tractor would be a profitable and economical investment if used for plowing alone.

To get the most out of his investment, however, a farmer should not be satisfied with maintaining a single-purpose machine, or one for which only a few uses can be found at the most. The tractor should be put to work at as many different tasks as it will perform quicker, easier, and cheaper than horses.

It is well before buying a tractor to make a study of the field and belt-power operations on one's farm that require power for their performance, and decide as far as possible how many of them can be done best with the tractor. With these different operations in mind, it will be easier to select a type and size of machine that will be suited to one's needs. As tractors become more generally used and as methods of using them are improved, the selection of the right machine will gradually become less difficult.

In view of the fact that farming with a tractor is very new, relatively speaking, the progress that many owners have already made in successfully putting their tractors to different kinds of work is remarkable. For example, a farmer in

Cherokee County, Iowa, says: "I have done everything on my farm with my tractor, except planting and cultivating corn, and the quality of work in each instance was better than that done with horses." A farmer in La Salle County, Illinois, uses his tractor for doing practically all kinds of field work (also excepting planting and cultivating corn), and he says that he can do it to far better advantage than he could with horses. His engine enables him to work faster, or to work on plowed ground with a disc and harrow, which is killing work for horses. This man has found by experience that there are some things a tractor is not capable of doing well. He considers it should be used only for doing those things which a horse should not be called upon to do anyway, that is, the heavy work on the farm.

Another Illinois farmer who has used a tractor 4 years, says: "The tractor is capable of doing a great deal more work than is commonly supposed. If it did nothing more than merely plow the ground, the great burden would be lifted from the horse, and would leave him much fresher and better able to do the remaining work of preparing the seedbed, planting, sowing, cultivating, and so on." Right along this line an Ohio power farmer says: "A farmer who does not own a tractor can scarcely have any idea of the many uses one can find for it. I would be completely lost now on a medium-sized farm without my engine."

Field Uses for the Tractor

Fitting and seeding. Some farmers after plowing their land with the tractor use their teams for disking and putting it in shape for seeding or planting, while others use the tractor for both fitting the ground and seeding. If there is other more important work, or work that the tractor can do easier and better than teams, or if there is not a great amount of seeding to be done, and if the teams need exercise and the work is not too hard, it may be more desirable to use them for fitting and seeding. Again, if the land is very hilly or if the fields are so small and irregular in shape as to require frequent turning, it may be preferable to do the fitting and seeding and the plowing as well with teams. As a general thing, however, land that can be plowed

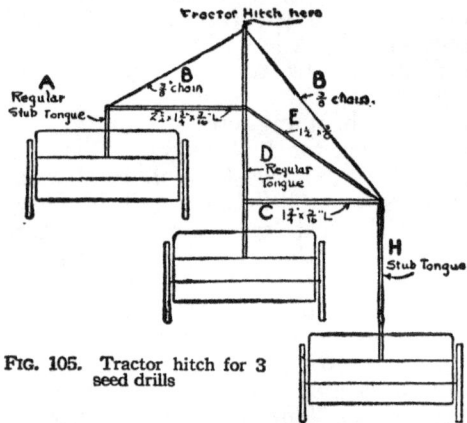

FIG. 105. Tractor hitch for 3 seed drills

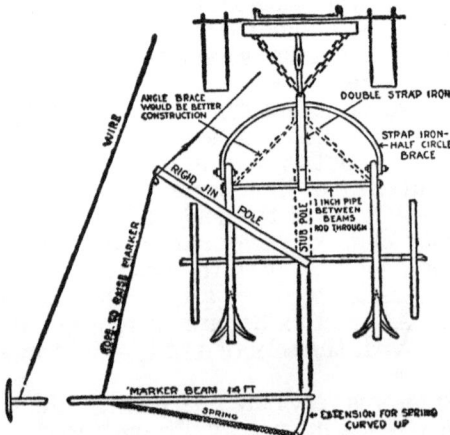

FIG. 104. Tractor hitch for a 2-row lister

more easily and economically with a tractor than with horses can usually be fitted and seeded more economically with the tractor. One of the principal objections at first to its use for this purpose was that it packed the plowed ground too much. It is true that it does this if the soil is too wet; but if the soil is not too wet to be worked, in 99 cases out of 100, no harm will result from the packing effect of the tractor wheels, and in a great many instances it proves actually beneficial. Even under unfavorable conditions, that is, in soil that is too wet, some have found that tractors produce no more harmful effects than horses.

An undoubted advantage which the tractor has over teams is its ability to cover more ground in the same length of time than the same amount of power in the shape of horseflesh; and if it can be used satisfactorily on plowed ground, without too great a power loss, its performance for fitting and seeding operations should at least compare very favorably with horses.

For fitting and seeding with a tractor, most owners aim to hitch two or more implements in combination behind it. On the larger farms where the fields are 20 to 30 acres in extent or larger, special hitches are used for spreading the implements out abreast, principally discs and drills. In this way a large acreage may be covered in a short time and 2 or 3 operations combined in 1.

On the small to medium-sized farms, where the smaller tractors are used in small fields, a common practice is to hitch a disc and smoothing harrow, or a disc, drill, and smoothing harrow or packer in tandem behind the tractor and combine the fitting and seeding in one operation. Another excellent method of preparing plowed ground for winter wheat, or, in fact, for any crop, is to follow the plow at least every half-day with a disc harrow and soil packer hitched in tandem behind the tractor. Because of the looseness of the soil this would be extremely difficult work for horses, but the tractor usually makes easy work of it.

If there is sufficient power in the tractor, it is a good thing in seeding to use the disc just ahead of the drill. It not only gives the ground an extra disking, but the seed is placed in freshly stirred soil where germinating conditions are more favorable to a good start.

One spring the writer used a tractor outfit for fitting and seeding corn stubble to oats. Directly behind the tractor, and fastened to it by means of cables, was hitched a large I-beam, which dragged along, crushed the corn stumps and leveled the ground. Behind this and hitched to the tractor was a double-throw disc harrow, and behind the disc came the drill, the tongue lying on the frame of the disc. The disc was weighted down somewhat, and as the soil was just right as to moisture, the disc prepared a fairly good seedbed. With the drill following the disc closely, the oats were put in under favorable conditions. As the teams on the farm needed exercise, they were put to work harrowing and cross-harrowing after seeding.

Many tractor owners in using an outfit of this kind build a platform on the disc to carry sacks of grain sufficient for half a day's run. If the construction and arrangement of the disc and drill will permit, a "gang-plank" may be built from the disc platform to the drill, so that the grain-box on the drill may be filled without coming to a stop. This saves considerable time in the course of a day, as otherwise it is not necessary to stop often, providing the tractor gives no trouble.

On the other hand, such an arrangement

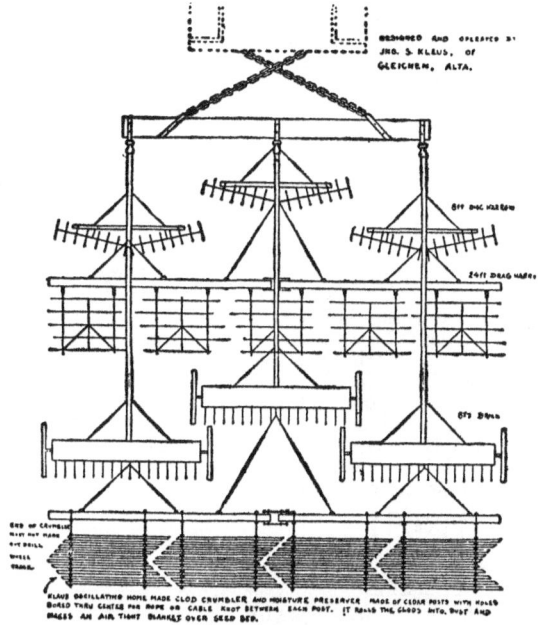

FIG. 106. Combination hitch for 3 double disc and 5 peg-tooth harrows, three seeders and a plank drag or soil crumbler

requires 2 men with the outfit, which often may be undesirable. However, to get the best results the outfit should be attended by 2 men. This allows the tractor operator to attend to guiding the outfit and watching the engine, and the other man to look after the implements handled. Herein lies one of the serious limitations of most tractors, which is, that, in order to drive them, the operator is compelled to sit on the tractor instead of on the implement that is being hauled. While this works out very well in plowing, in other kinds of work, more especially seeding and harvesting, it is a big advantage for the operator to be back where he can watch the entire outfit. Perhaps an even greater objection than this is the fact that in many cases 2 men are needed where, if teams were used, only 1 would be required. In this connection, however, it must be taken into consideration that the tractor will work much faster, as a rule, than horses, as well as a greater number of hours a day. On this account it may be more economical to use the tractor, even though an extra man is required.

Haying. The tractor is coming more and more into use for haying operations. No tractor as yet, however, is a complete success in hauling mowers, although 1 or 2 make fairly creditable work of it. These machines are so arranged that the operator can sit on the mower seat, where he guides and controls the tractor and handles the mower from the same position. Some owners have hitched two or more mowers behind a tractor by means of offset tongues. Such an arrangement is usually more expensive than mowing with teams, for, besides requiring a man on each mower, an extra man is needed to operate the tractor.

Whether or not it would be more economical to use the tractor for mowing would depend on conditions; ordinarily it would not. Mowing is comparatively light work, and, even in hot weather, the teams can rush it along almost as fast as the tractor. Besides, there is no question but that the horse is a more flexible motor for this work than is the tractor. The same thing holds true in the tedding and raking operations; for the most part teams will perform them best.

On land that is not too hilly the tractor will usually be found better suited to hauling hay wagons and loader than teams. A farmer in Wood County, Ohio, finds this very practical work for his tractor. The engine having more power than his team enables him to put on a larger load. It also travels much faster than horses, which is an advantage when hauling the load to the barn, or the empty wagon to the field. This man says that the men who do the loading find they can work with greater ease behind the tractor, as it produces a steadier pull than horses.

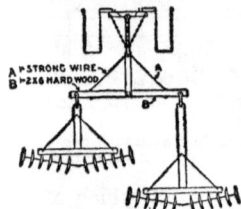

FIG. 107. A simple hitch for 2 disc harrows

Grain harvesting. Aside from plowing, and possibly preparing the seedbed, there is probably no more important operation on the farm for a tractor than harvesting grain. One of the serious objections in the past to its use for this purpose has been that 2 men were needed, while 1 could handle a 4-horse team and a binder. Some of the larger tractors have been used to haul 2 or 3, and even as high as 6 or 8, binders, by means of offset tongues, but a man or boy was needed on each binder. This objection is gradually disappearing with the introduction of new designs of tractors and improved attachments on older designs that permit the operator to ride on and control his engine from the binder seat.

Wherever it is practical to use the tractor for harvesting, it will usually be found superior to teams, principally because of its ability to work faster. The writer has found that it is not usually necessary to begin harvesting so early with a tractor as with horses. One can wait until the grain has reached the right stage of maturity to be harvested, and then keep his tractor and binder at work, if need be, 14 to 16 hours a day, or from dew to dew. Hot weather is not a hindrance to the tractor, the only stops necessary, if the outfit is in good condition, being for meals, to oil the machinery occasionally, fill the twine boxes, and make any minor adjustments, etc., when needed. In this way a large acreage of grain may be cut in a short time, if everything goes well.

However, for harvesting, as well as other farm operations, with a tractor the element of chance is a bigger factor than with teams. If the owner is not careful to see that the machine is in first-class condition before harvesting starts, and that it is kept in good order all the time it is working, a breakage or other mishap may occur that will seriously delay the work, even more so than would teams. Also because the tractor will work faster than horses, it will wear out the binder much faster. Where tractors can be speeded up to 3 or 4 miles an hour, a great many owners have attempted to haul binders at that speed. To say the least, this is almost without exception impracticable, and the delays caused by breakages usually offset the time saved. A speed of $2\frac{1}{2}$ miles an hour is quite fast enough for a binder.

Cultivating. One of the most serious limitations of most tractors, when it comes to putting them to work, is that they are not adapted to the cultivation of rowed crops. There are one or two makes, which come in the class of general-purpose tractors, that are fairly well suited for this purpose, but the rest are not. There have also been designed a few machines especially for cultivating, and

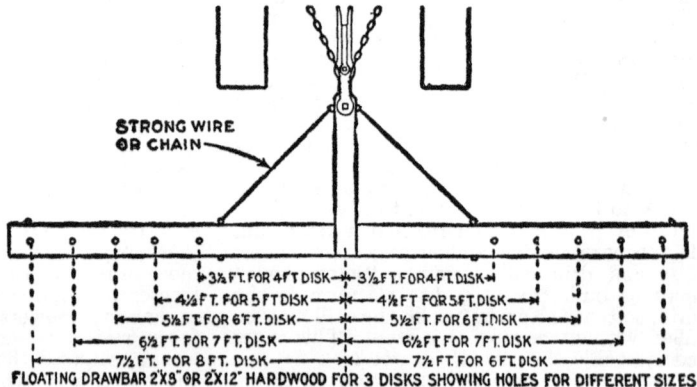

STRONG WIRE OR CHAIN

3¾ FT. FOR 4 FT DISK —— 3¾ FT. FOR 4 FT. DISK
4½ FT. FOR 5 FT DISK —— 4½ FT FOR 5 FT. DISK
5½ FT. FOR 6 FT. DISK —— 5½ FT. FOR 6 FT. DISK
6½ FT. FOR 7 FT. DISK —— 6½ FT FOR 7 FT. DISK
7½ FT. FOR 8 FT. DISK —— 7½ FT. FOR 6 FT. DISK

FLOATING DRAWBAR 2"X8" OR 2"X12" HARDWOOD FOR 3 DISKS SHOWING HOLES FOR DIFFERENT SIZES

FIG. 108. General purpose hitch and drawbar for tractor work. (This diagram, Fig. 95 and all between, courtesy of the International Harvester Co.)

success in this direction is very promising. These same machines without a great deal of change can be used, also, for planting and listing corn, planting potatoes, and other light work.

Belt work. There is quite a variety of belt work on the farm to which the tractor can be put. In general, this is easy as compared with some of the field operations, provided the tractor is well adapted for belt work. The table given previously will suggest quite a variety of tasks.

Speaking with reference to belt work on the farm, a tractor expert of the United States Department of Agriculture recently said: "Belt work is the largest item which can be included under one head; this represents, on an average, about 50 per cent of the work which the tractor generally does on farms. This, of course, includes many different kinds of work, but usually there are only two limiting factors involved: one is the amount of power available for the heavier operations such as cutting silage or running a separator; the other is the question of economy in doing the lighter jobs. Aside from these the nature of the belt work is immaterial; the tractor will take care of it."

Tractor Management

The "man behind" the tractor. Fundamental to the success of the farm tractor as a practical and economical investment is, as stated above, the human element—the "man behind." No matter how favorable conditions may be to its profitable use on any particular farm, unless the owner is a good manager, and operates and cares for his tractor properly, he must not expect to make his power-farming operations a success. Briefly, good management is even more essential when using a tractor than when farming with horses only, largely because a tractor will not stand abuse and bad management and still do good work, as well as teams.

While not heard so often now as in the past, a favorite claim of tractor salesmen has been that "anybody can run it." This has been the means of misleading a great many tractor buyers, making them expect more service from their machines than they could possibly deliver, and also giving them the impression that it was not necessary to study their operation very much to be successful. The conclusions reached on this point by the United States Department of Agriculture, after extensive tractor investigations by one of its experts, are as follows:

"It would appear that some manufacturers have felt that it was a discredit to their tractor to admit that a man need spend any time in learning to operate and care for it. Extravagant claims that 'anybody can run it' have resulted in many farmers feeling that it was an admission of a lack of even ordinary mechanical ability on their part to require instruction in the operation of a tractor or to ask for advice concerning it. This fact has been largely responsible for the need of so much expert service being required after machines have been in operation a short time.

"Actual experience in thousands of cases has shown so conclusively that running a tractor is not a job for either a boy or inexperienced man that it should be unnecessary at this stage of tractor development to have to take time to contradict the old statement that 'any boy can run it.' Every experienced tractor manufacturer knows better, and the new ones who still make such claims not only show their lack of experience, but are paving the way for service troubles at a later date by misleading the purchaser and preventing him from taking the trouble to inform himself fully regarding the operation of the outfit.

"The tractor is strictly a business proposition with the farmer. He cannot afford to risk delays with his work at critical seasons when a small amount of time and money spent in learning how to run the outfit will give a strong guarantee that such delays will be avoided."

Getting results. Good tractor management involves getting as much work as possible out of the machine to make it a profitable investment. As already pointed out, however, this does not mean using the tractor where it would be better to use teams, just to get work out of it. On the average, tractor owners do not get much more than about 50 days' service a year from their machines. In some cases the tractor may be profitable even at this rate; but, as a rule, it is more apt to pay if there can be found 100 or more days' work a year for it.

Getting the most work out of the small or medium-sized tractor on a medium-sized farm means using it for as many different kinds of work as is found practical. And to be more practical and economical than horses for any particular operation, it must be able to perform that operation quicker, easier, and cheaper.

Few tractor owners have used their machines for as great a variety of work as the tractor is capable of performing profitably. The principal reason for this is that the hitching to a tractor of implements designed to be used with horses is quite a problem. For the most part it has been left to the farmer to be solved as best he could solve it, little having been done during the early years of tractor building, by manufacturers, to develop suitable attachments for connecting various farm implements to the tractor.

It requires no little initiative and ingenuity on the part of the tractor operator to devise convenient hitches that will satisfactorily meet his particular needs, and with which he may operate his tractor outfit without bringing about unusual strains or unnecessary wear and tear on any of the equipment. Two fundamental aims to keep in mind when hitching implements to a tractor are: (1) A quick, easy method of connecting and disconnecting the tractor with and from the rest of the outfit; and (2) the greatest possible flexibility of operation, especially in turning. If the operator takes the pains to study this problem thoroughly and watch his outfit closely when at work, to discover ways and means of improving the hitching, he should not have much difficulty in getting satisfactory operation.

Tractor must be properly cared for. Good tractor management includes proper care of the tractor. The right kind of care is essential not only to keeping it in the best working condition at all times, and consequently avoiding expensive delays on account of breakdowns or neglect, but also to prolonging the life of the machine. And in this connection it cannot be too strongly impressed upon operators that they cannot expect to give the tractor the right care without a thorough study of the instruction book and following the manufacturer's recommendations. Many tractors have proved utter failures for no other reason than that they did not receive intelligent care.

On farms where the fields are badly cut up or irregular in shape, as is the case in so many instances, it is usually advisable to square them up wherever possible, even to the moving of fences, in order to get more satisfactory results in the use of a tractor. It is not only easier to handle a tractor in regular-shaped fields on account of the fact that it is not as flexible as a team, but it will do better work and work closer to the corners. Long, narrow fields are the most suitable for tractor operation, as a minimum of turning is required. It will no doubt be found that where a tractor will prove a considerably more economical source of power for the bulk of the work than horses, it would be profitable to relay out all or part of the fields on the farm to facilitate tractor work. Some tractor owners have done this and found it paid. As a rule, however, this change should be a gradual process and the result of careful study and planning.

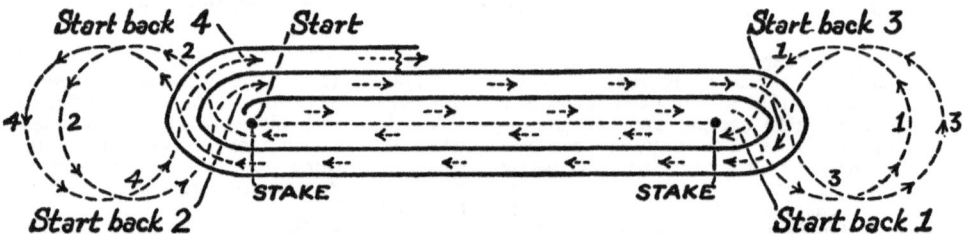

FIG. 109. How to begin a method of tractor plowing that does away with dead- and back-furrows. The arrows show the course of the machine and the figures the order in which the reverse loops are made. From here on the ends can be plowed around without a change of direction. (International Harvester Co.)

The Size and Type of Tractor to Buy

The right size of tractor to buy is naturally a difficult matter for the man who is about to buy his first. The smallest practical size on even the smallest farm is the 2-plow machine, and, according to an investigation conducted by the United States Department of Agriculture among 200 Illinois tractor owners, the minimum size of farm, in the opinion of these men, on which this size of tractor can be used profitably, is 140 acres. This does not necessarily mean that no farmer with less than this acreage should buy a tractor, for a great many farmers can and have made this size pay on even as low as 100 acres. But in considering the purchase of a machine for a farm of less than 140 to 150 acres, a great deal of thought and study should be given it in order to determine whether or not it would be a profitable investment.

Size. In making the purchase it is necessary to buy a tractor of the proper size in order that it may be most practical and economical. In the first place, one should get a machine that has plenty of power for all his work. A tractor of insufficient power to do the work required of it, will naturally prove unsatisfactory. Then again if one buys a machine with more power than he has any need of,

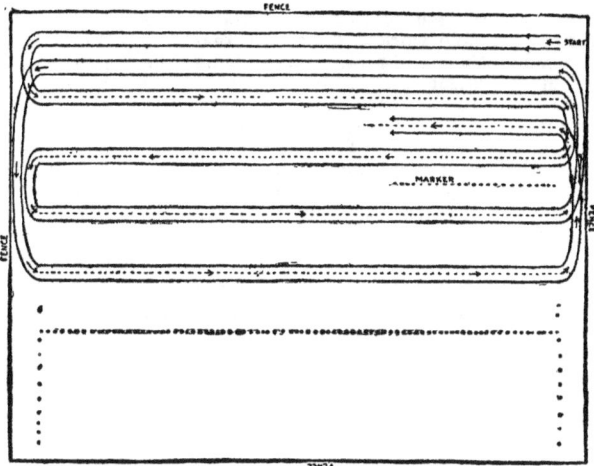

FIG. 110. Field laid out for listing corn with a tractor so as to avoid short turns. The dotted line shows course of the marker. (International Harvester Co.)

it may be too cumbersome and expensive to operate for a variety of tasks on the farm, for both light and heavy work.

In the Government investigation referred to, an important point relating to the best size of tractor, as brought out by the experience of the Illinois tractor owners, was that for farms having 200 acres or less of crops, a 3-plow tractor is considered the most desirable, while a 4-plow machine would probably be better on this size of farm than a 2-plow outfit. On farms of more than 200 acres and up to 300 acres of crops, a 4-plow tractor is generally believed to be the most suitable, while either a 3- or a 2-plow machine would be better than one pulling more than 4 plows. On farms of from 300 to 750 acres the 4-plow tractor was still favored.

Three- and four-plow tractors. Tractor manufacturers building a variety of sizes, as well as investigators who have made a careful study of this phase of the tractor question, are agreed that the 3- and 4-plow machine is by far the most economical size. Probably on most farms this size will give the best all-'round satisfaction. On the larger farms, where more power is needed, it is usually better to have two 3-plow machines than one 6. In a great many cases, farmers who have purchased 6- and 8-plow outfits have stated that it would have been better economy to buy two smaller outfits rather than the one large one. The reason for this is that very often, especially for the lighter work of harrowing, seeding, etc., one smaller machine would do the work just as well as the large one, and the expense of operation would be less. Another reason is that, if in plowing or work where both were needed, one should meet with an accident causing a delay, the other

would keep on working and the delay would consequently be less costly.

Recently a personal investigation was conducted among several tractor owners, owning different sizes of machines, to ascertain what it costs to do farm work with the different sizes. A careful analysis of the costs, including depreciation, repairs, and interest on the original investment, brought out the interesting fact that the cost of doing work with the 3- or 4-plow tractor was invariably less than with either a 2-plow machine or the larger sizes.

Information on this subject is sufficient to justify a definite statement that a 3- or 4-plow tractor will best meet the tractor requirements of the average farm. With this size one can plow much faster than with horses; besides he can handle 3 or 4 plows, whereas one man is needed with each 2 plows when horses are used. This size of machine has enough power to operate practically all belt-driven machines on the average farm, even a small-sized thresher or silage cutter. Also, it is not so large as to be too expensive for doing many kinds of work that do not require very much power.

The size of a farm does not, or should not, determine the best size of tractor to buy. One should outline carefully the different kinds of work for which he proposes to use his tractor, and then select a machine that will have plenty of power to do the most difficult tasks required of it. In this connection, if it is desired to have a tractor merely to furnish additional power during times when the work is rushed, in cases where there is only a small need of belt work and where most of the field work is done with teams, it is possible that a 2-plow tractor would be the most logical one to buy. Some have found this to be the case. A particular advantage of this size is that a smaller investment is required. This size is also recommended as a good one to buy at first, especially in cases of minimum power requirements, in order to gain experience in the handling and care of tractors.

The Department of Agriculture and other authorities are pretty much agreed that, generally speaking, it is better to make the mistake of buying too large a tractor than one that does not have power enough. This does not mean, however, that the purchase of the largest sizes is recommended, but it does refer to the rather too common practice of buying a 2-plow tractor to do the work that really requires a 3- or 4-plow machine.

Types of tractors. Tractors are variously classified, but the classifications of the greatest interest to farmers are based on (1) type of traction members; (2) number of wheels; and (3) motor.

Traction members. As to these there are two general types—the round-wheel and the track-laying or caterpillar. Until recently it had been generally considered that the track-laying type was more of a special-purpose machine, for use where it was necessary to travel on soft ground or to negotiate very wet or very sandy fields. This machine, however, is gradually coming more and more into favor as a machine for all conditions. In fact, at present it bids fair to become a strong competitor of the round-wheel type.

One of the principal disadvantages of the track-laying machine is the short life of the track, necessitating rather expensive replacements, and another is the much higher cost of this type, due to the expensive track construction and mechanism. One company claims to have solved the problem of excessive wear on the track; but with most machines of this type the track, especially under some conditions, is apt to be one of the main weaknesses. On the other hand, a particular advantage of this type is the fact that a larger percentage of the power of the motor is available at the drawbar than in a round-wheel tractor, because the slippage of the traction members in most cases is practically negligible. Especially is this true where the surface of the ground is inclined to be slippery, and also in plowed ground.

Number of wheels. As to this, the 4-wheel type is the most popular, the 2-wheel next, and the 3-wheel third and fast losing favor. The 2-wheel type is confined almost entirely to one or two makes. It is well adapted for use with quite a variety of farm implements, especially planters, cultivators, binders, etc., and has the particular advantage that it permits the operator to drive and control the tractor from the seat of the implement being hauled. Its particular disadvantage is the inconvenience of disconnecting it from one implement and hitching it to another.

All things considered, the 4-wheel type is perhaps better adapted than the others to plowing and other heavy work. Its principal limitations, in most makes, are that the operator is compelled to operate the tractor from the tractor itself, instead of from the implement it drives, and that the machine is not well adapted to cultivating.

The motor. As to the motor in a tractor, many manufacturers, and owners too, still favor the 1-cylinder and 2-cylinder types, although there is a decided tendency toward the use of the 4-cylinder motor.

The type to buy. In buying a tractor, price should be a secondary consideration; for it pays best in the end to add $200 or $300 or even more, and get a durable, reliable machine. It should not only be well designed and constructed with the best materials, but it should embody such features as ball and roller bearings, cut-steel gears enclosed and running in oil, and enclosed working parts; all these will add several years to the life of the machine. The ultimate type of farm tractor has not yet arrived, and it may be several years before it does; but, under favorable conditions and in the hands of a good manager, any one of several makes now on the market is practical and can be made a paying investment.

FIG. 111. Present day farm implements make the larger planting and better care of crops easier than they have ever been. All the world wants more food; shall it look to the farmer for it in vain?

PART II

Farm Implements

WHEN the aim and problem of man's existence consisted merely of the providing of his immediate family with food, shelter, clothing, and protection, there was no need for him to raise more crops or handle more food animals than he and his sons could look after with the help of whatever crude implements their stage of development happened to afford them. Modern illustrations of these prehistoric conditions are to be found in the undeveloped sections of the Canal Zone, of Africa, of South America and other unexplored or, at least, unmodernized regions. However, as men began to take up special lines of work, some found themselves confronted by the task of providing their industrially or artistically or commercially inclined neighbors with the products of the soil— that is, they became farmers. And the increased demands upon their industry gradually made it imperative that they multiply their productive ability. They had to have implements with which to handle larger fields, cultivate larger crops, harvest and dispose of larger yields. This, in briefest outline, is the history of the development of farm implements and machines; to what length it has been carried is suggested in the seven chapters forming this part of this volume.

It is easy to overlook the important part played by the manufacturers of farm machines in educating farmers to their use. Of course it has been good business policy, but it has been more than that; for the intelligent, effective use of labor-saving devices is one means of increasing the total yield of staple crops, and never has the need of increased production been as obvious and serious as it has become of late years. However, there are principles, comparisons, analyses, and descriptions that cannot find a place even in the excellent catalogs and instruction books issued by those who are best informed about the various implements; and which are none the less invaluable aids to a correct understanding of the place and purpose of the tools, and their successful operation. It is from this standpoint that the various tillage, seeding, harvesting, spraying, and other implements described hereafter, are treated. Other discussions of implements may be found in Vol. II, in the chapters dealing with the handling of the soil, the raising of special crops, etc.—EDITOR.

71

CHAPTER 7

Machines for Tilling the Soil

By F. H. Demaree, Farm Advisor of Grundy County, Illinois; who was born on a farm in Indiana. He graduated from Purdue University after having specialized in crop production and plant breeding, then went to the University of Missouri as assistant in crop production, becoming Assistant Professor within 3 years. He then became Agronomist and Advertising Manager for the J. I. Case Plow Works, in which capacity he became familiar with practically all types of implements used in nearly all parts of the country. As part of this work, he started a demonstration farm on which were tested all sorts of new machines and their parts. After 2 years at this, fearing that he might be "getting too far away from the farm," he spent a year with the Crop Improvement Committee of the National Council of Grain Exchanges, and then, in 1915, took his present position. His work has therefore brought him in touch with all kinds of farm implements, most of which he has operated either on his home farm or, in demonstrations, on other farms.—EDITOR.

Plows

WHEN man began to scratch the surface of the soil with a crooked stick, agriculture was born. Since that time the art of farming has been very slowly evolved to its present stage. Ages ago a type of wooden plow was designed, which was a great improvement over the crooked stick, but which was very cumbersome and easily worn out, to say the least. The past 75 years have seen the invention and development of the steel plow to its present high state of perfection and efficiency. This implement, more than any other, has made possible present-day farming methods.

Plow classifications. All plows may be divided in 2 general classes: those which turn the soil with a moldboard, and those which turn it with a revolving disk. Of these, the moldboard type is the original, and it still continues to find favor with the vast majority of farmers.

Moldboard plows. All plows of the moldboard type may be subdivided according to the shape or turn of the moldboard. This subdivision has been made necessary by the fact that soils differ widely as to the ease and manner with which they may be properly plowed. This difference may be further increased by the manner in which the land is farmed or by the type of crop grown on the land preceding the plowing. For instance, a heavy soil which has been growing a grass crop that has produced a sod cannot be properly turned with a short-turn or stubble moldboard.

On ordinary plowland the furrow slices should be turned over and, to a certain extent, lap one another. All weeds or trash will then be buried, and the ground can be easily leveled. On ordinary grass-sod land the furrow slice should be made with a medium lap, not flat, but far from on edge. Native sods which have never been broken should be turned down as flat as possible, owing to the fact that the sod is so stiff that it is exceedingly hard to work up a seedbed, and, unless the furrow slice is turned down flat, the sod will not rot, air spaces will be left underneath, and a poor crop will be apt to result. This is the only exception where a flat furrow slice is desirable. Ordinarily, a flat furrow slice beats down easily with heavy rains, and is, consequently, hard to get in planting shape. A furrow slice on edge is also hard to work down, and it leaves trash and sod on top of the ground.

In order to secure good plowing results, therefore, three distinct types of moldboard are made: sod, long-turn boards, sometimes called "sod" and "stubble boards," and short-turned boards, often called "stubble boards."

The typical sod moldboard is long and narrow, and it has very little turn. It is designed to plow native sods and thick grass

72

sods of heavy soils. It will turn a flat furrow slice or edge it up as lightly as desired.

The long-turn moldboard is designed to plow lighter sods, such as clover and timothy, and it will turn over even furrows, lapping them nicely without crinkling. It is also used almost exclusively in heavy clay soils, regardless of the crop previously grown on the land, because the long-turn board is lighter in draft on these soils.

The short-turn, or stubble, moldboard, should be used on all loose, easily plowed lands. This type of moldboard elevates the soil quickly, turning it almost completely over. It is very efficient in burying trash and in exposing the upturned soil to the sun. It will not do a good job of plowing in sod, and should not be used there if it can be avoided. When so used, the land looks very rough, with the furrow slices badly crinkled, and much sod exposed, which will certainly be dragged out on top when the land is being fitted for a crop.

Scouring. The different styles of moldboards mentioned above find their great adaptability to the kinds of work described, not only because good work is done, but because, in the main, the moldboards scour best under those conditions.

When a plow is scouring properly, the soil slips over the moldboard without any indication of sticking, leaving the moldboard bright and shiny, and giving a soil polish to the underside of the furrow slice as it slips from the plow. A plow will scour easily in a heavy clay soil, the particles of such a soil hang together with little tendency to stick to the plow. The long-turn board does good work with less draft in such soils. A sandy or gravelly soil, also, scours easily under most conditions; the grit keeps the finer portions of the soil from sticking. These soils are generally so loose in texture that a short-turn board will do the best job of turning. Loose loamy types of soil give the most difficulty in plowing; the soil particles do not hang together, and they stick to the moldboard badly. This is especially true when the soil is wet. To overcome this, short-turn boards are used. It has also led to the use of specially hardened, highly polished steel, so that the least friction with the soil is produced.

Under particularly aggravating conditions, a slatted or rod type moldboard may be used. In these, the upper end instead of being solid, is cut into narrow strips, or slats, which, owing to the lessened surface, do not cause as much friction as a solid board. In very sticky lands, this type of board should scour where others fail.

The walking plow. The walking plow is universal, one or more being found on practically every farm. It must be depended on exclusively where the land is very hilly or contains stumps and stones which will not allow the use of a riding plow. Under all

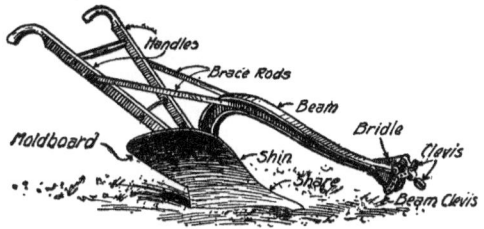

FIG. 112. A modern, general-purpose moldboard plow

other conditions, it has been practically discarded in favor of the lever riding plows, except for garden work and the plowing of smaller lands, ends, and corners. Walking plows are made with either a right- or a left-turn board. The right-turn boards are more favored in most localities.

In selecting a good walking plow, there are two important things to consider: the frog brace and the beam. The best brace is made of steel, thoroughly supports the moldboard, landside, and share, or lathe. If this is poor, the plow is short lived. Originally, all walking plows were made with wooden beams. A wooden beam can still be secured, but the use of steel beams has increased steadily. The great disadvantage of the steel beam lies in the fact that it can be sprung. When the set of the beam is sprung out of line in any manner, the plow never works properly; it should be sent back to the factory for adjustment. There is, of course, no danger in springing a wooden beam, but there is more or less danger of breakage, so that these two features about offset one another.

How constructed. The base of, or key to, the construction of the walking plow is the frog. This is a heavy steel or malleable iron piece curved on one side, to fit the shape of the share and moldboard, and perpendicular on the other, for the attachment of the landside. The beam is securely bolted to the underside of the frog. The share slips over the lower end of the frog and attaches to both the curved side and the square side, thus forming a very secure union. The moldboard proper is then fitted against the upper edge of the share and bolted to the frog at the lower end.

The care and adjustment of plows. But few adjustments are called for in the case of a walking plow, the main ones being at the clevis where the hitch may be raised or lowered to regulate the depth of the furrow, or moved from one side to the other to regulate the width of the furrow slice.

Good care of a plow always pays. Do not leave it in the ground at night, even though the job is not finished. Wipe off the moldboard and share so they will hold their polish. Grease these parts thoroughly during rainy spells and before putting the plow away for winter. A little such attention often saves hours of hard labor in scouring a rusty plow.

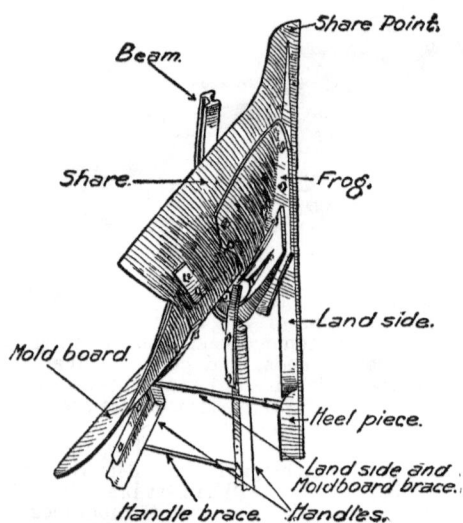

Fig. 113. Bottom view of moldboard plow indicating the essential working parts

One handle is attached to the moldboard, helping to brace the upper end of it and the other is usually attached to the beam high enough to avoid dragging against the furrow bank. The handles are separated and held in a position by brace rods, 1 being near the lower end and, generally, 2 at the upper end. The upper brace rods are the longer, thus giving the desired spread to the handles. Running between the landside and mold-board, there is also an additional brace rod, which supports both parts when the pressure of turning earth is thrown upon them.

Small moldboard plows. In the North, there is some demand for small plows for either 1 or 2 horses, for garden or small farm use. Many of such plows, however, are sold in the South and Southwest also.

Most manufacturers put these plows out in two types or series, the black-land type and the sandy- or loose-soil type. In both kinds the sizes range from a 5-inch to a 14-inch bottom. The black-land plow has a long, tapering moldboard with a narrow pointed share. The share is given plenty of suction to hold the plow in the ground. The sandy-land plow has a broader moldboard with a considerably quicker turn than is used on the black-land plows.

Although a sandy soil will generally scour easily, it is not advisable to use the long-turn moldboard, for this soil does not possess sufficient body to carry the length of the board and then turn into a clean furrow slice. It is apt to edge over leaving much trash exposed. A short-turn board, on the other hand, does good work under such conditions.

Middle breaker. The middle breaker or "middle buster" as it is commonly called, is a double-moldboard plow which throws a furrow slice to both sides, leaving a trench. Middle busters are practically always made in walking types only, as listers use the same construction, and middle-buster bottoms can be attached to riding plows. Middle breakers are used principally in the southwestern and western portions of our Great Plains states, where it is necessary to leave land in ridges and furrows in order to conserve moisture and avoid damage from the wind.

In construction, the middle breaker is very similar to the walking plow. The share is made in one piece, and in appearance looks like two ordinary plowshares, one a right- and one a left-turn, welded together and forming a sharp ridge down the centre. The mold-board wings, also, form the same sharp ridge where they join, and, when both share and moldboards are attached to the frog, this ridge is continuous. The moldboard wings may be in one solid piece, like the share, or in two separate pieces, so as to be easily removable.

Sulky plows. The original sulky plow was hardly more than a walking plow on wheels. It had no frame, and was controlled by hand levers. A large number of frameless lever-lift sulkies is still being manufactured every year, and these plows find a great deal of favor in practically all sections of the country. They are usually lighter in weight and less expensive than other kinds, and, being short-coupled, can be used to make a square turn when it is desired to plow ground in lands, going around the field to turn either in or out.

A more recent development of the sulky plow is the supporting of the beam in a frame, while the operation of the plow in raising and lowering it is controlled by a foot-lever lift. In plowing position, the plow is locked in the ground. This type of sulky requires more adjustment and is heavier and more expensive than the frameless type. On the other hand, it is much easier to operate and, generally, does better work, owing to the fact that the plow is held in the ground at a more uniform depth.

Construction of frame riding plows. The riding plow has all the main features of the walking plow, in so far as the plow bottom and beam are concerned. In addition, the beam is set in a frame supported by 3 wheels: the land wheel, front furrow wheel, and rear furrow wheel. The two furrow wheels are joined and held in line by a long connecting rod that extends from the upper end of the rear furrow wheelpost to the upper end of the front furrow wheelpost. This rod is adjustable at the rear connection, so that it may be shortened or lengthened, thus changing the angle of the furrow wheels.

Furrow wheelposts extend upward through the back portion of the frame and the front corner of the frame to which the axles are attached. In front, a ratchet and lever are attached to the front furrow wheelpost, which, together with the land-wheel lever, are used in leveling the plow and in changing the depth of furrow.

At the top of the front furrow wheelpost there is, usually, a little shift lever which is independent of the connecting rod and is used to change the angle of the front-furrow wheel. The pole is also attached to an iron extension with a hinge joint coming from the top of the front furrow wheelpost.

Most frame riding plows are now equipped with a foot-lever lift. This is a double lever within easy reach of the operator's foot, and is used to raise and lower the plow, after the desired depth has been secured, by adjusting the land- and furrow-wheel levers.

The plow bottom is swung under the frame by long supporting rods which are attached to both sides of the frame and to the plow beam. There are bearings in all 3 connections, so that the plow will work up and down freely. These supporting rods are known as bales. Some plows are made with a single bale; others, with a double bale, one in the front and one in the rear.

A rolling colter is usually placed on a riding plow. This is a heavy, sharp iron disc, attached by a yoke and spindle to the beam in such a manner that it runs just ahead of the inside edge of the moldboard, cutting both trash and soil so that a clean, even furrow may be turned.

Gang plows. The 2- and 3-bottom horse-drawn gang plows have been developed from the sulky. The usual gang plow is of the frame type, having a foot-lever lift and being locked in the ground, like the sulky, as described above. The 2-bottom gang is the one almost universally used. Especially is this the case since the advent of smaller tractors, which usually pull 3 bottoms. The 3-bottom horse gang is cumbersome, and requires so much horsepower that the teams are unwieldy.

Walking gangs are made for use in certain localities, such as the wheat country in the state of Washington, where large areas are to be plowed each year, but where the land is rolling, making the use of riding gangs more or less inadvisable.

Tractor gang plows. The tractor gang plow is a very recent development, and in the course of the last few years has seen many changes. The original tractor was designed to draw a heavy load. Consequently, heavy plows, ranging from 6 to 12 in numbers, were designed. Most of these plows have gone to the Great Plains states and Canada, where the land is quite level, and a tremendous acreage is plowed each year for seeding the following spring.

All heavy tractor plows are constructed with the beams attached to a heavy frame—the rear portion of which is set on a diagonal.

The beams are supported close to the moldboard by a guide wheel, and are held apart by some form of spacer. Originally, each beam was supplied with a hand lever and it was necessary for one man to ride on the plow to operate it. The demand for the use of less labor quickly caused the development of power lifts, so that at the present time practically all tractor plows are supplied with some sort of mechanical lift. Some of these lifts are so arranged that the plows are raised in order, starting with the one on the inside. Others lift two bottoms at once. They are all effective and answer the purpose for which they are designed, namely, in doing away with the necessity of a man on the plow. As indicated above, the general tendency is to

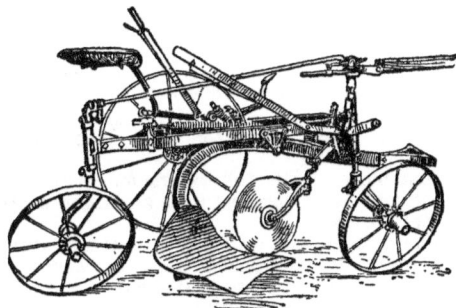

FIG. 114. Beam hitch, foot-lever lift, one-bottom sulky plow

reduce the size and weight of tractor plows. Even on large farms, smaller tractors are being used; consequently, lighter plows are necessary. The ordinary 3-bottom tractor plow is built with the beams securely fastened together instead of each one being free, as in the lever types, and the lift necessarily brings them all up together.

Reversible or hillside plows. The reversible, or hillside, plows, as they are variously called, are made both in the walking and the sulky types. The sulky is constructed with 2 plows, a right- and a left-turn, suspended in a single frame. The mechanism is so designed that first one and then the other plow may be lowered, so that the operator can cross the field, plow a furrow, turn around, lower the other plow, and come back, plowing a slice in the furrow just opened up. In this way, he can go back and forth across the field without making a dead furrow or plow in lands.

In the walking plows, the bottoms swivel for either right- or left-hand furrows, so that the land can all be plowed one way. They can be equipped with either jointer or colter which shifts automatically with the bottom. Such plows are used in experimental fields, where the presence of a dead furrow is undesirable, and on rolling and hilly land where it is de-

FIG. 115. Frameless, lever-lift type of riding plow. It has the advantages of light weight, low cost and short coupling, making sharp turns possible.

sirable to plow across the slope in order to prevent washing.

Disc plows. Disc plows are made in both sulky and gang sizes as well as larger units for tractor use.

The disc plow is a distinct departure from the ordinary moldboard plow so common throughout the country. Instead of a moldboard, a heavy steel disc is used, which is set at an angle and revolves around a centre pin or axle. As the plow is pulled forward, the disc both cuts and turns the earth in a well-defined furrow. The disc plow is invariably of the frameless type and is operated by hand levers. It is heavier and does not handle as easily as the moldboard types.

Plows of this type are used extensively in territories where the soil is hard and dry when plowing is to be done, or where there is difficulty in getting the moldboard plows to scour. There are many such places in Texas and northward over the Great Plains states. Farmers in Argentina, South America, also, are buying disc plows from the United States, particularly plows for use behind tractors.

Deep-tilling machines. The deep-tilling machine is a development of the disc plow. It is built like a gang disc, but with the back disc set directly behind the front one. It is so designed that the front disc throws its furrow slice and the back one cuts its furrow slice out of the bottom of the furrow just made. In this way very deep plowing can be done. Plowing in this manner to a depth of 12 to 18 inches seems to be most beneficial in regions of light rainfall. The loose, plowed soil collects and holds moisture much better than it will otherwise, thus practically insuring a crop.

In regions where rainfall is plentiful, deep tilling has not yet proved remunerative. While connected with the University of

Missouri, the writer made an unofficial test, comparing deep tilling done in both fall and spring with ordinary plowing also done in both fall and spring. The lands plowed were in long strips in the same field and as nearly uniform in fertility as could be selected. The whole field was planted to corn the following spring. Little difference could be observed, except that the corn on the land deep tilled in the spring was somewhat poorer than the rest. This would seem logical, as soil was thrown out on top that had never been disturbed before, and had not felt the action of those agencies that make plant food available.

The subsoil plow. Like deep tilling, the practice of subsoiling has shown itself to be most beneficial in regions of light rainfall. The subsoil plow follows directly behind the surface plow stirring the soil in the furrow bottom. Sometimes a small moldboard plow is used, or a single-shovel plow that stirs the soil in the bottom of the furrow without mixing it with the topsoil. Another type of subsoil plow has a long, narrow point with a thin cutting blade that simply splits the furrow bottom to a depth of several inches.

Subsoiling is done with three main things in view: to increase the water-holding capacity of the soil; to allow the air to enter the soil more freely; and to encourage deeper penetration of plant roots. As mentioned, this work is most practical in regions of light rainfall. Farmers in more humid sections find it more profitable to do their subsoiling with deep-rooted clovers.

Sizes and types of plows to use. The lay of the land, together with the size of the farm, will determine whether one should buy a walking, sulky, gang, or tractor gang plow. The larger types of implements do as good, and often better, work than the smaller ones. Their use invariably means cutting down the cost of crop production. When labor saving is greater than interest charges this is doubly appealing to the business farmer, because it means an actual saving of money and another link in the solution of the farm-labor problem. Data from the Ontario Experiment Farm show that the cost of plowing has been reduced there from $2 per acre, with a single

FIG. 116.   Deep-tilling machine built on the double disc principle

plow, to $1.25 per acre with a 2-furrow gang plow. This reduction is largely due to the fact that one man less was required to accomplish the same amount of work.

Sulky and gang-plow adjustments. The foot-lever-lift type of sulky and gang plows have special means of adjustment which should be thoroughly understood by every farmer, as they have much to do with the proper operation of the plow.

Both the front and rear furrow wheels of either a sulky or a gang plow should run straight ahead and toe squarely into the angle of the furrow. The rod connecting the two furrow wheels can be shortened or lengthened to overcome any difficulty with the rear furrow wheel. The angle of the front furrow wheel is controlled by the pole, or a special lever device at the top of the front furrow wheelpost. On practically all makes of plows, the iron to which the pole is attached is reversible, and can be shifted in such a manner as to make the pole stand square. If the front furrow wheel does not toe into the angle of the furrow, the wheel can usually be shifted on the axle until it is in the desired position.

In some makes of plows, the position of the rear furrow wheel can be changed by a simple adjustment in the rear, which causes this wheel to hug the furrow bank closer and in that way take off the pressure from the landside. If this is not done, the landside will drag and in a very short time will wear to a thin sharp edge on the lower side and will gradually wear completely out. Where this pressure is taken from the landside, the draft of the plow is considerably reduced.

If the plow does not take the ground properly, even though the levers are down, it should be given more suck. Practically all plows have an adjustment whereby the pitch of the beam may be raised or lowered, in order that this may be accomplished.

Other common causes of lack of penetration are dull rolling colters, dull shares, and a slipping of the frame on the rear furrow wheelpost.

Keep the colters sharp enough to cut through trash, stubble, or sod as the case may be. In dry ground, set the hub of the colter about 3 inches back of the point of the plow and in stubble land it is only necessary to run it from $1\frac{1}{2}$ to $2\frac{1}{2}$ inches deep. In sod, the colter should cut through the mass of roots, if a good job is to be done.

Shares should not only be kept as sharp as possible, but should be set as nearly like a new one of the same make as can be done. Various makers have ways of their own for sharpening and setting shares, the appearance of which should be maintained. To do this, it will be necessary to keep a new pattern share on hand to serve as a model for the blacksmith.

As mentioned above, the rear furrow wheel should take the pressure from the landside. In most frame plows, the frame is supported on the rear furrow wheelpost by a collar and set screw. If this adjustment slips, the frame sags down, and the plow is running like a sled with heel down and point up. In this case, raise the frame until there is a half inch clearance under the heel of the landside when the plow is standing on a level surface.

One of the great difficulties of most 2-bottom gang plows lies in the fact that they develop side draft. This is particularly true where 4 horses are hitched abreast. In order that a gang plow may pull straight ahead like a wagon, the hitch must be placed to pull from about the centre of the line of actual draft, then the equalizers must be so built that the pull of the 4 horses working together comes back to this point.

Many plows are so constructed that it is practically impossible to run them without side draft when 4 horses are working abreast. In that case, it is much better to use a strung-out hitch. All gang plows will work better with a strung-out hitch than where 4 horses are used abreast; and where 5 horses are used, this type of hitch should be the only one considered.

Where the horses in the team do not work at the same speed, or where there are one or

Fig. 117. A three-bottom tractor gang plow. As long as power is available to pull them every added bottom means one more man released for other work

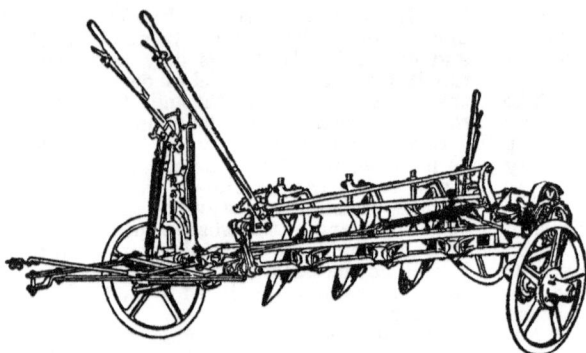

FIG. 118.   A three-disc tractor plow; a type especially fitted for use
in hard, dry soils or where moldboards fail to scour well

pull against the other and prevent lazy horses from shirking.

Correct engine plowing. Lay off the field in headlands from 200 to 400 feet wide, so that dead furrows will not be too frequent, and still the time consumed in traveling across the ends will not be excessive. Leave strips at each end wide enough to turn on with ease. Leave a strip of the same width on each side of the field. When the headlands have been plowed, start around the field, plowing out ends and side strips. In this way, the whole field can be nicely plowed out, except the corners.

Plowing in a circle leaves unplowed strips at first, when turning is short, and leaves too much unplowed and in the corners. Such plowing is also very hard on the plow, which is in a position of constant strain all the time. The life of a plow so used will necessarily be short and the quality of its work inferior

more lazy animals, it is a good plan in making a strung-out hitch to attach a pulley at the hitch on the plow, run a chain through this pulley, attaching one end to the rod running out to the lead team and the other end to the rear equalizer. This will make one team

Harrows

There are 3 general types of harrows in use: the drag, or spike-tooth harrows; the spring-tooth harrow; and the disc harrow. All of these have variations in design which make them adaptable for various purposes.

Drag harrows. There are 3 common types of drag harrows in use. The rigid wood-bar harrow which is made by driving the sharp spike teeth through a solid bar of wood and bolting these bars together to form a section. The bars are arranged in various ways, in the attempt to make a strong harrow, but, usually, the same purpose is served. These harrows have no levers, and the teeth are generally set at a slant which can never be changed. Their use is more common on level lands, where very little trash can accumulate on the surface after plowing is done.

The same construction is used to form another harrow in which the teeth are driven through wooden bars; but here the harrows are equipped with levers, and the teeth can be adjusted to practically any angle. This harrow is very similar in construction to the steel-bar-lever harrow; but, in the latter, the teeth are bolted firmly to steel bars, generally U-shaped, to secure greater strength. In both cases, the harrows are made up in sections and are sold usually ranging from 2 to 4 sections to the harrow. The choice between these 2 harrows is purely a matter of opinion. The wood-bar-lever harrow is a little lighter, and usually the teeth stay in position somewhat better than on the steel-bar types.

A variation of the drag harrow is found in a small harrow attachment for sulky and gang plows, which is attached to the plow and breaks up the soil at the same time it is turned over. Where soils are not too heavy, this attachment does exceedingly good work and can be universally recommended.

Spring-tooth harrows. A spring-tooth harrow is made with curved spring steel teeth which are firmly bolted to wooden or iron bars. The new types of spring-tooth harrows are fitted with levers, so that the teeth can be raised or lowered. Many of them have an iron side bar, which, when the teeth are completely raised, acts as a sled to transport the harrow.

The use of this implement is more general in the East than in other sections of the country. It works up the ground much deeper than the drag harrow, and is especially well adapted to stony or stumpy ground where the disc harrow cannot be used to advantage.

A variation of the spring-tooth harrow is found in the alfalfa cultivator. This tool was made by bringing to a point the broad ends of the spring teeth, so that, instead of being an implement that will thoroughly tear up the soil, the pointed teeth merely scratch and dig. This type of harrow is very efficient in digging out blue grass and weeds which come into alfalfa fields, but does no harm to the alfalfa plants themselves.

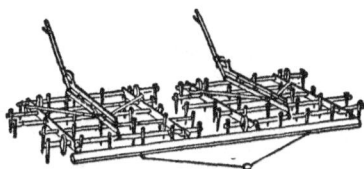

FIG. 119. Spike- or peg-tooth or drag harrow, one of the commonest of farm implements

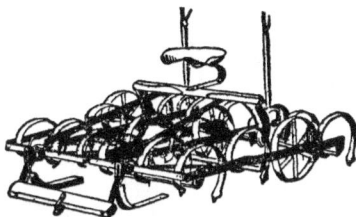

FIG. 120. Spring-tooth harrows are especially popular in eastern sections. This is a riding type with runners and wheels

It makes, therefore, an ideal implement for the cultivation of alfalfa.

The acme harrow. The acme harrow is constructed of long, heavy knife blades attached at the forward ends to a heavy bar. Each blade is notched about the middle, the front part curving slightly in one direction and the back part curving in the opposite direction. These blades are set about 8 inches apart, or close enough to thoroughly work the soil.

This peculiar construction allows the acme harrow to cut and slightly turn the soil, making it an efficient implement in the preparation of a seedbed. It both levels and pulverizes, thus doing the work of disc and drag harrow at the same time.

This harrow may be found in all sections of the country, but is especially esteemed in the East. In the corn belt the disc harrow has displaced the acme because of the ability of the former to prepare a good seedbed on stalk land for spring seeding without the use of a plow. The disc is also more effective as a pulverizer on plowed land.

Disc harrows. The disc harrow is a comparatively recent invention, differing so widely from ordinary harrows that it seems more akin to the plow than to the harrows. In action it both cuts and turns the soil, doing more to pulverize and compact it, as well as to kill weeds, than any other farm implement except the plow.

The disc harrow serves in more useful ways than any other tool on the farm. It has been found that stalk land when thoroughly disked is in splendid condition for seeding spring grain.

Land to be planted to corn, cotton, and other cultivated crops should be disked ahead of the plow, as loose fine earth is then turned down to the bottom of the furrow slice, where most of the roots of any growing crop will be found.

Plowed land should be disked to prepare for a cultivated crop, as weeds can be quickly and easily killed in this way. The ground is made fine and mellow on top and compacted somewhat beneath. This is the best kind of seedbed for any crop.

Hard, dry ground may be put in shape for plowing by first using the disc. Even a light mulch will cause the retention of enough moisture, as it rises from below, to cause the soil to become mellow enough for satisfactory plowing.

Where stubble lands are to be fall plowed it is an excellent practice to run a disc directly behind the binder; the soil then does not become dry and hard, and can be plowed to advantage any time after the crop is removed. When a tractor is used to pull the binder, the disc may be tied directly to the binder, as there is plenty of power to handle both.

Construction of disc harrows. The ordinary farm disc harrow may be made with solid discs, with cutaway discs, in which portions of the outer edge of the discs are cut out, leaving squared-off projections which are intended to dig a little better, or with spading discs. Instead of a solid wheel, the spading disc is cut into many sections completely down to the axle, the idea being to increase the power to penetrate the soil to greater depth.

Actual experience has shown, however, that the solid disc, when kept well sharpened, not only wears better, but usually does better all-'round work, than the other types.

Tandem disc harrows. The tandem disc harrow is formed by attaching a trailer directly behind an ordinary farm disc. This trailer throws the soil in the opposite direction to the first set of discs. When the ground is gone over once with this implement it is cut twice. The saving of time is an important item and the work is generally very satisfactory. Tandem disc harrows are made both solid in front and rear or cutaway front and rear, but the usual style is to use solid discs in front and cutaway in rear.

Reversible disc harrows. Another variation of the ordinary farm disc is what is known as the reversible harrow. This implement is made by attaching a gang of discs to the frame by means of an upright spindle,

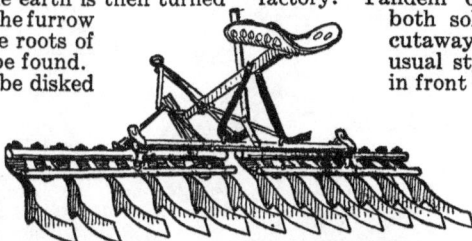

FIG. 121. The acme harrow leaves a splendid seedbed, but for best results it requires a soil in good condition in the first place.

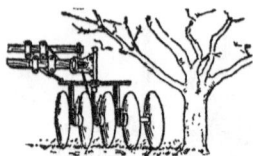

FIG. 122. Orchard harrow extension

set up close to the outer edge of the gang. When in one position, the implement looks like an ordinary disc harrow, but the gangs can be swung around so as to leave a wide space between the two. This type of harrow is found in common use in the South in preparing for cotton. It is of special advantage in throwing down the ridges made by the cultivation of cotton, corn, or sweet potatoes. The reversible feature allows this work to be done to the best advantage.

Orchard harrows. The orchard harrow is made on the same principle as the reversible disc, but the disc gangs are placed out on the end of an extension. This construction makes it possible to work under trees to good advantage with a disc harrow, which could not be done with an ordinary implement. These tools are coming into general favor in practically all sections of the country where orcharding is followed as a business.

Tractor disc harrows. Disc harrows, both single and tandem, are now being made for use behind tractors. They are built along the general lines of the ordinary disc harrow, but made heavier throughout. The style of disc is the same as for horse-drawn harrows, being either solid or cutaway, or both. They should always be equipped with tongue trucks and weight boxes, in order that sufficient weight may be put on to make them take the ground to best advantage. There is usually plenty of power when the tractor is used so that a good job of disking may be done. Double disking in this manner where oats, spring wheat, or barley is to be seeded on

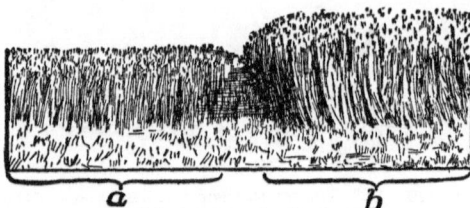

FIG. 124. Wheat plots comparing results of growing oats before seeding (*a*) and of summer cultivating or fallowing (*b*)

stalk land in the corn belt, is becoming a favorite practice and gives excellent results.

Disc harrow sizes. Disc harrows are made in a wide range of sizes, both as to length and height of discs. The 16-inch disc has proved the most popular, however, and the bulk of disc harrows sold are of this height. The length of the disc harrow depends on many things, principally the work to be

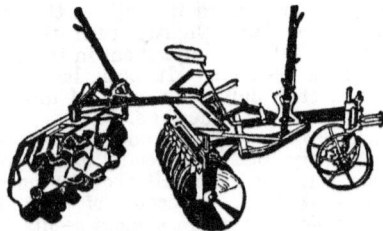

FIG. 125.   Combined disc and cutaway harrow

done and the horsepower available. In the corn belt, 2 sizes predominate—the 8- and 10-foot discs. The 8-foot disc will take 2 corn rows nicely when disking stalks, leaving a short lap. The 10-foot disc will just uproot 3 rows. The 9-foot disc will not reach the third row, consequently it leaves a wide lap. It takes more horsepower than the 8-foot, and is no more efficient. It is, therefore, being rapidly discarded in favor of one of the other sizes.

Tongue trucks. A tongue truck should always be used on a disc harrow. It steadies the disc, so that better work can be done, and at the same time takes the weight off of the horses' necks. It also eliminates most of the violent swaying of the pole which always annoys and often injures the animals next to it if the truck is not used.

Land Rollers and Packers

There are three distinct types of land rollers in use—the flat roller, either iron or wood, the corrugated iron roller, and the subsurface packer.

Flat rollers. Flat rollers are designed as clod crushers, and have

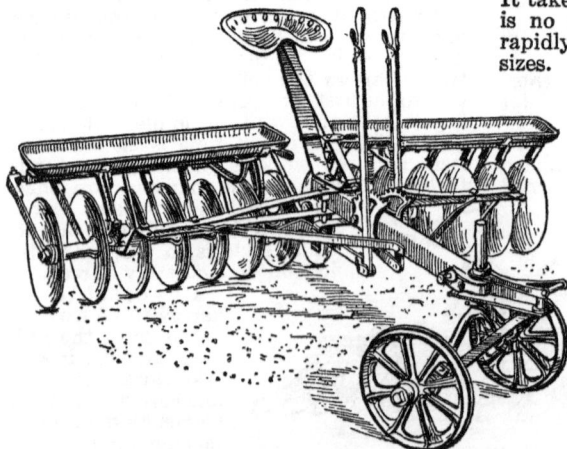

FIG. 123. A tongueless disc harrow, the truck serving to reduce the draft, which in any disc machine is considerable

been used for this purpose almost exclusively. It has been found, however, that land rolled down smoothly is crusted badly by a heavy rain. In addition, as capillary water rises to the surface, evaporation takes place much quicker from the smooth, compact surface.

Both of these conditions have been so universally noticed that the use of a drag harrow following a flat roller is always recommended.

Corrugated rollers. Corrugated-iron rollers are made with either solid or hollow iron wheels which come to a rather sharp edge on the outer circumference. This corrugation may be from 1 to 3 inches deep, curving out to form a flange on either side. The edges of the flange on adjoining wheels touch one another leaving the corrugations from 3 to 5 inches apart. It will be readily seen that not only will this type of roller crush clods, but that the corrugations will cut into the soil, causing some subsurface packing.

Corrugated rollers are made on a straight spindle or in sections. Those made in sections are more flexible and will do a better job where the land is uneven.

The heights of the roller wheels vary widely, most of them ranging from 16 to 20 inches. These higher wheels are used exclusively on the single rollers. Where a trailer is used, forming a double roller, the height of the wheels is greatly reduced, ranging from 8 to 12 inches.

The question is often asked as to which type of roller is to be preferred. Both single and double rollers do good work. The single roller is a little easier to handle in stalk land that has been disked up for oats or other small grain. Where the soil is very mellow and contains stalks or trash, this material is apt to push ahead to a certain extent of the small-wheeled roller. Other than this, little criticism can be made of either type. Every farmer who grows corn and small grain should have a corrugated roller.

Rolling spring-sown grain. Where oats or other small grain follows corn, the land is usually disked and the grain broadcasted or drilled. In either case, the corrugated roller can be used to great advantage as the last

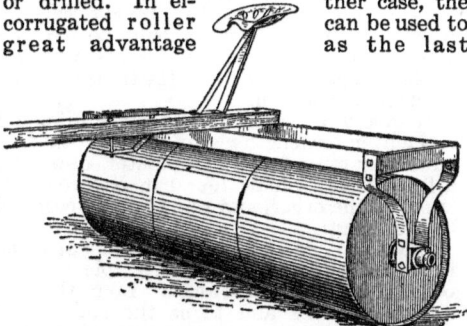

FIG. 126a. One form of soil packer, an implement particularly useful on light, loose soils

thing in the seedbed preparation. In Grundy County, Illinois, members of the Agricultural Improvement Association made tests showing the effect of the corrugated roller used at this time. Where oats were broadcasted, the stand was increased about 30 per cent. This was evidently due to pressing kernels, that had not been covered, into the soil, and also to the fact that the soil was pressed firmly around practically all the seed, holding the moisture and thus causing rapid germination. Usually, in broadcasting, many kernels lie near the surface, or in crevices where there is little or no fine dirt, dry out shortly, and never grow.

Rolling winter wheat and other small grain. Rolling winter wheat in the spring as soon as the ground is dry enough to work is an exceedingly profitable practice. The roller closes the cracks left by the freezing and thawing weather of early spring. This not only covers the exposed roots, but saves moisture. In a few days after rolling, a piece of wheat that has appeared quite dead, and which probably would never have come out except under the most favorable weather will begin to show life.

Oats, barley, and spring wheat can all be benefited by rolling in a dry spring, even after the plants are 3 or 4 inches high. It will be better, however, to roll at the time of seeding as explained above.

Rolling ahead of the corn planter. In any section where corn is grown, it is excellent practice to run a corrugated roller ahead of the planter. This firms the land sufficiently to enable the corn to be planted at a uniform depth; neither does the planter run too deep.

It is common practice in the corn belt to use the drag harrow as soon as planting is finished, usually crossing the rows. If the ground is very mellow and not rolled, the corn planter tracks are filled, burying the corn from 3 to 5 inches deep. Under this condition, even in favorable weather, the vitality of the seed is exhausted before the plants get out of the ground, and in cool, wet weather the kernels often rot.

Where the roller is used, the corn is planted shallow; and, if the soil is dry, the compacting of the roller causes enough moisture to rise and collect around the seed for germination. If the weather is wet, the seed is still close

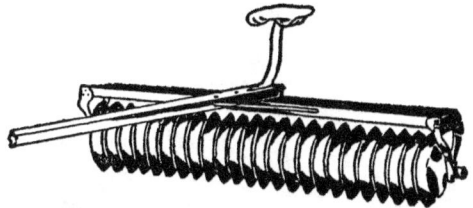

FIG. 126. Common, metal land roller, an implement that could and should be more widely used

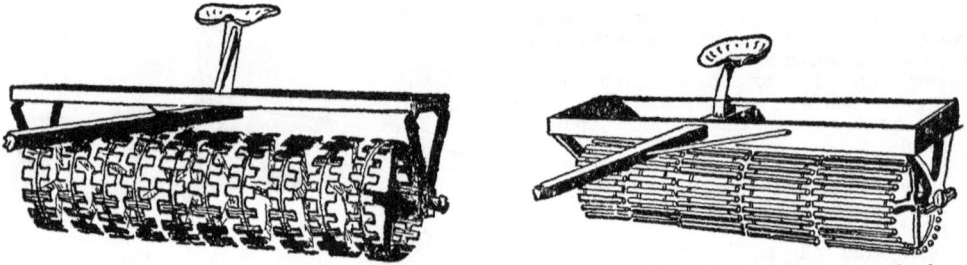

FIG. 127. Two modified forms of subsoil roller or packer. Implements of this type have been most developed and most utilized in the dry-farming operations of the semi-arid West

enough to the surface to get enough air to prevent decay.

The subsurface packer. The subsurface packer is more like a corrugated roller than any other tool. The wheels are farther apart, generally open, with a sharp rim. It is the object of this tool to cut rather deeply into the soil, splitting clods, and,

most of all, to firm the bottom of the furrow slice.

The subsurface packer has found great favor in regions of light rainfall, where a seedbed that is firm underneath and fine on top holds moisture to best advantage and presents the most ideal condition for the growth of crops.

Cultivators

There are a great many cultivators on the market and they present a large number of varieties of construction and manner of operation, most of which, however, are but different means of accomplishing the same end. All cultivators may be divided into 4 classes, according to construction—shovel, disc, spike-tooth, and surface—and into 2 general classes, in reference to soil topography, namely, those adapted to hilly fields, and those for use on level or slightly rolling lands.

Pivot-frame and pivot-axle cultivators. With hilly fields, the best practice is to cultivate them across the slope, to prevent washing. This procedure presents a difficulty at the outset in managing the cultivator.

For this kind of work there are 2 general types of cultivators now in use, the pivot-frame and the pivot-axle. Both of these implements allow the operator to incline the wheels up the slope, in order to keep the machine to its work. The pivot-frame accomplishes this by changing the direction of the pole, without interfering with the team, while the pivot-axle works directly on the wheels, changing their direction immediately, according to the way the operator pushes his foot. In both im-

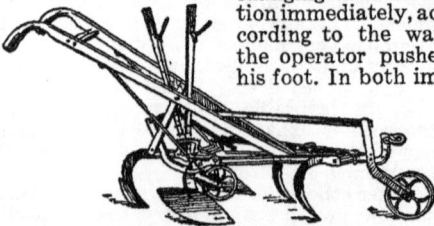

FIG. 128. The one-horse, single shovel plow was one of the first cultivators and the ancestor of our modern highly efficient soil-stirring implements.

plements the same device is used in dodging, whenever this is necessary.

The two-row cultivator. On account of the growing scarcity of farm labor, the 2-row cultivator is coming more and more into favor on the larger farms, where the land is not so rolling as to prevent its use. Many makes are now so improved that the work that can be done with them is equally as good as that of the 1-row machines, while the cost of the work is greatly reduced.

A good 2-row cultivator must have a strong, well-supported frame, to prevent sagging. A multiplicity of levers should be avoided. The dodging action should be positive, that is, the gangs should shift in the same direction as the foot is pushed.

It should be borne in mind, when using a 2-row cultivator lengthwise of the field, that the cultivator must follow the planter, otherwise, a little crook will be found in one row that is not in the other, and some hills of corn will be plowed up.

The disc cultivator. The disc cultivator is a distinct departure in principle from all other types of cultivators. Like the disc harrow, it cuts and turns the soil instead of tearing it. It is especially adapted to land infested with morning-glories, quack grass, smartweed and other strong-growing weeds that are hard to eradicate. The discs will not dodge around a tough root, and are not easily loaded up as is a shovel or scraper blade.

The gangs of a disc cultivator are reversible, and can be set to throw the dirt toward the crop or away from it. Each cultivator should be supplied with knife levelers which work in the opposite direction to the gangs, thus preventing hilling the corn or leaving a thin bare ridge when throwing the dirt from the crop.

The listed corn cultivator. This cultivator also is of the disc type, and is usually built as a 2-row machine. Listed corn is planted in the bottom of a trench; and it is the object of this implement to fill the trench gradually, as the corn grows, until the land is level enough to use an ordinary cultivator.

This cultivator has a low, heavy frame to which the cultivating gangs are attached. They have free side play through roller bearings. This enables the discs to follow the trenches, and avoids cutting out the growing crop. The success of this cultivator depends, therefore, on its ability to follow the inequalities of the trenches closely.

The surface cultivator. The surface cultivator is usually fitted with 2 scraper blades on each side, in the place of shovel gangs. Each blade is attached to a single standard, where adjustments to change the angle and pitch of the blade are provided.

The tending of all kinds of cultivated crops with this tool is rapidly coming into favor. Owing to the fact that the blades may easily get out of position and do little or no good instead of the first class work of which they are capable, the following specific directions are given for adjusting a surface cultivator.

FIG. 129. Steel frame disc cultivator with discs set to hill up along the row

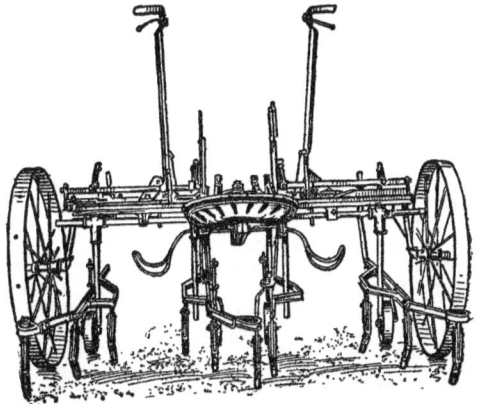

FIG. 129a. Two-row, riding cultivator, one of the means by which a farmer can win his battle against drought and weeds.

How to adjust a surface cultivator. Prepare the cultivator for the season by sharpening and polishing its blades.

Surface cultivators are usually supplied with 2 blades which are somewhat narrower at the heel and 2 which are of the same, or nearly the same, width throughout. Place the narrow row pair on the outside, and have the other pair trimmed down by a blacksmith to not more than 2 inches at the heel. Place these next to the row. Run the machine on a level surface—a floor or a plank will do—raising the tongue to the height carried by the team.

Loosen all set nuts, using kerosene and oil, and level all the blades with the surface on which the wheels rest; set them as nearly flat as the knuckles or blades will allow, and with as little angle to the row as will safely cover the surface. It will be found that the blades will clear themselves of trash best if the point runs slightly higher than the heel. This is particularly true in spring-plowed stalk ground. Use a rigid shield; if the machine is supplied with a floating shield, lock it. Place the blades about 4 inches apart and just high enough to permit enough soil to run under them and reach the row covering the small weeds that may have started at the hill. Blades should run only deep enough to keep soil flowing over them thus leaving a mulch between rows.

For the second cultivation, lower the standards carrying the outside blades a half inch. Move the gangs a little farther apart and turn blades to a slightly greater angle with the row. This will be the deepest cultivation. At any rate, the following cultivation should be more shallow. For the third crossing, lower the outside blades another half inch and set the gangs farther from the row.

The fourth cultivation should lay the corn by; set the outside blades flat once more, and raise the inside ones at the heel, so that they will only have the dirt run over them when crossing the ridges.

Don't stir loose dry soil, if not weedy; and don't get close to corn when ground is wet, thinking it will do less harm to the roots. Don't cultivate in wet soil anyway, if you can help it. Try to move just as little soil as possible. Have a good seedbed, and make it as level as you can. Don't be afraid to get close the first time, but keep away the third and fourth.

Remember the machine may get out of adjustment. Check up on the blades often by placing a straightedge under them. Never let any blade scrape the loose topsoil away, if it won't stay under, find out why.

The weeder. The weeder is really a spring-toothed cultivator with very narrow pointed teeth. These tools are built in sizes to take from 2 to 4 rows at a time. The 3-row weeder is the size usually found in the corn belt.

FIG. 131. Riding cultivator equipped with surface cultivator blades and, behind them, surface smoothing rakes to break down clods and ridges.

FIG. 130. Reversible riding disc harrow, with the discs set close so as to throw the soil away from the row; and (a) with discs spread and reversed so as to straddle a row

The most improved types of weeders are mounted on wheels and equipped with a spring-pressure lever to adjust the depth of the teeth.

A weeder can be used to advantage in tending practically any cultivated crop during the first period of its growth. Where the weeder is run over a field just as the crop is coming up, any crust that may have formed is broken and the sprouting weeds are destroyed in great numbers. This operation can be repeated in a few days and will both keep back the weeds and leave the ground in such good condition that the crop will generally get to a sufficient height to be easily cultivated before it is necessary to put a cultivator in the field.

Shovel versus Surface Cultivator

We usually expect to attain 3 objects by cultivation, namely: to kill weeds, to save moisture, and to aërate the soil. The University of Illinois has made repeated tests which show conclusively that killing weeds has more effect on the crop than any other factor, and that cultivation need only be deep enough to accomplish this. This being the case, the use of a shovel cultivator throughout the growing season, especially if run to any depth, is harmful since many of the roots of the crop are badly pruned by it while the weeds are being eliminated.

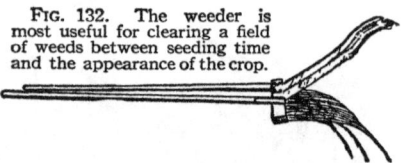

FIG. 132. The weeder is most useful for clearing a field of weeds between seeding time and the appearance of the crop.

Many of the best corn growers over the whole country are using surface cultivators altogether. Many more are using combination rigs, discarding the shovels after the second cultivation in favor of the scraper blades.

Practically all of the machines described above, except the disc, can be secured equipped with shovels or scraper blades. Owing to the importance of this matter of cultivation, it is advisable, when buying a new cultivator, to get one that is interchangeable for shovels and scraper blades.

Above, a light delivery truck well equipped with springs for the handling of easily damaged produce. Below, a powerful type equal to any task that a farm can impose

MOTOR TRUCKS ARE MADE IN ALL SIZES, TYPES, AND CAPACITIES. THE FARMER WHO REALLY NEEDS ONE HAS BUT TO CHOOSE THE KIND BEST SUITED TO THE MAJORITY OF HIS NEEDS

The old and the new in farm power. In the South "40 acres and a mule" is giving place to "400 acres and a tractor," and similar changes are occurring everywhere

A small tractor of the caterpillar type operating an ensilage cutter

THE SMALL FARM TRACTOR AS A COMBINATION MOVABLE AND STATIONARY SOURCE OF POWER IS ALMOST REVOLUTIONIZING CERTAIN PHASES OF FARM WORK

FIG. 133. Potato planter of the force feed type. The man in the rear feeds the seed pieces to the dropping mechanism

CHAPTER 8

Machines for Seeding and Planting Crops

By F. H. Demaree (see Chapter 7) who says of the implements described here: "As Agronomist with the J. I. Case Plow Works, I was afforded ample opportunity to study the construction and operation of all kinds of corn, cotton, and peanut planters and listers. I found, as most people finally do, that it is the successful operation of a machine that appeals most to the farmer. In most cases a great deal can be accomplished simply by following the directions given by those who make and know the machines. In following the work of drills and seeders in many sections of the country, I find that it is essential to get the style of furrow-opener that best suits local field conditions, then to standardize the drill every season. In starting my drill, I measure the correct amount of seed for 1 acre into the drill then set the drill as nearly as I can to sow that amount. I keep changing the feed until the right amount is seeded. This is the only way to get a proper stand." In view of the tremendous importance of proper seeding, which alone may mean the difference between a good crop and a poor one, the following information—based on practical tests and first-hand observations—should prove of very great value to the farmer.—EDITOR.

THE development of planters and seeding machines, as we have them now, is even more recent in point of time than that of plows and other tillage implements. In the early days of America corn planting was all done by hand, much of it on unprepared land. Kernels were pushed into the soil with a stick, or covered with a hoe, on a piece of clearing where the burning of stumps and brush had insured the land being free from weeds for a season or more. Even at the time the western prairies were being opened up planting was all done by hand. The land was plowed, furrowed out, then the corn dropped and covered with a hoe.

Later the planter, as we know it now in its essential parts, was developed. The first of these machines of the 2-wheeled type was the old "Step Drop." A boy sat between the wheels and worked a hand lever, dropping a hill at each step of the team. From then on, the drill and check-row types were rapidly perfected.

The use of drills and broadcast seeders is also very recent. Hand sowing had been the rule from time immemorial until the present day mechanical age brought the development of these

FIG. 134. Gravity-feed potato planter showing fertilizer attachment in front of driver

machines. The first seeders were all of the broadcast type, represented by a box mounted on wheels with an agitator inside the box and openings for the grain to fall through. This was far from a perfect seeder, but for many years it was the best available, for the force feed as described in this chapter has not been in existence much more than a generation.

Planters

CORN PLANTERS. Few farm implements offer such direct profit-making opportunities by correct use as do planters and other kinds of seeding machines. Corn planters may be roughly divided into two classes—the round-hole plate machines and the edge-selection planters. Machines of the former class may be subdivided into two classes—the single-cell and the full-hill drop. Both of the first two classes of planters mentioned may have either intermittent or continuous drive.

There can be little doubt that, so far as accuracy is concerned, the single-cell machines have outclassed those of the full-hill drop type; and it is, doubtless, for that reason that the great bulk of corn planters now sold are of the single-cell type, either round-hole or edge-selection plates. There may be considerable advantage in the better machines with intermittent drive, in that friction is considerably reduced and the life of the corn planter is prolonged. The drop mechanism is generally rather delicate, and a quick wearing of the parts is objectionable because it always means inaccuracy. Without proper adjustment throughout and a harmonious working of parts the highest accuracy cannot be obtained, and slight wear is often the cause of inaccuracy.

Value of individual hills. There are 3,556 hills of corn on an acre of land, planted 3 feet 6 inches apart. If each hill should produce 1 pound of corn, and a perfect stand be secured, the yield will be 50.8 bushels per acre, basing the field weight on 70 pounds per bushel. If this yield per hill can be doubled, then the yield per acre can be raised to 101.6 bushels.

As the yield increases, many other factors enter in to hold down the yield outside the question of stand, but for any good yield a first-class stand must first be secured. The illustration is introduced to show the real necessity of proper planting.

Careful examination of many corn fields in practically every corn state in the union has shown that often very good farmers have as low as 70 or 80 per cent of a perfect stand. In such cases the farmers have been plowing, planting, and cultivating from 20 to 30 per cent of their land with no prospect of return.

Round-hole versus edge-selection plate. Both of these types of corn planter have their advocates. The round-hole plate selects the kernels in their natural position. This is a strong argument in favor of the more accurate drop from this type of plate, as efficiency is bound to decrease when any artificial element, such as changing the position of the kernels in some way which is unnatural, creeps in. On the other hand, it is claimed that as corn varies considerably less in thickness than it does in width, the edge-selection plate is on this account more accurate than the round-hole type. Doubtless the different makes of both of these kinds of planters are accurate when properly operated.

In making a test with these two kinds of plates, the writer found a very interesting thing, which, to the best of his knowledge, had not then been made public. In both planters corn was used which had come out of a car of shelled corn and had only been graded once. There was, consequently, a rather large number of extra large and extra thick kernels in it. After a full planter box of corn had been run through each machine, it was found that the round-hole machine had planted every kernel in the box, whereas the edge-drop machine retained nearly a double handful of these extra thick kernels, which could not slide down into the cells. It became apparent that, if a farmer in using one of these machines did not have extra well-graded corn, after 10 or 15 acres had been planted, these "off-shape" kernels would collect in the bottom of the boxes, and, unless they were cleaned out every night, would seriously interfere with the accuracy of the drop of the machine.

The value of grading seed corn. First-class results with the edge-selection machine are practically dependent upon the thoroughness of the grading of the corn. This is not so true with the round-hole type, but even this machine gives better results if the corn is well graded. It may be mentioned that the efficiency of graded corn can be materially increased by shelling off by hand the butts and tips of the seed ears, and then throwing the ears into the different piles according to the width of the kernels.

Testing the planter. With every planter there is furnished a rather large number of plates, which are designed to cover such a wide range of corn that one set, at least, should give better results than any other with the particular corn to be planted. In judging results it is hardly enough to take a few kernels of corn and see whether or not they fit in the cells snugly or too loosely. The action of the planter while running has some effect in filling the plates. Neither is it enough to take the machine out of the shed, fill it up on the day you expect to plant corn, and then drive off down the road a few rods, letting it plant in the meantime, and then go back and count results. The jar on the machine is very heavy on the road, and more corn is apt to be sent through than will be the case in actual field conditions; hence, results are likely to be misleading.

It is much better to jack up the drive wheel of the planter and put in the plate which is thought best adapted to the corn to be planted; then run through at least 200 hills, if the corn is to be checked, counting accuracy on the basis of 2, 3, or 4 grains to the hill. If the corn is to be drilled, one kernel should come down at each click, which represents a cell as having passed the opening. Different plates tried in this way will often show surprising differences as to the degree of accuracy obtained by the machine, and will be found to have a direct bearing on the percentage of perfect hills that will be found in a field of corn at its maturity.

Insuring a good check. On lands that are level or only slightly rolling, the common practice is to check corn, in order that it may be cultivated both ways. Such cultivation aids materially in holding the weeds down. Often, however, the well-laid plans of the farmer are completely blocked by getting the check so poor that it is impossible to plow both ways. It is impossible to set a planter at the factory so that it will check accurately under all conditions. Just take hold of the pole of your planter and lift it up high. It will be seen that the shoes angle very little. The corn will drop almost straight down. Lower the pole, and the bottom of the seed can is several inches back of the seed can. The relative position of the shoe to the button

on the wire means a good check or a poor one.

In order to be sure that you have proper relation between these 2 parts, it is necessary, after you have planted the first 2 rows and gone back several rods on the second 2, to get down and dig up several hills of corn. The corn should be found 1 inch or $1\frac{1}{2}$ inches behind the button. If it is found that the corn is not being dropped quickly enough, that is, ahead of the points mentioned, then loosen the bolt that passes through the pole and uprights on the front frame and raise the pole higher. This will make the shoe stand back and allow the corn to drop before passing the button. If there is no such adjustment on your planter, shorten the neck straps which hold up the neckyoke. If the planter is dropping too quickly, that is, more than an inch and a half behind the button, reverse the pole adjustment.

The accompanying cut (Fig. 136) shows 4 rows of corn. The phrase "hill behind button" refers to the way the team is headed. Note the position of the kernels in reference to the button. After turning the team around, the wire is stretched at about the same tension as before, but the pull of the planter brings the buttons back past the spot where they were pulled by the planter on its opposite trip. Consequently, unless the corn drops behind rather than in front of the button each time, it will be badly out of line.

The clutch. When checking corn, the clutch controls the speed of the plates as well as the number of holes that pass the cut-off, thus determining the number of kernels planted. On most corn planters the clutch is found on the drill-shaft. Owing to the speed of the drill-shaft and the relative delicacy of the clutch parts, it is subject to a great deal of wear. Any failure to take hold, or slipping on the part of the clutch, is a contributing factor to uneven planting. In recent years one progressive manufacturer has taken the clutch from the drill-shaft entirely and

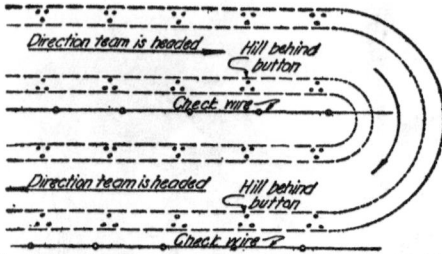

FIG. 136. Diagram to show good checking obtained **by** careful adjustment (see text, p. 89)

placed it on the axle. The parts are heavy and strong, yet they work with great precision and accuracy.

It is especially desirable to test the clutch of the corn planter thoroughly every spring. Do not take chances with worn parts; get new ones.

Fertilizer attachments. Attachments for distributing fertilizer are coming into increasing demand as the fertility of the land diminishes. These attachments are merely large cans, mounted on a support bolted to the main frame of the planter and driven from a sprocket on the axle. From the opening in the bottom of the can a hose leads to the heel of the planter shoe. The amount of fertilizer applied is controlled by the size of opening. The cans are equipped with a force feed device to ensure a constant delivery of fertilizer into the hose.

When using fertilizer for corn in this way, it is always advisable to drill it in the row. If checked in the hill, there is some danger of burning the young plants, or of causing too great a concentration of roots at one spot. Where drilled in the row, the roots spread in a normal manner. In addition, the cultivator usually spreads the material more or less, and the succeeding crop gets a greater benefit from the residue.

Pea and bean attachment. The practice of planting cowpeas or soy beans with corn is an excellent one. A special attachment for almost any planter can be secured for this work. These attachments consist of small cans, set immediately behind the corn boxes with hose leading into the planter shank. The gearing is attached to the drill-shaft, so that the plates operate in unison with the planter plates.

Cowpeas or soy beans planted with corn make an ideal combination for silage, hogging down, feeding-off with sheep, or to put in the shock for winter feeding. As near as can be determined, there has been practically no reduction in yield of corn due to the presence of the peas or beans.

Single-row planters. Many single-row planters of the walking type are used in the east, south and southwest sections of this country. The plates used in them are the same as in the ordinary 2-row machines.

One other feature has been subject to considerable change, that is, the device by which the plates are driven. Three methods have been used—the chain-drive, shaft-drive, and pitman-drive. The pitman-drive is the most recent and has become very popular, on account of its simplicity and durability. A pitman-rod is attached to each side of the drive wheel, so set as to work alternately. The plate is equipped with extension lugs and one pitman pulls it forward a notch then the other pushes a notch. Thus a hole passes the cut-off at every impulse, and a continuous rotary motion is given the plate.

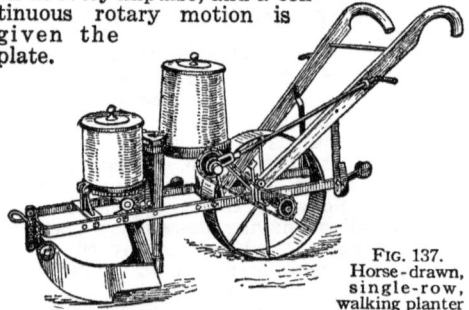

FIG. 137. Horse-drawn, single-row, walking planter

COTTON PLANTERS. Regular cotton planters are made both single- and double-row. They are equipped with shovel or sweep to open the row, and with shovels or discs in the rear to cover the seed.

The usual device for getting the seed out of the hopper into the hose leading to the ground has been a picker wheel. This little wheel with many sharp projections revolves in the opening from hopper to hose, tearing the seeds apart and forcing them in a steady stream into the hose. The use of the picker wheel invariably means thick planting, greater cost of seed, and increased cost of thinning.

Recently, there has been patented a cell drop that handles the seed with a plate much like corn. This planter bids fair to revolutionize methods of cotton planting. The cell-drop mechanism has a polished cell plate with sharp, hooked projections extending in the same direction that the plate is traveling. Above the cell plate is a retarding plate to prevent bunching of the seed, and above this an agitator plate which shakes and tears the seeds apart so that they will feed into the plate below. The cell plate is equipped with a cut-off, to insure greater accuracy. The cell drop is able to plant the cottonseed much thinner, thereby reducing the cost of seed and of thinning.

Cotton is also planted with the 2-row shoe type of planter, which is very much like an ordinary corn planter in appearance. These planters are commonly used when the land has been prepared in advance of the planting. The picker-wheel feed is used so far on these machines. They are interchangeable for corn by exchanging the cotton feed for corn plates.

Peanut attachment. Peanut attachments can be secured for most cotton and corn planters, either riding or walking type. The regular planter hopper is removed and the peanut hopper put in its stead. The nuts are taken out of the hopper by means of an endless-chain cup conveyor and dropped into the hose that leads to the furrow. Planting distances are controlled by the use of different sized sprockets which increase or diminish the speed of the conveyor.

Listers. A lister is a combination plow and planter, equipped with either corn or cotton plates. The plow is of the "middle-buster" type, that is, one having a double moldboard which in digging a trench throws the dirt both ways.

The middle-buster is set in a frame to which the planter box is attached. The whole implement must be strongly made and easily adjusted, to meet the adverse conditions under which it has to work. Both single- and 2-row types are made.

Why listers are used. Of more importance than the type of implement is the necessity for its use. Listing is a common practice in the central and southern portions of the Great Plains states. Where the annual rainfall is light, every method of conservation of moisture must be used.

Listed land is left with furrows and ridges at regular intervals. Land in this condition offers the best opportunity for rain to soak quickly into the subsoil, where it is protected from evaporation. Snow, also, is caught in the trenches and held far better than would be the case on an unbroken surface.

By planting in a furrow, weeds in the row are more easily killed; and as the furrows are filled the roots of the plants are left at a much

FIG. 139. Single-row, riding type cotton planter

greater depth than with surface planting. The crop stands dry weather better and is not so apt to blow down.

Methods of listing. Three methods of listing are used. Single listing is perhaps the most common. With this method the field to be planted is left untouched from the previous crop (except, perhaps, for the cutting of old stalks with a stalk cutter or disc) until the new crop is to be planted. Then the lister is put into the field and the seed planted in the furrows as they are made. The entire surface between the rows is covered with loose dirt, but a hard, unbroken ridge is left underneath. This method, while economical, should be used only on very deep, loose soils, if best results are to be expected.

Double listing differs from single listing in that the ground is listed twice. Usually the first job is done in the fall or in winter. At planting time the ridges are split and the seed is planted. This makes a better soil preparation, as all of the land has been stirred, more weeds are killed, and cultivation is easier.

Where land can be plowed either in fall or in winter and listed in the spring at planting time, better results can be secured than by either of the other methods. There is, however, a considerable increase in the labor cost. In thus preparing land the loose-ground lister is generally used. This machine is much the same as an ordinary corn planter, but is equipped with large double-disc furrow openers to form the trenches.

TRANSPLANTERS. Wherever tobacco, cabbage, strawberries, sweet potatoes, tomatoes, or other crops requiring transplanting are grown in commercial quantities, a transplanter is a labor saver and a money maker. The design of this machine is quite simple. It has a heavy triangular frame, low down, usually swung under the axle of the 2 main

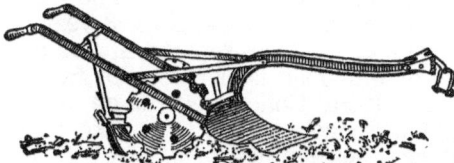

FIG. 138. Single-row, walking lister adapted to small scale corn or cotton planting

wheels. The point of the triangle is in front, supported by a pivoted truck. On top of the frame a large barrel is mounted for the water, and in the rear are a furrow opener and 2 seats for the planters. The barrel is equipped with a water valve which is tripped automatically. As the water is released into the furrow, one of the setters places a plant

FIG. 140. The riding lister does exactly the same things as the walking type but does them quicker and with less effort for the operator.

in it, one man setting every other plant. The coverers pull the loose dirt around the plant, compacting it slightly in the process. Distances between discharges of water are regulated by the use of different sprocket wheels.

The use of water allows plants to be set in the heat of the day without wilting, which is impossible with hand setting.

POTATO PLANTERS. Improved potato machinery is causing a steady increase in the interest shown in this important crop. Outside of planting and harvesting, the labor required to produce a potato crop does not differ materially from that in the production of a corn crop. A large quantity of seed is required, ranging from 8 to 14 bushels per acre. This quantity is heavy and cumbersome and is in itself a discouragement to the man who would like to grow potatoes.

The modern potato planter has overcome most of the labor difficulties in planting. The seed can be cut by machine, and the planter will put the pieces in the ground with remarkable accuracy.

How the potato planter works. The planter is provided with a large hopper in which the seed is placed. In some planters, the seed runs by gravity into the picking chambers; in others, by a feed wheel which is practically a force feed. The latter construction prevents the flooding of the picking chamber, and makes far more accurate planting.

In the picking chamber the seed is caught on the point of sharp, revolving pickers and carried over to the spout leading to the furrow opener. Planting distances are controlled by the use of different sprockets that change the speed of the pickers.

The furrow opener should make only a narrow trench into which the seed is dropped and immediately covered. The double-disc type of coverer, set to throw in, is universally used. This not only covers the seed, but starts a good ridge. Where a slight ridge is raised at planting time, a harrow or weeder can be used before the potatoes are up and the weeds practically eliminated between hills in the row. The cultivator will take care of those between the rows.

Single- and two-row styles. Potato planters are made in both single- and 2-row styles. The 2-row planter is, of course, heavier to handle; but it decreases the labor cost of planting, and is just as efficient as the single-row machine. In addition, a 2-row cultivator can be used to tend the crop by following the 2 rows of the planter at all times. Otherwise, a 2-row cultivator cannot be used in the potato field.

Drills

All small grain is now seeded in one of two ways—by drilling or by broadcasting with either an end-gate seeder or a broadcast seeder.

Practically all wheat, both spring- and fall-sown, is drilled in. The farmers have seemed to recognize the importance of using a drill to put in this crop. This is not so true, however, of oats and barley, particularly oats. In experiments conducted at various experiment stations, to determine the effect of drilling compared with broadcasting on the yield of oats, it has been found that a definite increase in yield can be secured by drilling instead of broadcasting. The Illinois Experiment Station secured, as a general average on 3 such fields for 3 years, an increase of 3.9 bushels per acre in favor of drilling. Kansas reports, after a long-continued experiment, an average increase of 5.3 bushels per acre in favor of drilling over broadcasting. The Ontario Agricultural College secured an increase of 4 bushels per acre as a result of drilling. Moreover, at the Iowa Experiment Station, a 4-year trial with winter wheat gave an average yield of 4.2 bushels more grain per acre for drilling as compared with broadcasting.

Additional advantages of drilling. Aside from the question of yield, there are the following advantages in drilling small grain: The seed is deposited in rows in firm, moist soil and covered at a uniform depth. An accurate amount of seed is distributed, less seed being required than in broadcasting. The ground is left with alternate small furrows and ridges. This favors the absorption of both heat and moisture, and gives some protection against drifting soil and winterkilling. Snow is held in the furrows to an appreciable degree.

Further, drilling is decidedly an advantage when clover or grass seed is to be sown with the grain. This work can be done at the same time the grain is seeded, thus saving labor. Owing to the more uniform stand and the unused space between the rows, the young seedling has a better chance to make a good stand than in broadcasted grain.

In sowing grass or clover seed in connection with a grain crop, one caution should be observed. Take the seed tubes off and let the seed broadcast ahead of the furrow openers. The dirt falling in together with the action of the drag chains will cover it sufficiently deep. Deep planting of small seeds is one of the main reasons for a poor stand. The seed fails to grow, or exhausts itself in pushing the young plant to the top of the ground. When using the drill for small seed alone, put the grain tubes on again, but set the furrow-openers to run very shallow.

Styles of furrow openers. From the farmer's standpoint, drills are usually classified according to the kind of furrow openers with which they are equipped. There are 5 kinds of furrow openers in use in various sections of the country, namely: hoe, shoe, single-disc, double-disc, and combination shoe and double-disc, commonly called the "open-furrow" type.

Hoe furrow openers. The hoe furrow opener was found on the early drills and is still popular in many localities. It does excellent work in a clean, well-prepared seed-bed. The hoe drill is at a disadvantage in trashy ground or where grain is to be drilled in stalk land that has only been disked. The delay caused by loading up has practically eliminated the hoe drill from this class of work.

One of the recent practical improvements in the hoe furrow opener is the attachment to each hoe of a spring trip. When any hoe hits an obstruction, it trips without breakage of any parts and swings back into working position.

Shoe furrow openers. The shoe furrow opener looks much like the ordinary corn planter shoe. This type of furrow opener, also, does excellent work in a clean, well-prepared seedbed. With the shoe, the depth of seeding can be very accurately adjusted, which makes it most desirable when seeding in alfalfa or other small seeds. In trashy ground or in stalk land, the shoes also load up, making them undesirable for this class of work.

Single-disc furrow openers. The single-disc furrow opener is by far the most popular type, serving the purposes of good seeding in the greatest variety of soil condition. Since half of the discs in the drill are set to throw in one direction and half in the other, this type of furrow opener has a decided cultivating effect that farmers have not been slow to appreciate. The single-disc drill will do good work in any type of soil that is not too stony, and it cuts through trash and corn stalks in an excellent manner.

The forward delivery of seed is an important improvement for the single-disc furrow opener. In the open-delivery style of disc furrow opener, the seed does not reach the bottom of the furrow before the upturn of the revolving disc catches it; thus the seed is often thrown on top of the ground. In the forward delivery the seed reaches the ground before the upturn of that section of the disc sets in and is consequently well covered.

Double-disc furrow openers. This furrow opener is formed by two straight, flat discs set at an angle toward each other, coming together where they enter the ground. This type of furrow opener is used extensively in heavy wheat lands of the Northwest. The double-disc has practically no cultivating action, but cuts through the soil like a shoe opener without turning it and without the consequent loss of moisture.

It also has an added advantage over the shoe

Fig. 141. Grain drill of the single-disc type with drag chains for covering the furrows

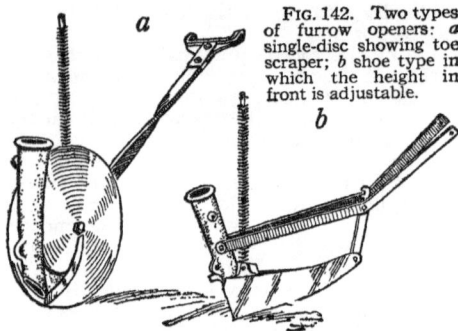

FIG. 142. Two types of furrow openers: *a* single-disc showing toe scraper; *b* shoe type in which the height in front is adjustable.

opener in that it will go through trash or stalks without clogging. In fact, it is practically impossible to load up a drill of this type, but the double discs do not cut through trash as well as the single-discs and under such conditions will not cover the seed quite as well.

Combination shoe and double-disc furrow openers. In this style of opener, the shoe is set between two concave discs and makes the first cut of the furrow. The discs are set at equal but opposite angles, and they throw the soil to each side, leaving a wide furrow from 2 to 6 inches deep, as desired. The dirt in the ridges on either side of the furrow gradually sifts back during the winter and spring until the land is nearly level again.

This style of furrow opener is made especially for the South for seeding winter oats and other small grain. The openers are from 14 to 16 inches apart and are often equipped with double seed spouts. This is especially desirable for winter oats, as they are invariably bearded, and do not feed through as rapidly as they should, to make a good stand, if a single-grain tube is used.

Force feed. There are two general types of force feed employed in grain-drill construction, the fluted-roll and upright-revolving-disc. The fluted rolls are attached on 1 long drill-shaft, and each roll is fitted inside a feed cup. The feed cups are usually fitted with an adjustable gate, which may be opened or closed for various kinds of seeds. The rate of seeding is controlled by a movable cut-off operated by a hand lever. This lever is set to act as a pointer for the scale which indicates the amount of seed being used. In the upright-revolving-disc type of force feed, the disc is placed at the side of the feed cup. Projections on the discs force the seed grain into the grain tubes. The rate of seeding is generally controlled by increasing or decreasing the speed of the discs.

Of the 2 types, the fluted-roll is the more popular and is used on the majority of drills now being manufactured.

Standardizing the drill. There is a wide variation in the size of seeds of the same kind, due to the variety, season, and soil

fertility. For instance, a kernel of Kherson oats is little more than half as big as one of Swedish Select. It is obvious, therefore, that a drill set to sow 8 pecks of large-berried oats will sow more than that amount of the small-berried kind.

The gauge or scale on a drill cannot be considered as more than an index. If correct seeding is desired, the drill should be checked up each year. For all practical purposes, the easiest way to do this is to measure into the machine the correct amount of seed per acre it is desired to sow. Set the drill for this amount, and drive ahead until the seed has been run out. Usually, the length of the field in rods is well known; the width sown can easily be measured. Multiply the length in rods by the width in rods, and divide by 160. This will give the area in acres or fraction thereof. If too little seed is being used, open up the drill; if too much, close the cut-off.

The condition of the grain at the time of seeding will also have considerable effect on the amount that will go through—whether it has been thoroughly recleaned, is absolutely free from chaff and trash, or whether these latter impurities are present. It should be especially noted that grain treated for smut with formaldehyde or hot water will swell considerably. This will affect the amount of grain put through to such an extent that in the latter case a standardizing of the drill similar to that described above will be essential.

Distance between furrow openers. There has been a distinct tendency in the last few years to close up the space between the furrow openers. Wide spacing gives weeds a better chance, and the seeds must be thicker in the row to sow the same amount per acre. This means crowding of plants, and it stands to reason that the grain will not do so well as when better spaced.

The most popular drills are now spaced only 6 or 7 inches apart. This close spacing has greatly increased the danger of clogging with stalks or trash. In order to offset this, most of the close-spaced drills are set zigzag, with every other furrow, opener set several inches

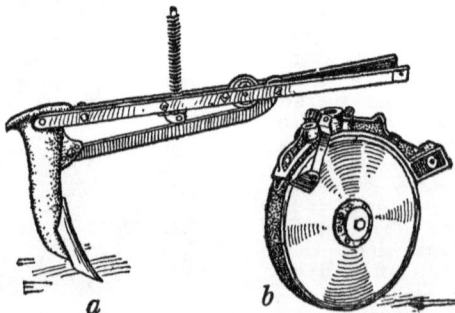

FIG. 143. Two more furrow openers: *a* simple hoe type; *b* double-disc type

ahead of those next to it. This in reality makes 2 rows of furrow openers.

Some manufacturers have gone so far as to set 2 drills in the same frame, so constructed that the back set of furrow openers split the middle left by the front ones. This makes very close spacing. The idea is excellent, but the machine is so cumbersome that it is not yet an unmixed success.

Fertilizer attachments. The fertilizer attachment for a grain drill is a secondary box next to the seed box, equipped with a force-feed device and spouts leading to the furrow openers. Owing to the fact that some fertilizers will "bridge," it is necessary to equip the fertilizer box with an agitator, to keep the material down to the force feed. The flat-disc type of force feed is the one in most common use. It gives positive results with a minimum of breakage.

Press drills. Press drills are made without supporting side wheels, but, following each furrow opener, there is a press wheel. This row of wheels is fastened on a long axle and set in a frame which in part supports the drill proper. Press-wheel attachments can also be secured for most types of grain drills. This equipment is used in sections where high winds blow the soil and seed badly and on light soils where compacting the dirt over the seed brings up the moisture and hastens germination.

The clover-and-grass-seed drill. The latest thing in seeding machinery for sowing clover, alfalfa, and grass seed is what is known as the clover-and-grass-seed drill. This machine is designed especially for sowing these seeds. It is built so that the seed is sown in furrows 4 inches apart, which is about the proper distance for a good stand. Another advantage of using this drill is that it permits the cultivation of the grain already above the ground and thereby tends to increase the yield.

The common practice is to sow 1 bushel of clover seed to 7 or 8 acres and 18 to 20 pounds of alfalfa to 1 acre. This is by the broadcast method. It is claimed and proved by tests that, with a clover-and-grass seeder, from 4 to 6 pounds of seed may be sown per acre, and that the stand will equal, if not surpass, that obtained by broadcasting from 2 to 5 times this amount of seed.

In 1 pound there are about 240,000 alfalfa seeds and 288,000 clover seeds. Since there are 43,560 square feet in an acre, then for every pound of alfalfa sown per acre there will be an average of $5\frac{1}{2}$ seeds per square foot, and for every pound of clover there will be $6\frac{1}{2}$ seeds per square foot. Farmers have been known to sow (broadcast) as high as 25 pounds of alfalfa to the acre. This is at the rate of 137 seeds per square foot, and no ground will support such a stand, if all the seeds grow. The Ohio Experiment Station found that 5 pounds of alfalfa seed, sown on an acre by a clover-and-grass-seed drill, gave a yield of 7,862 pounds of hay, and that 25 pounds, sown broadcast on the same acreage, gave a yield of 7,876 pounds of hay.

By the use of this drill the farmer is enabled to put the seed down in the moist earth at the right depth, where it germinates quickly. In this way, practically all the seed grows, and the tremendous loss resulting from broadcasting methods is eliminated.

The main disadvantages of the clover-and-grass-seed drill are that it means the purchase of extra equipment, and, where the seed is put in with spring-sown grain, the land must be drilled over again for the small seed alone. The value of saving seed and the better chances of securing a stand must then be weighed against increased labor and machinery costs. This is an individual problem.

One-horse drill. The 1-horse drill is the same in principle as the larger types just described, and it may be equipped with practically all the attachments. This type of drill is generally built with 5 furrow openers of either the hoe or the disc style. The advantages and disadvantages of each are the same with this drill as previously described. A drill with 3 furrow openers is also built for Southern use, where small grain is to be drilled between cotton or corn rows.

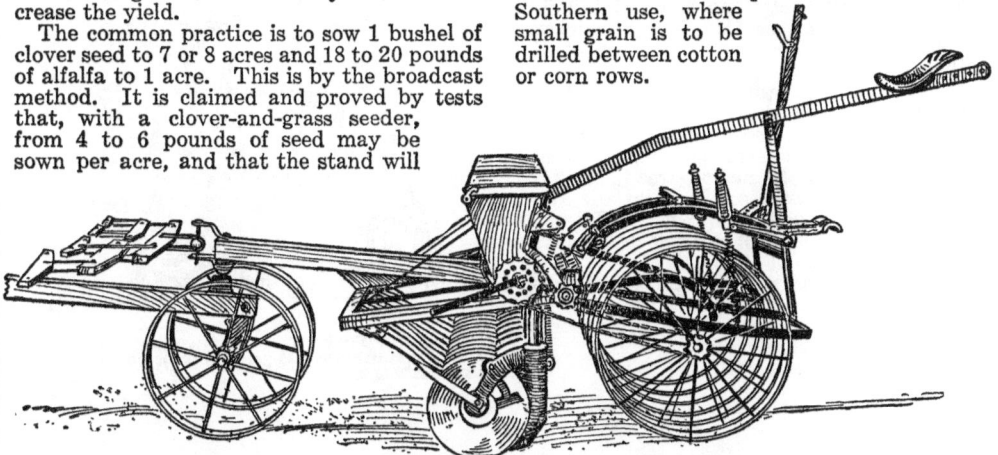

FIG. 144. Press drill. Note absence of regular wheels, their place being taken by the narrow ones, each of which firms the soil behind a furrow opener

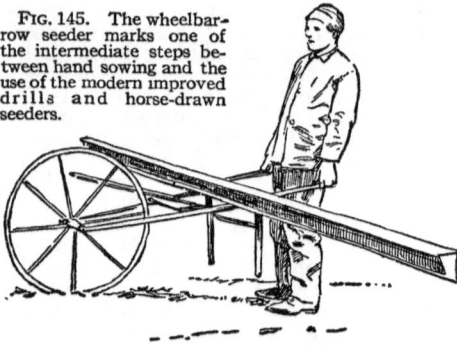

FIG. 145. The wheelbarrow seeder marks one of the intermediate steps between hand sowing and the use of the modern improved drills and horse-drawn seeders.

When drilling winter wheat or rye in standing corn, the drill should be run across the ridges, that is, in the opposite direction to that in which the corn was laid by. Otherwise, the outside furrow openers will go too deep and the middle ones too shallow. The matter of cross drilling should also receive careful consideration. In cross drilling, one half of the seed is sown when going one way; the other half, when crossing. This means twice the work but eliminates the unseeded portion left where the row of corn stood. In addition, the seedbed is put in much better condition. Since no previous preparation has been made, this extra work can well be afforded.

Broadcast seeders. Broadcast seeders are very similar to grain drills, so far as the seed-box and force-feed equipment is concerned. The seed, however, falls directly to the ground from the seed cups, and must be covered by other agencies. Some broadcasters are equipped with a cultivator attachment, which may be of either the shovel or the disc type. In the corn belt, however, farmers who broadcast their small grain prefer to work their land more thoroughly with disc harrows, so that if a broadcaster is used, it is merely for the purpose of distributing seed.

A good broadcaster must have a large, well-braced grain box. Sagging is very undesirable. The broadcaster is built in two styles: the wide-tread, having a wheel at each end, and the narrow-tread, in which the wheels are close together and the grain box extends beyond each wheel. There is practically no chance of sagging in the latter type.

End-gate seeders. The end-gate seeder is in common use throughout the corn belt particularly. It is used primarily for seeding oats, although wheat, barley, and rye are often sown broadcast with it.

In spite of data at hand showing the possibilities of increased production by drilling over broadcasting, the latter practice is by far the more common in nearly all parts of the corn belt. This is probably due to several factors. An end-gate seeder can be bought for less than one-fifth the price of a 4-horse drill. Seeding can be done considerably faster. Most farmers regard oats as a crop of secondary importance, and wish to get seeding out of the way as quickly as possible, in order to start work on their corn land.

The ordinary end-gate seeder is bolted on to a board that replaces the back end gate of the wagon, and is driven with a chain that extends from a big sprocket, bolted to one wagon wheel, up to the drive shaft of the seeder.

The grain is shoveled from the wagon into a large hopper, and delivered thence by a force feed into the distributing fans. The fluted-roll type of force feed, described above, is commonly used. An auger construction is also used to deliver the grain to the fans. Both types are efficient.

Double fans are now used in practically all end-gate seeders. They are set side by side, and run in opposite directions, both throwing away from the machine.

A clover-and-grass-seed attachment can be provided in the shape of a small extra hopper with seed spouts running down to the fans. This attachment can be operated either at the same time grain is being seeded or alone. Where clover or grass seed is put in with the small grain, it will rarely carry as far as the grain; consequently, bare streaks are left.

It is also not the best practice to disc down small seeds, as they may be covered too deep to germinate well. If the end-gate seeder is to be used, it is better practice to sow the grain; after this is disked in, sow the small seed and follow with a drag harrow or corrugated roller.

FIG. 146. For broad-casting grain, grass, or grass and clover mixtures, the end-gate seeder is an efficient and economical machine.

CHAPTER 9

Machines for Harvesting and Threshing Crops

By F. H. DEMAREE (see Chapter 7). In discussing plans for this chapter, Professor Demaree said: "It has been my privilege to have operated or worked with most of the implements described in this chapter. Here in Grundy County there is a growing interest in sweet clover. In the spring of the second year of its growth, the first cutting of this crop must be very high or the entire crop may be killed. Many plans were tried to hold the mower bar up to the desired height, until finally the plan described in this chapter was worked out. It has been uniformly successful.

"The seed-saving device is not original, but I have shown a number of farmers how to make the outfit which is an exceedingly valuable attachment when the binder is used to cut sweet clover. Many bushels of ripe unhulled seed will be saved in cutting even a small field.

"After having run one binder machine for 10 consecutive seasons, I can realize that it is an important and complicated machine. Mine is now 16 years old and still in use. Proper care, including housing, regular oiling, and keeping all parts tightened up, are all responsible for the long life this machine has enjoyed. My grandfather always said that oil was cheaper than machinery. I might add that a monkey wrench and a little time spent in tightening up nuts and bearings are both cheaper than a smash-up." It is practical knowledge such as this that counts whatever the type of farming and wherever it may be attempted.—EDITOR.

N O CROP can show a profit until it is finally marketed or utilized. The size of the profit (or loss) at that time depends on a good many factors, none of which is more important than the combined speed, thoroughness, and economy with which it is gathered. In this light, the tools with which this harvesting work is done, and their proper attention and handling, assume a correspondingly vital importance.

The mower. Owing to its labor-saving ability, the mower is an indispensable implement on practically every farm. It is not only used in the meadows at haying time, but it trims the roadsides, the pastures in the fall, and in a number of ways helps to avoid hand scything.

Taking everything into consideration, the mower, in spite of its usefulness, is not a very heavy-duty machine and should, if properly cared for, be made to last for a much longer period than the average farm tool.

In running a mower through heavy grass or clover, there is considerable strain on the sickle bar, as well as on the driving mechanism. When buying a mower, therefore, the principal features to take into consideration are strength of construction, fewness of gears with as direct drive as possible, and the necessary adjustments to keep the sickle bar in alignment.

97

FIG. 147. A modern, high-class vertical lift mower, the dotted lines showing the position of the sickle bar when raised and the insets showing how it may be adjusted to cut on sloping ground

Vertical or plain lift. The ordinary or plain lift allows the operator with foot lift or hand lever to lift the sickle bar several inches from the ground. This is for use in turning or transporting the mower short distances.

The vertical lift allows the operator to bring the sickle bar to an upright position without stopping the team. This feature is especially good when the mower is to be used in stumpy or stony land or where any obstruction more than a few inches in height must be passed. It will not only make the work easier, but will allow a much cleaner job of mowing.

High clipping. Where it is desired to clip higher than the ordinary set of the mower will allow, shoe runners can be used to raise the blades. In practically all shoe runners, the lower part is held in position by an upright piece of strap iron near the heel of the runner. This should have more than one hole in it; when fully extended it will raise the sickle bar considerably.

Sweet clover is a crop requiring very high clipping in the spring. The usual range is between 8 and 10 inches. For such work, it will be necessary to displace the short piece of strap iron just mentioned for one long enough to raise the shoe to the desired height. This adjustment should always be made when clipping the first crop of sweet clover. If it is not clipped high, the plants will be killed; and if the attempt is made to raise the sickle bar with the hand lever, without the support of the shoes, the weight of the sweet clover will probably spring the bar.

Bunchers. Two types of buncher at-

tachments for mowers are in use for cutting clover seed—the side-delivery and the rear-delivery. The side-delivery buncher delivers the seed behind the mower, so that it is never tramped on by the horses or run over by the machine. The seed is delivered in long rolls; and it is harder to turn or load for threshing than the flat bunches of the rear delivery.

The rear-delivery buncher is simply a large iron basket attached to the sickle bar. It has a foot trip, and can be dumped when full. The seed is delivered immediately behind the sickle bar, and the team and machine must pass over it the next round. If the clover is heavy, some of the seed will be shattered by such contact. Ease of handling the seed after being cut is the strong point of this type of buncher.

The dump rake. The dump, or sulky, rake is nearly universal in the hayfield. It is light, easy to operate, inexpensive, and can be used almost anywhere where a horse can be driven.

This rake may be either of the hand or of the self-dump type. In the latter type, a foot trip in front of the driver is given a push which causes a cam in each wheel to engage in a ratchet, thus bringing up the teeth.

The sulky rake does not leave the windrows in good condition for the use of a hay loader. Besides, if it becomes slightly overloaded between dumps, as it will in extra-heavy hay, it will not carry the load. This means hay strung along between the windrows, that is hard to pick up or must be reraked. The windrows themselves, too, are apt to be stringy, as it is nearly impossible to get the teeth down quickly enough to make a clean-cut windrow, especially where using a fast-walking team.

The side-delivery rake. The side-delivery rake is rapidly supplanting the dump rake where hay is grown in quantity and where quality is desired. It is now conceded that the way to make the best quality of hay is to follow the mower, either directly or within a few hours, with a side-delivery rake. This puts the hay in long, narrow windrows, protecting the leaves from the sun, and thereby aiding the evaporation of water from the stems. This will mean somewhat slower curing; but, on the other hand, the leaves will be retained, and the hay will all be soft and pliable, rather than harsh and sun-baked, as it is when exposed to the sun too long.

Side-delivery rakes are made in 2 general styles—the fork type and the cylinder, or reel, type. The fork-type rakes are made by attaching the raking forks to a crankshaft driven from the axle. The forks work in rotation, and move the hay out of the swath to one side, forming the windrow.

The cylinder- or reel-type side rakes have the teeth attached to long bars arranged in cylinder fashion. The cylinder is set diagonally in the frame and is driven from the main axle. The teeth on the various bars that form the cylinder are in constant touch with the hay, as the cylinder revolves, so that it is swept to one side with the least shattering.

Some cylinder and fork type side-delivery rakes may be make into tedders by shifting a gear. The hay is merely stirred by the rake teeth, in order that it may dry out better before raking. It is questionable whether this is as good practice as

to turn the windrow with the side delivery to hasten curing, if this seems necessary.

The sweep rake. When hay is to be stacked in the field, the most economical way to handle it is with a sweep rake and stacker. This type of rake has a set of long wooden teeth, mounted in a frame, and can be raised or lowered. The frame is supported by 2 forward wheels and either 1 or 2 rear wheels. The team is hitched to the rear of the rake, either directly or to one side, and pushes it forward. The hay may be gathered out of the windrow or the swath, but usually out of the windrow. When working, the operator lowers the teeth to the ground and drives ahead. The teeth slip under the hay gathering a load. When no more hay will stay on, the teeth are raised and the load driven to the stack. Then the teeth are again lowered and the team backed out.

The use of the sweep rake is typically Western; but it is rapidly finding favor over the whole country, where hay is put in stacks in the field.

Hay tedders. The hay tedder is a 2-wheeled machine with the tedding forks directly in the rear. These forks are clamped to a crankshaft, and work in rotation with a backward kick as each fork begins to rise. This picks the hay out of the swath, so that air can circulate freely through it, thus hastening curing. If a dump rake is to be used for windrowing, the hay tedder, run ahead of the rake a few hours, will improve the quality of the hay. With a side-delivery rake, the tedder is rarely needed except after rains.

Hay loaders. Like the side-delivery rakes, hay loaders are

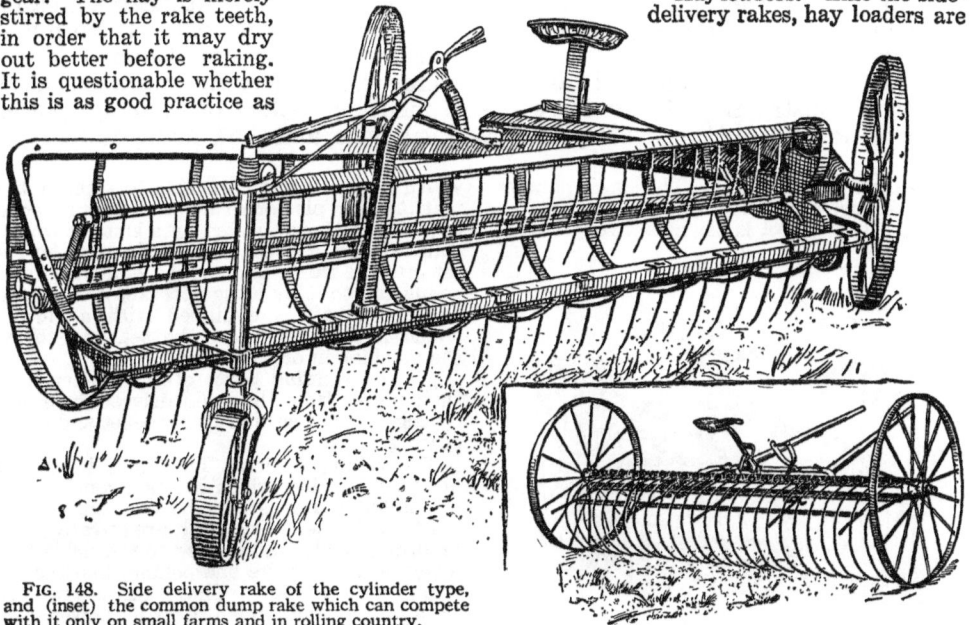

FIG. 148. Side delivery rake of the cylinder type, and (inset) the common dump rake which can compete with it only on small farms and in rolling country.

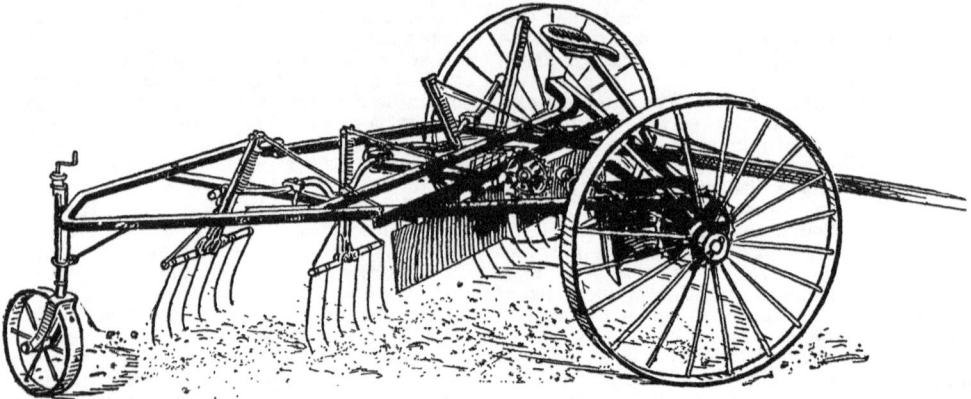

Fig. 149. Side-delivery rake of the fork type. Being likely to cause more shattering, this is better for grass than for leguminous hays

made in both raking fork and cylinder styles. In the first style, the fork bars are attached either to crooked crankshafts or rock shafts that give the forks a straight backward and forward movement. As the hay is pulled forward into the loader, it is caught by the conveyors and carried to the wagon.

The cylinder- or drum-type loaders are made with wire teeth attached to bars forming the cylinder. The drum is driven from the main axle and revolves continuously, lifting the hay on to the carriers or conveyors. The carriers may be either of the push-bar or slatted types. Both are efficient in delivering the load to the wagon.

Both the fork- and drum-type loaders are made to take the hay out of the swath or windrow, and both do good work. There is a growing tendency to use the windrow loader, owing to the better quality of hay made by windrowing it soon after being cut. These two types of loaders are in general use and favor.

Small-grain Harvesters

The grain binder. Aside from the plow, the grain binder has had a greater share in agricultural development than any other farm implement. It has made possible the harvesting of large areas of grain with a minimum of human labor.

The binder is a complicated piece of machinery, but there are several essential points of construction that deserve special mention. Chief of these are power transmission, the cutting device, the elevating device, and the binding device.

Power transmission. The grain binder has one large drivewheel, or bull wheel, as it is called. This wheel is set in a strong frame,

Fig. 150. The hay tedder or "kicker"

which is extended to one side to support the platform and platform canvas. The outer edge of the frame is supported by a smaller wheel, called the grain wheel. The main driveshaft is set into the frame back of the bull wheel and connected to it by a big sprocket drivechain. The driveshaft is equipped with a clutch, which is operated from the seat, to throw the machine in or out of gear.

All the running parts of the binder take their power from the main driveshaft, either by cog-driven shafts or by sprocket and chain gearing.

The cutting device. The grain is cut by means of a sickle, which passes through a row of pointed guards, set in the front edge of the platform frame. The teeth are serrated, like an ordinary file, and riveted to the sickle bar. They do not require sharpening, like the ordinary mower sickle. The sickle bar is driven by a pitman rod, which takes its power from the main driveshaft by means of a secondary shaft extended forward along the side of the frame and cog-driven.

Although not a part of the actual cutting device, the reel is used to aid this function. The reel has 2 important lever adjustments, one to raise and lower it, and the other to move it forward or back. Any one driving a binder should pay a great deal of attention to the reel. By lowering it, grain that is very short can be cut that would otherwise bend before the sickle and never be harvested. Tall grain should be cut with a high reel, in order to avoid straw wrapping around it and to avoid throwing the straw so far back on the platform canvas that it will not be properly elevated or properly bound.

Again, in cutting short grain, the reel should be carried well back, as well as low down, in order that the grain may be swept into the sickle and a clean job of cutting done. This, also, causes the short-strawed grain to fall farther back on the canvas, and aids in proper elevation to the packers. If the reel is not carried in this manner, short straw will have a tendency to fall close to the sickle bar, head first. It is practically impossible to make a good bundle when this occurs.

Just the reverse is true when cutting long-strawed grain. Carry the reel high and forward. If it is swept too far back on the platform canvas, it will be elevated in the same relative position and bound too close to the butt of the bundles. Such bundles will be hard to shock and will take water badly.

The elevating device. The ordinary grain binder has 2 elevating canvases with which to deliver the straw to the packers. The platform canvas runs the full length of the binder platform, which varies from 5 to 8 feet in length. This canvas is driven by 2 rollers, 1 at each end of the platform. Its movement is toward the body of the machine, where the straw is delivered to the elevating canvases.

There are 2 elevating canvases, set at an angle of about 90 degrees, one above the other. These canvases are worked by rollers, the same as the platform canvas. The lower roller of the bottom elevator canvas is close

to the inside roller of the platform canvas, which allows the grain to feed up on to the elevator. The 2 elevators run in opposite directions, thus between them lifting the grain and delivering it directly to a sheet-iron sloping platform, where the packers feed it into the binding device.

The binding device. The packers are curved iron fingers working on a crooked crankshaft which gives them a forward and backward movement, lowering each packer as it moves back, and raising it as it comes forward.

The grain is pulled forward until it is in position on a steel spring trip, where it is held by an iron upright. As more grain is fed forward, the weight on the trip increases until the spring is tripped. At this point a large iron needle comes forward, carrying the twine. This is passed over the bundle, the knot is tied, and the twine cut, all in one operation. At this point, the binder head is so timed that it starts to revolve, and the long iron fingers or rods push the bundle on to the bundle carrier.

A good bundle should be tied around the middle or slightly above the middle. Two adjustments must be watched by the operator. The butter board, which keeps forcing the butt ends of the straw into place, can be moved backward and forward, and has a great deal to do with the place at which the bundle is tied. In addition to this, the binder head can be shifted back and forth by a hand lever. When shifted forward, the bundles will be bound closer to the heads; when shifted back, they will be bound nearer the butts. This shift is a necessary one to watch, if a good job of binding is to be done.

Do not oil the knotter jaws. If they are tight or rusty, loosen them up with kerosene before starting. After that, the twine will keep them polished. Oil causes dust to settle on them, and also allows the twine to slip off too easily. If the bundles are not tied, the knotter jaws may be either too tight or too loose; there is a spring tension adjustment for this. Or the trouble may be in the tension of the twine at the box, or in an improperly threaded needle. Be sure the twine goes through all the little eyes before reaching the needle provided for it.

Grain and seed savers. When cutting some crops, especially over-ripe ones with a binder, there is more or less shattering of seed. This is particularly true of sweet clover. In order to save

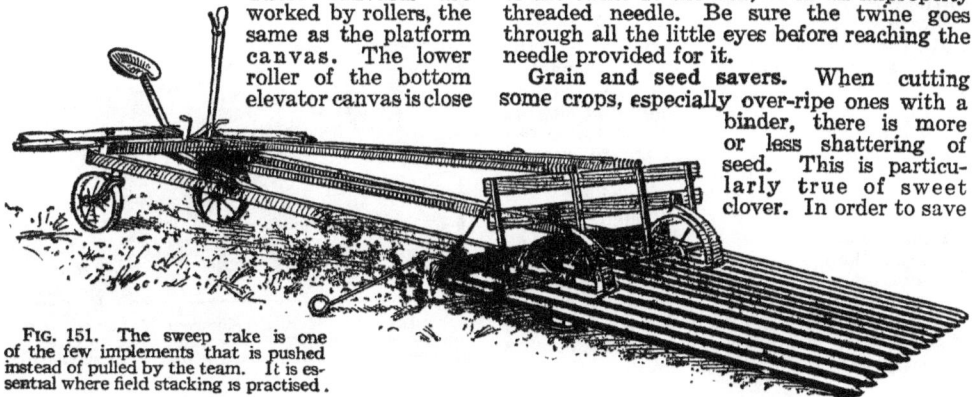

FIG. 151. The sweep rake is one of the few implements that is pushed instead of pulled by the team. It is essential where field stacking is practised.

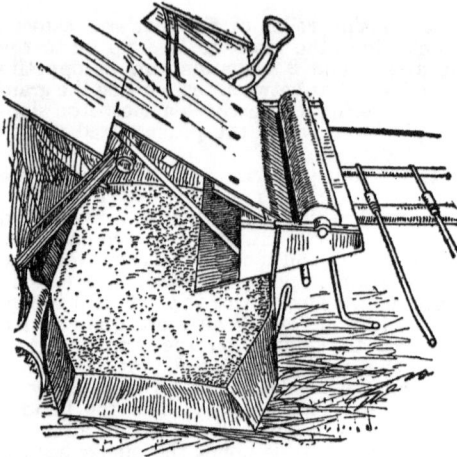

FIG. 151. Front view of grain saver or binder. Front end of pan is unhooked and let down

this seed, it is necessary to catch it at the points at which it will fall from the binder to the ground. There are 2 main points to be protected—under the packers and between the platform canvas and the elevator canvas.

A grain-saving device attachable to a binder is now on the market; but, if a farmer is so minded, he may make a good one himself at small cost. The materials and special tools needed are: three 3 by 8 foot pieces of galvanized iron, 3 dozen 1 by $\frac{3}{16}$ inch stove bolts, a vise, iron drill, snip shears, 3 dozen iron rivets, about 30 feet of $1\frac{1}{2}$-inch strap iron; and a hack saw or other instrument for cutting iron.

There are 2 main pans or troughs. One is attached between the platform and bull wheel, fastened to an angle iron on the under edge of the platform, just beneath the lower roller of the elevator canvas, and to the frame of the binder next to the bull wheel. The other trough is attached under the packers, between the bundle carrier and bull wheel, and fastened to the binder frame and to the bundle-carrier bolts.

Two iron straps are used for each trough. Have the troughs extend the full width of the machine, with sloping ends. In cutting tin, be sure to make an allowance of 10 inches or more for each end. One end should be sloping to such a degree that seed may be raked out easily. The pans are fastened to straps, which, in turn, are fastened to the binder. Stove bolts may be used for doing this.

Straps for fastening the troughs to the machine should be cut so that they will be 36 inches long, exclusive of end fastenings. The bottom of each trough should be about 9 inches wide, and the sides about 12 and 15 inches. The sides should be bent at an angle of 45 degrees. Of course, these measurements will vary with different machines.

An apron should be placed under the elevator canvas, attached to and a few inches under the top edge of the packer table, hanging out over the platform trough. The lower end hangs loose, resting on brace irons. It should be the full width of the binder. A strip of tin should be placed under the lower edge of the packer boards, extending a few inches out, and sloping back over the bundle-carrier trough. This should be an inch or two longer than the width of the machine, curving in, so that all material will run into the trough. Extension tins are placed on the back side of the binder, running from the lower edge of the elevating canvas, up under the seat, and over the top roller cogs, down along the edge of the packer table, connecting with a strip under the latter. At the end of the packer table, this extension tin should be curved in, so that it will funnel over and into the back trough.

The small-grain header is built on the same general lines as an ordinary binder except that the platform can be carried very high, so as to clip off only the heads of the grain with a very small section of straw. This machine is pushed ahead of the team, to avoid running down the grain, and to simplify construction. The reel sweeps the grain directly on to the platform canvas, which delivers it to an elevator that extends upward from the side of the machine to reach a grain rack that is driven alongside. A combination or inter-changeable header and binder is also on the market.

Where the ordinary header is used, the grain must be well matured at the time of cutting. It is taken directly from the machine and built into large stacks, to stand until threshing. This practice can only be followed in the drier sections of the Great Plains and Pacific Coast regions, where the rainfall is sufficiently light to allow it. A still larger combination header and thresher is sometimes used, which not only heads the grain, but threshes and sacks it at the same time.

FIG. 152. A modern development in harvesting machinery is the employment of an engine to drive the machinery, leaving only the traction to be supplied by the team

Results: small in amount, poor to medium in quality

Results: good and moderately large, but measured after all by the physical limitations of the animal body

Results: excellent, limited only by the restrictions imposed by nature upon the operator

THE PROGRESS OF FARMING IS MARKED NOT ONLY BY THE INCREASED AMOUNT OF WORK DONE, BUT ALSO BY THE GREATER EASE WITH WHICH IT IS ACCOMPLISHED

Where conditions are right the deep-tilling machine (in this case of the disc type) does a work that can be done in no other way

Because haying is so nearly a universal farm practice, improved hay-harvesting machinery is among the most valuable of modern agricultural inventions

THE OBJECT OF A FARM MACHINE IS TO SAVE TIME, SAVE LABOR, INCREASE THE WORK DONE AND DECREASE ITS COST. DO YOUR IMPLEMENTS DO THIS?

FIG. 153. The reaper does good work so long as binding and shocking are not required

back to the conveyor, they are caught by the revolving points of a star-shaped knife and cut off, then elevated into the wagon box.

The bean harvester. The early method of harvesting beans was to pull the mature plants and cure them in piles or on stakes. Of recent years, a bean harvester has been developed, and its use is widespread. Essentially, this implement consists of a frame on wheels and carries 2 knives. These blades slip along just beneath the surface of the ground, cutting and throwing together 2 rows of beans at the same time. Harvesting should be done when the plants are mature, but should not be delayed until the pods are too ripe as considerable loss may be caused by shattering.

Corn Harvesters

The corn binder. Corn is a difficult crop to handle with machinery on account of its size and weight, and also on account of the loose soil in the cornfield, which makes it very difficult to secure sufficient traction to operate a machine. To overcome these difficulties and to make a successful binder, it was first necessary to make a heavy machine. The drivewheel is of good width and equipped with lugs to grip the soil. A short, heavily trussed main frame extends out from the drivewheel and is supported on the other side by a lighter grain wheel.

From the outer portion of the frame 4 steel beams or tubes are carried forward several feet. Two come together, forming a point on the outside, and the other 2 come together on the inside, forming another point. These are the gathering points of the machine. The 2 inside bars converge, almost coming together at the point at which they reach the main frame. At this point, a sloping or curved knife is riveted to each beam; and directly behind these stationary knives is a large single-sickle section that works back and forth, driven by a pitman rod.

At the outer end of each gathering point is attached a shield board, which extends backward and upward to the main frame of the binder. Under each shield board are located 2, and sometimes 3, sets of conveyor chains. These chains have extended lugs, which reach out into the narrow passageway between the 2 shield boards.

When the binder is being operated, the driver lowers the 2 gathering points sufficiently to catch any down stalks and drives down a row so that the corn stalks must enter the passageway between the shield boards. The conveyor chains hold the corn upright, and each stalk is sheared off by the stationary knives or the sickle. It is possible to lift and cut any stalk of corn that is not lying in exactly the same direction the binder is moving.

After the stalks are cut, they are shifted

The reaper or self-rake. This machine is designed to meet the needs of farmers who prefer to leave grain in gavels (small bunches) on the ground, rather than to bind and shock it. For such crops as flax, buckwheat, and clover seed the implement is ideal; but it is doubtful whether it should ever be used in place of the grain binder for wheat, oats, and other small grains which sprout and discolor easily when left on the ground.

The reaper is built on the order of the mower, but with only one, large drivewheel, from which the power is taken in about the same manner as in the mower. The sickle is run by a short pitman rod from the main shaft. In addition, there is an upright shaft with a heavy housing, equipped at the top with a wide collar which has a deep depression on the outside next to the platform.

Four large rakes or sweeps are pivoted at the top of the upright shaft, and as they revolve are each supported by a roller that runs on the collar. As each roller comes to the depression, the rake dips down and sweeps the platform clean, depositing the bunch behind the reaper, out of the way of the team for the next round.

The platform is usually about 5 feet wide and is supported on the outer edge by a small grain wheel. A circular rail is put on the back side of the platform, to conform to the swing of the rakes and to aid in guiding the bunches as they are delivered to the ground.

The kafir header. Kafir corn, milo, maize, and similar crops may either be cut with a corn binder and shocked or it may be headed, leaving the stalks in the field.

The kafir header is a simple machine, attachable to the side of a wagon box. It has a long, narrow platform, equipped with an endless-chain conveyor. On the outside of the platform is a high shield which tapers to a point, to act as a gatherer.

The machine is attached to the wagon at a decided angle, with the gathering point down. It is driven by a sprocket and chain from the rear wheel of the wagon. As the heads are fed

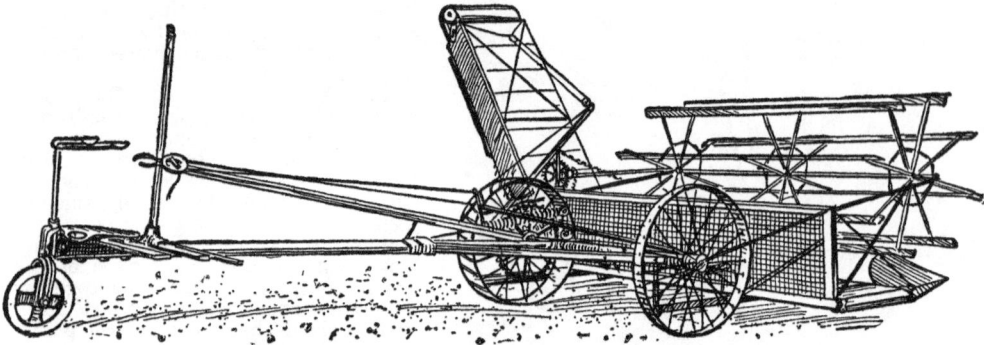

FIG. 154.   The grain header used mainly on the sorghum crops does a similar work but is an entirely different implement from the grass seed stripper (Vol. II. Fig. 274)

back to binding position. Some binders lay the stalks down, and others stand them up in a butt pan for binding. The latter device seems the more practical for all kinds of corn, as this pan can be raised and lowered to suit the height of the corn.

The binding mechanism. The binding mechanism is the same in principle as that for small-grain binders. The stalks press against a spring trip rod as they are fed in by the packers. When this rod is tripped by the weight of the corn, a needle carrying the twine comes forward, passing the twine over the knotter, and the bundle is tied. When the trip is sprung, the discharge arms are also released and force the bundle out on to the bundle carrier or into the elevator.

The bundle carrier is at the rear of the machine and usually somewhat to the right, so that the bundles may be delivered far enough to one side to be out of the way of the team when cutting the next row. From 3 to 5 bundles may be carried; they are dropped by means of a small foot treadle under the foot of the operator.

The elevator attachment is an important addition to the corn binder, when filling the silo. The bundle carrier is taken off and the elevator attached to receive the corn directly from the discharge

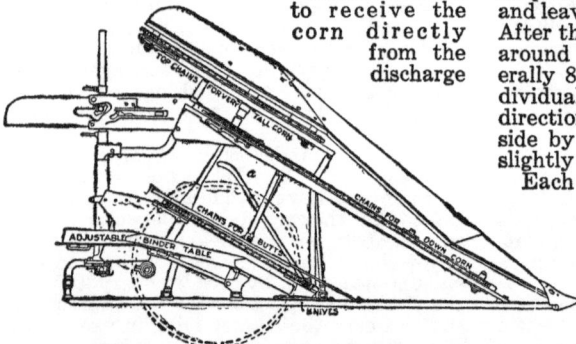

FIG. 155.   Diagrammatic skeleton side view of a typical corn binder (Office of Experiment Stations, Bulletin 173.)

arms. Each bundle is then delivered directly to the wagon, as it is driven alongside of the binder. The elevator frame is adjustable, so that it may be raised or lowered to suit the height of the load.

Every corn binder should be equipped with a tongue truck. Without a truck, the heavy binder is very hard on the necks and shoulders of the horses. Where a truck is used, it is also much easier to turn at the end and make a square corner. Some trucks are now built so that the wheels turn at an even greater angle than the pole, thus making a square turn exceptionally easy.

The corn picker. The corn picker, or field husker, is a comparatively recent invention and is still in the process of development.

In design it looks somewhat like the corn binder, having the same long gathering points that run close to the ground. Just inside of these points, gathering chains with extended lugs are usually placed. These lugs assist in picking up down corn and feeding it back.

As the corn leaves the gathering chains, it is caught between 2 snapping rolls. These rolls may be made with ribs that start at the bottom and wind toward the top, corkscrew style, or they may be corrugated. The rolls snap the ears from the stalks, leaving the stalks and leaves in a pretty badly mangled condition. After the ears are snapped off, they are carried around to the husking rolls. These are generally 8 in number and work in pairs, the individual rollers in each pair running in opposite directions. In some machines, the rolls are set side by side; in others, one roll of a pair is slightly higher than the other.

Each set of rolls is equipped with a husking device, according to the manufacturer's idea. Some have raised shoulders and husking pins, while others have alternating sections and cylinders. As the corn is delivered from the snapping rolls, it is put in a lengthwise position by means of agitators. This gives the husking rolls a chance to tear away the husks. Any corn that may be shelled off, together with the husks, falls through the rolls to a cleaner chain or conveyor. A screen is provided to catch the corn, which is carried back to the wagon elevator while the husks fall back on the ground. The wagon elevator is so adjusted that it may be raised or lowered to suit any wagon. Some machines are now made so that the elevator may be thrown out of gear independent of the husking part. This arrangement is very convenient for turning, as there is often not room for the wagon to turn with the picker. When the wagon is in position again, the elevator is once more thrown into gear and the accumulated ears pass on to the wagon.

Like the corn binder, a picker can get all of the down corn except that lying in exactly the same direction in which the picker is moving. The machine is very heavy, and usually requires 5 horses to pull it. Besides this, an extra wagon must be driven alongside to take care of the husked corn. Where labor is scarce, some men make a practice of taking 3 or 4 wagons to the field and husking them full before unloading any. In this way 2 men can husk and crib about 5 acres in a day, provided there is an elevator at the crib. The average corn picker is still far from perfect; but the machine is correct in principle and will undoubtedly soon be in common use on farms which are large enough to warrant the employment of such equipment.

Potato Diggers

The walking digger. This implement looks much like a walking plow, except that it is fitted with a broad, bill-shaped shovel or scoop, instead of a moldboard. Attached to the back of the shovel are iron rods. The potatoes are lifted out of the ground by the scoop and pass back over the rods, which sift out the dirt leaving the potatoes uncovered in the furrow.

The riding digger. This digger has a long, sloping frame supported by the drivewheels. Each wheel is equipped with soil lugs, to insure plenty of traction. At the lower end of the frame there is a pointed scoop. The bottom of the frame is a slatted endless-chain conveyor.

The potatoes are lifted out of the ground by the scoop; and, as they are carried back, the dirt falls through the slats, while the potatoes are delivered over iron bars at the rear to the ground again.

FIG. 156. An improved potato digger carrying a gatherer which does all the picking and delivers the crop a boxful at a time.

On one digger, a side elevator has been added that places the potatoes in a basket or hopper instead of letting them fall to the ground. The potatoes may be dumped in piles on the ground or put in baskets, crates, or sacks. This attachment does away with a great deal of hand labor, and should be extensively used in potato-growing sections.

The potato digger is a heavy draft machine and must be strongly built to resist the wear. Where the digging season extends over a number of days, a small binder engine may be mounted on the digger to run all parts except pulling the machine forward. This will greatly lessen the draft of the implement.

Cotton Pickers

An efficient cotton-harvesting machine would be a boon to a huge army of American farmers. Unfortunately no such implement is yet found in the field as a practical, thoroughly reliable, satisfactory piece of farm equipment. Nevertheless the following report from Mr. Arthur Johnson of the Experimental Department of the International Harvester Company, is interesting both as a statement of conditions in 1917, and as a forecast of possible developments in the future:

"There have been efforts for the last 25 years to produce a cotton picker; the efforts have been divided mainly into 2 classes. The first and oldest type of picker is one in which there are a large number of spindles located on a revolving cylinder, 2 cylinders being used, one on each side of the plant. The spindles revolve on their own axes, and protrude into the plant as the machine is hauled along the row of cotton and as the cylinder revolves, there is mechanism on the opposite side to doff the cotton and elevate it into a sack.

"With this type of machine the cotton picked is dirty, and loses some of its value in grading on that account. Moreover, as you doubtless know, the cotton ripens gradually, the bolls on the lower part of the plant ripening first, and as the season progresses those bolls higher up in the plant gradually come to maturity.

"In using a machine of this kind there is a chance for some of the immature plants to be spoiled. Then again, in spite of the fact that some of the men engaged on the development of this type of machine have been able to apply an unusually large number of spindles and place them as close together as mechanically possible, they often miss some of the bolls of cotton and do not do a clean job. Our observation on this type of machine leads us to be very skeptical as to any possible success.

"The other type of machine is the pneumatic type, where a man has a number of nozzles operated in connection with a blower and the nozzle used to suck the cotton out of the boll as it is presented to each boll, and is used in lieu of the fingers of the hand picker, so to speak.

"This type of machine does a clean job, the cotton grades well. The machines are so far along that they operate without trouble. The grave question about this type of machine is whether it is any faster or more economical than hand picking, as to make it a practical machine. This scheme has possibilities in it, in our opinion, depending upon the skill of the designer in getting a nozzle with the flexible pipe connected back to the blower in such shape as to be handy to the operator."

Grain Separators

The successful threshing of grain depends first on getting the grain out of the straw, then separating it from the chaff, and finally delivering it to the wagon or sack. There are, of course, many features about a thresher that make for convenience and labor saving that are not essential to the main task of producing clean grain from the bundles.

Feeding device. Old-style threshers and many of the smaller rigs used to-day are fed by hand. The bundles are pitched to a platform on either side. The bands are cut by hand, and one man feeds the grain into the cylinder.

A self-feeder equipment is now used in most places. This feeder is simply a platform hinged to the thresher at the cylinder, and equipped with an endless conveyor which carries the bundles to the cylinder. There is a guard board on either side, and a dividing board down the centre; this is to keep the bundles coming straight to the cylinder, where the machine is being fed from both sides.

The most important thing about the self-feeder is the governor. This is so timed with the cylinder that, when the cylinder speed is reduced to a certain point, the feeder stops automatically, thus keeping the machine from becoming choked up and wasting grain. It is very important that the governor device for the feeder should always be in first-class condition.

In the way of saving labor, the self-feeder does away with 2 band cutters and the feeder. It would soon save its cost on the saving of labor.

Cylinder, concaves, and band knives. As the grain leaves the feeder, it is caught by the band knives, and the bands are cut. These knives may be of the revolving type or may work back and forth on a bent crankshaft. The straw then passes between the cylinder and the concave teeth, where most of the grain is knocked out. The new types of cylinders are made of iron bars with rows of teeth bolted in place. There is an open place between the bars, so that no grain can lodge. The number of bars used is variable; but, for successful threshing, the fewer the number of bars, the greater must be the speed of the cylinder, and vice versa.

The concaves are in rows directly under the cylinder, the teeth being so set that there is a little clearance between them and the cylinder teeth. Adjustment for this distance can be made as required. The number of concave bars used is also variable according to threshing conditions. The concave bars are usually divided in one or more places by a blind bar that is grated, so that the grain will fall through on to the grain pan.

Grates and beater. Directly back of the concaves and extending backward and upward is a grate. This grate is provided with large openings, so that most of the grain falls directly through to the grain pan.

The beater is stationed just above the grates. It has much the same action as a flail, beating grain out of the straw that escapes the cylinder and concave teeth, and forcing the shelled grain through the straw so that it will drop between the grate bars on to the grain pan.

Straw conveyors and racks. After the straw has passed the beater, it is taken by a conveyor to the racks. Some conveyors are of the open-web type, operated by sprocket and chain. Others are of the push-bar type, made of notched wooden bars that move back and forth on a bent crankshaft. Both types allow grain to fall through to the grain pan, if it has been able to pass over the grate.

The straw passes directly from the conveyor to the racks. These racks are the width of the machine, and extend back to the straw carrier or blower. Racks of the newer type are designed to have a circular or rotary motion, that is, up, forward, down, and back. This keeps the straw from bunching, and passes it to the rear in a steady stream. Any grain that may still be caught in the straw is shaken out, and drops through the racks into a return pan, which works it forward to the grain pan or conveyor.

Stackers and blowers. After the straw reaches the rear end of the rack, it is passed on to the stacker or blower, as the case may be. The stacker is merely an outside elevator attached to the rear end of the machine. It is adjustable up and down, and most stackers may be extended in length as the stack goes up.

The wind stacker, or blower, is most commonly used. As the straw leaves the rack, it is forced through the blower pipe by

Fig. 157. A modern grain separator set up and ready for business as soon as power is supplied. The stacker has yet to be elevated and directed to where the chaff is wanted.

a wind blast from a fan in the lower part of the rear end of the machine.

Grain conveyor and sieves. The grain pan or conveyor extends the length of the machine from front to rear. The rear end of this conveyor is known as the chaffer sieve. It is full of oblong holes, with the punched-out metal turned up to form rows of tips down the length of the sieve. A chaffer extension reaches back to the fan chamber, so that much of the chaff, short pieces of straw, and sticks are worked over to the blower.

A cleaning fan is placed under the machine somewhere near the centre. A wind board is so located that the blast from this fan strikes the conveyor about half-way back. The strongest part of the blast will then pass through the shoe sieve near the front end, giving it cleaning capacity its entire length.

The grain passes through the shoe sieve, falling on a solid surface, called "the shoe." This slopes down to the grain auger. As the grain slips down the shoe, it must pass over a screen at the back of the shoe, where most of the weed seeds not previously removed are taken out.

Tailings auger. Under the chaffer extension and back of the shoe sieve and shoe is the tailings auger. This catches unthreshed heads, grain that is blown over, and trash from the conveyor that has not passed out to the blower, and brings all of this material back to the cylinder again.

Few tailings should be returned. If much good grain comes back, it may come either from the conveyor or from the shoe sieve. If it comes over the shoe sieve, there is probably too much chaff on the sieve. To remedy this, close the chaffer sieve above slightly. If, however, the good grain is coming through the chaffer extension, then the chaffer should be opened still more.

When the clean grain gets to the grain auger, it is elevated to the weigher or wagon spouts or sacking spouts, as the case may be.

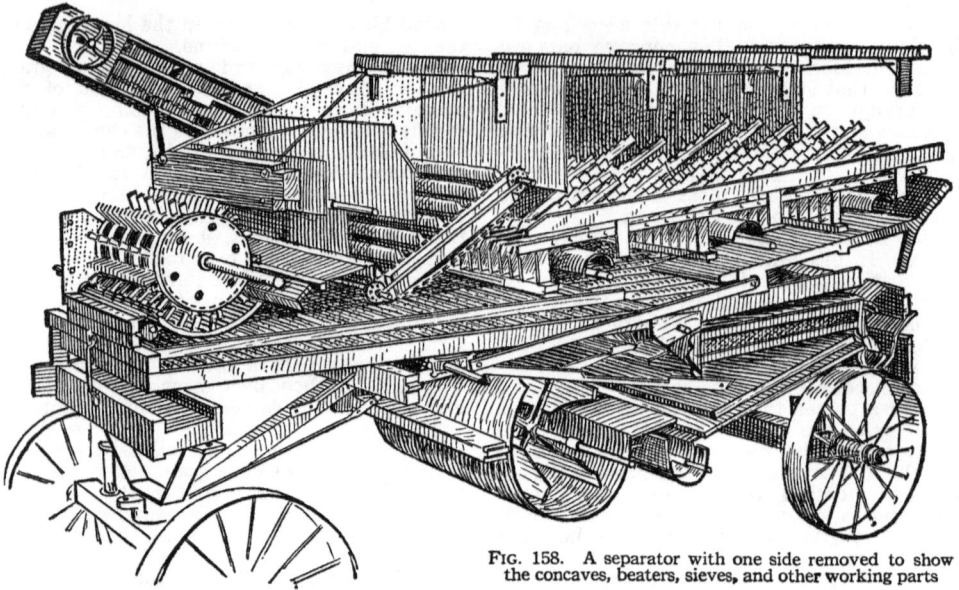

FIG. 158. A separator with one side removed to show the concaves, beaters, sieves, and other working parts

The Clover Huller

The clover huller is much the same in principle as the grain separator, but essentially different devices are used to accomplish the work.

As the clover is fed into the machine, it passes between the cylinder and concave teeth, where the heads are knocked from the straw. All of this material passes back on a shaker floor or separating table. The floor of this table is slotted so that the chaff and heads will fall through; but the heavy straw is worked back to the blower, or stacker, by saw-toothed strips which run the length of the table.

The chaff and heads may pass through one or more shakers, according to the make of the huller, and are then delivered to the hulling cylinder.

This cylinder and the concaves may be equipped with rasps or steel beads. Both do effective work; but it is claimed that in damp clover the rasps are more apt to gum up.

After leaving the hulling cylinder, the hulled seed with the chaff is elevated to a chaffer sieve, where most of the stems and coarse material are taken off to the blower, as in the thresher. The seed and fine chaff fall through the chaffer on to a finer-meshed shoe sieve. Underneath this sieve one or more fan boards are placed. The blast from the fan comes up through the sieve and floats the chaff toward the blower.

The seed and chaff that fall through the shoe sieve pass over a sand screen to the seed auger and are elevated to the recleaner. The unhulled pods and the heavy chaff fall into the tailings auger, and are returned to the hulling cylinder to pass through the machine again.

The recleaner contains an independent fan and sieves. As the partially cleaned clover comes from the main shoe, it is deposited on the front end of the upper sieve of the recleaner. The blast from the fan, together with the shaking motion of the sieves, accomplishes the cleaning. Before it reaches the bottom of the sieve chamber, the seed must pass over another sand screen, where particles of sand and soil are removed. It then runs into the seed spout and thence into the bag.

FIG. 159. The pea and bean thresher is rapidly extending its field of usefulness as cowpeas, soy beans, and other legumes are more widely grown throughout the South

Hulling with a separator. Clover may be hulled with a grain separator by inserting a set of special concaves filled with corrugated or rasp teeth. It will also be necessary to close up the shoe sieve if it is adjustable, and better work will be done if a regular clover sieve be placed under the shoe sieve.

To secure good, clean seed, a recleaner, like the regular recleaner on a clover huller, can be secured for most separators. This is attached to the side of the machine, and operates as described above.

The clover-hulling attachment will do very good work when hulling sweet clover, owing to the fact that the seed is so easily knocked off the stems. With other clovers, however, there is a considerable waste of seed, owing to the fact that there is no hulling cylinder in a separator and the rasp teeth fail to knock all of the seed out of the chaff.

Pea and bean threshers. The successful hulling of peas and beans presents a more difficult problem than one would naturally suppose. These crops absorb moisture readily after being harvested, and there are invariably a number of pods that are tough, even though the straw seems dry. In addition to this, there are always on the vines late pods that are hardly mature when the earlier pods are beginning to shatter. This variability in maturing also makes for difficult threshing.

Regular pea and bean hullers are built on the same general lines as a grain separator, but are equipped with two cylinders instead of one. A slow speed must be maintained to avoid cracking beans, particularly the ripe ones. For this reason the front cylinder is run slower than the back one. Most of the ripe beans are knocked out of the pods by the front cylinder, and fall through the grates and racks to the grain pans before reaching the second cylinder. The second cylinder runs at a higher rate of speed than the first, which tends to knock the beans out of tougher pods. All of the coarse material passes from the second cylinder to the straw racks. These are equipped with lifting rakes which keep the straw well agitated. Hulled beans and unthreshed pods fall through to the grain pan. Most of the unthreshed pods go over the chaffer to the tailings auger and are returned again to the second cylinder for hulling.

The cleaning of peas and beans is not essentially different from that of grain, as previously described. A special grader or delivery spout is used, however, to aid the work. This grader has a vibrating motion. A small screen at the bottom permits sand, dirt, and small stones to fall through to the ground. A coarse screen above separates sticks and pods that may have escaped the sieves. The peas or beans fall through this screen on to the bottom one and are delivered from one spout while the trash falls from another.

Peas and beans may be hulled with fair results by an ordinary thresher if it is properly equipped. The cylinder speed must be decreased; the chaffer and shoe sieve should be adjusted and opened a little wider than for wheat; and the extension beyond the chaffer must be opened wide enough to permit unhulled pods to drop through to the tailings auger.

If hulled beans are returned with the tailings to the cylinder, there will be more cracked beans with the cleaned ones. It is a good plan, therefore, to open the bottom of the tailings elevator and let them run out on a canvas, then put this material through the separator when cleaning up.

Corn Shellers

Hand shellers. The ordinary hand corn sheller consists of a narrow steel or wood-frame box, with legs sufficiently long to make the sheller easy to operate.

Two styles of shelling devices are in general use—the picker-wheel and the cylinder sheller.

In the picker-wheel type, an iron chute leads down into the machine, the upper part being adjustable with sharp iron points, called the rag iron. This iron is held in position by a spring, and brings the pressure on the ear of corn. At one side of the chute is a large upright revolving plate with projecting points, called the picker wheel. These points shell the corn from the cob, giving the ear a revolving motion in the process and thus insuring clean shelling.

The shelled corn falls through to the delivery spout, being subject, as it descends, to a wind blast which blows out the chaff. The cobs are carried upward by the picker wheel and delivered to the outside.

The cylinder shellers have a conical iron grate for a chute, in which an iron cylinder studded with many projections is revolved. Since the size of the grate decreases toward the lower end, increasing pressure is put on the corn. The projections on the cylinder shell the corn from the cobs, giving each ear at the same time a revolving motion, just as in the picker machines. Similarly the shelled corn falls through the grate bars and is delivered below, while the cobs are pushed on through to a separate delivery spout.

FIG. 160. The hand-power corn sheller has long been one of the most widely used of all American farm implements.

Power shellers. Power shellers are designed for commercial work or for large farms where much shelled corn is used. In most communities where corn is sold in quantity, it is shelled on the farm before being hauled to the elevator. This practice is a distinct advantage in many ways. There is less bulk to haul; the grade of the corn can be readily established, thus insuring an equitable price to the producer; and the farmer can usually shell his corn cheaper per bushel than the elevator man is willing to do the same work.

Shelling device. Practically all power shellers are built with either a picker-wheel or cylinder shelling device. In principle, these two devices are the same as described under "Hand shellers"; but they are built larger and stronger to suit the requirements of heavier work. Owing to the fact that considerable trouble arises from breakage of the springs that put the pressure on the rag iron in the picker wheel, the cylinder shellers are growing in favor for commercial work.

Self-feeder. For both styles a self-feeder is now invariably used. In the picker-wheel type, there are several holes, each leading down to an independent wheel, the number of holes depending on the size of the sheller. The self-feeder must be designed to shake the ears down into grooves, so that each groove will deliver its ears lengthwise into a sheller hole.

Separation. When the ears of corn are forced through the shelling device, the grain falls downward, but the cobs and pieces of husk are caught by a conveyor and carried to the outside of the machine, usually being delivered to a V-shaped cob stacker equipped with an elevating chain. The conveyor has a set of agitators which keep the cobs and husks stirred up, so that any loose kernels that have caught in this material will fall back to the rear shoe and thence down to the grain auger.

Recleaning. The bulk of the grain falls straight down from the shelling device, striking a front shoe that is directly over the fan. As the corn comes over the edge of this shoe, it is subject to an air blast, which removes the chaff and stray bits of husk. The grain then falls into the grain auger and is elevated directly to the wagon or to sacks.

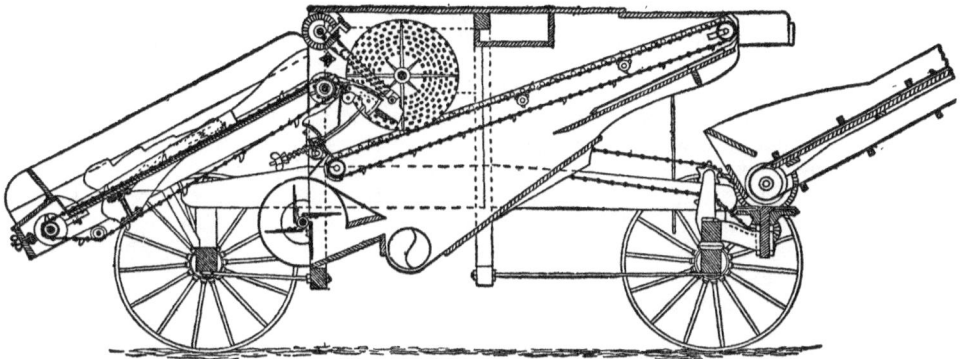

Fig. 161. The power driven corn sheller, a product of corn belt needs and tendencies

CHAPTER 10

Machines Used in Preparing Crops for Use

*By F. H. DEMAREE (see Chapter 7). The easiest possible way to dispose of a crop is to pasture it; where this is impossible machine harvesting must be resorted to. Some crop products go still farther and require an additional handling; for example, baled hay, ensilage, etc. The machines and implements used in thus working over or "changing" the important crops, and the right way to use them, form the subject of this chapter.—*EDITOR.

MANY of the machines described in this chapter are complicated and expensive. The observations of the writer point plainly to the fact that neither the individual farmer nor one engaged in custom work will be successful with many of these machines without careful study of each one of them and the functioning of its parts.

Not only may an expensive machine be quickly ruined by improper management and want of care, but the chances of serious accidents are great. While this chapter was being written, a friend of the writer's caught his hand in the gear of an ensilage cutter and had a finger torn off—and he does not yet know how it happened. Manufacturers are taking pains to make their machines as safe as possible: enclosing gears, extending oil pipes to points of safety, using high-class materials, etc. Consequently, most accidents are traceable to negligence or ignorance on the part of the operator. Intelligent care cannot, however, be exercised without knowledge of the machine.

The Hay Press or Baler

Baled hay or straw occupies only about one fifth of the space required to store the same material when loose. This fact is in itself the chief reason for baling. If barn room is inadequate for loose hay, baling will generally be found cheaper than additional buildings. The bulk of hay and straw is so great that it must always be baled for shipment. Bales for the market should be tight, neat, and trim; the successful operation of the baler is, therefore, important.

The hopper. The main part of a baler consists of the hopper and the baling chamber, which latter is merely an extension of the hopper, made of heavy angle-iron bars. The hopper is solid on both sides, but opens directly into the baling chamber on the inner end and has a false end on the other, which is pushed forward to compress the hay or straw. This is known as the plunger.

The hopper sides are extended upward and outward above the level of the baling chamber, to secure greater capacity. An extension from one side forms the feed table.

Feeder and plunger. As the material to be baled is pitched into the hopper, it must be forced far enough down for the plunger to push against it. In old-style balers, the feeder does this with his foot. In modern balers, a self-feeder has been devised that works alternately with the plunger, forcing the charge of hay down between strokes.

The self-feeder consists of a heavy steel arm with a right-angle bend. It is hinged at the upper end and geared with the plunger, as indicated above. The other end of the arm is equipped with a broad feeder head with sufficient surface to force the hay to the bottom of the hopper.

As the hay reaches the bottom of the hopper, the plunger starts forward. Two styles of plungers are used: the pitman, which pushes the block, and the toggle-joint plunger, which pulls instead of pushing. When the pitman type is used, the power must be located back of the hopper. Where the toggle-joint plunger is employed, the power is in front of the baling chamber, the plunger rod running underneath the chamber floor.

The pitman plunger works on the same principle as any ordinary pitman rod. Where the toggle-joint plunger is used, at the beginning of the stroke, the rear toggle link is

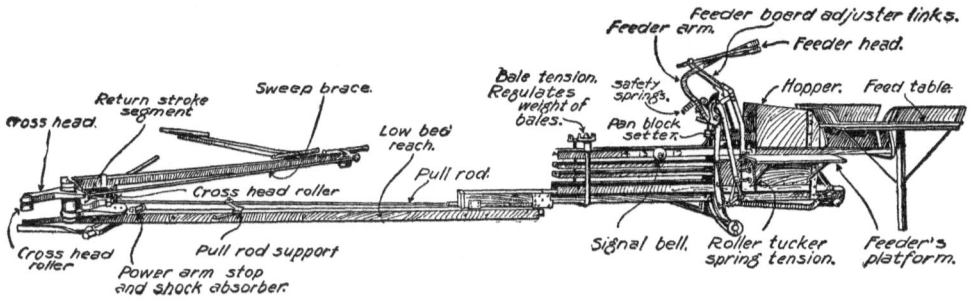

FIG. 162. The essential working parts of a horse-operated hay baler

pulled upward until the stroke is half completed and the link is in a vertical position. As the stroke advances, the pull at the joint begins to straighten out the links, and increased pressure is the result. There is some danger of buckling with a pitman-style plunger, but none with the toggle-joint type.

Bale tension. The outer end of the baling chamber is fitted with a tension rod which encircles both sides and bottom of the chamber near the rear end. Across the top is a tension tie, through the ends of which the tension rods pass. The tension-rod ends are threaded and fitted with a hand wheel or other tightening device. As these are screwed up, the tension on the end of the bale increases. Some tension rods are constructed to permit tightening from the side as well as from the top and bottom. This is known as the double-bale tension. Proper tension insures a neat, compact bale.

Blocks and block setter. The proper length of the bale is indicated by a signal bell or is ascertained by watching a measured scale on the bale chamber. When a bale in process of formation reaches this point, a heavy block, usually of wood, is inserted in the bale chamber, to cut off the finished bale and allow a new one to start. In some balers, the block is inserted by hand; but in the modern power types an automatic block setter is used. A special case for the block is located under the feeder arm. When the block is set, the case is tripped by a hand lever, and as the feeder arm goes on the downstroke, the block case is carried forward and downward to the bottom of the hopper. A bracket on the feeder head pushes the block into the bale chamber, and the case returns with the upstroke again.

The blocks are so constructed as to leave 2 large grooves on both faces. These grooves are about 6 or 8 inches apart. As the new bale is pushed forward, a wire is inserted through each groove next to the straw. As the next block appears, the same wires are inserted through the grooves of that block, so as to encircle the bale. The ends of each

wire are spliced loosely and, as the bale leaves the chamber, the expansion of the straw draws them tight.

Power. Balers are operated by both horsepower and motor power. Owing to the size and capacity of the baler, a 1- or 2-horsepower sweep is furnished. Those operated by motor power are of various sizes and may be driven by stationary gas or kerosene engines, mounted on special trucks for the purpose, or by small tractors, steam engines, or unmounted engines which have sufficient power to carry the load.

Where horsepower is used, the sweep is attached to the outer end of a heavy beam extended from the baler proper. Underneath the sweep is a sprocket segment with an extended arm called the power arm. The plunger rod is attached near the outer end of this arm. A heavy crosshead is attached to the inside end of the sweep, with a roller at the other end and on the under side. As the sweep comes around, this roller engages the power arm, causing the plunger inside the hopper to move forward. The pressure increases until the roller reaches the other

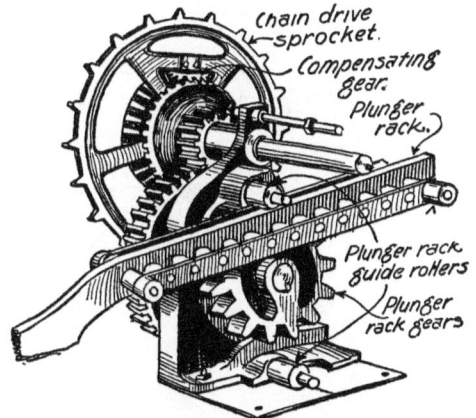

FIG. 163. Details of the apparatus by which the hay baler is operated by engine power.

end of the arm, which causes the plunger rod to be released and spring back for the next stroke.

Where motor power is used, a power jack takes the place of the sweep. The end of the plunger rack is made to mesh with the driving gear of the jack and is held securely in place by heavy rollers. It is driven forward to the end of its stroke, then automatically released and allowed to return for the next stroke.

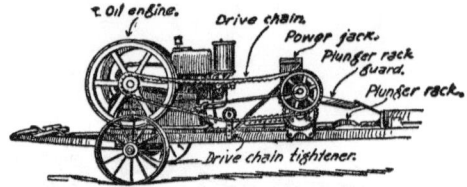

FIG. 164.   The driving equipment of a power hay baler. The rest of the machine is the same as in Fig. 162

Silage Cutters

The use of silage is becoming more widespread each year. As corn is the main crop used for silage purposes, the cutter must be designed with special reference to this crop.

In design, the silage cutter is a long, narrow conveyor attached to a circular fan case. In some machines the cutting knives are on the fly-wheel inside of the fan case; in others, the knives are bolted on a cylinder and are outside of the fan case. Machines of the first type are made with the fan case set at right angles to the conveyor. In the others, it usually sets parallel to the conveyor, the cut silage being fed directly into the fan from the cutting knives.

A long, adjustable pipe extends from the upper side of the fan case into the silo. The machine is mounted on trucks, front and rear, so that it can be transported like a wagon. Strong, heavy materials must be used if this machine is to stand up under its work.

The conveyor. The conveyor is of the endless-chain type, equipped with either steel or wooden slats. It is sufficiently long to accommodate a bundle of stalks lying flat, and it has guard rails on both sides. At the inner end of the conveyor a force-feed device is provided. This is usually of the fluted-roller style. The rollers are both above the conveyor and flush with its end. They are held under spring pressure, and revolve in the direction of the knives. As the cornstalks come between them, they are flattened out and held rigidly as they are being cut. The knives work against a heavy steel angle bar, known as the cutter bar, which is located just inside the feed rollers.

Cutter knives. As mentioned above, two different styles of knives are used in ensilage-cutter construction. Where the knives are bolted to the flywheel a simpler machine is produced.

In this type of cutter, a heavy, solid flywheel is placed inside the fan case. It revolves on the main driveshaft and just clears the

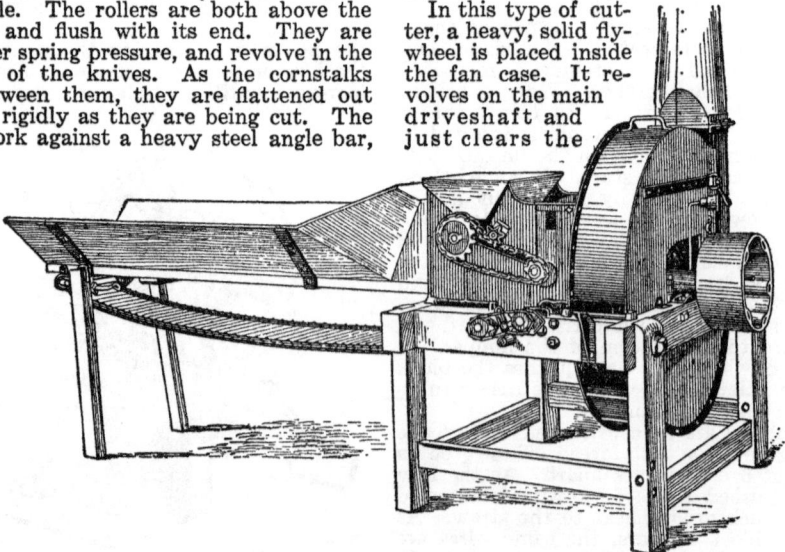

FIG. 165.   Ensilage cutter complete, showing feeding platform at left, cutting knife compartment in centre, fan box and drive pulley at right, and base of blower stack in upper right hand corner

cutter bar. The shaft extends to the outside of the fan case, where the drive pulley is attached, and through the fan case to a small gear house, where power is transmitted to the conveyor and means provided to change the conveyor speed.

The flywheel has openings cut through, to conform to the opening from the conveyor. Above each opening a heavy knife is securely bolted. Below each opening there is a fan blade extending from the driveshaft to the outer edge of the flywheel and practically as wide as the fan case, allowing for clearance.

Where the knives are bolted to a revolving cylinder, they are generally somewhat curved and set spiral. This is an advantage over straight-set knives, as they slice instead of making a square cut. The knives on the cylinder work against a cutter bar, as described. The cut pieces of silage fall into a trough and are then delivered into the blower fan.

The bundles should be fed in butt end first, and, as the stalks are forced past the cutter bar, the ends are clipped off by the swiftly revolving knives. The small pieces are thrown and blown upward through the pipe to the silo.

Care and adjustments. Many accidents occur in operating silage cutters. Some of these are unavoidable, and others should never happen.

Do not lean against the conveyor. The clothing may catch and pull a person to the machine in that way. It is dangerous to attempt to remove an obstruction in or near the self-feed rollers while the machine is in operation. Throw it out of gear first.

It is also a wise precaution to tighten up the knife and fan-blade bolts at least once a day. If a knife blade gets loose, it is very apt to strike the cutter bar. In this case, something is bound to break. Flying pieces have often caused injury to the workmen as well as great damage to the machine.

The knives should also be kept sharp. Usually, two sets are furnished with each machine, so that one set can be sharpened while the other is in use. Dull knives do poor work and require more power. Some machines furnish a grinding attachment, run from the gear chamber, so that a set of knives may be ground while the machine is in operation.

Every silage cutter should be equipped with a clutch and a handy shifting lever. This lever should throw the machine in and out of gear, forward and reverse.

Shock corn silage. Corn that has had to be shocked and left to cure in the field may be cut with the silage cutter at any time during the winter. This is an especially good way to handle shock corn for winter feeding. The grain and roughage are thoroughly mixed, and the stock eat a greater portion of the stalks than would be the case if fed whole. In addition to this, the short pieces of stalk and pith that are not eaten make fair bedding, and are much easier to work up into manure than whole stalks. Unless it can be siloed (p. 465) do not pile up too much of this material early in the fall, as it is apt to ferment in warm weather.

Corn Huskers and Shredders

The value of shredded corn stover is a moot subject. Some farmers rank it equal to timothy hay, others not so high. Feeding experiments have not shown that shredding adds enough to the value of whole corn stover to pay for itself, whereas the ensiling of stover if rightly done as mentioned above, is profitable.

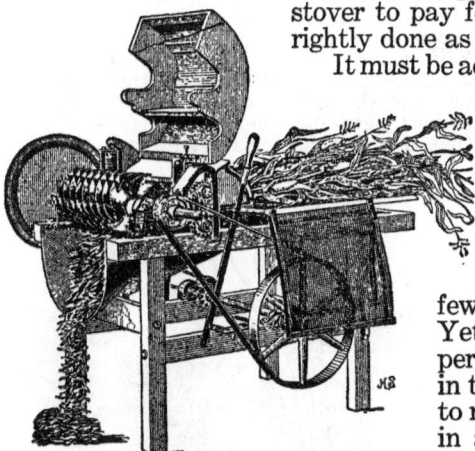

It must be admitted, however, that shredding is gradually growing in favor, so that some factor other than cost must enter in the proposition. It seems to the writer that that factor is the production of easily-handled roughage.

It is a hard job to dig shock fodder out of the snow and hack the bundles loose from the frozen ground, and few farmers will undertake it on a large scale. Yet it is thoroughly recognized that fully 30 per cent of the food value of the corn crop in the stalks and leaves is lost if these are left to rot in the field. Shredding puts this food in an available shape, so that, although the cost may be increased over that of shock fodder, it is money well spent,

FIG. 166. Corn shredder with hood lifted to show course of fodder through knives (Office of Experiment Stations, Bulletin 173.)

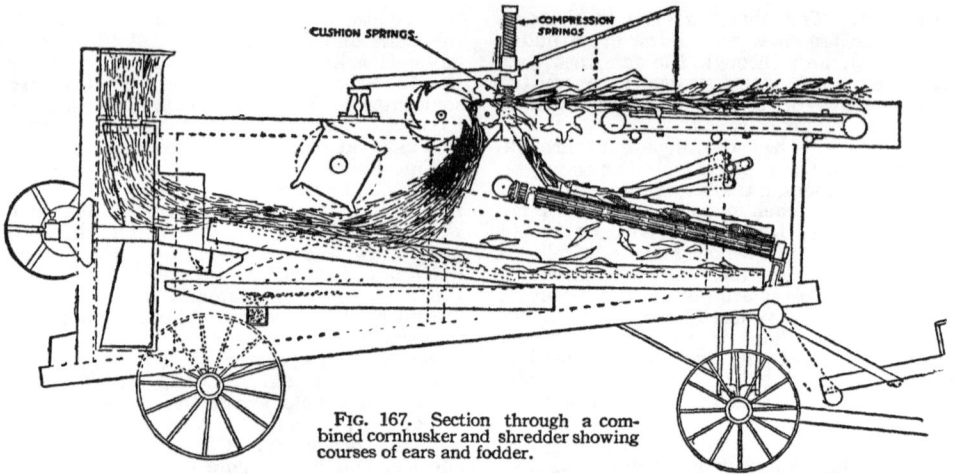

FIG. 167. Section through a combined cornhusker and shredder showing courses of ears and fodder.

Design. The husker and shredder is built along the lines of an ordinary thresher, but with special equipment for the work to be done. Many points of its construction are the same in principle as those of the separator, while many others correspond to those of the corn picker. Both of these machines are described in Chapter 9.

Self-feeder. The original shredder was made to be fed by hand; but so many accidents occurred that a self-feeder was designed, and this should be used on all machines. This feeder is simply an endless-belt conveyor, long enough for the bundles of fodder to be thrown on to it. At the inner end of the feeder is a separate revolving shaft equipped with spikes or knives. This is known as the feeder head, and assists in passing the stalks on to the snapping rolls. A sloping hood is placed above the feeder head, so that the ends of the stalks cannot fly upward and get crosswise of the feeder.

Snapping rolls. When the stalks leave the feeder head, they pass between 2 large snapping rolls, set crosswise of the machine. These rolls are very similar to those used in the corn picker, and answer the same purpose, namely, that of removing the ears from the stalks. In the shredder, the stalks pass completely through, while in the picker they do not. Snapping rolls of the better type are corrugated, so as to pinch off the ears with the least possible shelling. The rolls are held together under spring pressure. The stalks can pass through, but the pressure is so great that the ears are immediately broken from the shanks as they attempt to pass.

Husking rolls. The unhusked ears, after being snapped, fall directly down to the husking rolls. These rolls are in pairs, and in construction and operation are very similar to those used on a corn picker. The same variations in styles are found here as in the

picker, and there seems to be practically no difference in their efficiency. The number of husking rolls found in a shredder increases with the size and capacity of the machine.

Agitators that work back and forth are placed over the husking rolls to keep the ears moving parallel with the rolls. This is essential to clean husking. As the ears move down the length of the husking rolls, the husks are stripped from the ears by the picker projections and passed downward to a shaker rack. The husked ears go on over the ends of the rolls into a conveyor, which carries them to the wagon box or crib.

Shredder head. When the stalks pass through the snapping rolls, only the ears are removed. The stalks are then immediately engaged by a shredder head or a set of revolving knives. The shredder head is a heavy iron cylinder equipped with close-set knives or plates which tear the stalks and leaves into small pieces. Many kinds of knife equipment are in common use. Some are saw-toothed and set in the shredder head on a spiral. Others are set square, but have a short right-angle bend. Any type of knife that is strong enough to stand up under this

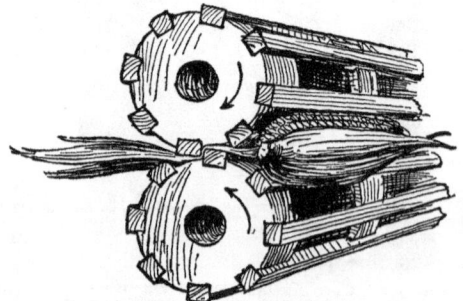

FIG. 168. Detailed view of husking rolls in a husker and shredder

heavy work and that does not powder the leaves too badly, is a good shredder.

The shaker rack. The shaker rack is placed just underneath the husking rolls and extends from the forward end of these rolls back to the entrance to the blower. The rack is perforated or slotted, so that any shelled corn or foreign material may be shaken through to the sieve below. The rack bars are saw-toothed, with their projections toward the blower.

The shredder head is directly above the shaker rack, so that the shredded stover is delivered directly down to the shaker. The husks from the ears are dropped on the forward end of the shaker by the husking rolls and shifted back to the blower with the shredded material by the back-and-forth movement of the rack.

Saving the shelled corn. The shoe is equipped with a sieve and screen the action of which takes out the dirt, snow, weed seeds, and other foreign material. The shelled corn is passed on to a conveyor, which delivers it to the sack. Usually, an air blast is taken from the blower fan to expel small pieces of stalks or leaves that may have remained with the shelled corn.

If shredded stover is to be kept any length of time, the removal of the shelled corn is important, as this will ferment during warm weather if left in the shredded material in any quantity.

Feed Crushers and Grinders

There are 3 important objects to be attained in grinding feed, namely: to increase its palatability, to increase its digestibility and to make available food material which in its natural state animals will not masticate. Much has been said and written about the value of ground feed as compared with feed in its natural state. Some experiments show that grinding pays well, others do not. This question of relative values, however, depends largely upon the first cost of the feeds under consideration. It is a safe assumption that, with the increasing cost of both grain and rough feed every process will be used extensively that will make available food materials that are now largely wasted.

The sweep-power mill. Probably the most common type of feed grinder found on the average farm is the one operated by a horse-drawn sweep. On account of the space required for the sweep, this grinder is usually set out in the barnyard. The mill is mounted on a base which catches the ground feed as it comes from the burrs. The mill legs, or supporting frame, are bolted to the base, which, in turn, must be anchored to the ground by stay rods and stakes. At the top of this supporting frame there is a large ring gear.

The upper part of the mill consists of a hopper, to which another ring gear is attached on the lower side. The upper gear is placed directly over the lower one, separated from it by small rollers that mesh in each. The sweep is attached to the upper gear and hopper, and when the mill is in operation these parts revolve on the lower gear, thus producing the power.

Heavy arms extend upward inside the hopper. They are attached at their outer ends to the main frame and to an iron collar in the centre of the hopper which they support. An iron spindle runs through this collar, supporting on the under side a set of revolving arms and a cone studded with heavy projections. When ear corn is being ground, the revolving arms crush the ears against the stationary ones, and the cone reduces the crushed portions to still smaller portions. The bottom of this cone has a corrugated section which acts as a grinding plate or burr.

It works against a similar face on the inner side of the supporting frame, which is, of course, stationary. Since the grinding burr is cone-shaped, the opening between the 2 faces is wider at the top than at the bottom, so that increasing pressure is put on the material as it is forced through. Two sets of burrs are generally furnished: one with coarse corrugations for corn, and the other with finer corrugations for small grain.

Fig. 169. Typical power-driven feed-grinder. Hand lever at right adjusts burrs and determines fineness of feed delivered.

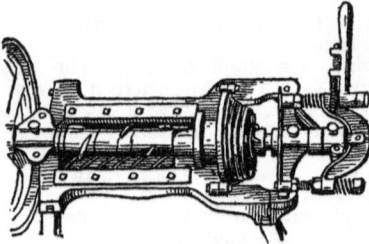

FIG. 169a. Top view of crusher and burrs, hopper and cover having been removed

FIG. 171. A combined roughage cutter and feed grinder, by means of which a balanced ration can be prepared in one operation.

The capacity of such a mill will range from 20 to 40 bushels per hour for ear corn and 10 to 20 bushels for shelled corn or small grain.

Single and duplex grinding plates. The ordinary belt-power grinder is made with circular opposed grinding plates or rings, either single or double, instead of the cones used in the sweep-power mills.

In single-plate mills one grinding ring is bolted to the housing which forms the grinding chamber. It has a specially corrugated face, which is turned to the face of the companion ring bolted to the driveshaft but adjustable there so as to give desired pressure.

In the double or duplex mills, the ring on the driveshaft has a grinding face on each side, and there are two stationary rings, one on either side of it. It can readily be seen that if the same grinding surface is to be secured with the single rings they must be of larger diameter. But with plates of larger diameter the grinding pressure is farther from the driveshaft; consequently, the power required to operate such a mill of equal capacity is necessarily larger.

Feed-grinder construction. The ordinary feed grinder consists of a hopper mounted on a sturdy frame with legs long

FIG. 170. A bagging elevator, that can be attached to any power-driven mill, saves much shoveling and prevents much waste

enough to give good clearance for the ground feed spout. The drive shaft passes through the bottom of the hopper and the exposed section is generally used as a crusher. This may be in the form of a spiral knife or of extended lugs set in spiral order, or it may consist of a combination of knife and lugs.

The crusher is used to reduce ear corn and other very coarse feed and also to act as an auger to force the reduced material into the grinding rings or burrs which may be set at varying distances apart so as to deliver feed of the required degree of fineness.

Regulating fineness. All grinders, whether of the single or duplex type, have a pressure regulating device whereby either fine or coarse feed may be produced by shifting the revolving burr on the shaft. All mills have a screw or lever device whereby this shift is made almost instantly and while the machine is in motion.

Most manufacturers make more than one set of plates for each mill. These are designed for either fine or coarse grinding, and will produce different grades of feed, according to the pressure put on them. The capacity of a mill is greatly increased by using coarse rings.

Bagging elevator. The bagging elevator is simply an enclosed narrow box with a dividing board in the centre. An endless-chain conveyor runs around the board, gathering the feed that is delivered to the bottom entrance of the elevator from the feed spout of the mill, and elevating it to the upper end, where the bag is attached. The elevator is supported by legs that can be adjusted to suit the height of the bags being used. A longer elevator may be secured to deliver the feed directly to a wagon box, if desired.

Crushers. It is often desired to feed corn merely crushed, not ground. Machines to meet this requirement are made by omitting the grinding plates and, usually, the lower crusher. The machine then consists of a hopper set on legs. One of the spiral crushing knives becomes the driveshaft as well; the other knife is geared to it; and the operation is the same as previously described.

Since a grinder is generally desired on a farm where any of this work is done, most machines are constructed so that the grinding rings may be separated far enough for the crushed material to pass through without further reduction. In this way the grinder may be used as a crusher without the expense of buying another machine.

Grinding mixed feed. Two different kinds of feed may be ground and mixed together at the same time by attaching an extra hopper for the small grain to the edge of the main hopper. This secondary hopper is equipped with a slide in the bottom, by which the amount of grain fed in may be regulated at will. The upper and lower crushers thoroughly mix the two kinds of grain by the time they reach the grinding rings.

Fig. 172. The simplest type of hand-operated feed cutter

Cutters and Grinders

Roughage cutters. The simplest form of rough-feed cutter is a knife hinged at one end to a cutting table and working against an iron cutting bar. The material is held in position by hand, and is cut in short lengths by bringing the knife down with the other hand. This is a very slow process.

An excellent cutting box, operated either by hand or by power, may be secured consisting simply of one or more pairs of knives set on a cylinder. As they revolve, they work against an iron cutting bar located at the inner end of a feed table from which the material is delivered to the cutting cylinder.

Combined cutters and grinders. A combined cutter and grinder can be secured that will cut and grind corn fodder, alfalfa, clover, pea hay, sheaf oats, or other rough feeds, together with ear corn or with ear corn and another small grain, if a secondary hopper is used.

This machine has an upward extension of the grinding chamber in which the cutter is located. A feed trough delivers into the cutting cylinder the material to be cut. For best

results, the feed trough should be equipped with a force-feed device that will assist in holding the roughage for the cutter knives.

The small pieces of cut material are delivered to the lower crusher, together with the crushed grain from the upper crusher. All the material is mixed together by the lower crusher and forced into the grinding rings, where it is reduced to the desired fineness.

A combination cutter and grinder may be used either as cutter or as grinder, as desired. In producing cut feed alone, it is only necessary to open the grinding plates, so that they will not function. The feed will then be delivered in the condition in which it leaves the cutting knives. In case it is desired only to grind grain, the cutting device may be disconnected, and will remain idle during the process.

Such a combination machine means not only a saving in the first cost, but also a reduction of the amount of labor and space required for its operation thereafter. Another advantage is that, with such a machine at hand, the farmer will more readily utilize corn stalks and other roughage instead of wasting them. Fodder so prepared will be more nutritious, more palatable and more generously consumed by the animals. And they, in turn, as the result of the choice of the right machine will make increased growth and become more profitable thereby.

Fig. 173. A rotary cutter, even though hand-driven, is a great improvement over the type shown in Fig. 172

Wood Saws

In timbered sections of the country, the farm woodlot generally furnishes its share of the farm income. Saw logs, cordwood, mine props, poles, etc., are all produced, according to the demand in the community. In the ordinary woodlot where the big timber has been removed, cordwood is usually the chief source of income.

Most of such wood may be worked up to advantage with the least labor by using a power saw run by a gasoline engine.

Design. The ordinary power saw consists of a heavy driveshaft mounted in a frame of convenient height. At one end, usually the left, if not otherwise specified, is bolted a circular saw. At the other end is the drive pulley and counterbalance or flywheel. Most saw blades are shielded with a hood or iron strip as a safety device.

Tilting or sliding frame. A section of the frame is designed to hold within itself the length of wood being sawed. In some cases this section is hinged from below, so that it can simply be lifted up and the wood pressed against the saw for cutting.

In other saws, the table slides back and forth between the saw blade and flywheel. Stop blocks are provided to hold the piece of wood steady. In sawing, the piece is laid on the table with one end extending beyond the saw the desired distance; the table is pushed forward, and the saw engages the wood in the process.

The pole saw. When it is desired to saw timber of considerable length, a saw should be secured which has the flywheel hung under the frame, out of the way of the table. This is accomplished by using a secondary shaft for the flywheel, below the main driveshaft. Extra pulleys and belting or gearing are required to deliver the power to the flywheel. The flywheel is stationed on the opposite side of the frame from the saw, to give the necessary counterbalance.

This saw may be used for cordwood as well

Fig. 175. Pole saw. The underslung fly wheel makes possible the cutting of poles of any desired length

as for sawing poles or other longer sections of wood. For the ordinary farm it is the best owing to its greater adaptability.

Portable saw rig. For commercial use, a portable saw rig is a good investment. This rig has both engine and saw mounted on a heavy frame, the engine forward and the saw at the rear, with table dropped to a convenient height. In principle and operation, the saw is the same as an ordinary unmounted saw. The outfit is easier to transport and set for sawing, thus saving time.

Fig. 174. Tilting frame wood saw, for cutting short lengths

Fig. 176. Sliding frame saw with especially well-protected blade

CHAPTER 11

Garden Implements
and
Hand Tools

By F. F. ROCKWELL, *practical farmer, gardener, author and contributor to agricultural and horticultural press. Though born in the city he spent many summers in the country even before he took up practical farming and greenhouse management in Connecticut in 1900. From 1906 to 1909 he attended Wesleyan University and was in the publishing business. In 1909 he returned to farming, practising also as consulting agriculturist. In 1917 in this latter capacity he was in charge Farms and its orchards. Later he became Manager of the Nurserymen's National Service Bureau.*—EDITOR.

THE importance of tools and machinery in the production of all kinds of garden and field crops is continually increasing. As the price of labor and the difficulty of obtaining skilled labor increase, the importance of an adequate equipment of those things which make labor more effective increases in proportion.

While the gardener and the farmer must depend more and more on the use of machinery, nevertheless there are two general conditions under which machinery will not and cannot pay, namely (1) insufficient use for it and (2) lack of attention to its proper care. These are so widely encountered and so serious that a word of warning in regard to them is in place. The mere fact that a machine will save labor is no guarantee that it will pay. *There must be sufficient use for it* to justify making the investment. A mechanical weeder for 10 rows of onions in the garden, or a potato digger for half an acre of potatoes would be an absurdity. Good judgment at this point is a most important factor in successful farming.

Still more important, however, is *the proper care of machinery*, once it is bought. It has been estimated by government experts that the farmer gets fewer hours of actual use out of many complicated machines than the manufacturer spends on them in their construction. Field-crop and garden machines of most kinds are adjustable or convertible for various operations, and are designed for very close and accurate work. Unless given excellent care, however, they will do poor work, will be much harder to use, and will soon become worthless. Every tool should be kept under cover when not in use, should be wiped and dried off immediately after use, and should always be kept well oiled.

Tools for Working the Ground

Besides the standard plows of various types, there are a number designed particularly for garden work, either in addition to the others, or for turning the ground in very small areas where an ordinary plow is not available.

The 1-horse swivel plow (Fig. 177) is particularly useful for putting in succession plantings in small areas where a small block of some vegetable has been cleared off. It is good also for opening furrows or trenches, preparing ridges, etc. Another type of light 1-horse plow is furnished with a set of different moldboards for deep or shallow work, plowing growing vegetables, hilling up, and so forth.

123

A very convenient tool for small places or for garden work on the farm, to save the use of a plow and team, is the plowshare attachment for a regular 1-horse cultivator. This makes a substantial, small, reversible steel plow which will cut a furrow 4 inches deep and 8 to 10 inches wide.

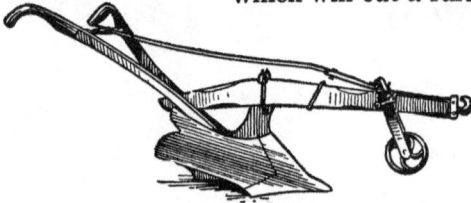

FIG. 177. One-horse, reversible or swivel plow, especially adapted to small garden use

The small garden outfit motor-driven and guided by a man walking behind, may also be utilized for light plowing. For the small place, devoted to poultry and "specialties," where a horse is not kept, this implement will do such plowing and cultivating as may be necessary, throughout the season, after the regular deep spring plowing has been done.

Tools for Fining the Soil and Preparing the Seedbed

While the standard disc, spring-tooth, and spike-tooth harrows are used in the preparation of the soil in the garden and in intensive field operations or on truck crops, there are a number of special harrows made to meet certain conditions which are very widely used in making specially prepared seedbeds for potatoes, onions, tobacco, and the like.

Harrows. One of the most popular and generally useful of these harrows is the acme (Fig. 121), a pulverizing harrow and clod crusher. This is particularly valuable for use just before planting, as it turns up a fresh surface, the blades being so constructed that they form practically a miniature gang plow which thoroughly churns and turns up the upper 3 or 4 inches of the soil. It also does much more leveling than the ordinary harrow. Further, it can be set to be used as a smoothing drag. It is one of the most useful implements which the market gardener or truck farmer can possess. Care should be taken to renew the teeth often enough, so that it will continue to do as good work as when new.

The Morgan spading harrow, which is popular in some sections, is constructed somewhat upon the principle of the cutaway disc harrow; but the indentations are so deep that they leave a series of sharp teeth which cut deeply into the soil, chopping it up very fine, and also loosening it up. It is good on hard, compact soil such as may be deficient in humus, and where not much trash has been plowed under.

The Meeker smoothing, or small disc, harrow and leveler (Fig. 178) is used in the preparation of seedbeds that are required to be extremely fine, such as those for planting the seed of onion, lettuce, celery, carrots, and other fine crops. It practically takes the place of hand raking for finishing off the seedbed, one man with a horse and a harrow being able to do as much as 10 to 20 men could do by hand. The small revolving discs push under all small stones, trash, and other similar material—most of which a rake would pull out—and the adjustable leveling

board smooths and levels the soil, leaving it almost like a table top. If possible, the last harrowing given with this harrow should be at right angles to the direction in which the rows will run, so that the lines made by the drill marker will be more distinct. This harrow is useful also wherever the creation of a fine dust mulch to retain moisture is wanted, and for covering in broadcast seed, grain for winter covered crops, and so forth. The chain, or link, drag harrow is also used for smoothing seedbeds and covering grains; but it does not do such thorough work as the Meeker; also it uncovers or collects trash and small pieces of manure which the Meeker harrow would push down out of the way.

FIG 178. The Meeker smoothing harrow leaves the soil in the best possible shape for the sowing of vegetable seeds.

Fertilizer Distributors

Intensive feeding of all garden and truck crops is one of the great factors in their successful cultivation. Sometimes, all the plant food to be used can be put into the ground before planting; but, in many cases, it proves more profitable to supply some of the plant food in the form of quick-acting fertilizer as top- or side-dressing in the drill or row with the seeds or plants, or as a top- or side-dressing during growth. The old method of doing this by hand was disagreeable and slow. A number of practical machines designed for this purpose are now on the market at reasonable prices. In the ordinary market garden or on the truck farm, a machine of this kind can be used frequently throughout the year, and, in addition to its use on vegetable crops, it may be employed to advantage on small acreages of potatoes or corn or root crops for truck for which alone its purchase might not pay. With a machine of this type it is possible to sow dry commercial

FIG. 179. Wheelbarrow type of fertilizer distributor.

fertilizers, of any kind and in any desired quantity, in open furrows prepared for planting or along the sides of rows, several rows at a time. The fertilizer is carried out of the hoppers steadily and evenly on endless canvas belts, and the spouts, which are adjustable, may be set to distribute it in narrow or in broad rows or broadcast. One of the larger machines will side-dress 5 rows any distance apart up to 2 feet, 4 rows up to 32 inches, 3 rows up to 4 feet, or 2 rows up to 8 feet, and will broadcast over a strip $8\frac{1}{2}$ feet wide, the capacity being 400 pounds of fertilizer at a time.

For smaller operations, where only a limited acreage of truck crops is grown in connection with the general farming, a man-power distributor (Fig. 179) is a practical and rapid-working machine which will, with care, save a great deal of time and disagreeable work. It will, moreover, give better and more uniform results than hand distributing. It is particularly useful in growing strawberries and truck crops like cabbage and tomatoes. It can, of course, be employed in many places where a horse machine cannot be used. It will apply from 3 to 40 pounds of fertilizer to 100 yards of row, distributing between, or on both sides of, 1 to 4 rows at each passage; and it is not hard for the operator to use.

Some of the leading makes of wheel hoes also have fertilizer-distributing attachments which may be used with the seed drill or separately for top-dressing or side-dressing.

Marking and Spotting Machines

Marking or "striking out" rows or check rows for planting seeds or setting out plants is one of those small jobs which cut down the total of a day's work; nevertheless, marking out must be done carefully and accurately, or it will interfere with the cultivation, whether by horse or by wheel hoes, throughout the season. Even with crops which are not usually planted in check rows, such as lettuce, early cabbage, and so forth, because of the extra time required to crossmark the rows when planting, some time may be saved in hoeing and wheel hoeing, if all the plants are accurately spaced.

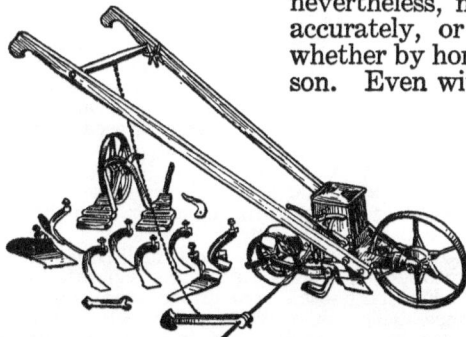

The most complete and up-to-date tool of this kind (Fig. 181 c) admits of the following combinations: 5 rows 8 to 12 inches apart; 4 rows 14 to 16 inches apart; 3 rows 12 to 24 inches apart;

FIG. 180. A wheel hoe with complete set of attachments, including seed drill (shown ready for use), an extra wheel, and various types of cultivator teeth.

and 2 rows 25 to 48 inches apart. Not only are any of these combinations of rows marked off with accuracy at one passage, and a line marked for the return trip, but by the use of removable plugs on the circumference of the broad wheels, each row may be "spotted" for plants to be set 3, 6, 12, 15, 24, or 48 inches apart. For small gardens, a simpler machine is made, which will mark and spot 2 rows at a time; or one wheel may be used as an extra attachment to an ordinary wheel hoe. (Fig. 181 a and b).

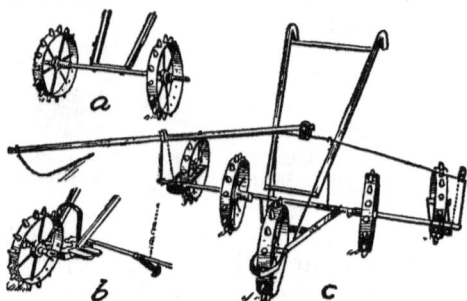

FIG. 181. A row marker adjustable for laying out from one to five rows, and for "spotting" at any desired distance in each.

There are adjustable markers which mark out the rows only; with these, however, as with the common homemade drag markers, it is much more difficult to make a straight row, as they are pulled behind the operator. Where there is not enough marking to justify the purchase of a regular marking tool—a device which does not cost very much and which with care will last for a generation—it will, however, pay to take the time, say, on some stormy day, to make a set of markers covering all the row widths which are likely to be needed. If these markers are made with a *detachable* handle, which will serve for all the "heads," they can be stored away in a small space and will be much less likely to get broken than when constructed in the ordinary way.

Various types of markers are used for field crops; but the best of these have discs instead of runners, and can be used as coverers and furrowers as well as markers. The ordinary "wing" marker is too widely known and employed to need description here.

Planters and Transplanters

Accuracy in the sowing of seed is of the greatest importance. The yield with any crop depends directly upon the fullness and evenness of the stand obtained. It costs just as much to prepare the ground for, fertilize, cultivate, spray, and supervise the growing of, a poor stand as a full one. Careful work on planting may mean all the difference between profit and loss on the entire season's work. It pays to have the best of planting machinery. But planting machines are, as a rule, more complicated and more finely adjusted than those for other purposes, and they should be given particular care. *Never allow unused fertilizer or seed to remain in the machine,* and keep all planters not only under cover, but in a dry place and off the ground.

Seed drills. There are, in general, 3 types of seed drills, each of which has its advocates. The first of these, and the kind most commonly in use, has a gravity feed with a brush or a metal agitator working just above the seed opening in the hopper, to prevent packing or bridging of the seed. In the second, or force-feed type, the seed is carried out of the feed hopper by a revolving drum, in which there are small cavities into which a certain amount of the seed falls, being then dropped into the delivery tube as the drum revolves. Planters of this class do very accurate work; but, as they require separate cylinders or sizes of drums for different sizes of seed, and take longer to change, they are better adapted to large gardens or to work where a considerable amount of each kind of seed is to be put in than where changes will have to be made frequently. The third type operates on the "snap" principle, a spring-and-cam or trip mechanism being used in addition to the gravity feed. The modern drills are readily adjustable for the different sizes of seed, widths of rows, and so forth, and will sow in hills as well as in continuous drills. Except where there is a great deal of planting to be done, it pays to get a seed drill in combination with a wheel hoe. Market gardeners and others who have use for a seed drill frequently throughout the season prefer a separate drill.

For large-scale operations, the multiple seed drills, which will sow several rows at a time, are coming into more general use. Their advantage lies in the fact that not only is much time saved in the work, but each unit of rows is spaced with absolute accuracy, even if the rows themselves are not perfectly straight. Machines for cultivating 2 or more rows at a time can, therefore, be used with much better results than where the rows are planted individually.

For very small gardens, and for use in greenhouses and frames, there are a number of small seed sowers (Fig. 182) designed with a handle for use like a slide hoe, or with a short handle for use on raised benches. Some of these open the furrow and cover the seed, others merely drop the seed; but they all save a good deal of time compared with hand sowing, and sow more accurately than an inexperienced operator can sow by hand.

For very small plantings, of either potatoes or corn, hand planting machines (Fig. 183) can be used; and they will do the work much more accurately and very much more quickly and easily than it can be done in the old-fashioned way with a hoe. Machines of this

type cost very little, are simple in operation, and, with ordinary care, will last indefinitely. Even where one has but a small patch to plant, the cost of one of these handy implements will be saved quickly. They are also particularly useful in going over the fields as soon as the plants are up, to fill in skips, which latter will occur, even in careful planting. For this purpose, they are invaluable supplements to the larger machines, and offer a means of stopping serious crop leaks which are often overlooked.

Plant-setting machines. One of the slowest and most tedious jobs connected with gardening and trucking has always been the transplanting or setting out of plants from the seedbed or coldframe. For a long time this work was done entirely by hand. At length, the plant-setting machine (Fig. 183) came to the rescue of the large planter. More recently a practical "back and time saver," in the form of a small hand plant-setter, has been invented, to serve the smaller planter and for jobs where the big machine would not pay.

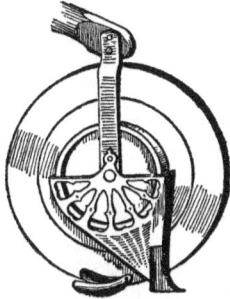
FIG. 182. The smallest type of seed drill consisting of wheel, hopper, and furrow opener.

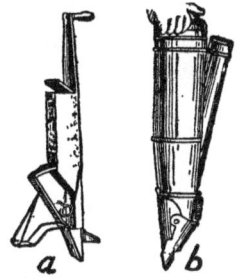
FIG. 183. Hand corn planter (a) and plant setter or transplanter (b)

Weeders and Weeding Machines

One of the most important additions made within recent years to the general line of farm tools is that of the spring-tooth weeders, which are widely used for corn, potatoes, and many other rowed crops. Like most other agricultural tools, to do their best work, *they must be used at the proper time.* Intelligently handled, they are capable of saving, or rather of preventing, a tremendous amount of hand work. The old style of field weeder with 3 rows of spring-steel, pointed teeth was fixed to a light, rigid frame. The newer forms are adjustable in width and have removable teeth, so that, where desired, a tooth or two immediately over the row may be removed. A riding form, with adjustable spring tension on the teeth and with an extra feeding attachment for grass or grain has proved very popular in some sections. In operation, these weeders pass over the growing crops—corn, potatoes, and so forth—which are strong enough to withstand the light harrowing or raking given by the flexible teeth, while all small weeds, up to an inch or two in height, are pulled from the ground and shaken loose from the soil by the vibration of the weeding fingers. Besides removing the weeds that are close to the plants better than can be done with any form of cultivator, they loosen up and aërate the soil, and create a fine

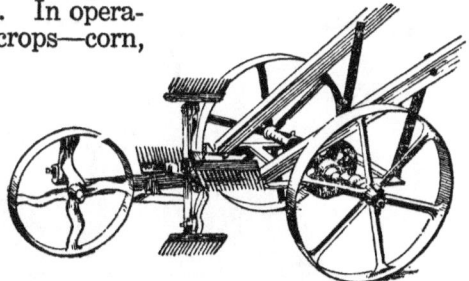
FIG. 184. One of the recent garden weeders in which the teeth revolve at right angles to the direction in which it moves.

dust mulch which is of great service in retaining moisture. They are especially good for getting over the surface of large areas in a short time after a rain, during the early stages of growth, preventing the crusting and baking of the soil.

Weeding machines, that is, machines for the actual removal of the weeds in a manner similar to hand weeding, as distinct from the tools, are a very recent development. While they are still being improved and have not yet come into general use, nevertheless they are being used extensively by a number of large growers and have apparently come to stay. They are peculiarly suited to work in light loam and in muck soils comparatively free from stones. They have proved of value for thinning rowed crops, such as carrots, as well as for weeding.

There are 2 general types of these weeding machines, one of which operates with wire fingers or weeders, and the other with a weeding blade or comb. In either case, the fingers or combs are passed through the row with a rotary motion by the mechanism of the machine. With the first type, the weeds are scratched or raked out; with the latter, they are caught, held in the comb, and pulled out. The comb machines will work, of course, only when there is enough difference between the form or size of the vegetables to be weeded and the weeds to be removed to allow the former to slip through the combs while the latter are held and pulled out.

Cultivators and Special Tools for Cultivating

The riding and walking cultivators used for garden and truck crops are for the most part identical with those which are in use for ordinary farm crops. For

garden and truck crops, however, it is especially important to have a machine that may be readily adjusted for width of row and depth of cut, and that will do fine and close work. A cultivator with *lever* adjustment for both width and depth should, therefore, be selected. In one of the latest types of the 1-horse cultivator, or horse hoe, the levers are out of the way and are easy to adjust while the machine is in operation, the frame is light but exceptionally strong, and the wide, flat back sweep leaves a smooth surface in-

FIG. 185. For gardens more than 50 by 100 feet in size, a one-horse, adjustable cultivator is almost a necessity.

stead of a deep furrow as with the old-style, wide-back tooth. An important extra attachment for machines of this kind is the vine turner or leaf guard, which

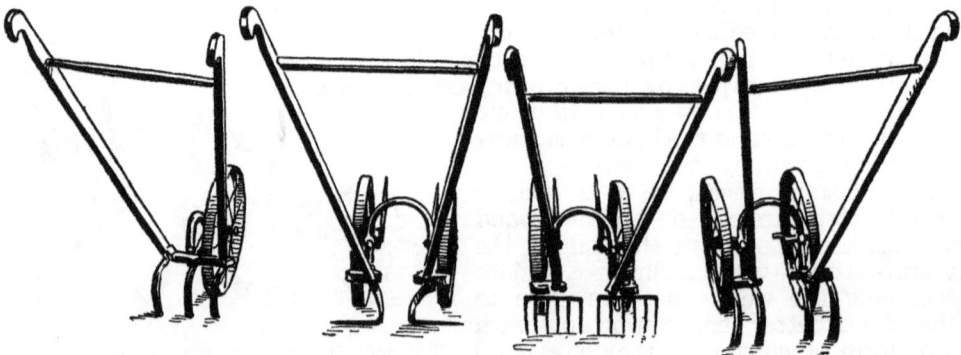

FIG. 186. Four wheel hoes showing single- and double-wheel types and three styles of cultivator blades

can readily be put on and which makes even closer work possible without injury to the foliage.   The solid front wheel has this advantage over the old spoke style: it does not sink so deeply into the soil when the latter is very wet or very dry, nor does it get stuck by small stones or trash.   One of the most popular tools with the market gardener, strawberry grower, and so forth is the 12-tooth, or diamond-tooth, cultivator.   One of these machines, with the pulverizer attachment, which can be had to fit on the back end of the frame, will take out practically all small weeds, will work almost as close to the plants as one could with a hoe, and will leave the ground as finely pulverized and smooth as if an iron rake had been employed.

Hand cultivators. Hand cultivators, or wheel hoes, are so universally known and used that little need be said here concerning them, except that with a machine of this kind, used constantly throughout the growing season, and for many different kinds of work, it pays to get a tool of the best quality that can be had, even though it may cost several dollars more than an inferior one. Of course, such tools should be well cared for, and kept so free from dirt and rust that all changes and adjustments may be made easily and quickly. Very often this is not the case. While the single-wheel type costs less than the double-wheel, it usually pays best to get the latter, as it can be used to straddle the row, doing much closer work, especially while the plants are small, than the single-wheel. Also, for work in crops that are nearly grown, the double-wheel may be converted into a single-wheel in a few minutes. Among the most important of the many extras or attachments for standard wheel hoes are the "onion" or extra-high-guard hoe blades, the advantage of which is that, in close work, they prevent lumps of soil, small weeds, and so forth from falling over on to the seedling plants. This allows both the wheel hoeing and the hand weeding to follow it to be done more rapidly. Other particularly good attachments are the wide or flare-bottomed cultivator teeth, instead of the old standard narrow cultivator teeth. The advantage of these is that, while they break up the crust and pulverize the soil, they cut off small weeds instead of allowing the latter to pass around them as the narrow teeth often do. Plowing, raking, hilling, ditching, and furrowing attachments all have their special uses and, under many circumstances, will prove profitable investments.

Machines for Special Purposes

In connection with some of the garden and truck crops, special machines have been developed for use in harvesting, preparing for market, and so forth.   The most generally used, perhaps, of these is the potato digger.   While there are a number of types of this machine on the market, it is doubtful if any are really satisfactory except the high-grade standard elevator diggers which are now universally used wherever potatoes are grown on a commercial scale.   The recent addition of a small gas engine, to operate the mechanism of the digger, enables work which formerly required 4 horses to be done easily by 2.   Moreover, it is done better, because the elevator, separating apparatus, and so forth is run at an even rate of speed, no matter how fast or how slow the machine may be moving ahead.   The latest development is the picking attachment, which saves the labor of picking up from the row after the digging.

A new machine, which many large vegetable growers have found to be one of the biggest time savers ever introduced, is the "bundle-tying" machine for such crops as green onions, carrots, beets, rhubarb, asparagus, and so forth.   This machine may be operated either by foot, somewhat like a sewing machine, or by power.   Its tension may be varied to give moderate or tight tying; and it is adjustable for tying bunches and so forth.

Hand Tools

The small, or hand, tools which are used with garden and truck crops are so numerous and so well known that they do not require any extended mention here.

Nevertheless, there are a few general considerations in connection with them which are too often overlooked. Such common tools as spades, digging forks, rakes, trowels, and hoes, while standard in general design or shape, vary widely in the quality of the material and in the kind of workmanship that goes into them. With these simple things, as with larger implements, it always pays to get the best, and then to take the best care of them.

The difference between the work that can be done with a suitable, sharp hoe or weeder, and a poor-shaped or dull one, is considerable, even on a few rows of work; and, in the aggregate for the season, this may easily amount to the entire cost of a new implement. One marked feature, always to be sought in tools of this kind, is the *solid one-piece shank*, instead of the old-fashioned ferrule, which may wear or work loose. The selection of these tools always depends largely on personal preferences, as well as on local conditions. But there is often danger of preference becoming prejudice. The progressive grower should always be ready to give up something good for something better, in the little things as well as in the large.

FIG. 187. The machine buncher and tier is an invaluable labor and time saver for the market gardener who considers appearance worth while.

The spade and spading fork. The spade and the spading fork, used for turning over small plots of ground, finishing up corners, and around irrigation posts, in the cold-frames, greenhouse beds, etc., are, perhaps, the simplest tools made. But even here quality counts. A poor spade or fork, soon bent out of shape and becoming loose in the handle, is a constant source of annoyance and of inefficient work. Buy the best, because it pays the best, and see that the handle is strapped with metal up the front and back, protecting it against breakage at the weakest point. The tool should, if possible, be selected personally, so that one with a suitable "hang" for the user may be obtained.

Rakes. The iron rake is one of the most constantly used tools in any kind of garden work. The size, number of teeth, etc., required will depend on the kind of work to be done. In preparing seedbeds, raking out weeds, and the like, a large-sized rake might often be used with a considerable saving in work where a smaller one is employed simply because it "will do." The "bow" rake, attached to the handle from the *ends* of the head instead of at the middle, is not so likely to break or bend at the middle as the ordinary kind; it is also better for leveling soil, in which the back of the head is used. A straight, strong, well-balanced handle is also an important item in securing good work with a rake.

Hoes. Most farmers are stingy with hoes. It is not at all unusual to find half a dozen old, round-cornered, loose-shanked, broken-edged hoes—some of them out in the fence corners—on a place where the gang plow and the riding cultivator and the cream separator are in prime condition and the mowing machine and manure spreader are under cover. The old-style, deep, heavy hoes were made before the days of modern cultivating machines. For most work with hoes nowadays, in garden and truck crops, the wider, lighter hoes will be found easier to use and more efficient. The very small, narrow-bladed, or "onion," hoe, is little known in many sections

FIG. 188. One of the modern improvements over the old time hoe

where it could be used to great advantage. Of course, these lighter hoes are not adapted to "chopping" a thick sod of half-grown weeds; but, for stirring the soil around and close to plants, when the weeds are small, much easier and quicker work can be done than with a large, heavy hoe. An important thing to remember is, to *buy hoes with solid shanks*— ferrule and head in one piece.

For use in closely-planted crops, after they are well grown, the scuffle, or slide, hoe often has to be used to get through when the wheel hoe can no longer be used. A new type of slide hoe, which is proving very popular, has round iron runners in front of the blade, which is thereby held at a uniform angle, and does not glance to the side and thus injure the crop. Also, the weight of the tool is supported, so that it is less tiresome to use. A number of different blades, adapted to different kinds of work, give it a wide range of usefulness.

The smaller hand tools—trowels, transplanters, weeders, etc.—are, on the basis of the work they save in proportion to their cost, among the most important on the place. For this reason it will pay to see that they, also, are of the best quality when bought, and that they are taken care of afterwards. These small tools are so easily misplaced that a special spot should be assigned to them, and their return to that spot *immediately after use* insisted upon. A new type of trowel is found in the long, heavy-bladed one with the handle at right angles to the blade. This is especially useful in the transplanting of tomatoes, cabbage, etc.

When even a few of such rowed crops as onions, carrots, beets, etc., are grown, enough small hand weeders to equip as many hands as are likely to be working in them at one time should be provided. There are probably no other tools which will more quickly repay

their cost in time saved, when conditions make their use necessary.

There are a number of new tools, such as adjustable weeding rakes, long-handled weeders, etc., on the market, many of which are excellent for general work or for special purposes. Before investing in any of these, however, be sure that they are substantially built to stand actual service. Many a dollar idea gets on to the market first in a thirty-five-cent dress.

All edged tools should be kept sharp. There is just as much waste of energy in using a dull hoe as in using a dull mowing-machine knife. They should also *be kept free from rust*, which greatly decreases both their length of life and their efficiency. A piece of an old bur-

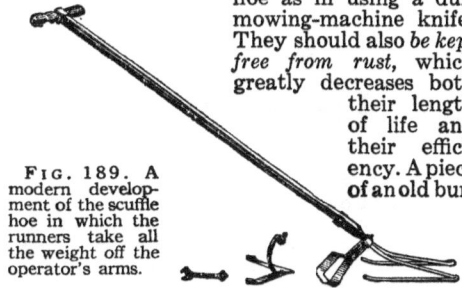

FIG. 189. A modern development of the scuffle hoe in which the runners take all the weight off the operator's arms.

lap bag and a bottle of kerosene (kept tightly corked to prevent evaporation), to be used on the tools after they have been in use, will prove one of the most profitable small investments that can be made about the place.

Above all, the grower who realizes how much of his success must depend on the machines and tools he has to work with will provide room to house them from the weather. He will have a place for each, and will see that each tool after the season of its use, is put back in its place, and that any broken or worn-out parts are replaced *before the season of its use.* Though such advice may seem old-fashioned and trite, it is as true to-day as when first given.

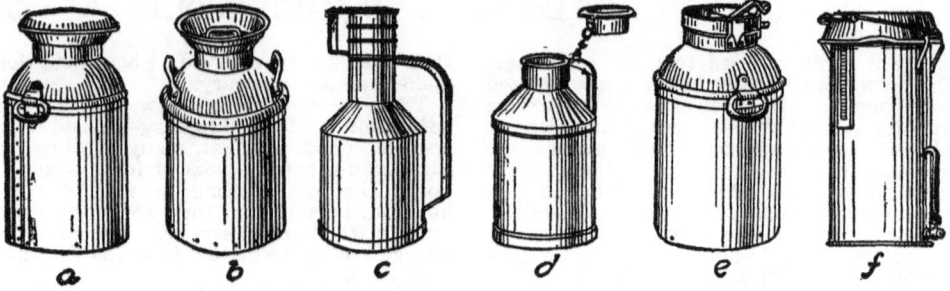

FIG. 190. Types of milk cans for various tastes and purposes. See text (pp. 139 and 140) for descriptions

CHAPTER 12

Dairy Machinery for the Farmer

By W. A. STOCKING, Chief of Dairy Department and Professor of Dairy Bacteriology, New York State College of Agriculture since 1908. He grew up on a large Connecticut dairy farm and graduated from the agricultural colleges of that state and of New York. From 1899 to 1901 he was Farm Superintendent and Instructor in Agriculture at the former institution; from 1901 to 1906, Professor of Dairy Bacteriology there and Dairy Bacteriologist of the Connecticut (Storrs) Experiment Station; and from 1906 to 1908 Assistant Professor at Cornell. He is author of a "Manual of Milk Products" and a number of bulletins.—EDITOR.

CHAPTERS 40, 41, 42, 43, and 44 of Volume I include some discussion of dairy machinery as related to the different lines of dairy work. This chapter treats many of the same tools and machines but more from their mechanical point of view. In this case it is the machine itself rather than the work it does and the reason for its use that is considered first. There are, of course, many makes and styles of these articles. Only the farmer himself can say which is best for his purposes; but this discussion will help him to decide carefully, wisely, and well.

BABCOCK TESTERS AND EQUIPMENT. The butter-fat testers now on the market vary from a little 2-bottle machine up to those handling 40 bottles and include types operated by hand, steam, and electric power. The smallest hand machine (Vol. I, Fig. 505) is useful when only a few samples are to be tested. It is light, well made and can be easily attached to any table or bench.

For testing more samples the 8-bottle type is very satisfactory, turns easily and runs smoothly and quietly, but for a larger amount of testing a power machine is better than a hand one. An attached hot-water tank for filling the bottles is a great convenience.

Accurate glass-ware is one of the most important features of a testing outfit. The best is that made in accordance with

FIG. 191. Milk samplers. (See text)

specifications formulated and adopted by the Federal Bureau of Standards at Washington and the American Dairy Science Association and known as "Standard" Babcock test bottles and pipettes. The milk bottle is graduated to one tenth per cent, and the cream bottles to five tenths per cent. The straight sided cylinder with a lip is the most commonly used acid measure (Vol. I. Fig. 500 e), but some persons prefer other forms such as the dipper or, the device that measures the right amount of acid automatically.

Milk and cream samples for testing must be kept in bottles or jars with tight seals—either snap tops, screw tops or glass stopples—so that no moisture can evaporate.

Milk samplers. The McKay sampler (Fig. 191a) consists of 2 nickel-plated, brass tubes one of which fits inside the other, and both of which have milled slots that coincide

FIG. 192. Combined acid bottle and pipette.

when the handles stand together. The slot so formed is closed by turning the handles at right angles to each other. The Scoville sampler (Fig. 191b) has a check valve which is closed by pressing the sampler down against the bottom of the can or vat. The milk thief (Fig. 191c) is simply a hollow tube in which the column of milk is held by pressing the thumb over the top. Each of these types of sampler is designed to take a column of milk the entire depth of the can or vat.

Cream to be tested by the Babcock method

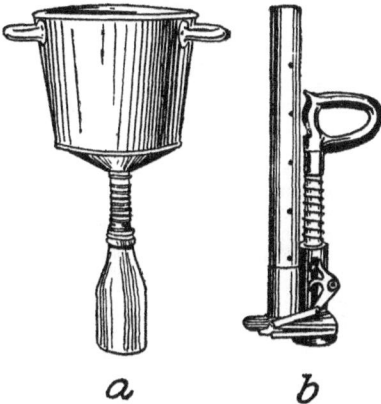

FIG. 193.   Single-bottle hand filler (a) and hand capper (b).   The proper caps are stocked in the cylinder

must be weighed on a sensitive balance. One of the simplest, least expensive, but wholly satisfactory forms on the market is shown in Fig. 212. It is fitted with agate bearings and is sensitive to one thirtieth of a gram. The empty cream bottle is balanced by a sliding poise with a balance nut; after the cream is added it is balanced by means of the hanging poise placed on the notched beam, the balance being clearly indicated by the pointer beam and indicator plate.

BOTTLE FILLERS AND CAPPERS. During recent years great improvements have been made in milk bottling and capping

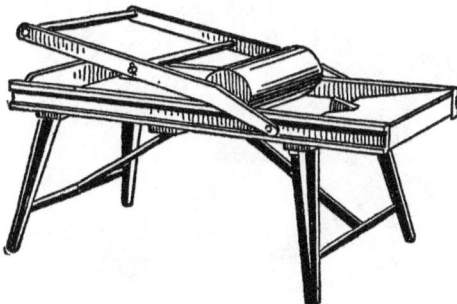

FIG. 194.   Simple farm butter worker.   (See text)

machines so that the dairyman may now select, from an almost endless variety, the outfit best suited to his special needs. For dairymen who fill not more than 100 or 200 bottles

FIG. 195.   Butter lost in separating when machine is level (A) and out of balance (B)

per day the single-bottle hand machine (Fig. 193a) will prove very satisfactory, easy to use and efficient. It is automatic, filling every bottle to the same point without overflowing. As the filler is placed on the bottle, the valve opens allowing the milk to flow and as it is removed the valve closes completely so that there is no loss by dripping. When considerable quantities of milk are to be bottled, a machine which will fill several bottles at once (Vol. I, Fig. 537) will be better than the hand filler.

The hand capper (Fig. 193b) provides a very satisfactory method for capping a limited number of bottles. When the capper is placed on the bottle, a thrust of the handle forces the cap firmly into position; as the handle returns to its position the next cap (from the stack in the cylinder) comes automatically into place. A capper of this sort can be worked rapidly and is more sanitary than handling the caps by hand.

BUTTER WORKERS. For use in connection with any small churn. There are several types of butter worker. That shown in Vol. I, Fig. 544 is an old standby in which the butter is placed in the tray, sprinkled

FIG. 196.   Butter lost in separating one cow's milk for a year when the inflow of milk to machine is correctly A and B and incorrectly C adjusted.

with salt, and worked by turning the crank which moves the wooden roller over the butter. This process is repeated until the desired amount of working has been secured.

Another of the oldest styles is shown in Vol. I, Fig. 546. The butter is placed in this worker and subjected to the desired pressure by means of the lever which is pivoted loosely at the lower end.

In a third type (Fig. 194) the butter is rolled out into a thin sheet by means of the large wooden roller, sprinkled with salt then folded over with a hand ladle and rolled again. The proper amount of pressure can be obtained by bearing down on the lever, the lower end of which slides in grooves at the sides of the worker table.

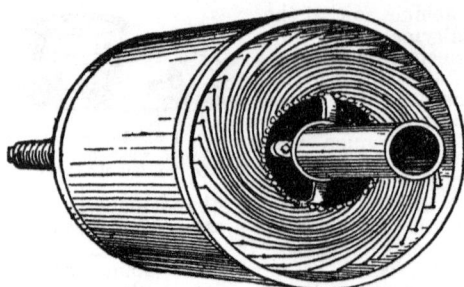

FIG. 197. View into the top of a separator bowl of the link blade type

CENTRIFUGAL CREAM SEPARATORS.

The old-time methods of raising cream in shallow pans or by setting cans in cold water have given way very largely to the use of the centrifugal separator. The chief advantages of the latter, which are responsible for the change, are: (1) more complete recovery of the butter fat; (2) the richness of the cream may be controlled to suit the use that is to be made of it; (3) the cream is obtained fresh and is therefore of much better quality than when obtained by the slower methods; (4) the fresh skim-milk is at once ready for feeding calves and pigs; and (5) the time and labor saved.

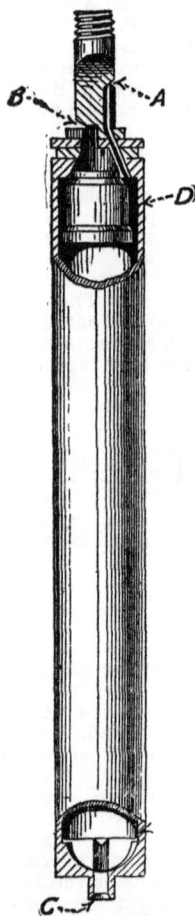

FIG. 198. Hollow bowl used in one type of separator. *a* skim-milk vent; *b* cream vent; *c* milk inlet; *d* dividing wall.

The modern cream separator was invented about 1879 on the principle that the fat globules in milk are lighter than the skim-milk, and that therefore, as the milk is revolved rapidly in the separator bowl, the skim-milk tends to flow toward the outside, and the fat toward the centre, of the column each running out through holes provided for it. The many different makes and styles of separators developed and placed on the market of late years, are all based on the same principles as the earlier ones, but differ widely in their mechanical details. The devices that separate the cream

and skim-milk as they pass through the bowl may in general be divided into 3 distinct groups: (1) The disc type; (2) the link-blade type; (3) the hollow-bowl type.

A representative of the disc type is shown in Fig. 200. In this particular separator the milk enters through the centre and flows between the leaves or discs of the distributing device where the separation takes place, the cream moving inward and upward at the centre while the skim-milk moves outward and upward at the outside of the discs next the inner surface of the bowl. Both cream and skim-milk pass

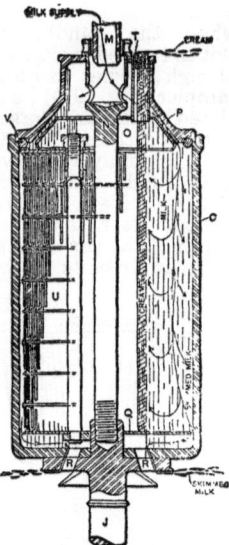

FIG. 199. Section through link-blade bowl as shown in Fig. 197.

out through holes at the top to the delivery spouts.

In link blade machines (Fig. 197) the separation takes place as the milk travels between the curved blades of the separating device, the cream passing up and out at the centre and the skim-milk up along outer ends of the blades next the inner wall of the bowl.

In the tubular or hollow bowl type (Fig. 198) the milk enters at the bottom. The bowl revolves at a high rate of speed and as the milk moves upward the skim-milk moves out

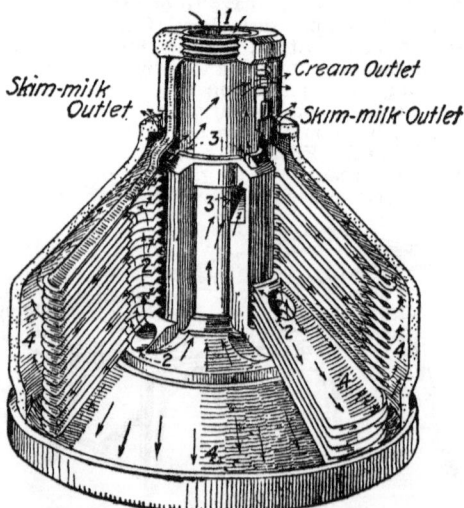

FIG. 200. Section and interior view of a separator bowl of the disc type

toward the wall while the cream forms a column in the centre. The chief features of this type are the long, slender bowl and the absence of an inner separating device.

CHEESEMAKING EQUIPMENT. For the dairyman who wishes to make small amounts of the soft cheeses such as cottage,

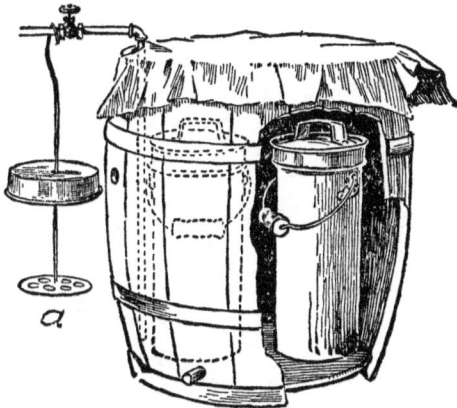

FIG. 201. Pasteurizing outfit that any farmer can make. A stirring device to be used in the can to insure even heating.

Neufchâtel, or cream, the following equipment will be found convenient and desirable.

Experience has shown that cheeses of better quality can be made if the whole or skim-milk is pasteurized. A simple, convenient outfit for pasteurizing small quantities consists of a barrel sawed off at the proper height with a steam pipe extending into it so that the water can be brought to the proper temperature (Fig. 201). Milk can be pasteurized in this barrel, in either the shot-gun or the 40-quart can; care must be taken to see that the milk is stirred so as to be uniformly heated. After the heating process has been completed, the same outfit can be used for cooling the milk by replacing the hot water with cold.

A convenient draining and curd table (Fig. 223) has racks in the bottom that allow the whey to run off through the outlets at one end. Pressure can be obtained as shown by wrapping the curd in the draining cloths, laying a board over the surface and setting upon it cans full of water. The sides of this table are removable for convenience in operation.

Convenient molds for making soft cheeses are shown in Fig. 202.

CHURNS. On many farms it is occasion-

FIG. 202. Two types of soft cheese mold

ally desired to make a small amount of cream into butter. For this purpose the little glass churn (Fig. 203) is very satisfactory. It consists simply of a heavy glass jar with a smooth steel screw-top and a well-made dasher operated by gears and crank handle much as in an egg-beater. Such churns are made to hold from 2 to 5 pints.

FIG. 203. Glass churn for kitchen use.

Three of the older types of hand churn are shown in Vol. I, Fig. 543. This swing churn is extremely simple with no floats or paddles inside, the rounded ends and slanting top producing the desired agitating of the cream. The fact that the opening is always on top is a favorite with many persons.

The barrel churn, usually made of hard wood, well bound with metal hoops, is one of the old standbys. The lid contains a sight glass so that the progress of the churning may be observed without removing the cover. The smaller sizes of this type of churn work easily by hand and give good results, but for the larger sizes it is well to have some kind of power.

The rectangular or box type is also well known on farms. It works easily, bringing a complete separation without injury to the

FIG. 204. Two sanitary modern churns. Stoneware type at left, steel barrel type with geared handle at right.

grain of the butter. The opening is large making the churn easy to clean and there is no mechanism inside.

One of the newer types of churn is made of fine glazed stoneware (Fig. 204) which will not absorb liquids, with a cover of clear glass through which the process of churning may be observed and which rests upon a rubber ring giving a tight joint. Its chief advantage is the ease with which it can be thoroughly sterilized and kept in perfectly sanitary condition.

Another attempt to develop a strictly sanitary churn has produced the steel barrel heavily tinned inside. The cover is cork lined making a perfectly tight joint.

In the power-driven combined churn and butter worker the processes of washing, working, and salting are done in the one container. The churn is provided with 2 rolls between which the butter is worked, and special gears which are thrown into place after the churning is completed and the working process is commenced. While exhibiting the good features of the modern creamery churn and butter worker, it is reasonable in price and is giving good satisfaction on many farms.

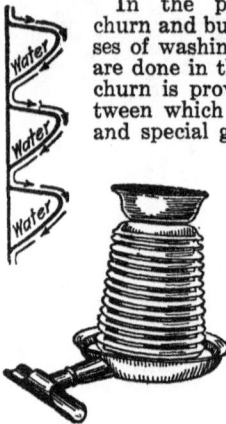

FIG. 205. Combined milk cooler and aërator. *a* section of spiral coil containing cold water.

COOLING TANKS AND COOLERS.

In the handling of milk and cream no one thing is of greater importance than proper cooling, for which every dairyman should provide suitable equipment. The cooling may be done by setting the cans in cold or ice water or by running the milk over one of the many forms of coolers now on the market.

For a small amount of milk or cream, good results may be obtained by using a pickle or alcohol barrel, either full height or sawed off at the proper point to prevent the water getting into the cans, and either cold running water or broken ice. It is always well to use a thermometer for testing the actual temperature of the milk.

For larger amounts of milk, a homemade tank of either wood or concrete may be used. Such a tank as shown in Fig. 207 will give good service. It may be made of 2-inch No. 1 clear cypress, all joints tongued and grooved or with slip tongue joint and the corners carefully rabbeted (Fig. 206). If desired it may be painted with a good oil paint down to the water line, but it is better not to paint below that point. Several styles of wooden tank may be obtained from dairy supply houses.

In many respects a tank made of concrete is the most satisfactory. If it is sunk into the floor so that only about a foot extends above

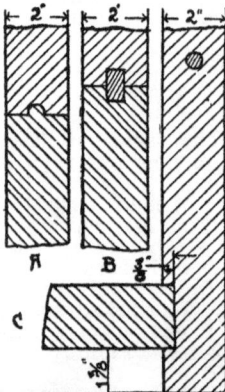

FIG. 206. Joint construction in wooden cooling tank. *a* and *b* tongue and groove; *c* rabbeted corner.

FIG. 207. Wooden cooling tank such as can be bought ready made or built at home. Inset shows popular conical metal cooler (see text below).

it, much labor in lifting the cans in and out will be saved. The top of the concrete should be faced with band or angle iron to protect it against wear in lifting the cans out and in, and there should be an outlet in the bottom so the tank can be easily cleaned, since more or less milk is sure to be spilled making the water foul and unfit for use.

Under some conditions a galvanized-iron tank which is cheap, quite durable and easy to keep clean, will give good results. Its chief disadvantage is the rapid loss of cold by radiation through the thin metal walls. In a more expensive, well-insulated metal tank built much like a refrigerator, there is very little loss of cold through radiation.

When milk is cooled in cans in any style of tank it is necessary to stir it frequently till the desired temperature is reached. This means added labor which is often a serious obstacle. When rapid cooling and less labor are desired, one of several forms of aërating coolers now on the market may be used. One of the simplest and cheapest is shown in Fig. 205. The open-topped conical drum may be filled with cold water or ice water or it may be connected to a running water supply by means of a piece of hose. When cold or ice water is used, it should be stirred frequently to prevent the layer next the cooling surface becoming warm. The milk tank, which rests on the cooling drum, has a circular row of fine holes at the base through which the milk flows, spreading out as it strikes the cooling drum, in a very thin, even filin.

In the form shown in Fig. 205, running water

While the man works, the horses stand still; while they move his labor ceases. This is both a tiresome and an unprofitable method

With a spreader the operation is continuous and the distribution uniform; a maximum area is covered and with a minimum of human effort

THE TEST OF A SUCCESSFUL FARMER IS NOT HOW HARD HE WORKS BUT HOW MUCH HE ACCOMPLISHES ECONOMICALLY. WHAT GOOD ARE PROFITS IF ONE WEARS HIMSELF OUT IN SECURING THEM?

Threshing scene in **Egypt,** one of the few countries where native farming methods have remained unchanged for centuries

Threshing scene in our own wheat belt where science and practical experience combine to increase production and efficiency

FARMING GIVES MANY MEN A LIVING IN SPITE OF, RATHER THAN BECAUSE OF, THEIR METHODS. BUT THIS IS NO REASON FOR STICKING TO OBSOLETE, INEFFICIENT PRACTICES

enters at the bottom and circulates upward through the spiral coil till it reaches the overflow pipe at the top. The milk does not drip from one ring to the next but follows the entire surface, which becomes colder and colder toward the bottom. This cooler is made of copper heavily tinned both inside and out.

Where economy of space is important a very durable style of cooler shown in Vol. I Fig. 533 may be desirable. In this also running water enters at the bottom and flows out at the top; the milk flows in a thin film over the outer surface, the wavy construction giving a very long surface for the amount of space which the cooler occupies.

MILK PASTEURIZER. When it is desirable to pasteurize milk on farms, a tank of the style shown in Fig. 208 will be found convenient and satisfactory. With this it is possible to control the heating and holding temperature of the milk very accurately. The milk can also be cooled in this same tank by replacing the hot water with cold, or ice water. This pasteurizer is made in a number of sizes for handling different quantities of milk.

MILK BOTTLE CASES AND DELIVERY BASKETS. There are many styles of cases for carrying bottles on the delivery wagon. Some are made of wood with metal dividers, while others, entirely of metal, are somewhat more expensive but will outwear the wooden ones, and are also easier to keep in a clean and sanitary condition. When cracked ice is used on the milk during delivery, it is best to have cases with water-tight bottoms. When cracked ice is put over the bottles and the cover closed tight, the milk will remain at a low

FIG. 209. Common and uncommon containers for retailing milk

temperature for several hours even in hot weather.

The dairyman who retails his milk needs some kind of a delivery basket or rack. One open wire style is very light but is well made and durable; another has a similar frame covered with a close mesh woven wire that makes it serviceable for carrying bottles of any size. A third style is made of tinned steel, heavier and able to stand much hard wear.

MILK BOTTLES. Most dairymen have their milk bottles made with some special identification mark, but aside from minor modifications, they are all made after one of the styles shown in Fig. 209. Any of the ordinary flat paper caps can be used with (a), while (b) requires the "beer-bottle" style of metal cap; (c) and (d) show 2 styles of individual service fibre milk bottles which, under certain conditions, are very desirable since they do away with all cost and trouble of collecting, washing, and sterilizing used bottles.

MILK CANS FOR SEPARATING AND SHIPPING. Among the different methods of raising cream by gravity, the Cooley-can system is generally recognized as the best. In the can used (Fig. 190f), the cover extends down over the sides forming a water-tight seal so that the can may be entirely submerged in cold water. A glass gauge on the side shows the depth of the cream. The skim-milk is drawn off from the bottom, the siphon preventing the cream from running out.

There are a few quite distinct types of shipping can aside from details of construction and size. The principal point of difference in the different styles is the form of the cover. Fig. 190a shows a can with flat top cover the edge of which extends well beyond the lip of the can thus protecting it from dust and dirt, and also the drop style of handle; (b) illustrates a style with hollow top cover and rigid strap handle; (e) shows a type in which the cover may be locked, closing the can air-tight, making it impossible to disturb the contents and preventing evaporation (this form is often desirable when cream is to be stored or shipped).

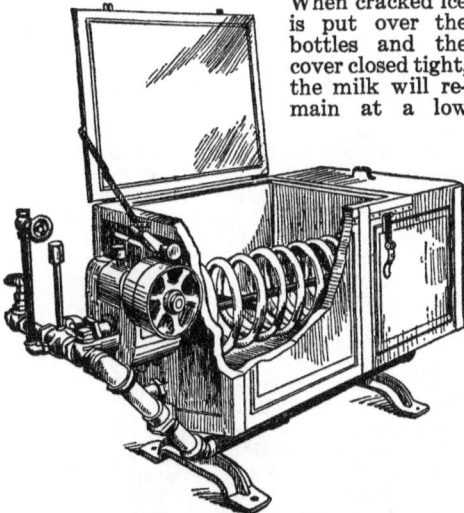

FIG. 208. Power-driven pasteurizer which may also be used as a cooler by running cold water or brine through the revolving spiral coil, instead of steam.

FIG. 210. These are both good pails, but the small-topped form is by far the better.

(d) shows an entirely different type fitted with a metal cap or plug attached by a short piece of chain, and commonly used in the New England milk districts for market milk.

Each of these styles is made in several sizes. The last one described usually holds $8\frac{1}{2}$ quarts, the other 3 types are made to hold 20, 32, or 40 quarts.

A convenient can sometimes used for peddling milk from a wagon has a 1-quart measure for a cover (Fig. 190c). This arrangement protects not only the milk can but also the measure from contamination with dust and dirt while on the route.

MILK PAILS. The milk pail is an important factor in determining the sanitary quality and the market value of milk. A good milk pail should (1) be easy to use; (2) be durable; (3) be easy to keep clean; and (4) as far as possible, prevent the entrance of dust and bacteria into the milk.

FIG. 211. Good (A) and poor (B and C) seam construction in dairy tinware.

To meet the first requirement a pail must be of the right form and size to hold easily between the knees and below the level of the cow's udder. For most men a 12-quart pail of the form shown in Fig. 210 is the most desirable; if too flaring it is difficult to hold, and if too high it is inconvenient to use.

To meet the second need the pail must be made of metal, heavy enough to stand use without binding or denting. Pails made from heavy tin, well reinforced at top and bottom either with band iron or wire, are good and the lowest in price. A galvanized iron bottom adds somewhat to the cost, but a pail so strengthened will wear much longer and in the end be just as cheap. Pails made of heavy, pressed steel, well tinned inside and out, are strong and durable, greatly outwear the thinner ones, and on many farms are giving excellent results.

Because of the great importance of cleanliness only metal pails should be used. Furthermore, there should be no open seams or cracks into which milk can get and which can not be easily reached with the wash cloth or brush. All seams and corners should

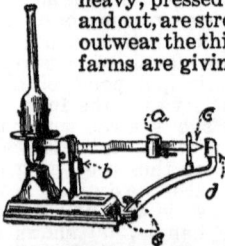

FIG. 212. Scales for use in testing dairy products. a sliding poise; b hanging poise; c pointer beam; d indicator plate; e levelling screws. (See text p. 134.)

be well flushed with solder as shown in a Fig. 211; such conditions as are shown in B and C should never be allowed in pails or other utensils used for handling milk. From the point of ease in cleaning, pails made of pressed metal without seams are especially desirable.

Keeping quality is a very important factor in the market value of milk, and since keeping quality is largely dependent upon the number of bacteria in it, it is desirable that as many as possible be prevented from entering the milk. Many of the bacteria in fresh milk fall in from the body and udder of the cow during milking, hence it is important to use a pail made to keep out as many as possible of these organisms. A hooded or covered pail (Fig. 210) is very desirable for this purpose, being easy to use and able to exclude approximately two thirds of the bacteria which would fall into an ordinary pail. The crown in the hood makes every part of the inner surface visible and easy to wash. A covered pail costs a little more than the ordinary sort, but the hood adds stiffness and makes it wear longer.

MILK SCALES. A good spring balance for weighing the milk from individual cows when records are kept should be provided with a loose pointer which can be adjusted by means of a thumbscrew to offset the weight of the empty milk pail so that the exact weight of the milk may be read direct. The scale that reads in pounds and tenths of a pound makes the calculations much simpler than one showing pounds and ounces. Such a balance is made in sizes weighing up to 30 and 60 pounds respectively. For weighing milk in cans, a platform scale with the beam graduated to half pounds is desirable.

MILK SEDIMENT TESTERS. The amount of insoluble dirt in a particular lot of milk may be ascertained by the use of a sediment tester of which there are several styles on

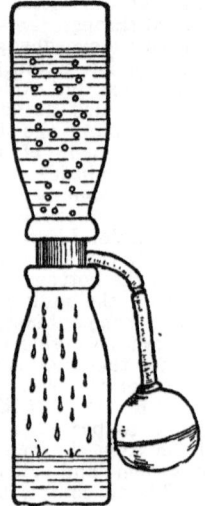

FIG. 213. Diagram to show action of sediment tester.

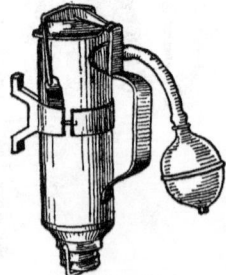

FIG. 214. Metal type of sediment tester

FIG. 215. Sanitary metal milk stools. (See text)

the market. In using one of the cheapest and simplest, a pint bottle is filled with the milk to be tested; the tester containing a disk of filter cotton is placed over it then turned upside down over a second, empty pint bottle. A few squeezes of the bulb give enough pressure to force the milk through very quickly (Fig 213).

In another style (Fig. 214) the cotton filter is easily clamped in place on the bottom, the chamber filled with milk, the air-tight cover clamped on and the required pressure obtained by means of the bulb. This may be used as a portable tester by the milk inspector or, in the factory or creamery, may be attached by means of the bracket over the weigh-can for testing the milk of different farmers as their milk is received.

A style of filter is made for use where a large number of samples are to be tested. A number of filter tubes are placed in a hot-water jacket which hastens the filtering process, especially in cold weather. Each of these devices uses a standard 1-inch filter disc, and holds a pint of milk.

MILK STOOLS. The usual wooden milk stool soon becomes very dirty, is not easily cleaned, and is usually quite short-lived. To meet the present day demand for a more permanent stool which could be kept clean, the stools shown in Fig. 215 have been placed on the market and are now in use on many dairy farms: *a* is a single-piece stool, the seat, leg, and foot being welded together and the entire stool then given a heavy enamel coating, making the surface smooth and easy to wash and keep clean; *b* is of the same size, but has 3 legs which are securely riveted to the seat and stiffened by a brace about half way down; the whole stool is then heavily enameled making it practically a one-piece stool; *c* is made of malleable iron cast in one piece, galvanized, and very strong and durable.

MILK STRAINERS. Whatever differences exist in different methods of handling milk, one operation is rarely omitted—that of straining. Of the many forms of strainers on the market, many are too complicated for practical use. A good strainer should be simple in structure, easy to use and capable of removing all particles of dirt. Fig. 216*a* shows

FIG. 217. Complete two-cow unit of one type of mechanical milker.

one of the simplest and cheapest styles. The one or more layers of strainer cloth are placed over the bottom of the cylinder and held in place when the loose collar is pressed into place. In another type (Fig. 216*b*) the milk passes through 2 layers of metal gauze of different mesh which are held in place by a wire spring, and can easily be removed for washing. If a heavier, more durable strainer is desired, the one shown in Fig. 216*c* in which the milk passes through 2 sets of strainers will give good results. It is made of heavy tin, each section being a single seamless piece.

MILKING MACHINES. Since about 1910, a number of different styles of milking machines have been placed upon the market. In this development, many manufacturers have endeavored to imitate the action of either the calf or the hand milker. In all the machines, the milk is drawn from the teat by the creation of a vacuum in the teat cup, or by pressure outside, or by the combined action of a vacuum in the teat cup and pressure applied to the side of the teat. Some of the chief differences between machines now on the market are in the form of the teat cups and in the principle by which they work. On this basis there seem to be 3 general types. In one the teat cup consists of a rigid metal cone with a flexible rubber cup at the top, which forms a close ring around the teat. The body of this teat cup is not flexible and has no inner lining; the milk is drawn by creating alternately a vacuum and a normal atmospheric pressure in the cup. This form of cup is made in 3 sizes to fit cows with large, medium, and small teats. It is shown on the machine in Fig. 217.

In a second type of cup, (Fig. 218), the outer wall is rigid but there is an inner flexible lining. The milk is drawn by creating first a vacuum in the inner tube and then pressure in the space between the inner lining and the outer wall, which exerts a squeezing effect upon the teat. This action, in which the periods of the vacuum and pressure alternate automatically, is supposed to have the effect of preventing the accumulation of blood in the blood vessels of the teat.

FIG. 216. Three types of strainer as described above

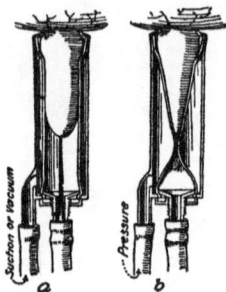

FIG. 218. How teat cups of the second type (p. 141) work.

The third type of cup (Fig. 219) also has a double wall. One side of the outer wall is rigid while the opposite one is soft and flexible, the flexibility being greatest at the top. When the vacuum is applied, the soft side is drawn in, causing pressure on the teat first at the top and gradually extending downward. This pressure is intended to imitate the action of the hand in hand milking.

Many variations of these types are found in the machines now on the market.

For creating and controlling the vacuum and air pressure, many devices have been developed but as in the case of the teat cups, there appear to be three general types after which the other makes are patterned. In one of these (Fig. 217), the automatic control is placed upon the cover of the milk pail, and is operated by means of a vacuum pump connected with the air pipe line placed over the cow stanchion. This regulator or "pulsator" is automatic and can be adjusted to give any desired number of strokes per minute; it is easily removed from the pail for cleaning and sterilizing. In this form of milker, there is continuous vacuum in the pail which seals it tight and avoids any possibility of the cover falling off during the milking process.

A second type of pulsator (Fig. 220a), instead of being attached to the milk pail, is applied at the pipe line above the stanchion, and is operated by compressed air. It requires the use of 2 rubber tubes from pulsator to milk pail and also from the pail to the teat cups instead of one tube as in the case of the first style. This pulsator does not come in contact with the milk and therefore requires no washing; it also maintains a constant vacuum in the pail which seals down the cover.

In the third type (Fig. 220b) the vacuum is created only in the tube system so the cover of the pail is not sealed down during the milking. The vacuum and release in the teat cups are caused by the alternate opening and closing of the valve at the bottom of the vacuum chamber in the lid of the pail. Like the other styles, this machine is operated by means of a vacuum pump and some means of power which may be located at any convenient place, outlets for the attachments of the rubber tubes being distributed along the pipe line above or beside the stanchions.

STERILIZERS. In the handling of milk and other dairy products, nothing is more important than the thorough sterilizing of all the utensils. In the case of a few utensils this may be done by thorough scalding with boiling water or by submerging them in boiling water. If there are many, the use of steam will be more satisfactory, both from

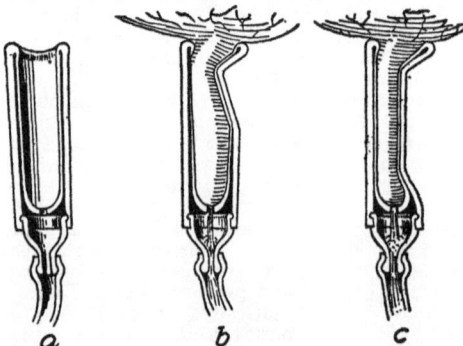

FIG. 220. Pulsating mechanism of two types of milkers. a is attached to pipe line, b forms cover of the pail.

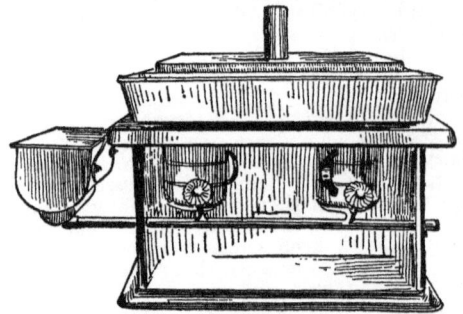

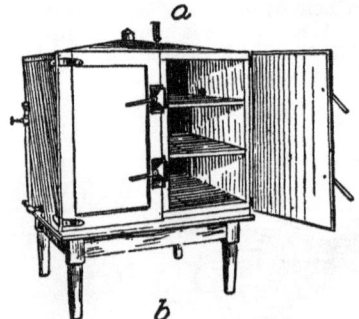

FIG. 219. Three stages in the operatio of the third type of teat cup

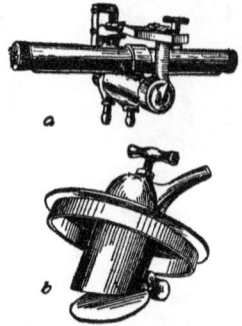

IG. 221. Two sterilizers suitable for house and farm use, described on the following page

the standpoint of complete sterilization and the saving of labor. On many dairy farms large wooden chests or chambers built of concrete or hollow tile are giving excellent results. Fig. 221a shows a small sterilizer devised by the U. S. Department of Agriculture (see Farmers' Bulletin 748) for use over a 2-burner oil stove in small dairies, and which can be made by any good local tinner. A galvanized iron box confines the steam generated from water held in an ordinary roasting pan, the steam then escaping through the tube on top. Pails and cans are sterilized by being inverted over this exhaust pipe.

Where a supply of steam is available the heavy galvanized iron sterilizer (Fig. 221b), in which the steam enters through perforated pipes in the bottom makes a very useful sterilizer, since all the utensils used during the day can be treated at once. When the steam is turned off the door should be opened for a few minutes to allow the steam to escape so the utensils can dry off. They can then be placed on racks or left in the sterilizer till needed for use. Larger, higher grade machines, insulated and filled with loose shelves, trucks and metal rails, drainage devices, etc., are made for use in large dairies.

WASH SINKS AND BOTTLE WASHERS. A good wash sink is a very important part of the dairy equipment. The old-style wooden sink is hard to keep clean and is being rapidly replaced by the metal tank type. A round-bottomed sink is especially good for washing

FIG. 222. A good metal double wash sink with bottle brush operated by a foot treadle

milk cans. If it contains a centre partition, one side can be used for washing and the other for rinsing. Such a tank costs somewhat more, but really serves as two sinks, and is very economical and serviceable.

A small number of bottles may easily be washed with a regulation bottle brush, but if 100 or more are to be handled daily a foot power-washer (Fig. 222) will be a great time saver. For still larger numbers of bottles a power-driven washer run by steam turbine, belt, or electric power, is desirable.

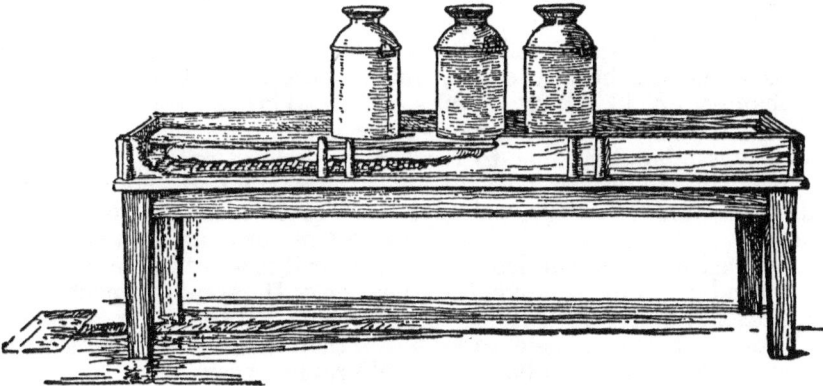

FIG. 223. Farm-made apparatus for pressing cheese (see p. 135)

FIG. 224. A triumph of modern invention in spray machinery. A horse-drawn, engine-driven machine covering twelve rows with sixteen nozzles

CHAPTER 13

Machines for Spraying Crops

By F. F. ROCKWELL, (see Chapter 11). The greatest development of spraying apparatus has been along the lines of vegetable and fruit growing activities, and it is these to which Mr. Rockwell has given a large share of his attention. As in the case of the garden implements he describes in an earlier chapter, his statements and recommendations are invariably based on personal practical trials and experiences. That he is equipped to put his ideas in interesting, convincing form has already been proved by his books which include "Around the Year in the Garden," "The Key to the Land," "Vegetable Gardening," "Gardening Indoors and Under Glass," and others.—EDITOR.

SPRAYING has become, during recent years, an acknowledged factor in the successful culture of many of the most important crops of the farm, garden, and orchard. Many of the large growers would sooner think of attempting to get along without their harrows or cultivators than without their spraying apparatus. Spraying is coming into such universal use, not only because it is a protection against possible loss from insects and disease, but because it *increases profits,* in normal as well as in exceptional years. (For detailed directions as to when and how)to spray, what to use for different pests, etc., see Vol. II, Chapters 32, 33, 34 and articles on special crops.)

Wonderful advances have been made in the apparatus for spraying of all kinds, from the smallest hand sprayer to the largest orchard outfit. This has made spraying a much less disagreeable task than formerly, and, more important, it has made it much more certain and effective. It has also greatly decreased the cost of applying the spray material. The cost of spraying, however, which is, after all, the factor which will determine whether it is a profitable thing to do or not, will depend very directly upon the efficiency of the spraying machinery used. Further, efficiency is not determined solely by large capacity. A 2-row machine which *will keep working* will be cheaper to use than a 6-row outfit which continuously gives trouble and has to be replaced after a few years' use.

The requisites of a good spraying machine, of no matter what type, are as follows: reliability, good pressure, accessibility, convenience, and facility of repairs.

Reliability. This point, important with any machinery, is relatively more important in the case of the sprayer than with most other farm machines. The results of the work done, the cost of applying the spray, and the completion of the work *on time*—which is more important in spraying than in almost any other farm operation—depend directly upon reliability, upon having a machine so practical in design that it will *keep working.*

Good pressure. The results of spraying depend directly upon the thoroughness with which the work is done. A steady, high pressure that will break the spray up into a fine mist, reaching every part, penetrating crevices, and adhering when larger drops of the spray would merely strike and roll off or run together, is absolutely essential for good work. A constant, *even* pressure, as well as high pressure, is of equal importance.

Accessibility. The best of pumps will have to be cleaned and regulated occasionally, and will sometimes need repairs. Construction or design which will admit of such work being done quickly and easily is of great importance. A sprayer that requires a machine-shop equipment to get at its interior is not a good pump to have to depend upon. The working parts should not be made accessible, however, at the cost of risk either to the operator or to the machinery. Working gears and so forth should be inclosed or protected by guards so that they will be "foolproof"; and the whole apparatus should be protected by screens, side curtains, or similar guards to prevent the possibility of limbs, broken-off twigs and so forth from getting caught in any of the working parts.

Convenience. Time is an important element in spraying, because it affects not only the cost, but also the results, of the work. A spraying equipment should be such that the work can be done as expeditiously as possible. Small things count. A hinged tower that can be folded down, instead of a rigid one; an equipment of tools on the spray rig for making all adjustments and slight repairs; angle or other special nozzles carried *with* the outfit; a rig that will turn short, but with the centre of gravity low down, so that there is no danger of tipping over; an adequate supply of such extra parts as are most likely to wear out or give way, such as washers, gaskets, hose connections, and so forth; in a power machine, a good, reliable engine; a tank that can be conveniently filled, but will not slop over when it is full—all these things, and many others, are minor points perhaps; but in the aggregate they make a

FIG. 225. Spraying melon vines with double nozzles on two lines of hose, the machine being of the horse-driven traction type.

great deal of difference in the actual efficiency of the outfit in use; and they should be borne in mind when selecting a machine.

Repairs. After determining the type of machine best suited to your purposes, get the best of that kind there is to be had. You may find a wide range in prices, but in spraying equipment, if anywhere, "the best is the cheapest." Many of the materials used in spraying are highly corrosive. A brass tank or brass working parts may be more expensive than those of some other metal, but the additional expense will pay in the long run, because the weakest part of the machine will be the one to determine its working life.

Good modern spray machinery is made to give long, reliable service without constant tinkering and repairing. When it finally gives out it is generally worn out and should be replaced. But even a good machine sometimes needs repairs in a hurry. It is well to buy your equipment, therefore, from an established concern from which you can always get advice or repair parts without any trouble. Other things being equal, buy the machine for which you can get this service in the shortest possible time.

These general principles apply to all spraying equipment, and should be taken into consideration when one is making a selection of any kind of spraying apparatus, from a hand syringe to a high-powered orchard outfit.

General Types of Sprayers

With the very wide variations that exist in the forms, sizes, and mechanical construction of pumps for many different purposes, the simplest basis on which to make a classification of the different kinds of pumps is the kind of power by which they are operated. There are three general types: those worked by hand power; those driven by traction from the wheels of the rig on which they are mounted; and those driven by engine power.

The power-driven outfits are, of course, the most expensive, but their capacity is so much greater that they are coming into more and more general use, even for those purposes for which the other two types have been generally used. The traction type has the disadvantage of being more uneven in the pressure supplied than either of the other two, and it is on the whole more likely to get out of order.

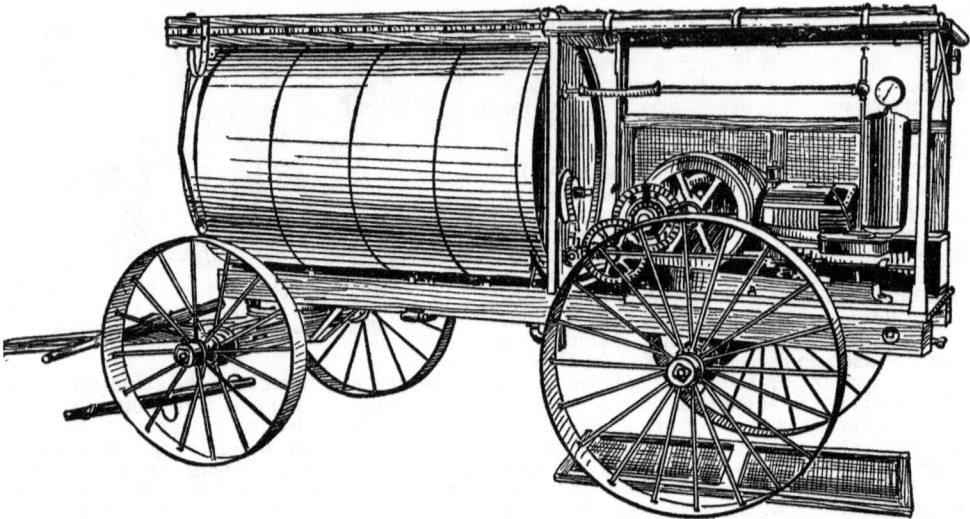

FIG. 226. A power sprayer suitable, and indeed essential, for large commercial orcharding operations. A tower can
be erected on it for reaching tall trees

Frequently outfits of this type cannot be used under unfavorable conditions, such as on wet ground or over steep grades, where either of the other types could be used.   Where, as is usually the case, there is a good deal of other work for a small engine on the farm, the spraying-outfit engine can be used for that purpose and need not be charged up entirely to the spraying equipment.

The disadvantage of the hand-operated pump, compared with either the traction- or the power-driven type is, of course, that it is either very limited in capacity or will require the services of an extra man in addition to the applying of the spray material.   The larger hand outfit now, however, is entirely practical and good for commercial work, where the amount of spraying done is not too great.   It can be utilized for many purposes, being readily adapted for work in the orchard, on garden or field crops of limited area, and for spraying, disinfecting, and so forth.

Another general distinction among pumps of various kinds is based on the manner in which the spray material is applied—whether by compressed air or directly by the force of the pump.   The former method is more flexible and gives a more constant and even pressure; and it is under more accurate control, with less danger of breakage, if anything goes wrong.   Both types, however, are used successfully, and neither is best for all purposes.   In small hand sprayers, the compressed-air type has the advantage of allowing the pumping to be done in advance of the actual spraying, so that one's whole attention can be given to the latter while the material is being applied.

Outfits for Orchard Spraying

Orcharding, like the growing of any other commercial crop, is being done on a larger and larger scale.   Spraying in commercial orchards is done mostly by power-driven outfits of 100 to 250 gallons capacity.   The size of the orchard is not the only factor in determining which size of tank will be the most economical to use.   On hilly or rough land, a tank of relatively small capacity, when full, may be all that a pair of horses can handle.   Here, as in selecting the make of your machine, it is important to get one that will *keep working*.   Delays and breakdowns are expensive; and the bigger the machine, the more expensive they are.

Where spraying is one of the important operations on the farm, it will pay to have the spraying rig complete in itself, rather than to have to depend on using another wagon gear. Spraying outfits can be bought, however, that are mounted on a substantial base, ready to use on a farm gear, where that seems desirable. One of the handiest outfits for a moderate-sized orchard is a small engine pump and tank mounted as a unit on a platform which can be slipped in to a suitable farm wagon, or placed upon a flat truck. Some of these outfits are designed to work without even being secured to the wagon in any way; but, of course, bolts or lag-screws, to prevent possible slipping on hillsides or rough ground, will be only a reasonable precaution. Such outfits may be had to carry 50 to 100 gallons of spray, and weighing, complete, 500 to 700 pounds. An outfit of this kind, especially if there is available a suitable place to store it during the periods when it is not in use, so that when wanted it may be let down on the wagon body by means of a block and tackle, makes an ideal 1- or 2-man outfit for the small orchard. While it costs a little more than a hand pump and a barrel outfit for a wagon or cart, it has several times the capacity, will save one man's time, and is thoroughly practical in every way.

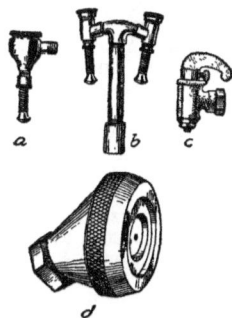

Fig. 227. Popular types of nozzles: *a* and *b* single and double Vermorels; *c* bordeaux; *d* a recent improved development.

Fig. 228. Extension rods for orchard spraying, one of bamboo, the other of metal with sliding grips

For the home orchard or the small commercial orchard of only a few trees or few acres, where it is often operated in connection with poultry or intensive truck gardening, a barrel-pump sprayer is the most economical and will answer its purpose. One of the objections to this type of sprayer is that the small quantity of material carried makes it necessary to come back to the base of supplies very frequently. Even when only a 1-horse cart or wagon is used for transporting the spray outfit, much of this wasted time may be saved by carrying 1 or 2 extra barrels of the spray material with tight covers along with the barrel with which the spraying is being done. When this extra supply is needed, it can be simply siphoned into the barrel into which the pump of the spraying outfit is attached, care being taken, of course, to keep it agitated while it is being drawn off, so that the mixture will remain uniform. With an outfit of this kind, at a very low, initial cost, two men will cover a great many trees in a day's work.

Equipment for orchard work. The larger spraying outfits are usually equipped with an elevated platform or stage with a substantial railing around it, to enable the men who are handling the hose to be in a better position to do their work, to reach the tops of tall trees, and to save spray material. Under some conditions, however, the spray tower may not be needed, or may even be in the way. It should be easily removable, and, in the most convenient type of apparatus, it is hinged on one side, so that it may be folded over for going under large limbs or wires on the road, farm gateways, and so forth.

One of the most frequent causes of annoying, awkward, and costly delays in spraying is worn-out or dried-out hose and hose connections. An ample supply of all the minor repair accessories—hose fittings, washers, connections with bands, etc.—should be carried along with the machine. Also, a lead of hose should be kept available in case of accident. With the high pressure used,

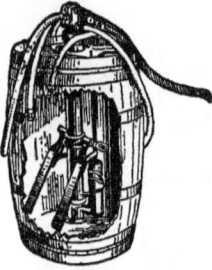

FIG. 229. Barrel spray pump, an excellent all-round farm outfit, partly cut away to show agitators

and the strenuous treatment necessary in spraying, one can never be sure when a hose is going to "go."

For most orchard work, extension poles for use in connection with the hose are required. These are made either of steel or of bamboo with an inner tube. The former, with adjustable wooden hand grips, are coming into more general use. Drip guards, formerly used and recommended for the operator's comfort are being rendered unnecessary by improved pole construction and materials. Special strainers for filling the tanks may be had for most outfits and are well worth their price in time saved both in excluding foreign matter from the tank and in preventing clogging of the nozzles. Where there are large acreages to be taken care of, a pump for filling the tank will be a paying investment; for even a slight delay at each trip, this point will cut down the day's work from a tankful to several tankfuls.

A new system in orchard work uses a nozzle which throws one big sheet of mist for 15 feet or more, so that fruit trees of ordinary size may be covered clear to the top branches from the ground. It enables one man to apply as much spray as two or three could with the ordinary pole-and-small-nozzle equipment. A good-sized tree can be covered in half a minute. The spray "guns" may be used singly or in pairs, and if desired, in connection with the regular spraying apparatus. Any one interested in getting a new orchard outfit would do well to investigate this new system and, if possible, to see it in actual operation before

he decides to purchase the standard spray equipment.

Another type of the power sprayer for orchard work is the compressed air sprayer in which *only the air* is carried with the outfit, the engine and compressor being located at the central point, where the tank is filled. Where it is possible to get air under sufficient compression without going to too great an expense, this system has some advantages, especially in very hilly orchards, or where the work is being done on a very large scale. The weight of the outfit to be transported is, of course, very much reduced, and therefore more of the spray material can be carried. The danger of breakdown or injury to the equipment in operation is also much less. The equipment, if used, should be such as to make it possible to keep up the pressure required for good work until all the contents of the tank are emptied—a petering out of air pressure means poor results. Under the right conditions, the work done is as good as by the regular equipment. It is merely a question of which will pay the best, and that depends principally on local factors.

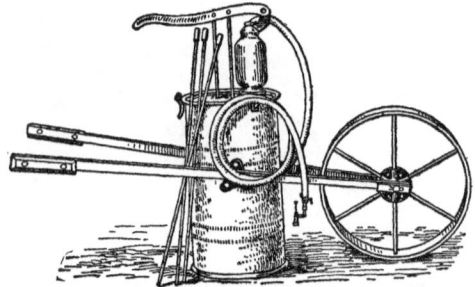

FIG. 230. Wheelbarrow type of hand sprayer suitable for whitewashing as well as crop spraying. Tank can be removed and barrow part used for other farm work

Spraying Equipment for Small Fruits and Field Crops

For spraying field crops, small fruits, vineyards, and so forth, the traction type of sprayer is used more generally than either the engine-driven or the hand-power type. Spraying of this kind can be done automatically at a uniform rate of speed, and the traction-power is, therefore, sufficiently steady to answer the purpose. With machines of greater capacity, and the recognized necessity of using high pressure to get a mist spray, traction as a source of power becomes less practical and is being gradually replaced by lightweight but highly efficient gasoline engines.

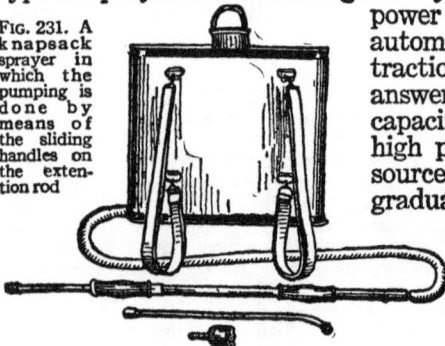

FIG. 231. A knapsack sprayer in which the pumping is done by means of the sliding handles on the extention rod

Much of what has been said in regard to orchard outfits applies to the equipment for field sprayers. The main things in selecting a pump for this purpose, are to make sure that you are getting a practical, substantial pump, and to obtain an equipment in the way of spray arms, nozzles, and so forth, that will be exactly suited for the purpose for which you wish to use it. In this connection, it should be kept in mind that for most crops it has proved most effective so to concentrate the nozzle fire that it will strike the row or crops from the side or even at an upward angle as well as from the top. This is especially true in the case of fungous diseases and of such insects as the melon louse, which congregate on the under surfaces of the foliage. With the increased pressure obtained in using an engine-driven sprayer for field work, however, resulting in a fine spray or mist that completely envelops the plant, it is not so necessary to have part of the nozzles ar-

Fig. 232. A convenient and popular knapsack sprayer—

ranged for going sideways or at an upward angle. In fact, in the latest type of power field sprayers, there is no attempt to hit the individual rows with the spray. The nozzles are so arranged that the entire area covered, some 25 feet in width, receives a dense, foggy spray under high pressure which penetrates to all parts of the plant (see Fig. 224). The engine upon this machine may be used to

Fig. 233. To which an extension rod can be attached for spraying small trees or the under surface of leaves.

operate other portable farm machines such as potato diggers, binders, etc. There is little doubt that the power sprayer will replace the traction sprayer for field operations, where planting is done on a large scale.

Hand-power sprayers. The hand-power sprayers, while not practical for work on a large scale, are, however, of the greatest use for taking care of small patches of truck of various kinds, garden crops, and fruit trees, for home use, and the spraying and disinfecting of animals and buildings. A small hand sprayer has its place in the regular equipment of practically every farm. Unfortunately, such poor care is taken of these small machines that usually they wear out within a few years, not from use, but from neglect.

There are, in general, 4 types of these small sprayers: (1) the bucket or ordinary force or suction pumps; (2) the barrel sprayers which are very similar, on a little larger scale, but are of large enough capacity to be of practical use for commercial field crops or for orchards of small acreage; (3) the knapsack sprayers, which, while of small capacity, are readily portable and are useful in small market gardens and about the place generally; and (4) compressed-air sprayers, which cover the same field as the knapsack type, but are, in many ways, more convenient, as they are easier to carry, will not slop over, and furnish a spray under higher pressure more completely under the control of the operator.

A more recent modification of the barrel type is the small portable tank sprayer, which has as wide a range of use as the knapsack and compressed-air types with a much greater capacity. This type has proved the efficiency of its work and become popular very rapidly. It fills a long-existent definite need of the smaller grower, who has wanted a sprayer that could be transported conveniently by

hand and yet have enough capacity to be of practical use for commercial work. Some sprayers of this type have been equipped with traction drives from the front wheel and, under good conditions, are practical for field work.

For the very small place, for use in the greenhouse and frames, and where a few hens, or a cow or two are kept, the simple small hand sprayer, with glass or brass tanks, with a capacity for a quart or so of spray material are of real service; with these, as with any spraying equipment, however, it decidedly pays to get the best. There are very many of these small sprayers on the market, very flimsily made of tin

Fig. 234. With a good pump and nozzle the bucket outfit is satisfactory for work on a small scale.

FIG. 235. The smallest type of outdoor sprayer for either wet or dry preparations.

and solder that will go to pieces almost before you can get them home from the store. There are a few, however, that are more substantially made, and, if used with care, will give years of service. They will cost two or three times as much as the makeshift affairs that are usually offered for sale, but are well worth not only the difference in price, but any trouble one may have in hunting them up when one is ready to buy. The best of these sprayers have 2 nozzles, with one of which the under sides of the leaves of vegetables or ornamental plants may be reached. It is often worth while to have a little sprayer of this kind on hand, even on a large place, for occasional use and to give prompt attention to the first insects which may appear before it would pay to spray the entire crop. Around the lawns and borders such an implement is especially effective since even the children can operate it. By keeping flowers, vines, and shrubs free from insects and disease one gives added protection to the farm crops as well.

Equipment for Dusting or Dry-Spraying

Within the last few years, the application of spray materials in the form of dust or powder has been developed very considerably and is now being used by many large commercial growers in place of the wet sprays. The saving in transportation, and in the application of these spray materials is, of course, very considerable. The gains now being made by this method, in actual field practice, are due both to the advances in the chemical preparation of the spray material, and to the equipment for applying it. It is perhaps too early to predict that the dry or dust sprays will replace wet spraying to any great extent, but such may quite possibly be the case. In orchard work the dust spray has, in many instances, proved as satisfactory as the wet, in controlling the codlin moth; in general it has been less successful against scab. As late as 1918 the cost of dust spraying was considerably the larger. Any one who is buying new spraying equipment will do well to investigate carefully the possibilities of this method for the crops and the conditions he has to deal with. Many of those who have changed to the dry system seem to feel that they are getting just as good crop protection as formerly at very considerably less cost per acre. The principle, of course, is not new, and this kind of spraying has given good results on small areas for many years.

FIG. 236. Four-row traction dust spray machine, convenient and highly efficient under certain conditions

PART III

Power on the Farm

IN 1915, after a careful investigation, an engineering authority of national prominence estimated that the power supplied by horses, mules, windmills, steam and gas tractors, and gas engines on farms amounted to 23,905,000 horsepower. This total, compared with figures taken from the last census, represents 5,149,714 more horsepower than was then being used by all the manufacturing enterprises in the country! This suggests what power on the farm means. Yet it is only a partial suggestion, for, in the first place, it leaves out all water- and electrical-power plants; and, in the second place, the use of power on farms has increased tremendously even since the above estimate was made.

The chapters that make up this part of Farm Knowledge undertake to explain some of the principles of power and its application to farm needs; and also to describe some of the means by which it can be created and utilized to increase the farmer's efficiency, his productive ability, and his chances for success. What he has already done in this direction has helped greatly to increase the Nation's per capita productions of staple crops, sufficiently, indeed, to balance the steady decrease in the number of persons engaged in agriculture.

Yet even more may confidently be expected in the future. Dr. B. T. Galloway, of the U. S. Department of Agriculture, once said: "The farm of the future will so utilize modern labor-saving devices and efficiency methods that human labor will be reduced to a minimum and the farmer and his children will have time and opportunity for and means of living a satisfactory life." Perhaps the following verses, published in the *Kansas City Star*, will ere long express as much of truth as they originally did of humor, and refer as accurately to other sources of power as they do to the undeniably versatile tractor.—EDITOR.

The tractor on the farm arose
  Before the dawn at four;
It milked the cows and washed the clothes,
  And finished every chore.

Then forth it went into the field
  Just at the break of day;
It reaped and threshed the golden grain
  And hauled it all away.

It plowed the field that afternoon,
  And when the job was through,
It hummed a pleasant little tune
  And churned the butter, too:

And pumped the water for the stock,
  And ground a crib of corn;
And hauled the baby round the block
  To still its cries forlorn.

Thus ran the busy hours away,
  By many a labor blest;
And yet when fell the twilight gray
  That tractor had no rest.

For while the farmer, peaceful-eyed,
  Read by the tungsten glow,
The patient tractor stood outside
  And ran the dynamo,

CHAPTER 14

Power and
Power Machinery on the Farm

By R. P. Clarkson, Professor of Engineering, Acadia University, Nova Scotia, since 1912, and before that, Assistant Manager, Worcester (Mass.) Boiler Works; Designer of Special Machinery, United Shoe Machinery Co.; Instructor in Mechanical Engineering, University of Vermont; Examiner of Patents and Electrical Engineering Expert for the U. S. Government; and Consulting Engineer of various engineering projects, agricultural and otherwise, in Canada and the United States. Notwithstanding this extensive experience along technical engineering lines, Professor Clarkson has given much time to the study of the application of mechanical and engineering principles and methods to practical agriculture. He has spent a good many summers on farms and for a number of years has answered engineering inquiries from readers for the "Rural New Yorker." He has also contributed many papers to other farm journals and is the author of "Practical Talks on Farm Engineering." * Consequently his contributions to "Farm Knowledge" reflect an understanding of just what farmers want to know and in just what form they want to have their questions, answers or unfamiliar subjects presented to them.—EDITOR.

Some Mechanical Terms and What They Mean

SUCH terms as energy, work, and power are so frequently used in the following chapters that a brief statement of their exact meaning is necessary.

Energy is capacity for work or ability to do work. Coal has stored-up energy —capacity for doing work by heating water and changing it to steam by means of which huge engines may be operated. Gasoline, gunpowder, dynamite, the wound-up spring of an alarm clock—all have the same stored-up energy ready for use when released. This type of energy is called *potential energy*, the term "potential" in this connection having exactly the same meaning as when we speak of "potential" life in the tiny seeds or the "potential" value of a great organization or a great idea.

On the other hand, the falling weight of a pile driver, the cannon ball flying rapidly through the air, the kick of a mule, the wind and the swiftly rushing stream all have energy due to motion. They represent tremendous forces— not stored up, but actually being exerted. This is *kinetic energy*. All potential energy must become kinetic before it can be utilized, just as a gold dollar must be spent before it is of real value. Stored in the pocketbook it represents "potential" buying energy, but it cannot purchase anything until actually put into circulation.

Work is done only by expending energy. Lifting a weight, hauling a loaded cart, the action of the brakes in bringing an automobile to a stop, the action of steam or of the gasoline explosion on an engine piston—each of these represents the effect of a force acting through a distance. While we ordinarily say and assume that work is done whenever effort is put forth, technically and in the sense of the following chapters, there must be motion when work is accomplished. The force acting must act through a distance. If, for example, a horse pulled strenuously on a load without moving it, he might exhaust himself through the effort

* Doubleday, Page and Co., Garden City, N. Y. $1.20 net.

put forth, yet in the technical sense no work whatever would have been done. If the load moves, the work done is the *average pull measured in pounds multiplied by the distance the load moves in feet.* It is expressed as so many "foot pounds" of work.

Power is the rate of doing work. Here the element of time enters in. A

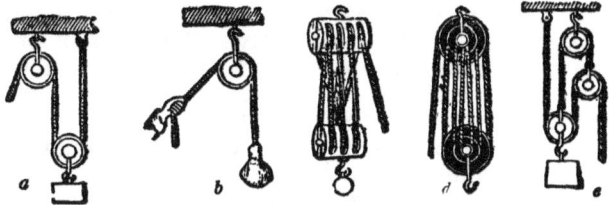

FIG. 237. Pulleys enable us to do heavier work than we otherwise could, but the greater the gain in force the greater the loss in time and distance. Thus in *a*, a 5-pound pull will lift a 10-pound weight, but 10 feet of rope will have to be pulled to lift it 5 feet; in *b* the pull and weight are equal, the pulley being merely a convenience; in *c* a pull of 10 pounds will lift 70; in *d* a pull of 10 pounds will lift 60; in *e* a 4-pound pull will lift 16.

2-horsepower engine will fill any silo if you give it time enough, but a 15-horsepower engine will fill the same silo in much less time. *The amount of work done per second, per minute, or per hour measures the power of any machine.* Here is the reason for the difference in size between the stationary farm engine and the automobile engine capable of exerting the same power. The stationary engine is of slow speed, so that the piston moves through a comparatively short distance in a minute and each explosion represents considerable force requiring large and heavy construction. The automobile engine, on the other hand, is a high-speed machine with less powerful explosions, occurring much more often. Its piston moves through a much greater distance in a minute; hence it may be made small and light. Both engines do the same amount of work in a minute, if operating at the same horsepower, only the smaller engine does a little at a time and does it oftener.

Work, as mentioned above, is measured in *foot pounds.* Multiply the force in pounds by the distance in feet through which it acts and the result is the work done. In raising a weight of 12 pounds 2 feet from the ground, the work done is 24 foot pounds. If the average total pressure of steam on the piston of a steam engine is 500 pounds and the length of the stroke is 2 feet, 1,000 foot pounds of work are done every stroke. If there are 150 strokes per minute, the work done per minute by the steam on the piston would be 150,000 foot pounds.

Power is measured by the amount of work done per minute, the unit being 33,000 foot pounds per minute or 1 *horsepower.* This number 33,000 is merely an accepted standard and does not represent the power of an average horse, although at one time it was thought to do so. In the engine spoken of above, the work done per minute by the steam on the piston is 150,000 foot pounds. As 33,000 foot pounds per minute represent 1 horsepower, this engine has an input of a little more than $4\frac{1}{2}$ horsepower (150,000 divided by 33,000 equals $4\frac{54}{100}$).

Some of the input (total power) of every machine is used in running the machine itself, largely in overcoming friction and in moving heavy parts. For example, in the engine above probably 1 horsepower is required out of the $4\frac{1}{2}$ to operate the engine. The remaining $3\frac{1}{2}$ horsepower would be available for operating a sawmill, a silage cutter, a pump or any

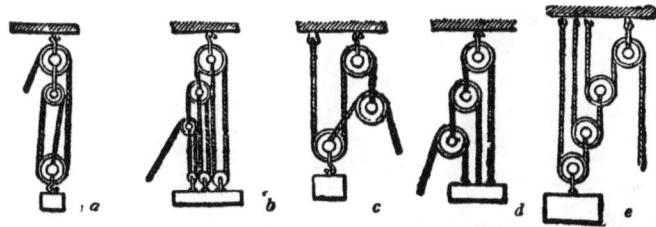

FIG. 238. Types of pulley combinations. In *a*, a pull can lift three times itself; in *b*, six times itself; in *c*, five times itself; in *d*, seven times itself; and in *e*, eight times itself.

other piece of machinery around the farm. Thus the engine is not 100 per cent efficient. The engine gives the value of only $3\frac{1}{2}$ horsepower output for every $4\frac{1}{2}$ horsepower input and has an efficiency of nearly 78 per cent, which is found by dividing the output by the input ($3\frac{1}{2}$ divided by $4\frac{1}{2}$ equals .78 multiplied by 100 equals 78 per cent).

Efficiency of machines. No machine known has 100 per cent efficiency. Every machine has to have more power put into it than it delivers, whether water wheel, electric generator, steam engine, potato digger, pump, or other apparatus. Some machines are more efficient than others, usually because of better design and better workmanship, not because of any difference in the process which goes on within the machine. Some processes are extremely inefficient, and so we think of the machine itself as being inefficient. An instance is the steam engine. Without doubt 80 to 90 per cent of the power exerted on the piston by the steam is delivered to the driving pulley; that is, a well-made, carefully designed steam engine has an efficiency equal to that of most well-constructed machinery. As a means of getting mechanical power out of steam, however, it is inefficient, for the process which goes on inside the engine is astonishingly wasteful. The steam contains a certain amount of energy when it enters the engine, but it still contains much of that same energy when it is finally exhausted. The engine is able to use but a small part of the steam's energy, but what it does take is usually delivered with but little loss to its driving shaft.

However, the efficiency of any machine is not necessarily the most important factor in its *value* or importance. Many machines could be made more efficient by refining them; but the refinements would cost so much that the added value of the output would be much less than the interest charge against the investment. The real desirability of any machine must take into account both its cost and its efficiency, and a nice balance between the two must be preserved. In most cases dealing with farm machinery and power, the first cost as well as the operating expense is a deciding factor. Increased efficiency at extremely high cost is desirable only in rare instances.

Simple Machines for Increasing the Power of Men and Animals

The work that a man can do armed only with a simple machine, such as a crowbar, a winch, a block and tackle, or a lifting jack, is surprising to people not familiar with these tools. It is quite true that a man's strength cannot be increased by these machines nor by any form of machinery. He can do only just so much work anyway; but various devices, including levers, pulleys, inclined planes, etc., enable him to exert his force in such a way that it is multiplied. Tests have shown that a well-developed man can deliver about one eighth horse-

FIG. 239. A stump puller worked on the horse-sweep principle illustrates the use of both the lever and the pulley or windlass

Under stress of war conditions, farming sometimes returns to prehistoric conditions and methods. This is an actual farm scene as photographed in France in 1917

The upper picture shows what sometimes has to be done; this one shows what inventive genius and mechanical skill have developed for use under more favorable conditions

MAN POWER IS NO LONGER THE MOST ABUNDANT AND CHEAPEST FORM FOR FARM USE. IT MUST BE CONSERVED IN EVERY POSSIBLE WAY

Natural power, such as that from wind, water, and the heat of the sun, is the most abundant, but not always the cheapest or most efficient.

Horse power is reliable, flexible under varying conditions and does not require mechanically trained operators. Its greatest disadvantage is its overhead cost

THE FARMER'S POWER SUPPLY IS SAFEGUARDED BY THE FACT THAT IF HIS ENGINES BREAK DOWN HE CAN FALL BACK ON NATURAL SOURCES IN THE EMERGENCY

power continuously, but that for a minute or two he can put forth something more than half a horsepower.

Suppose, for example, he develops one fifth horsepower, that is, does 6,600 foot pounds of work per minute on a machine, say on a lifting jackscrew. The jackscrew does no more work on its part than he does. However, the man may exert a force of only 66 pounds through a distance of 100 feet in pushing the bar around, while the jackscrew may raise the weight on it only 2 feet, in which case it will have lifted with a force of 3,300 pounds—far more than the man could have exerted. The object of the jackscrew, the crowbar and other similar helps is, then, to multiply the force that can be exerted on any object; in such a case the distance through which such object can be moved in a given time is always less than if the force could be applied directly.

The lever. Suppose a large stone is to be moved. It cannot be lifted directly and a crowbar is necessary. This is put in the position shown in Fig. 240 with the fulcrum placed just as near the load as possible. Why? Because, as a result, less force is required to lift the load. The work done by

FIG. 240.   How the lever works

the man at one end of the bar, that is, the force he exerts times the distance through which it is exerted is the same as the weight lifted times the distance through which it is moved, just as in the case of the jackscrew. The nearer the fulcrum is to the load, the longer the sweep moved through by the man and the shorter the distance the load is moved. In the same proportion, however, the load moved is greater than the force exerted.

The inclined plane in many forms is also a machine with which we can perform a task which would be too much for us unaided. Very few of us could roll a barrel of sugar right up the side of a building (from A to B, Fig. 241), but most of us could roll it up a slope (as from C to B) if it were not too steep. In either case the same amount of work is required; but in the first case a large force must be exerted through a short distance (A B), while in the second the distance

is greater, but the force required is correspondingly less. In most cases, distances over which to work are practically unlimited whereas our strength is decidedly limited.

Another way of lifting the barrel would be by means of a lifting jackscrew as already described above. In fact, this machine is merely an adaptation of the inclined plane. To show this, cut out of a sheet of paper a 3-sided figure representing an inclined plane. Wrap this paper around a lead pencil or broom handle (Fig. 242), so that the edge B C lies along the pencil and the edge A C overlaps itself. The edge AB now forms a screw thread around the pencil just like the winding thread on a jackscrew. If you cut the paper so that the distance A to C is just equal to the distance around the broom handle, then BC is just equal to the "pitch" of the thread,

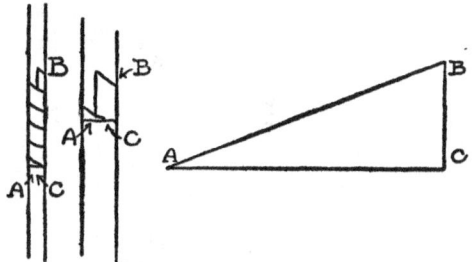

FIG. 242.   How the screw works upon the inclined plane principle.   (See text)

that is, the distance the screw moves up or down in making 1 revolution. Operating a jackscrew is, therefore, like pushing a load up an incline—the less the slope, the easier the task. In choosing a jackscrew, then, remember that the less the pitch for any given diameter and the greater the diameter for any given pitch, the easier the jackscrew will operate under a load, because the shorter BC is compared with AC, or the longer AC is if BC is unchanged, the less the incline or slope of AB. Further ease is obtained in operating a jackscrew by using a long bar as a handle, since this makes use of the leverage principle, as discussed in the case of the crowbar above.

FIG. 241.   The inclined plane and what it does. (See text)

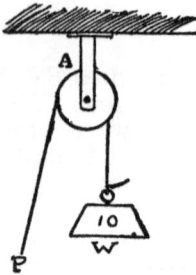

FIG. 243. Simple pulley

Pulleys. With a series of pulleys known as a block, a tackle, or sometimes as a block and tackle, similar principles are involved. Almost unlimited multiplication of force is possible by increasing the number of pulleys. A rope passed over a single pulley and pulled on one end will exert a similar pull on the other (Fig. 243). To balance a 10-pound weight (W) a pull or force at P of 10 pounds is needed. It makes no difference whether a pull of 10 pounds is applied at P or a weight of 10 pounds is hung there—in either case the weight (W) is balanced. Then the hook (A) supports 20 pounds with both weights on, and this strain would still be 20 pounds if a 10-pound pull were substituted for the weight (W).

This suggests a way of multiplying force. Suppose a horse can pull 200 pounds. Then if we hitch up our single pulley as shown in Fig. 245, with the supporting hook (A) hitched to a load; the stump S (corresponding to W in Fig. 243) will exert a force of 200 pounds, the horse will exert a pull of 200 pounds and we will be moving 400 pounds. For every foot the load moves, however, the horse must move 2 feet, taking up a foot of rope on each side of the pulley. Again, in exerting a large force through a short distance a lesser force has to be exerted for a greater distance.

A series of pulleys, either separate or, more often, in a single or double block, is used for the same purpose—that of letting a small force, acting through a long distance, move a large force over a correspondingly small distance. In Fig. 244 (in which the pulleys in each block could just as well have been side by side instead of one below the other), a force of P pounds is being exerted on the rope. Then every part of the rope pulls on each block with a force of P pounds, whether P equals 10 or 100. The lower

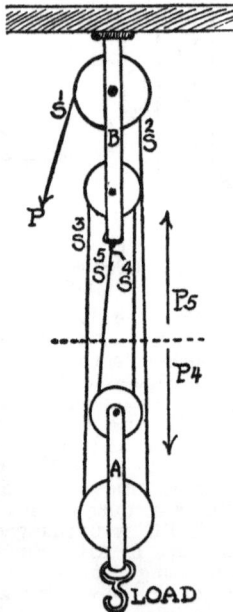

FIG. 244. Compound pulley

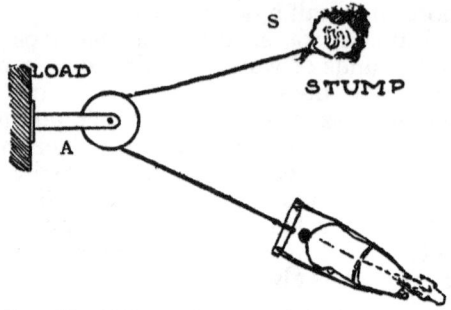

FIG. 245. Pulley applied to land-clearing operations

block (A) is pulled on by 4 ropes so the whole pull on it is 4P and this is the load it will hold. If the pull (P) on the rope is 100 pounds, then the pull on the lower block will be 400 pounds. The upper block (B) has 5 ropes (S1, 2, 3, 4, 5) each pulling P pounds on it, or all together pulling 5P pounds. If the upper one is held fast, the lower block will move only 1 foot for each 4 feet the end of the rope is pulled; if the lower block is fastened, the upper one will move only 1 foot for each 5 feet the end of the rope is pulled.

The author is familiar with a tremendous job of land clearing done very rapidly by means of a rig like this. The block A was anchored to a large stump, the end of the rope hitched to an automobile, and the block B fastened to a whole group of young saplings, which were yanked out of the ground at a very satisfactory rate. Of course, horses could be used in place of an automobile, but less speed would be obtained in clearing.

The wheel and axle, especially in the form of a winch, a capstan, a horse sweep (Fig. 239) or a treadmill, involves the same principle as the lever. A leverage is obtained whereby a small force works through a long distance and in course of time accomplishes a large task. The horse sweep is made for any number of horses, even 12 or 15 being sometimes employed. It is not very efficient, although satisfactory as a cheap method of getting a small amount of power for a considerable period.

A treadmill is more frequently used to get power from animals. It consists of an endless apron carried over 2 cylinders and supported by rollers on a platform, the arrangement resembling a belt over 2 long pulleys. A shaft passes through one cylinder and from this power is taken to the machine operated, either directly or through belts or gears. In such a treadmill the work done is the force exerted by the horse in lifting a portion of his weight up the incline. The power developed depends on the slope of the tread, the speed with which the horse moves, and the friction of the machine (which is lessened as far as possible by careful and complete lubrication). For best efficiency and continuous action the horse should not travel faster than 2 or $2\frac{1}{2}$ miles per hour. The slope of the tread-mill should not be greater than 1 in 4 (1 foot rise for every 4 feet of length) and this is very likely to overwork a horse unless care is taken to keep him at it for only a short time. A slope of 1 in 6, or 1 in 8 will enable an average horse to work a reasonable day without becoming overtired.

At a slope of 1 in 8, a horse lifts one eighth of his weight and transmits this force to the apron at the rate he walks. If he weighs 1,200 pounds and walks 2 miles an hour (176 feet a minute), he does $\frac{1200}{8} \times 176$ = 26,400 foot pounds of work per minute and so delivers four fifths horsepower continually. This is about the maximum rate at which a horse of this weight should be worked for any continuous period. Twelve-hundred-pound horses have been known to deliver 3 and 4 horsepower for short periods, while in a very brief, horizontal pull a 1,500

FIG. 246. In this homemade power plant the dog continually ascends an inclined plane; but this, being movable, exerts a lever action around the axis of the wheel the motion of which is conveyed to the grindstone by the rope belt.

pound horse has been known to exert an effort greater than his own weight. The proportion of the power delivered by the horse available for useful work depends upon the efficiency of the machine; in most cases it should be from 60 to 80 per cent.

Power Machinery on the Farm

The great change in the condition of life throughout the country during the past century has been due in a large measure to the substitution of power and machinery for hand and animal labor. The growth in population and wealth is due mainly to this cause. Starting with the invention of spinning and weaving machinery, followed successively by the steam engine to furnish power and trans-portation means and methods to suit the changed conditions, every step has seemed to bring together and bind more closely the various sections of the coun-try, to promote the standard of existence, and to make possible the great demo-cratic movements of the age. No better example could be given of the results of power substitution on existing conditions than the influence of the invention and marketing of the automobile and its effect upon the life of the farming com-munity. Distance is removed, the country is but a part of the town, and all residents of the county are neighbors and ready to join in the many neighbor-hood and community movements, whether for commercial, industrial, or social betterment.

Choosing and Purchasing Machinery

Recognizing the value of power machinery, the problems which present themselves are: (1) What are the main functions or purposes of machinery? (2) What machinery will best serve me? and (3) What can I afford to buy? An understanding of the true function of power machinery is a real aid in any proper purchasing decision. All power machinery does 1 of 4 things: (1) it saves time in accomplishing certain results, usually by permitting the application of greater force or by allowing force to be applied in a more efficient manner; (2) it saves effort or force, usually at the expense of time; (3) it saves cost, usually by per-mitting the use of less expensive sources of force or less expensive materials; or (4) it saves neither time, force, nor cost, but accomplishes the desired result in a neater, better, or more desirable way than that possible with hand labor.

The real value of power machinery. We are all too apt to measure the value of new apparatus by immediate returns to us in dollars and cents. Nothing is so

FIG. 247. The treadmill (see also sketches at head of chapter) is a further adaptation of the inclined plane

certain as this to narrow our viewpoint and hinder our progress. A saving of time in accomplishing a certain object may or may not result in a saving of cost, but frequently it will make all the greater the enjoyment of what one has. A saving in force may not result in a direct saving in money; but if it makes things easier to do, it makes possible the doing of more things or the greater enjoyment of life while those things are being done. If the end accomplished is in any way better or more desirable than that which could be attained without power, it is worth while and should not be at once condemned because of its greater cost. Not only is the thing worth doing surely worth doing well, but the care and desire to do the thing well has a much deeper effect than can be measured by mere material standards.

The real cost of machinery. In any purchase, several questions must be taken into consideration. In regard to the purchase price: (1) Does the price asked include all necessary attachments or accessories required to get the full benefit of the machine? (2) Is the machine *better* than some cheaper competing machine, or is it merely *different* or fitted with nonessential additions which the cheaper machine lacks? (3) If the machine is guaranteed, just what good is the guarantee; does it mean that defective parts will be replaced free, or that the machine may be returned and the purchase price refunded if the machine is not satisfactory; and is the guarantee for a long enough period so that the machine may be tried out under varying conditions? (4) Will the machine prove worth the purchase price? In answering this last question you must consider the result to be accomplished. Will the machine do the work you wish to accomplish with some kind of saving?—not necessarily some saving in cost, but one either in time, effort, convenience, or cost.

Cost of machinery operation. The items of machinery operation are many. There is always the human element—the number of attendants necessary and, sometimes more important yet, the degree of skill or intelligence required of them. Obviously, if 2 laborers may be replaced by 1 man, and that one must be a highly skilled mechanic, hard to get in the first place, and requiring a salary away out of proportion for the work desired, the apparent saving in attendance is not at all real. On the human side, 4 things must be considered: (1) How many men are required? (2) What degree of skill or intelligence is required? (3) How easy is it to secure men of the required standard in the locality and at the times when the work is to be done? (4) To what extent will any other help or operations be affected?

Outside of the human element, there are 3 things to be taken into account: (1) the material required for the machines; (2) the auxiliary machinery necessary; (3) the deterioration or depreciation in the machine as affected by its method of operation, foolproofness, complicated nature, and materials of construction. The materials required for operation will come under the heads of: (a) lubricants, (b) substance operated on, (c) waste, (d) fuel, and (e) upkeep material, such as paint or other coating, wrapping, packing, or repairing. For example, a water pump will possibly require under (a) both oil and grease. Under (b) it must be considered whether the pump is for sea water or fresh water, whether the water is clean or sandy or full of suspended matter, and whether it is acid or injurious in any way in its effect on metals. Under (c) there would be no consideration; while (d) would bring out the method of operation, whether by direct steam action, by belted drive from some other source, or by direct connection to an outside motor or engine. Under (e) would be considered paint for the exterior, packings and washers for stuffing boxes and valves, possible protection against unusual temperature changes, possible duplicate parts such as bolts, washers, nuts, valves, fittings, oil and grease cups, etc.

The auxiliary machinery necessary would be for 1 of 3 purposes: (1) to drive or supply power to the machine in question; (2) to be operated by the machine in question; or (3) to supply some operating need of such machine. For example, the auxiliary machinery required for a mowing machine is (1) some driving power, such as a horse or tractor, and (2) a grindstone or emery wheel to supply an operating need, viz., the sharpening of the knives. This, in turn, requires a method of driving, a means of lubrication, and, usually, a water supply to the stone.

Depreciation of machinery. The depreciation of the machine should not be considered along with its repair. Depreciation is brought about by more than mere wearing out. An automobile, for example, after running its first thousand miles or so should be a much better machine, run more smoothly, have greater speed and flexibility, be more easily handled and better adjusted than when new, and yet its depreciation is a very heavy percentage. Probably no other thousand miles causes greater depreciation, yet no other thousand miles does the car more good, if properly handled. Depreciation really means change in selling value under conditions of forced sale. All farsighted men allow each year for the depreciation of their belongings and outfits and, in one way or another, make arrangements so that, when required, the machinery may be completely marked off the books. Changes in styles or fashions, possible changes in law or in standards of living, new inventions which render the old machine obsolete—the knowledge of the possibility of any or all of these things causes every prudent man to charge up against operating expenses every year a more or less definite fixed percentage of the value of the machine, which is called "depreciation." This amount should be such that, if continued right along through

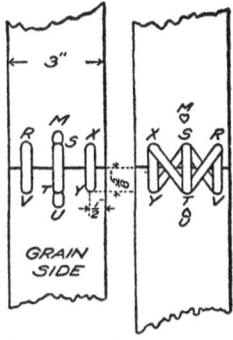

FIG. 248. Lacing a three-inch belt. With grain side down, put one end of lace down through S and up through M; then lace other end as follows, passing downwards through the first hole, up through the second, down through the third, etc.: V, R, V, R, T, S, Y, X, Y, X, T, S, U.

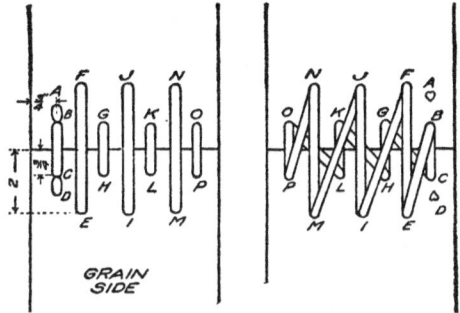

GRAIN SIDE

FIG. 249. Lacing a six-inch belt. Fasten one end of lace down through C and up through D. Run the other (long) end alternately down and up through holes B, C, G, H, K, L, O, P, O, P, N, M, J, I, J, I, F, E, F, E, B, C, B, and A. The grain side runs on the pulleys. (This and Fig. 248, from Farmers' Bulletin 638)

the probable useful life of the machine, the whole sum charged up, together with accrued interest, would be sufficient to replace the machine.

In most cases, the depreciation when a machine is new is much greater than for the same machine when 3 or 4 years old. Also, the percentage depreciation on a second-hand machine is not so heavy as on a new machine, not because of the less cost, but because of that peculiar psychology in men which causes them to consider a machine purchased from the original buyer much more desirable than when purchased from the second owner, yet there is no choice whatever between the second, third, and fourth owners. A second-hand machine is second-hand, whether it has changed hands once or a dozen times. This is particularly true of farm machinery as a general rule.

Repairs. Repair of the machine, altogether apart from its depreciation, is a yearly item of expense. The more complicated the machine, the greater the repair bills and the more likely that delay will occur because of the need for getting a skilled workman to make the repair. The more nearly foolproof the machine is, the less the repair bill, no matter how skilled and intelligent the operator may be. Repairs are affected in various ways, for example, by the nature of operation, whether inspection can be made while operation is going on, whether quick shutdowns can be made in emergencies, and whether the load must be thrown on with a jerk or may be gradually applied without jar. Last but not least, the materials of construction affect the repair bills. Particularly is this true of any wearing part, such as a friction or rubbing surface, of any part exposed to undue heat, and of any part exposed to undue strain or shock. Such parts should be specially designed for their purpose, and the materials specially chosen to meet the condition existing.

Classification of Power Machinery

Most of the machinery used on the farm is power machinery, particularly as the tractor invades more and more the province of horse and cattle. Yet all farm machinery may be classified under, perhaps, 4 heads: (1) machinery affecting land for crops; (2) machinery affecting the care of stock; (3) machinery affecting the preparation of farm products for market or home consumption; and (4) machines affecting general living conditions on the farm. In Class 1 would be placed the tractor, plow, harrow, ditch digger, etc.; in Class 2, litter carriers, milking machines, pumping apparatus, sterilizing machines, etc.; in Class 3, wood saws, harvesting machines, corn shellers, cold-storage machines, etc.; and in Class 4, running-water apparatus, electric lights, heating apparatus, vacuum cleaners, dish washers, laundry apparatus, automobiles, etc. Most of these having been discussed in other chapters, it will only be necessary here to treat of two or three, such as the power ditch digger and the various pumps.

The distinguishing feature of the modern farm is the inclusion in its machinery of some of the items of Class 4. The other apparatus is largely money-making. The machinery of Class 4 does more than make money: it makes life, helps to remove drudgery, and brings an amount of joy and happiness into the life of the farm housewife that is more than desirable—it is a necessity, a right which she may properly claim as a copartner in the operation of the farm as a business unit.

Machinery for Crops

One of the most important recent developments in farming is the growing adoption of the farm tractor on the smaller-sized farms. The tractor having been treated in Chapter 6 of Part I of this volume, only a few features will be mentioned here. Some of the advantages in particular may well be recapitulated. They include the low cost of maintenance when not in use; possible 24-hour use when desired; flexibility in operation in the field or for road hauling and as a stationary power plant about the farm, movable by its own power; low cost of operation per unit of work and, lastly, elimination, to some extent, of the labor demand at various seasons of the year.

Agricultural machinery is taken up in detail in Part II, and the function of this article is merely to call attention to the different types of the same machine for different purposes. The small farmer must not, and cannot economically, have plows and harrows each specifically designed for special service in some part of his field. On the other hand, the big farmer, to achieve success, must have this sort of equipment. Each must consider all types; and, where 1 or 2 are sufficient, the right compromise must be studied out from an intimate investigation of all.

Power ditching machine. A machine not in wide use, but which is extremely desirable where extensive ditching is done, is the power ditch digger. There are several types, all different in details, and, of course, in quality of workmanship. There are 2 functions which these machines have: (1) excavation to required dimensions, width, and depth; and (2) the delivery of the excavated material to some point on one side or the other or on both sides of the trench. A necessary feature of operation is the ability of the machine to propel itself forward as the ditch is dug. So far as the writer knows, in no type of digging machine is any provision made for filling in the trench. Where this is desired, it is done with horse shovels, scrapers, and by hand shoveling. On the farm, the ditches are more often left open for drainage and irrigation purposes.

The excavating mechanism is frequently a bladed wheel about 10 feet in diameter, chain-driven, and pivoted on the end of an arm which moves up and down, hinged on the main body of the machine. This up-and-down adjustment permits the regulation of the depth of the trench and, also, a gradual start of operations from the ground surface, the wheel being fed slowly

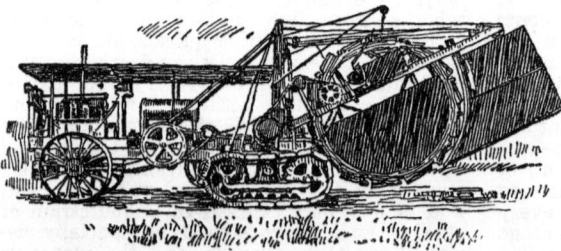

FIG. 250.   Gasoline-driven, caterpillar-tread ditching machine, showing digging buckets around wheel at right, also apron and travelers for removing earth to side of ditch.

into the ground to the proper depth, and feed of the wheel being regulated by the propulsion of the whole machine forward. An engine, mounted on the body of the machine, is connected to a gear which is in mesh with a jackshaft on which a clutch is mounted. This clutch is so arranged that the engine may be connected through a chain drive to the rear wheels to furnish power for propelling, or to the digging-wheel, through a chain drive, for excavation, or connected to operate both wheel and propulsion at the same time.

The digging blades or shovels around the circumference of the wheel are bucket-shaped and so located as to retain the excavated material until, in the turning of the wheel, the buckets at the top are upside down and the material is dumped on to a canvas belt conveyor which passes through the centre of the wheel. This belt conveyor is arranged to dump the material on either side of the wheel, as desired. If the material is wanted on both sides, as would be apt to be the case in a deep or wide trench, 2 belts, turning in opposite directions, are used. The speed of the machine is from a half mile to 2 miles of trench per day, the trench being 4 or 5 feet deep and 2 or 3 feet wide in one cut.

Machinery Affecting the Care of Stock

PUMPS. Power pumps are properly placed in Class 2, as affecting the care of stock, although, to some extent, overlapping class 4, if they are used for house water supply and, perhaps, Class 1, when they are used for irrigation purposes. They are of 3 distinct types: piston, or plunger, pumps; centrifugal, or turbine, pumps; and rotary pumps. The differences between them are in cost, speed, efficiency, and space required. Both piston and turbine pumps will force water to any height met with in practice. Rotary pumps are not good for really high heads. While

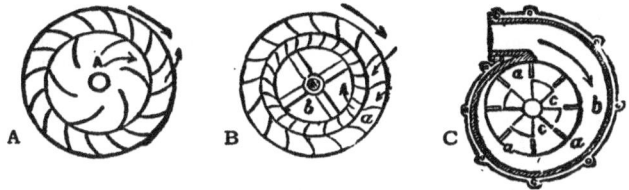

FIG. 252. Three turbine water wheels. In A the water enters at the centre, and is directed by the guides against the buckets of the outer wheel which it forces around before escaping at the edge; in B it enters from the edge and revolves the wheel inside, escaping at the centre; in C the casing is shaped to make the water act against the buckets all the way around, with a final thrust on the interior buckets (c) through which it escapes.

piston pumps will operate at nearly the same efficiency regardless of variation in speed, turbine pumps must operate at high speed and any variation causes a large change in efficiency. Constant or nearly constant speed is desirable.

Plunger pumps. These are of the same type as the usual hand pumps. A piston moves back and forth in a cylinder, on one stroke drawing in the water through a check valve and on the next stroke forcing it out through a similar check valve operating the other way. These check valves operate in 1 direction only, so that the intake valve opens only on the suction stroke and the delivery valve opens only on the delivery stroke. In a double-acting pump, both sides of the piston are in operation, the stroke on one side drawing in the water through one set of valves, and the stroke on the other side of the piston forcing out the water through another set of valves.

The speed of such pumps is usually low, say from 40 to 60 revolutions per minute on large pumps, and somewhat more on small sizes. The action is positive, the water being forced out against any pressure that the pump is built to withstand.

This type of pump is most efficient, as it can be driven by any kind of engine; but it is being very widely displaced by centrifugal and rotary pumps because of the comparatively low cost of the latter.

In some kinds of plunger pumps, used for city waterworks and boiler-feed purposes, the water plunger and a steam piston are connected together on a single piston rod, and the steam acts directly through the pistons on the water. This is a sort of combination engine and pump; but the steam is not allowed to expand at all, only the initial pressure being used. This results, of course, in a great waste of steam. On this type there is no flywheel, and the speed can be regulated very nicely by means of the steam throttle.

Another type of direct pump, with flywheel, has a steam cylinder arranged like a steam-engine cylinder, expansion of the steam being permitted. This is widely used for elevators, boats, and condenser purposes.

The ordinary farm pump is operated by belting or gearing from some form of separate

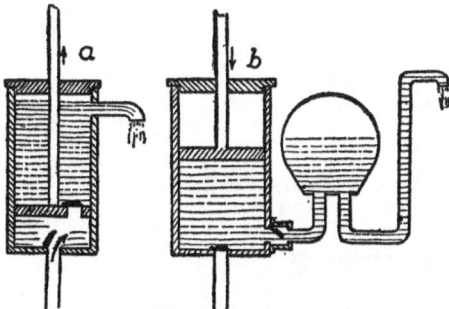

FIG. 251. Diagram to show working of single lift pump (a), and force pump with air dome to give a steady flow of water (b).

power machine such as an electric motor, gasoline engine, steam engine, or water wheel.

Rotary pumps. Pumps of this type give about the same efficiency as centrifugal pumps under the same operating conditions. They may be operated at low speed, are fairly positive in their action, are of low cost as compared with piston pumps, and take up very much less floor space. They are objectionable at high speed because of the noise they make. As compared with centrifugal pumps, they need not be regulated so closely in speed, a variation of 100 per cent not making much difference in efficiency. They take up more room than a centrifugal, and cannot be driven over 700 to 900 revolutions, even in small sizes.

A flat, circular casing contains 2 intermeshing impellers, each shaped like the figure 8. Each is on a separate shaft, and the two turn in opposite directions, being geared together, and the power being applied to one shaft only. The water is drawn in at one side of the casing and propelled out of the other. A variation in speed changes the amount of water delivered in about the same proportion. Rotary pumps are used in small sizes for light duty, where a cheap, dependable, variable speed, but positive-acting pump is desired. They are used only for force-pumping against a short head or low pressure, although, as ordinarily constructed, adaptable to heads up to, perhaps, 50 feet or so.

Centrifugal pumps. Centrifugal and turbine pumps are very much alike, the latter being a refinement of the centrifugal as regards the elimination of cross currents in the water within the casing. There are 2 essentials in each: (1) a casing of somewhat similar shape to the rotary casing, and (2) a single impeller of such a shape as to draw the water through the centrally located intake and force

it to the circumference of the casing and out of the discharge opening.

In the centrifugal pump the casing is spiral, like a snail's shell, with the suction inlet at the centre and the full-sized discharge opening at the edge of the main part of the casing. There is only 1 impeller, on a single shaft turning at a high speed. The action on the water is similar to that of a wagon wheel running in a rain-filled rut and throwing mud and water off the tire with considerable force.

In the pump the force is directed; and a single pump, properly designed, is capable of raising water to a height of several hundred feet with an efficiency of over 80 per cent. By arranging several impellers side by side in separate casings, 1 impeller is made to deliver water to that next to it, and a greater range of usefulness is developed. Two, 3, and 4 such impellers are frequent, the pumps being called 2-, 3-, or 4-stage pumps, as the case may be. A stock type of 3-stage pump will deliver 200 gallons per minute against a head of 750 feet running at 2,500 revolutions per minute. Usually, however, the pumps are designed for a lift of about 125 feet to each stage.

Centrifugal pumps in their widest field are found raising huge quantities of water short heights. For irrigation and drainage purposes, they are especially valuable, single units raising up to 100,000 gallons per minute for a lift of from 4 to 10 feet. They may be obtained in small sizes for ordinary farm use and, where they can be driven at high and reasonably constant speed, they are most satisfactory and by far the cheapest, lightest, and smallest pumps that can be obtained. They are nearly automatic in action, having no valves and but few parts; their depreciation and wear are thus reduced to a minimum. The cost of centrifugal pumps runs from 12 to 15 cents a pound for the weight of the pump.

Air compressors. In many cases, the use of compressed air is very advantageous. In underground pneumatic tank systems, air compressors are almost essential. A small compressor for tire filling or for blowing purposes, where a blower or fan cannot be used and the jet of air is substituted, is frequently very handy. These compressors are usually low-speed, they operate like a single-cylinder piston pump, and are dependable. There is usually an unloading valve, to throw off the load automatically when the air pressure reaches any predetermined point. Unlike pumps, air compressors must have a water-cooling jacket, like that on the gasoline engine, because the action of compressing air creates considerable heat. Small compressors cost from $30 to $100.

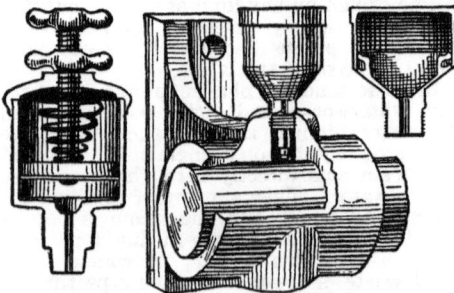

FIG. 253. Spring-equipped grease cup (*left*); ordinary grease cup in section and on a bearing (*right*)

Care of Power Machinery

In all types of pumps, as in other apparatus, proper and sufficient lubrication must be provided. Lubrication is the one thing which means a long life to any machine. Without it, most machines will be ruined in a very few minutes of operation under load. It is particularly true that the importance of lubrication be impressed on the farmer operating power machinery, because, while ordinary farm machinery should be well lubricated at all times, it is not usually built so carefully and accurately that lack of lubrication will cause immediate disaster. With power machinery, however, the careful fitting and adjustment of parts, the tremendous forces called into play, the very high speeds usually called for, and the special materials used, all require special care; and the first step in proper attention is lubrication.

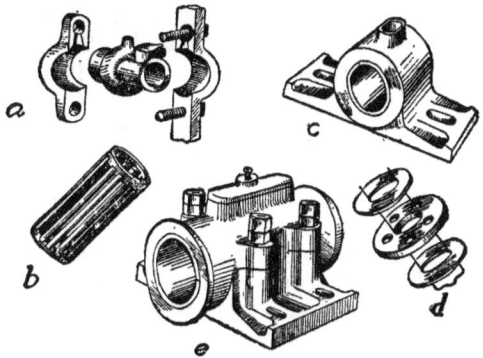

FIG. 254. The bearings are the centre of machine efficiency: *a* self-aligning bearing; *b* roller bearing; *c* plain bearing; *d* ball bearing (the balls fitting in the small openings in the centre plate); *e* split bearing (Ore. Bulletin, 133).

Next to lubrication, care must be exercised by the operator to keep all nuts and bolts tight. Large strains develop, and vibration occurs which rapidly loosens all parts. Periodical inspection during operation and a thorough tightening after every shutdown is desirable.

Bearings. The features of all power-operated machinery which should be carefully looked into at purchase and carefully cared for afterwards are the bearings—the portions of the machine which hold the moving parts in place. They are all of plain, babbitt or similar metal, roller, or ball type. The plain bearings are usually carefully drilled holes in proper metal or hardwood, the moving part, which is usually a rotating shaft, being held in place by the material. Such bearings must be well lubricated. The babbitt bearing is a plain type lined with soft metal, so that the friction created is very slight. These bearings, also, require lubrication. In both of these types, the friction of the shaft turning in the bearings is a rubbing or sliding friction.

Roller and ball bearings substitute rolling friction for the rubbing friction of the other types. This requires very much less effort to overcome it and eliminates a large part of the friction in any machine. These bearings are necessarily very well protected from dirt and dust, being packed in grease, which not only protects, but lubricates. The main objection to ball and roller bearings is their high cost. Ball bearings are especially adapted for end thrust conditions, whereas roller bearings are used particularly for radial loads as on axles and horizontal shafts. The diameter of the balls used varies from a very small size up to 1 inch on very heavy bearings. The rollers used most generally run from three eighths of an inch up to a half inch in diameter. The radial bearing in length should be from 3 to 4 times the shaft diameter.

The difficult feature about all bearing settings is to get perfect alignment. To aid in overcoming this difficulty, a type of bearing called "self-aligning" is made, the main bearing box being pivoted in the frame so as to be slightly movable.

Machine materials. A complete study of, or even a complete glance at, the multitudinous materials used in machine construction would fill many volumes. Steel and iron, with alloy metals, form the most important group. Steel in particular has received a great deal of study. By combining very small amounts of various rarer metals, such as nickel, tungsten, and titanium, with steel, we get a tremendous increase in the testing strength of the steel and considerable change in other characteristics. While ordinary steel shows a strength of, say, 60,000 pounds per square inch, alloy steels show up to 200,000 pounds per square inch and sometimes more. This permits lighter and cheaper construction.

Steel differs from iron only in its purity and smaller percentage of carbon. Cast iron is pig iron cast in sand molds. It cannot be forged or welded. It is used for loads in compression. When a particularly hard, difficult surface is wanted, part of the mold is made of iron in place of sand and the cast iron runs against it, becomes chilled, and the surface gets very, very hard.

Cast steel is cast iron with some carbon removed. It is more dependable and stands shock better than cast iron. Gears are frequently made from it.

Wrought iron is almost the pure metal. It is easily welded and worked. Steel is, of course, easily worked, but is hard to weld except by the electric-welding process.

FIG. 255. A portable wood-sawing outfit run by a stationary gasoline engine. With such a machine the farmer can get out his own lumber and often do a good deal of profitable work for his neighbors

CHAPTER 15

Windmills

By PROFESSOR R. P. CLARKSON (*see Chapter 14.*)   *The windmill is so familiar an object on the farm that its importance, value, and needs as a piece of machinery are apt to be overlooked until a breakdown occurs.   At this point the farmer may blame the weather, the mill itself, the makers, etc.; or he may have an awakening and see himself and his neglect as the real causes of the trouble.   With knowledge of just what kind of a machine it is, what it does and how to care for it, he might have prevented the trouble in the first place and also assured himself of longer service from the mill.   This chapter provides such knowledge.*—EDITOR.

THE source of power to which the farmer has naturally turned in the past, and to which he still rightly clings, is the wind.   For ages wind power has been employed to drive the vessels of the sea, and for almost, if not fully, as long the windmill has been used for grinding grain.   During the last decade or two, serious thought has been given by scientists and engineers to the project of using the force of the wind indirectly through the novel form of wave motors.   Wind and falling water together with the tides and the heat of the sun are the four so-called "natural" sources of power to which, in ever-increasing degree, the minds of the power engineers are turning.   Undoubtedly, the windmill, even in its most highly developed form, is the least efficient of all our power apparatus, while the hydraulic turbine is by all means the most efficient of our prime movers.   Almost all inventions and changes in design and construction of our prime movers have but one aim—the increasing of efficiency.   The increased utilization of water power and the study which that involves are responsible for the great advances shown in the modern turbine over the old grist mill wheel.   The same amount of study has not, of course, been devoted to the windmill, and as a result the improvement over the old-fashioned Dutch mill is not great.

Advantages.   The value of any source of power to the farmer is in direct ratio to (1) reliability; (2) cheapness of installation; (3) cheapness and ease of upkeep; (4) durability; and (5) simplicity.   In all these features, the windmill stands well.   It is reliable in its action, and its general average of operation in any locality is a thing easily ascertained.   It is cheap to install, its cost depending on what is desired.   It requires but little care, and the upkeep cost is extremely low.   Its life is long, if reasonable attention is paid to rust prevention.   It is easily understood, and no part of the machinery is at all complicated or easily put out of order.

A disadvantage in utilizing the breezes, as in all natural sources, lies in the continual variation in their intensity and direction, as well as in their complete cessation at times.   The changes in direction may be momentary, if induced by surrounding local conditions; or they may be very decided, if due to the shifting of pressure centres and the resulting change in the point of the compass from which the wind blows.   Modern mills are arranged to turn with the more decided shifts in direction, thus presenting at all times a proper surface to the wind action.   So far as possible, an effort is made to avoid local effects by raising the fan portion of the windmill to a considerable height, and by placing the structure in a clear, open position.

167

The following table gives as closely as possible the velocity and pressure of the wind for ordinary breezes.

WIND TABLE

KIND OF WIND	VELOCITY MILES PER HOUR	VELOCITY FEET PER SECOND	PRESSURE EXERTED LBS. PER SQ. FT.
Light breeze .	3 to 4	4.4 to 5.86	.03 to .05
Gentle breeze .	6 to 8	8.8 to 11.7	.11 to .19
Brisk wind .	15	22	.68
High wind . .	30 to 35	44 to 51.3	2.7 to 3.66
Gale. . . .	45	66	6.1
Hurricane . .	80	117.4	19.2

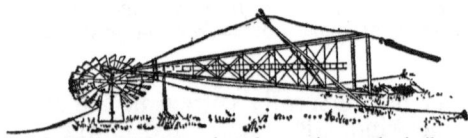

FIG. 256. Steel mills of moderate size can be built on the ground, then hoisted into place, fastened down, and hitched up to the pump.

Wind velocities. The engineer is, of course, chiefly concerned with the force of the wind, and with the pressure which will be exerted by it against any wheel or other structure placed in its path. A hasty examination of the few figures given above shows that the pressure exerted increases very, very rapidly as the velocity of the wind increases. When the velocity increased 20 times, from 4 miles to 80 miles per hour, the pressure increases 384 times, from .05 pounds to 19.2 pounds per square foot. The effect of this enormous increase in pressure is seen directly in the output of any mill. With a mill having a capacity of 8 to 10 bushels of grist per hour under ordinary brisk wind conditions, the possible output would be doubled if a high wind prevailed.

Old-type mill. The old type of Dutch tower mill is still seen, with its foundation of stone surmounted by a frame tower having latticed sweeps or blades over which sails can be drawn. The tower is movable by a "veering mechanism," to bring the blades into direct relation with the breeze. Some of these mills are of several stories and are often 100 feet high or more. The blades are frequently 25 to 30 feet from shaft to tip of blade; and they are so geared to the grinding stone that the stones make from 10 to 12 revolutions to each turn of the sails.

The modern windmill. The modern windmill is built for a much wider range of work. Its chief field is, apparently, pumping; but it has many other uses, including in some localities that of operation of the farm electrical plant. It differs from the old windmill in being made of steel, with a light turbine-shaped wheel, usually 10 to 12 feet in diameter, and never in the largest mills over 40 feet in diameter. The steel towers are from 40 to 100 feet in height. A ladder provides access to the mill wheel for repairs.

The wheel. The wheel itself is a very important part of the machinery, although in the old mills, a large proportion of the power was doubtless lost in transmission through the gears and shafts from the wheel to the millstones. The most modern wheels have a steel framework with radially placed sails of sheet metal, shaped somewhat like the blades of a screw propeller. They resemble, also, the blades of some forms of fans. The exact form of curvature adopted is not mathematically determined, even as much as the shape of water turbine wheels; but, like them, it is an outgrowth of experiment and test. Obviously, with the varying velocities of wind, there must be a compromise in the shape of sails, to give good results with each velocity. There must be a best shape for any one velocity; but, where a compromise is made, it can readily be seen that there is room for considerable difference of opinion, and indeed the manufacturers differ quite decidedly among themselves. Experiments have shown, however, that a twisted sail has but slight advantage over a surface which has a flat constant angle presented to the wind. To put it crudely, the fancy and complicated shapes occasionally marketed are not so desirable as the simple plain forms of sails resembling the old flat Dutch sails, except that the present form is curved radially from shaft to outer ring.

With even the best wheels, a large proportion of the air passes through between the sails and, of that which does strike the sails, only a portion of the energy is absorbed. The result is an over-all efficiency of windmills of only 8 to 15 per cent.

Veering devices. Various

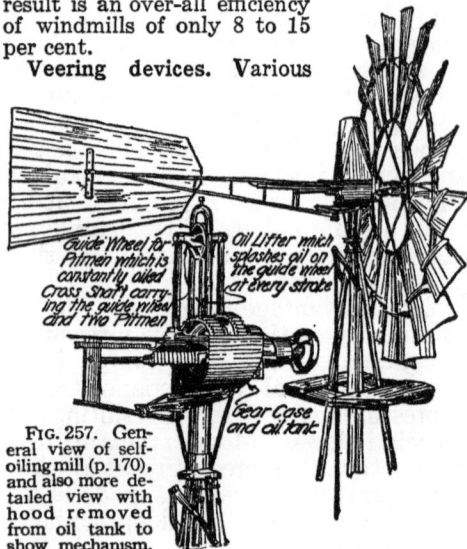

FIG. 257. General view of self-oiling mill (p. 170), and also more detailed view with hood removed from oil tank to show mechanism. (Aermotor Co.)

FIG. 258. Small, farm-made mill in which fan, wheel, and tower are all of wood.

veering devices have been employed in the various types of wheels, to keep the sails headed up into the breeze. At present, the tail vane stuck out behind from an extended arm is almost universally adopted. The best mills are mounted on roller bearings and easily moved, so that the size of the tail vane must be carefully determined by experience. If too small, the mill does not respond quickly to slight variations in direction; if too large, the mill is continually moving or oscillating in a stiff breeze. The vanes, like the sails, are made of sheet metal.

Transmission gear. The transmission gear in modern mills is the device for changing the rotary motion of the wheel to a reciprocating motion suitable to operate the pump plungers. The change must be made from the horizontal rotating shaft of the wheel in the air to the vertical plungers extending down to the ground. This gear is the main point of difference between the various windmills; and in most trials, such as used to be held at the various fairs, this gear has been the main reason for the superiority of certain makes of wheels over other makes, the difference in the mills being much greater than those made possible merely by the slight differences in sail design.

Although in all forms of engines the force end has a reciprocating piston stroke, which is changed to a rotary shaft motion by means of a crank and connecting rod, this same device is not used in any of the larger types of windmills to

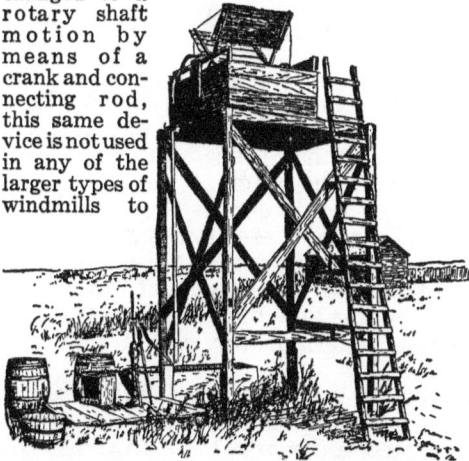

FIG. 259. A small jumbo mill in Nebraska supplying a small hotel and stable, and which cost $3.70 (Cornell Countryman IV-4.)

reverse the operation and change rotary into reciprocating motion. It is used, however, in many, or possibly all, of the smaller and cheaper mills, the alternate form of construction being a flat pinion arranged to mesh alternately with 2 parallel racks, one of which thus provides a downstroke and the other an upstroke to the pump plunger.

Automatic regulating device. Most wheels in this country have some sort of automatic regulating device, to prevent the wheel from gaining too high a rate of speed. The device, of whatever form, is arranged to change the sail adjustment, and, frequently, it also throws the wheels out of gear with the pump. This is done because of the fact that the pumps used will not work properly at a high speed. The mill itself should stand a gale of wind and operate satisfactorily. In many of the better types, a device is provided which, if anything breaks, automatically throws the mill out of the wind direction.

The storage of water. With any type of pumping apparatus so intermittent in its action as the windmill and so variable in its output, a form of storage must be used. An elevated tank or reservoir is most common. Occasionally, the site is so arranged that a small reservoir may be built on a hillside, usually a masonry or concrete structure partly dug into the hill and partly walled above ground level. More frequently, an elevated tank is built on the same tower with the windmill, or, possibly, in the attic or in the barn loft. In all these cases, gravity pressure is relied on to give a flow of water.

FIG. 260. The water tank may be supported on the mill tower, on a separate tower of its own, in the attic of the dwelling or the loft of the barn.

In some installations, the water is pumped into a pneumatic tank, buried underground below the frost-line or placed in the cellar. By creating an air pressure from the compressed air in one end of the tank, a flow of water is established. This air pressure may be pumped in by means of an air compressor, or the water itself may be depended upon to create sufficient pressure by sealing in the air already in the tank. In very few cases is this type of installation desirable, unless the pumping is done by some type of engine.

Tanks. Both wooden and metal tanks are used, of all sizes from 150 to 150,000 gallons. Usually about a 4- or 5-day supply is arranged for. For wooden tanks, pine and cedar are regarded as desirable in the East.

while cypress is most favorably regarded in the South and Southwest. Metal tanks are occasionally of cast iron, but more often of sheet metal, either iron or steel. They cost nearly twice as much as wooden tanks. In both cases, care must be taken to keep the tank painted, clean, and tight.

Wooden tanks are generally thought to be better than metal ones. They are easier to erect; they neither sweat nor rust; they do not follow temperature changes so readily; and they are less likely to freeze. As they cost much less, wooden tanks are much more common, even in the very large sizes.

Whether the tank should be inside or outside, depends largely on local conditions. The inside tank is restricted in size, both because of the limitations of the space avail-

able and, also, because of its great weight and the necessity of making special provision for its foundation support. On the other hand, it can be more readily protected from freezing in winter and overheating in summer. The life of wooden tanks is found to be seldom less than 15 years, and it is frequently 25 years or longer. Under inside conditions, the higher figure may well be reached.

Usually, the outdoor tower with tank is the most expensive construction. Placing the tank on the windmill tower is just as satisfactory, if a sufficiently high mill be used; and the cost is less than for a separate tower or for indoor construction. Frequently, however, the height necessary to give water everywhere cannot be reached except by means of a separate tower.

Characteristic Features of a Desirable Mill

The power of any mill depends on a multitude of variable features, not only of construction and operation, but of atmospheric conditions. The size of the wheel does not necessarily indicate the power. The larger the wheel, the slower the speed at which it operates. For example, a 10-foot-diameter wheel has a rated speed of about 70 turns per minute; a 16-foot wheel 60 turns; an 18-foot wheel 45 to 50 turns; a 25-foot wheel perhaps 35 turns. Some of the mills run faster than this. Under average conditions, the horsepower developed may be from one sixth horsepower for the smaller wheel in a stiff breeze up to, perhaps, $1\frac{1}{2}$ horsepower or more for the larger. This is only suggestive, however, because in actual tests of 2 types of mills having wheels of the same diameter, operating in the same wind, one has been found to deliver twice the power of the other.

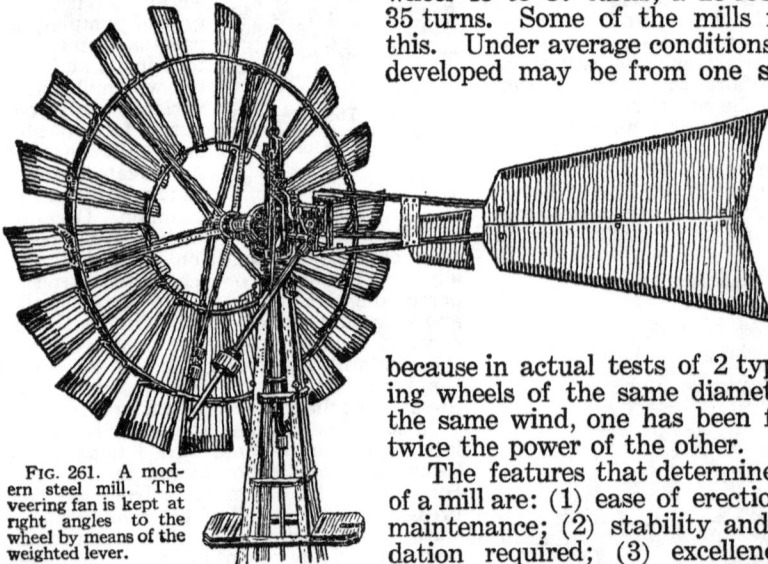

FIG. 261. A modern steel mill. The veering fan is kept at right angles to the wheel by means of the weighted lever.

The features that determine the desirability of a mill are: (1) ease of erection, handling, and maintenance; (2) stability and nature of foundation required; (3) excellence of design as regards both engine and pump; (4) regulation and degree of automatic action; (5) efficiency as determined by size and output; and (6) cost.

In connection with the first feature, low towers, say 40 feet and under, can be assembled on the ground and raised complete, including sails, while larger towers require piece-by-piece construction. In any case, a concrete base, to which the tower may be anchored, is very desirable.

In a brisk wind, using a 2-inch water cylinder, a 10-foot wheel of average manufacture, geared 3 to 1, will make a lift of 250 feet; with a 3-inch cylinder it will lift 150 feet; and with a 4-inch cylinder it will lift 70 feet. A 16-foot wheel with a 3-inch cylinder will lift 500 feet and with a 4-inch cylinder will lift 300 feet. A 20-foot wheel with a 3½-inch cylinder will lift 800 feet; with a 5-inch cylinder it will lift 375 feet; and with an 8-inch cylinder it will lift 125 feet. The actual amount of water delivered depends on the number of strokes per minute of the pump and its efficiency (Chapter 14).

There is no doubt that the windmill in the central and western states, with a reasonably constant supply of wind, is the cheapest power on the farm. It must be utilized in small units. Even for power mills, as distinguished from pumping mills, sizes never run above a few horsepower output, while, in pumping mills, it is unusual to find even 1-horsepower capacity. The cost of such power is very difficult to determine with any degree of accuracy. It is not sufficiently low to justify any attempt to use it for continual power output in localities where some form of storage must be provided in combination with the windmill. It is best adapted and is cheapest for intermittent work. This includes pumping water, grinding grain, irrigating, and even wood sawing, if the wheel is large enough.

Protection in winter. There is need for considerable care in installation, where the windmill is to be used throughout the year. Particularly is it desirable to see that the pump cylinders and pipes are protected from freezing during very cold weather. This is best done by keeping all pipes below the frost line and by sinking the pump to a corresponding depth. Where any part is exposed, it is desirable to use a wrapping of felt or heavy stuff, or to fashion a casing which may be filled with sawdust, rags, crumpled paper, or similar substances. In the case of long, exposed pipe lines, it is desirable to place valves so that the water may be shut off and the pipes emptied on specially cold nights.

Cost. The cost of the various types of windmill varies widely, from $60 in normal times for a small mill on a 30-foot tower to, perhaps, $2,500 for one of the finer English mills of 30-feet diameter on a 40-foot tower. Such short towers are usually not desirable, except in open country; and even then dissatisfaction is apt to result from their purchase. Not only is the velocity of the wind less uniform near the ground surface, but it is actually very much less than the velocity higher up. At 100 feet, the velocity is normally nearly twice that at 25 feet. Therefore, a mill on the higher tower may give 4 or 5 times the power output of the same wheel on the short tower. With the higher towers, the normal cost of a substantial windmill would be from $250 to $500. This is, of course, a very considerable investment for so small a power plant. It is feasible, therefore, only where the prevailing winds are sufficiently strong and constant to give fairly constant power at the times when power is required. When this is true, the cost of the power is the interest on the investment plus depreciation of the machinery, plus the cost of oiling, painting, and occasional inspection. This will, probably, amount to about $60 or $70 per year for a $500 mill. From a knowledge of windmill performance it is conservative to assume that this outfit in the Central or Middle West would give a yearly output of at least 3,000 horsepower hours, and in favorable localities, would approach 5,000 horsepower hours. In the East, not more than 1,200 to 1,500 horsepower

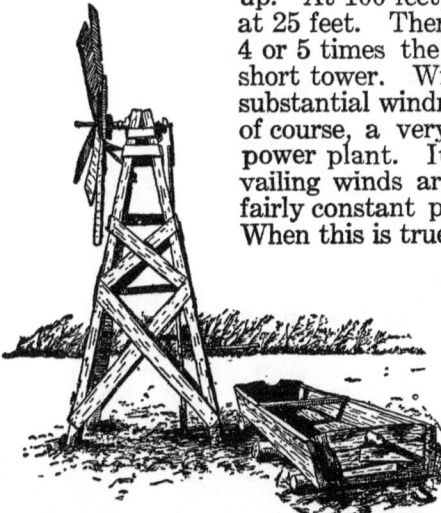

FIG. 262. Side view of a small turbine windmill successfully used on the Great Plains. (Cornell Countryman IV-4.)

hours could be estimated. At this figure, the cost of windmill power would vary from a minimum of perhaps 1¼ cents per horsepower hour in the West to perhaps 5 cents per horsepower hour in the East. This would indicate that in the western windy states, the cheapest form of power is the windmill, as nothing else, except, perhaps, a water-power installation, can approach 1¼ cents per horsepower hour for small power users. In the East, however, the cost is approximately the same as for gasoline and steam power, though somewhat greater than the cost of kerosene power. The choice, then, is a matter of convenience. If the larger power plant is in use anyway, the windmill may be dispensed with. If a small power is needed intermittently and without constant attendance, the windmill may work out to the best advantage.

Homemade windmills. To give a small amount of power and for a corresponding small outlay of money, the homemade windmill is worthy of consideration in states where the prevailing winds are pretty definite in direction. The jumbo type and the battle-ax type of windmill are both usually built fixed in direction, but the latter type may be mounted with a vane to veer the head into the wind. The construction is all wood, and, depending on size, may cost anywhere from $2 or $3 up to $30, not counting labor or such material as is usually found on a farm. The battle-ax type may be extended into a straight vane turbine, a cheap type of which is shown.

The shafts of these mills are usually wooden, some smooth pole serving the purpose or, if available, a short rod or piece of pipe may be driven into the ends of the shaft proper to serve as bearings. The wheels or sails and tower structure can be made of poles, with sides of old boxes for vanes.

All of these types must be built fairly large, to give appreciable power, but, with diameters from 8 to 12 feet, they will show results. Such a mill should pump water sufficient for 50 to 100 cows to a height of 10 or 15 feet, if there is sufficient storage capacity to permit the mill being fully utilized.

Care of the windmill. In the so-called self-oiling types of windmill, oil magazines or reservoirs are provided which, when filled, will supply sufficient lubrication for from 4 to 6 weeks during the usual operation of the mill. If the mill is not self-oiling, it should be lubricated after each 8 or 10 hours of use. If a considerable period of time elapses before this amount of use, oiling should be done about twice a week. At each oiling, go over the mill, wrench in hand, and tighten any loose nuts or bolts. Note the condition of the structure, and, when rust appears, use paint.

Fig. 263. Giant battle-ax windmills used in irrigating a 15-acre Nebraska orchard. Costing $25 each to make, they will average about 1,000 gallons an hour each in a 15-mile wind. (Cornell Countryman, IV-4.)

CHAPTER 16

Water Power on the Farm

By PROFESSOR R. P. CLARKSON (see Chapter 14). The farmer who has on his property, or within convenient reach of it, a permanent stream has at hand a source of power that is constant, uniform, and easily harnessed either for the driving of machinery or the production of the more easily carried electric power (Chapter 19). Water has thus been used by man for many ages, yet even to-day, millions of units of power are being wasted because of insufficient knowledge as to how to capture and apply them. The data given here will enable the farmer to do a good deal; it should also stimulate him to learn and to do still more.—EDITOR.

O N ANY farm where there is a flowing stream, a pond, or a lake, the question of water-power possibilities should be carefully investigated. The requirements for power are, first, a supply of water continuous during the time power is needed, and, second, a drop or fall between the level where water is available and the place where it must be discharged. This is called the "head" of water. A very small head, or difference in level—even a few feet—is all that is necessary. The power obtainable depends upon three factors: (1) the weight of water which can be used per minute; (2) the fall, or head, available; and (3) the efficiency of the water wheel used. While it is well for the farmer himself to make preliminary investigation into the possibility of power, good engineering advice should be employed before any considerable amount of money is spent.

Head. This can usually be measured directly without difficulty, by means of the farmer's level. If the water supply is a lake which can be discharged through a wheel into a stream at a lower level, the head is measured between the two water surfaces. If the supply is a stream, the measurement of fall may be more complicated. A dam will probably be necessary to back the water up and form a storage pond. The head is then measured between the water surface of this pond and the surface of water in the stream at the point where the wheels discharge.

Usually, if a stream is being considered, there must be a rather quick drop, as indicated by swiftly flowing water, or the power probabilities will not be large. If there is a considerable drop, 2 points are chosen between which the main fall takes place. Call the upstream point A and the downstream point B. A dam is erected at B, sufficiently high to back the water up to the level at A and make available the fall or head from A to B.

Water flow. After head is determined, measurement must be made of the water flow between the two points. Unless measuring apparatus, such as a current meter, is available, the simplest way is as follows: Pick out a stretch of stream which does not vary greatly in width. Say the stretch is 200 feet long. Measure the width of the stream in feet at 10 stations along this stretch. At each station take 10 measurements in feet of the depth of the stream, spacing the measurements equally from shore to shore. Average the depth measurements at any one station and multiply the average depth by the width

FIG. 264. An automatic measuring instrument being used to determine the depth of water on a weir. The inset shows one form of current meter sometimes used in place of the weir system of measuring stream flow.

FIG. 265.   In measuring the flow through a weir with a simple straightedge, take the depth far enough back of the edge to avoid the surface curve of the water.

of the stream at that station. This will give the cross section in square feet. Get this cross section at each of the stations and average these 10 cross sections. This will give an approximate average cross section of the 200 feet stretch of stream chosen. If this is multiplied by the average velocity of the stream in that stretch, you will have the approximate flow of the stream in cubic feet per second. This velocity may be observed by noting the time taken for a chip to float the stretch of 200 feet, or, perhaps, an easier way would be to notice how far a chip will float along the stretch in 1 minute or in half a minute. Make 15 or 20 trials with floats, and get the average velocity of the stream—that is, the feet per minute covered by the floats. To be safe, take two thirds of this float velocity and multiply by the average cross section of the stream, to give the discharge in cubic feet per minute. The following table shows the amount of water required theoretically to give 1 horsepower with various heads. Twice

the quantity will give 2 horsepower, 3 times the quantity will give 3 horsepower, and so on.

WATER REQUIRED THEORETICALLY FOR ONE HORSEPOWER

HEAD IN FEET	CUBIC FEET PER MINUTE
5	105.6
10	52.8
15	35.2
20	26.4
25	21.1
30	17.6
35	15.1
40	13.2

Practically, the horsepower will be only a fraction of that indicated, as some power is used up in the machinery and some is wasted entirely. The exact proportion of theoretical power obtainable depends upon the type of water wheel used.

To calculate quickly the horsepower of a stream or fall, engineers are in the habit of using a short and quite accurate method, as follows: *Divide the flow of water in cubic feet per second by 11 and multiply by the head.* The result gives the horsepower which may be developed, using a wheel of 80 per cent efficiency. If, for example, a head of 9 feet is available with a water flow of 22 feet per second, the horsepower which may be developed with a turbine of 80 per cent efficiency is found by dividing 22 by 11 and multiplying by 9, the result being 18. To find the horsepower of falling water exactly, multiply the weight in pounds of water falling per second (a cubic foot of fresh water weighs about 62½ pounds) by the head in feet. This gives the foot pounds per second. Divide the foot pounds per second by 550 to get the theoretical horsepower available.

Water Wheels

For small, low-priced installations the gravity type of wheel is frequently used. Wheels of this type are operated directly by the weight of falling water exerted through the falling distance or head. Such are the breast and overshot wheels illustrated. The undershot wheel is operated by the current and involves no dam or other works. It is the old-fashioned mill wheel. The Pelton wheel is an impulse type, being operated by a jet of water forced against the blades of the wheel. Turbines of various kinds are also used. They are the most efficient type of wheel for low heads and are most generally used in commercial plants. They may be of either the impulse type, in which the direct force of the water against the moving part is used, or the reaction type, in which the back "kick" of the water as it leaves the vanes, furnishes the driving force. This kick is commonly used in rotary lawn and garden sprinklers, which are really reaction turbines.

The efficiency of wheels may be stated as follows: Undershot, 15 to 25 per cent; breast, 55 to 65 per cent; overshot, 65 to 75 per cent; Pelton, 75 to 80 per cent; turbine, 80 to 90 per cent.

A fall of from 6 to 8 feet and upward usually makes a turbine installation advisable, if considerable water is available. Under very small heads, of only 2 or 3 feet, undershot and breast wheels may be desirable, or, if conditions permit, an overshot wheel may be adopted. Where the undershot wheel is used, the stream is frequently narrowed to about the width of the wheel, thus giving the wheel the benefit of all the water in the stream running at a somewhat higher velocity than in the open stream. This type of wheel is rapidly disappearing altogether. One of its strong features, however, is its comparatively low cost, no dam being required.

Recently an entirely new type of turbine, known as the Clarkson current turbine, has been invented. This is operated by the current flow and requires no

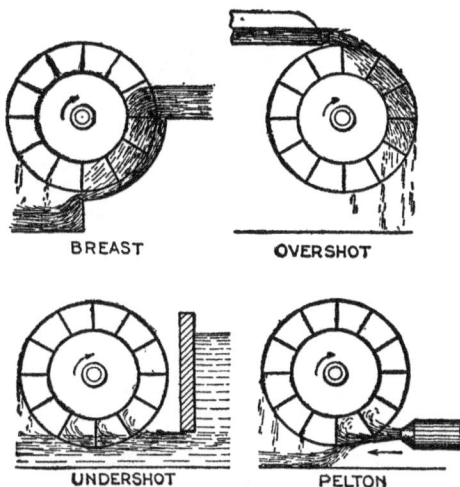

Fig. 266. Types of water wheels. The overshot is the most efficient of the simpler forms, but where there is much fall the necessary size (the diameter equalling the fall) may make it impracticable.

dam, thus having these advantages in common with the undershot wheel. Its efficiency, however, is very much greater than that of the undershot wheel, and may run up to 60 per cent or more.

Just what installation is best in each case, and the cost involved depend upon local conditions. Before determining the size of wheel to install, the conditions of water supply at all seasons of the year must be taken into account. Some streams, for example, have large quantities of water available for a few months of the year, and for the remaining time are either frozen or completely dried up. Such streams are obviously unsuited for continuous year-round power purposes. In commercial installations, the flow of such streams is often equalized to some extent by the utilization of big storage reservoirs which can be drawn upon in time of drought; but for small installations, this arrangement can usually be only sufficient for day-by-day regulation, and cannot take care of seasonal water shortage.

Cost of water-power plants. The cost of a small water-power turbine plant may range from $50 per horsepower, where conditions are unusually favorable and the power is used directly, as for sawing, grinding, or other purposes, up to several hundred dollars per horsepower, where a large dam is required and electric generators are installed to give a supply of electricity for lighting and other purposes. The following table* gives some figures about successful farm water-power plants in various parts of the country. All of these are turbine plants.

While nothing very definite can be said about cost, it is usually true that, with other conditions unchanged, the higher the head, the less the cost. The larger the plant, the less it will probably cost per horsepower.

*From "Practical Talks on Farm Engineering," by R. P. CLARKSON. Doubleday, Page and Co., Garden City, N. Y.: 1915.

FARM WATER-POWER PLANTS

HEAD	POWER DEVELOPED	LENGTH OF DAM	COST OF PLANT
6	17 hp.	36 feet	$1,000
11	8 hp.	350 feet	1,000
15	5 hp.	(used old dam)	225
17	15 hp.	200 feet	700

In most cases under normal conditions, from $100 to $125 per horsepower will cover the cost of the entire installation, including electric machinery, wiring, lights, etc. The cost of operation is the interest and depreciation on the plant, the expense for oil and repairs, taxes, and such supervision as is necessary.

On the other hand, plants using the gravity

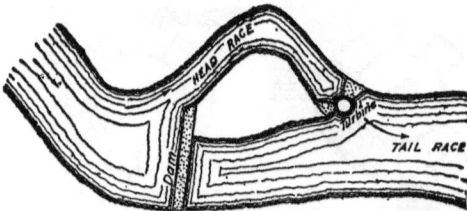

FIG. 267. Stream viewed from above to show where dam and turbine should be located. With sufficient water available, a supplementary, simple water wheel could be located at the dam.

or current types of wheels, where they are suitable, may cost only a third as much as a turbine plant or even less if constructed by the farmer himself. They involve, usually, only wooden construction or, at most, wood and concrete.

When a simple, inexpensive outfit of this kind has proved the importance and value of power development to you and your household, then you may feel more inclined toward the expenditure required for a good, permanent turbine plant, if conditions permit. It should be remembered that additional convenience and the saving of unnecessary labor and trouble are all sufficient to warrant considerable outlay and are truly as valuable as the more direct saving of money.

It must not be assumed that a water-power plant is necessarily the cheapest way to get power. It may be neither the cheapest in first cost nor the cheapest to operate. In fact, water-power plants are usually more expensive to build and install than steam plants or gas and oil engines. Water-power installations are sometimes so very costly at the beginning that the very heavy interest and depreciation charges, taken together, with the maintenance and operation costs, make the cost of the power produced greater than that of such power when supplied from other sources. These items of interest and depreciation should properly be considered always as part of the cost of power.

The thing to keep in mind is the cost per horsepower hour for the power you can use. This must be modified somewhat by the feature of convenience. Simple farm water-power plants require almost no continual attendance. No damage, except to the immediate machinery, can result even from gross neglect. The installation may be practically noiseless. These are some of the points—frequently deciding factors—which differentiate water-power plants from either steam, gas, or oil installations. The discussion of these alternative types of power plants may be found under their respective headings later on in this volume.

The Hydraulic Ram

One of the more important problems of country life is the supplying of water in plentiful quantities for use in the house and barn. Laborious hand pumping and frequently, also, the carrying of water by hand are resorted to on many farms. Windmills are commonly used for pumping purposes and make very satisfactory pumping engines. Of course, any form of power can be utilized to run a pump, but one of the cheapest and most satisfactory, as well as almost foolproof, machines for pumping is the hydraulic ram. It will operate at high efficiency continuously without any attendance.

To the average person, the hydraulic ram is a mysterious thing. Working day and night for years without attention and without rest, it is the farmer's most dependable friend for pumping water. The efficiency of the ram, when used for lifting water only 4 or 5 times as high as the fall, is as great as that of the best pumps, and is much better than that of most pumping apparatus. For other ranges, where the lift is small up to 25 times the fall, table A gives the efficiency of a ram.

The efficiency of a ram falls off so greatly as the delivery height increases that rams are seldom used where the lift is more than 25

FIG. 268. Rough stone dam (*in background*) creating a head with which to operate ram (*at left*), the delivery pipe from which is seen in the foreground.

TABLE A

Lift divided by fall	2	3	4	5
Per cent efficiency	90	85	80	75
Lift divided by fall	10	15	20	25
Per cent efficiency	57	42	30	23

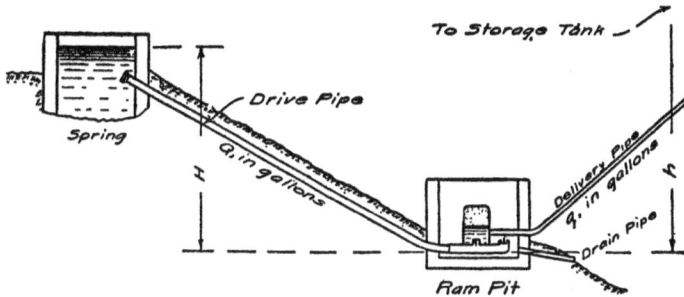

FIG. 269. Diagram to show relative positions of spring or supply tank, drive pipe, ram in ram pit and delivery pipe. The letters are the same as those used in the efficiency formula as given below.

times the fall. For what are known as "common rams," the general rule for calculation is that one sixth of the water supplied to the ram will be lifted to a height 10 times as great as the fall. Exact calculation may be made for any ram by using the following formula:

$$q = \frac{Q \times H \times e}{h}$$

where q equals the quantity of water raised, in gallons, Q is the quantity supplied to the ram, in gallons; h is the lift from ram to storage tank, in feet; H is the fall from supply down to ram, in feet; and e is the efficiency of the ram taken from Table A, page 176, where h divided by H is the lift divided by fall.

For example, there is a fall of 10 feet, and ram can be supplied with 25 gallons of water per minute. The storage tank is in the attic 40 feet above the ram. How much water per minute will be supplied to tank? From Table A, the ratio of 40 feet lift to 10 feet fall will permit an efficiency of 80 per cent. Then, using the figures given and substituting them in the formula:

$$q = \frac{25 \times 10}{40} \times 80 \text{ per cent} = 5 \text{ gallons per minute.}$$

It is apparent that if 25 gallons of water are delivered to the ram and only 5 gallons reach the tank, there must be a great waste of water. The water is wasted, but the energy of its fall is utilized in lifting the remaining quantity to the greater height.

A diagrammatic form of ram is shown on page 178 (Fig. 271). There are five main parts: the drive pipe A, the waste valve B, the delivery pipe D, the air chamber C,

and the admission valve E. The water flows down A and out of the waste valve B, when the ram is first started. When sufficient velocity has been gained by the water, it closes valve B suddenly. This confines the water in the casing and, as the movement of such a large bulk of water cannot be stopped instantaneously, the valve E is dealt a hammer blow which opens it and allows a small amount of water to flow into the air chamber. The valve E then falls shut again and, then, as the water has slowed down, the waste valve B again opens, the water flows out, gains velocity, shuts the valve B again, opens valve E, and more water is forced into the air chamber. This action continues indefinitely as long as water is supplied to the ram.

TABLE B: SUPPLY REQUIRED TO DELIVER ONE GALLON PER MINUTE

Ratio of lift to fall . . .	2	3	4	5
Gallons per minute required to operate ram .	2.22	2.47	5.00	6.67
Ratio of lift to fall . . .	10	15	20	25
Gallons per minute required to operate ram .	17.54	35.91	66.67	108.70

The presence of air in the chamber C is necessary; for it compresses when the sudden blow is struck on the valve E, and this allows

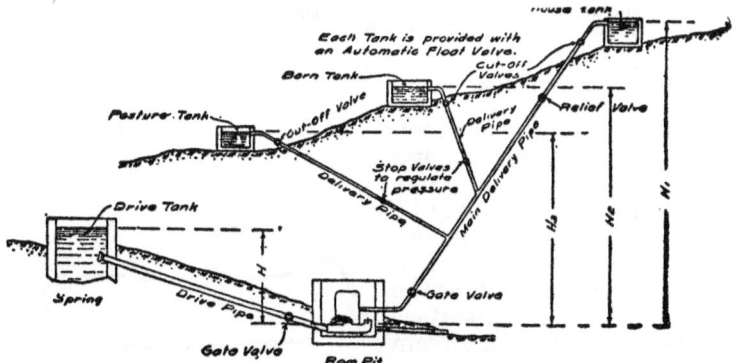

FIG. 270. Showing how one ram can be used to supply three different tanks at three different levels

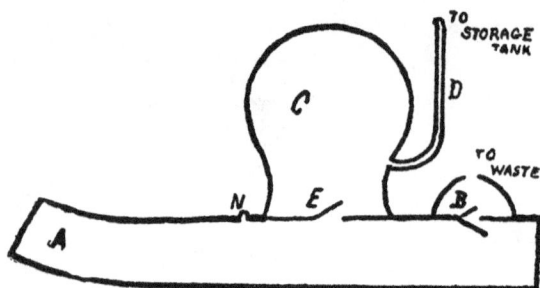

FIG. 271. Diagram made as simple as possible to show parts and operation of the hydraulic ram (see text)

that valve to open. Of course, the water will absorb a little of the air, and after a time the air in the dome will be exhausted. This will cause the ram to stop; and, to prevent such stoppage, there must be a way of admitting more air into the air chamber. This is done by boring a small hole at N. The water, rushing into chamber C, sucks in through the hole N just a tiny bit of air, enough to prevent the exhaustion of the air chamber. On many of the higher-priced rams, a "sniffer" valve is located at some such point as N, to serve the same purpose as the tiny hole here recommended.

The ram, as described above, will raise a portion of the water supplied to it to any

FIG. 272. Small concrete spillway to take care of the surplus water above that needed to operate a ram or wheel.

desired height. If, however, it is desired to pump clear water from a brook or spring by means of undesirable water from some pond or stream, it may be done with safety by using a ram-pump. This resembles the ram shown except for the addition of the parts H, K, V, and S, as shown in Fig. 273. As before, the water to operate the ram comes through the drive pipe, but the water to be pumped enters through the small pipe K,

and passes through the valve E, when the latter is opened. A check valve at V prevents the clear water being forced back up the pipe K, while a stand pipe at S keeps sufficient water pressure on the pipe at H to fill the righthand end of the casing at all times and even to allow a little to leak through the waste valve B. Thus, none of the impure water gets near enough to the valve E to be in any danger of being forced into storage. The ram-pump is best used where the supply of pure water is decidedly limited in quantity.

Rams and ram-pumps are usually placed at the bottom of pits dug into the ground, the head being increased in that way while the waste water flowing from the waste valve is easily drained from the pit through open-joint tiles or through a drain pipe laid from the pit to a lower level.

Rams are commonly made in 6 sizes, from that requiring only 1½ gallons per minute to operate it up to one requiring 25 gallons per minute. The price ranges from $5 up to $25 for these sizes. Larger sizes are made, and often a whole battery of rams is installed where the supply of water is large. Rams so combined may deliver through a common pipe or through individual pipes leading to as many tanks. Similarly one ram may be made to supply more than one container as shown in Fig. 270. Ram-pumps are slightly more expensive. If possible, the ram to be purchased should be provided with an adjustable arrangement on the waste valve, so that the latter will not stick if a higher head is used than was at first thought to be possible. If this is not done, care must be taken that the ram bought is workable on the highest head of water that is likely to be employed.

The Siphon

Occasionally, it is necessary to carry water for use up over the top of an intervening ridge. This may be done without continuous pumping by means of a siphon, provided (1) that the lift required from the surface of the supply pond or stream to the top of the ridge is not more than about 25 feet, and

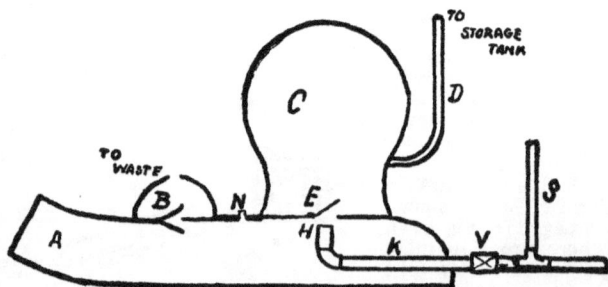

FIG. 273. Simple ram varied by the addition of parts H K V and S so that a large head of impure water can be used to elevate pure water from a smaller supply.

(2) that the discharge point of the siphon is below the level of the supply pond.

In theory, the siphon may be made so that the lift from the surface of the supply pond to the topmost point of the siphon would be 34 feet, but in practice it is very difficult to get anything like this height without special care and precautions. The large number of joints, their possible leakage, the friction loss of the water flowing in the pipe, all combine

FIG. 274. Two arrangements by which the head and the length of the drive pipe can be brought into correct relationship. It is the height of the water surface, not its extent, that determines its pressure.

to lower the possibilities. Also, in high altitudes the atmospheric pressure, which causes the siphon to operate, is much less, and the height to which the water can be raised is less in proportion.

For temporary use, any sort of flexible pipe as, for example, a garden hose, makes a very satisfactory siphon. The siphon principle may be used to empty barrels or tubs where it is not desirable to use a spigot; but the barrels should be raised, so that the discharge end of the hose will be kept below the surface of liquid in the barrel. Similarly flooded cellars and basements may be emptied without pumping, if the discharge end of the pipe can be put below the surface of the water in the cellar, as is frequently possible if the house stands on a hill.

Often a siphon, ingeniously used, will do away with an expensive pumping outfit; and it not only costs less to install, but will operate at practically no cost and without any attendance.

A successful siphon must be air-tight, and the supply end of the pipe must be kept below the water surface. Preferably, there is a valve at each end of the siphon, the one on the entrance end being below the water surface and that on the discharge end, which may indeed be a faucet, being materially below the level of the other one—5 to 10 feet or more is advisable. At the very topmost point of the siphon, there should be a short standpipe connected with the siphon; and it, too, should have a good valve in it.

The diagram (Fig. 275) shows the arrangement for supplying a barn from a pond located on the other side of a hill. To start the siphon, close the valves at the ends of the pipe A and C, open the valve at B, and fill the pipe completely until it overflows the standpipe. Then shut valve B as tight as possible. The

siphon is now full of water, and if the valve or faucet at C be opened and that at A opened at the same time—not before C—the water will flow continuously from A up over the hill and out at C. The valve A and B must not be touched; but the flow must be regulated entirely by C, which may be opened and closed like any faucet. If at any time there is leakage of air sufficient to stop the action of the siphon, careful repairs must be made, and the siphon must then be started as before by filling through B.

In most cases, the distance from the surface at A to the top point below B cannot be much more than 25 feet, but the distance from the top point below B down to the discharge at C may be anything. The difference in levels between A and C should be at least 4 or 5 feet and preferably more. A very small pipe should not be used for siphoning, as the friction is greater in proportion than with large pipes, and a very small leakage of air which might not affect the operation of pipes of considerable diameter would operate to stop the smaller siphon entirely.

FIG. 275. Diagram to show working of siphon (see text)

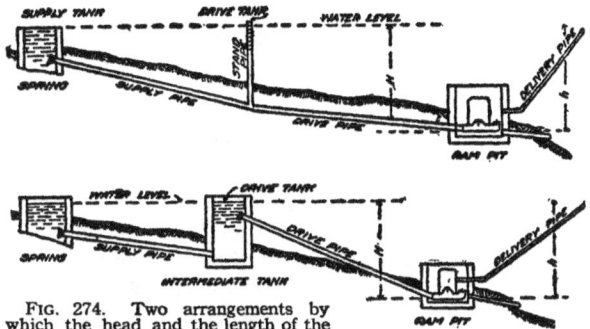

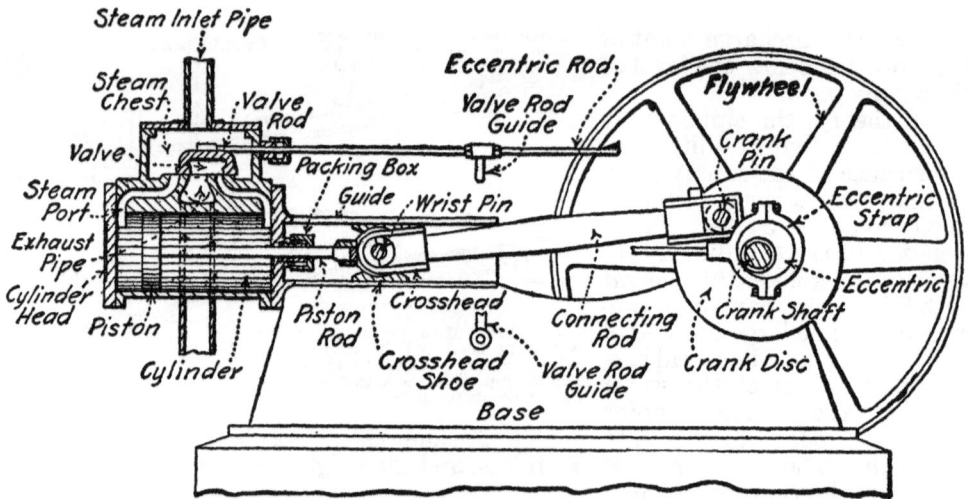

FIG. 276. Sectional diagram of a horizontal steam engine with important parts named

CHAPTER 17

Steam Engines

By PROFESSOR R. P. CLARKSON *(see Chapter 14). The most extensive use of steam on the farm to-day is probably in connection with the heating of greenhouses and the washing and steriliz-ing of dairy utensils, etc. It should not be overlooked, however, as a source of power (preferably stationary, but also, where conditions are favorable, in tractor form) because of the simplicity of the principles by which it operates, and of the comparative cheapness of the maintenance of a steam power plant. Moreover, when such an outfit is operated, there remains the exhaust steam with its value almost undiminished as far as heating the house, work buildings, greenhouses, etc., is con-cerned. For some years the steam tractor hauled, unchallenged, the huge gang plows and the mighty seeding and harvesting implements that paved the way to the profitable cultivation of the western plains. Now the lighter, more easily fed internal combustion tractors have stepped in and largely displaced the pioneer. But it still has its place, both on large farms and small, and the farmer can well afford to familiarize himself with its requirements and its fields of service.—*EDITOR.

IN ANY discussion relating to farm power, it is necessary to consider not only the very common internal-combustion, or oil, engine, but, also, the less fre-quently seen steam engine. The great advantage of the steam engine lies in its flexibility of power output. By increasing the boiler pressure of steam, you make possible a very great overload or increase in engine power. A so-called 10-horse-power farm engine may thus be made to give 25 or 30 horsepower for a time, with, of course, a corresponding increase in steam consumption. This cannot be done with any other type of engine or motor. The output of kerosene and gasoline engines is definitely limited by certain features of their construction.

Another and, in some cases, an important advantage in the use of steam power is the possibility of using the exhaust steam about the farm to heat water for wash-ing or other purpose or for the heating of the house, shop, or other buildings. Where these things are desirable, steam power must be considered.

It is seldom advisable to purchase small-sized steam plants—2- or 3-horse-power—as they are almost invariably carelessly and poorly made and, at best, extremely uneconomical. Even comparative economy of operation of steam equipment cannot, in fact, be reached in anything short of 20 or 30 horsepower;

and even this size of
plant suffers by com-
parison with a kero-
sene-engine outfit, ex-
cept in the special cases
mentioned above.

In attendance and
care required, the steam
plant is more exacting
than any other source
of power.   In cost of
operation, including la-
bor, it is usually more
costly.   A great disad-
vantage lies in the fact
that initial labor and
considerable fuel con-

FIG. 277. The steam tractor is powerful, reliable and simple to run and care for. Its chief disadvantages, compared with the gas and oil types, are its frequent need of water and the bulkiness of its fuel.

sumption take place before the plant is ready for work in the morning; and, if it
is worked up to its proper capacity, the end of the day finds a boiler full of
steam and a fire pot full of partially consumed fuel.   All this is wasted unless there
is some secondary use for steam from the power plant.

The steam plant, although requiring a great deal of care and attention because
of the possibility of explosion owing to too great a steam pressure, or of destruction on account of lack of water in the boiler, is not, as built for farm use, a very
complicated outfit.   There are several types, differing largely in degrees of porta-
bility.   In the usual form, the engine is mounted directly on a horizontal boiler
which, in turn, is on a light-wheeled frame.   With this it is not possible to use
economically the many refinements available for stationary plants, especially
those used in large-sized stations, such as are found in our towns and cities for
electric lighting.

The essential parts of any steam-power plant are: (1) a boiler or container in
which water can be heated to steam and (2) an engine in which the steam can be
made to expand and exert pressure continually.   These are the only fundamental
parts.   There are all sorts of additional arrangements and devices used with each
part of the plant, either to save labor or to insure safety of operation.   On the
boiler, for example, there is a gauge glass to show directly the height of water,
and try cocks to indicate by trial the presence of water in its proper place and
steam in its place.   There is always a pressure gauge to show what the steam
pressure is, and a safety valve which operates and releases the steam, if the pressure becomes too great for safety.   Usually, an additional safety device, called a
fusible plug, is installed.   This plug is arranged to melt, if the water gets too low
and the boiler thus becomes too hot.   When the plug melts, the steam blows on
the fire and puts it out.   Some boilers are fitted with a feed-water heater which
heats the water before it goes into the boiler, the heat required being usually
furnished by the exhaust steam from the engine and occasionally by the hot gases
passing up the smokestack.   The latter arrangement is called an "economizer,"
and is seldom found in any but large stationary steam plants.

Boilers

Water-tube and fire-tube types.   The
steam boiler, which is usually cylindrical in
shape for small installations, may be either
vertical or horizontal.   The vertical type

saves floor space and is most used indoors.
Sometimes the water circulates through a
network of pipes which are surrounded by
hot gases from the fire.   This is called a
water-tube boiler.   Very often the water is
in a large container through which a number

of tubes or flues pass, these tubes carrying the hot gases through the mass of water on their way to the smokestack. This is the fire-tube type. In this type, also, the fire box is almost entirely surrounded by water.

Locomotive type. The locomotive type of boiler is most used for portable purposes and is always used on steam tractors. It is a horizontal, fire-tube type, with an enlargement of the shell of the boiler at one end to form the fire box. The water surrounds the fire box. Above the water space and attached to the top side of the cylinder is a small chamber, called the steam dome, from which the steam is drawn to the engine. This arrangement is made to prevent the steam, as it is drawn off, carrying water spray with it. The steam dome thus helps to make available dry steam. Occasionally, the steam drawn from the boiler is also passed through a collection of highly heated tubes, called a superheater, which raises the temperature above the normal boiling point of water.

Grates. The grate design depends on the fuel to be used. Most farm boilers have interchangeable grate riggings. The grate with smaller openings is used for cobs and wood and other similar fuels. Where straw is used, still further changes are made. The straw, being light in weight, is quick-burning, and flashy. To prevent its waste, deflectors are frequently used to prevent the draft carrying the flaming straw through the tubes, and a supplementary grate is used below the main grate to catch such particles as fall without being entirely consumed. Combustion of these particles is completed on this grate and thus a considerable heat saving is effected.

Traction engines. In many cases, particularly with traction engines, there is some way provided for increasing the draft and thus forcing the fire to burn more rapidly. The usual way is to exhaust the steam from the engine up the stack. The steam rushing up the stack under pressure creates a suction and draws air rapidly through the grate. Of course, when the engine is not running, live steam

from the boiler must be substituted. When this is done, only a small jet is used. With forced draft on traction engines, some form of spark arrester is needed in the stack. This is usually a wire screen formed into a cone and filling the stack. Occasionally, a sharp turn in the smoke path is provided; and this separates the heavy sparks from the smoke. The heavy particles strike the side of the bend, lose their inertia, and fall into a receptacle placed to receive them. The smoke, however, being of a gaseous nature, makes the turn.

Injectors. To deliver water to the boiler without the use of a feed-water pump, the steam injector is sometimes used. It is practically universal in locomotive work; but in stationary plants it is not so frequent as a main method of delivering water. Small boilers are frequently equipped with some type of injector; and it is a desirable thing to know something about its use. In a sense, it acts like the atomizer spray. There are 2 intakes: one for steam under pressure and the other for water. The steam rushing in over the water intake partially exhausts the air and draws water up the tube. The steam condenses on meeting the water, and gives to the water considerable motion. The water is thus forced through a check valve into the boiler. This is the simplest form, and illustrates the action of all types. The troublesome features about the use of the injector are: (1) its inability to handle hot water because the vacuum action causes the water,

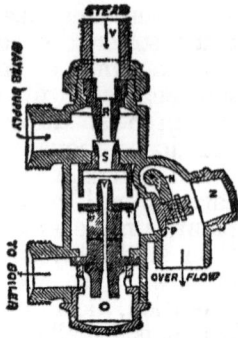

FIG. 278. Sectional view of one type of injector. When steam is turned on through V, it creates a suction in the narrow parts R and S, which draws water in through the supply pipe and starts it on its way into the boiler.

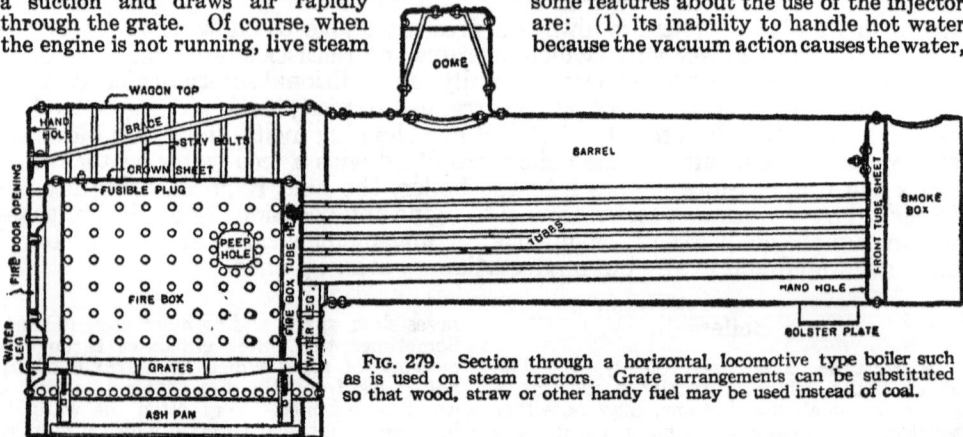

FIG. 279. Section through a horizontal, locomotive type boiler such as is used on steam tractors. Grate arrangements can be substituted so that wood, straw or other handy fuel may be used instead of coal.

if hot, to flash into steam; (2) the uncertainty of operation except under proper working conditions; and (3) its lack of flexibility, that is, the difficulty of maintaining it in operation under varying conditions of load. On the other hand, the injector is self-acting by means of the boiler pressure, and the steam used serves to heat the water as it is fed into the boiler. Nevertheless, even with these advantages, the injector is a great waster of heat as a mere pump. Its efficiency lies in its combination action as pump and feed-water heater. For these purposes, it is a proper device for small single-unit boilers, being less expensive, usually, than the apparatus which it is designed to replace.

Care of the boiler. Proper care of the boiler means attention to only 5 main points: (1) there must be a proper amount of clean water in the boiler; (2) there must be a deep, clean fire on the grate; (3) the pressure of steam must be kept reasonably constant at the proper gauge reading; (4) the boiler itself must be kept clean and in good repair; (5) all accessories should be kept in perfect working order.

Where water is dirty or impure, foaming is likely to result. The water will rise and fall in the gauge glass, the engine will lose power, and water is very likely to be carried over with the steam into the engine, causing a sort of clicking noise. If this condition persists, damage will result.

Almost all water has in solution something which is detrimental to the best operation of the boiler. Soft water is apt to eat into the boiler and accessories because of its probable acidity, while hard waters are sure to cause a scale. Continual cleaning of the boiler and the careful use of soda ash with soft scale is usually advisable. If the scale becomes hard, one of the many boiler compounds made for this special purpose is necessary.

To keep a deep, clean, bright fire all through, continual watchfulness is essential. Clinkers must be kept loose from the grate and the furnace cleaned entirely once in a while. The method of firing is important also. The fresh coal is usually placed near the door and spread back after coking. In this way a bright fire is maintained even when a fresh charge is put in, and the gases of the fresh coal are entirely consumed in passing over the bright coals in the rear. The principal disadvantage in this method of firing is the maintenance of an open door during the consider-able time necessary for spreading; but quick and careful work will cut this time down to a minimum.

The two processes "banking" and "drawing" should be mastered. If a boiler is used constantly every day, it is usually considered more economical to bank the fire overnight, rather than build a new fire each day. Frequently, too, fires must be banked to prevent a rise in pressure during the day if, for example the fire is bright and the engine is shut down for a while. Banking is usually accomplished by covering the fire with ashes or screenings or fresh coal, and closing all drafts. Sometimes the fire-pot door is left ajar. Drawing a fire should never be attempted without first lowering the heat of the fire by smothering with dirt or ashes or, if need be, fresh coal. After smothering to some extent, the fire may be drawn rapidly without the terrific flare which would result if it were drawn while glowing. The fire should always be drawn to cool the boiler quickly or to remedy any condition caused by extremely low water.

The size of a boiler is usually stated in terms of capacity and the quality or economy of action. The former is given in boiler horsepower; the latter, in pounds of steam evaporated per pound of coal. The standards used vary considerably and are most unfortunate in many respects. The horsepower standard adopted by one leading engineers' society says that the equivalent of 1 boiler horsepower is the evaporation of 30 pounds of water at 100 degrees under a gauge pressure of 70 pounds. Another way of rating is to take from 10 to 14 square feet of heating surface of the boiler as the equivalent of a boiler horsepower. Still a third approximate method of rating, only occasionally used, is to consider one half a square foot of grate surface as equivalent to 1 horsepower. Except in the case of marine boilers, this rating has nothing whatever to do with the horsepower of the connected engine. In marine work, however, the horsepower of the boilers is uniformly spoken of as the horsepower of the engines which they serve.

The economy of boilers depends so much on the handling of them that little can be said about this feature. The amount of coal consumed in the average farm boiler per horsepower of engine output varies from 60 to 80 pounds per day with ordinary firing. With expert firing this could probably be cut down to perhaps 40 pounds. Such firing, however, is almost never available.

Engines

Next to the boiler which furnishes the steam, the important link in the power chain is the engine which utilizes the steam. The former is essentially very simple. The latter, although it looks complicated, is in reality just as simple. In a cylinder of smooth bore a piston fits and slides freely to and fro. This piston

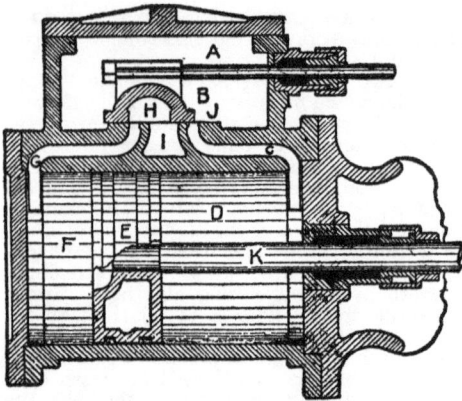

FIG. 280. Section through a steam engine cylinder and valve. Steam from the chest passes through the port (C) into the rear end of cylinder (D) and against piston (E) which it forces toward the front end (F). At the same time the steam previously used is forced out through port (G) into the exhaust chamber (H) and through port (I) into the air. By this time the valve (B) has shifted and begins to admit steam through (G) and let it out through (C), thus driving the piston back again. The piston rod (K) acts on the fly- or driving-wheel

is actuated by the live steam from the boiler, which is, of course, at a pressure, usually considerable; in stationary boilers of small size, it runs from 69 to 150 pounds gauge. This live steam is admitted to one side of the piston by a valve mechanism which, at the same time, opens the other side of the piston to the atmosphere or, in condensing engines, to more or less of a vacuum. Consequently, the force of the live steam moves the piston and thus operates the flywheel and the power shaft. The valve mechanism is operated by an eccentric on the power shaft or connected with it, so that when the piston reaches the end of its stroke the valve shifts to reverse the side on which the steam pressure acts. This reversal together with the momentum of the flywheel forces the piston back on the return stroke.

Various refinements of this valve mechanism are available, so that the steam may be admitted during varying lengths of time and then allowed to expand in the cylinder, thus reducing its pressure as it acts on the piston, and permitting exhaust at a considerably lower temperature and pressure. This results in much greater engine economy than if

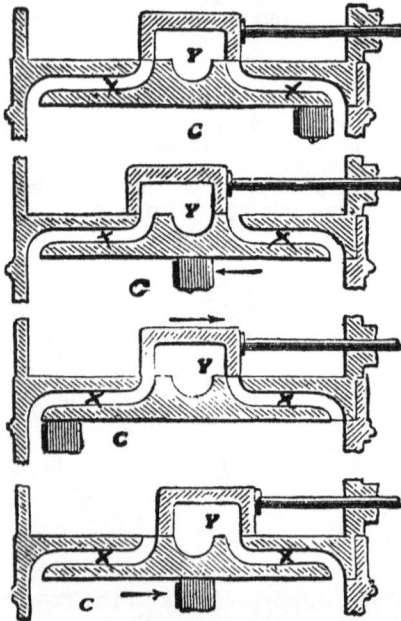

FIG. 281. The valve and ports of a steam engine illustrating the four positions referred to under Fig. 280. Here C is the cylinder, Y the exhaust chamber, and X the ports.

the steam were used directly at boiler pressure throughout the stroke. In this connection, it should be stated that the correct setting of the valves on an engine is probably the most important thing about the economical running of this mechanism. There are 4 points involved: (1) the opening of the port for the admission of live steam immediately before the piston reaches the end of its return stroke; (2) the cutting off of the live steam at the proper point for true economy in the work being done. This may be at one fourth or one third or one half of the stroke and does vary from time to time; (3) the release of the exhaust steam at the proper point, which is just before the end of the forward piston stroke; and (4) the closing of the exhaust port sufficiently early to retain a small amount of steam in the cylinder for cushioning purposes. This steam is compressed as the piston nears the end of the back stroke, frequently making the pressure against which the live steam is first admitted a considerable amount.

This process of obtaining power is very simple and, of course, very old; and it is, in fact, very wasteful. In some engines, less than 2 per cent of the heat supplied is available for useful work. In the best engines with all refinements, not more than 18 per cent is available. The great loss of heat is in the exhaust steam, which is discharged from the engine and can do no more useful work so far as the engine is concerned. If this source of heat can be adequately utilized for any other purpose, as described at the beginning of this article, the lack of engine economy is not so discouraging.

FIG. 282. Flyball type of governor by which the admission of steam into the cylinder is automatically regulated.

The governor. While in most steam engines the control of variations in speed during different parts of a single revolution is taken care of by the action of the flywheel, yet the regulation of the speed, revolution after revolution, must be handled by some control of the steam pressure. The supply from the boiler at all times must be just enough to handle the load without slowing down under heavy loads or speeding up under light loads. This is arranged in any one of a number of ways on various types of engines apt to be found on the farm.

The governor may operate the throttle and act equivalent to changing the boiler pressure; it may act to cut off the live steam at some point in the stroke, and thus determine the expansion.

The common form of governor is the fly-ball type. The centrifugal action of rotation throws the balls out, and in so doing raises the slide and operates some type of valve. Promptness of action and delicacy of adjustment are essential features which must be looked after in design. The resisting force against which such a governor acts is the weight of the balls, although some types use a spring in place of depending on gravity action.

The location of the governor is not important, except as it avoids complication of machinery and action. If designed to regulate the pressure, it is usually placed on a vertical shaft, balls being suspended from the shaft by hinged arms or rods sloping to either side with the balls on their extremities, connecting links extending down to a sleeve which slides up and down on the shaft and operates directly or through mechanism to a valve on the steam line. In some cases the governor is placed on the flywheel, the weighted or spring-controlled arms being pivoted to the wheel spokes and operating directly a rod which affects the eccentric mounted on the same shaft. This eccentric determines the position of live steam cut off in the steam chest. The governor thus regulates the amount of steam admitted and the ratio of expansion in the cylinder. This type is called the automatic cut-off. It is perhaps more often found on high-grade engines which operate at high speed.

Classifications of engines. There are a number of classifications for engines which should be mentioned. The kind of work to be done determines whether the engine should be (1) stationary; (2) portable; (3) traction, or (4) marine. Their speed divides engines into (1) high-speed, and (2) low-speed. Whether the engine exhausts into the air or into some sort of condenser, determines it to be (1) non-condensing, or (2) condensing. Whether the steam goes on both sides or on only one side of the piston makes it (1) double-acting, or (2) single-acting. If the steam exhausts from the cylinder of one engine to a larger cylinder of another engine, built to utilize low-pressure exhaust steam, it may be a compound engine. Most farm engines are high-speed, non-condensing, single- or double-acting, and may occasionally be compound, though oftener they are single 1-cylinder rigs. Of course, stationary, portable, and traction types are frequently used.

Measuring horsepower. There are 2 values which are known as the "horsepowers" of any engine. One is called "indicated horsepower," usually written I. H. P., while the other is the "brake horse-

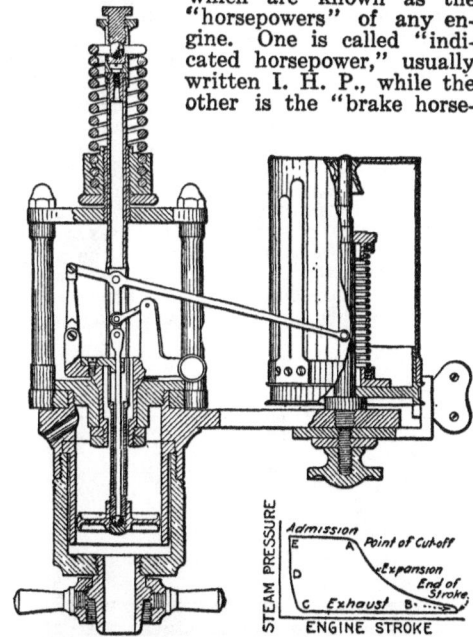

FIG. 283. Sectional view of an indicator for measuring the horsepower of an engine. The inset shows a typical indicator diagram or curve.

power," written B. H. P. The indicated horsepower of a steam engine is the mechanical work done in a certain time by the steam acting on the piston. Some of that work goes to run the engine itself, overcoming the friction of the bearings and the drag of the moving parts, so that only a portion of the force exerted can be delivered to the belt pulley. The work which can be done by this portion at the pulley in a certain length of time is the brake horsepower.

To measure the indicated horsepower of a steam or oil engine, an instrument known as the "indicator" is used (Fig. 283). There is a cylinder to which steam or explosion pressure is admitted from the engine cylinder. The pressure forces the piston back against the resistance of a coiled spring, which has been experimented with previously, so that the pressure exerted by the steam on the little piston is known from the amount the spring is compressed. As the area of the small piston is usually just one square inch, the pressure indicated by the compression of the spring is the pressure per square inch of the engine piston. So, if we multiply this indicated pressure by the total area of the engine piston, the result obtained is the total steam pressure on the engine piston. This varies continually, on account of the movement of the piston and the expansion of the steam.

Indicator diagrams. There is, also, on the indicator a rotating drum which turns through a distance proportional to the stroke of the engine piston. A pencil is so arranged that it moves up and down with the indicator piston; and as the drum rotates beneath the pencil the latter draws a diagram with its length proportional to the engine stroke, and its height proportional to the pressure on the engine piston. In Fig. 283 is shown the shape of such a diagram for a steam engine, which is known as an "indicator diagram." Oil-engine cards are similar, but narrow. Mathematical calculations show that the area of such a diagram as this is proportional to the product of the average pressure on the piston during the stroke and the length of the stroke. In other words, the area of this diagram is proportional to the work done on the piston of the engine by the steam during one stroke, so that, knowing the number of strokes per minute made by the engine piston, we may easily find the work done per minute. This, divided by 33,000, gives the indicated horsepower (I. H. P.) of the engine, because 33,000 foot pounds per minute equal 1 horsepower.

The prony brake. To measure the brake horsepower (B. H. P.) of any engine, an instrument known as the "prony brake" or, more technically, "the absorption dynamometer," (Fig. 284) is used. This consists of a band which may be tightened around the engine pulley, creating great friction on the pulley and requiring constant force acting to overcome this friction. As this force is acting constantly on the rim of the pulley, in one revolution of the pulley the force acts through a distance equal to the circumference of the pulley. The circumference is $3\frac{1}{7}$ times the diameter. The product of the length of the circumference and the force of friction acting will give the work done in one revolution. Then by counting the number of revolutions per minute and multiplying this number by the work done in one revolution and dividing by 33,000, we get the brake horsepower.

It is difficult to measure the force of friction directly, so that it is measured by suspending weights on the end of a long arm (Fig. 284). By the principle of the lever, the force acting at the circumference of the wheel is to the force exerted by the weights at the end of the arm as the length of the arm measured from the centre of the shaft is to the radius of the wheel or pulley. That is,

$$\frac{\text{Friction force}}{\text{Weights}} = \frac{\text{Length of arm}}{\text{Radius}}$$

and hence

$$\text{Friction force} = \frac{\text{Length of arm x weights}}{\text{Radius of Pulley.}}$$

And the horsepower, as stated above, being friction force times the circumference times the number of revolutions per minute (written R. P. M.) divided by 33,000 will give the following value for brake horsepower (B. H. P.) by substituting the value of the friction force found above:

B. H. P. =

$$\frac{\text{Length of arm x weights x } 3\frac{1}{7} \text{ x pulley diam. x R. P. M.}}{\text{Radius of wheel x 33,000.}}$$

equals

$$\frac{\text{Length of Arm x weights x } 3\frac{1}{7} \text{ x 2 x R. P. M.}}{33,000}$$

because the diameter is twice the radius and we may divide them.

If, now, we take pains to have the length of the arm measured from the engine shaft to the weights, just $3\frac{1}{2}$ feet long, the formula becomes simplified and we obtain, by dividing:

$$\text{B. H. P.} = \frac{\text{Weights x R. P. M.}}{1,500}.$$

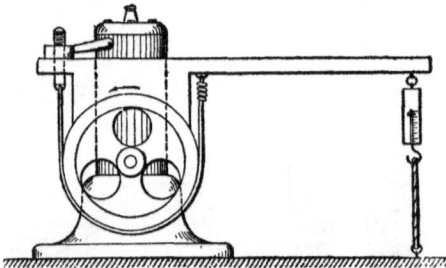

FIG. 284. One form of prony brake. The pull or weight on the arm, caused by the friction of the wheel against the belt is read on the spring balance shown attached between the lever arm and the floor at the right.

The belt which creates the friction is usually made of heavy canvas, held by springs at one end, while a turnbuckle is used at the other end, in order that the belt may be tightened at will and the force of friction increased. In long-continued tests, it is frequently found necessary to throw water over the belt and pulley to keep them cool. In place of the weights, a spring balance may be used, care being taken that the turning direction of the pulley is such as to pull against the balance. With small engines, up to 10- or 12-horsepower, a balance reading to 25 pounds is large enough.

Gasoline and other oil engines for farm use are usually rated at their tested brake horsepower, but the power of steam engines, unfortunately, is not so accurately stated. The commercial power rating of steam engines is ordinarily only one half or one third of what they actually will do under test. The custom in making calculations is to assume that a steam engine of any specified rating will give the same power as a gasoline engine of twice the rating.

An engine is usually rated by the amount of work it will do continuously. This is true of all stationary steam engines. Traction steam engines, however, are given their commercial rating by manufacturers, this being usually only about a third of their actual horsepower, as shown by tests. It is obviously not satisfactory to compare traction engines rated at any given horsepower with other engines, either stationary steam or any kind of kerosene or gasoline types.

Lubrication. The essential points in operating a steam engine are the bearings and lubrication in general. The bearings must be kept sufficiently tight though not tight enough to cause heating. As the bearing wears, it must be taken up. Each bearing must be kept fully lubricated with the proper kind of lubricant. Any old thing will not do for the lubrication of a steam engine. The oil used for the cylinder is very heavy, as it must stand the high heats and pressures of the cylinder and steam chest. For the places requiring oil outside of the cylinder, a light oil is satisfactory, but it should be of good quality.

Owing to the pressures on some bearings and the pressure within the cylinder and steam chest, some form of device for forcing oil into these has been found necessary. Usually, some form of steam-operated oil cup is used. In the most common type, outside of a direct steam-pressure-operated force feed, the steam passes in one side of the oil, condenses, and forms a partial vacuum which permits the pressure on the other side to force oil along the lubrication pipes.

Prices. It is difficult to say much about prices because of the wide variation between manufacturers and between types and sizes of engines. Boilers erected will run from $10 to $15 per horsepower of engine which they serve. The engine will cost from $15 to $25 per horsepower, depending on whether high- or low-speed, whether simple or compound, and whether small or large, well-built or just fairly built. The life of good equipment should be 15 or 20 years at least.

Starting up and stopping. In operating a farm steam engine, the greatest care should be taken in starting up. With cold cylinders into which steam is turned, there is bound to be considerable condensation of the steam on the cylinder walls. The operator in starting, therefore, should be careful first to open the cylinder cocks, allowing any water collected to flow out. They should be kept opened as long as the engine is being warmed up. The drainage cocks on the steam chest should also be opened. Then the throttle should be opened slightly, and the engine allowed to warm up slowly before throwing the throttle wide open. If possible, both ends of the cylinder may be warmed at once. When the engine reaches full speed, close all cocks, open up the lubricating devices, and the engine is ready for work.

In stopping, reverse the operation. Close the throttle and open up the various cocks. Close the lubricators. Wipe off each working part with a bunch of waste, watching carefully for a loose nut, a worn bearing, a scored wearing surface, or other trouble. Constant vigilance and care mean the prolongation of the life of the engine and thus pay bountifully in actual cash saved.

Steam-plant troubles. In the above discussion the various essentials of the steam-plant operation are described. It is lack of attention to these essentials which causes trouble to develop. With most difficulties, the result is serious as regards the plant, because of the temperature and force which are handled. The only troubles likely to occur are owing to (1) loose parts or (2) improper lubrication.

Looseness of parts may be detected by ear and by careful inspection when shutting down and before starting up. It must be remedied at once, or serious smash-ups may occur.

Lubrication must be properly carried out according to the manufacturer's instructions, or the parts will heat, expand, and bind. The result is a shutdown, a wait for cooling, and thorough lubrication after inspection. If, on inspection, injury to the parts is found to exist, new parts must be provided or the old part repaired. Lack of lubrication with first-class oils and greases always ultimately costs a great deal more than proper attention in the first place.

CHAPTER 18

Electricity on the Farm

By W. K. FREUDENBERGER, *who, though an engineer by profession, has never lost touch with the farm conditions and environment amid which he was born and brought up. Until entering college he lived on a farm in Moniteau County, Mo. Since graduating from the University of Missouri, he has practised electrical engineering. For about 10 years he was in charge of the light and power work with the U. S. Steel Corporation, the American Smelting and Refining Co., and the Colorado Fuel and Iron Co. Then, for six and a half years, he was chief engineer of the Public Service Commission of Nevada. At present he is with the similar commission of his home state. Meanwhile he has found time to complete a special course in agriculture at the University of Missouri and to manage, with marked success for 14 years, a large farm near Columbia.*—EDITOR.

E LECTRICITY is the principal source of light and power for most classes of people in this country—except farmers. However, they, also, should make more general use of electricity for these purposes, because (1) it produces a much better, safer, and more convenient light; and (2) as power for driving the numerous small machines about the house, barn, and dairy, it saves much time and hard work. Its use for cooking, ironing, and other household tasks and in the incubator and brooder, also, is very desirable.

The average farmer knows how to manage his farm and make of it a profitable business enterprise; but many farmers who are well-to-do and know how to make money do not know so well how to use it in the best way to obtain for themselves and families those conveniences, comforts, and niceties which alone make it worth while to accumulate wealth, and which make life pleasant and worth living.

Rural life can be made more attractive than city life. In the city, the struggle for existence is incessant, and home life falls far short of life in the country with its pleasant and healthful and beautiful surroundings. Less hard labor and more recreation are needed, however, to make the average farmer's life pleasant and attractive. Nothing can bring about these results so surely as labor-saving and time-saving electric appliances. Many city people recognize the advantages of living in the country, and have country estates or country homes. They surround themselves with all the modern conveniences, however; for, without these, country life would lose some of its attractions.

Advantages of the electric light. Electricity produces a white, soft light which is pleasant to the eye and which enables one to do any kind of work at night almost as well as in the daytime. It is much more convenient than oil lamps, which have to be moved about, because chandeliers, brackets, or cords may be suspended from the ceiling or fixed to the wall wherever needed. Plugs, cords, and drop lights provide table and desk lamps when desired. The oil lamp requires frequent cleaning and filling, and must be lighted with a match, whereas electric globes require nothing more than an occasional dusting and a washing, at the regular house-cleaning time, and the

FIG. 285. Electricity in the home makes for health, hospitality, and happiness. The farmer is just as entitled to these as the city man.

If the farmer would give his implements the care that a carpenter gives his tools, he would find their cost lessened, their life lengthened, and their usefulness greatly increased.

Equipment promptly and carefully repaired loses little if anything in efficiency and saves both time and money for its owner

EVERY TOOL AND MACHINE THAT THE FARMER USES DESERVES SHELTER WHEN NOT IN USE, INTELLIGENT HANDLING, AND FREQUENT OVERHAULING

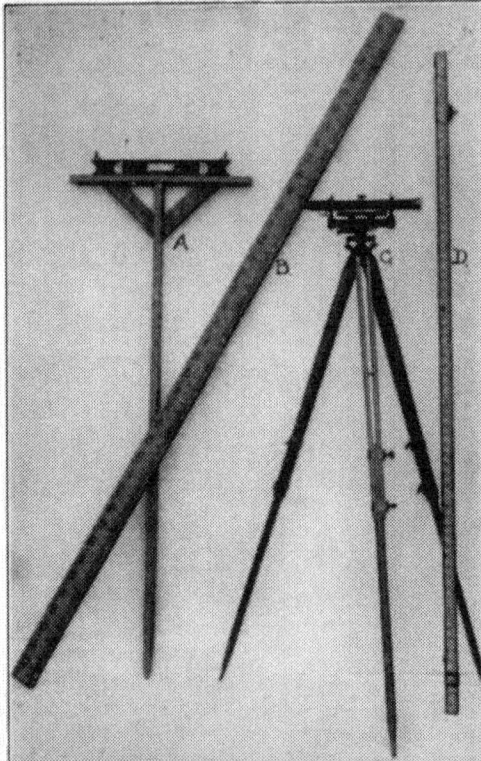

These instruments, a knowledge of some fundamental principles, and a little practice are all the farmer needs for most of his engineering tasks. A, mounted carpenter's level; B, sighting rod made of a flexible tape tacked on a 1 x 4-inch board; C, engineer's level; D, self-reading rod. (See Chapter 21.)

Concrete and other modern building materials enable the farmer to do most of his construction work without outside help

PROBABLY NO INDUSTRY, NOT PRIMARILY ENGINEERING OR MECHANICAL IN NATURE, MAKES AS MUCH USE OF THE PRINCIPLES OF THESE SCIENCES AS DOES FARMING

light is turned on and off by simply pushing a button or turning a switch. Oil lamps give off an offensive odor, burn up the oxygen in the room, are frequently smoky and sooty, and smudge the fingers, while the electric light is clean and pleasant to use, gives off no smoke and very little heat, and does not throw off any poisonous gases.

There is still another and a most important advantage possessed by the electric light: it is perfectly safe to use and produces no flame, whereas there is always danger of explosions and fires with oil lamps and lanterns from the use of matches in lighting them, or from overturning by accident, or by stock, if at the barn.

On every general farm, there is a large amount of work to be done about the farmstead that requires much time and hard labor to accomplish, such as pumping the water for house use and for stock, grinding corn and other grain, chopping hay, turning the grindstone, running the cream separator, churning, washing and ironing, cooking meals and washing dishes, sweeping and housecleaning, and many other tasks. On account of a shortage of farm hands, increasing wages, and shorter hours of labor on the farm, it is becoming more and more necessary for most farmers to do all of their own work; and, as practically all of their time is needed in the fields during the cropping season, very little can be spared for work around the house, barn, and dairy. It is very necessary, therefore, that such work be done quickly and with little labor. These results can be accomplished with electric power.

How electricity may help farm women. It has been the custom of many farm women in the past to employ house servants to help with the really enormous amount of work that has to be done in and around the house, but it is now practically impossible to secure house servants on farms. Under these conditions, it becomes almost imperative that farm women use electric power to do this hard work and thus to save much time and to conserve their strength, in order that they may be able to do all the work without hired help.

How time and labor may be saved by using electricity in the home, may be demonstrated in many ways. When, for instance, an electrically-driven pump is used to supply water for house use and for stock, neither the labor nor the time of the farmer is required for this purpose to any great extent. The time used need not be more than a half hour per week. Stock can then be provided with good, clean well water instead of hot, contaminated water from muddy ponds. Water can be easily stored under sufficient pressure to provide fire protection for the farm buildings, and for the convenient use of the family in the kitchen, laundry, bath, and closet from faucets in any part of the house.

An electrically driven cream separator and churn will not only save time and labor, but no cream will be left in the milk, as is the case when the separator is operated by hand, and more butter, of better quality, will be obtained from the cream.

When a grindstone is driven by electric power, the time of one man is saved, and the work of grinding ax, sickle, or other tool is done much quicker and better.

A feed grinder, for grinding corn or other grain, may be run by electric power and the time required for hauling to town for grinding

be saved; also, the cost of grinding may be much reduced.

Farm women should be provided with

FIG. 286. The washing machine, electrically driven, conserves the time, strength, and good nature of the housewife.

electric washing machines, wringers, irons, vacuum cleaners, kitchen ranges, dish washers, and other electric equipment, and be free from the drudgery of their work.

The electric washing machine with wringer attached takes away practically all of the hard work of washing clothes, makes the time required for a family wash much shorter, reduces the wear and tear on the clothes, and leaves a woman free to attend to other duties much of the time.

Electric irons are used almost as universally by townspeople as are electric lights. When once tried, they are soon found to be indispensable. In the old process of ironing, one must do the work near a hot stove, and be continually walking back and forth between stove and ironing table, changing irons; whereas, with an electric iron, a cool place may be selected in which to work, and the entire washing may be ironed without moving from one position or changing irons a single time.

By using an electric vacuum cleaner, the daily sweeping and dusting can be much better accomplished, and in one operation, without hard labor and without having to breathe into the lungs germ-laden dust. Much of this work can be accomplished at the same time that the churning is being done or the clothes are being washed when these operations are performed with electric machines. So, also, the electric dishwasher can be put to work 3 times a day, helping greatly to reduce the load on the overburdened housewife.

In addition to the foregoing illustrations of the desirability and necessity of using electricity on farms, the following important reason for its use may be advanced: Many farm boys and girls, who are now leaving good farm homes, could, it is safe to say, be kept from doing so if conditions were reasonably improved at their homes, as they might be by the use of modern electrical equipment.

Sources of Electrical Supply

Farmers may choose between 2 principal, sources of electrical supply for their farms. First, they may obtain it from the electric-service companies operating in practically every city and town in our country. Second, they may purchase small electric plants and obtain their electricity from this source. In general, it is advisable for farmers to secure their electricity from the first-mentioned source, if possible, if fairly reasonable rates can be obtained and the cost of the electric lines for carrying the electricity from the company's plant to the farms is not too great. In case electricity cannot be obtained from the electric-service companies, farmers must purchase their own plants. In that event, each may have an individual plant of his own, or a group of farmers may coöperate and purchase a plant in partnership large enough to supply them all.

It is advisable for farmers to purchase their electricity from the electric-service companies, if it can be obtained at a price no greater than it would cost to produce it on the individual farms with private plants. It must be evident to any one that it would cost much less to generate electricity in very large quantities at one central plant than it would to produce the same quantity in hundreds of small plants, because the fuel and other supplies needed could be purchased at lower prices in large quantities, and the same quantity of fuel would produce a much larger quantity of electricity in the large plant because the large plant is much more efficient.

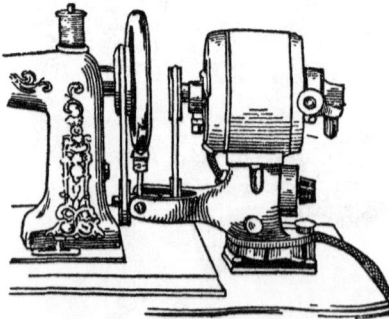

Fig. 287. The sewing machine is only one of the various household machines to which a small motor can be attached.

Furthermore, experience has shown that the capacity or size of one large plant, used to serve a large number of customers, need be only about one third the aggregate capacity required if each customer has his own plant; because they would never all be using large quantities of the electricity at the same time. Therefore the large plant would cost very much less than the group of small plants. In fact the cost of the electric-service plant together with all of the transmission lines which are required to carry the electricity from the towns to the farms, should not cost more than the group of small plants, if the same high grade and quality of equipment is used in both cases.

As a general rule, it is possible for the electric-service companies to supply electricity at a lower cost than farmers are able to produce it on their own farms from private plants. This is the principal reason why it is advisable to purchase the supply. Another reason is that these companies have made a special study of the electrical business and thoroughly understand every detail of it. They employ experienced electricians who know how to handle the machinery, the electricity, and every part of the plant to the very best advantage. On the other hand, farmers could not be expected to handle their private plants so advantageously, they would

FIG. 288. One way to grind an ax. Compare Fig. 288a.

have breakdowns and other annoyances, and their plants could not operate with such regularity and such freedom from interruptions of service as the large plants.

There are other advantages in purchasing the electrical supply. Farmers would not have to make large investments, which would be necessary if they purchased their own plants. They could get a large amount or small amount of light or power from the electric company as needed, whereas, if they generated their own supply, the plants would have to be large enough to produce the largest quantity that would ever be needed at any time.

Electrical Terms and Expressions

In dealing with the subject of electricity it is necessary to use a number of terms and expressions the meanings of which are not generally understood. A few of the more important ones will, therefore, be defined here.

Electric current; ampere; ammeter. The electricity flowing through a wire, lamp, motor or other apparatus is called an *electric current*. The *ampere* is the unit of measurement of an electric current. An *ammeter* is an instrument used for measuring the number of amperes of current flowing at any time.

Voltage; volt; voltmeter. The *voltage* of an electrical supply is the electrical pressure. The *volt* is the unit of measurement of voltage or pressure. A *voltmeter* is an instrument which is used to measure the voltage.

Watt; kilowatt; horsepower. Electric lamps and small apparatus are rated in watts, the *watt* being the unit of measurement of capacity or size. We have 15-watt lamps, 20-watt lamps, and other small sizes, also 100-watt lamps and large sizes. Large generators and transformers are rated in kilowatts. The kilowatt is usually written K. W. One *kilowatt* is equal to 1,000 watts. One *horsepower* is equal to 746 watts, or nearly three-fourths of a kilowatt. One kilowatt is slightly more than 1⅓ horsepower.

FIG. 288a. The modern, electrical way saves one man's time, the other one's back and keeps the edges keener because more easily and, therefore, oftener done.

How Electricity is Measured and Sold

Electricity is measured for sale in kilowatt-hours (usually written KWH) by a device called a watt-hour-meter or recording wattmeter. One kilowatt-hour is 1 kilowatt working 1 hour. A 50-watt lamp uses 50 watt hours in one hour. In 20 hours, it would use 20 x 50 or 1,000 watt hours, which is 1 kilowatt hour. If the price is 10 cents per kilowatt hour, then the cost of burning such a lamp 20 hours would be 10 cents. The wattmeter measures and records with great accuracy the amount of electricity used, whether for lighting, power, or heating.

Many farmers using electricity. In all sections of this country large numbers of farmers are already using electric light and power, obtaining electricity from private plants or from electric-service companies. A San Francisco, California, company supplies more than 8,000 farmer customers with electricity; one at Fresno, California, more than 1,600 farmers; and there are many thousands of farms supplied by other electric companies in California, Washington, and Oregon. The greater portion of this electric demand on the Pacific Coast is for operating electric pumps for irrigation. In the central sections, especially in those states included in the Corn Belt, Wisconsin, Indiana, Illinois, Iowa, Missouri, Ohio, Pennsylvania, Nebraska, Kansas, Minnesota, and other states many thousands of farms are supplied from electric companies operating in nearby towns. Large numbers of farmers and ranchers in the Rocky Mountain states, the southern states, and the Atlantic Coast states are also using electric light and power.

Rates and charges for electricity. The terms and conditions under which transmission lines are built to farming sections by electric service companies, and the rates charged for the service vary widely in different places. The following examples of typical arrangements are cited as illustrations, all these being taken from the "Corn Belt":

Case 1. A company has several lines extending into the country wherever a small group of farmers who want electric service is favorably located. Ten or more farmers can usually be served from one 10-kilowat transformer and a pressure of 2,200 volts. Each farmer owns a transformer of from 1 to 2 kilowatt capacity which reduces the pressure to 110 volts for use in lamps or motors. The farmers furnish the poles and set them in the ground, and pay 10 cents per kilowatt-hour, with a minimum charge of $2 per month. This is more than city patrons served by the same company have to pay, as is usually the case, because of the extra investment in the long transmission lines and the loss of current in them and in the transformers.

Case 2. A company serves a number of farmers within a radius of 4 miles from town. The farmers paid the entire cost of the electric line, which amounted to $425 for each farm. Their average electric bill is from $4 to $6.50 a month, for which they do their washing, ironing, water pumping, cream separating, and feed grinding, and light all their houses, barns, and outbuildings.

Case 3. A company serves farmers from

FIG. 289. The kind of power plant that is run by stream or impounded waters and that supplies electricity to farmers of certain sections at a remarkably low figure

a 2,200-volt line, furnishing the line and equipment and making a flat charge of $4 per month per farmer.

Case 4. A company serves farmers at 12 cents per kilowatt-hour with a minimum charge of $5 per month. It furnished the transmission line, but each farmer erected or paid for his own service line.

Case 5. A company serves farmers from a line which runs to a nearby town. The line is owned by the town, the farmers paying a rental of 50 cents per month each for the use of it. The company's charge for service is 12½ cents per kilowatt-hour and the average bill per farmer is about $20 a year for service plus the $6 for line rent.

Case 6. A company has 275 farmer patrons. In order to obtain service, a farmer is required to pay the company $75 for extending its lines along the public highway to his farm. The farmer builds his own service lines from the company's line on the public highway to his home. The $75 payment is later returned to the farmers in small monthly amounts, in the form of discounts on service. The rates paid are 10 cents per kilowatt-hour for the first 25 kilowatt-hours per month; 8 cents per kilowatt-hour for the next 25 kilowatt hours per month; 6 cents per kilowatt-hour for all in excess of 50 kilowatt-hours per month. A discount of 5 per cent. is allowed for prompt payment. The minimum monthly

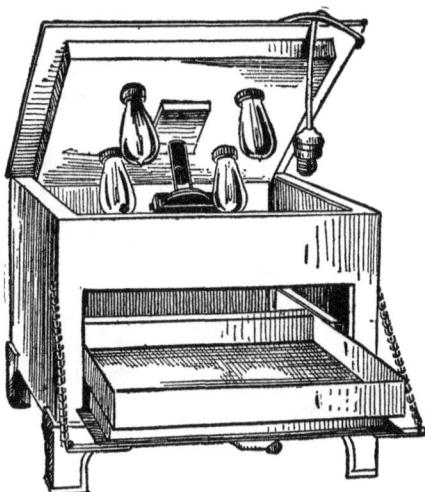

FIG. 291. The electrically heated incubator is no longer a doubtful experiment, but an established fact

charge is $1 and the average bill paid per month is from $1.50 to $1.75.

Rates on the Pacific Coast and in other sections where water power is obtainable are lower than in sections where fuel is required. The Fresno, California, company referred to above makes the following rates for farm service: 4 cents per kilowatt-hour for the first 200 kilowatt-hours per month; 2 cents per kilowatt-hour for all in excess of 200 kilowatt-hours per month. The minimum charge per month is $2.50.

There should be a more uniform system adopted for serving farmer patrons and of charging them for current and service. The service company should own and operate all lines, in the interest of uniformity, economy, and good service. A proper minimum charge should be established, and this should seldom be lower than $2 or $2.50 a month. Under some circumstances, it might be necessary to make it somewhat higher.

For the reasons given above, it is necessary for an electric company to make a higher charge for farm service than for city service either in the rate per kilowatt-hour or in the minimum charge per month, or in both.

Coöperative and Private Electric Plants

FIG. 290. An electric range, clean, efficient, cool, easily controlled and, where current is reasonable, decidedly economical.

There are, of course, many communities where it is impossible for farmers to obtain electricity from a city plant or a commercial electric light or power company. In such cases two courses are open to them: They may individually install private, individual plants, or they may form a coöperative company and install and operate a central plant to supply them all. The latter plan often provides the better service, and generally at a lower cost per farmer, thus

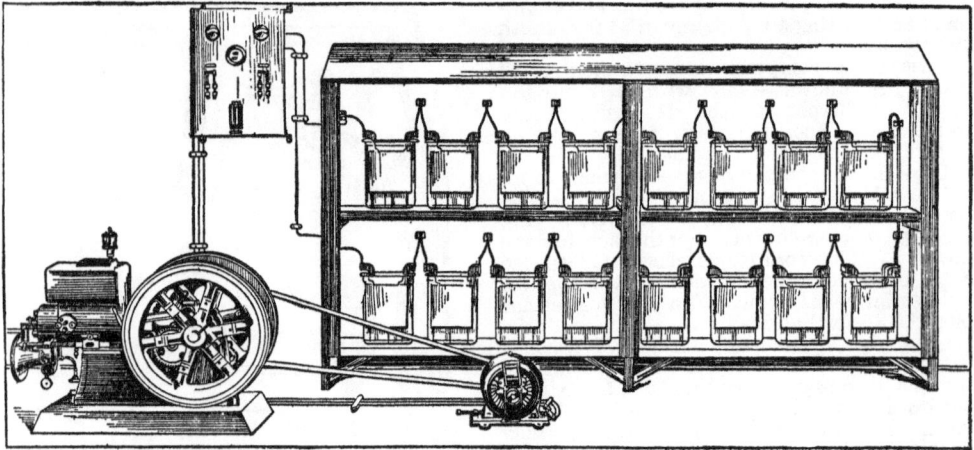

Fig. 292.   A complete lighting plant for the farm house and barns.   It consists of the source of power (gas engine) at left, the electricity maker or generator in foreground, the storage batteries at right, and the switchboard carrying the current gauges, etc., in rear at left.

bringing electricity within reach of many who might be able to invest a small amount in the company and pay a regular charge per year or per month when they could not afford to buy even a small individual plant outright.

A farmers' coöperative electric system could be operated somewhat on the plan of a farmers' telephone line. Under such a system, small, medium, and large farms should be able to get reasonable service at an expense not exceeding $3.50, $5, and $7.50 per month respectively, including the depreciation of the plant. The investment would be about $400, $600, and $800 respectively.

When coöperative companies cannot be formed, farmers who can afford to do so will find it highly profitable to install private plants.   There are a large number of manufacturing companies which make a specialty of furnishing complete plants suitable for such use.   The better-grade plants are fairly easy to operate, so that by following the instruction book furnished with each plant, a farmer will soon learn how to obtain good results, especially if he has had experience in operating a gasoline engine or an automobile.

These small plants usually consist of an electric generator, a gasoline engine to operate the generator, a switchboard with switches and electric meters to measure and control the electricity, and a storage battery.   A storage battery is used in connection with farm plants for the following reason: If no storage battery were used, it would be necessary to run the engine and generator all day in order to have power or light at any time that it is needed.   This would be objectionable on account of the excessive cost for gasoline, and the machinery would wear out rapidly. By using a storage battery, the engine and generator may be run for a few hours every 3 or 4 days to generate electricity and store it up in the storage battery, and then the battery is always ready to provide either electric light, power, or heat at the turn of a switch, day or night, while the engine and generator are idle.

It is very important that the manufacturers' instructions be carefully followed. The bearings of the engine and generator must be oiled at proper intervals, and the storage battery will require a little pure water occasionally as well as other attention.   Solid foundations should be provided for engine and generator, to prevent vibration.   The plant must be enclosed in a tight building and carefully shielded from rain.   The machinery should be kept clean.

In many cases, it will be found cheaper to purchase an engine large enough to drive the generator and some of the larger farm machines besides, such as pump, churn, washing machine, and feed grinder. The engine could be belted to a jackshaft, and the individual machines driven from this jackshaft by the use of proper pulleys and clutches. When this plan is adopted, the electric plant need not be so large nor so expensive. It would be advisable to erect a building large enough to accommodate the electric equipment and the other machines, with plenty of room so that they can all be operated conveniently.

Farmers should buy their electric plants from reliable dealers and require proper guarantees. A local dealer should be required to furnish an experienced electrician to install the machinery, put the plant in smooth running order, and give the farmer careful instructions on how to operate it and care for it.

In any case it is well to have the advice and assistance of such an expert when installing an electric system. Time is usually saved by such a procedure, and sometimes the farmer is prevented from making mistakes which might result in damage which it would cost ten times as much to repair as the services of the electrician would cost in the first place.

A good idea may be obtained of the sizes and prices of plants for farm service usually supplied by the dealers, and the amount and kind of service such plants will provide, from the data given in Table I.

TABLE I.—FARM ELECTRIC PLANTS, 30-VOLT SYSTEM
(AUTOMATICALLY REGULATED)

SIZE OF DYNAMO IN KILOWATTS	STORAGE BATTERY AMPERE HOURS	LIST PRICE WITH SWITCH-BOARD AND METERS	SIZE OF ENGINE AND LIST PRICE	LIST PRICE OF COMPLETE PLANT
.5	50	$260	1½ H.P.—$ 78	$338
.5	90	310	1½ " — 78	388
.7	90	360	1½ " — 78	438
.7	180	455	1½ " — 78	533
1.0	90	375	3 " — 132	507
1.0	180	470	3 " — 132	602
1.5	90	550	Direct connected sets	550
1.5	180	645		645

The .5-kilowatt plants are recommended by the manufacturer as being of suitable capacity for lighting a house of 6 or 8 rooms, an average-sized barn and outbuildings, and for operating electric fan, vacuum sweeper, cream separator, churn, washing machine, and other apparatus with motors not larger than one-fourth horsepower.

The .8-kilowatt plant will light a house of 8 to 10 rooms, good-sized barn and outbuildings, and operate almost all the usual electrical devices and electric machines with motors not larger than ½ horsepower.

The 1-kilowatt plant will light a home of 10 to 12 rooms, large barns and outbuildings, operate almost all the usual electrical devices and electric machines with motors not larger than 1 horsepower.

The 1.5-kilowatt plant will light a large house, very large barn and outbuildings, operate all the usual electrical devices and

electric machines with motors not larger than 1½ horsepower.

Standard voltage. Many electric manufacturing concerns supply 30-volt plants; some of them also supply 110-volt plants. This raises the question as to which voltage is best to use.

Electric-service companies supply lighting service at 110-volts in nearly all cases, and that has become the generally accepted standard. For other purposes they usually furnish either 110-volt or 220-volt service.

Those who have private plants installed may use either the 30-or the 110-volt systems. The 110-volt plants are preferred by many for the following reasons: When the buildings to be supplied are at considerable distances from the electric plant and much current is used, the cost of wire for carrying the current is excessive when 30-volt plants are employed, which is not the case when 110-

FIG. 293. Three desirable conveniences that electricity makes possible in the farm home: toaster for breakfast table, vacuum cleaner and fan.

volt plants are used. Motors and heating appliances have been standardized for 110 volts and are more readily obtainable for that voltage than for 30 volts; also they are somewhat cheaper.

The 30-volt plants are satisfactory when the electric plant is located very near the buildings to be served and large motors are not required.

In Table II, an estimate is given of the size of engine, generator, and storage battery to be used for various quantities of electricity required per month, the cost of good-quality plants of the various sizes, and what it will cost to operate them. From a study of this table, it will be seen that these plants are not so very expensive in first cost, an average-sized plant costing about $500; but if we assume that such plants will not give more than 10 years' service, and charge a proper proportion of the first cost in with the operating expense each month, the total expense

will run from about $5 per month for a very small plant to about $12 for a large plant, or perhaps $8 per month for a plant of average size. This does not take into consideration interest on the plant. Almost any electric-service company within reasonable distance should be able and willing to furnish service at a much lower cost than this. This, however, is no argument against the wisdom of purchasing private plants. As before stated, if it is impossible for farmers to obtain service from electric companies, they will find it profitable and desirable to install private plants.

The engines are somewhat larger than necessary for the electric plants and they might be used to drive other machines, in addition, if desired.

Portable storage batteries. For those who cannot afford to install private plants, portable storage batteries will service very well to light the home. Five cells of the modern Edison battery, weighing about 100 pounds, will light a small farmhouse very well for one week on a single charge. Such a battery has a capacity of 1,395 watt-hours. It would have to be taken to a garage or to an electric plant about once a week to be recharged. Such a battery would cost about $100. It would require nearly 2 kilowatt-hours for charging each week, which should not cost more than 50 cents. Lead cell batteries of similar capacity would be heavier, but less expensive.

FIG. 294. A safe, handy electric lantern for use in unwired barns, cellars, and outdoors

Electric lanterns and flashlights. Electric lanterns are now being supplied by dealers similar in size and shape to the well-known railroad lantern. They are perfectly safe to use and cannot be put out by wind or rain.

TABLE II.—SIZE AND COST OF FARM PLANT AND COST OF OPERATION FOR VARIOUS QUANTITIES OF ELECTRICITY REQUIRED PER MONTH
(30-VOLT SYSTEM)

Kilowatt-hours per month.	10	20	30	50
Size of engine (horsepower)	1.5	1. 5	3	3
Size of generator (kilowatts)	$\frac{1}{2}$	$\frac{3}{4}$	1	$1\frac{1}{2}$
Size of storage battery (ampere hours)	50	90	90	180
Cost of plant installed	$365	$465	$535	$675
Operating expense per month	$1.95	$3.00	$4.40	$ 7.00
Depreciation per month	$3.05	$3.87	$3.62	$ 5.62
Total expense per month	$5.00	$6.87	$8.02	$12.65
Cost per kilowatt-hour	$ .50	$ .34	$ .27	$ .25

They never get sooty, hence seldom need cleaning. They are lighted by simply moving a contact lever, thus doing away with the trouble and danger of lighting with matches. Two sizes are made, one with a 1½-candle-power lamp and one with a 2-candle-power lamp. A battery for the small size costs about 45 cents and will operate continuously for 10 or 15 hours; one for the large size, operating for the same period costs 75 cents. If these lanterns are used for operating an average of 15 or 20 minutes a day, the cost of batteries would be about 35 cents per month for the small size, and 55 cents for the larger.

Large flashlights of the usual tubular shape are also supplied with 2-candle-power lamps. These have strong reflectors and produce a bright light. Batteries for these flashlights cost about the same per month for similar use as do those for the large-sized lanterns. The dimensions of the flashlight are: diameter 1¼ inches, length 13 inches, and the cost is $4. Various smaller sizes are also made.

The Various Uses of Electricity

Electric lighting. Tungsten lamps are now used almost exclusively for lighting all kinds of buildings. Carbon-filament lamps, so popular for many years, have gone out of use because they require about 3 times as much electricity to produce a certain amount of light as do the tungsten. After-the-war list prices for the usual sizes of tungsten lamps are given below in Table III.

TABLE III.—TUNGSTEN LAMPS

	SIZE IN WATTS.	SPHERICAL CANDLE POWER	LIST PRICE CENTS
110-VOLT LAMPS	10	6.	35
	15	10.	35
	25	17.7	35
	40	30.	35
	50	38.	35
	60	45.8	40
	100	80.	85
30-VOLT LAMPS	5	3.4	35
	10	7.3	35
	20	15.5	35
	40	32.2	35
	50	58.8	70
	75	93.7	80
	100	133.3	$1.20

Size of lamps required. In the early days of electric lighting, 16-candle-power lamps were used quite generally for lighting an average-sized room. At the present time,

40-watt lamps, giving 38 candle power or more are the most popular. For a room 14 by 16 feet, 50-watt, 60-watt, or 100-watt lamps are more satisfactory. Two or more lamps of smaller size may be used instead of one large one, and better results obtained in many cases. Small lights may be used in closets, hallways or porches, and in outhouses and barns. Table IV gives the size of lamps generally used for different rooms under various conditions.

TABLE IV.—SIZES OF LAMPS USED ON VARIOUS FARMSTEADS

LOCATION	FARMSTEAD		
	SMALL	MEDIUM	LARGE
Dining room	25-watt	40-watt	60-watt
Living room	50-watt	60-watt	100-watt
Kitchen . .	25-watt	40-watt	60-watt
Front hall .	....	25-watt	50-watt
Front porch.	25-watt	25-watt	40-watt
Rear hall .	....	25-watt	40-watt
Bedrooms .	25-watt	40-watt	60-watt

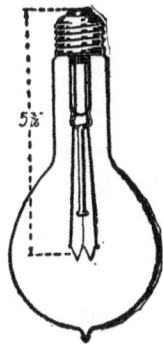

The estimated amount of electricity required per month for lighting a farmstead, including barns and outbuildings, may be given as 5 kilowatt hours for a small farm, 10 for a medium sized one and 15 for a large one. In estimating the size of plant required it must be remembered that all the system will not be in full use at any one time. While the chores are being done, but little light will be used in the house; in the evening when more rooms are lit, the barns will be dark, and so on.

Installation should be substantially done. An experienced electrician should be employed to do the wiring of the house and other buildings and to install the electric fixtures.

FIG. 295. Improved types of incandescent lamps for farm lighting: 100-watt (*left*), 50-watt (*right*) with sizes indicated.

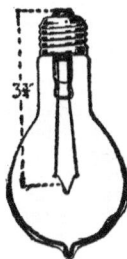

For large, well-built farmhouses the electric wiring should be done in as neat and substantial a manner as is done in similar houses in cities. Switches should be placed in each room, for greater convenience in turning lights on and off. For the average farmhouse, however, switches for individual rooms may be omitted, and pull-sockets used instead. Pull-sockets are provided with short chains, which hang just beneath the lamps, within easy reach. A slight pull on the chain lights the lamp, and another pull turns it

FIG. 296.
10- and 20-watt lamps are made in this style.

off. By using pull-sockets instead of switches, the cost of wiring may be very much reduced.

FIG. 296a. The 40-watt lamp is the size most generally used.

Generally, the lamps are suspended from the ceiling on lamp cords. They should be suspended as low as possible for best results, but not low enough to be in the way. Six feet from the floor is about the right distance. They should not always be placed over the centre of the room, but where the light is most needed, in each case.

The cost of wiring a farmhouse for lighting might be as low as $2 for each room for small one-story buildings, and 4 or 5 times as much for large, expensively-furnished houses, including cost of fixtures.

Shades and reflectors. There are many varieties of somewhat gaudy and highly colored lamp shades on the market. Although such varieties are frequently low in price, they are neither the most ornamental nor the most efficient. In fact, they are sometimes detrimental, rather than beneficial. The best shades and reflectors for electric lamps are those made from clear transparent glass, such as the Holophane reflectors.

Pumping the water supply. When it is necessary to pump from a deep well, and considerable power is required, it is sometimes advisable to locate the electric plant near the well, so that one engine may be used both for the plant and for pumping water. A separate motor is preferable, however, in most cases.

One very popular water supply system is that in which the water is pumped into an elevated tank where it is under sufficient pressure to provide flowing water at faucets in any part of the house, provide water under pressure for fire protection, and supply water to tanks and troughs for stock. The water tanks should be elevated 50 or 60 feet, or somewhat higher than the buildings to be served.

The same results are obtained with another much-used system, in which a large air-tight water tank is placed in a building or underground, and the water supply pumped into it under a pressure of from 25 to 50 pounds. The air originally in the tank is compressed and acts as a pressure regulator, forcing the water out, when faucets are opened, at practically the same pressure at which it was forced in by the pump.

In either case the electric motor which drives the pump may be made to operate automatically. That is, a float is rigged to stop the motor when the tank is full; then, when the water in the tank falls to a certain point, the float starts the motor again and the tank is soon refilled. In case the pressure tank is used, a pressure apparatus performs the same duty. When these systems are used, very little time is required to look after the water supply.

A nonautomatic system is preferred by many. That is, instead of the pump motor being started by a float or pressure apparatus, it is started and stopped by hand. The latter system is simpler, somewhat cheaper, and less liable to get out of order. It is an easy matter to start the pump before sitting down to a meal, and to stop it when the meal is finished and very little time is required.

The cost of such systems is from $150 to $200 and upward, and the power required would usually be from 2 to 4 kilowatt-hours per month.

Washing and ironing. The proper solution of the washing and ironing problem for farm women, as for city women, is to send clothes to the steam laundry to be washed and ironed where it is done by machinery. If this method is considered too expensive, farmers should establish coöperative laundries. Failing these methods, the only remaining course is to use electric washer, wringer, and iron. Electric washers and wringers are usually run by 1 motor of about $\frac{1}{4}$ horsepower. Manufacturers of these machines claim that there is not half as much wear on clothes washed by machine as there is when rubbed on the washing board. The big incentive to buy an electric washer and a wringer, however, is the fact that it requires only 1 or 2 hours to do a family wash in this way, and all of the hard work is done away with. The electricity required is about 2 kilowatt-hours per month, and the cost of washer, wringer, and motor is from $75 to $150.

Ironing with an electric iron makes this part of the work much less disagreeable. One need not work in a stove-heated room, but may select as cool a place as can be found,

FIG. 297. The electric flat iron is especially appreciated in hot weather.

and connect the iron to the nearest lamp socket. The ironing can be done much quicker, also, because it is not necessary to stop to change irons; the iron keeps at the proper temperature so that the ironing can be done to the best advantage all the time; and there is no walking back and forth from ironing table to stove. The cost of an electric iron is from $3.50 to $7 and the electricity required to operate it is from 5 to 10 kilowatt-hours per month.

Cream separator and churn. A cream separator and a churn can be arranged for operation from a single motor, which may be one-sixth or one-fourth horsepower in size for the average farm. Dairy farms should have larger machines, also electric milking machines. The power required for churn and separator for an average farm would be 2 or 3 kilowatt-hours per month. By using electric power all the cream can be taken out of the milk, when the speed of the separator is properly adjusted. When the separator is turned by hand, it is impossible to regulate the speed so that all the cream will be taken out. The electric motor takes the hard work out of churning and of separating the cream, also out of milking, when electric milkers are used. Farmers having 10 or more cows to milk will find that it is economical to install electric milking machines. Those of small size will use from 3 to 10 kilowatt-hours of electricity per month.

FIG. 298. Why turn the separator and churn by hand when a portable motor will do it?

Refrigeration. Farmers recognize the necessity of having a cool place in which to keep dairy products, fruits, and vegetables during the summer months. Many farmers build icehouses, and put up ice during the winter for use in summer. Others have no suitable bodies of water within a reasonable distance from which to obtain ice, or cannot afford to build icehouses. In any case, a small farm refrigerating plant will provide the low temperature required for preserving farm products as cheaply as can be done by storing winter ice for the purpose. For example, a test was made with a small outfit in 1915, in central Illinois during the month of July, in which a refrigerator box 16 x 36 x 50 inches was used. The power required to operate the refrigerator was less than ¼ horsepower. The motor was run an average of 5¼ hours a day for the 31 days. The temperature of the refrigerator box was kept between 40 and 45 degrees F. In the 31 days, only 36.2 kilowatt-hours were used, which, at 10 cents per kilowatt-hour, would cost $3.62. Of course, the cost of ammonia which is necessary for the refrigerating machine, and of oil for lubrication, would add materially to the total cost. But it would only be necessary to operate the machine from 4 to 5 months of a year, depending on the climate.

Feed grinder, fanning mill, corn sheller, wood saw. A feed grinder requires considerable power, and, therefore, some farmers prefer to locate it in the building with the electric plant, so that it may be run from the same engine. If one can afford the additional investment, somewhat better results may be obtained by using an electric motor to run the feed grinder; but the other method would be quite satisfactory. A size requiring a 2-horsepower motor to drive it should be large enough for the individual farmer. It requires about 40 kilowatt-hours to grind 100 bushels of corn, but only about 8 kilowatt-hours for cracking the same amount.

The fanning mill, or seed cleaner, could be driven from the same motor that runs the feed grinder, or a portable motor should be rigged up for various uses, including the work of running a wood saw and corn sheller.

When electric power can be purchased from an electric company, such work as silo filling and threshing grain can be profitably done with electric motors; but it would not be economical to install a private plant large enough for such work, except on very large farms.

Many farmers are now buying farm trac-

FIG. 299. Electric soldering iron.

FIG. 300. An important advantage of the electric washer is that it can be used in the kitchen but kept elsewhere out of the way when not needed.

tors. Such work as feed grinding, corn shelling, silo filling, threshing, sawing wood, and other heavy, intermittent work can be done with the tractor power.

Electric range, hot plate, and toaster. To do all of the family cooking on an electric range during the summer season would require from 100 to 200 kilowatt-hours per month. Only fairly well-to-do farmers could afford to purchase a range, which would cost from $45 to $100 and pay $5 to $10 a month for current. But farm women who have electric ranges are exceedingly fortunate and well favored. If current can be gotten from water power on the farm at very low cost they can prove economical as well as convenient.

Those who must be more economical can purchase two disc hot plates and prepare two meals a day on them, using the wood range for the heaviest meal and for baking. The cost could be reduced about one half in this way.

There are numerous other uses to which electricity may be put on a farm, such as house cleaning and sweeping with a vacuum sweeper, running the sewing machine with a small motor, dish washing, running ice-cream freezer, electric fan for ventilating and cooling a room, running the grindstone, heating pad, and heating the incubator and brooder.

In Table V are given a list of machines and apparatus that can be operated electrically, the size of motor generally used, and the amount of electricity required per month.

TABLE V.—SIZE OF MOTOR REQUIRED AND ELECTRICITY REQUIRED PER MONTH FOR VARIOUS MACHINES

MACHINE	SIZE OF MOTOR	USE PER MONTH	KILOWATT-HRS. PER MONTH
Churn	$\frac{1}{6}$ to $\frac{1}{4}$ Hp	6 to 8 hrs.	1.0 to 1.5
Cream separator	$\frac{1}{8}$ to $\frac{1}{4}$ Hp	10 to 15 hrs.	1.0 to 2.0
Washer and wringer	$\frac{1}{6}$ to $\frac{1}{4}$ Hp	5 to 10 hrs.	1.0 to 2.0
Milking machine	$\frac{1}{2}$ to 2 Hp	30 to 60 hrs.	2.0 to 10.0
Water supply pump	$\frac{1}{4}$ to 2 Hp	10 to 20 hrs.	2.0 to 8.0
Vacuum cleaner	$\frac{1}{6}$ to $\frac{1}{4}$ Hp	30 to 40 hrs.	3.0 to 8.0
Refrigerator	$\frac{1}{4}$ to $\frac{1}{2}$ Hp	90 to 180 hrs.	20.0 to 50.0
Range	4,000 watts		75.0 to 150.0
Iron	500 watts	8 to 15 hrs.	3.0 to 7.5
Sewing machine	$\frac{1}{30}$ Hp		
Buffer and grinder	$\frac{1}{30}$ Hp		
Dish washer	$\frac{1}{8}$ to $\frac{1}{4}$ Hp		
Ice-cream freezer	$\frac{1}{6}$ Hp		
Electric fan	45 watts		
Toaster	500 watts		
Feed grinder	2 to 3 Hp		
Grindstone	$\frac{1}{6}$ to $\frac{1}{4}$ Hp		

CHAPTER 19

Internal-Combustion Engines

By PROFESSOR R. P. CLARKSON (*see Chapter 14*). *With the greatly increased use of the automobile, the principle and management of the stationary gas and oil engines (which are considerably simpler) have lost most of their mystery. To-day there are probably more farmers who can talk intelligently about ignition, compression, magnetos, etc., than there are men in other lines of work not primarily connected with engines who can do the same. A full course in the subject requires considerable time, study, and experience. All this chapter attempts is to tell the novice the essential things that he should know and to remind the experienced man of some of the important details that he may already know but may have forgotten.*—EDITOR.

GASOLINE, kerosene, and crude-oil engines of every kind belong to the internal-combustion class of motors. They are primarily heat engines, just as are the steam engine and the hot-air engine; and the principle of operation does not materially differ in the various types of heat engines. In every case, there is some working substance which, through the application of heat in some manner, causes expansion and contraction in the working space of the engine and, usually, causes a piston to slide back and forth. In this way, the heat energy of the working substance is transmuted into mechanical energy and made available for doing useful work of some kind.

In the steam and hot-air engines, the heat is supplied by the combustion of fuel outside the engine, this combustion acting to heat up the working substance. In the oil and gasoline engine, there is combustion of fuel, but it takes place right in the cylinder of the engine. Engines of the former class—steam and hot-air—are thus called "external-combustion engines," because the fuel action is outside, while those of the latter class—oil and gasoline—are termed "internal-combustion engines," because the fuel action is entirely inside the engine.

Efficiency of engines. The theoretical study of engines has shown very clearly that only a certain amount of the heat energy which is supplied to an engine can be converted into mechanical energy. The higher the temperatures that can be reached with the working substance, the greater the efficiency which it is possible to obtain. Internal-combustion engines are so designed that the explosion temperatures are very high; and, with the explosion taking place right in the engine, there are no losses of heat in transmission. Therefore the resulting efficiency may be made very much higher than that of any steam type. In efficiency, the internal-combustion engine has reached more than 50 per cent and runs frequently as high as 40 per cent, while 15 per cent is high for a steam engine, and 20 per cent is seldom reached. The figures, of course, apply to the different engines as means for getting mechanical energy out of fuel.

FIG. 301. Typical water-cooled, stationary gas engine with spark plug ignition system, the current being supplied by dry batteries in the box at left.

203

How an engine works. In all engines of the internal-combustion type apt to be found on the farm there are 5 main operations: (1) the fuel must enter the engine; (2) it must be compressed in one end of the cylinder; (3) it must be ignited (4) then expand against a piston, and (5) the burned gases must be got rid of by being forced out of the engine entirely. All this must be done automatically and very rapidly. In the very highest speed engine, this entire series of operations is completed 30 times in every second. In most stationary engines of small size, the speed is not more than one fifth of this; but it is fast enough to necessitate the most careful timing of every part of the action.

The fuel is stored in a tank, from which it flows by gravity to a device called a "carburetor," or "vaporizer." This is nothing more nor less than some form of atomizer in which conditions are made favorable for the vaporizing of the liquid fuel, and the thorough mixing of the vapor with air, thus forming an explosive mixture. This is drawn into the engine by the suction of the moving piston in the cylinder. The time of drawing in the charge is regulated by the inlet valves, which are operated by cams on the engine cam shaft this being geared directly to the crank shaft itself. After the mixture is drawn in, the valves are all shut, the piston comes up and compresses the mixture considerably. Then, at the proper

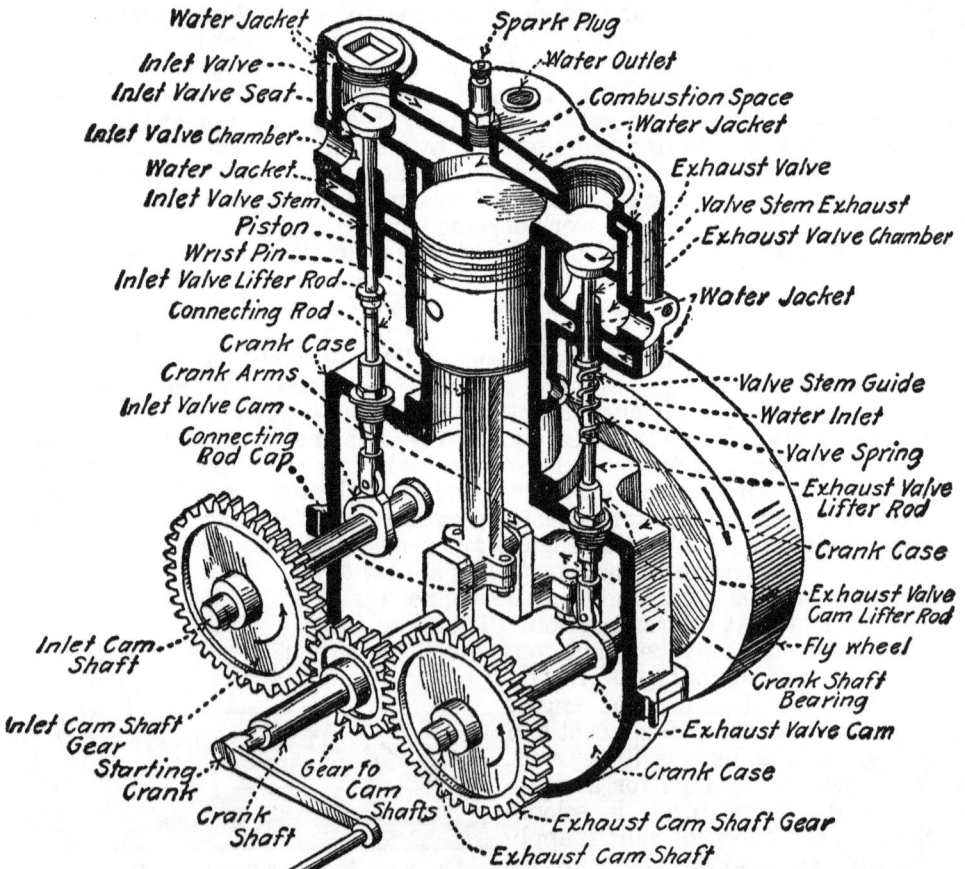

FIG. 302. Diagrammatic view of one cylinder of a gas engine with connecting parts, partly in section to show internal arrangement

moment, a spark ignites the mixture, which explodes and, by expansion due to the heat, forces the piston down again. On the return stroke, the piston drives out most of the burned gases. They are still very hot, of course, and the heat they carry away is one of the large sources of lost efficiency in oil-engine work.

Fuels

The determination of the best fuel for economical use in small engines has occupied a great deal of thought during the past few years, because of the constantly increasing price of gasoline and the very much cheaper cost of kerosene in many parts of the country. Crude oil for engines of about 50 horsepower and over has also found strong advocates, and it is used to a considerable extent. The problem presented is very much complicated by the incidental points connected with the use of the fuels.

Kerosene is a heavier distillate than gasoline, both being obtained from crude petroleum. The crude petroleum itself is in some form used in the larger engines. Naturally, it is cheap, as there has been no material work done on it from the well to the consumer. Both kerosene and gasoline are products of "cracking" the petroleum, and are incidents in the continuous succession of products from crude petroleum. This material not only gives us our various fuel oils, but also the many mineral lubricating oils, both light and heavy—the greases, dye and munition products, and, not least valuable, the well-known vaseline, or petroleum jelly.

Theoretically, kerosene has a higher heat value than gasoline, in the proportion of 11 to 9; but it is not possible to get the full value of kerosene in a gasoline engine. A much higher temperature is required to vaporize it than gasoline, and more evaporating surface is required. Then, within the engine, combustion is not apt to be complete, so that a deposit of carbon is left on the cylinder walls, piston head, and spark plugs, thus requiring frequent cleaning. Any gasoline engine will run after a fashion on kerosene, if started and warmed up on gasoline. Some adjustment of the carburetor may be, and usually is, required. Some kerosene engines may be started by heating the ignition tube

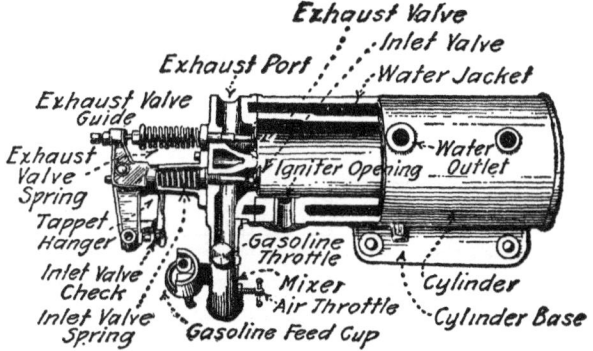

FIG. 304. Top view of a gas-engine cylinder, partly cut away to show water jacket and arrangement of valves

with a blow-torch, but this is not recommended for farm use.

The amount of kerosene usually required is over 1¼ gallons to 1 gallon of gasoline, because of the incomplete combustion. In engines specially built to run on kerosene, trials have frequently shown an actually smaller consumption of kerosene than of gasoline in a gasoline engine doing the same amount of work in the same time. In general, the amount of fuel consumed per horsepower

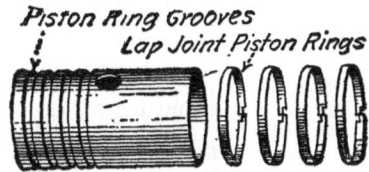

FIG. 305. A piston removed from a cylinder to show the rings which create the compression and the grooves in which they fit.

per day of 10 hours is about 1 gallon, sometimes more, but seldom less, even with expert attendance or supervision.

There are many good engines built to run on kerosene fuel that give more or less satisfaction. There is every reason why that type of engine should be purchased. There is an economic importance in the use of kerosene in place of gasoline as fuel; for it will tend to lower the price of gasoline by lessening the demand, and will, at the same time, leave this more refined product for special purposes in which kerosene will not serve as a satisfactory substitute. While, of course, by increasing the demand for kerosene, the tendency will be to increase the price, there is a plentiful kerosene supply due to the large production of it as a

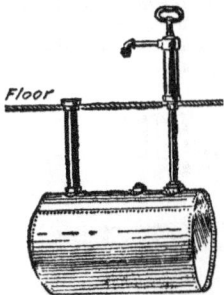

FIG. 303. Fuel for gas engines should be kept in an underground metal tank, preferably well away from all buildings, and equipped with tight supply pipe and delivery pump.

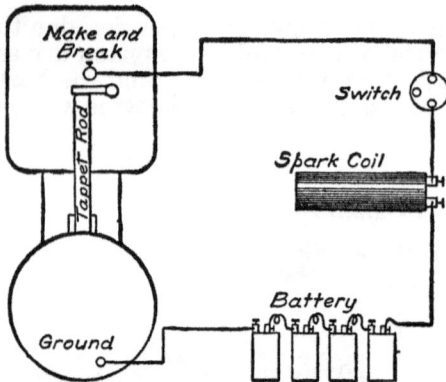

FIG. 306. Diagrammatic wiring plan of a make-and-break ignition system in which a spark coil is used. The batteries are connected in series and with the engine where the wire is grounded. The make-and-break mechanism is operated by the tappet rod and this by moving parts of the engine

by-product in refining to produce gasoline and other oils.

All these things being taken into consideration, there is no doubt as to the superiority of the kerosene engine over any other type of power plant on the farm. This is being demonstrated by the constantly decreasing use of steam and other engines in proportion to the total power used.

Carburetors

The first action on the fuel is to prepare the explosive mixture. This is done mainly in the carburetor. The common form of carburetor has 2 chambers, one being a float chamber in which the fuel is kept at a constant level by the action of a float—either of coils, hollow brass, or aluminum—this float rising as fuel flows in and shutting off the supply at a valve when the proper level is reached. The fuel from the float chamber then flows to a second chamber, through which air is brought by the suction of the engine. This air is frequently warmed by some auxiliary connection, as by an intake situated near the exhaust pipe or manifold. The air passes through a constricted tube and thus acquires a high velocity. In the smallest part of this tube, where the velocity is highest, the rushing air passes over the surface of the fuel oil, picks up some of it, and mixes with it on the way to the cylinder.

There are many different types of carburetor, but the general action here outlined fits all of them more or less. The tendency is to effect in some way a very perfect and intimate mixture of the gas and air before they enter the cylinder. With the engines of very high speed, now in use to a large degree where light weight is essential, sufficient time is not given between explosions for complete combustion of the fuel. As explained below, this combustion is not instantaneous.

The kerosene vaporizer differs from that for gasoline in that it involves some device, usually a hot tube or hot-water jacket on the carburetor for warming the fuel and helping to atomize it. There is also a heating device on the intake manifold to vaporize the kerosene. These help on the familiar principle that liquids vaporize more readily when heated.

Ignition System

Two thirds of the troubles encountered in gas-engine operation are due to defects in the ignition system or to lack of knowledge of proper ignition control. The ignition system is the vital part of the oil engine, and it must work properly and be controlled in the correct manner.

There are 2 divisions of ignition systems under which all designs may be properly classified: (1) the make-and-break, or low-tension, and (2) the jump spark or high-tension. These names refer to the particular method by which the spark in the cylinder is made.

Make-and-break system. With the make-and-break design, there are 2 contact points in the cylinder, one of which is movable and may be turned away from the other suddenly by a spring-trigger arrangement, after having been in contact for a very small interval of time. The 2 points are connected in circuit with a battery and a coil of wire wound about an iron core. When the points are separated, the momentum of the current causes it to jump the gap created between the points, thus giving the required spark. The purpose of the coil used is to increase this tendency of the current to continue to flow even after the circuit has been broken. The coil itself consists merely of a few turns of insulated copper wire wound about a soft iron core. Such an arrangement as this has been used for many years in electric gas-lighting systems, and is there known as a spark coil. It is commonly referred to in connection with gas engines as a "make-and-break coil" or "non-vibrating coil."

The make-and-break system, because of the

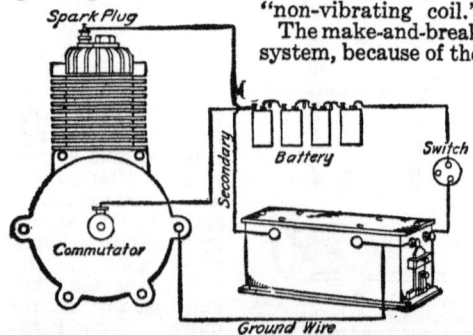

FIG. 307. Diagrammatic wiring plan for an ignition system using dry batteries, vibrating spark coil (in box) and spark plug, the timing of the spark being effected by the commutator

A piece of level bottom land on which water stands for several days after a heavy rain.   One line of tile would promptly remove this surplus moisture

Whether done by hand or machine, the construction of underdrains is a cheap and simple operation as measured by its results

DRAINAGE IS A MEANS OF RECLAIMING POOR LAND, IMPROVING GOOD LAND, AND BENEFITING THE FARM AND THE FARMER IN MANY WAYS

Sometimes Nature provides irrigation water at the mere cost of tapping the supply. *At left,* a shallow flowing well in Colorado; *in centre,* a typical artesian well in Mississippi; *at right,* water from an underground river fork in Kansas.

Man's task is the distribution of the water where it is needed, with the least effort and waste

IT WOULD SEEM AS THOUGH THE EARTH HAD PREPARED A SUPPLY OF MOISTURE IN THOSE
REGIONS WHERE THE HEAVENS DENY IT. THE RESULT IS—IRRIGATION

difficulties of mechanical design, cannot be used on high-speed engines but is used on a large proportion of stationary farm engines. It has many advantages and many disadvantages over the jump-spark system. A much hotter spark can be obtained with the make-and-break, because of a greater flow of current; there is not so much leakage of current; and the system is not so readily put out of service by dampness and dirt. On the other hand, good contacts are required all through the system, and, particularly, the contacts within the cylinder must be kept clean. This is sometimes difficult because of the presence of soot and oil. Another disadvantage is the larger number of moving parts and wearing surfaces.

The jump-spark system. The jump-spark design is that in which a spark plug is used in the engine cylinders. Here the spark points are stationary (but adjustable), with a fixed distance between them. They are in circuit with the secondary winding of an induction coil, commonly referred to as a "jump-spark coil," or "vibrating coil." It consists of 2 windings. The primary has a few turns of comparatively large copper wire and is connected to the battery. The secondary has many thousands of turns of fine wire, the fine wire being used solely to allow the coils to be crowded close to the core and to save space and cost. Owing to the large number of turns in the secondary, the voltage, or "pressure," of that circuit is higher than that of the battery circuit, and so it can force a flow of electricity across the gap. The current flowing in the secondary is less than that in the primary, and it cannot be measured easily and directly by convenient instruments. The current in the primary, however, may be measured by means of a pocket battery ammeter, and should not exceed one fourth or one half an ampere, if the circuit is in proper condition.

The principal disadvantage of the jump-spark design is the high tension or voltage used, because of the difficulty with which proper insulation is obtained. The least dirt or moisture is fatal to its workings. The

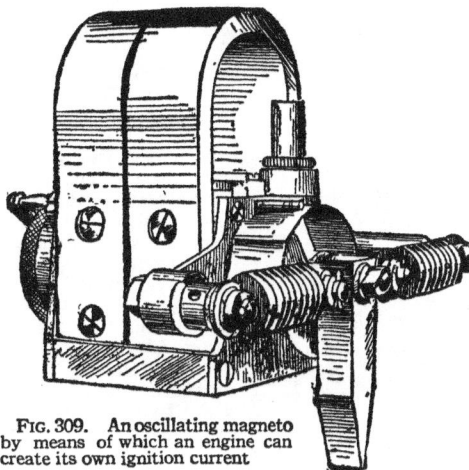

FIG. 309. An oscillating magneto by means of which an engine can create its own ignition current

vibrator in the primary circuit, used to open and close the circuit rapidly, sometimes gives trouble. This system is generally adopted for medium and small-sized engines.

Magnetos are common sources of electrical energy in ignition systems, being really a special type of generator or dynamo generating alternating current at either a high or a low voltage. The low-tension magneto replaces the battery in the make-and-break system and, occasionally, in the primary circuit of the jump-spark design. Frequently there is a special spark coil built in low-tension magnetos designed for automobile ignition but this is not common in farm-engine magnetos. The high-tension magneto, when used, takes the place of the whole jump-spark system, if desired, the spark plugs being connected directly to the magneto terminals.

Another type of magneto used on stationary engines is of the oscillating make-and-break type shown in Fig. 309. To use this type an engine must be equipped with a cam to work the oscillator. This is controlled by a spring which, when it is released, draws it back breaking the contacts and thereby making a spark. Often a double spring oscillating attachment is arranged with the springs set horizontally thereby giving a somewhat better balanced magneto and insuring a more even spark.

In all of these systems, the electrical action is practically instantaneous; but although combustion in the engine cylinder is extremely rapid, there is a definite lapse of time between the closing of the electrical circuit and the point of maximum pressure set up by the explosion of the gases. The exact length of this period depends upon the proportions of air and vapor in the mixture. The combustion period with a mixture of 1 part gas to 4 parts air is four hundredths of a second, and that with a mixture of 1 part gas to 14 parts air, thirty hundredths.

On this account, the spark circuit must be

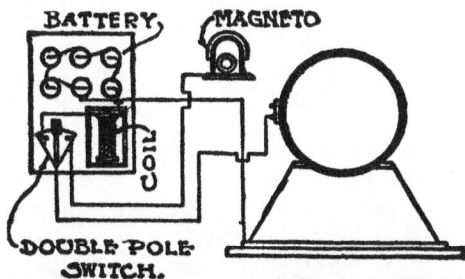

FIG. 308. Wiring plan when batteries and low tension magneto are used in a simple make-and-break system. Actually the magneto is attached, metal to metal, to the engine base so as to complete the circuit

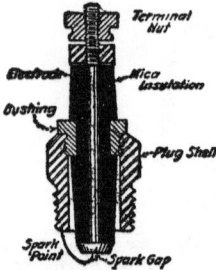

FIG. 310. Section of a spark plug. The metal thread screwing into the cylinder head makes one connection; the central electrode, joined to the spark-coil wire by means of the thumb nut, makes the other.

closed a little while before the piston gets to the exact point where it is desired that explosion shall take place. This may even be before the piston reaches the end of its compression stroke, for even so, the force of the explosion does not occur until after the maximum compression has taken place and the piston started back.

The richness of the mixture varies from time to time, and so, of course, there must be changes in the point of ignition. This variation in the mixture is due to changing the throttle, opening and closing it from time to time as the load varies. Then, too, with an increase in the speed of the engine, the spark must be advanced, because the circuit must be closed earlier in the stroke, to allow the same period of time to elapse before the piston reaches the end of stroke, the piston traveling so much faster than before. On the other hand, if the engine is being started, the piston is traveling slowly, and so the spark must be retarded. That is, the circuit must be closed when the piston is at the end of the compression stroke, or after it has passed the end of stroke usually the latter. In either case, the maximum force of the explosion will occur after the piston

has started back. Care should be taken that explosion shall not occur when the piston is exactly at the end of the stroke, because that causes bad knocking, owing to the full force of explosion being transmitted directly to the crank and crankshaft bearings.

If explosion occurs before the piston reaches the end of the compression stroke, when the engine is being started, it may turn the crank forcibly in the reverse direction and so injure the operator who is trying to turn it the other way. If the explosion occurs too early, when the engine is running, there will be a loss, of power, because the force of the explosion will oppose the motion of the piston. Then, too, combustion is slower with the gas under less pressure, so that the engine will become overheated, if run continually with a much-retarded spark.

These facts underlie 3 rules of spark control, which should be memorized and understood by every engine operator. They are:

(1) Always retard the spark before starting the engine.

(2) Always advance the spark as the engine picks up speed.

(3) Always retard the spark when the engine slows down under a heavy load.

In every case when the engine is running, the object of spark control is to secure an explosion at the moment when the crank has passed the dead centre and the piston has started back on the return stroke. This will give the maximum power and the most economical operation. An explosion at any other time in the stroke wastes fuel and injures the engine—from undue strain, if before the piston reaches the end of stroke, and from overheating, if after.

Types and Parts of Internal-Combustion Engines

Four-cycle engines. The entire operation, from the intake of the mixture to the exhaust of the burned gases, is a cycle of operations. In one class of engines it takes 4 strokes of the piston, the first stroke sucking in the charge, the return stroke compressing it, the explosion forcing the piston out on the third stroke, and the piston pushing out the burnt gases on the fourth stroke. Then it is all done over again. This kind of engine is a four-stroke-cycle type, usually called a "four-cycle type." This is to distinguish it from the two-stroke-cycle engine, the details of valve construction necessary to make the two-stroke cycle possible being very different.

Two-stroke-cycle engines. In the two-stroke-cycle type, the explosion forces the piston out, this movement of the piston compressing the mixture in the crank case. The compressed charge flows up into the cylinder as the piston, toward the latter part of its outward stroke, uncovers an inlet port in the cylinder side. Just before the inlet port is uncovered, an outlet or exhaust opening on the opposite side of the cylinder is uncovered, so that the

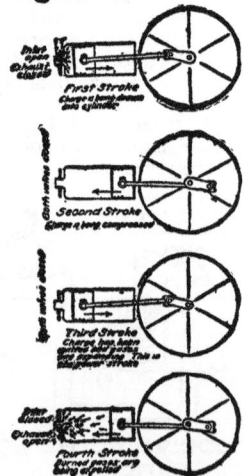

FIG. 311. Diagram showing the successive operations in the working of a 4-cycle engine.

exhaust gases start out and are helped somewhat in the scavengering process by the inflow of the new charge. This inflow is stopped as the piston returns on the instroke, the mixture is compressed, and ignition takes place, followed by the explosion.

In this case, the piston really acts like a valve, allowing both intake and exhaust at a certain fixed time in the stroke. As both inlet and exhaust ports are open at the same time, there is some mixing together of the fresh charge and the burned gases, thus cutting down the explosive force on the piston. If it were not for this fact of a weaker resulting mixture and, also, the loss of some of the fresh charge through the exhaust port, the two-cycle

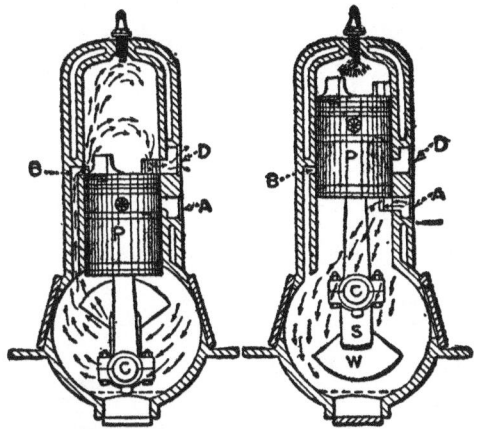

FIG. 312. How a twocycle engine works. Note that the crank case is enclosed and takes part in carrying out the series of operations.

engine would give twice the power at the same speed as the four-cycle and with a similar weight and bulk of material. As a matter of practice, the weakening of the charge and the nature of the valve openings make the two-cycle type a slow-speed engine, and with the same size of engine only about 30 per cent. power is obtained, using perhaps slightly more fuel in proportion than with a more similarly powered four-cycle machine.

Diesel engines. Engines of the Diesel type (Fig. 313), more or less modified, are a modern development for farm use and have attained considerable popularity because they start and run on any of the cheap, low-grade fuels such as crude oil, fuel oil, kerosene, tops, distillate, etc. They are of the four-cycle type but require neither carburetor nor electric ignition system. On the suction stroke of the piston pure air is drawn in through the intake valve (a). During a part of this stroke the fuel required flows by gravity into a fuel cup (f) through a separate needle valve (d) the amount of fuel admitted being regulated by the action of the engine governor on this valve. On the compression stroke the air is compressed very highly so that it becomes extremely hot, ignites a part of the fuel in the fuel cup and the pressure created by this burning portion forces the remaining fuel oil as a spray (h) out of the cup and through the spray openings (g) into the cylinder (k) where it burns and creates an explosive pressure, forcing the piston (l) out on the expansion stroke. At the end of this stroke the exhaust valve opens and the piston traveling back on the exhaust stroke forces the burned gases out through the muffler.

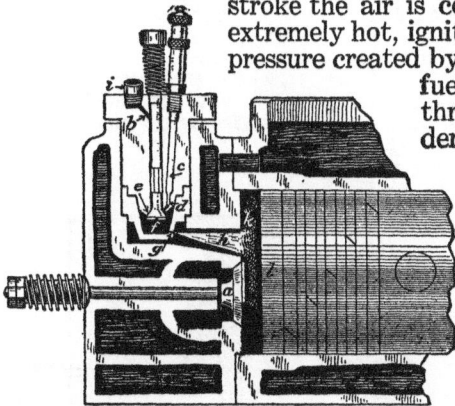

FIG 313. Section through cylinder head of Diesel type engine. *a* air intake valve; *b* fuel passage; *c* needle valve stem; *d* needle valve; *e* valve into *f*, fuel cup; *g* spray opening; *h* spray of vapor; *i* fuel supply pipe; *k* cylinder; *l* piston.

Valves. The important part of engine design is in the size, location, and operation of the valves. They determine to a large degree the possible speed, power, and efficiency of the engine. In the four-cycle engine they are usually operated by a series of cams on a shaft geared to the crank shaft and running at half

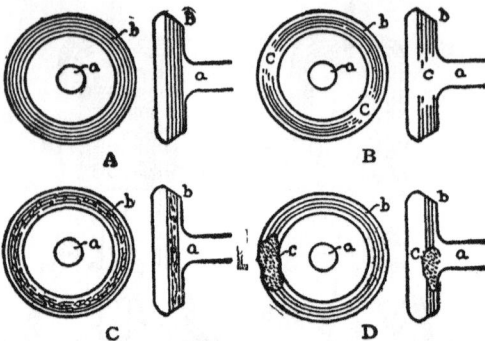

FIG. 314.   End and side views of valves (*a* stem, *b* face) showing in A normal condition and, in B, C, and D various troubles. B illustrates worn spots (*c*) which cause the escape of gas; C shows a pitted surface; D shows a burned spot (*c*) which cannot be repaired, but which fortunately is not common. The other troubles can usually be corrected by grinding.

its speed. These cams open the valves at the proper points in the stroke, and coiled springs close them. When carbon accumulates in the cylinder, it is apt to be found on the valves and on the valve seats. This prevents tight closing, allows leakage, and causes loss of power. It is, therefore, a good plan to grind the valves occasionally with a fine pumice-stone paste, so that they will always seat tightly.

Governors. Many stationary oil engines have governors of one kind or another. The

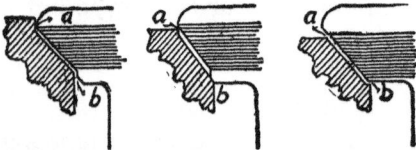

FIG. 315. One type of valve trouble comes from the wearing of the seats and valves until shoulders are formed, which prevent tight closing and therefore complete compression.

hit-or-miss type is very common. In this, the exhaust valve is held open and the inlet valve closed until the engine slows up, then the exhaust valve closes, and proper admission of gas takes place. Sometimes the governor operates so as to admit gas only once in so many strokes, depending on the speed of the machine. In all cases, the governor itself is some modification of weighted levers which are forced out by centrifugal force as the speed of rotation increases. They are like steam-engine governors in principle.

Cooling systems. Two general methods of cooling are practised—air cooling and water cooling. Occasionally, an oil-cooling method is also employed. Cooling is necessary in the interests of efficiency to increase compression, speed of explosion, and so forth, and in every way to make the temperature of

explosion as high as possible. Such high temperatures as are obtained would cause expansion of all parts, increase friction, cause binding, and, in some cases, melt the metal, if it were not for some means of carrying off the surplus heat from the parts as rapidly as it is created. This, of course, causes a tremendous loss of energy, and constitutes a big problem in design. Any engine will operate better hot than cold or warm.

The air-cooled engine has flaring ribs or flanges on the outside of the cylinder casting, in order to increase the radiating surface. Air-cooled engines are lighter and usually run better than water-cooled ones. They are always small units. Water-cooled engines are more generally used, though not always the best. In this type, the cylinder has a double wall through which water circulates, from a tank or radiator, around the explosion cylinder and, in large engines, through the piston which is made hollow. Both types sometimes have auxiliary fan equipment to blow cold air over the engine, or, more often, to draw air through the radiator and increase the radiation.

The water is sometimes circulated by a small pump but more often, as a result of the change in temperature setting up currents which flow completely around the system. Occasionally, water, from some plentiful source just flows continually through the whole water jacket. The trouble most often encountered with water cooling results in the winter from forgetting to empty the radiator or tank and water jacket. Such forgetfulness frequently results in a cracked cylinder or broken pipes. Oil, alcohol and special preparations are frequently used in freezing weather to prevent this. However, as a rule it is best to drain the water-cooling system whenever it is necessary to leave the engine idle for any length of time in a place where the temperature is likely to go much below the freezing point.

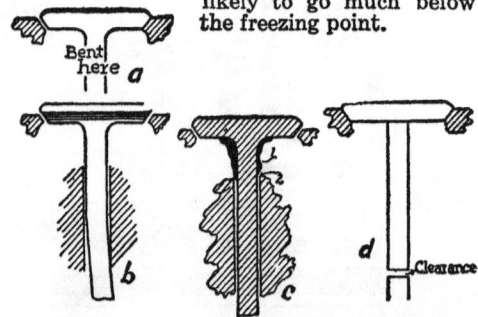

FIG. 316.   Other valve troubles: *a* and *b* bent stems; *c* accumulation of carbon preventing tight closing; *d* incorrect adjustment of stem so that valve is not lifted when it should be.

Care and Management

Cleanliness is the first essential in the proper care of engines. The cylinder, valves, ignition parts, and the machine generally should be kept free from dirt and other foreign substances. Otherwise, short circuits in the electrical system and loose parts in the machinery will result.

Oiling is another essential. All cups should be kept full and running at all times when the engine is operated. In starting, the engine should be turned over by hand to see if it is "free"; it should then be started, and the oil system put into operation at once. A high-grade oil is required for the engine cylinder where temperatures are very high, while medium or light-weight oils are used for bearings and the like. In winter, lighter oils can be used than in summer. Proper oil circulation should be insured by the use of an oil pump in the system.

The fuel oil must flow freely, and admission of air into the storage tank must always be provided for, in order to force the fuel feed. Usually, there is a tiny hole in the cover, which takes care of that point.

The water jacket must always contain water, as a few minutes' run without water cooling would be very likely to destroy the engine.

The ignition system must be kept in operating shape and given careful inspection involving the tightening of all connections, the keeping of all wires dry, the maintaining of the voltage of the battery used, and certainty that the ignition points of the spark plug or other electric igniter are clean.

Except in cases of overheating and binding parts, an engine will stop only when there is either lack of the explosive mixture or a broken electrical connection. An intermittent engine action may be due to a poor electrical connection, leaky valves, or poor mixture.

Sources of current. The electric current furnished to the ignition system must come from (a) dry batteries, (b) storage batteries, (c) magnetos, or (d) generators. All are perfectly good sources. In first cost and ease of handling, nothing is so good as the dry battery. While they last, dry batteries give entire satisfaction. One of the difficulties in using them is the need of electrical instruments to indicate when they require replacing. When used for starting only, they are fairly economical.

Storage batteries, while expensive in first cost, make a very cheap source of power in operation. They must be watchfully cared for, the liquid maintained at the proper specific gravity, the amount of liquid kept to a certain point, and they must be continually recharged as they run out. Some source of direct current for charging is required, and this is very seldom readily available on the farm.

Magnetos and dynamos will furnish electricity so long as motive power is applied to them. Their power is usually furnished by the engine itself, starting usually being effected by dry batteries. The difference between the magneto and the dynamo, or generator, is one of make-up and not of prin-

FIG. 317. Horizontal type of internal combustion engine especially adapted for farm use. Horse power may range from 1½ to 12 or more. Ignition is effected by means of oscillating magneto shown in place at the side of the cylinder

ciple. The magneto is a whirling coil in a permanent steel magnet, while the generator is a whirling coil in an electromagnet. They are the most desirable sources of current for farm-engine purposes.

Valve action. The timing of valves to open and close at predetermined points in the stroke is a very difficult matter. There are 2 valves—inlet and exhaust. The inlet valve must be opened during the early part of an outstroke. Occasionally, it is operated by the suction of the piston, but more frequently by cams. The exhaust valve opens before the end of the explosion stroke and remains open until the end of the exhaust stroke. Directions for timing and for setting are always provided for each individual engine by the manufacturer, but, when received from the factory, the engine should be properly timed. There is, therefore, rarely any necessity of timing it again unless the engine should be taken apart for cleaning and repair. In this case it would be advisable first of all to mark with a prickpunch the matching points of the various parts, so that in reassembling the engine they can be fitted together again just as they were before being taken apart.

Judging and Buying an Engine

Size required. For most indoor operations on the farm, such as running a churn, washing machine, grindstone, corn sheller, pump, etc., an engine will need up to 2 or 3 horsepower. Feed mills and elevators will use up to 5 horsepower. An ensilage cutter should be from 6 to 30 horsepower, while a threshing machine usually takes somewhat more than that. The best plan where all of this work is to be done is to get 2 engines: one of, say, 4- or 5-horsepower, and the other larger and big enough to take care of the heaviest work you have to do. A heavy engine running under a small load of 2- or 3-horsepower will take as much fuel and oil as a 5-horsepower engine running at full load. The saving effected by using a small engine for light work will pay for the engine in a short time.

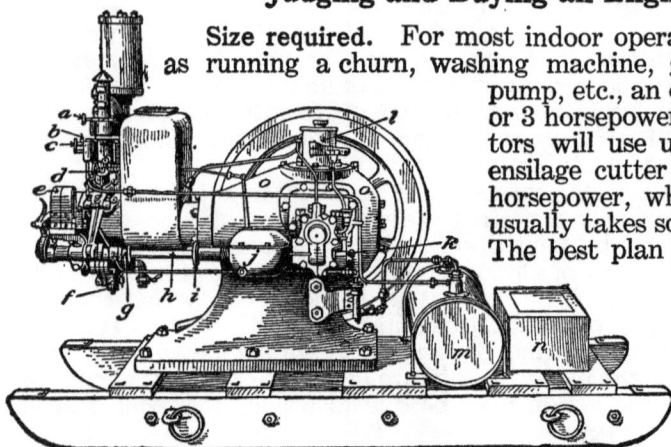

FIG. 318. Horizontal, internal combustion engine of a standard farm type mounted on skids for easy transportation about the farm. A truck may replace the skids. One fly wheel has been removed to show parts as follows: *a* gasoline valve; *b* water valve; *c* kerosene valve; *d* intake valve; *e* magneto; *f* exhaust valve; *g* compression relief cam; *h* side shaft; *i* speed changing device; *j* governor; *k* fuel pump; *l* mechanical oiler; *m* fuel tank; *n* tool box; *o* crank case (enclosed)

Sources of Electrical Supply

Determining the horsepower of an engine. The use of the indicator and prony brake for finding the power of an engine has been discussed in the preceding chapter. Gasoline and oil engines are generally rated at their brake horsepower, but it is also possible to obtain a theoretical figure on the basis of the engine's dimensions. Thus the horsepower of an engine depends on its speed, the diameter of the cylinder, and the length of the stroke. Of course, the fuel and mixture have a great deal to do with it. Four-stroke-cycle engines will give power approximately in accordance with the following formula:

$$\text{Horsepower} = \frac{\text{Diam. x Diam. x Stroke x Speed}}{18,000}$$

The diameter of cylinder and length of stroke must be measured in inches; the speed, in revolutions per minute.

Essential points in buying an engine. In buying an engine, see that it is large enough for the work, yet not too large. In construction, it must be heavy enough to "stand the racket." Look to the thickness of the frame for this. The bearings must be easily accessible; any adjustments should be positive; and the bearings themselves should be large, the thickness of metal being about one-half the shaft diameter, and the length of bearing being $2\frac{1}{2}$ to 4 times the diameter of shaft. The lubrication system should be positive in its action. The general construction should be well done. Consideration should be given to the questions whether a stationary or a portable type is wanted and whether of high or of low speed. The high-speed type is very much lighter, but not so favored for farm work. The governor should be carefully looked over, to see if it is simple, satisfactory, and not likely to get out of order readily. The ignition system should be complete and well protected from the weather, the oil, and continual handling.

Internal-combustion versus steam engines. There are many points in favor of the oil engine, such as less chance of explosion, less complication, and less knowledge necessary to operate it. The interest on the first cost of the plant will be less, and the depreciation should be slightly less. Yet the steam engine has a number of points in its favor, the most important being that it can be over-loaded. The steam pressure may be increased to the strength capacity of the boiler, and in that way the 10-horsepower engine may be made to give 25 or more horsepower for a short period, if an emergency should arise. Of course, this would mean increased coal consumption. Then, too, the exhaust steam may be used, to some extent, about the farm for all manner of purposes—heating, ster-ilizing, etc. In fact, it is found that exhaust steam has really greater heating value than live steam of the same temperature and pressure, probably because of the slight pulsation or throbbing from the action of the engine.

The steam boiler requires constant attendance in keeping up the fire and keep-ing the water level neither too high nor too low. It requires attendance in getting up steam long before it is time to start using it; and, when the engine is not used, if the fire is kept up, care must be taken of the outfit.

Against this may be put the fact that the oil engine may be started at a moment's notice and, when done with, may be shut down and no attention what-ever paid to it until the next time it is needed, provided ordinary care is taken of it.

Fig. 319. Farm type of gasoline engine mounted on low metal truck to which is fitted a removable sliding-table saw rig. With this detached the engine may be used for any other kind of belt work.

CHAPTER 20

How to Care for Farm Implements

By F. H. DEMAREE (*see Chapter 8*). *Every farm implement should have various cost items charged against it so that the owner may know whether or not it is giving service in proportion to its value. The most important of these items are: the original cost; the interest charge for the money invested; the yearly repair cost; and the annual depreciation, which is based on the useful lifetime of the machine. An investigation in Ohio showed the average investment in machinery on a number of farms, averaging 160 acres in extent, to be $774. Another investigation gave 3 per cent as an average yearly repair cost on such machines. Adding 6 per cent interest, such a farm has a charge of nearly $70 to make against such machinery each year. If it lasts 10 years, the total cost of principal (purchase price), interest, and repairs will by that time be $1,470.60. If the average 160-acre farm has 120 acres of crops, each must be charged with its share of this cost or slightly more than $12 in the period, or $1.20 per year. If the machines last but 5 years, the cost will jump to $1.87 per acre per year. While such figures are but averages, they indicate one result—the financial one—of making implements last and keeping them in good condition. There are other results, such as easier and quicker work, better-satisfied help, horses less pulled down, etc., which, though harder to put on a cash basis, are no less important. This chapter tells how to insure these highly desirable results by giving farm machines the right kind of care.*—EDITOR.

IN ORDER to decrease the cost of his implements, the farmer must either make them last a longer time or operate them over a greater number of acres. Since the size of the farm is a fixed thing, and since it may be impossible or inadvisable to coöperate with the neighbors in the use of important tools, then the one great means of decreasing cost is to extend the life of the tools.

Causes of Depreciation

Normal wear. There are a number of important factors entering into the depreciation of farm machinery of which the chief, probably, is use. That is what the implement is for, to use, but it is naturally desirable to extend the service over as long a period as possible.

In order to do this, all tools must be kept in good repair. Broken parts, dull working points, and loose connections all tend to destroy the usefulness of an implement far earlier than its normal time.

Plow shares, cultivator shovels, disc harrows, scraper blades and other portions of tools that stir the soil must be kept bright and sharp for efficient work. An extra share (lathe) should be kept on hand for each plow to be used in case of emergency, or valuable time will surely be lost.

Before putting any implement in the field after a period of idleness it should be gone over thoroughly and all bolts, bearings and connections tightened up. See that broken parts, even if minor ones and not seriously interfering with the operation of the machine, are properly repaired. One of the big advan-

tages of a tool shed and shop on the farm lies in the fact that rainy days can be well employed in looking over and repairing the farm tools.

Rust and weathering. When a piece of iron rusts, small scales form on it; later these drop off leaving small pits. Rusted plow shares, moldboards, cultivator shovels and other parts that work in the soil are full of these pits. When thoroughly covered with a coat of rust they never scour properly again. Each succeeding coat of rust makes them worse and makes them require more time to take on a soil polish again.

FIG. 320. Do you wear out your implements or rust them out? You wouldn't leave a piano out in the snow, yet it costs no more and is no more useful than a binder. Why is it?

216

FIG. 321. Well-cared for, nicely adjusted tools do the best work with the least effort. This plow is hitched and balanced just right or it would not steer itself.

The remedy for rust is oil or grease. Every tool that requires a polished surface for proper work should be kept well greased when idle. It is very much worth while to keep a grease rag with every plow, planter, seeder and cultivator. At night, clean the dirt away from scouring parts carefully, then smear them well with the grease rag. When one unhitches at night, he never knows but that a sudden shower will keep him from the field the next day, and maybe longer. In that case, if the scouring parts are not greased, much delay will be caused in getting them polished again.

When an implement or tool of this description is put away for the season, the scouring parts should be given a thick smear of axle grease. This will prevent the rust and, when the tools are to be used again, it will only require a few minutes to scrape off the grease and wash the scouring surface with kerosene. The tool is then ready for business.

The action of sun, wind and rain will soon cause the best of paint to blister and crack. It does not take long for unhoused implements to look old; as the paint scales from iron parts, rust scales form and fall away, the piece attacked gradually weakens until it suddenly breaks under the strain. Wooden parts may become water-soaked and start to decay after the paint peels off, or become attacked by boring insects of various sorts. All of these things shorten the life of the tool. *Good housing is the remedy.*

Lack of adjustment. One of the chief causes of undue strain and wear on farm tools when in operation is the lack of proper adjustment. Almost all implements of modern make have some places of adjustment. These may be for the purpose of taking up wear or of adapting the tools to different conditions. The man who pushes right ahead until the implement absolutely quits is merely inviting trouble and expense.

Every implement of importance carries with it directions from the factory. If not supplied, insist on getting such directions. The makers of farm tools prepare direction sheets with a great deal of care after testing out the tools under various working conditions.

Wherever possible, paste the direction sheet on the implement where it will not be destroyed or disfigured. If this is not possible,

get a cheap letter file, and file all such sheets away alphabetically, so they can be found when needed.

When using a new implement, study it until you know how it ought to operate. Study the direction sheet carefully, then see that the various points of adjustment are kept in proper position. This is what makes an old implement *run like new.*

Lubrication

Farm machinery is of necessity subject to much abuse. Tools that are put in the field will always be called on to work under adverse conditions that can not be much improved. For this reason manufacturers are doing much to protect bearings with proper housings and in using better grades of wear-resisting materials. It remains for the farmer to do his share in the matter of proper lubrication.

Kinds of lubricants. There are 3 kinds of material used for lubrication: oil, grease, and dry materials, such as graphite and mica. Of these the liquid oil, hard oil, and grease are in common use on most farms.

The lubricant must be adapted to the machine. Oiling a machine does not necessarily lubricate it. All wearing parts are more or less rough. It is the function of the lubricant to fill in the unevennesses of the wearing surfaces in order that they may not touch, and, at the same time, to have "body" enough to form a film with a tension great enough to keep the surfaces coated.

Light machinery with highly-finished bearing surfaces require a lighter-bodied oil, than heavier, more roughly-finished tools. A cream separator is one of the light-running machines in common use.

For heavy field tools, such as binders, mowers, rakes, planters, parts of threshing machines, etc., a medium heavy oil will give best results.

For bearings where dirt and dust are apt to accumulate badly, hard oil or grease supplied through pressure cups is most desirable. These materials not only do a good job of lubrication, but the pressure keeps the lubricant constantly working outward which, of course, pre-

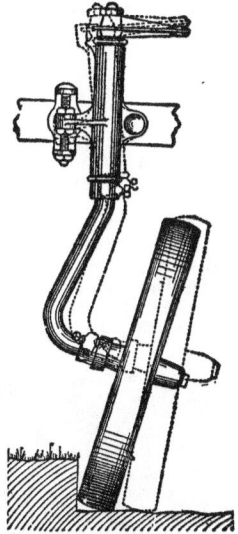

FIG. 322. Modern machinery is made adjustable to different conditions and to the effect of wear. The dotted lines show how the position of this furrow wheel of a sulky plow can be varied.

vents much dust or grit from working into the bearing.

The speed at which bearings are run is also a factor in good lubrication. A slow-speed bearing carrying a great deal of pressure requires hard oil, grease or a very thick oil with plenty of body. On the other hand, a high-speed bearing requires a thin, light-bodied oil. If a thick oil or grease is used in such bearings, the friction in the lubricant itself will be sufficient to cause the bearing to heat.

Graphite and mica are not commonly used on the farm for lubrication. Their higher cost compared with the other materials just mentioned hinders their use. A little powdered graphite dusted into bearings, however, helps smooth them up wonderfully. It will not gather dust as will hard oil or grease, so that its use should receive more consideration. A stick of ordinary lamp black can be used to advantage on drive chains of all kinds. Simply hold the stick to the chain as the machine is in operation.

Shelter

Need of shelter. In discussing the subject of shelter for farm tools, the author of Bulletin 338 of the U. S. Department of Agriculture, makes the following statement: "Much has been written and said about the waste incurred by lack of housing for machinery on farms. Many large and successful farmers do not shelter their machinery at all. The principle which guides them is that, if a machine not housed at all will wear out before it is injured by exposure, there is no need to shelter it. The larger the amount of work that can be done with an implement annually, therefore, the less need there is to house it. The waste, by depreciation, of capital invested in farm machinery is caused primarily by the inability of the smaller farms to wear out their implements with profitable use. An economic remedy is partial reorganization of their business and coöperation with neighbors so that more work can be done annually with the machinery equipment."

It is true that a great many farmers seem to go on this theory, at least judging by the

FIG. 324. A sensible, economical implement shed. It pays to make a building worthy of the equipment it shelters. This would be improved by the addition of doors.

number of unhoused implements one sees over the country. It is also true that there is no available data on the relative life of machinery when housed and when not housed. Practical observation, however, will not permit of any other conclusion than that *housed implements last longer, give less trouble and look decidedly better* than unhoused ones.

Furthermore the economic remedy suggested above fits too few cases to be effective; a farmer cannot change his land acreage at will. Moreover, the use of the various farm tools is practically universal on every farm in the community at any given time. For instance, when ground should be plowed, every farmer is at it; when corn is small, every farmer should be cultivating. Practical necessity rather than the lack of desire to coöperate has driven each farmer to maintain a rather complete line of machinery according to his system of farming.

It is very true, however, that the small farmer—the man with few acres—has a machinery investment proportionately larger than the man with more land. The small size of his business will then be a disadvantage and without working more land, his only recourse in reducing the cost of his tools is to practice as much coöperation with neighbors as circumstances will permit, and to insure *greater length of life for his tools.* As before, this means careful handling and adequate housing.

Storage places for implements. It is good business to reduce storage charges on farm tools as much as possible. Crib driveways and lean-to sheds are the most common places that the average farm presents for storage. It must be admitted, though, that implements so stored are nearly always in the way at one time or another, so that a machine shed on every farm is highly desirable.

A good machine shed. A farmer with 160 acres of land that is mostly tillable and growing corn, oats, wheat and clover will have about the following equipment: Walking plow, gang plow, drag harrow, 2 disc harrows, corrugated roller, wheat drill, corn planter, 2 cultivators, mower, side-delivery rake, hay loader, grain binder, manure spreader,

FIG. 323. Such protection as this is a little—but only a little—better than none at all. Why not go all the way?

wagon and handy wagon. In normal times this equipment, when new, will have a value of about $1,000.

This machinery can all be housed in a building 24 by 36 feet. In making such a shelter, it is not necessary to have a foundation except for the posts; these supports are best made of concrete. The floor should be of earth or cinders, slightly raised above the level of surrounding land. A drain around the outside of the building is necessary to prevent water from soaking up through the floor.

For greater convenience and at little extra cost, both sides of the building can be constructed with sliding doors the entire length of the shed. The tracks should be laid so that one door will slide by the one next to it. The doors should be about 10 feet wide so as to permit driving into the shed with any implement. Where doors are on both sides they afford an opportunity to drive into the building, unhitch and drive the team out on the opposite side. This will do away with a great deal of hand pushing and lifting when storing the tools for winter.

The Farm Shop

Some sort of farm shop, where tools or other farm equipment may be taken for all but the more complicated repairs, is indispensable to a well-organized farm. Such a farm will save both time and money in its ability to have this work done at home. If the proper equipment is at hand, many repair jobs can be done before the broken parts could be taken to town, let alone repaired there or replaced.

Locating the shop. If a machine shed has been built somewhat along the lines sug-

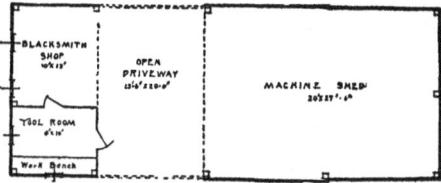

FIG. 326. Another type of shop and machine shed combined. Plenty of light and some heat are needed to make winter repair work both comfortable and efficient.

gested, an addition 14 or 15 feet wide at one end will make an excellent shop, large enough for all ordinary purposes.

The shop should be well lighted with windows on 3 sides; if it is a part of the machine shed, a wide door should connect the two.

A concrete floor is preferable, owing to the ease with which various pieces of equipment may be bolted down securely to it. There is also the advantage of less danger from fire and the ease with which the shop may be kept clean and tidy.

The following general arrangement and equipment for a farm shop to be located in a special building has been suggested by Prof. I. W. Dickerson, formerly of the University of Illinois. Such a building should be about 18 feet square to give ample space for work. The plan is to run most of the equipment by power, but in case this is not done and some of the equipment is, therefore, not installed, the layout will still prove valuable in helping one make a plan for a good, convenient shop.

Shop equipment. Fairly complete equipment for carpenter work, iron work, soldering and harness repair can be bought in normal times for about $175, not including a gasoline engine. Suggested equipment for a complete shop is as follows:

FIG. 325. Floor plan of a combined tool shed, shop, garage and wagon shelter. The latter has only a roof, but would be better if walled in on at least two sides.

For Carpenter Work

1 bit-stock or brace
8 auger bits, $\frac{1}{16}$, $\frac{1}{4}$, $\frac{5}{16}$, $\frac{3}{8}$, $\frac{7}{16}$, $\frac{1}{2}$, $\frac{5}{8}$, $\frac{3}{4}$ inch
1 claw hammer, $1\frac{1}{2}$-pound
1 carpenter's square
1 try-square, 8-inch
1 marking gauge
1 ripsaw, 26-inch
1 handsaw, 26-inch
1 keyhole saw
1 jack plane, 14-inch
1 smoothing plane, 8-inch
3 firmer chisels, $\frac{1}{4}$, $\frac{1}{2}$, $\frac{3}{4}$ inch
1 level, 25-inch
1 draw knife, 12-inch
1 dividers, 8-inch
1 wood rasp, 14-inch
1 screw driver, 10-inch
1 wood bench and vise
1 screw driver, 6-inch
Nails and screws

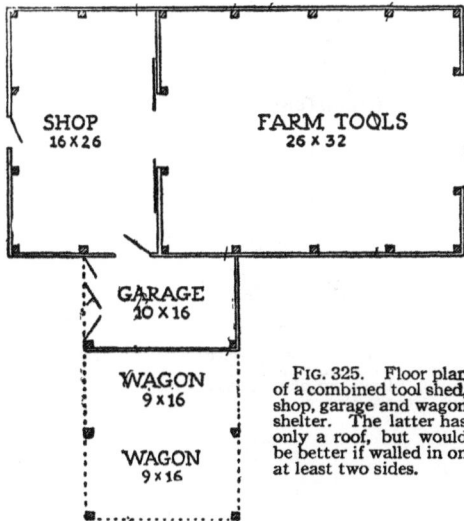

For Iron Work

1 blacksmith's sledge, 10-pound
1 anvil hand hammer, 3-pound
1 machinist's ball-peen hammer, $1\frac{1}{2}$-pound
2 cold chisels, $\frac{1}{2}$ and $\frac{3}{4}$ inch
5 punches, $\frac{1}{16}$, $\frac{1}{8}$, $\frac{3}{16}$, $\frac{1}{4}$, $\frac{3}{8}$ inch
1 centre punch
1 adjustable hacksaw frame
1 dozen hacksaw blades
12 twist drills, $\frac{1}{16}$, $\frac{3}{32}$, $\frac{1}{8}$, $\frac{5}{32}$, $\frac{3}{16}$, $\frac{7}{32}$, $\frac{1}{4}$, $\frac{5}{16}$, $\frac{3}{8}$, $\frac{7}{16}$ $\frac{1}{2}$, $\frac{5}{8}$ inch
6 assorted files with handles
1 screw-cutting outfit consisting of two stocks and tap-wrench, and 7 sizes taps and adjustable dies, $\frac{1}{4}$, $\frac{5}{16}$, $\frac{3}{8}$, $\frac{7}{16}$, $\frac{1}{2}$, $\frac{5}{8}$, $\frac{3}{4}$ inch
1 straight hardy, 1-inch
1 cold-cut, $1\frac{3}{8}$ inch
1 hot-cut, $1\frac{3}{8}$ inch
1 straight lip tongs
2 bolt tongs, $\frac{3}{8}$ and $\frac{7}{8}$ inch
1 forge with hand blower
1 anvil, steel-faced, 100-pound
1 iron bench and vise

For Soldering

1 tin snips
1 square-pointed soldering copper, $1\frac{1}{2}$-pound
1 bar half-and-half solder
Large crystal sal-ammoniac
Commercially pure hydrochloric acid
Powdered rosin

For Harness and Leather Work

1 hand-belt punch, 4 sizes
1 hollow-drive punch
1 belt awl
1 coil belt-lace wire
1 bunch cut laces, $\frac{1}{4}$ inch
1 box copper rivets and burrs, assorted
1 lever riveting machine and box hollow steel rivets, needles, wax, thread
Iron repair stand with three lasts

Miscellaneous Tools

1 monkey wrench, 12-inch
1 monkey wrench, 8-inch
5 double end S wrenches
1 button wire-cutting plier, 10-inch
1 pipe-threading outfit, with pipe-stock and 6 adjustable dies for $\frac{1}{4}$, $\frac{3}{8}$, $\frac{1}{2}$, $\frac{3}{4}$, 1, $1\frac{1}{4}$-inch pipe
1 single-wheel pipe cutter
2 Stillson pipe wrenches, 10 and 14 inch
1 open-hinge pipe vise
1 pinch-point steel crowbar
1 trowel for concrete work, 10-inch
1 pointing trowel
1 sidewalk edger
1 sidewalk groover
1 putty knife
1 glass cutter
1 melting ladle for Babbitting
$2\frac{1}{2}$ pounds Babbit metal
Assorted paint brushes

Machinery equipment. Where power is desired, a 3- or 4-horsepower gasoline engine will be needed, unless electricity is available. In addition, an emery wheel, grindstone, post drill-press, main shafting, pulleys, hangers and belting will complete the equipment. In figuring sizes of pulleys, it will first be necessary to know the number of revolutions per minute the engine runs normally, then to find at what speed the various other tools should run and belt them up accordingly.

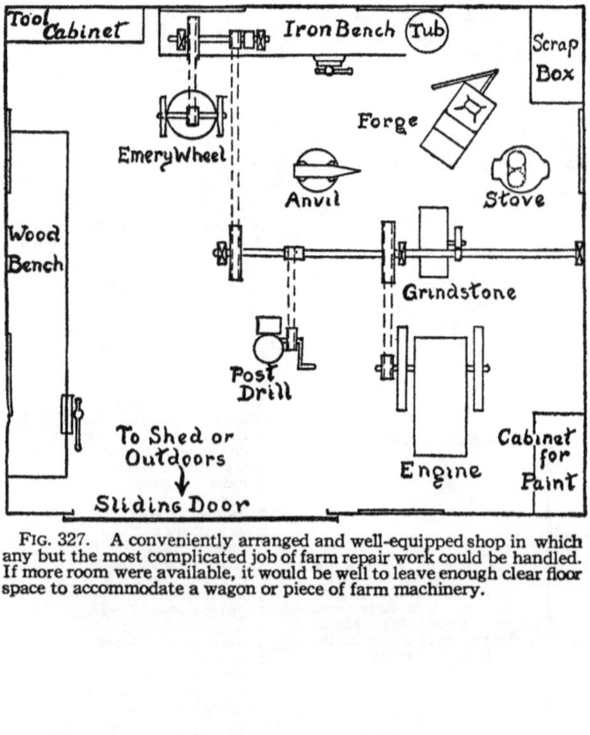

FIG. 327. A conveniently arranged and well-equipped shop in which any but the most complicated job of farm repair work could be handled. If more room were available, it would be well to leave enough clear floor space to accommodate a wagon or piece of farm machinery.

PART IV

Farm Construction and Engineering

WHEN the city man, the business man or the manufacturer has a piece of constructive work before him, whether it is the paving of a backyard, the building of a flight of stairs, the planning of a factory or the layout of a new equipment of machinery, he usually calls in a professional engineer either to advise or to take over the whole job. This expert may restrict his activities to but one branch of engineering, such as mechanical, electrical, civil or architectural, so that the completion of the task may call for several consultations and the combined services of half a dozen skilled workers.

When, on the other hand, the farmer is confronted by a correspondingly important task, he usually, almost invariably, attacks and completes it alone. This is partly because he is more or less out of reach of, or at all events out of touch with specialists in the line of work at hand. It may also be due to the fact that their number is relatively small, even though every year is seeing more technically trained engineers apply themselves to the opportunities offered in agricultural lines. In any case he solves the problem in one of two ways. Either he goes ahead with what skill his experience has given him, and by rule of thumb and the help of fortune turns out something that is more or less satisfactory and efficient; or else he seeks out what information he can, that is in usably simple form, and applies the rules and principles that he is able to make out, to the task at hand.

The latter is, naturally, the safer course, since the principles of any operation constitute the foundation upon which its successful practice rests. However, there has long been a lack of data and advice in these connections, sufficiently detailed and accurate to be of value, and sufficiently elementary to meet the average farmer's needs. It is with the hope of meeting these requirements that the chapters in this volume have been prepared.

Those immediately following are concerned with what may be called the Civil Engineering of the farm, or, from another point of view, farm construction and maintenance. Their applications to the handling of soils and of crops have been treated in Volume II, just as have the uses of farm machinery in their relation to soil management and crop production. The wonders that are being accomplished by means of irrigation and the millions of acres that merely await the construction of drains before becoming profitable, suggest the practical value of farm engineering operations, to mention only two of them. Moreover, in many respects, we are only at the threshold of the possibilities they offer. The farmer of the next quarter century who prepares his way with the help of the printed information that is within his reach and carries out his plans with the help of modern mechanical and scientific helps, will find his occupation not only the most useful and noble of occupations, but one of the most remarkable and inspiring as well.—EDITOR.

CHAPTER 21

Practical
Farm Surveying

By E. W. LEHMANN, *Professor of Agricultural Engineering in the University of Missouri, whose practical experience began in southern Mississippi, where he was born and lived on a farm until he went to the A. & M. College of that state. Since then he has gone back and worked on the farm awhile each summer, and at present he has a half interest in a farm adjoining the old home place which his brother and he are improving. He has done all kinds of farm work "from chopping cotton to harvesting corn, and from milking to dipping the cattle for the Texas fever tick." While his actual farm experience has been gained in the South, he feels that he has the northern viewpoint. His college training brought him degrees from the Mississippi A. & M. College, the Texas A. & M. College, and the Iowa State College, and also took him for shorter courses to the University of Wisconsin and Cornell University. He taught physics 3 years at the Texas A. & M. College before taking up agricultural engineering work at the Iowa State College where for 3 more years he taught a practical course in farm surveying and drainage as well as courses in farm sanitation, machinery, and concrete construction. While in Iowa he looked after several drainage problems. In 1916 he left to take charge of the agricultural engineering work at the Missouri University. Since taking the position he has supervised some tile drainage work in that section.*

The reader will find it of interest and value to refer to this chapter in connection with the article on Farm Arithmetic in Chapter 18 of Volume IV.—EDITOR.

MEASURING and mapping land. Surveying is the operation of determining the dimension, position, volume, or area of any part of the earth's surface. A survey includes maps as well as data and notes obtained in the field, such as the description and location of points, corners, and monuments. Every farm has its engineering problems, many of which are classed as problems for a surveyor. The great need, however, is not so much for more surveyors as for more farmers with the ability to do their own surveying. The farmer as a surveyor should be able (1) to measure accurately the area of his fields; (2) to subdivide them into equal parts, if need be; (3) to locate corners, roads, buildings, fences, and tile drains; (4) to make maps including these objects; (5) to lay out tile drains, roads, and terraces, and (6) to establish the proper grades for the men to work by in constructing them.

The present system of farm management (Volume IV, Chapters 1 and 2) makes it necessary that the farmer should know the exact acreage of each field and the area occupied by each crop, so as to be able to estimate the amount each piece of land should produce. He should know the space occupied by each road, fence, and ditch, and the location of each line of tile. He should know the distance between his buildings and have them arranged so that the work of doing the chores is reduced to a minimum.

Instruments Used in Surveying

The equipment used in farm surveying work need not consist of high-priced instruments. Quite often the work at hand can be done by means of inexpensive instruments and simple devices that can be constructed on the farm, a few of which will be discussed below in connection with the regular surveying equipment.

The chain and tape are used in measuring horizontal distances. Land surveys were originally made almost entirely by means of the Gunter's chain, which was often referred to in deeds, conveying property from one party to another. This chain is 66 feet long divided into 100 links of 7.92 inches each. The reason it is such a convenient length for measuring areas is that one square chain is equal to one-tenth of an acre. The objections to the chain are that (1) it is heavy and (2) the wear at the link ends will tend to make it inaccurate.

SURVEY OF FIELD WITH STEEL TAPE				
A B	482			
B C	472			
C D	279			
D E	282			
E A	272			
E B	553			
B D	337.6			

Head Chainman   John Smith
Rear Chainman   Henry Brown
September 3, 1917.

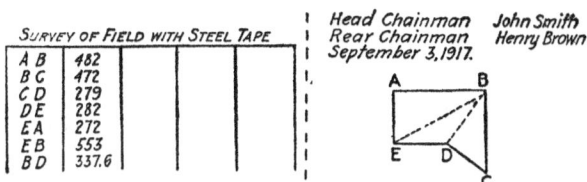

FIG. 328. Part of two pages of a notebook showing how to record data in measuring a field with a tape

Table of Linear Measure Using Gunter's Chain

7.92 inches (in. or ")	make 1 link (li)	
100	links	make 1 chain (ch)
80	chains	make 1 mile (mi)

Equivalent Table

Mile	Chains	Links	Inches
1	80	8,000	63,360
	1	100	792
		1	7.92

Table of Surface Measure With Gunter's Chain

625 square links (sq. li.)	make 1 square rod (sq. rd.)
16 square rods	make 1 square chain (sq. ch.)
10 square chains	make 1 acre.
640 acres	make 1 square mile or 1 section

Equivalent Table

Acre	Square Chains	Square Rods	Square Links
1	10	160	100,000
	1	16	10,000
		1	625

Table of Linear Measure (in feet)

12 inches (in. or ")	make 1 foot (ft. or ')	
3 feet	make 1 yard (yd.)	
5½ yards or 16½ feet	make 1 rod (rd.)	
320 rods	make 1 mile (mi.)	

Equivalent Table

Mile	Rods	Yards	Feet	Inches
1	320	1,760	5,280	63,360
	1	5½	16½	198
		1	3	36
			1	12

Table of Surface Measure

144 square inches (sq. in.)	make 1 square foot (sq. ft.)
9 square feet (sq. ft.)	make 1 square yard (sq. yd.)
30¼ square yards	make 1 square rod (sq. rd.)
160 square rods	make 1 acre.

Equivalent Table

Acre	Sq. rods	Sq. yards	Sq. feet	Sq. inches
1	160	4,840	43,560	6,272,640
	1	30¼	272¼	39,209
		1	9	1,296
			1	144

Tapes. The tapes suitable for use in surveying work are the metallic and the steel tape, the latter being used almost altogether at the present time. The metallic tape is of cloth reinforced with brass wires to prevent its being stretched when in use. The steel tape is the more accurate. It is marked either by etching or by stamping, and is graduated into feet and inches or feet and tenths of a foot.

The usual width of steel tapes is either one fourth or five sixths of an inch. They can be obtained in almost any length, the 50- and 100-foot lengths being the most common. They are arranged to be carried in metal and leather cases or on a reel. In many cases 100-foot lengths and even greater ones are carried in coils which can be easily wound with a little practice.

The tape has a more permanent length than the chain, because of its lack of wearing surfaces. It is light and its smoothness makes it easily handled. The fact that it is light is sometimes a disadvantage when the wind blows.

Marking or chaining pins. Chaining pins made of stout steel wire are used in marking, temporarily, the end of the tape or chain while measuring. Eleven pins make a complete set, and are carried on a ring made of spring steel wire with a catch. A set of pins can be easily made of No. 6 wire by any blacksmith.

Range or flag poles. Range or flag poles are used in establishing a line in surveying or locating a fence. They are usually about 8 feet in length, painted with alternate foot lengths red and white so they can be easily seen at a distance. The lower end of each flag pole is shod or spiked with metal. A good flag pole can be easily made of 2 x 2 inch scantling by beveling the corners and painting it.

How to measure with a tape or chain. The line to be measured is first marked with range poles; the head chainman then takes the 11 pins, marks the starting point with one and leads off with the zero end of the tape and the other 10 pins toward the point to which the distance is to be measured. Just before the full length of the chain has been drawn out the rear chainman signals to the head chainman by calling "halt," "chain," or "tape." As the tape is stretched to its full length by the head chainman, he is lined up by the rear chainman who calls "stick" when the chain is properly lined and drawn taut. The front chainman then sticks a pin, being careful to place it so no error will be made, and calls "stuck." The rear chainman then pulls up the rear pin and both men move ahead and repeat the operation. This process is repeated until the head chainman has set his tenth pin, when he calls, "out" or "tally," at which the rear chainman walks forward and gives 10 pins he has collected to the front chainman.

In the use of either tape or chain in measuring distances, care must be observed to see that the tape is kept horizontal. This caution must be kept in mind, especially when measuring on a slope. When measuring down a slope the front chainman has to use a plumb-bob to determine the point at which to stick the pin.

Errors made. In the use of the tape or chain the most common error is due to lack of sufficient pull on tape. For accurate results the tape should be tested between 2 points a measured distance apart to get the necessary pull. Careless plumbing and incorrect alignment are also causes of error. The effect of the wind in causing the tape to sag should be avoided. Care should be taken to observe the zero point on the tape and to take readings carefully. Errors are sometimes made by omitting whole chain lengths.

The compass is used to establish new lines, to determine the bearing of lines already established, and to retrace old lines which are lost. The essential parts of a magnetic compass are the line of sight, the graduated circular box, and the magnetic needle. The line of sight is attached over the N and S points on the circular box. The E and W points on the box are reversed and the north point of the needle can be easily designated from the South point. This makes it possible to make all readings direct. The compass box is attached to the tripod by a ball-and-socket-joint, and leveled by plate levels. In reading the compass, point the north end of the compass box along the line and read the north end of the needle. For example, when the box is pointed to the northwest, the needle still points to the magnetic north and falls in the part of the circle marked with the N and W. The reading would be N so many degrees W. Such a reading would be a magnetic reading and not true.

The difference between the magnetic north toward which all compasses point and the true north is called the *magnetic declination* or *variation*. It varies from place to place and is not constant (the same) for any one place at different times. Since the governmental surveys are based on *true north* readings it is important to know the variation when making a survey with a magnetic compass.

There are positions on the earth's surface where the magnetic variation from the true north is zero, that is, the magnetic readings and the true north readings are the same. Such a line extends across the United States, beginning in Michigan and passing to the south through the Carolinas. At all points west of this line the compass shows a variation toward the east; at all points east of the line it has a variation toward the west. This variation is as much as 20 degrees in the extreme eastern and western parts of the United States. In all government surveys, the variation at the time of the survey is recorded, so it is an easy matter to determine this value when new surveys are made.

Care in using the compass. The magnetic needle is a fine, hardened piece of steel, carefully balanced, hung on a delicate pivot. Both are protected by a device which lifts the needle from the pivot when not in use. This must be done each time the compass is moved from one position to another. Care must also be observed when taking a reading to see that the needle is not affected by local attraction. If too near a wire fence or a pile of old iron, or if the instrument man carries a heavy bunch of keys, the reading may be incorrect.

The plumb line is the simplest and most universally used of all surveying instruments. The finest transit requires a plumb line so that it may be located over a given point. For farm work the difference in elevation between points fairly close together can be determined by means of a combination of plumb line and carpenter's steel square. For short distances the grade for drains, roads, and terraces can be established with this instrument. Another device for leveling work where the plumb line is used (Fig. 330) is nothing more than an A-frame constructed of 1 x 4-inch lumber with 1 x 2-inch braces of light, well-seasoned wood. An instrument made to span 12½ feet is a convenient size. The legs should be cut 9 feet 2 inches long and nailed together at right angles at the top of instrument. They are then spread until they measure exactly 12½ feet, when the crossbar is nailed fast so it will be exactly 3½ feet from the ground. The plumb line is suspended from the top of the A-frame and allowed to swing freely until it comes to rest. The crossbar is marked where the line crosses it. The A-frame is now turned end for end on the same support and the bar again marked where it was crossed by the line. The point midway between these 2 marks is distinctly marked and called zero. When the instrument is in use and the plumb line falls on zero, the points at which the legs of the instrument rest are on a level line. If a grade is to be established and the fall is to be 1 inch in 12½ feet, a 1-inch block is put under one of the legs when in level position and a mark made on the crossbar. This instrument cannot be used to advan-

SIGHTING N.E.   SIGHTING N.   SIGHTING N.W.   SIGHTING S.

FIG. 329. Compass readings and what they mean (see text)

FIG. 330. A-frame carrying plumb-bob and line, a simple but efficient leveling device for the farmer

tage when the wind is blowing. If a carpenter's level is available, the A-frame with plumb-bob should not be used.

Bubble tube. The bubble tube is an essential part of all surveying instruments. By means of it the level, compass, and transit can be adjusted so that the line of sight will be in a definite direction or plane exactly at right angles to the plumb, or vertical line. The bubble tube is a curved glass tube partially filled with alcohol or ether. The sensitiveness of the bubble depends on the curvature of the tube. The carpenter's level is made by attaching a level tube to a block or frame and adjusting it so the bubble will remain in the same position when the level is reversed on a perfectly level surface. A point on the tube is then marked to indicate its position when the instrument is level.

The carpenter's level for surveying. While the carpenter's level is used almost altogether for leveling buildings and in construction work, it can also be used to advantage in farm leveling, the same method being observed as when using an engineer's level. Figure 331 illustrates carpenter's level mounted and with sights attached. Readings with the carpenter's level must be made over comparatively short distances. An A-frame, as discussed under plumb line (p. 224), with a carpenter's level attached to the crossbar makes a very serviceable farm level. In making this instrument take care to have the horizontal bar to which the level is attached a definite distance above the bottom of the A-frame. To use this in establishing a definite grade, one leg is shortened a little with a saw or the other is lengthened by means of a block of wood nailed under it. To make a fall of 1 foot to the 100 feet, a block 1½ inches thick would be placed under one leg of an instrument spanning 12½ feet, since 8 "steps" with such a frame would make 100 feet. Where this kind of an instrument is used for laying out terraces, the terrace is marked off by walking the instrument around the hill. In working from the outlet the short leg is kept

in front and moved up and down the hill until the instrument is level. It is then carried forward and the back leg is placed where the front leg rested. Every few steps the points where the legs rested should be marked with stakes.

The engineer's or surveyor's level. This consists of a bubble tube, a line of sight and a vertical axis. The line of sight is attached parallel to the bubble tube and at right angles to the vertical axis about which they revolve. A very satisfactory farm level can be secured at a cost of about $20. The uses of such a level on a farm are many, including: (1) cross-section work to determine the necessary excavation in ditching and road building; (2) determining the difference in elevation between 2 points, as in finding the total fall in a stream to see if a hydraulic ram can be installed; (3) establishing grades for drains, roads, terraces, feeding floors, and walks, etc. To make it possible to take readings at some distance, a telescope is provided for a line of sight. Cross-hairs are set in the telescope to be used as sights.

Level rods. Leveling rods are graduated in feet, and tenths and hundredths of a foot. For very accurate work a target is used by which the readings may be taken to one-thousandths of a foot. Rods known as speaking or self-reading rods are read direct from the instrument and are best suited for farm leveling work. A graduated tape tacked on a 1 x 4-inch board 12 feet long makes a very satisfactory rod.

Setting up the level. In setting up the level the work at hand will to a great extent determine its location. Always select a point from which readings can be taken for equal distances in both directions. Set up on firm ground if possible with plates about level. Spread the tripod enough to make it solid, placing 2 tripod shoes or feet parallel to the general line of levels. Level the instrument up by bringing the telescope directly above one set of foot screws. When one of the foot or leveling screws is tightened the other must be released to prevent it binding; thus they must be turned in opposite directions at the same time to keep them to a snug

FIG. 331. Carpenter's level mounted on a standard and equipped with sights, as used in simple surveying. Inset shows the bubble tube (enlarged and with the curve somewhat exaggerated).

bearing. When turning these screws, the bubble will always go in the direction of the movement of the *left thumb*. When the bubble indicates that the instrument is level over one set of screws, revolve the telescope to stand over the other set and make the same adjustment. Continue this process until the instrument is level in whatever direction it is pointed. An instrument may be perfectly level, but it will not give accurate results if out of adjustment. The bubble tube and the line of sight must be parallel and perpendicular to the vertical axis. The instructions for testing and making adjustment which accompany each instrument are better than any general statement that can be made here.

To take readings, focus the telescope so that the rod can be seen clearly and adjust the eye-piece to see the cross-hairs. Be sure that the rod is vertical and that the bubble is in the centre of the tube before taking a reading. The telescope must be readjusted for readings when the rod is taken farther away or brought nearer the instrument. Do not disturb the tripod after the instrument has been properly set up and adjusted.

The **water level** (Fig. 332) consists of 2 glass tubes fastened upright 3 feet or more apart to a board and connected by a pipe or piece of rubber tubing. The whole thing is fastened to a staff or tripod for use. Water is poured into one tube until it can be readily seen in both. Since water seeks its level, the height of the water in the 2 tubes will give 2 points in a line of sight for leveling work. If a small quantity of ink is added to the water, the line of sight will be more readily seen. Corks should be put in the tubes when the instrument is being moved about the field.

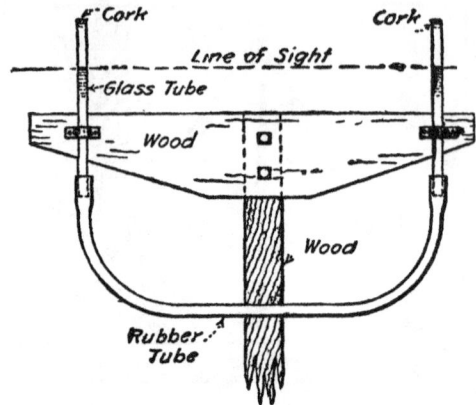

FIG. 332. How a home-made water level is constructed. When in use the corks are of course removed from the tubes.

Transit. The complete transit has been called the universal instrument because it can be used for so many purposes. It differs from the plain transit in that it is equipped with a vertical arc, and has a bubble tube attached to its telescope. It is adapted to the same uses as the compass and level, and measures horizontal angles with great accuracy; but, in addition, it can be used in determining vertical angles and distances, and in numerous other ways which need not be mentioned here. It requires a great deal of practice to become proficient in the use of the transit. It is not an instrument a farmer would be justified in buying unless he expected to undertake a great deal of reclamation work. In most cases of that sort the services of a trained engineer would be needed.

Systems of Land Surveys

The two systems of land survey in the United States are: (1) the survey by metes and bounds, and (2) the rectangular system. The *survey by metes and bounds* was the original method of surveying land and still exists in that part of the United States that was first settled. In this system the boundary line of each tract of land is fully described. To illustrate, a certain farm is described as "comprising all land that is included within boundary line beginning at bald rock at the southwest corner of the farm running north, 45 degrees east 80 chains to the twin elms on hill, thence due east 60 chains to cedar stump, thence due south 60 chains to concrete block on road, thence, south, 45 degrees west 80 chains to cedar stake in ground 4 feet east of birch tree, thence north, 45 degrees west 84 chains to starting point, containing 852 acres more or less."

In order to simplify the making of surveys, to reduce the litigation of land, and to make locations more easily designated, Congress, in 1785, adopted a system since known as the United States Rectangular System of Public Land Surveys. The fact that the earth's surface is a sphere made it difficult to lay out its surface into rectangular areas. However, this difficulty was successfully overcome. The system has as a basis the true meridians, which are lines radiating from the north pole, and standard parallel lines running east and west, often called correction lines.

In each land district a *principal meridian* is chosen, and from this line an east and west line is run called the *base line*. Their intersection is the initial point of the survey. The townships are numbered north and south of the base line and the ranges east and west of the meridian. In each township there are 36 sections, each 1 mile square, containing 640 acres more or less. Each section in the township is numbered beginning in the northeast corner going west then east in the second row, etc., putting number 36 in the southeast corner. Each section is subdivided into halves, quarters, quarter-quarters, and even smaller units. A 40-acre farm may then be designated as being the N E $\frac{1}{4}$ of N W $\frac{1}{4}$ of Sec. 1, T. 3 N., R. 2 E. of some particular meridian. An important feature of this system of surveys is that all corners and lines established and approved by the government are unchangeable.

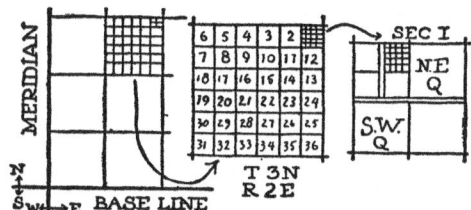

FIG. 333. The rectangular survey system showing how the ranges are divided into townships, these into 36 sections, and these again into quarters, etc. The farm described in the text is shown by the shading on the right hand figure.

The work of making a survey by metes and bounds or the reëstablishment of lines should be put in the hands of a competent engineer. It is to the farmer's interest to have permanent monuments established at all corners. Many costly lawsuits are the results of temporary monuments being placed with the idea of putting in permanent ones later. All monuments described in the sale of a piece of land acquire a perpetual and controlling significance; if recognized in a deed their position controls the location absolutely. Lost monuments can be reëstablished only by consent of parties concerned or by judgment of the court. A surveyor or engineer has no power to reëstablish a monument; he can simply act as an expert witness.

Practical Problems for the Farmer

Using chain or tape. The need of erecting a line at right angles or perpendicular to another is a common everyday problem often met with in laying out fences and buildings. The simplest method is by the application of a right-angle triangle. For ordinary conditions, use 6, 8, and 10 feet as the sides of the triangle. Measure 8 feet along the line from the point A where the perpendicular is to be erected to the point B. With 6- and 10-foot chords, strike off arcs from each end of the 8-foot base A and B. The point of intersection C will give the perpendicular as the line A C. For more accurate results use a larger multiple of 6, 8, and 10 feet, as 24, 32, and 40.

Laying out the foundation of a building. Nearly every building is laid out with reference to some other building, road, or boundary line. If it is to be erected parallel to a road, it is necessary to establish a line on the center of the road to be used as a reference line. If there is an old fence along the roadside it would probably make a very accurate base line. Erect a perpendicular to the base line, extending it to the location of the building. In case of a square or rectangular structure, two corners are fixed on this line and the positions of the other corners determined by erecting perpendiculars at the first two corners. Each corner should be marked on a stake with a nail driven exactly at the intersection of the lines. Six or 8 feet from

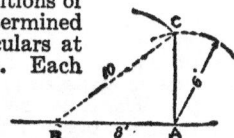

FIG. 334. How to lay off a right angle (see text)

the corner drive 3 large stakes, and on these nail strong braces. Points on the braces are determined where the lines of the building, if continued, would cross, and at these points notches are cut, giving an easy reference in excavating.

Laying out the orchard. There are two general plans for laying out an orchard: the rectangular and the alternate, the latter having several variations. The rectangular system is most used; in it each tree is set at the corner of a rectangle, usually a square.

FIG. 335. The practical application of Fig. 334 in
building a barn foundation

In the alternate system the trees in the even rows are placed midway or "staggered" between the trees in the odd rows.

To locate the trees in the rectangular system, lay out a base line along one side of the field at the proper distance from the fence, and lay out other lines across the ends of the field perpendicular to the base line. Set stakes along these lines at intervals equal to the distance between the trees. To locate other trees, set a range pole on the opposite side of the field at second stake on opposite end. Before measuring across toward this range pole set a pole at the third stake to measure back to when through with the second row. Before measuring the third row set a stake for the fourth, and so on. By having a pole at each end of the field the work is greatly facilitated.

A second method is by lining the trees in. A line of stakes properly spaced is established along each side of the field forming a perfect rectangle, then 2 additional rows are staked through the centre of the field at right angles to each other. With these stakes in place, the man setting out trees would have 2 stakes in each direction by which to line the trees in. The alternate system may be laid out in the same way as the rectangular.

Surveying the farm layout. As already mentioned, the present system of farm management necessitates a close study of the farm layout. The farm that is divided up into a great number of small fields cannot be farmed at a profit. The whole farm should be surveyed and mapped and plans developed for a future ideal arrangement. To make such a survey with a tape, if the fields are irregular in shape, it is necessary to divide them into rectangles and triangles so the areas may be easily obtained. The notes taken in the field should be complete in every detail. Sketches should be made with corners lettered, all distances tabulated, and all important points described in a well-bound field book. It is impossible to make an accurate map with a poor set of notes and sketches made in the field. In making such a survey no angles are measured, so all lines should be checked for accuracy. Extreme care must be observed in measuring over hills. To chain over a hill between 2 points not visible from each other, set a range pole at each point, then let the chainmen with range poles take positions on each side of the hill from which each can see over the hill and past the other chainman to the range pole beyond. They should then range each other in until they are on line. In measuring either up or down a slope the lower chainman should use a plumb line so the chain can always be kept horizontal. The chain is pulled taut in a horizontal position and the plumb line dropped to locate the correct position for the pin.

In chaining between two points where the view is obstructed by woods or other objects, run a trial line as near as possible toward the given point, leaving fixed points at known distances. Upon finding the error at the end of the line, correct all other points into line a proportionate amount. For example, if the random line is 1,000 feet in length and the error at the end of the line is 20 feet, then the error at the 100-foot point would be 2 feet, at the 500-foot point 10 feet, at the 750-foot point 15 feet, and so on. After points have been measured in from the random line, the desired line can be measured by these points.

To measure a line beyond a house or other obstacle there are several methods. Probably the simplest is to erect a perpendicular to the line by the triangle method, extend it out beyond the obstruction, erect another perpendicular to this line, and carry it until the obstruction is passed, then by means of two more right angles the original line can be continued.

Mapping out the farm. To make a neat, accurate map, a set of drawing instruments should be available. However, a usable map may be made with pencil, rule, straight-edge, and triangle. The first step in map-making is to select a suitable scale; 100 or 200 feet to the inch is usually convenient. If a 66-foot chain has been used in measuring, either 2 or 4 chains to the inch would be better. If the farm is of regular dimensions, lay out the boundary line first, then fill in details. Always remember that any point can be located by measuring from two known points. In the case of a field that is irregular in shape, one side can be mapped as a base line and other points located from it. A compass is a convenient instrument for doing this. The diagonals of all fields should be measured so they can be used in checking the corners on the map. All objects, such as drains, ditches, fences, and outbuildings should be suitably designated, and each field should be numbered and number of acres in it given on the map.

Computing the area of fields. To determine the area of different shaped fields, it is neces-

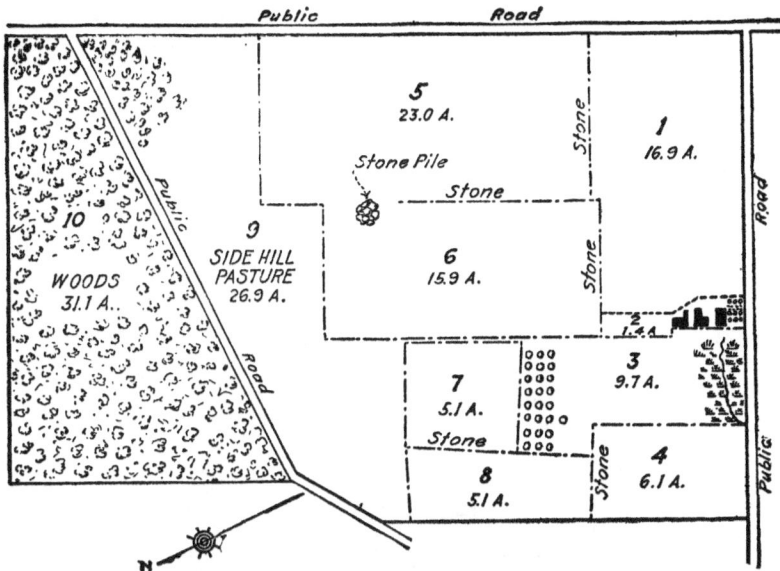

FIG. 336. Typical map of a farm, giving the most necessary facts and providing a good place on which to note plans, changes, cropping systems, etc.

sary to be familiar with the use of a few simple mathematical formula. (See also Volume IV, Chapter 18, D).

Rectangles. The area of a rectangular field is obtained by multiplying the length times the breadth (Fig. 337)

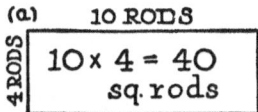

(a) 10 RODS

$$10 \times 4 = 40 \text{ sq. rods}$$

FIG. 337

Triangles. There are several methods of determining the area of a triangle: (1) If the length of one side, the base of the **triangle,** and the perpendicular distance from this side to the opposite angle, or height of the triangle are known, the area is equal to one half the product of the base times the height (Fig. 338). (2) If the

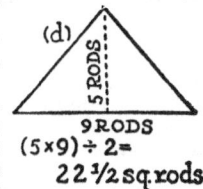

(d)

9 RODS
$$(5 \times 9) \div 2 = 22\tfrac{1}{2} \text{ sq. rods}$$

FIG. 338

three sides of a triangle have been measured, add the length of the three sides and divide the same by 2; from this result subtract each side in turn; multiply these three remainders and the half sum together; the square root of the product equals the area of the triangle (Fig. 339)

(a) $(5+6+7) \div 2 = 9$

(b) $\begin{aligned} 9-5 &= 4 \\ 9-6 &= 3 \\ 9-7 &= 2 \end{aligned}$ (e)

(c) $4 \cdot 3 \times 2 \times 9 = 216$

(d) $\sqrt{216} = 14.7$

FIG. 339

Parallelogram. The area of a parallelogram, a 4-sided figure with opposite sides parallel, is equal to the product of one of its parallel sides and the perpendicular distance between it and the opposite side (Fig. 340).

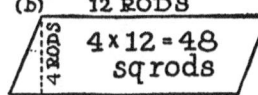

(b) 12 RODS

$$4 \times 12 = 48 \text{ sq rods}$$

FIG. 340

Trapezoid. The area of a trapezoid, a 4-sided figure with two sides parallel, is equal to the product of one half the sum of the two parallel sides times the perpendicular distance between them (Fig. 341).

Nearly all fields can be divided into suitable triangles, rectangles, and trapezoids, the areas of which can be determined. The area of a many-angled field (or regular polygon) can be found by dividing the figure into triangles by radii drawn from any point within it.

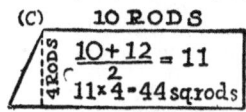

(c) 10 RODS

$$\frac{10+12}{2} = 11$$
$$11 \times 4 = 44 \text{ sq rods}$$

12 RODS

FIG. 341

Area of a field with a curved or irregular boundary. In determining the area of such a field, a base line is established across the field and perpendiculars erected at intervals, thereby dividing the area of

FIG. 342

the field into a number of trapezoids, the area of which can be obtained as outlined above (Fig. 342).

Problems in Using a Level

To determine the difference in elevation between two points. The simplest problem of this kind is when a reading can be taken on each point when the instrument is about midway between them. Assume that the rod reading on the first point is 10 feet, and on the second point 5 feet. These signify that the first point is 10 feet below the line of sight and the second point 5 feet below the line of sight. Hence, the difference in elevation between the two points is the difference between the two rod readings, or 5 feet. To determine the difference in elevation between two points at some distance apart the following method is followed: Set up the level and adjust it as previously outlined (p. 225) at a convenient distance for reading on the first point and in the direction of the second. As a matter of convenience assume the elevation of the starting point to be 100 feet, which we call bench mark (B M) 1. The first reading will be to determine the height of the instrument; all readings on the known or assumed elevation are for this purpose and are called back-sight (B S). The back-sight reading on B M 1 added to the elevation of B M 1 equals the height of the instrument. With the height of the instrument known, the elevation of a reference point, called a turning point, can be established nearer the point of unknown elevations by taking a fore-sight reading; a reading taken on a point of unknown elevation is always called a fore-sight (F S). Establish the turning point by driving in a stake so that the fore-sight distance will be approximately the same as the back-sight distance. The fore-sight reading subtracted from the height of the instrument equals the elevation of the turning point. The level is then moved to the second set up and a B S taken on the turning point to determine the new height of instrument. A fore-sight then establishes a new turning point and the process is continued to the end of the line. The difference in elevation between the points as found may be checked by the difference between the sum of the fore- and back-sights.

Profile leveling. In establishing the grade for a tile drain on a road, it is not only necessary to know the elevations of the lowest and highest points but also those of all intermediate points. The process of determining the elevation of these points is called profile leveling. The actual procedure is the same as in determining the difference in elevation between any two points except that a number of fore-sights are taken from each set-up of the instrument. Where leveling work of this character is done, it should be based on some point of permanent elevation, such as a concrete walk, a concrete floor, or a large stone. The elevation of this point is assumed and is used as a bench mark (B M) to determine the height of the instrument at the first set-up.

It can also be used in checking at any future time. If such a point is not available a solid stake should be driven into the ground to be used as a reference point.

To run a line of levels for a tile drain, set the instrument up at a point where a reading can be taken on the bench mark and the outlet of the drain as well as 400 or 500 feet along the drain. A back-sight is taken on the bench mark and the height of the instrument is established. A fore-sight is then read on a point at outlet of drain and at intervals of 50 feet up the drain. When as many readings have been taken for the first set-up as is convenient, a turning point is established, a new set-up is made and the work is continued. All points where readings are taken are numbered as follows: Station 0, Station 0 + 50, Station 1 + 00, Station 1 + 50 and so on; Station 2 + 93 would thus be 293 feet above the outlet. The readings are not only taken at regular intervals but at other points where there is a decided break in the surface of the ground, where there is a change of direction in the line of tile or where a lateral tile enters. At each of these stations a short stake should be driven nearly flush with the surface of the ground; also, alongside it, a long stake for a marker or guide on which the station number and depth of cut is written. The short stakes can be made of most any pieces of solid wood and the long stakes of plaster lath or similar boards.

Establishing the grade. There are many drains with a decided slope where the grades can be easily established with a line and gauge rod without an instrument. This method can nearly always be used in the drainage of seepy spots on hillsides. Begin by driving a 4-foot stake at each station. Then, starting at the outlet, fasten a line a definite distance above the proposed bottom of the ditch; if a $6\frac{1}{2}$-foot gauge rod is to be used and the ditch is to be 4 feet deep at the outlet, then measure up $2\frac{1}{2}$ feet from the top of the grade to locate the line on the long stake. Draw the line up to each of the other stakes, being sure always that there is ample fall toward the outlet. To determine the depth to dig at any point, it is only necessary to measure from the line down to the surface of the ground and subtract this distance from $6\frac{1}{2}$ feet. If there is a uniform fall along the line of the drain, the depth can be kept practically the same at all points.

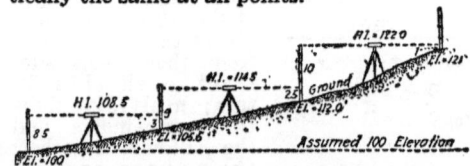

FIG. 343. Diagram showing the process of profile leveling. El is elevation; H. I. is height of instrument; and the figures give the readings of both back- and fore-sights. (See text and Fig. 344.)

To compute the amount of excavation in digging a ditch for drainage or the amount of excavation or filling necessary in road work, a grade line must be established. To establish a grade for a tile drain, the elevation of the highest and lowest points must be considered. The tile must not be too deep nor too shallow. Where the land is uniform the elevation of the outlet and the highest point will usually control the grade. The best method to follow is to select trial grade lines along the line of drain until the grade and depths at the various controlling points are satisfactory. This is done by assuming the grade line to be a certain depth below the grade stake at the outlet and a certain depth below at some other controlling point, say Station 5. It is found that the elevation of Station O, the outlet, is 95 feet and the depth

at that point is assumed to be 4½ feet; the elevation on the grade line would then be 90½ feet. The elevation of the grade stake at Station 5 is 96½ feet assuming the depth of drain to be 3½ feet; the elevation in grade line would be 93 feet or 2½ feet higher than at Station O. A rise of 2½ feet in 500 would be equal to a rise of 6 inches in 100 or 3 inches every fifty feet. When the elevations of the intermediate points have been determined the cuts can be computed as follows: Add the rise in 50 feet to the elevation of the grade line at Station O to determine the elevation of grade line at Station 0 + 50. The elevation of all other points on the grade line may be found in the same manner. The difference between the grade-line elevation and the surface elevation at any station is equal to the cut at that station.

LEVELMAN John Smith
RODMAN Henry Brown
September 3, 1917
Clear and Warm

Survey
between concrete slab at well
and bottom of concrete tank on hill

STA.	B. S.	H. I.	F. S.	ELEV.	(Description of concrete slab at well)
BM1	8.5	108.5		100	Turning point 400 ft. East of BM1
TP 1	9.0	114.5	3.0	105.5	" " 800 " " - " "
TP 2	10.0	122.	2.5	112.0	Tank 1000 " " " - " "
Tank			1	121.0	

FIG. 344. Notebook page showing how to keep a record of a job of leveling as pictured in Fig. 343

CHAPTER 22

Practical Farm Drainage

By PROFESSOR E. W. LEHMANN *of the Agricultural Engineering Department of the University of Missouri, whose experience in this field has been referred to in Chapter 21. Drainage is not like plowing and cultivating, a work that every farmer must make an almost daily feature of his activities. But it is one that a great number of farmers have found profitable and one that a great many more would do well to study and undertake. Its relations to soil quality and condition have been discussed in Volume II, Chapter 3. In these pages Professor Lehmann takes up the actual details of planning and carrying out a drainage project, large or small.—EDITOR.*

PERFECT underdrainage has been one of the chief factors in making the land of the Middle West the highest-priced general farm land in the United States. While it is the fertility of the soil that produces the big crops, it is the indirect results of drainage which makes the production possible. This is true not only in the Middle West, but in every other part of the United States where complete drainage is practised. Good drainage is one of the first essentials of maintaining a soil for permanent agriculture. If drainage is lacking, the greatest benefit derived from growing legumes, from applying manure, and from thorough tillage will not be realized. In a poorly drained soil nitrogen-fixing bacteria do not thrive, very little of the total fertility is made available for plant growth, and thorough tillage is impossible. As far as the physical condition of the soil is concerned, the three essentials of crop production are a soil of the right temperature, right condition of moisture, and sufficient ventilation. All of these are affected by underdrainage.

Benefits of drainage. Drainage makes the soil firm, so that the entire field can be cultivated. The areas of waste land so noticeable in poorly drained fields are replaced by growing crops. It makes possible early cultivation in the spring as well as early cultivation after rains. It warms the soil by removing the cold ground water, by eliminating a great deal of surface evaporation, and by making possible the entrance and downward movement of warm rains. It allows better aëration of the soil, which increases the decomposition of organic matter thereby making more plant food available. It makes possible a deeper root development

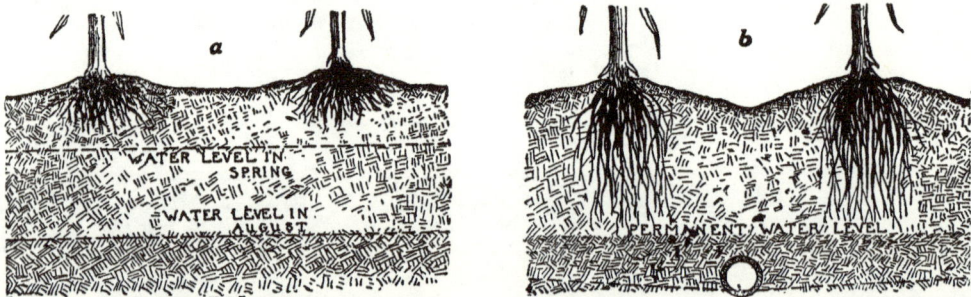

FIG. 345. Drainage stimulates deep root development. Roots held back by 'high water in spring cannot reach the supply in summer (*a*). When a drain keeps the water table at one place, the roots reach down to it and never want for moisture (*b*).

in plants by lowering the water table or line of saturation. Plants with a deep root development are better able to resist the effect of a drought.

Drainage increases the storage capacity of the soils, allowing the rains to soak in instead of washing over the surface. Thorough drainage prevents heaving due to frost. Less winter wheat is killed on well-drained land than on poorly drained. It improves the sanitary conditions of the surroundings and, lastly, it increases crop production enough to pay for the installation in one or two years.

Land needing drainage. There are a great many conditions that make drainage necessary. The most common are: Where low lands are flooded with water from higher surrounding land; where there are seepy or spouty spots on hillsides, due to a tight sticky soil underneath the surface; where nearly level fields or rolling tracts of upland do not have adequate natural surface or underdrainage; where there are marshes and swamps, and in irrigated sections where there are swamps and sloughs due to over-irrigation. Besides the 70 or 80 million acres of swamp and overflowed land that need to be reclaimed by drainage, it has been estimated that there are between 100 and 150 million acres now classed as cultivated land that could be tile-drained with profit. There are few farms that have perfect drainage. A heavy subsoil or other natural conditions may interfere with the natural tendency of the free soil water to pass downward by gravity. The result is a wet slough or seepy spot filled with marsh grass. Proper drainage is needed to make such land productive.

Kinds of drains. Farm drains are usually classed as open ditch or tile or a combination of the two. The vertical drain is another type which has been quite satisfactory in certain sections.

Surface Drainage

Open ditches. There are two kinds of ditches used in drainage work, the surface relief ditch and the outlet ditch. The surface relief ditch is a wide, shallow ditch used in connection with a tile on nearly level bottom land that receives the surface run-off from a large area of hill land. It removes a part of the heavy rains that occur occasionally and which would otherwise cause injury. A relief ditch should be so broad and shallow that it will not be a hindrance in cultivating the field and not have a tendency to form a gully by washing. It usually follows the line of natural drainage, and may be made directly over the main tile or to one side of it. When made directly over the main tile there is some danger of washing due to the soft earth above the tile. Where the surface water from a large field is drawn into the relief drain by

dead furrows or terraces it should be well sodded. Nearly all tile systems should be provided with relief ditches in connection with the mains.

Hillside ditch. The hillside ditch is a type of surface ditch used in preventing the flow of water from hills directly upon bottom land.

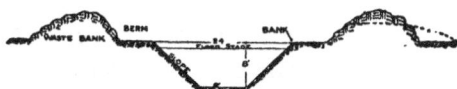

FIG. 347. Cross section of drainage ditch giving names and dimensions of parts. The right bank would be better if sloped as shown by dotted lines.

It is carried around the slope on a slight grade, thereby breaking the flow of water and reducing soil washing. This type of ditch is built by plowing several furrows down hill making it of sufficient depth to carry the surface water.

Terraces. The terrace is an outgrowth of the hillside ditch and is devised as a means of draining the water from hills in a satisfactory way and at the same time preventing soil washing, which is a big problem in hillside drainage. The Mangum terrace is about the best type. It is laid out with a very slight slope, usually about 6 inches fall per 100 feet. They are spaced on the slope so that the vertical distance between them is not more than 6 or 7 feet. As an outlet for a terrace, a well sodded surface relief ditch may be used if it is not convenient to discharge the water into a sink hole or wooded strip of land. A convenient method of

FIG. 346. Surface relief ditch combined with under drain. (This and Fig. 347, Wis. Bulletin 229.)

FIG. 347a. A large outlet ditch such as this is often a necessary evil. But it should be kept outside the cultivated area.

bu lding a terrace is to back furrow and drag the dirt toward the centre with a drag or grader. The terrace when finished should be 16 to 20 feet wide with the crown about 12 to 16 inches higher than the drain above.

Outlet ditches. An outlet ditch is often a necessary evil in draining swamps, bayous, lakes, and in straightening winding sluggish streams. While a ditch hinders cultivation and requires a great deal of maintenance it is the only satisfactory means of removing the water from large areas. In some sections relatively small outlet ditches used in draining a few thousand acres are being replaced by large tile drain. In localities where the value of the land does not justify an expensive system of drainage, open ditches are essential and very effective in soils of an open character. The nature of the soil allows the water to pass downward through it, thence laterally into the ditch. The soil becomes saturated before there is much flow into the ditch. In heavy clay soils open ditches afford very poor drainage. The soil absorbs the water very slowly and the greater part of it flows directly into the ditch, carrying a great deal of soil particles with it. The result is the water is not given an opportunity to replenish the moisture in the soil which is needed when rainfall is deficient.

Design of an outlet ditch. An outlet ditch must have sufficient capacity to carry the surface water in time of heavy rains. It should be deep enough to be an adequate outlet for tile mains when there is normal flow in the ditch. It must not be too deep in proportion to its width. A depth of 6 to 10 feet is desirable. This will be governed by the topography and type of soil. A ditch cannot be made as deep in a soil that caves easily as one that does not. The necessary capacity of a ditch determines its width. In no case should the width be less than twice the depth of the ditch. The actual dimensions have to be determined for each individual problem. If a very large area is to be drained the services of an engineer should be secured. Quite often a local man who has studied and made note of the conditions of flow in the natural drainage lines during high water is a good judge as to the necessary size of outlet ditch.

TABLE I.—NUMBER OF ACRES DRAINED BY OPEN DITCHES

DEPTH OF WATER 5 FEET.   DEPTH OF DITCH AT LEAST 6½ FEET

GRADE		AVERAGE WIDTH OF WATER						
Per Cent	Feet Per Mile	6 Feet	8 Feet	10 Feet	15 Feet	20 Feet	30 Feet	50 Feet
0.02	1.0	980	1,470	1,900	5,000	7,150	23,800	43,800
0.04	2.1	1,390	2,090	2,800	7,200	20,400	33,500	62,500
0.06	3.2	1,710	2,560	5,100	17,600	24,700	40,800	75,500
0.08	4.2	1,980	2,980	6,100	20,400	30,000	48,800	88,000
0.10	5.3	2,220	5,010	7,600	23,400	83,400	54,500	98,000
0.15	7.8	2,720	6,300	17,100	28,700	40,500	66,700	120,000
0.20	10.6	4,820	7,300	19,500	33,000	47,000	77,000	139,000
0.25	13.2	5,370	16,300	21,900	37,500	53,000	86,000	155,000
0.30	15.8	5,900	17,900	23,900	40,700	57,000	94,000	170,000
0.40	21.1	6,830	20,600	27,700	47,000	67,000		
0.50	26.4	7,600	23,000	31,000				
0.60	31.7	16,700	25,200	33,900				
0.70	37.0	18,100	27,300					
0.80	42.2	19,000						
0.90	47.5	20,500						

TABLE II.—NUMBER OF ACRES DRAINED BY OPEN DITCHES

DEPTH OF WATER 7 FEET. DEPTH OF DITCH AT LEAST 9 FEET

| GRADE | | AVERAGE WIDTH OF WATER | | | | | |
Per Cent	Feet per Mile	8 Feet	10 Feet	15 Feet	20 Feet	30 Feet	50 Feet
0.02	1.0	2,300	4,700	16,600	28,000	48,000	88,500
0.04	2.1	4,850	6,740	23,400	35,400	58,000	106,000
0.06	3.2	5,920	17,000	29,600	43,400	72,000	129,000
0.08	4.2	6,940	19,100	34,200	50,000	83,000	150,000
0.10	5.3	7,720	21,800	38,400	56,000	92,600	167,000
0.15	7.8	19,400	27,000	47,200	68,500	112,000	202,000
0.20	10.6	22,400	31,300	54,200	78,700	130,000	235,000
0.25	13.2	25,000	34,800	60,500	88.000	146,000	
0.30	15.8	27,400	38,200	66,200	96,500		
0.40	21.1	31,700	44,100				
0.50	26.4	35,400					

There are a number of formulas developed by the use of which, when the controlling factors are considered, the area which a ditch will drain can be computed. The accompanying tables were computed by means of Kutter's formula for average drainage conditions and are a help in determining the proper size of ditch to drain a certain area. For relatively small areas, a ditch should have a greater capacity than a large area, due to the fact that the water reaches the ditch more quickly, and less water has a chance to soak into the soil or evaporate. For this reason, for the areas above the heavy lines in table the ditches are large enough to carry off three fourths inch of water in 24 hours; for the areas between the heavy lines, the ditch will remove half an inch in 24 hours; and for the areas below the heavy lines, the ditch will remove one fourth inch in 24 hours.

To prevent caving, the sides of a ditch should be sloping. Under average conditions a 1 to 1 slope is sufficient. The areas in table above are computed for ditches with this slope. If the soil caves very badly a greater slope would be necessary, while in a stiff clay soil the banks may be left steeper.

Location and construction of outlet ditches. All outlet ditches should follow the lines of natural drainage as near as possible. The ditch should be staked out and grade established as outlined in Chapter 21, page 230. Nearly all outlet ditches at the present time are constructed by means of some kind of ditching machinery of which there are a great many types, designed to be used under nearly every condition. The cost of excavation by machinery is much less than by hand method. In some places ditches have been blasted very economically by means of dynamite. At the Montana Experiment Station, ditches were blasted at much less expense than they could have been dug by hand. In blasting, 2 sticks of 60-per cent dynamite were placed in holes 22 inches apart. The holes were made 2 feet deep by driving in 2-inch steel bars. Due to the soil caving, 1-inch galvanized pipe was used for loading the holes. The dynamite was put in through the pipe which was removed. The holes were filled with water over the charge, necessitating no other tamping. About 25 holes were fired at a shot, using the middle hole as a primer. Into one of the sticks of dynamite in the priming hole was inserted a blasting cap attached to a fuse, all parts being waterproofed. On the explosion of the priming charge, all other charges are exploded by concussion. After the blast the ditch is straightened up by hand.

There are many small open ditches constructed by plow and scraper. Such ditches are usually so small that they are not very satisfactory as outlets.

Maintenance of open ditches. The efficiency of every tile drain depends upon the condition of its outlet. There are a great many outlet ditches that are neglected, and a few years after they have been built, they are filled with mud to such an extent as to be practically worthless. The waste banks become rough and irregular, straw and cornstalks collect in the bottom of the ditch and are covered with silt and clay which gradually fills the channels. All ditches should be regularly inspected each year. All obstructions, such as trash and drift wood, should be removed. Bushes and weeds which would hinder the flow of water should be cut out. The deposit of silt along the sides and bottom of a new ditch furnish an ideal seedbed for the growth of weeds and willows, and if not removed immediately they form a lodging place for silt for future years. The method to pursue in cleaning out ditches depends on local conditions. Where ditches are spanned by bridges the equipment possible to use is often limited. In small ditches that practically dry up in the summer the team and scraper method may be used. In some cases in larger ditches silt banks and drift may be dynamited and the material carried away by the stream. A method that has been successfully used for breaking up the silt in ditches that contain 8 or 10 inches of water is to have a small flat-boat on which is installed a force pump driven by a gasoline engine. The inlet pipe to pump should be equipped with a strainer. The discharge pipe should be slightly smaller than inlet pipe and equipped with suitable fire nozzle that will deliver sufficient stream of water to break up the silt beds. Another method is to use a centrifugal pump and pump the silt over the water bank. Such a pump will remove 25 to 40 per cent solids with the water.

Underdrainage

Tile drainage. An ideal system of drainage is one that is adequate, permanent, not a hindrance to cultivation, and uses the least possible land. A tile-drainage system, if properly designed, meets the requirements of an ideal system.

History. The pioneers in underdrainage used various methods to provide a channel for the water to discharge as it seeped through the soil. Ditches were dug and brush and twigs from trees were piled into them and then covered with earth. An .improvement over this method was three straight poles, one resting on the other two at the bottom of the ditch. Some of these drains worked for a great many years. Flat stones were used, a row upright on edge along each side of the ditch with a flat stone on top. Many drains of this type are still giving good service. The first tile drains used in the United States were installed by John Johnston, a Scotchman, of Geneva, New York,

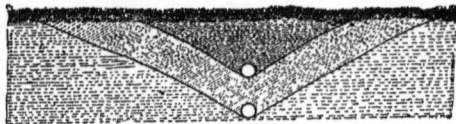

Fig. 349. The relation of the depth of tile to the width of the surface area drained (see text)

in 1835. The crude, hand-made tile used were shipped from England. A machine for making better tile was soon introduced, which greatly stimulated the installation of this type of drain.

Principles of tile drainage. There are certain essential principles to be kept in mind in considering a tile drainage problem. The main line which acts as an outlet for the laterals should follow the line of natural drainage as near as possible. All drains should be laid as straight as possible and in the direction of the greatest slope, an exception being where tile are used to intercept seepage water. Avoid abrupt turns and short laterals.

Spacing and depth of tile. The distance which the laterals are spaced apart will depend on a number of facts, chief of which are the type of soil and the depth of lateral. The depth is also determined by the type of soil. In sandy loam with a sandy subsoil where there is not an appreciable surface flow or seepage of ground water from higher land during and after storms, the laterals which are 4 feet deep may be spaced 125 feet apart. A great many people make the mistake of placing the tile too shallow in an open soil. Tile placed 4 feet deep will draw water from a greater distance each side of the line than when placed only 3 feet deep. This is illustrated in Fig. 349. In the prairie sections it is rather general practice to space the tile 100 feet apart, on an average of $3\frac{1}{2}$ feet deep. On close, stiff soils the tile should be placed to a depth of 2 to 3 feet and an average of 50 to 60 feet apart. It makes little difference whether the depth is greater than 3 feet unless the tile are covered by soil which does not allow the water to seep through readily. There is really little information on the proper spacing of tile in gumbo soils. This type of soil is heavy and sticky when wet and it would seem necessary to place the tile very close together to get any results. However, reports from one experiment station state that the tile need not be laid closer than 150 feet. Other information would indicate that it is better practice not to place the tile farther than 100 feet apart in the gumbo soils or in the type of soil usually found along river or creek bottoms.

Size of tile. The size of tile for mains can be obtained by referring to Table III. This table is worked out on a basis of the main tile removing one fourth inch of water from the area in 24 hours. To use this table it is necessary to know the acreage to be drained

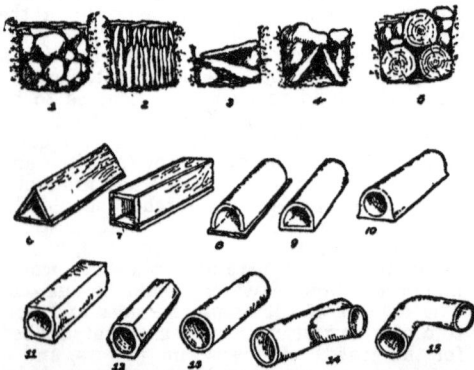

Fig. 348. Types of drain, old and new: 1–4, stone drains; 5, pole drain; 6 and 7 board drains admitting water at bottom edge; 8, horseshoe tile on board; 9–13, tile of various shapes; 14, Y-junction; 15 elbow. (Cornell Reading Course).

TABLE III.—NUMBER OF ACRES DRAINED BY TILE REMOVING ¼-INCH OF RAINFALL IN 24 HOURS*

Inside Dia. Inches	Fall per 100 Feet in Inches.													
	¾	1¼	1½	3	4½	6	8¼	10½	12	1' 6"	2' 0"	4' 0"	7' 6"	10' 0"
4	4	7	8	12	14	16	19	21	22	28	31	44	60	69
5	7	12	14	19	25	28	32	37	39	47	55	77	105	122
6	10	19	21	30	39	43	51	58	61	74	86	122	166	194
7	15	28	32	45	56	63	75	85	90	109	126	179	244	282
8	22	40	44	62	78	88	104	118	124	152	177	250	340	394
9	29	53	59	83	106	118	140	158	167	204	236	334		
10	38	69	77	109	137	154	181	206	217	267	308			
12	59	109	121	171	217	244	287	326	342	418				
14	92	159	176	251	318									
15	104	190	212	300										
16	121	222	248	325										
18	164	298	325											
20	213	389												
21	241													
22	270													
24	336													

*Table computed by J. H. Ames, Iowa Highway Commission.

as well as fall per 100 feet. For example, to drain 150 acres with a fall of three eighths inch to each 100 feet, it would require an 18-inch tile while a 9-inch tile would drain the same area with a fall of 10¾ inches per 100 feet. To be safe in selecting proper sized mains, it is well to increase the area by 50 per cent before using the table. In this case, a tile would be selected for 225 acres instead of 150 which would necessitate a 20-inch or 10-inch tile, depending on whether there was seven-eighths inch fall per 100 feet or 10¾ inch fall per 100 feet. The latter size tile would be best to use. The size of sub-mains may be determined in the same manner.

In selecting the proper size laterals, one should never use less than 4-inch tile. Where the land is rather flat with a fall of 4 inches or less per hundred feet, 5-inch tile would be better. Five-inch laterals are not as liable to be stopped up as smaller sizes. It has been found that it is not good practice to have the laterals more than one fourth mile in length; nor is it economical to make the laterals short.

How the tile works. Contrary to the common idea, the water does not enter the tile through its walls but at the joints between the ends of the tile, although tile should be placed as closely together as possible. The water enters the tile due to the force

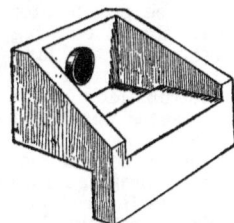

FIG. 350. Concrete outlet bulkhead, simple and cheap but permanent and efficient. A grating is sometimes needed to keep out rats, etc. (Farmers' Bulletin 805.)

of gravity, and it is this force that carries the water through to the outlet.

Tile drain outlet. The possibility of an outlet or a place for the water to flow eventually is the first thing to look for in draining a piece of land. It is the most im-

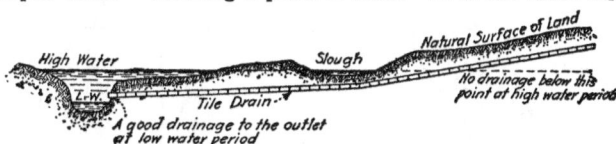

FIG. 351. The elevation of the outlet and the drain as a whole should be based on the high-water, not the low-water, level. (Minn. Bulletin 15).

portant feature of an effective drainage system and must be properly built and protected. When the outlet tile receives little or no protection, the earth around the tile is washed away and one tile after another then drops down and is washed out, causing a gully to be formed. Cattle and hogs in tramping around such an outlet will knock the tile out of place and partially or entirely stop it up. When soft tile are used at an outlet freezing will break them. The proper protection for an outlet should be a concrete bulkhead constructed as shown in Fig. 350.

A glazed sewer tile is sometimes used as an outlet. A galvanized iron pipe makes a very satisfactory outlet into an open ditch which is liable to be deepened or widened at any future time. This type of outlet is not permanent but can easily be replaced. The concrete bulkhead should be used when the outlet is into some permanent stream. A good outlet is enough lower than any point in the field so that there is no danger from back water or silt. Many tile outlets are submerged sometimes during the year, but this does no damage if the tile are prop-

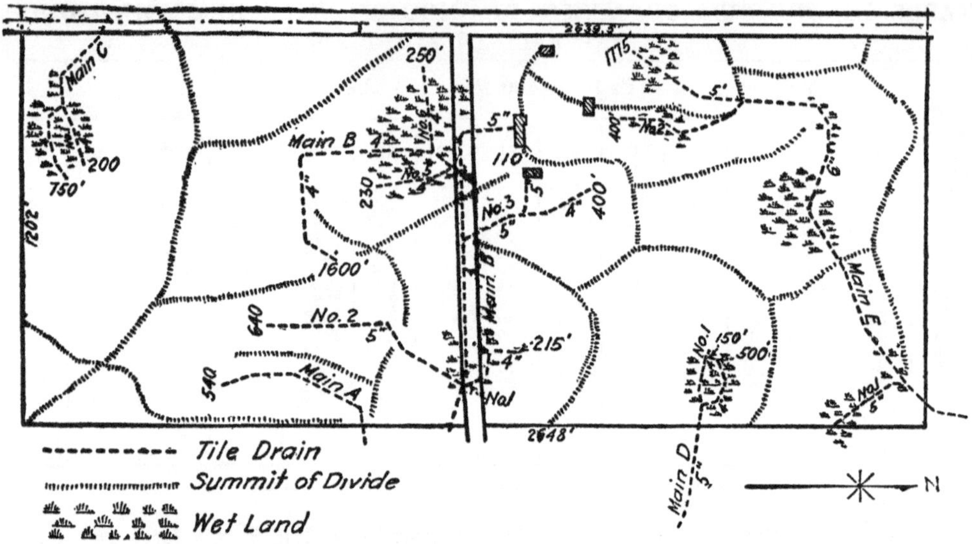

FIG. 352. Map of a drainage system showing lines of tile following the low places between the divides. Figures show size of tile and length of each run. (Minn. Bulletin 15.)

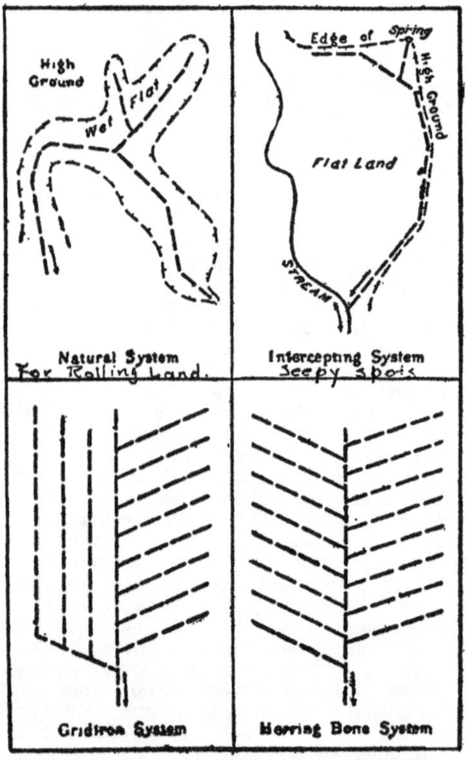

FIG. 353. Systems of tile drainage for special conditions

erly laid. To be effective as soon as possible after heavy rains, the outlet must be at a higher level than the normal water line of the ditch so as to allow free flow from the drain. In some cases this necessitates deepening the stream or grading it for a certain distance below the outlet. In many cases crooked, meandering streams can be made into effective outlets by straightening them and removing logs and brush. In some cases it is necessary to secure an outlet by draining through a neighbor's field or across a public highway. Proper coöperation between parties concerned is the best way to handle such cases.

Draining a farm. The first thing any one should do in preparing to drain a farm is to make a preliminary investigation to determine the best method to follow. Most farmers know the type of soil, how fertile it is, and are able to make a fair estimate of the value of crops that can be grown after drainage. Such an estimate should be a big factor in determining whether a piece of land should be drained or not. The farmer also knows the parts of the bottom land which are wettest and where the seeps appear on the hillsides. He also knows how high the water stands in the creek under normal conditions and where the natural drainage lines are in the fields. All of these facts should be considered in the preliminary investigation. The type of soil should be examined carefully, both the surface and the subsoil. The presence of a tight clay or a sandy soil would affect the arrangement of the drains very materially.

FIG. 354. An economical method of draining bottom land. The fewer junctions and outlets the better (Minn. Bulletin 15).

Planning the system. A system of tile drainage is composed of laterals, submains, and mains. The laterals are the branch drains that empty into the mains or submains. The submains empty into the mains and the mains carry all of the water to the outlet. Quite often a single line of tile is all that is necessary. Plans for complete drainage should be made at the outset. All mains should be large enough to carry the water from the laterals. On rolling land the natural system of drainage is used. The mains and submains follow the lines of natural drainage and lead to the laterals located in the small sloughs and draws as found on such land.

An economical system of tile drainage suitable for bottom land is one with long laterals with as few junctions and outlets as possible (Fig. 354.)

Draining a seepy hillside. A seepy spot on a hillside is due to the surface soil being underlaid with a tight heavy clay which outcrops somewhere down the slope. The excess water in the soil as it pours downward reaches this layer of tight dense soil and follows it to the surface. This seep water will quite often ruin the soil for cultivation for several rods down the slope. In draining such a spot the tile must be laid across the slope on the upper side of the wet outcrop and deep enough to intercept the water flowing along the tight soil. These can best be located while the ground is wet, or have the spot marked out during the wet period.

The method of staking out lines and establishing a grade is discussed in Chap. 21.

Digging the ditch for tile. After the lines have been staked out and the grade has been properly established excavation can

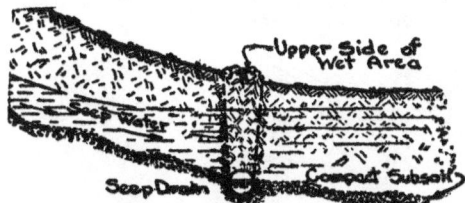

FIG. 355. Location of tile placed to drain a seepy hillside.

begin. The tools used in this work for the ordinary type of soil are shown in Fig. 357. Where very hard soils or roots and stumps are encountered a pick and grubbing hoe are needed.

In heavy, sticky soil the open spade can be used to advantage. The round-nose shovel is used to remove the crumbs after the first and second spading. The tile scoop is used for cleaning the bottom of the ditch to receive the tile. The scoop is made in different sizes to accommodate the different sized tile.

The excavation should begin at the outlet or lower end of the drain. The topsoil, or first spading, is usually placed on one side of the ditch and the bottom spading on the other. The ditching must be carefully and accurately done. Each part of the ditch should be tested with a gauge rod to see that it is to grade at all points, being careful to shape the bottom of the ditch to receive the tile. Proficiency in digging a ditch for tile comes only with practise. All lines should be thoroughly checked.

On large jobs a ditching machine (Fig. 260) can be used economically in excavating for tile. Where a machine is used, sight bars are established a definite distance above the proposed bottom of the ditch and are sighted on in regulating the cut of the machine.

Laying the tile. The tile are laid as fast as the ditch is completed. A

FIG. 356. Bringing the ditch to grade by means of line and gauge rod.

great many tilers place the tile in the ditch by hand; others prefer to use the tile hook. It requires practice to get the tile to fit closely in the ditch. It is usually done by quickly revolving the tile until the ends make a good joint. If a space of one fourth inch or more is left between the ends of tile there would be danger of entrance of silt. In laying large tile that cannot be handled easily with hook or by hand, a derrick is used. At the end of string of tile and at sharp curves where spaces between the ends of tile are likely to occur, broken pieces of tile should be placed to keep the line from filling.

Where quicksand is encountered in digging, extreme care must be observed to keep the tile to grade. Boards are sometimes placed in the bottom of the ditch. To prevent the sand from entering the tile at the joints they should be surrounded with coarse hay or grass,

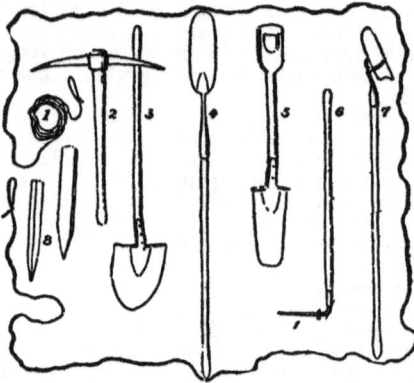

FIG. 357. Hand tools for drainage work: 1, line; 2, pick; 3, round shovel; 4, scoop shovel; 5, narrow ditching spade; 6, tile hook for lowering pipe; 7, bottom smoothing scoop; 8, gauge pegs.

FIG. 359. A ditching plow is sometimes used in stiff soil, but it leaves a ragged top edge

burlap or tarred paper sometimes being used. If a supply of coarse gravel is at hand, it should be packed around the tile. It will be more permanent and very effective.

Proper caution must be observed in joining the laterals to mains. It is best practice to make a smooth curve so that the flow of water from the lateral into the main will be down stream. With small mains a "Y" connection is best but a little more expensive. In joining a lateral to a main it is best to connect it with the top of lateral on level with the top of main. The hole in the main may be started with a hammer or pick and enlarged by breaking off small pieces with a monkey wrench.

FIG. 358. One way to join lateral and main lines

Place broken pieces of tile around junction to prevent entrance of silt.

Blinding the tile. The tile should be covered with a small amount of dirt just as soon as they are placed in the ditch. This is called blinding or priming. The tile are usually blinded by spading off the sides of the ditch, care being taken not to knock the tile out of line. The purpose of blinding is to hold the tile in place until the ditch is filled.

Filling the ditch. The trenches are not usually filled until after all tile are laid and blinded. One common method of filling is to use a walking plow with a long double-tree having a horse on each side of ditch. An "A" shaped drag can be used to advantage and light graders and scrapers are also used. Either of these methods is much more economical than to fill by hand. Where tile are laid in tight soils it is good practice to place the top soil directly over the tile, and the soil from the bottom of the ditch in last. In other words, the back-filling is just the reverse

of the excavation. Under certain conditions gravel or sand should be placed over all the tile to allow the surface water to pass through more quickly. This is advisable in draining an old pond or barn yard where the top soil is puddled to such an extent to prevent the passage of surface water.

Care of tile drains. After the tile are installed they should be watched to see that they are working properly. The condition of the soil a few days after a rain is the best indication as to how the tile are working. If a wet, seepy spot is found over a tile line it is a sure indication of an obstruction. This may be due to a broken tile or filling in of silt where the tile are off grade. Sometimes roots from trees and plants will stop the tile. No willows should be allowed to grow along a tile line unless the sewer tile are used and the joints closed with cement. Alfalfa, though a deep-rooted plant, has given very little trouble in stopping up tile. One condition where alfalfa roots would probably go into a tile is where a line passes through a dry, well drained piece of land carrying water from a wet, seepy spot. Tile lines draining a seepy spot or spring should be carefully inspected each year.

Cost of draining. The cost of drainage work consists primarily of the cost of excavation and cost of materials. In large drainage projects the cost of engineering service, legal proceedings and organization become rather large items. In open ditch work the cost of excavation may be more than ninety per cent of the total cost. In tile drainage the cost of excavation varies from fifty to seventy-five per cent of the total. All of these items vary greatly in different localities.

Cost of tile. Drain tile, if bought in carload lots directly from the factory, are much

FIG. 360. Two other methods of joining mains and laterals. At left, the better; and the commoner of the two

The furrow method of irrigating, usually practised on tilled land, but here on a field of timothy

A large market garden establishment with many acres under irrigation by the pipe-line system

Furrow irrigation of an orange orchard. Thorough preparation oī the surface before planting is essential

THERE IS AN IRRIGATION SYSTEM FOR EVERY SET OF CONDITIONS UNDER WHICH ARTIFICIAL WATERING IS NEEDED

The standpipe system, with either stationary or revolving sprinklers, is excellent for permanent flower and vegetable gardens

Lawn perfection can be maintained by means of a portable nozzle-line system with an automatic pipe-revolving device

OVERHEAD OR SPRINKLER IRRIGATION, AN INVENTION OF THE TWENTIETH CENTURY, IS OPENING THE WAY TO NEW POSSIBILITIES IN HUMID REGIONS

FIG. 361. Filling a ditch with a plow and long doubletree

cheaper than when purchased from local dealers. Table IV is representative for tile delivered at middle western railroad stations.

TABLE IV—COST AND WEIGHT OF TILE (1919)

SIZE	FEET IN 15-TON CARLOAD	PRICE PER 1,000 FT.	PRICE PER ROD
4 inch	5,000	$ 36.00	$ .59
5 "	3,000	49.00	.81
6 "	2,500	66.00	1.09
8 "	1,500	117.00	1.83
10 "	1,000	175.00	2.90
12 "	850	200.50	3.30

Prices per rod and per 1,000 ft. are on the same basis.

Cost of excavating and **laying.** The cost of excavating the ditch and laying the tile depends upon labor conditions, type of soil, depth of drainage, and size of tile. The cost of complete drainage is often as much as $25 to $50 an acre.

The cost of excavating increased practically 75 per cent. during the years 1916 to 1919. But land values have also increased and owners of land that needs drainage cannot well afford to hold back on account of high prices. The following figures give costs per rod for digging trench, laying the tile, and back filling where the wages for good diggers are 40 to 50 cents an hour.

SIZE OF TILE INCHES	DEPTH IN FEET		
	3	4	5
4 to 6	$ .60	$.95	$1.45
7 to 8	.76	1.20	1.75
9 to 10	.87	1.36	2.10
12	1.00	1.55	2.04

The cost of back filling alone is not more than 3 to 5 cents a rod, if a team and plow are used, but prohibitive if done by hand.

The data on this page make possible an approximate estimate of the cost of the laterals of a system. To this must be added the cost of mains, submains and back filling.

Feet between laterals	50	66	80	100	150	200
Feet of tile per acre	872	660	545	436	291	218
Rods of tile per acre	53	40	33	26	18	13

Kinds of tile. Burned clay and concrete tile when properly made give equally good results. A poor quality tile of either kind will disintegrate when in use. Contrary to the general idea clay tile should be hard burned, not porous. A vitrified clay is more durable and better than the salt glazed variety. The following general requirements are from Bulletin 31 of the Iowa Engineering Experiment Station: "All drain tile shall be good, sound tile, of first class quality. They shall be entirely free from cracks and fire checks extending into the body of the pipe in such a way as appreciably to lower its strength. No pipe shall be accepted having pieces broken out in such a way or to such an extent as appreciably to affect the strength of the pipe, or to permit the entrance of soil.

"The pipe shells shall have uniform, strong, dense structures throughout, without serious flaws or weak spots.

"All pipe shall give a clear ring, when stood on end or laid on one side, evenly supported at the lower end, or along a line of one side, and free elsewhere, and tapped with a light hammer while dry.

"All pipe shall be regular and true in shape. The average diameter shall not be more than 2 per cent. less than the specified diameter. No two diameters of the same pipe shall differ from each other more than 7 per cent. nor shall the average diameters of adjacent pipe differ more than 4 per cent.

"Pipe may be furnished in lengths of 1, 2, 2½ and 3 feet, but 1 foot lengths shall not be used for sizes more than 15 inches in diameter. No pipe, designed to be straight, shall vary from a straight line more than 1½ per cent. of its length.

"If cement tile are used, they shall show a uniform, dense structure, with clean aggregates, well graded as to size of materials, and with the grains and pieces of aggregate well coated and the pores well filled with good Portland cement. There shall be no spots of specially great porosity. Fractured surfaces shall show broken pieces of aggregate firmly bedded in the concrete. The general appearance of the material shall be at least equal to that of first class gravel concrete, in these proportions: 1 part first-class Portland cement; 2 parts clean, coarse sand; 1 part pebbles, from 1 inch to half the thickness of the tile wall in diameter."

Drainage records or maps usually show all lines of tile. Such a record is important

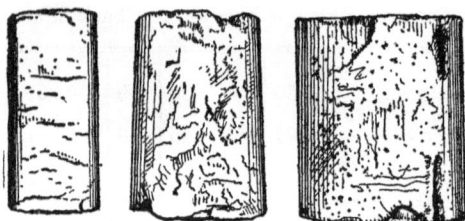

FIG. 362. The kind of tile *not* to use. It is poorly made and will soon break down and cause trouble and expense.

in the location of lines of tile when the system is extended. It is also of value if the farm is sold. Fig. 352 is an example of a good drainage map.

Drainage of irrigated land. It has been estimated that more than one million acres of land in the irrigated sections of the United States need drainage. The greater part of this need of drainage is brought about due to over-irrigation, also to waste from canals. Quite often the lower lying lands are injured by the seepage from higher land. It is necessary to find the source of the damaging water to be able to locate lines of tile intelligently. The drains must be located to intercept the water before it does any damage. After the intercepting drain is installed and the seepage from outside sources cut off the natural drainage of the area usually is sufficient to take care of the water applied.

Drainage laws. Nearly every state has a general drainage law which gives the land owners authority to form drainage organizations of a coöperative character. Laws are always necessary in the construction of large drains in which a number of people have a common interest. While the laws vary in detail in the different states, they are practically the same. The essential features of a drainage law according to C. G. Elliott, formerly of the United States Department of Agriculture, are: "First, the right given to property owners under certain prescribed conditions to petition the proper authority for the construction of drains which will be of public benefit; second, provision for making and collecting assessments to defray the cost of the work, and also for the appraisement and payment of damages to property incident to such construction; third, the establishment of the perpetual right of land owners included in the district to use the ditches or drains which are constructed; fourth, authority under proper legal regulation to incur debt and sell bonds for obtaining money with which to perform the public part of the work."

Special Drainage Systems

Surface inlets or vertical drains. Surface inlets are used to allow the surface water to flow directly into the tile without seeping through the soil. For this reason, they are of great value in dense soils in depressions where the water collects and would drown the plants if not removed. In draining barnyards and feeding lots where the soil is of a clay formation and is puddled by tramping, surface inlets should be used. The simplest type of surface inlet is to have a part of the trench above the tile filled with rock or broken stone. A convenient form of surface inlet is a large sewer tile with side connections for the drains and provided with a grate at the top. A concrete box also makes a very satisfactory surface inlet. A surface inlet made of either a concrete box or sewer tile should be made deeper than the drain so that any sand or silt that is washed in will not be carried down the tile.

The silt basin. The silt basin is a small well for collecting the sand and silt that is carried along in the tile. It may also be used as an observation well. The silt basin is needed most in sandy soils where there is a long drain or where the grade decreases, thus tending to deposit silt. The combined inlet and silt basin are often used at the upper end of a line. The silt basin is constructed of either sewer pipe or a concrete box. The bottom of the basin should be carried down two or three feet below the outlet tile.

Collecting basins or relief wells. Relief wells are used in connection with a tile drain to intercept seepage water which percolates down a slope. They are very useful in the drainage of irrigated land where the seepage water flows 8 or 10 feet under the surface. The tile line is installed to the ordinary depth and is supplemented with wells bored at intervals in the bottom of the trench to a sufficient depth to provide a passage through which the water may rise to the drain.

Drainage wells. In limestone areas there are many sink holes that furnish outlet for drainage systems. In the same type of soil wells are bored and

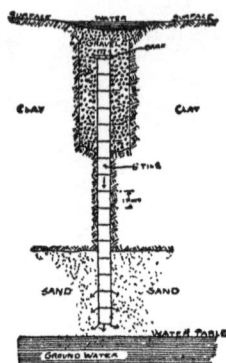

FIG. 363. A vertical drain that may serve as an outlet for a lateral line or, by itself, drain a pothole. (Wis. Bulletin 229.)

FIG. 364. A combined open catch basin, silt basin and inspection basin. (Kansas Report.)

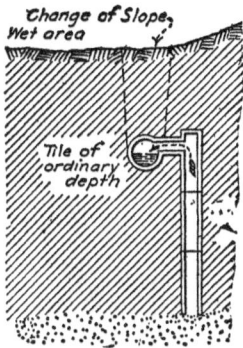

FIG. 365. Relief well for bringing water from below up to the drain line.

used for the same purpose. There are single wells in use that are an outlet for a system of drainage for more than 1,000 acres. Such a well must be properly curbed to prevent it washing. This type of outlet for drainage water is not always advisable. This is especially true in closely settled communities where the water supply for the home is used from the same type of wells and would be liable to contamination from the field drainage.

Drainage of the home surroundings. One of the essentials of a good house location is to be well drained. Where the natural drainage is not perfect, artificial drainage should be provided. This is necessary not only from a standpoint of health, but also of convenience and a means of providing better growth of plants in the yards and gardens. A sanitary barnyard requires that it be well drained. Eaves troughs and down spouts should be pro-

FIG. 366. Blind inlet or drainage well sunk below the surface to permit cultivation.

vided for all the buildings. Where a cistern is not used they should be connected to a tile drain. A greater part of the mud common to the average barn-

yard is due to the water flowing from the roofs of the barns and sheds. Ample drains with surface inlets should be provided for the feed lots. Surface relief ditches should be used to divert any water

FIG. 367. Stone drainage well and screened wooden box that serves as a filter. (Cornell Reading Course.)

from higher land from flowing into the barnyard. The foundations of all the buildings should be well drained, especially the house. Drains extending underneath the cellar wall will prevent water entering the cellar.

Road drainage. Proper drainage is one of the first requirements of a good road. There are three forms of water that must be removed from a road to drain it properly. The surface water on the top and sides of the road, the ground water in the road bed, and the flood water.

To secure good surface drainage the road must be properly graded and crowned, having the crown not less than 24 to 30 inches above the flow line of the side ditches. The crown should be kept smooth by dragging, keeping all ruts filled. The side ditches should be properly maintained to prevent them from forming gullies. The ground water should be removed or lowered by tile drainage. Often one line of tile laid under the shoulder of the road on the upper side is all that is necessary. Where the road is flat a line on each side is usually better than one under the centre of the road. To take care of flood water by conducting it across a road, proper sized culverts should be selected for the area drained. Quite often a relatively small concrete culvert can be substituted for a long dilapidated wooden bridge which is always a source of danger.

Profits from drainage. The profits from drainage are manifold. The increase in production and the rise in value of land are the most noticeable. There are thousands of acres of once practically worthless land that have been transformed by drainage into productive farms that sell for $200 an acre. Many acres that have been cropped at a loss for years double their production after drainage. Every farm that has a few wet acres is farming at a loss until they are drained.

Drainage is also profitable from a standpoint of health. By drainage, the mosquito pest is partially or entirely eliminated, reducing the malaria in the community. In one district malaria was reduced 70 per cent after drainage. The health condition of the farm animals is affected by drainage. An important feature in combating hog cholera is to have well-drained lots. Sickness and disease are very costly, so an investment that will improve health conditions will pay big dividends.

FIG. 368. Using the scraper crosswise of the furrow in making an irrigation ditch (see text, p. 252)

CHAPTER 23

Practical Farm Irrigation

By PROFESSOR R. P. CLARKSON (see Chapter 14), who discusses the engineering details of ordinary irrigation on and under the surface, and F. F. ROCKWELL (see Chapter 11), who takes up overhead irrigation and its equipment. When we speak of irrigation work we think first of the immense projects of the West, which impound millions of gallons of stream water and provide moisture for the needs of many settlers and their many cultivated areas. The details of these enterprises are, however, of no practical interest to the small, individual farmer. He wants to know (1) how to carry out irrigation plans and projects on his own place, when this can be done; and (2) how to handle and distribute the water supplied to him at the edge of his fields by the local large-scale project upon which he depends. These problems are the ones treated in this chapter.—EDITOR.

THE value of irrigation, its purposes and its various relations to soil management principles, are treated in Volume II, Chapter 3. This article treats of only the engineering side of the subject, the practical details of things to do, how to do them, and what to do them with.

A good general lays out his plans carefully before starting anything. The first step in irrigation should not be taken until the entire campaign is planned out. Determine the needs of your land, the nature of the crop, the soil conditions at the time, the water available, the money and time at your command, any expectations of the season that can be estimated in advance, in fact every feature that is likely to affect results; and then determine the method of irrigation to be adopted.

Principles of Irrigation Practice

The first step in any case is to study the slope of the ground and carefully prepare the surface. It must really be quite well graded, gulleys and hollows must be filled in, while ridges and humps must be leveled.

Leveling the surface. This is best accomplished in 2 operations: (1) Go over the land and grade roughly any particular spots that need attention. (2) Follow up with plow and harrow over the field, and, if necessary, with some type of drag or scraper, the split-log drag (p. 272) being useful and easily made. A light U-shaped scraper formed of a 2-inch plank placed on edge will prove even more satisfactory. The dragging should take place in both directions, the breaking up of the field by means of plow and harrow need, of course, be done but once.

To be successful the land levels must

FIG. 369.   Leveling the surface with a home-made plank scraper, preparatory to irrigating

FIG. 370. Finishing a small ditch with an A-shaped plank scraper

not be determined solely by the eye. Stretch a cord between stakes and grade to the cord. The surface must not be horizontal, indeed this condition could not be attained on many farms, but the surface of any particular field should be flat and even, not broken by hillocks which would remain dry, or by hollows and dead-end ditches which would receive more than their share of water.

Preparing side hills. On side hills of any considerable steepness of grade, the land must be terraced to accomplish anything like even distribution of water. This terracing is done by first breaking the field with plow and harrow to make it workable. The best plan is then to decide on the distances between terraces. They should be small and close together so that the field has in reality a sort of saw-tooth surface down hill. Oftentimes the whole work required can be done with a plow turning out furrows the proper distance apart, say 4 or 5 feet, and then grading the intervening land to a gentle slope from the top of one furrow to the bottom of the next lower furrow. This does away with the necessity of removing the topsoil during grading operations and restoring it afterward. The conservation of the topsoil is, however, sometimes required where a thin topsoil and poor subsoil are encountered.

Methods of Irrigating

Furrows and flooding. With a terraced hillside and, in fact, with practically all steeply sloping land, the flooding from a canal at the top of the slope is the easiest and cheapest method of getting water on the surface. If it is a long hill several canals may be laid out at intervals of perhaps 50 to 100 yards down the hill, depending on the nature of the soil. The water is led into each of these canals, the canals are dammed up by methods described later, and the water rises and flows over the side of the canal. The grade of the canals must be carefully established so as to give a gentle slope with a fall of, say not more than 1 foot in 1,000 to allow the water to flow properly. The water, when it rises in the canals, must flow over the bank in a more or less unbroken sheet. To assure this the edge of the bank over which flow takes place must be leveled accurately. At the bottom of the hillside a canal should be placed to catch the surplus flow and direct it to further irrigating purposes on the lower lands.

Bedding layout. If the ground is practically level a bedding method may be employed. The land is plowed in ridges from 12 feet wide for hand cultivation to 25 feet or more wide for horse machinery. The result is really a series of wide but very shallow ditches, the earth being all worked in so as to give merely a smooth surface everywhere. Along the crest

of each ridge a small distribution canal may be run from the main supply canal at the head of the ridges. These supply canals are about 100 yards apart, and divide the field into plots of that length, the distribution canals and ridges being parallel to the length of the plots. At the bottom of the depression between the ridges may be placed open drainage canals or tile underdrains the method of laying the tile being described in Chapter 22.

Orchard irrigation is becoming more or less extensively practised, the furrow and flood system being very common and simple in this connection. Occasionally the bedding layout is used, the distribution furrows being run between the trees and so spaced as to throw the rows of trees into the depressions. A modification of this, however, requires only a single distribution furrow alongside each tree row, the land being sloped slightly from the furrow to the opposite side of each tree where

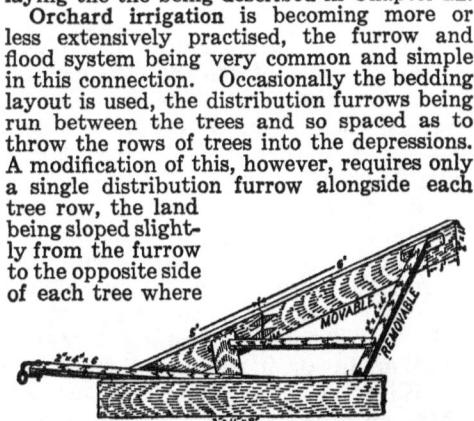

FIG. 371. Terrace drag or A-shaped scraper used in hillside irrigation and in making small ditches (Farmers' Bulletin 882).

the earth is piled into an embankment. This sloping can be done with a hoe at each individual tree, the furrow side being cut down so that flooding over the side is permitted only at the tree point. Thus there is a little bed or check at each tree which may be flooded as desired.

Vineyard irrigation is practised in both of the ways outlined above.

The irrigation level and its use. The most convenient form of cheap level for use in almost any farm operations which do not extend over any great distance is made by fastening a good carpenter's level to a well-made straight edge 20 feet long. At each end fasten and firmly brace a leg say $3\frac{1}{2}$ feet long. One leg is best made adjustable to slide so that grades may be determined with it. The shorter leg is placed on the surface of the ground at the start of a ditch or grade line, and a hole is dug for the longer leg to such depth that the straight edge is parallel. The difference in length of the legs then gives the fall in grade for the 20 feet in the direction of the longer leg; multiplied by 50, it is the total fall per 1,000 feet.

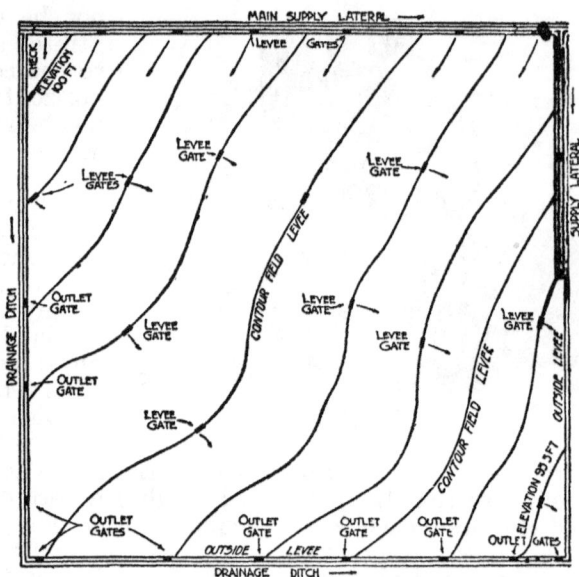

FIG. 372. Rice field laid out for irrigation by flooding. Note that there is only about 7 feet difference in elevation between the upper left-hand and the lower right-hand corners. This makes less than a foot of fall between each two levees. (Calif. Bulletin 279.)

When one hole is dug and levels obtained, the short leg can be placed in that hole and a

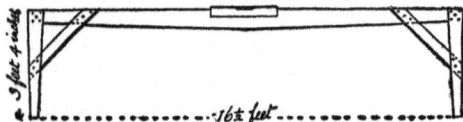

FIG. 373. Irrigation level. The dimensions ($16\frac{1}{2}$ x $3\frac{1}{4}$ feet) are in practically the same proportions as those given in the text (20 x $3\frac{1}{4}$ feet).

second hole dug for the longer leg until a level is again obtained. This operation can be repeated to the end of the line, then the ditch excavated to the grade joining the bottoms of the holes, stakes driven, and a cord stretched, to make sure there are no intermediate humps and hollows. It is a good plan to check the grades over again with the level after the ditches are opened up. The flowing water will give a good indication of the evenness of the grade.

Cost of preparing land. Dr. Elwood Meade, a well-known authority on irrigation subjects, states that the cost of preparing the land for irrigation is from $3 to $30 per acre. The flooding method is, of course, usually the cheapest method and frequently costs considerably less than $3 per acre irrigated. Unless some particularly expensive obstacle is encountered, the cost will seldom go over $4 or $5 per acre.

Pumping irrigation water. In many places irrigation is impossible without pumping; this type is always very costly, especially when compared with some neighboring project in which water is obtained from reservoirs by gravity flow. Low-head, large-volume pumping has been worked out by means of the development of centrifugal or turbine pumps and on a big scale it is found to be economical. Small volume pumping, however, depends almost entirely on the windmill or oil engine or, in rare cases, the farm steam engine.

With the many disadvantages attached to the pumping of a water supply, there go also some advantages among which two in particular stand out: (1) There is always a high degree of control of quantity of water and period of distribution. (2) There is no controversy over sharing the use of the supply with

FIG. 374. Many irrigation projects in the Southwest are supplied from flowing wells such as this

others. In other words, your pumping plant is independent and in control entirely of your operations.

Frequently underground or well water is the only source of supply for pumping, and through the Middle West the windmill adapts itself particularly to these conditions where the operation of irrigation is on a small scale. However, the small size and power output of windmill units make them a highly expensive and more or less undependable source of power for acreage irrigation.

The first cost of pumping plants for about 15 installations in the South and Middle West where the head of water pumped was about 20 feet and the average acreage treated was about 200, was about $15 per acre, while the annual cost per acre was slightly under $7. With plants having heads averaging 40 feet and supplying on an average only 100 acres, the first cost was about $25 per acre and the annual cost about $12 per acre.

Practical considerations. It would not be advisable to irrigate all land which requires it. A practical consideration of costs is necessary in each case. Some fields are so low or the soil is so heavy that a system of irrigation would, of necessity, require also a system of drainage to prevent the fields being turned into swamps and marshes. In this case the total cost of the irrigation and drainage might be very much greater than the

FIG. 375. Where the going is firm a wheeled scraper will do more and quicker work in leveling and ditch-making than the ordinary drag type (Fig. 369).

possible value of the additional products obtained from the land as a result of the treatment. To take another case, if the land is situated far above the water supply, the cost involved in raising and storing the water may be prohibitive.

Cost, not physical conditions about the farm, is usually the only obstacle in the way of irrigation. Usually where it pays at all, it pays well, for it makes possible a crop which should return in a very few years the total outlay involved.

Source of water supply. The supply of water involves the most serious part of any irrigation project. For small garden and market farming, wells can as a rule be depended upon. For fairly dry regions, however in the case of irrigation over a number of acres, wells are not dependable, except for partial service.

The requirements for a project of considerable size are such that only streams, lakes or springs can be relied upon, unless unusual conditions as to flowing or artesian wells are available. Even then, after a period of use they have been found to require expensive pumping to keep up their supply. In some cases as in the West, large, artificial storage basins or lakes are created by proper engineering measures, but as a rule this form of water supply cannot be afforded unless a very large project is contemplated. A small, artificial pond, unless continually fed by streams or springs would not be a dependable source of supply.

Quantity of water required. A number of factors enter into a study of the amount of water required to irrigate each acre properly. The nature of the crop is important as some plants require more moisture than others, which results in either a larger amount being used at each operation or more frequent operations. The climate affects the evaporation and thus the total amount of water that must be applied in order to give the crop a certain amount. The nature of the soil treated affects the absorption and retention of the water. Lastly, the subsoil affects the water requirements through absorption or drainage action. An examination of each of these controlling factors must be made before the layout of any irrigation system, in order that the water requirements may be met not only by the source of supply, but by the various ditches and canals whose size and grade are designed with a view to a definite water-carrying capacity.

The following table indicates the effect of different types of soil, the absorption figure being for either surface or subsoil:

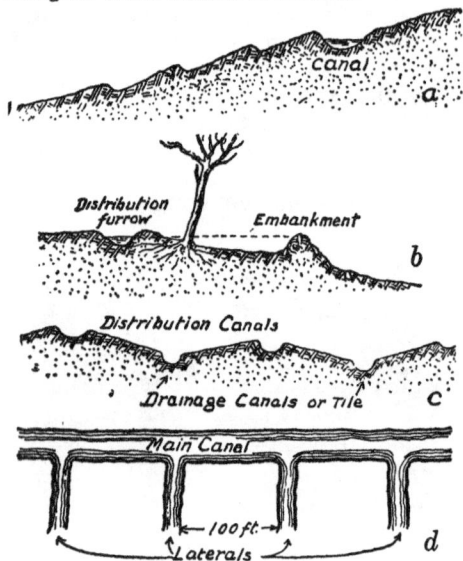

FIG. 376. Diagrams of different irrigation methods: a, section of hillside terraced for furrow and flooding system; b, section showing orchard irrigation details; c, section and d, ground plan of the bedding layout. In each case the slopes are exaggerated to illustrate the principles. (See text p. 247).

SOILS	WATER	
	ABSORBED	EVAPORATED IN 3 HOURS
Sandy . .	25 to 30%	55 to 70%
Loam . .	50	25 to 35
Heavy Clay	60 to 80	20 to 25

This table shows that (a) the soils which absorb the most water retain the most; and (b) the finer-grained soils such as loam and clay absorb the most water. From study along these lines, it is suggested that a soil which contains 20 per cent sand should be irrigated once in 15 days, while a soil having 80 per cent sand requires an application of water every 5 days.

Conveying the water. The following table gives the amount of water required per acre,

for various depths, together with the flow required, assuming all the water is put on the land in 3 hours.

AMOUNT	FLOW IN	
	SECOND FEET	GALLONS PER MINUTE
1 acre inch . . .	$\frac{1}{3}$	$151\frac{1}{4}$
1½ acre inches . .	$\frac{1}{2}$	$226\frac{7}{8}$
2 acre inches . . .	$\frac{2}{3}$	$302\frac{1}{2}$
3 acre inches . . .	1	$453\frac{3}{4}$
4 acre inches . . .	$1\frac{1}{3}$	665

An acre inch is the amount of water required to cover an acre uniformly 1 inch deep. It equals 3,630 cubic feet; or 27,225 gallons.

One second foot is a cubic foot per second. It equals 1 acre foot in 12 hours and 6 minutes; or 1 acre inch per hour; or 7½ gallons per second.

The size of ditches required depends on 3 things: (1) the amount of water which must pass a given spot; (2) the seepage and leakage losses, and (3) the greatest velocity of the water which will not cause erosion. This velocity depends to some extent upon the slope of the ditch. The usual maximum velocity allowed in irrigation canals is 3 feet per second. The aim is to give as high a velocity as will not cause serious erosion of the soil and yet will prevent the deposit of silt in the main canals. The grades allowed vary from an inch or two per thousand feet to 1 foot in 100 feet. With the slighter falls, large canals must be used, while with the greater slopes the cross section can be correspondingly smaller. The carrying capacity of any ditch is its cross section in square feet multiplied by the velocity of the flowing water in feet per second.

Where wooden and iron or stone aqueducts or pipes are used the grade is made as steep as practicable, so as to make the necessary construction as small and thus as cheap as possible. All wooden or other construction in connection with running water must be made very heavy and substantial. Water is a very heavy substance and in motion can exert tremendous force. Wherever possible, bolts and lag screws should be used in place of nails or spikes.

Measuring water. The simplest and most accurate device for general use in water measurement is a weir specially formed and set to make calculations easy. It consists of a flat board or panel set on edge vertically and extending across the stream. A rectangular notch is cut in from the top edge of the board, the edges of the notch being beveled to flare downstream. The edge of the weir must be horizontal. The width of the notch and the height of the surface of the water above the horizontal edge of the notch are the two measurements which must be carefully observed. The following weir table gives the discharge from such a weir in cubic feet per minute per inch of length for varying depths of water above the edge of the notch in the weir:

DEPTH IN INCHES	DISCHARGE	DEPTH IN INCHES	DISCHARGE
1	0.4	8	9.1
2	1.1	9	10.8
3	2.1	10	12.7
4	3.2	11	14.6
5	4.5	12	16.7
6	5.9	13	18.8
7	7.4	14	21.1

It is important to obtain the depth of the water by measuring from its surface, not directly over the notch but some distance back of it because the water surface will slope downwards as the weir is approached from upstream, because of the velocity of the water as it runs over the weir. The weir is usually placed just below the head gate and thus measures the total intake. Its construction is so simple and cheap that it can be used in any number to check up water quantities.

The canvas dam. It is convenient to have a movable arrangement with which to dam up a ditch, and to meet this need the canvas dam has proved very satisfactory; it may be moved along a ditch or transferred to another at will. It consists of a canvas square large enough for the ditch, and usually about 4 feet on a side. One edge of this is tacked between 2 wooden strips which project a foot or two on either side so as to rest over the edge of the ditch. Near the top edge of the canvas and just below the strips, in the centre of that edge of the canvas, is an opening, perhaps 8 inches square, covered with a canvas flap one edge of which is held between

FIG. 377. A hillside orchard terraced so that the bedding system may be used, three or more lateral ditches being run along each terrace

FIG. 378. Large alfalfa field irrigated by means of lateral furrows supplied from the main canal in foreground. The water is being turned into them by the canvas dam in the left foreground.

the strips. When in use the dam is laid in the furrow with the canvas extending up stream, the flap side up, and the strips resting across the furrow. The bottom and sides of the canvas are loaded with earth to hold them down. Thus placed it effectually blocks the furrow and raises the water. By means of the flap and opening, the level can be fairly well regulated.

Sometimes a flat piece of thin metal with a handle on one side, or a contrivance of 2 pieces hinged, with handles on each (Fig. 146, Vol. II) is used as a dam; for small furrows either is quite satisfactory.

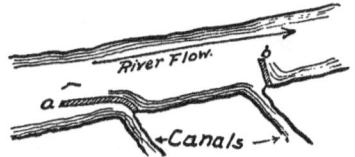

FIG. 379. Wing dams for tapping a large stream. a, longitudinal, b, cross stream type.

Wing dams. For the extensive use of water from large streams, some convenient method of tapping the stream has been found necessary. The usual plan, if the level of the stream is to be raised only a few feet, is to build a wing dam of either the longitudinal or cross-stream type. A point is picked out where the greatest fall of the stream occurs, and a dam is built parallel with the shore from the head of the fall down to the entrance to the irrigation canal. Below this entrance the dam curves to shore. This kind of dam may be simply built and forms a sort of flume in which the level of the water along approximately its entire length is the same as at the flume intake. It is not always essential that it be absolutely tight. Often an earth and stone dam is sufficient.

Where the fall is more rapid the cross stream wing dam is used. This is a simple dam extending straight out into the stream, the canal entrance being, of course, up stream from the dam. Such a dam

Fig. 380. A canvas dam in detail and in place. The water reaching it will be forced to overflow the ditch on the lower side.

must be very strongly built if the stream is very large.

Ditch drops. To lower the water in a ditch without serious erosion of the soil is a special problem on steep hillsides, and has been solved by the use of a series of wooden steps within a flume. At both ends of the flume, the sides are flaring to the width of the ditch, the step part of the flume being narrow to save lumber and to give strength to the construction. The flare at the discharge end must be sufficient to let the water out without great velocity into the lower ditch which, otherwise, would be deepened considerably at that point by the action of the water.

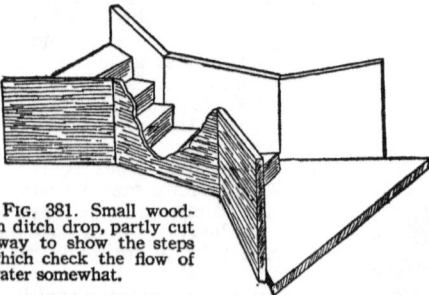

Fig. 381. Small wooden ditch drop, partly cut away to show the steps which check the flow of water somewhat.

Division boxes. Where the side branches or laterals join any main canal or ditch, the water is apt to wear away the corner of the outlet as it turns sharply from the direction of the canal to the lateral. At such places division boxes should be placed, the ditches lined with boards and regulating gates placed at each opening. The wooden flume should

always extend a considerable distance down stream from the gate to provide for times when the gate is opened under a considerable head of water. Under these circumstances the water would rush with high velocity from the partially opened gate and very seriously affect any type of earth conditions.

Water gates. A single gate should be used wherever water is diverted or enters any canal of size. The construction should be permanent. A good type is a simple panel sliding vertically between strips on the inside of the flume. The gate should be loose, and directly above it there should be some construction to hold the gate up at varying openings. With gates of any considerable size, arrangements must be made for some mechanical device to aid in lifting them as the pull required on a gate having even 8 or 10 square feet exposed to a head of several feet of water is very considerable.

Reservoir storage. The importance of a regulating reservoir is not generally admitted. The necessity in irrigation where the supply of water is limited can be very well pointed out. In the case of a small spring making available only 2 quarts per second, saturating

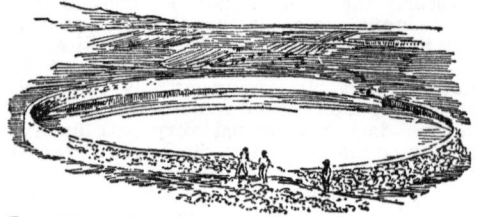

Fig. 382. Rubble stone storage reservoir in pr cess of construction

the surface and being drained away by the subsoil, all the water it supplies may be readily taken up by the soil around about within 50 yards. By arresting this flow and accumulating it in a reservoir, there may be stored in 24 hours about 43,200 gallons of water, equivalent to one quart for each square foot over 4 acres; a week's storage would be sufficient to supply 12 acres with the equivalent of an inch of rainfall. This indicates clearly the importance of the conservation of water in every step of irrigation from the intake to the drainage canal. It also points out the reason for considering the use of drainage water from upper levels to be distributed on the lower fields. Every tiny stream is valuable in this type of work. Continuous flow is far more essential than volume in every phase of the work.

Constructing ditches. Large ditches are best made with the scraper, after plowing. The land is plowed deeply along the direction of the furrow so as to make it workable. The scraper is used across the direction of the ditch, starting on one side, traveling down the

FIG. 383. Wooden division box used to divert the water from a main canal to a lateral without cutting away the banks. (Farmers' Bulletin 864.)

PRICES, WEIGHTS AND COST OF TILE LAYING:

DIAM-ETER	PRICE PER 1,000 FT.	COST OF LAYING ER ROD 3 FT. DEEP	WEIGHT PER FT. POUNDS	AVERAGE CARLOAD
4 inch	$ 22.00	$0.36	6	6,500 ft.
5 "	32.00	.36	8	5,000 "
6 "	43.00	.36	11	4,000 "
7 "	55.00	.38	14	3,000 "
8 "	75.00	.44	18	2,400 "
10 "	100.00	.50	25	1,600 "
12 "	150.00	.55	33	1,000 "

dip of the furrow and up the other side where the material is placed as an embankment. The depth of the ditches is usually made about one sixth the width, the depth being measured from the original surface of the land. Of course, the embankment adds considerably to the finished and usable depth of the ditch. A canal 10 feet wide is thus about 1½ feet deep at the centre below the original land level; a ditch 3 feet wide is perhaps 6 inches deep.

The smaller ditches are often made directly with a listing plow having a double moldboard which throws the earth out on both sides. One or two passes usually gives depth enough. To finish such a ditch, an A-shaped scraper may be run through it. This scraper is made of two 2-inch planks fastened together at one end and braced apart at the other.

FIG. 383a. Long ditch drop passing under road in foreground.

On a hillside, the ditches should be excavated in such a manner that all the earth is thrown out on the lower side; this is usually necessary in order to utilize the full cut.

Drainage and Subirrigation

A method of subirrigation practised to some extent consists in laying tile with broken joints in a manner similar to that followed in drainage operations. Water is supplied to and leaks out through the joints for distribution. On the other hand, low lands and fields with a dense subsoil, yet requiring irrigation, frequently need proper underdrainage to keep the ground water level down and to prevent a surplus moisture supply from turning the fields into swamps. In laying tile for drainage or for subsurface irrigation, similar precaution must be taken with the

exception, perhaps, that in drainage operations, very special care must be taken in establishing and completing the grade or slope of the drains.

Subirrigation methods follow those of drainage except as to depth of laying. There are 4 distinct steps: (a) Laying out the direction and slope of lines; (b) digging the ditch; (c) laying the tile and making connections; (d) filling the ditch.

The first step should include making a sketch, however rough, of the job complete as planned. This sketch should be kept correctly, and as fast as any changes are decided upon, they should be plainly marked with distances and other data and filed for future reference as to the location of the lines.

The ditch can be dug with the plow, the depth being usually between 1 and 2 feet below the surface. The grade should be well established with the level and should not be more than 15 or 20 inches per 1,000 feet. The tile is laid in the ditch, with broken joints perhaps a quarter of an inch wide. Each joint can well be covered with a piece of stone or broken tile before the earth is carefully shoveled back and the ditch filled.

FIG. 384. Wooden water gate. Note apron at front end which is sunk in the soil to prevent excessive erosion. A smaller one should be affixed to the opposite end. (U. S. Dept. Agr. Bulletin 115.)

Spray or Overhead Irrigation

The mechanical problem in applying water by the spray system involves a number of points, for example: (1) A minimum of labor; (2) breaking the water up into drops fine enough so that they will not injure plants or pack the soil; (3) uniformity of application; (4) perfect control of the water at all times; and (5) a minimum of expense for equipment. As a result of the efforts of inventors and manufacturers to achieve these ends, 3 distinct types of irrigation equipment have been developed—the nozzle line, the rotary sprinkler and the stationary sprinkler.

The first consideration in irrigation equipment is, of course, the water supply. This is most commonly a running stream. In using this source of supply, the first step is to provide means by which to make sure that the water, before it gets into the main supply line, is free from sand, mud, or other impurities which might clog up the system. There are strainers in the pipes

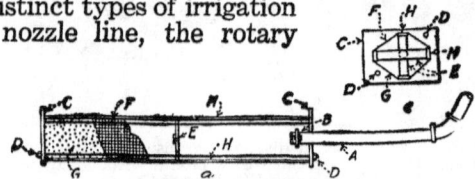

FIG. 386. Lengthwise section and (above) cross section of homemade strainer with which to obtain clean water for spray irrigation. A, suction pipe to pump; B, locknut to hold strainer on pipe; C, end pieces; D, iron brace rods and nuts; E, wooden braces to hold lengthwise bars (H); F, heavy galvanized wire screen tacked to H; G, perforated sheet brass tacked and soldered in place around wire netting.

at the beginning of each distribution line, but these can be counted upon to take out only a comparatively small amount of impurities; otherwise they would themselves clog up so quickly as to require frequent cleaning. The most convenient way of cleaning the water is to strain it at the source of supply. A fine perforated brass "foot strainer," may be used, but this has not sufficient area under most conditions, and consequently has to be cleaned too frequently. An inexpensive strainer may be constructed easily as shown in Fig. 386. If a small stream is to be used, it will probably be necessary to dam it and to make a reservoir to store the water for use in drought periods. Even if the flow of the stream is sufficient to furnish the water required to supply the crops for, say, a week's time, it may not be sufficient to keep up with the demands when the system is in operation. The water reserve should be big enough to supply the pump at full capacity for 8 or 10 hours' run. If water is scarce, it may be necessary to put in a concrete reservoir that will prevent waste by leakage and drainage.

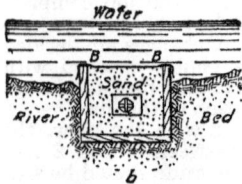

FIG. 385. Increasing the efficiency of a strainer by sinking it in a wooden or concrete box of sand and covering with a sheet of canvas or burlap (B B).

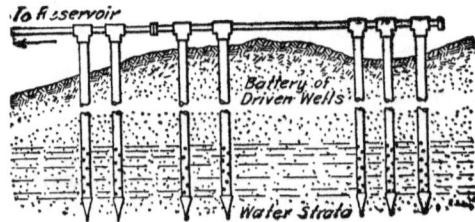

FIG. 387. How a battery of driven wells may be connected up to provide an increased supply of water

Fig. 385 shows a method of straining the water by the use of sand in addition to the wire strainer. This may be necessary in shallow streams with fine sediment or slime on the bottom. From ponds, it is usually not so difficult to get a supply of water already clean enough to use. The water from driven artesian wells is usually clean. Water from a battery of shallow driven wells which would not be sufficient to meet the pumping capacity of the plant may be collected and stored in a reservoir at a slightly lower level. (Fig. 387.) Or water may be lifted by a ram and stored in a cistern or tank of suitable size for use when needed.

Water Pressure

To operate successfully any one of the systems of overhead irrigation the water should be under considerable head or pressure, from 30 to 60 pounds per square inch for the nozzle line type, and 20 to 80 pounds for the spray and rotary sprinklers, according to conditions. The amount of water which will be required should be accurately figured out, as this will form the basis upon which to determine the size of the main supply line and the capacity of pump and engine or tank and elevation as the case may be. If water will be required once a week at the driest season, the pumping outfit should be capable of supplying water for about one-fifth of the total area to be irrigated every working day.

Where conditions are such that a reservoir or tank can be established without much cost, it may be cheaper to install a ram to elevate the water automatically rather than to put in a pumping outfit. Usually, however, it is not possible to get sufficient elevation without going too far from the tract to be irrigated. One may estimate roughly on a little less than half a pound pressure for every foot of elevation or fall—that is the difference, in a vertical line, between the tract to be irrigated and the water supply.

For systems of moderate size— requiring up to 100 gallons or so per minute for all the lines it will be necessary to operate at one time, the one-cylinder, double-action pump, with a good capacity air chamber, will usually be the most satisfactory. The importance of the air chamber is that it equalizes the pulsations or variations in the water pressure between strokes, so that practically an even pressure is maintained. A centrifugal pump may be used where conditions are suitable; it costs considerably less for the same capacity. Either type of pump used may be either belt or gear driven, or connected direct to the engine. However, when the double-action pump is geared direct, there should be a check valve or bypass set to open at a suitable pressure so

FIG. 388. A centrifugal pump connected directly to an electric motor is efficient and inexpensive for lifts of less than forty feet. 1, Discharge pipe; 2, suction pipe; 3, pump case; 4, stuffing box; 5, coupling; 6, motor; 7, base.

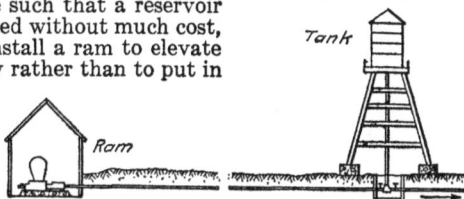

FIG. 389. A ram and elevated storage tank can be used to get the necessary pressure

that if all lines are turned off, or if anything happens to check the system while the pump is in operation, there will be no breakage in the machinery. There should also be a check valve in the main line just above the pump to hold the water between the times when the system is in operation, and a foot valve on the suction pipe, or some other convenient arrangement for priming the pump when starting it.

Pump. The type of engine used is not material, but it should, of course, be one that delivers power economically and is capable of running for a long time with little attention, for a system may be in operation on one field or crop for from 1 to 12 hours. Since in times of severe drought, it may be necessary to operate the system almost continuously night and day, it is always advisable in installations of any size to have an auxiliary pumping outfit that can handle at least part of the work if anything happens to the main pump. Many installations have both a gas engine and an electric motor; where a system is extended and a larger pump put in, the old one is usually kept as an auxiliary.

Main or supply pipes. The next consideration, is, of course, the distribution of the water to the field or fields where it is wanted for use. The system of mains or supply pipes should be worked out carefully so that the small-

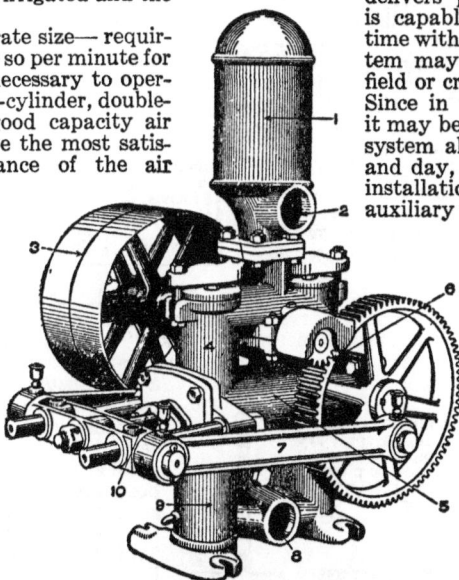

FIG. 390. A double-acting displacement pump is the preferred type for irrigating a few acres under average conditions. 1, Air chamber; 2, discharge; 3, light and loose pulleys; 4, valve chamber; 5, cylinder; 6, gear; 7, connecting rod; 8, suction opening; 9 base; 10, packing gland. (U. S. Dept. Agr. Bulletin 495.)

est possible amount of the larger sized pipes need be used; at the same time, sufficient water and pressure at every point where it is wanted must be assured. Table, below, gives the amounts of water carried in different sizes of pipes.

There should be an expansion joint in the main line to take care of expansion and contraction of pipe as a result of temperature changes.

If there is much sand or sediment in the water, in spite of the strainer at the source, there should also be a "trap" or "bucket strainer" in the main line. Of course, sharp angles and numerous bends should be avoided as much as possible as they greatly increase the friction of the water in the pipes. Unless the whole pipe line is down below danger of frost, it should be laid to a slight grade so that it can be drained out clean in the fall at convenient points. Of course, the size of the main line may be decreased as branches or laterals for irrigating are taken off it. Valves should be placed wherever there are branches and at other convenient points for controlling the water.

Equipment for Distribution

Each of the systems of spray irrigation—nozzle line stationary, and rotary sprinkler—and each of their various modifications, requires a system of laterals or small-sized branch pipes running off from the main supply line at right angles to feed the nozzles or sprinklers that distribute the water over the soil.

Laterals. In the nozzle line type, these lateral lines must be above the soil. They are placed at from a few inches to 10 feet or so above the ground; the usual height is 6 to 7 feet. The laterals for the sprinkler types may be placed either above or below the ground. Usually they are put below, because this plan is easier and keeps the pipes more out of the way. Vertical pipes or risers are distributed at the proper intervals, and at the tops of these the sprinklers are placed.

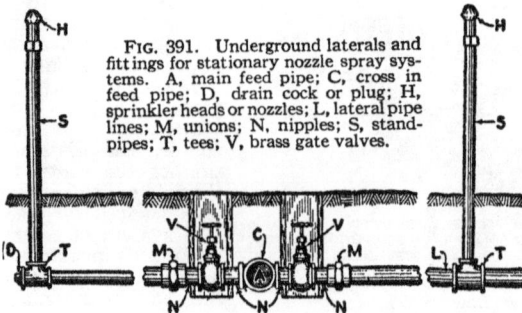

Fig. 391. Underground laterals and fittings for stationary nozzle spray systems. A, main feed pipe; C, cross in feed pipe; D, drain cock or plug; H, sprinkler heads or nozzles; L, lateral pipe lines; M, unions; N, nipples; S, standpipes; T, tees; V, brass gate valves.

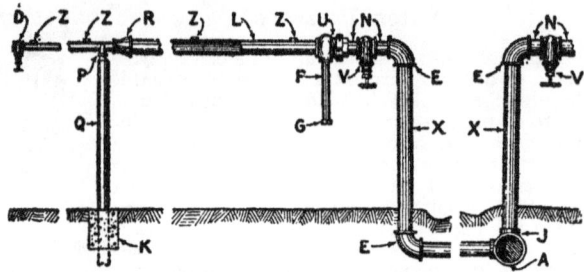

Fig. 392. Fittings for overhead nozzle line system: A, main feed line; D, drain cock; E, elbow; F, handle for turning nozzle line; G, cap on handle; J, side outlet tee; K, concrete base for pipe pact; L, nozzle-line pipe; N, nipples; P, pipe hanger or support; Q, pipe post; R, reducing fixture; U, turning union; V, gate valve; X, risers to nozzle line; Z, nozzles.

The standard distance between the lateral lines for the nozzle line system is 50 feet. If water has to be used at less than 30 pounds pressure, they may be placed nearer than this; and under higher pressure they may be placed as far apart as 56 feet. The length of the nozzle lines may be anything that will suit conditions; 600 feet, however, has been found a desirable maximum length in large installations. The distributing nozzles are placed 2 to 4 feet apart along the nozzle line; 3 feet is the standard distance.

TABLE 1—AMOUNT OF WATER FOR SPRAY IRRIGATION DIFFERENT SIZES OF STRAIGHT IRON PIPE WILL CARRY WITHOUT EXCESSIVE FRICTION

DIAMETER OF PIPE	QUANTITY PER MINUTE	DIAMETER OF PIPE	QUANTITY PER MINUTE
	Gallons		Gallons
½ inch . .	1 to 2	3 inches .	75 to 125
¾ inch . .	3 to 4	3½ inches.	125 to 175
1 inch . .	5 to 8	4 inches .	175 to 250
1¼ inches .	9 to 16	5 inches .	250 to 400
1½ inches .	17 to 25	6 inches .	400 to 600
2 inches .	25 to 45	7 inches .	600 to 900
2½ inches .	45 to 75	8 inches .	900 to 1,200

The nozzle line may be either supported upon posts or suspended from overhead cables. The latter system leaves the ground much clearer for working, there being only a few posts to the acre; but it is, of course, more expensive. For commercial installations, either cedar posts or second hand iron pipe posts set in concrete are usually used. The supporting posts are set at intervals of 15 to 20 feet according to the size of the pipe used for the nozzle line (see Table 2, p. 257). To facilitate the turning of long nozzle lines, roller-bearings or other supports are set on top of the posts or suspended from overhead cables.

Usually a valve is placed on each lateral near the main feed pipe so that its flow can be controlled as a unit. In

TABLE 2—SIZES OF PIPE FOR MAIN SUPPLY LINE

FLOW (GALS. PER MIN.)	LENGTH OF LINE IN FEET				
	50	100	200	400	600
30	1½	2	2	2½	2½
75	2	2½	2½	3	3
100	2½	2½	3	3	3½
200	3	3½	3½	4	4
300	3½	3½	4	4	5
500	4	5	5	6	6

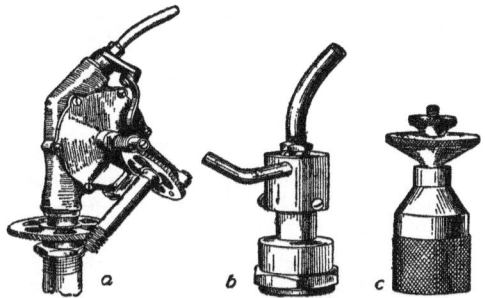

FIG. 394. Three types of sprinklers for the upright pipe system: *a* and *b*, rotary type; *c*, stationary, mist-producing type.

the nozzle line equipment, a device called a "turning union" (Fig. 393) is installed at the beginning of each nozzle line. This contains a strainer and has a hollow pipe handle to be used in turning the nozzle line from one side to the other. By removing the cap from this handle and turning the water on, loose scale, sand, etc., can be flushed out of the strainer chambers.

With the rotary sprinkler type of irrigation, where the individual sprinkler covers a large area—that is, a circle 35 to 90 feet in diameter— the price of the equipment may be lessened considerably by getting only as many sprinklers as may be operated by the system at one time. By having caps on the risers, these sprinklers may be changed from one section of the field to another, wherever most needed. While this cuts down the cost of installation it means, of course, some additional expense in operation.

TABLE 3—SIZES OF PIPE FOR NOZZLE LINES WITH NOZZLES 4 FEET APART. IF CLOSER, LARGER PIPE IS REQUIRED

NOZZLE	LENGTH OF LINE FEET	NUMBER OF FEET OF PIPE				
		¾-IN.	1-IN.	1¼-IN.	1½-IN.	2-IN.
No. 1 4 Feet Apart	150	150				
	200	130	70			
	250	100	150			
	300	100	150	50		
	400	90	160	150		
	500	90	160	150	100	
	600	90	160	175	175	
	700	90	160	175	175	100
No. 2 4 Feet Apart or No. 1 3 Feet Apart	150	115	35			
	200	100	100			
	250	90	100	60		
	300	90	100	110		
	400	80	100	120	100	
	500	75	100	120	120	85
	600	75	100	120	120	185

Portable Sprinkler Systems

In small gardens, and for the occasional irrigation of field crops during extreme drought, it is often desirable to irrigate, although a permanent installation would be objectionable or would involve too heavy an investment. The nozzle line systems may be bought in portable form as complete assembled units. By the use of special lever couplings, it is possible to put a nozzle line system up in sections so that it can readily be dismantled and re-assembled in exactly the right positions. Portable posts may be used or short temporary posts may be driven in the field. An extension to the main pipe line can be made either under ground or, if wanted only temporarily, by laying the pipe along the ground. With a sprinkler system, laterals and risers for the temporary work may be supplied and the sprinklers borrowed from the regular installation, or some additional ones may be kept on hand for this purpose. Temporary extended use of a system in this way, often means crop insurance, and as the pumping plant and main supply line

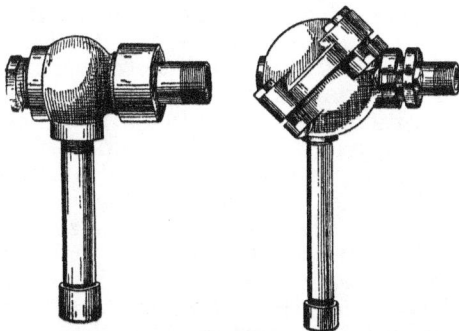

FIG. 393. Two types of handles for turning the nozzle pipe lines

are already in use, it may frequently be accomplished at a very low cost per acre.

Automatic Turning Devices

While small systems of the nozzle line type are usually operated by hand, it is possible to provide for the automatic operation of the system by mechanical turning devices. These are of two types: (1) a gear system operated by a belt from the pumping engine, and (2)

a form of water motor or tilting device operated by the flow of water on its way to the distribution lines. The advantage of an automatic turning system of this kind is not merely in the saving of labor but also in the fact that the distribution is usually much more even than can be or, at least, usually is, obtained by hand turning. More water can be applied without saturating the surface of the soil, as the application at any particular point is more gradual.

TABLE NO. 4—SHOWING NUMBER OF GALLONS OF WATER REQUIRED TO COVER AN ACRE AND TIME REQUIRED TO APPLY THEM, USING THE STANDARD NOZZLES NOS. 1 AND 2, SPACED 4 x 54 FEET APART, OR 200 NOZZLES TO THE ACRE

NO. 1 NOZZLE

POUNDS PRESSURE		10	15	20	25	30	35	40	45	50	
FEET ELEVATION		23.1	34.7	46.2	57.8	69.3	80.9	94.2	104	115.5	
One Acre 1-inch Deep	27,152 Gal.	18	15	13	11	10	9	9	8	8	Hours
		51	17	18	54	52	42	23	54	26	Min.
One Acre ½-inch Deep	13,576 Gal.	9	7	6	5	5	4	4	4	4	Hours
		26	39	39	57	26	51	42	27	13	Min.

NO. 2 NOZZLE

POUND PRESSURE		10	15	20	25	30	35	40	45	50	
FEET ELEVATION		23.1	34.7	46.2	57.8	69.3	80.9	94.2	104	115.5	
One Acre 1-inch Deep	27,152 Gal.	15	12	10	9	8	8	7	7	6	Hours
		18	26	49	40	50	10	37	12	51	Min.
One Acre ½-inch Deep	13,576 Gal.	7	6	5	4	4	4	3	3	3	Hours
		39	13	25	50	25	5	48	36	26	Min.

Installing a system. If the work of installing a system is to be done in a hurry—as frequently occurs—it may be begun from both ends. However, the installation of pump, strainer, engine and so forth, usually represents the longest part of the job and should be attacked first. Second, the main supply line into the various fields should be assembled and put into permanent position, care being taken to knock off scale, etc., on each length of pipe before it is put in place, and to use red lead (on the thread ends of the lines only) in making up the joints. The main pipe should be so distributed through the various fields that the nozzle lines, as suggested above, will not be over 600 feet in length. These and risers for the nozzle lines being put in place, the exact positions for the lines of post may be deter-

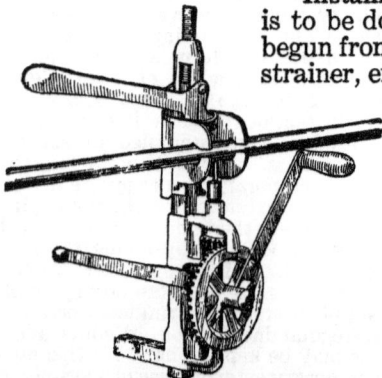

FIG. 395. Hand drill for boring nozzle holes in horizontal pipe line. These must be in a perfectly straight line and of uniform size.

Above, the problem; *in the centre,* the solution; *below,* the problem solved

POOR ROADS ARE AMONG THE COMMONEST CAUSES OF FARM FAILURE. BUT THEIR IMPROVE-
MENT IS LARGELY IN THE FARMER'S HANDS

The farmhouse needs light, air, shade, and protection; so does the garden. Note how the needs are here supplied for both

The farm should centre around its main feature. Here this is clearly the dairy herd

THE WELL-PLANNED FARMSTEAD PLACES ALL RELATED FIELDS AND BUILDINGS AS NEAR TOGETHER AS POSSIBLE AND THE DWELLING IN A CONVENIENT LOCATION WITH RESPECT TO ALL OF THEM

mined and the posts put in place. The nozzle lines themselves may now be assembled, and drilled and tapped for the nozzles. One of the most important points in the installation is to make the holes for the nozzles at regular intervals and in an *exact, straight line*, along the shell of the pipe. Otherwise the distribution of water will not be even. This alignment is accomplished by the use of a special drilling machine, so constructed that it can be clamped to the pipe rigidly, the exact position wanted being determined by a small spirit level. A special drill bores the holes and threads or taps them at one operation so that the nozzles can be put in quite rapidly.

Before inserting the nozzles the pump should be run for a while, with the ends of all nozzle lines *open*, to allow the dirt, scale, etc., which may have accumulated in the pipe, to be washed out.

FIG. 396. Nozzles in different lengths of pipe line *must* be kept in line; and the way to accomplish this is by means of pins and socket holes and the unions of such sections.

CHAPTER 24

The Building and Care of Farm Roads

By CURTIS HILL, *City Engineer of Kansas City, Mo., since 1913, but, in spite of his urban activities, a man with the training and the experience that fit him to write of farm conditions. He was born and raised on a farm in Jackson County, Mo., attended school there and at Independence, then took the engineering course at the University of Missouri, following it with a year in the Engineering College of Cornell University. His wide practical experience has included railroad construction; the making of field surveys and estimates for canal work for the Government in New York; highway bridge erection around St. Louis; and the following official positions: With the city engineering department of St. Louis for 6 years, State Highway Engineer of Missouri for 6 years, and the position he now holds. He is a member of the American Society of Civil Engineers.*—EDITOR.

WHEN a good, hard-surfaced road runs by a farm, giving the farmer means of easy access to central points, such as railroad station, village, and church, and to neighbors, it is a further pleasure, comfort, and saving to have easy access from central points on the farm to this main traveled road. If it pays the farmer to have a good road passing the farm to town, will it not also pay, in some degree, to provide means *on* the farm to reach that main road with the same ease and the same sized loads that he can haul over it? A good road means cheaper transportation, whether it be a "through-the-farm" road or a main county road. The through-the-farm road is here discussed from the constructional point of view only.

Travel on these roads is comparatively light, and they do not call for the type of construction required on roads traveled by many vehicles per day. This applies not only to a farm road, but also to one over which several farmers travel, a side road for a small community, or a country lane. Such a road must be simple in construction and inexpensive. A few farms may need a higher class of road; but the ordinary driveway, as herein outlined, will answer every purpose for the great mass of farms. Furthermore, the limited financial ability of many farmers is a condition to consider. The higher principles of road building, which is a science, are left for application to the main roads and to other roads which many vehicles use and where the demand is for a better and more costly construction.

Kind of road and how to choose it. The first step in building a road is to

FIG. 398. Scratching up the surface of a road with a heavy spike-tooth harrow in readiness for a new surfacing and rolling.

262

FIG. 399. This kind of bridge looks bad and is dangerous; it hurts the farmer's pride, his reputation and his pocket book and may do him serious personal injury.

study the local conditions and to choose the methods, plans, and materials most suitable to the needs and purpose for which the road is to be built. No one method, plan, or material will be equally adaptable in all places, but must be varied to suit local conditions. Use should be made of the natural material, placed in a way to give good results. For example, shale or slate will not wear well in contact with wheels, while gravel will. If there is a local supply of shale and but little gravel, the base of the roadbed may be made of shale and the surface of gravel.

Therefore, the determination of the kind of road to build, the width and thickness of roadbed, and the kind of material, depend upon the demands of travel to be made upon the road, the means for construction, and the accessibility of material. Local conditions should always govern, whether use is to be made of natural material as found or of artificial material, that is, natural material broken up or treated. One section may not have material, the best use must then be made of earth; in another, straw, sawdust, or sand may be mixed with clay or gumbo; in still another, shale or slate may be used to advantage. In some sections of the country, near the coast, shells are obtainable from which the road may be made; in others, there are banks, or ridges, composed of a chert-bearing soil which packs hard and becomes watertight under the pressure of travel, and here the chert road prevails. A special feature in many places is the gravel or the broken-rock road. So that, depending upon local conditions, we may have a farm road of any of the following materials: earth, straw, sawdust, or sand with earth, clay, or gumbo; shale; slate; chert; gravel; rock or other prevailing material.

Equally with the character of the road, the cost of building, also, will depend upon local conditions, upon the nearness and quality of material, width and thickness of roadbed, price of labor and teams, location, and topography.

Drainage

The three principal divisions of actual road making are location, construction, and maintenance. With the exception of drainage, all details pertain to one or other of these. Drainage is involved in all three. The all-imperative, dominant, and paramount question is that of good drainage. If there is not good natural drainage, an effective system of artificial drainage must be installed; for, if the drainage be not good, the road will not be good. Drainage may be conveniently discussed under two heads: (1) surface drains, and (2) underdrains.

Surface drains. Surface drains are the open ones, as crowns, ditches, and culverts. The object of the crown is to make a drained roadway for the travel and to get the water quickly off this traveled roadway into the side ditches. Make the side ditches to drain entirely off the rights-of-way at every opportunity. Where a ditch must cross the road,

FIG. 400. This kind of culvert is easy and cheap to build; once built it is permanently safe, efficient and attractive.

FIG. 401. Every time a farmer drives over an eyesore like this without deciding to fix it—*and then doing so*, he goes one step lower along the road of carelessness, inefficiency and failure.

take it under the roadbed and not over it, and keep it on the side of the road and not down the middle of it. Side ditches are also made to protect the roadbed from water from adjoining lands; therefore, if the road be on a hillside, have a ditch on the upper side of the road. Culverts are of sufficient importance to constitute a subject in themselves. The cheapest in first cost are not always the inexpensive; for, as a general rule, the weaker the structures, the greater will be the cost for repairs and maintenance. There are some exceptions—where ready money is not available and plenty of timber is—where the wooden culvert is best; but one of metal, clay tile, stone, or concrete masonry is more often the economical construction.

Make the culverts long enough and with good head walls and wing walls, to throw the water into the barrel of the culvert and to prevent it from getting through the road alongside of the culvert or from washing out the fill and sides of the road. A good rule-of-thumb method to determine the size of opening is "1 square foot area in opening of culvert for every 5 acres of drainage area." This rule may be varied for a smaller culvert area, if the country is flat and the flow sluggish, and may be increased where the hills are steep, with much fall to the branch and a rapid run-off.

Underdrains. Underdrains are constructed from 2 to 4 feet deep. They may be made of buried logs; of a trench half-filled with stone, hay or straw being thrown in over the stone, and the remainder of the trench filled with earth or gravel; or of clay tile or concrete pipe; in fact, of anything along or through which the water can find its way freely. They must be so placed as to cut off the underflow in the ground before it reaches the road surface—to take the ground water out of the roadbed. Roads which dry out slowly after a rain, or in the spring of the year, need underdrainage to carry out the water and dry them out more quickly. Wet-weather springs or "spouty" places often occur on hillsides, where the surface drainage is good; but they never dry out. Under-drainage is the proper treatment for such places. These drains may be made directly across the roadway or laid along under the centre or side, leading out into the side ditch or a culvert or to low ground on one side of the road.

The drainage question may be summed up as follows: The roadbed must be protected from water, water which comes up from beneath as well as that which falls from above; and, in order to have a good road, the water must be gotten *out*, *off*, and *away*—out of the road by means of the underdrains, off the road by means of the crown, and away from the road by means of the ditches.

FIG. 402. Stone dam in a surface road drainage ditch, built to stop the flow of water and direct it through the culvert. Water should always cross a road *below* the surface.

Location

The question of location enters more largely into consideration in a country of broken topography than in a level section. Where the position is not determined by the nature of the farm improvements, the choice of a suitable location for a road may eliminate a culvert, provide a good creek crossing or a good foundation for the roadbed, give good drainage, provide sunshine and air for the road-bed, lower the construction cost, or reduce the grades.

The reduction of grades (not grading) is a very important matter. One can afford to go around a hill, rather than over it, with a loaded wagon. Very often "the longest way round is the shortest way home." The term "grade" is not the cut, or fill, but is the rise and fall in vertical feet in each 100 feet in horizontal

PER CENT OF GRADE	POUNDS WHICH A TEAM CAN HAUL
0 per cent, or level	4,000
1 " " " 1 foot in 100 feet .	3,600
2 " " " 2 feet " " " "	3,240
3 " " " 3 " " " "	2,880
4 " " " 4 " " " "	2,160
5 " " " 5 " " " "	1,600
10 " " " 10 " " " " .	1,000

FIG. 403. How grade affects the load that can be hauled. This is on a smooth, hard road; on dirt the ratio is slightly less. Thus what one horse could pull on a level country road, would take two on a 5-per-cent, three on a 10-per-cent, and four on a 15-per-cent grade. (International Harvester Co.)

length of road. This is spoken of as the "percentage of grade" and is based on the 100-foot length. Therefore, a 2-per-cent grade is one of 2 feet rise or fall in each 100-foot length. A 5-per-cent grade (up) is a rise of 5 feet in 100 feet, or 25 feet rise in 500 feet. The accompanying table shows the importance of a grade (steepness) to traffic and the additional force, or pull, required by a team on different grades.

Construction

Under the head of construction is included the work of actually building the road, such as grading, ditching, culverting, hauling, and placing the material.

The farmer, of course, is more or less confined to the class of tools and implements available on his farm; but, if he has much road work, he will save both time and money by obtaining those best suited for the work. If he has chert or other hard-compacted soil to loosen and haul to place on the road, he will find the "rooter plow" the best implement. Gravel may be sorted with forks made for the purpose, and a manure spreader may be used for spreading sand. In grading work, allowance should be made for shrinkage. Usually about 10 per cent is added to the height, so that the correct height will obtain after the fill has settled. The side slopes should be given the natural slope, about 1½ horizontal to 1 vertical for earth, clay, or gravel, and about 2 to 1 for light loamy, or light sandy, soils. The drag scraper is best for moving earth up to about 100-foot-length haul; wheel scrapers, from 100 to 1,500 feet; and wagons, for greater distances.

FIG. 404. Upkeep is as important as construction. This drainage tile will soon be washed bare and rendered useless.

The regulation dump wagon or home-made slat-bottom wagon bed is suitable for hauling earth, gravel, and broken rock. A plow, a grader, a harrow, and a drag are necessary implements for shaping up and finishing the roadbed.

Earth roads. Earth-road building consists in grading, draining, and shaping the ditches and roadbed (Fig. 405). The earth road must be drained and the surface become compact and solid before it is a good road. The supporting power of earth, when thoroughly wet, is only about one-fourth of that same earth when just sufficiently wet to pack without yielding. If the drainage, then, is properly attended to, the rest will follow.

Simple crowning and smoothing on all earth roads with grader and drag cost about 3 to 5 cents per cubic yard, or $30 to $60 per mile. When grading is to be done, moving earth, for the various hauls on the average road, costs about 20 cents per cubic yard.

Sand-straw roads. These should be kept rather flat, to retain some of the moisture in the road. The straw method of treating a sand road is to cover the traveled roadway to the desired width with 2 or 3 inches of woody straw, uniformly spread, and maintained by keeping the straw raked into the ruts. In some instances, the right of way outside the traveled roadbed has been planted with clover, rye, or some grass or grain having a woody

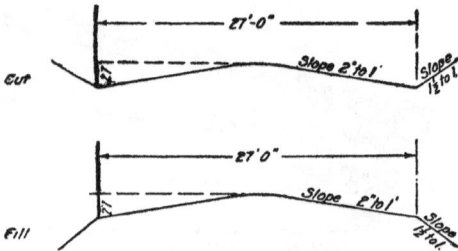

FIG. 405. Graded earth roads showing desirable relation between width and crown

fibered stem. This is cut and raked upon the centre strip of roadway each season. Sawdust (which is bad for an earth road) may be used instead of straw. The straw or sawdust treatment will help retain the moisture and in time will change the nature of the soil, making it firmer.

Sand-clay roads. The sand-clay road is one composed of a mixture of sand and clay. If, instead of clay, gumbo be used, the road may be termed a "sand-gumbo" one. Other kinds of cohesive soil of a plastic nature may be mixed with sand, to make a more compact and impervious roadbed. If the natural soil be sand, clay must be added; if clay, then sand must be added. The roadbed is prepared and shaped in the same manner as with an earth road (Fig. 405). The added material should be spread and mixed on the prepared roadbed with plow, harrow, disc and grader to the uniform thickness and width desired. All material dumped on the roadbed should be evenly spread as it is deposited. Each succeeding load should be hauled over the preceding one, after the first has been spread. The clay should be well pulverized and disintegrated and the materials mixed while comparatively dry. Enough clay, usually about one-third, must be used to fill the "voids" or interstices, of the sand. If there should be too great a proportion of one material, the opposite material must be added until the whole mixture shall be of the proper consistency. This may be done by hauling the opposite material from that of the natural roadbed or by adding the roadbed material by running the plow a little deeper or by bringing such material in from the sides or side ditches. The whole of the materials should then be mixed wet and puddled. Upon completion of the mixing, the roadbed should be reshaped and dragged while yet damp enough to pack well. If a roller be available, the roadbed should then be well rolled. No soft or uneven spots must remain, and the finished roadbed must present a true and even surface.

It may not come up to expectancy at first; but keep going with the drag as you would on an earth road, adding such material as is needed, and it will ultimately make a good farm road.

The machinery required is a No. 10 plow, drag scrapers, disc harrow, spike-tooth harrow, road grader, drag, and 2 x 4 slat-bottomed wagon beds. With a mile haul for the added material and $1.75 for labor and $4 for teams, a 10-foot strip of sand-clay will cost about $1,000 per mile.

Foundations. Other kinds of road, now about to be discussed, require more or less preparation of subgrade or foundation to receive the material. A rock road, for example, is an earth foundation covered by a compact crust of small broken stone. The earth foundation supports this crust of broken stone as well as the loads which come upon it. This crust of rock placed upon an undrained, wet, or boggy earth will gradually work down into it until the wheels cut through. The foundation should be well drained, shaped, and compacted before it receives the wearing surface material.

After the drainage has been well provided for, excavate a trough-shaped section of the width and depth of the proposed road material. This trough should be crowned and shaped to the desired finish, that is, to conform to the proposed cross section without unevenness or soft spots and should be well compacted before the surfacing material is put on. If a roller is not available, let the subgrade become settled before adding the material. Where the material is to be put on an earth roadbed of approximately the desired shape, which is already firm and well compacted, it is better not to break up the old roadbed, since a solid subgrade, or foundation, is the thing desired. Put the material on this firm old roadbed, and draw in earth for the shoulders from the sides or ditches. This constitutes the first step. (See Figs. 408 and 411.)

In placing the material on this prepared foundation, spread it evenly by using a slat-

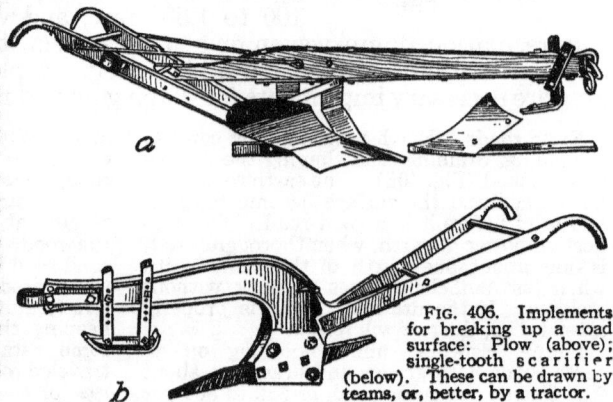

FIG. 406. Implements for breaking up a road surface: Plow (above); single-tooth scarifier (below). These can be drawn by teams, or, better, by a tractor.

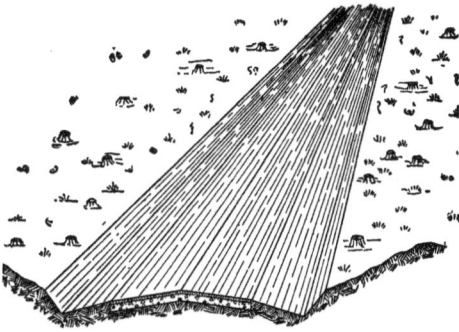

Fig. 407. Cross section of a road showing crown, ditches and relative location of foundation and surfacing.

bottomed, or dump, wagon and dump with the team slowly moving; otherwise, dump on a platform and shovel on to the roadbed, in order to spread the material evenly. If the material is dumped on the roadbed in a heap, it compacts more at that point than between the loads, and the road may settle in a wavelike surface. When dumping from wagons, dump the material where the wheel tracks will come and spread it each way; for the greatest wear upon a road falls within a strip under each wheel track, and the greatest compactness will then be on these lines. Let each load slightly overlap the preceding one.

Shell, slate, shale, and chert roads. In some of the coastal sections, where oyster shells are available and other materials are not, shell roads are built. The shells are

spread evenly upon the prepared foundation, where they are compacted by the traffic. They are soon ground into a powder and blow or wash away, so do not make a lasting road, especially under heavy traffic.

Slate and shale roads are both inferior kinds, the materials grinding easily into dust and powder under traffic, or disintegrating from atmospheric conditions. The material is broken up and placed on the roadway in a manner similar to that described for shell roads.

Chert, to which reference has already been made, is also one of the inferior road-building materials, but is better than either shells, slate, or shale. It is placed in the same manner as the others, but wears and weathers better and also binds together and consolidates more effectively.

Other kinds of material of a quality equal to any thus far described may be found and utilized. It is not worth while to put the expense upon them required for a high-grade road. Any of them might be used for the base under a broken-rock or gravel road. (Fig. 409.)

The cost of a road built from one of these materials for a 10-foot road will be about $800 to $1,000 per mile, exclusive of grading, but including shaping the roadbed and a haul of 1 mile for material.

Gravel roads. For gravel roads, the gravel should not run too unevenly (this can be governed by forking in loading), and should contain 15 to 25 per cent of fine material, to act as a filler for the gravel. A gravel which nature has packed and cemented in the bank, for example, often indicates the

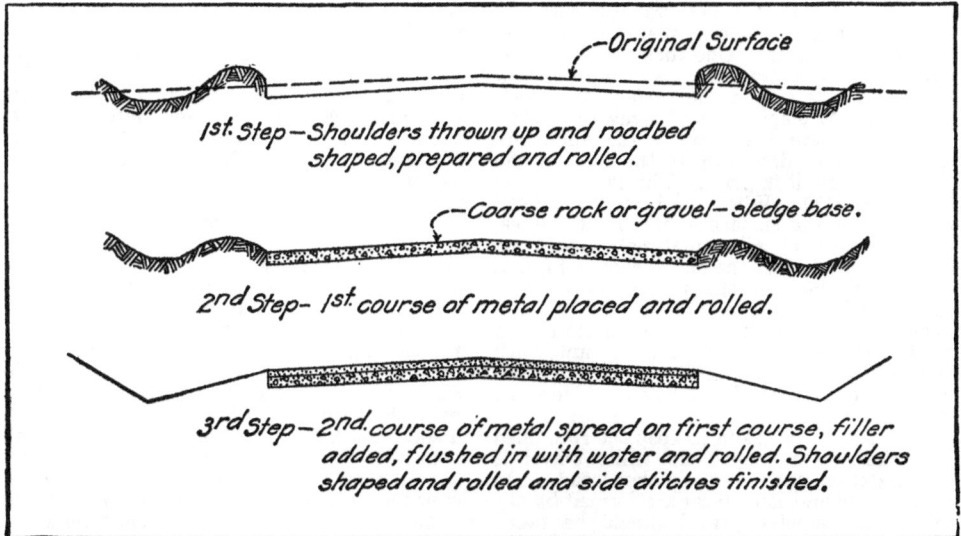

Fig. 408. Three steps in making a rock or gravel road of the higher grade or Class A type

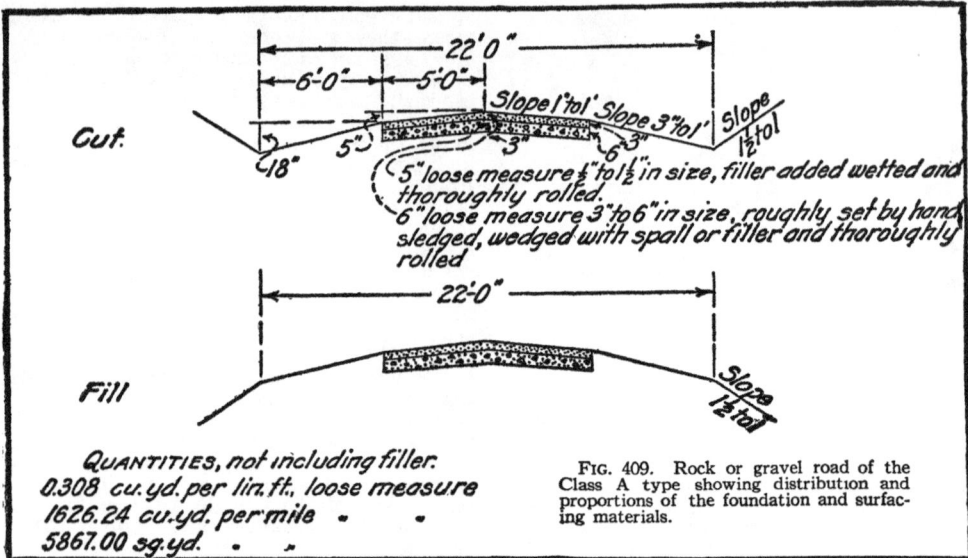

Cut.

22'0"
6'0" 5'0"
Slope 1"to1' Slope 3¾"to1' Slope 1½to1
5" 3" 3"
18" 5"
5" loose measure ¾" to 1½" in size, filler added wetted and thoroughly rolled.
6" loose measure 3" to 6" in size, roughly set by hand sledged, wedged with spall or filler and thoroughly rolled

Fill

22'0"
Slope 1½to1

QUANTITIES, not including filler.
0.308 cu. yd. per lin. ft., loose measure
1626.24 cu. yd. per mile
5867.00 sq. yd.

FIG. 409. Rock or gravel road of the Class A type showing distribution and proportions of the foundation and surfacing materials.

presence of lime, clay, or other cementing medium and will always recement in the roadbed. If it is a clean gravel, it is hard to consolidate without some smaller material, and this is called the "filler."

Gravel roads, like rock roads, may be divided into two classes: Class A, of the better class of construction; and Class B, of the cheaper kind (Figs. 410, 411). For a Class B road, the gravel should be placed on the prepared earth foundation and between the shoulders of earth thrown up from the side ditches to receive it. The gravel should not be larger than 3 inches, measured on the greatest dimension. One inch of filler, preferably limestone screenings, should be spread evenly over this, and the surface then well harrowed with a spike-tooth harrow (Fig. 398).

At the same time, the earth shoulders should be kept drawn up to the sides of the gravel, to hold it in place. This is the second step (Fig. 411). When the harrow has been passed over the surface a number of times, another inch of filler material should be spread over it, and the sides, shoulders, and ditches finished to the proper slopes and shape. Watering and rolling may be dispensed with, so that, with the addition of a grader, this class of road can be built with the ordinary farm implements. In some soils, a drag may be substituted for the grader.

Care should be taken to keep the ruts filled and the crown shaped up while travel is consolidating the newly finished road. As fast as ruts and low places are formed by the wheels of vehicles, gravel should be raked into them (the drag or harrow can take the place of hand work), and this work should

be kept up until the roadbed is solid. Let nature do the sprinkling, travel do the rolling, and yourself do the dragging; and in a year's time the road will be pretty well consolidated.

The cost of a 10-foot gravel road for a 1-mile haul of gravel, exclusive of grading, will be about $1,000 a mile (Fig. 409).

Rock roads. Rock roads of Class B, or the cheaper road of rock is constructed in a manner similar to that described for a gravel road of the same class. Rock from the crusher, taking what is termed the "crusher run," is substituted for gravel, but, otherwise the details of gravel-road construction apply here also. The rock should be selected with a view to its hardness, toughness, and cementing value. Where, as is generally the case, there is but the one kind available, there is frequently a choice between the different ledges, between ledge and field rock or between solid and shell rocks.

The broken stone is spread on the roadbed to the required width and thickness. It should be broken to such size that no stone shall be larger than 3 inches, measured on the greatest dimension. It must be spread evenly upon the foundation, to present a true surface to the finished road. Stone screenings, or crushed rock less than one-fourth inch in size, or other suitable filler, should be spread over the surface and then be harrowed with a spike-tooth harrow as long as the teeth will penetrate the surface. Another inch of filler should then be evenly and uniformly spread over the surface, and the shoulders and ditches finished. The finished road should be given the same attention while consolidating under traffic as that described for gravel roads.

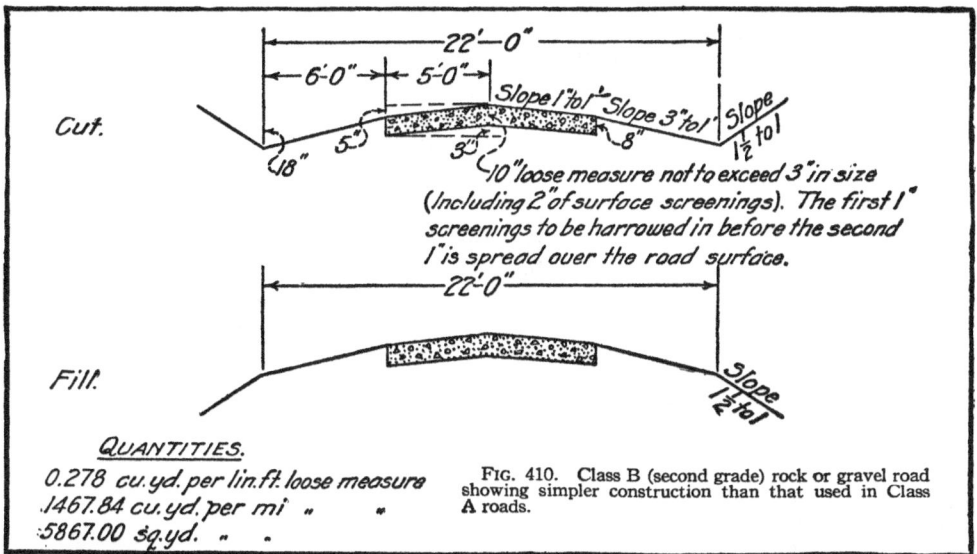

Cut.

Fill.

QUANTITIES.

0.278 cu.yd. per lin. ft. loose measure
.1467.84 cu. yd. per mi "   "
.5867.00 sq.yd. "   "

FIG. 410. Class B (second grade) rock or gravel road showing simpler construction than that used in Class A roads.

The cost of a rock road, exclusive of grading, calculated for a 1-mile haul of rock is about $1,500 for a 10-foot road (Figs. 410 and 411).

Roads of higher grade. The roads thus far described, and designated Class B, in which road metal is used, represent a kind of construction more often within reach of farmers for roads and farm lanes than the more expensive and better kind, which we have termed Class A. If duly cared for and properly maintained, they make good farm roads.

The construction of roads of the higher grade, or Class A, may be divided into 3 steps or stages: (1) the subgrade, or foundation; (2) the base, or first course of metal or hard road material; and (3) the wearing surface, or second course of metal. The material, or metal, is put on the prepared foundation in courses, usually 2, but never less, and neither course should exceed 8 inches in thickness, as the complete effect of the roller extends only through 6 or 8 inches. The lower course of material may be of an inferior quality to that of the top course. The interstices, or voids, of the stone or gravel should be filled with a suitable binder (preferably limestone screenings in water-bound construction) well dampened ahead of the roller. As the top course will be in contact with the wheels, it should be of a lasting and wearing quality; and water should be used to flush the filler into the voids, since road material will compact better damp than dry. The foundation should be shaped out, all soft spots removed and replaced with good earth and thoroughly and evenly rolled ahead of the material. Care must be taken to get the metal on evenly and uniformly and the

whole well rolled. The crushed rock is separated into the required sizes by means of screens at the crusher, and care should be taken to see that the size is regular and within the specified limits.

The subgrade, or foundation. The complete subgrade should be of an even finish, shaped to a true surface, rolled until thoroughy compacted and with no soft or uneven spots remaining. This is the first step (Fig. 408).

Ditches should be excavated of such widths and depths as the drainage requirements demand, and with true grades and sufficient incline to furnish a free and uniform flow of water. Whenever necessary, subdrainage should also be provided.

A shoulder should be left when preparing the subgrade on each side of the roadbed, and be rolled with the final rolling. Rolling should be commenced at the sides and on the shoulders and, working toward the centre of the road, be continued until the material is well consolidated. All broken rock, screenings and filler should be spread evenly upon the roadbed.

The base. A layer of broken stone should be spread on the prepared subgrade to the desired width and thickness. The stone for this course must be broken to such size that no stone shall be less than 3 inches, nor larger than 6 inches. It should be placed upon the road to present a true surface to the finished road. Stone screenings, broken rock, and spalls should be sledged and worked into the crevices, making what is termed a "sledge" base, and the whole should be thoroughly rolled. This is the second step (Fig. 408).

The wearing surface. The second course

of stone, the wearing surface, should then be added. This is the third step (Fig. 408). This stone should be not less than one-half inch nor larger than $1\frac{1}{2}$ inches. It should be evenly and uniformly spread and rolled to partial compactness, after which sufficient screenings (rock less than one-fourth inch in size) to fill all voids should be spread over the surface and flushed in with water. The whole should then be rolled until thoroughly compacted, further screenings and water being added during the rolling until the voids are thoroughly filled with screenings, and the surface becomes smooth and rounded, and no further compacting is possible. No material should be used which will produce mud or dust. Neither should clay or any sticky material be used, because when it is picked up by the wheels the fine surface rock will be picked up with it.

The same specifications will apply to this class of gravel road, as the former except that the gravel of the first course may run from 2 to 6 inches in size and contain from 20 to 30 per cent of filler (smaller than 2 inches) reasonably well distributed throughout the mass. The second course of gravel should be from half inch to $1\frac{1}{2}$ inches in size and should contain from 20 to 30 per cent of material of less size. It is better if this last course of gravel does not contain the finer material and that limestone screenings to that extent be added, flushed in, and rolled. The finished rolling, watering, and shaping is the same as that indicated for a rock road.

The telford road. The telford road belongs in this class of construction and is one with a base course of heavy stone placed athwart the road.

The stone is broken to suitable size and set by hand with the broad end, or edge, down and in courses at right angles to the centre line of the road and firmly wedged by hand. All irregularities of the surface should be reduced with a hammer until no stone has a maximum projection or depression exceeding 2 inches, and the whole should then be thoroughly rolled. The crown to the base may either be formed in the foundation or may be made by using stone of increased depth toward the centre in the base. The rest of the construction is the same as that described for this class of regular rock or gravel road.

The cost of this class of road is nearly double that of Class B (Figs. 410 and 411), for the same width, length haul, and other conditions. The itemized cost for average country conditions is about as shown in the accompanying table.

NATURE OF WORK	COST	
	PER SQ. YD.	PER CU. YD. (loose)
Quarry rent . . .	$ .013	$ .05
Quarrying. . . .	.100	.40
Crushing	.075	.30
Hauling (1 mile) .	.075	.30
Shaping roadbed. .	.035	.12
Spreading material .	.03	.11
Rolling	.015	.06
Sprinkling. . . .	.014	.05
Superintendence . .	.013	.05
Incidental. . . .	.015	.06
TOTAL	$ .385	$1.50

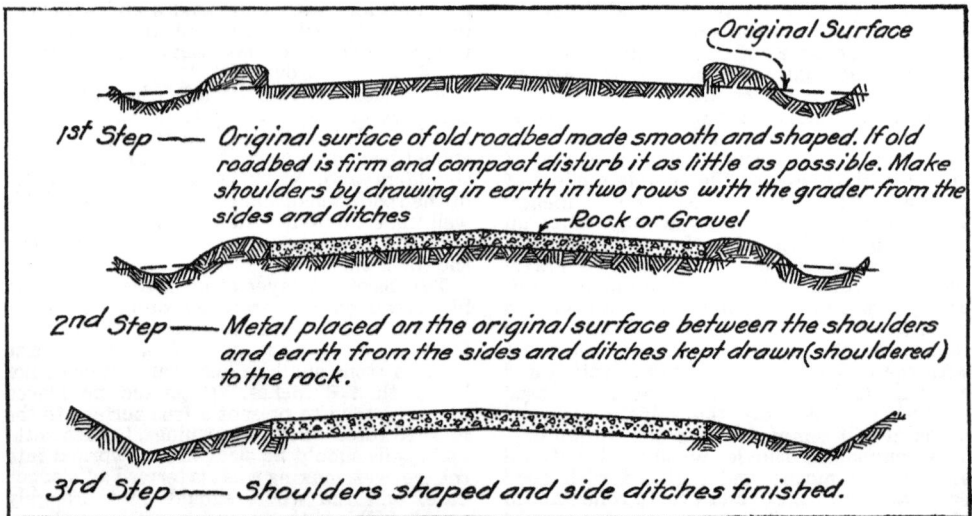

1st Step —— Original surface of old roadbed made smooth and shaped. If old roadbed is firm and compact disturb it as little as possible. Make shoulders by drawing in earth in two rows with the grader from the sides and ditches

2nd Step —— Metal placed on the original surface between the shoulders and earth from the sides and ditches kept drawn (shouldered) to the rock.

3rd Step —— Shoulders shaped and side ditches finished.

FIG. 411.　Three steps in the making of a class B rock or gravel road

Bitumen in road making. The roads thus far described are of what is termed "water-bound construction." In these, water is a vehicle to carry the filling material into the voids, and also to dampen the material, so that it will be more firmly compacted. This construction depends more upon consolidating than upon cementing the particles together. In "bituminous-bound" roads, bitumen is substituted for water, to get a more tenacious cementing of the parts together. This bitumen is especially prepared for road purposes from tar, asphalt, or asphaltic oil. For construction purposes, it is generally applied in what is termed the "penetration method," which requires some changes in method of construction from any thus far described. In fact, a greater degree of care and slightly different methods are used to develop a still better class of roads than those hitherto described, in both water-bound and bituminous-bound types.

Concrete or brick might be recommended for some localities, but both of them, as a rule, are too costly for application to farm roads, country lanes, and side roads. If one intends to embark on this class of construction, he should employ a road expert.

Sprinkling method. Some use may be made of the "sprinkling" method. This consists simply of an application of road oil by sprinkling. It is principally a dust preventive, but anything which prevents the surface of a hard road from grinding into dust adds to the life of the road. When uniformly coating the stone, it forms a bond which prevents the rapid wear and disintegration of the roadbed surface. The surface of the road should be placed in good condition, cleaned of all foreign substances, and be perfectly dry, and the application of oil should be made in dry, warm weather. The oil is applied by means of a sprinkling wagon and allowed to penetrate to a depth of, usually, 1 or 2 inches. It may be applied to an earth surface as well as to one of rock. This method requires from one-fourth to one-third of a gallon of oil per square yard of surface per application with 1 or 2 applications per year, and costs about 2 cents per square yard.

Earth-and-oil roads. The ordinary method of constructing an earth-and-oil road is to plow and pulverize the earth about 4 inches deep. While applying the oil, the earth is cut, rolled, and turned with plow, disc harrow, and grader until it becomes well saturated with the oil. It is then ironed out, with a light roller. The oil and mixing can not be well done in 1 application of oil, but the oil is gradually applied while mixing. It takes about $2\frac{1}{2}$ gallons per square yard of surface and will cost about 15 cents per square yard. Road oil of from 40 to 50 per cent asphalt is used. Oils of over 50 per cent asphalt are too heavy for this class of work, and will not penetrate unless heated.

Maintenance

Traffic and the elements are continually tearing the roads down, so we must be continually building them up, A small hole soon becomes a large one, and a roadbed can be kept smooth only by constant care and attention. It is this constant care and attention in the proper use of the drag which makes this implement one of the best for maintaining earth roads. It is also a good maintenance tool for the sand-clay road and for some gravel roads (Fig. 413).

Drags. The King, or "split-log," drag (Fig. 412) is made from the 2 halves of a split log 10 or 12 inches in diameter and 8 or 10 feet long. The halves are fastened about 30 inches apart with strong 3-inch wooden dowels, or stakes, driven into 2-inch auger holes; and they are set with their flat sides to the front and staggered about 1 foot. Five or 6 feet of the ditch end of the front slab is faced with iron for a cutting edge. This iron face is 2 or 3 inches below the lower edge of the slab at the ditch and cutting end, and runs out flush with the lower edge of the slab at the other end. A plank platform, for the driver to stand on, is laid across the dowels. The long end of the hitch chain, 7 or 8 feet in length, is fastened over the top of the front slab and to the cross-stake, or dowel, at the spillway end of the drag. The short end of the hitch chain, about 3 or 4 feet long, is

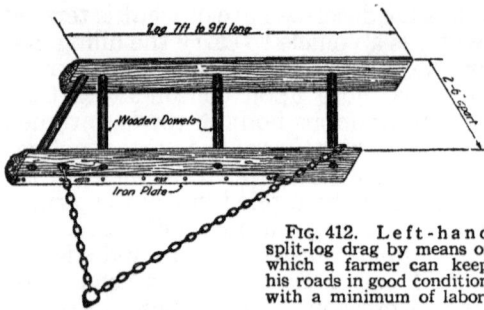

FIG. 412. Left-hand split-log drag by means of which a farmer can keep his roads in good condition with a minimum of labor.

passed through an auger hole near the centre line and from 6 to 8 inches from the end of the cutting, or ditch end of the front slab.

A plank drag (Fig. 413) is made in a similar manner, with 2 x 12 timber for each slab, reinforced by a 2 x 6 of the same length and bolted along the centre line to each 2 x 12.

The length of the hitch chain, slipping it backward and forward through the hole in the ditch end of the slab, regulates the hold taken on the earth. The position of the snatch hook to the doubletrees regulates the cutting effect. The shifting of his weight, or position, by the driver on the platform of the drag also regulates the cutting, carrying, and spilling effect of the drag. The drag is hauled at an angle of about 45 degrees with the line of the road.

The time to use the drag is when the road surface is moist, but not wet. The effect of the drag is to put on just enough soil to fill all holes, ruts, or hollows, pressing the water out of them and, by a smoothing and puddling effect, leaving the surface in better condition to sustain loads and shed rains. Some soils stick to the drag if too wet, others will stand wetter dragging. Different soils require different treatment. The rate of dragging should be about 1 round-1 mile-1 hour.

Keep roadbed and side ditches clean. On any kind of road, the sides between the traveled roadbed and the side ditches should be kept clean, so that the water may be shed freely to the ditches. The ditches and culverts, also, should be kept open and clear of obstructions. It is also a good principle of construction always to *repair a road with the same material of which it is made*.

Where ruts, holes, or defects appear in gravel, rock, or other hard-surfaced roads, use the hand rake to fill them with small loose stone raked in from the high points or the edges of the road. If additional material be needed, haul on small-sized stone and keep it raked into the defective places. On the better class of roads, the spot to be

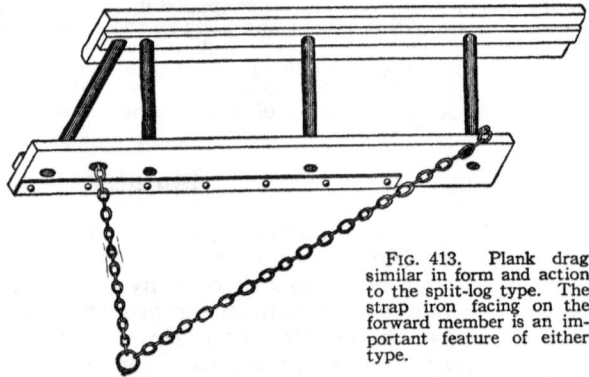

FIG. 413. Plank drag similar in form and action to the split-log type. The strap iron facing on the forward member is an important feature of either type.

repaired is cleaned before depositing the material, and either tamped or rolled.

Defects and wearing away will develop in any road. Repairing these defects and replacing this loss *as soon as possible*, is maintenance.

FIG. 414. Frostproof root cellar

CHAPTER 25

Concrete and
Its Use on the Farm

FIG. 415. Substantial, fireproof farm garage.

By H. COLIN CAMPBELL, Sanitary Engineer and Director of Publicity and Editor, Portland Cement Association. He was born and raised on a farm and has owned and operated a profitable Virginia farm since 1909. However, he also graduated from the Columbia University School of Mines, and from the College of Physicians and Surgeons. For twenty-nine years he has practised engineering, having for the last fifteen identified himself with the development of concrete construction. He prepared the courses of concrete construction for the International Correspondence Schools, has contributed to or served on the staffs of several farm papers and is the author of various text books on the use of concrete.—EDITOR.

CONCRETE is an artificial stone. Its ingredients are portland cement, sand, pebbles or broken stone, and water. Portland cement, when combined with water, undergoes certain changes which make it harden; and in this hardening process the portland cement acts as a binder to unite the sand and pebbles or broken stone into a firm mass which, in the course of a comparatively short time, becomes very hard and eventually, is to all intents and purposes, stone.

Portland cement. Portland cement is a manufactured product. It was given the name "portland" because of a resemblance which the first manufactured cement when hardened, had in color, to stone quarried at Portland, England. The name, therefore, means nothing except that such cement is a manufactured one. There are so-called natural cements, but these are comparatively little used in building construction, and hereafter in this chapter, when cement is referred to, it should be understood that portland cement is meant.

Portland cement consists principally of limy and clayey materials. There are several other substances in cement, but their presence is due principally to the fact that they are contained in small quantities in such limy or clayey materials.

When the correct quantities of raw material for portland cement have been selected, they are burned in a kiln at a temperature which converts them into a sort of slag or clinker. This clinker is afterward ground in what are known as tube mills, these being large revolving steel drums partly filled with flintlike pebbles which are tumbled about as the drum revolves, thus reducing the clinker to a fine flourlike powder. This powder is portland cement.

Being a manufactured product, the quality of portland cement can be controlled with practically any degree of exactness required. Many persons are under the impression that portland cement varies greatly in quality. Some people refer to "good" portland cement and likewise to "bad" portland cement.

Portland cement is made to meet specification requirements which have been found necessary to make it suited to the uses to which it is put; namely, to make mortar and concrete.

The only thing that can happen to portland cement to make it unfit for use is exposure to damp while in storage. Portland cement is purposely made sensitive to water. If it is stored where it can be affected by damp, it undergoes gradual and progressive hardening which will make it unfit for use. Therefore, portland cement, before use, should be stored in a tight, dry shed and piled on a board floor sufficiently above ground to prevent absorption of damp. If, when the sacks are opened, there are no lumps which cannot be crushed by light pressure between one's fingers, the cement has suffered no injury. If, however, there are lumps which require more than such light pressure to crush them, it is an indication that damp has acted upon the cement, and any such lumps should be discarded. Portland cement is now almost invariably shipped in cloth sacks or paper bags containing 94 pounds net. It is usually billed by the barrel, a barrel being figured as 4 sacks. Manufacturers and dealers make a charge for the cloth sacks, because, if they are taken care of and returned to the dealer or manufacturer in good condition, they will be redeemed for the price at which they were billed. Sacks should, therefore, be opened carefully and kept dry. Sacks which have been wetted or torn will not be redeemed.

Aggregates. "Aggregates" is the name given to the sand and pebbles or broken stone which are combined with the portland cement in the manufacture of concrete.

Sand. Aggregates are referred to as "fine" and "coarse." Sand is called "fine aggregate." Not every kind of sand will do for concrete. Concrete sand must be graded from fine to coarse, and in such grading the volume of coarse particles should exceed the fine ones. Usually concrete sand should range from quarter-inch particles downward to the finest permissible ones, exclusive of dust.

Sometimes, stone screenings from hard rock, such as granite, are used in a concrete mixture in place of sand. They, also, must meet the same specifications; that is, they must range from quarter-inch downward in size and must be free from dust.

Pebbles or broken stone. Pebbles or broken stone are referred to as "coarse aggregate." By "pebbles" is meant coarse material ranging in size from quarter-inch up to 1 or 1½ inches or, in some cases, 3 inches, depending upon the kind of work for which the concrete is being made. The same specifications would apply to broken stone, when used as coarse aggregate instead of pebbles.

Many persons think that natural bank-run material, which is sand and pebbles as taken directly from a gravel pit, is just the same as properly graded sand and pebbles. This is not true, because in a concrete mixture it is very necessary that the materials be proportioned so that voids, or air spaces, will be reduced to the lowest possible minimum. This cannot be done without using an excess quantity of cement where natural bank-run gravel is used instead of definite volumes of sand and pebbles or broken stone. The reason for this will be made clear by the following: Practically every gravel pit contains twice as much fine material (sand) as coarse material (pebbles). In most concrete mixtures, proportions for which are given below, it will be seen that correct proportions involve a ratio between fine and coarse materials practically the reverse of the ratio between such materials which exists in the natural gravel bank. No matter how solidly the materials may seem to lie in the deposit, there are from 33½ to 45 per cent or more of voids or air spaces. This can be proved by taking from the pit enough material to fill a tight box having a capacity of 1 cubic foot, packing the material in the box as firmly as possible, and then pouring water into it until the box just overflows. An experiment of this kind will prove that, in addition to the compacted material from the gravel bank, there may be added to the box from one-third to nearly half its capacity of water. Bankrun material must, therefore, be screened so as to separate the fine material from the coarse to enable correct proportioning of the two when preparing a concrete mixture.

Screening materials. For screening bankrun material, a suitable screen may be made by building a frame from 2½ to 3 feet wide and 6 feet long from 2 by 6-inch lumber and nailing over this screen-wire netting having quarter-inch square meshes, or slotted screen wire having three-eighths inch slots with cross wires from 4 to 6 inches apart. When such a screen is set at an angle of about 45 degrees, and the bankrun material thrown at its upper end in shovelfuls, the material rolling down the face of screen will be separated into fine and coarse particles. It may be necessary to pass the coarse material over another

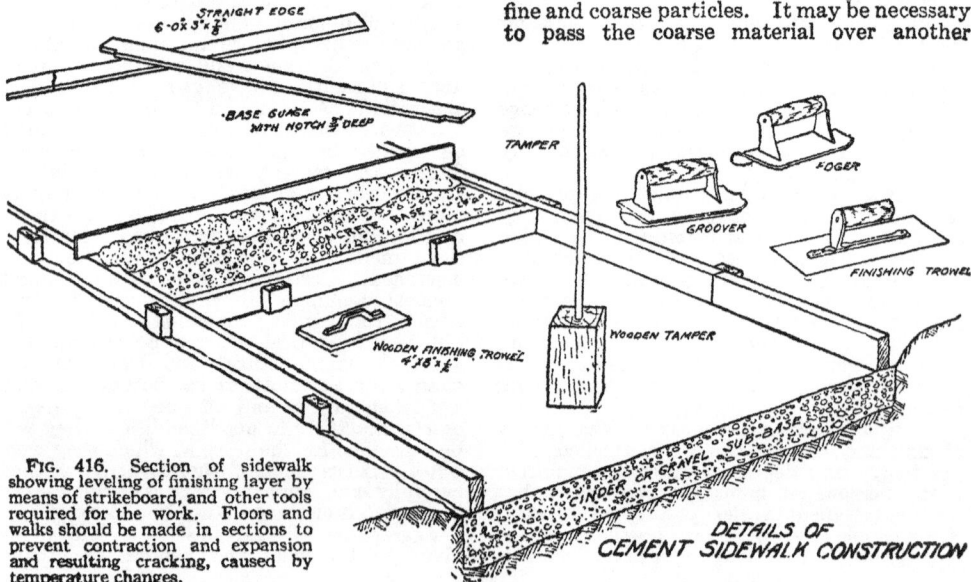

Fig. 416. Section of sidewalk showing leveling of finishing layer by means of strikeboard, and other tools required for the work. Floors and walks should be made in sections to prevent contraction and expansion and resulting cracking, caused by temperature changes.

screen having 1¼-inch or 1½-inch meshes, so as to exclude particles that would be too large. The maximum size of particles that can be used in any concrete mixture is governed entirely by the nature of the work. In reinforced concrete work, the largest particles should not exceed 1 inch in greatest diameter; otherwise, the concrete cannot be solidly compacted in the forms nor brought in intimate contact with the reinforcing material.

Use clean materials. Several things must be borne in mind when selecting aggregates for concrete making. The materials must be clean. They must be free from dust, loam, clay, or other foreign material. If there is much of such material present, it will prevent the cement from coming in contact with the surface of the aggregate particles, and this, in turn, will prevent the cement from acting as a binder to unite the mass. A very small quantity of foreign materials such as those mentioned, say, not more than 3 or 4 per cent, is not objectionable, if the concrete is thoroughly mixed. If foreign material is present in excess of 4 or 5 per cent, it must be removed by washing.

Washing aggregates. Small quantities of aggregates may readily be washed in an inclined trough by throwing them in at the upper end of the trough and introducing water there in sufficient quantity to cause the materials to roll down the trough. This tumbling about will loosen, or wash out, clay and loam which will be carried away by the water as it leaves the trough at the lower end. Such a trough may readily be arranged with a screen near its lower end, so that the sand and pebbles will be separated as well as washed. Another simple way of washing materials is to build a large tray, say, about 4 feet wide, 10 feet long, and 8 or 10 inches deep, prop up one end slightly, and throw the materials in at the higher end. Then apply water and keep the contents of the box stirred up, so that the overflow water will carry off the silt or loam.

Quality of aggregates. Aggregates that are soft, such as would come from shale rock, are not suited to concrete. Sand should be what is known as siliceous; that is, it should consist of particles of quartzlike rock. Pebbles should be hard and round or egg-shaped, rather than flat and long. As pebbles of the latter shape will not compact solidly, they will not make dense concrete.

When stone screenings are used in place of sand, they should be the product of crushed trap rock or granite. If broken stone instead of pebbles is used for coarse aggregate, it should have been obtained by grinding granite, hard limestone, or trap rock. Sometimes slag is used for coarse aggregate. There are various kinds of slag, however; but only slag which is the refuse from blast furnaces in smelting iron ore is suited for concrete aggregate. Slags that come from other ores usually

contain free chemicals which may act injuriously on the cement. One exception to the foregoing statement is material known as "chats." This is the waste material obtained in the process of smelting, or reducing, zinc ores. It makes a good aggregate, but cannot be obtained in graded sizes, as it almost invariably runs from half inch downward in grading.

Fire-resisting aggregates. Another point should be borne in mind in selecting aggregates, and that is their ability to resist fire. Concrete is recognized as a highly fire-resisting building material. The degree of fire resistance it has depends largely upon the aggregates. In the process of manufacture, portland cement is subjected to very high heat, so that in itself it is highly fire-resisting. Trap rock and slag are fire-resisting rocks, because in nature they resulted from a transformation in which high heat was present. Granites, although hard, tend to burst, when exposed to high heat, while most limestones disintegrate or crumble when exposed to heat, because, as nearly every one knows, quicklime is made by burning limestone. Therefore, for high fire-resisting qualities, aggregates must be selected with the above consideration in mind.

Adaptability of concrete. Concrete is a most adaptable building material. This is due to the fact that, when the mixture is finished, it is in what we might call a plastic form. If immediately placed in forms or molds and then allowed to remain undisturbed, it will harden and assume the shape of the mold, or form, in which it has been placed.

The advantages of concrete are that it is a strong material, may be easily prepared and used, and, when properly employed, makes to all intents and purposes a permanent fireproof structure—more nearly so than can be secured with any other building material. Also, it makes a structure that is water-tight, and one that can readily be kept clean, which means sanitary.

Most of the materials of which concrete is made may be obtained anywhere. There is hardly a farm that has not on it or near by a

FIG. 418. The essential tools for concrete work are easy to get on the farm: *a*, rake for spreading; *b*, metal wheelbarrow; *c*, wooden float; *d* and *e*, shovels, the first better for mixing; *f*, pail; *g*, water barrel; *h*, sand screen; *i*, iron tamper for foundation.

gravel pit from which the necessary sand and pebbles may be obtained. Cement is sold in practically every town. The little reinforcing steel that may be needed is about the only thing that has to be shipped in; while, under careful supervision, common labor can do the greater portion of the work. Any intelligent farmer who is willing to observe a few simple rules that mean success with concrete can build almost any farm structure with the help of his farm labor.

Tools required. The tools required are few and simple. Practically all of them are already a part of the farm equipment or may be home-made. A screen for separating sand from pebbles has already been described. A mixing platform may be made by nailing 1 x 4-inch tongued-and-grooved boards on 2 x 4 stringers placed 2 feet apart. Tongued-and-grooved boards are specified, because it is necessary that the platform be tight, so that, when water is added to the other materials,

the cement will not be washed away. The platform should not be less than 8 by 10 feet square; and there should be nailed 2 x 3 stringers around 3 sides, so that, when turning the materials in shovel mixing, they will not be thrown off the platform.

Concrete is usually proportioned by volume, and in such proportioning 1 sack (94 pounds net) of portland cement is considered 1 cubic foot. Therefore, the cement need not be measured unless the quantity of concrete required calls for less than a sack of cement. On most work, however, the job will call for more than a 1-sack batch; therefore, it is always well to mix batches that require 1 sack of cement. This does away with the necessity of measuring that ingredient.

For measuring sand and pebbles or broken stone, a bottomless frame will serve as a measuring box. This may be of any capacity from 1 cubic foot upward; not larger than from 3 to 5 cubic feet is best. This box should have, marked around the interior, lines indicating capacities of 1, 2, 3, etc., cubic feet, so that, if it is desired to measure any quantity less than the full capacity of the box, definite known volumes may easily be obtained. In use, the box is set on the mixing platform and the required quantity of materials placed in it; it is then lifted, and, if necessary, again filled.

Square-pointed shovels are needed for mixing the materials. A hose, or in its absence, a water barrel and pails will be required for handling water. A wheelbarrow may be needed, to move the concrete from the place where mixed to that where deposited. A wheelbarrow with a steel body is best. It should be cleaned out each time after use, so that it may not eventually become filled with particles of hardened concrete.

About the only other tools necessary are a strikeboard, to level off the concrete when being used in forms for walk or floor construction; a wooden hand float, to finish the surface; a steel trowel for smoother finishing, if needed; and, when walks, floors, and similar pavements are being made, a groover or jointer and an edger, to finish properly the joints and edges of slabs. These last-mentioned tools may be obtained in any hardware store.

Concrete Mixtures and Their Preparation

For very accurate results, concrete is proportioned after quite extensive tests have been made on the sand and pebbles or broken stone, to determine the percentage of voids, or air spaces, in their volume. For most work, such scientific or exact methods are not necessary. The concrete is, in most cases, proportioned after what are known as arbitrary methods. This practice is most common on the average run of work, and is satisfactory because long experience has shown that certain mixtures, presuming that the materials are well graded, answer for certain classes of work.

FIG. 419. Screening sand. Clean, sharp aggregates are as necessary as good cement

Mixtures are described 1:2, 1:3, 1:2:3, 1:2¼:4, 1:3:5, etc. A 1:2 mixture means that the concrete is composed of 1 part, or 1 volume, of cement and 2 parts, or 2 volumes, of sand. In other words, a 1:2 or a 1:3 or a 1:1½ mixture means a sand-cement mortar. The first figure stands for cement; the second, for sand. In a 1:2:3 mixture, for instance, we have a third figure. This refers to the volume of coarse aggregate—pebbles or broken stone; so that a 1:2:3 mixture means 1 part or 1 sack of portland cement, 2 cubic feet of sand, and 3 cubic feet of pebbles or broken stone.

Tables of mixtures. In the following table of so-called arbitrary mixtures are given the ones commonly used and recommended for the classes of concrete construction indicated for each. From this table the reader will be able to select mixtures suited to almost any class of work, if he will bear in mind that walls, floors, and pavements, for example, belong to one construction; that fence posts, hitching posts, and clothesposts make up another class; that watering troughs, feeding troughs and tanks are similar, and so forth.

TABLE OF ARBITRARY MIXTURES

1:2 mixture.
Used for wearing course of 2-course floors.
1:2:3 mixture.
Used for concrete roofs; 1-course concrete walks, floors, driveways, barnyard pavements, fence posts, watering troughs, and tanks, such as cisterns, reservoirs, etc.
1:2:4 mixture.
Used for concrete walls and reinforced concrete work in general.
1:2½:4 mixture.
Used for silo walls, grain bins, building walls when stucco finish is not to be applied, manure pits, dipping vats, hog wallows, and similar classes of construction.
1:2½:5 mixture.
Used for building walls that are to be stuccoed; base of 2-course walks, feeding floors, barnyard pavements; basement walls and foundations for ordinary conditions where water-tightness is not essential.
1:3:6 mixture.
Used for heavy foundations and footings.

MORTARS

1:1½ mixture.
Used for inside plastering of tanks, silos, and bin walls, if necessary.
1:2½ mixture.
Used for fence posts, when fine aggregate only is used.

Mixing concrete. Concrete may be mixed either by hand or by machine. Good concrete may be mixed either way. If only small quantities are to be made, hand mixing will probably be found most convenient; but, if the work is of any considerable volume, machine mixing will insure a more thorough mingling of ingredients and will considerably reduce the labor of mixing. Small power-operated mixers in almost endless variety, and all of them efficient, may be obtained at relatively low cost. They are to be recommended wherever the amount of work to be done warrants obtaining one. Frequently, farmers in a community unite and purchase such a machine and use it jointly as needed, thus making its cost practically nothing to each of those who are interested in it.

FIG. 420. Water-tight mixing floor with bottomless measuring box in foreground, and water barrel in rear. The edging around the floor is handy, but not essential.

Hand mixing. If concrete must be mixed by hand, the following method should be used: Measure the required amount of sand. Spread it out in a thin layer on the mixing platform. Dump on this the correct quantity of cement. Turn the cement and sand thoroughly 3 or 4 times, using a square-pointed shovel, until the entire mass is of a uniform color, absolutely without streaks of brown or gray. Measure out the required quantity of pebbles or broken stone. Sprinkle it so that everywhere the surface is wet. Add it to the mixed sand and cement. Turn the entire mass once or twice. Then add water, preferably from a hose spray or from a sprinkling pot, while some one keeps turning the sand and pebbles, so as to prevent the cement from being washed away. Turning should be continued until the mass is of a quaky, or jellylike, consistency.

Quantity of water. Too much water is as bad as too little. The right consistency in concrete mixtures is very important, for several reasons. Correct proportions and correct consistency insure water-tight concrete. A mass that is mixed too dry cannot be compacted to the utmost density in the forms, and will not make water-tight concrete. A quaky, or jellylike, consistency will not flow. If placed in a pile, it will remain as placed until slightly disturbed, when it will tend to flatten out. It can be carried on a shovel without spilling, while a wetter mixture will flow and will cause the sand-cement mortar to separate from the pebbles.

Machine mixers. Practically all makes of mixers to-day are of what are known as the "batch type." This means that separately measured materials must be placed in the drum thus insuring that each batch is properly proportioned and not the result of guesswork. Mixers come in various capacities. Small ones for farm use may be had for $75 and upward, in most cases complete with gasoline engine to operate. Several manufacturers make small mixers in which the drum, or mixing receptacle, consists of a barrel which has vanes or blades mounted on the inside. The barrel is revolved until the concrete

is mixed, and is then tilted, to dump out the contents. Other mixers have cylindrical or cube-shaped drums; but all have certain fundamental principles in common.

The choice of a mixer for farm use is largely a matter of deciding upon the capacity which it is desired the machine shall have, rather than upon its make. In this connection it may be well to mention that many farmers make their own patent fertilizer, and that those who have been most successful have mixed the ingredients in a medium-sized batch concrete mixer. This illustrates the point that a power-operated mixer can be made to earn its cost in another way than in using it for concrete construction.

Time of mixing. When mixing concrete by hand, the tendency is to slight the operation somewhat, because there is some work about hand mixing and the men on the job get tired toward the end of the day, and later batches are likely to be not so well mixed as earlier ones. No definite time for mixing can be stated, when hand mixing is being followed. Then it is simply a question of turning the materials until the mass is of a uniform color and consistency and the various materials have been thoroughly combined with one another. A machine mixer does not get tired, and the time of mixing can be definitely stated. After all materials have been placed in the drum, it should be revolved for a definite number of revolutions or a definite period of time. Manufacturers base the number of revolutions on the time required to make them, and this usually amounts to at least 1 full minute of mixing. One minute and a half, however, is better, since it has been proved that the time occupied in mixing has great influence on the strength of the resulting concrete.

No more concrete should be mixed at any one time than can be used within about 30 minutes. Neither should sand and cement be combined any considerable time in advance of adding the pebbles or broken stone and mixing water. Dampness of the sand will cause the cement to commence hardening, and thus destroy part of its effectiveness when the remainder of the materials are added and the batch of concrete completed. Loam and clay are more likely to remain moist than is clear sand, which is one reason why the latter should be carefully chosen and, if necessary, washed before it is used. The hardening action that takes place when cement and water are combined commences very soon after the combination. This will have progressed far enough within a period of 30 minutes, so that after that time any unused concrete should be thrown away. Neither concrete nor mortar that has commenced to harden should be retempered, as it is called, by adding more water and remixing. Such concrete or mortar will not have full strength.

FIG. 421. A small engine-driven concrete mixer, good for the farmer who does much construction work. It is tipped and emptied by means of the crank.

Forms

What they are. Forms, or molds, are the receptacles in which concrete is placed immediately after mixing, so that, when hardened, it will take the shape desired. For most work, forms are made of wood. Where a particularly smooth surface is desired as on an exposed wall face, forms should be of planed lumber. A still smoother finish can be obtained by lining the forms with sheet steel. The concrete will stick less to forms made of planed lumber or lined with metal forms than to those made of rough lumber. Also, the concrete surface resulting from the use of well-made forms is much more pleasing.

In foundation work, forms are often unnecessary for that portion of the work below ground; that is, when the earth is so firm that the walls of the trench are self-sustaining. When concrete has been placed to ground level, forms will, of course, be required for any further work. For a basement or cellar wall, it may be that only an inside form will be needed. It is necessary, of course, where the foundation is to serve as an inclosure for the basement or cellar, that the interior face of the work be smooth and well-finished. It is also necessary that the foundation wall be water-tight, and in such cases the concrete can be placed more satisfactorily by using forms than by depending solely on the earth trench.

Form construction. For small buildings, form studs, or uprights, to which the sheaths, or form boards, are nailed, may be 2 x 4's or 2 x 6's, placed from 18 inches to 2 feet apart. The sheathing boards must be either tongued and grooved or tightly jointed on the edges by being planed, so that cement and water will not run through joints while the concrete is being placed. Only ordinary carpenter skill is required to build forms for most concrete work; but all concrete form work requires careful thought, to plan for window and door-

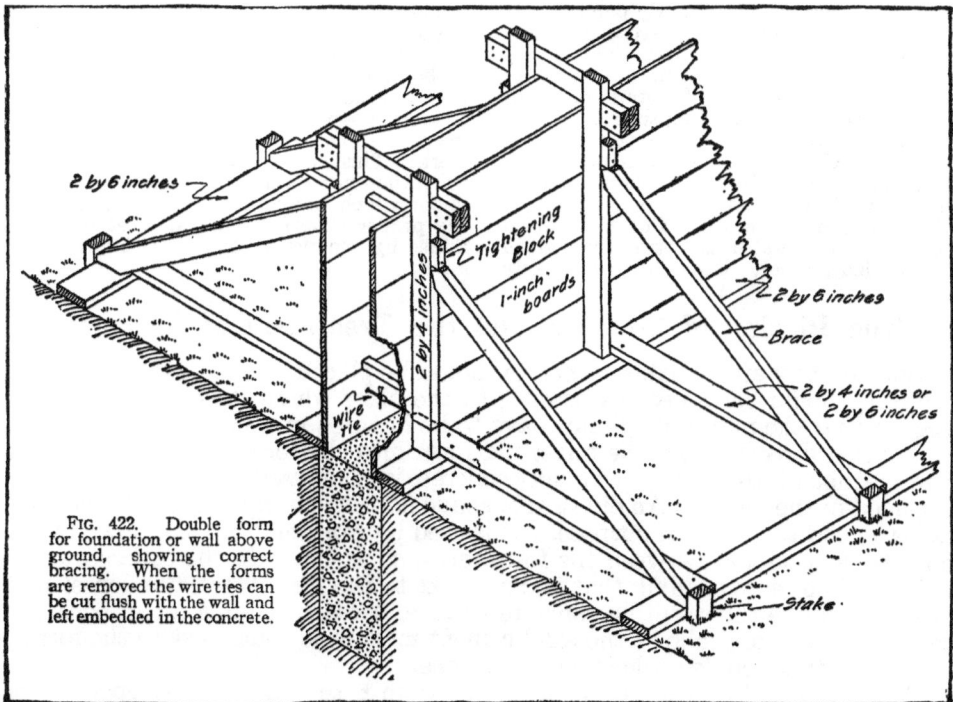

FIG. 422. Double form for foundation or wall above ground, showing correct bracing. When the joints are removed the wire ties can be cut flush with the wall and left embedded in the concrete.

2 by 6 inches

Tightening Block

2 by 4 inches

1-inch boards

wire tie

2 by 6 inches

Brace

2 by 4 inches or 2 by 6 inches

Stake

way openings and for any other departure from a plain, straight surface in the structure being built.

Setting up forms. Forms must be set up plumb and straight to line and be well braced with stiff braces, to prevent them from bulging out of shape, from ramming or compacting of the concrete when placing. They must also be set at the correct distance apart and tied or clamped, so that they will remain the required fixed distance apart while the concrete is being placed, in order that the concrete section wall, or whatever it may be, may have the required thickness. Wire ties, or bolts, drawn against wood spacers in the forms are used to hold them a fixed distance apart. As the concrete is placed up to the wood spacers, the latter are removed.

Ties and spacers. When wire ties are used, they are usually cut when the forms are removed, the main piece of wire is left in the concrete, and the projecting ends are cut off, the wire ends being hidden by pointing up or plastering the holes with a little cement-sand mortar. If bolts are used as form ties they must be greased before the concrete is placed, so that they can be knocked out of the concrete afterwards. The resulting holes are then pointed or plastered up with cement mortar.

Wetting or oiling forms. Before concrete is placed, the forms should be well wet down, to prevent the concrete from sticking. Sometimes, forms are wiped or lightly painted with a mixture of linseed oil and kerosene, to prevent the concrete from sticking. This is very effective, but not so economical as wetting them down. However, if forms must be used a number of times, oiling them before use each time makes them last longer, because the lumber absorbs the oil, thus preventing absorption of water from the concrete and, in turn, preventing warping and bulging.

Kind of lumber to use. For most form work, air-dried lumber is better than green or kiln-dried lumber. Green lumber is likely

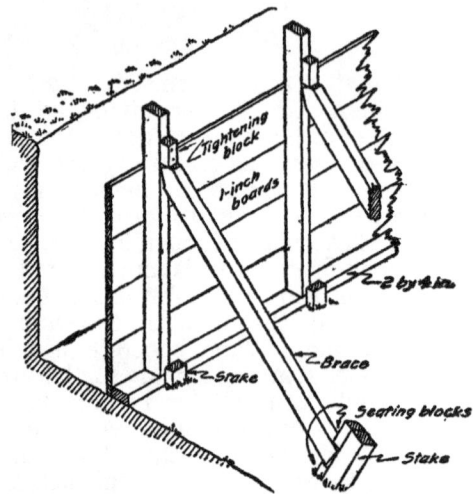

Fig. 423. Simple form for a retaining wall or foundation where the soil is firm enough to provide one side

to dry out, causing cracks between sheathing boards, while kiln-dried lumber is likely to bulge and swell through absorbing water from the concrete.

Form removal. A very important detail of concrete work has to do with the time of form removal. Concrete hardens best in moderately warm weather, provided it is not allowed to dry out rapidly. Forms can often be removed from wall sections in from 24 to 36 hours after the last concrete is placed. In cold weather, concrete hardens very slowly. It is not possible to say definitely when forms may safely be removed from any concrete work. This is a matter dependent entirely upon the temperature and weather conditions under which the concrete has hardened, and the proper time for removal must be determined by experience coupled with good judgment.

The Placing of Concrete and the Treatment of Surfaces

Immediately after mixing, concrete should be placed. The mixing platform, or mixer, should be as near as possible to the place where the concrete is to be used, so that the material will not have to be handled unnecessarily. It should be shoveled directly from the mixer platform into the forms, or dumped direct from the mixer into them. If a quaky concrete is being used, it should be puddled or spaded in the forms. An old garden spade flattened or a hoe straightened out makes a good tool for this purpose. It is used by working it up and down in the concrete next to and between the form faces, so as to settle the concrete thoroughly. The more spading that is done next to the form faces, the denser and smoother, will be the resulting concrete surface, because spading forces back the coarse particles, and allows the sand-cement mortar to come next to the forms, thus preventing pebble pockets on the surface.

Concrete should be placed in layers of from 6 to 8 inches. Deeper layers cannot be properly spaded so that they will unite with the concrete previously placed. Also, concreting operations should be carried on as continuously as possible, so that there will be few construction seams or lines of cleavage in the work. This is especially necessary when tanks, troughs, and similar receptacles that must be water-tight are being built. Sometimes, however, it is not possible to continue the work without interruptions; in such cases, when leaving off work, the concrete should be roughened in the forms. Then, when concreting is resumed, this roughened surface should be scrubbed with a broom and water and painted with a cement mortar paint made by mixing portland cement and water until a creamlike mixture results. This paint is then applied to the scrubbed surface in the forms and fresh concrete placed at once. If properly done, this will prevent a leaky construction joint.

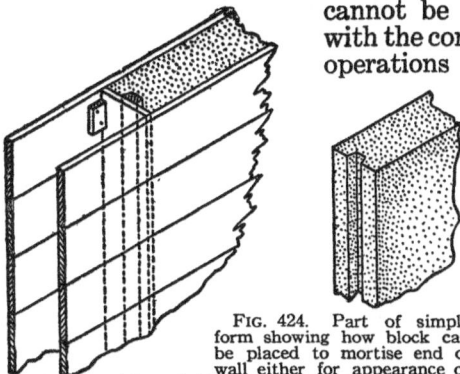

FIG. 424. Part of simple form showing how block can be placed to mortise end of wall either for appearance or to admit a sliding division board, a door frame, or the tongue of another section of wall with which a strong joint is desired.

Protecting finished work. The action which takes place when cement and water are combined is a chemical one. The process is technically called "hydration." It is commonly referred to as "a process of crystallization," and this accurately describes it. A certain amount of water is necessary to complete this chemical change, or complete crystallization, which leads to hardening, in other words, to make the cement set. This setting of cement is what causes the hardening of concrete. It cannot take place properly, naturally, and completely unless all of the water which is used in making the concrete mixture can be retained in the concrete until hardening is complete. To insure this, it is necessary that when the work is finished it shall be protected in some way to prevent loss of water. Many persons have the wrong impression that the hardening of concrete is a drying action. This is exactly the reverse of what should take place. If the concrete is allowed to dry out, it will not attain the same degree of hardness as when prevented from drying out. Therefore, when concreting has been finished, some means must be taken to protect the work. What this protection must consist of depends upon the nature of the work and the time of year. Sometimes, the only protection required is that secured by leaving the forms in place for several days and keeping the work thoroughly wet down. In other cases, the work must be covered with wet burlap, hay, straw, or similar material, and this covering be kept wet down for several days. Many leaky tanks, troughs, walls, etc., have resulted from failure to protect the concrete after placing. This one detail is very important; indeed, it is almost equal in importance to the correct proportioning, mixing, and placing of materials.

Finishing concrete surfaces. A concrete surface may be given several different kinds of treatment, if it is desired to vary the appearance from that obtained merely by contact with the forms. If forms are well made and the concrete is properly placed, no after-treatment is needed for the average concrete surface; but, as concrete is very faithful in reproducing any irregularity of form work, it may be necessary to go over the surface in some way, to remove unpleasing irregularities.

Rubbed surfaces. Probably the simplest treatment that can be given a surface is to go over it with a brick or carborundum stone after the forms are removed, keeping the surface wet, and rubbing all over with the brick or stone, to grind down form markings. Sometimes, a cement water paint is applied, but as a rule such applications peel off after a while and leave the surface more unsightly than were no such treatment given.

Colored surfaces. Concrete can also be

colored by combining with the mixture a small amount of some mineral coloring matter. Only mineral pigments should be used, because in time other colors fade and they usually fade unevenly, so that the resulting surface is much more displeasing than it would have been if no attempt had been made to color it.

Slap-dash and pebble-dash. Concrete surfaces that are to be plastered or stuccoed can be varied in a number of ways. Usually, a thin mortar is thrown against the concrete surface from paddles. After a little experience, very pleasing surfaces can be produced in this way, the mortar being colored with mineral pigments; and, if the work is properly done, the surface is referred to as having a "slap-dash" finish.

A "pebble-dash" finish is secured by plastering the surface with cement mortar as soft as can be worked under the trowel and, before this mortar has commenced to harden, throwing against it clean, hard, uniformly graded pebbles. If before they are thrown against the surface, the pebbles are wet with cement water paint, they will adhere more firmly to the surface.

Tooled surfaces. Sometimes, a concrete surface is tooled, this usually being done by going over it with a bush-hammer or pein hammer, such as stonecutters use. A concrete surface cannot be finished in this way, however, until it is 36 days or more old, so that the hammer or cutting tool will not dislodge or loosen particles of aggregate from the surface. This manner of finish is laborious and somewhat expensive and is not often used, except on buildings that require some treatment to produce a desired architectural effect.

Washing and scrubbing the surface. Other surface finishes are secured by washing or scrubbing off the film of cement from the surface of the aggregate particles. Washed surfaces, however, are usually confined to more or less ornamental objects such as vases, lawn benches, sundial pedestals, etc. Preparation for securing such surfaces is usually made when the concrete is mixed, and consists principally of using selected aggregate for some desired color that it is intended to expose on the concrete surface by the scrubbing or washing process. If forms can be removed within 24 hours after the concrete has been placed, the aggregate surfaces can usually be exposed by scrubbing down the concrete with a stiff brush and water. If, however, the concrete has hardened too much to allow this, it is necessary to use an acid wash. A common one consists of 1 part muriatic acid to 3 or 4 parts of water. Scrubbing must be

done quickly, the work watched carefully, and the surface drenched with water immediately the surface film of cement has been removed, so that the action of the acid will be stopped. Otherwise, it would go on acting with the cement and would finally loosen the aggregate particles from the surface.

Painting. Very rarely is a concrete surface painted. If, however, it is desired to paint the concrete, a paint prepared especially for that purpose must be used and the concrete must be dry and thoroughly hardened before the paint is applied. There are in concrete some free lime and other chemicals that act on ordinary paints and prevent them from adhering permanently to the concrete.

Waterproofing. No waterproofing treatment is necessary with concrete that has been properly proportioned, mixed, placed, and protected after finishing. If, through faulty workmanship, however, the concrete is not water-tight, there are several treatments that can be applied to render it so. One of these is what is known as the Sylvester process, which consists in the application of alternate solutions of soap and alum to the surface. Details of the treatment are as follows:

Two solutions are used in this treatment. The first consists of three-quarters pound of soap to 1 gallon of water; the second of half pound of alum to 4 gallons of water.

The surface to be treated should be perfectly clean and dry.

The soap wash should be painted on when at boiling heat. This should remain 24 hours, after which the second, or alum, wash should be painted on in the same manner. These coats may have to be repeated alternately a second time, until the walls are made impervious to water. The alum and soap thus combined form an insoluble compound, filling pores in the concrete.

Another treatment for the same purpose consists in painting the surface several times with a solution of water glass in water. Water glass is chemically known as silicate of soda. This chemical, also, reacts with free chemicals in the concrete and closes the pores. Several applications may be necessary to produce the required effect.

Various compounds are sold as waterproofing mediums. Some of these are mixed with the dry materials when the concrete is proportioned; others are applied to the finished work; and, while all of them have something to recommend their use, none of them will take the place of good workmanship. In other words, the workmanship cannot be slighted merely because some one of these waterproofing compounds is being used.

Reinforcing Concrete

Probably the most notable property of concrete is its compressive strength; that is, its ability to carry heavy loads placed directly upon it. In fact, the first use that was ever made of concrete was to build mass work or heavy foundations. Experiments with concrete proved, however, that it was possible to extend its use to practically every part of the building by imbedding in it reinforcing material in the shape of steel-wire mesh or steel rods. Concrete is not very strong in tension, that is, in resisting loads that tend to bend it or pull it apart. But, if suitable reinforcing material be placed in it, it can be used for beams, roofs, columns, and other parts of structures, just as any other building material may be used, and with decided advantage, as compared with other materials, because concrete is fireproof and, unlike other materials, grows stronger as it grows older.

Where reinforcing is necessary. As a rule, concrete foundations do not require reinforcing. Building walls, however, over a certain length should have suitable rods or mesh fabric embedded in them, to take care of stresses or strains resulting from changes of volume in the concrete under different temperature conditions, because concrete, like other material, expands and contracts as the temperature rises and falls. Reinforcing is necessary in some other uses of concrete, such as beams, columns, fence posts, etc. Probably no better example of the principle of reinforcing concrete can be chosen than the ordinary fence post. If a concrete fence post were made without reinforcing, it would probably break in the fence line very quickly because of the pull of the fence wires on it. Or, if not from this cause, if stock tried to break in or out of the inclosure, it would not take much of a blow to break the post off at the ground. If steel rods are properly placed in the concrete, such breaking cannot occur. But the rods must be properly placed. Many persons think that a single rod or piece of gas-pipe at the centre of the post accomplishes all that is necessary in the way of reinforcing. This can readily be proved incorrect by reference to the two accompanying sketches.

The first (Fig. 425) is supposed to illustrate a beam which has been cut in two at its centre and joined by a hinge. The two ends of the beam rest on supports, as shown. At the top of the hinged joint is supposed to be a block of rubber; at the bottom, a coiled steel spring. It requires no imagination to see that, if a load is placed on top of the beam, it will bend at the hinged joint; and, supported at the two ends as it is, this bending under load will tend to close the gap at the top, where the piece of rubber is inserted, and, at the same time, make the gap at the bottom, where the coiled spring is, wider. In other words,

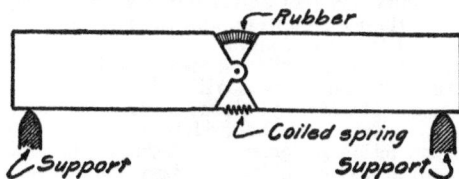

FIG. 425. Diagram showing the effect of the strain on a supporting beam held at the ends (See text)

the rubber will be squeezed together, or compressed, while the spring will be pulled, or lengthened; that is, the spring will receive the pulling strain and, at the same time, will tend to resist the bending. Suppose now that, instead of being broken and hinged as shown, the beam is made solid and a steel rod is

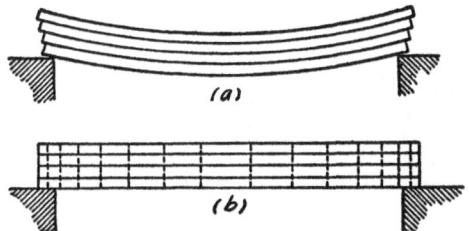

FIG. 426. Diagram showing how and why vertical reinforcing is effective

imbedded in the concrete, say, half inch or more from the lower face of the beam. The adhesion between the concrete and steel will compel the steel to resist the tendency to bend; that is, it will take up the tensile strain brought about by the load on the beam.

Another example is given in the second illustration (Fig. 426). This shows the manner in which the bending strains on a concrete beam act and may be resisted. Figure *a* is intended to show several planks or boards, laid one on top of the other and separated at their ends. If they are long enough, they will naturally bend from their own weight, while, if a load is placed upon them, they will bend still more and in bending will slip past or along each other, thus causing their ends to become uneven, as shown. If these planks are bolted together, as shown in Figure *b*, the slipping will be prevented and the planks will lie practically straight.

The principle of reinforcing concrete is applied by every farmer who has strengthened his singletrees to withstand the pull or strain of the team by bolting a piece of strap iron along their back face.

Steel is to be preferred above all other materials for reinforcing concrete. In fact, steel is the only material used. The reason for this is that steel is the only material which expands and contracts under changes of temperature in the same degree or ratio as does concrete. This is very necessary, as it prevents the breaking of the bond, or adhesion between the concrete and the reinforcement; and this bond, or adhesion, is necessary to secure the full effectiveness of the reinforcement.

Installing fittings in concrete. Anchor bolts, hinge eyes, and various other fittings that are to be attached in concrete for setting machines or for hanging gates or doors should be arranged for when the forms are made. Anchor bolts can be suspended from crosspieces resting on forms (Fig. 427).

Anchor bolts for setting gas engines are placed in a similar manner. In such cases, however, care must be taken to make an accurate template, bored so that the position of holes in it will correspond exactly to the position of holes in the machine or other device to be set.

Concrete can, of course, be drilled and bolts be leaded in place after the work has been finished, but it is much more satisfactory to provide for them at the time the work is going on.

The desirability of using concrete as largely as possible in farm buildings has often been urged elsewhere in this work. In the following paragraphs, the places where it may be advantageously employed by the farmer are more specifically set forth together with some helpful information in reference to dimensions of various structures and to the materials for concrete and reinforcement.

Some Things a Farmer Can Make With Concrete

Floors, walks, driveways, etc. Floors, walks, and driveways are similar classes of concrete pavements. A description of one serves as a guide for the construction of the others, with but a few minor exceptions.

Perhaps the most profitable kind of concrete floor that the farmer can lay is a feeding

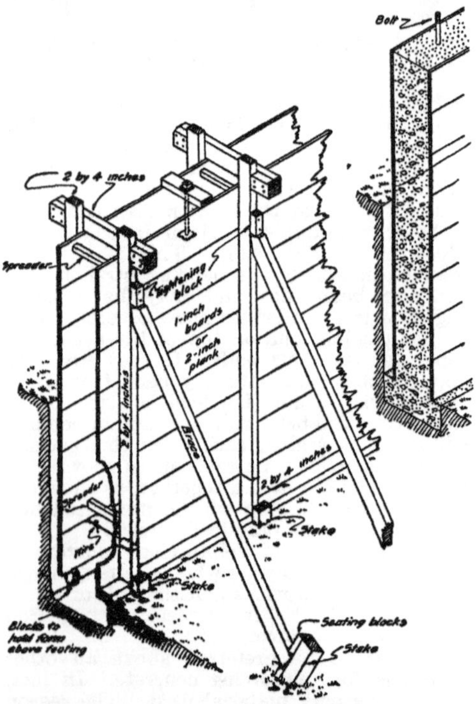

Fig. 427. Double forms for foundation extending both below and above the surface, and with a footing. Note how bolts or other fittings are installed by being hung in place before the concrete is poured.

floor or barnyard pavement. On such floors no grain or fodder fed to animals is lost, nor is any trampled in the mud. Floors in dairy barns or cow stables and in hogpens are desirable, profitable, and sanitary improvements. All of these classes of floors are usually laid on the ground. A feeding floor is built just like a sidewalk. In fact, a feeding floor is nothing but a series of sidewalks laid side by side.

Feeding floors, barnyard pavements, and dairy-barn floors must be laid on firm, well-drained soil. If the area where the floor is to be laid is not firm, it should be made so by thorough rolling and tamping. If the soil is not well drained, then tile drains should be laid at the edges of the floor, these drains leading to some suitable outlet, so that water will be kept from beneath the concrete. Feeding floors are usually made 5 inches thick and of a 1:2:3 one-course concrete. This means that the same mixture of concrete is used for the entire thickness of the floor. Two-course floors or pavements consist of a $1:2\frac{1}{2}:5$ base on which is laid a 1:2 cement-mortar top, or wearing course. The base is usually 4 inches thick, and the top, or wearing, course, 1 inch. Where floors, walks, driveways, or similar pavements are to be used by loaded wagons, they should be not less than 6 inches thick.

Feeding floors and barnyard pavements are usually built in slabs from 5 to 6 feet square, but in no case exceeding 10 feet square.

Concrete walks about the farm are rarely more than 3 feet wide, and the slabs may be from 3 to 5 feet long. Walks are sometimes made no more than 4 inches thick. This is all right where there is little freezing and expansion of wet soil; but, in general, a thickness of 5 inches is to be preferred. One-course construction is better than 2-course, because there is no possibility of separation between the base and the top course, as sometimes happens in 2-course work when the top course is not laid until after the base has commenced to harden.

Finishing concrete walks, floors, and pavements with a wooden float is better than finishing them with a steel trowel. A wooden float gives an even, yet gritty, texture to the surface, that provides a good foothold for man as well as for beast.

Protection of concrete floors and pavements immediately after laying is very important. As soon as the concrete is hard enough to resist the pressure of one's thumb, it should be covered with a 2-inch layer of moist earth. This covering should be kept wet by sprinkling several times daily for a week, before the walk or floor is put into use. If the pavement is to be driven over, it should remain unused for at least 2 weeks after laying.

Concrete feeding floors are usually built with a curb and apron around them. The curb prevents animals from shoving food off while eating, while the apron, extending down 12 inches or more below the base of the floor, prevents hogs from rooting underneath.

Concrete feeding floors and barnyard pavements are easily kept in sanitary condition. It is customary to slope such floors and pavements slightly in 2 directions and at the low point to connect with a tile drain leading to a concrete manure pit. This arrangement saves every bit of fertilizer and adds to the profit of the improvement.

Foundations. For most buildings that require strength rather than water-tightness, foundations are made of a 1:2½:5 or a 1:3:6 concrete. For small farm buildings like poultry houses and hoghouses, which are rarely more than 1 story high, foundation walls need not be more than 6 or 8 inches thick. It is difficult, however, to dig such a narrow trench to proper bearing soil; consequently they are often made thicker, especially when the concrete is to be deposited without forms. If, however, excavating would have to be continued too far to reach firm bearing soil, then the trench is made wider and the foundation started by first laying a footing (that is, a layer of concrete 6 inches or more high and 10 inches or more wide), to distribute the load of the building over a greater area of soil, thus helping to prevent settlement. No fixed rule can be given for the thickness of foundation walls nor for the width and thickness of footings, as these depend on the size of the building and hence the load that the soil is to carry.

For poultry houses, hoghouses, and dairy buildings, footings may be made 6 to 8 inches high and from 10 to 12 inches wide. On top of this the foundation wall proper is started. Eight inches is usually thick enough for such small buildings. When ground level is reached, the thickness of the wall may, in most buildings of the kind mentioned, be reduced to 6 inches.

For 2-story houses, foundation walls should be about 10 inches thick and should start on a footing 10 inches high by 15 inches wide. For heavy barns, foundation walls may be from 10 to 12 inches thick and should start on a footing 10 to 12 inches high and from 20 to 24 inches wide.

In each of the foregoing cases, the dimensions for footings are only approximate, as the bearing capacity of the soil and the weight of the building to be carried are the final determining factors.

Troughs, tanks, manure pits, and hog wallows. Among the most common uses of concrete on the farm is its application in the building of watering troughs or tanks. Manure pits, cisterns, hog wallows, etc., are merely other forms of tanks and, like the first-mentioned, must be water-tight. Watering troughs, tanks, and cisterns should be made of a 1:2:3 concrete, while manure pits and hog wallows may be made of a 1:2½:4 mixture. In tank construction, it is important that the work be carried on continuously, if possible, to prevent seams in the work through which leakage may occur. However, there is a way to prevent leakage in such seams, if directions are carefully followed, and these have been given above.

Manure pits are usually built partly beneath ground, for convenience in throwing manure into them. Often they are built so that one end has a sloping pavement leading into it, for convenience in backing a wagon into the pit for loading. Manure pits are sometimes connected with cisterns for the purpose of draining into these receptacles the liquid content of the manure. This is pumped from the cistern from day to day and either sprinkled over the manure in the pit, to regulate decomposition, or is hauled in tank wagons to the fields and there sprinkled on the ground. The cistern for the manure pit should be of 1:2:3 concrete, to insure that the walls are thoroughly water-tight, thus preventing seepage of the contents and possible contamination of the domestic water supply.

All kinds of tanks, manure pits, and hog wallows must be properly reinforced. No table of reinforcement can be given for such structures, because each change in shape and size

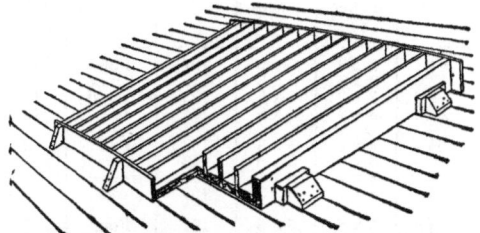

Fig. 428. Series of fence-post forms partly cut away in front to show simple construction. Reinforcing rods should be used in these.

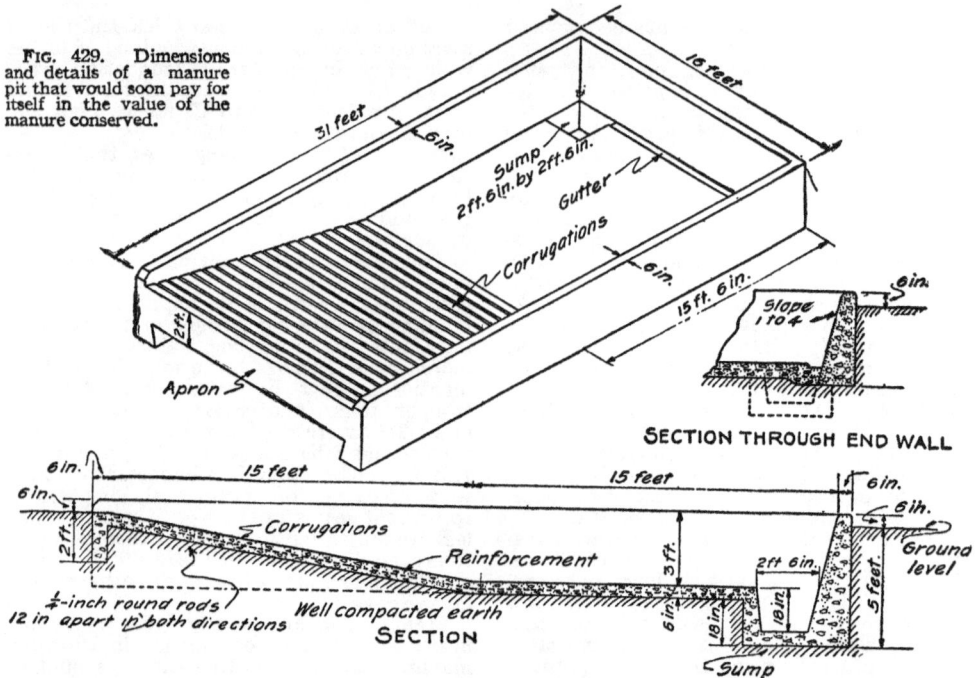

FIG. 429. Dimensions and details of a manure pit that would soon pay for itself in the value of the manure conserved.

SECTION THROUGH END WALL

SECTION

makes a different structure for which reinforcement must be specially calculated.

Concrete water tanks in the barnyard or pasture lot should have a concrete pavement laid around them or, at least, on the side from which cattle approach the trough to drink, so as to prevent the surroundings from becoming a mudhole.

The inside face of small watering troughs should be battered, or sloped, so that the walls of the tank or trough will be thicker at the bottom than at the top. This detail of construc-

tion helps to relieve the tank from pressure due to ice, when water in the tanks freezes.

Posts. Posts of various kinds—clothesposts, gateposts, etc.—can be readily made in concrete during spare hours on the farm. For ordinary fence posts, gang molds can be made of wood, the capacity of such molds being based on one filling with a 1-sack batch of 1:2:3 concrete. If any considerable numbers of fence posts are to be made, however, some one of the many good types of metal commercial molds is to be preferred, because a

TABLE I—DIMENSIONS OF CONCRETE LINE POSTS, AND MATERIALS NEEDED

DIMENSIONS			Weight of post in pounds	Amount of reinforcing metal required	MATERIALS						
					1 part Cement to 3 parts Sand			1 Cement, 2 Sand, 3 Pebbles or Stone			
Length in feet	Top in inches	Bottom in inches			No. posts per barrel cement	FOR 10 POSTS		No. posts per barrel cement	FOR 10 POSTS		
						Sacks cement	Cubic feet sand		Sacks cement	Cubic feet sand	Cu. ft. pebbles or stone
6½	3 x 3	5 x 5	107	Four	15.1	2.6	7.9	21.1	1.9	3.8	5.8
7	3 x 3	5 x 5	115	¼-inch	14.0	2.8	8.5	19.5	2.1	4.2	6.2
7½	3 x 3	5 x 5	123	round	13.2	3.0	9.1	18.4	2.2	4.4	6.6
8	3 x 3	5 x 5	131	rods	12.3	3.2	9.7	17.1	2.4	4.7	7.1
6½	4 x 4	5 x 5	133	Four	12.2	3.3	9.8	17.0	2.4	4.8	7.2
7	4 x 4	5 x 5	143	₇⁄₁₆-inch	11.3	3.5	10.6	15.8	2.6	5.1	7.7
7½	4 x 4	5 x 5	153	round	10.6	3.8	11.3	14.7	2.8	5.5	8.2
7½	4 x 4	5 x 5	163	rods	9.9	4.0	12.1	13.8	2.9	5.9	8.8
6½	5 x 5	6 x 6	197	Four	8.2	4.9	14.6	11.4	3.6	7.1	10.6
7	5 x 5	6 x 6	213	⅜-inch	7.6	5.3	15.8	10.6	3.8	7.7	11.5
7½	5 x 5	6 x 6	228	round	7.1	5.6	16.8	9.9	4.1	8.2	12.3
8	5 x 5	6 x 6	243	rods	6.6	6.0	18.0	9.2	4.4	8.8	13.2

TABLE II—DIMENSIONS OF CONCRETE CORNER POSTS, AND MATERIALS NEEDED

Dimensions		Weight of posts in pounds	Amount of reinforcing metal required for each post	PROPORTIONS AND AMOUNTS OF MATERIALS						
				1 Cement, 3 Sand			1 Cement, 2 Sand, 3 Pebbles or Stone			
				No. posts per barrel cement	For 1 Post		No. posts per barrel cement	For 1 Post		
Length in feet	Size in inches				Sacks cement	Cu. ft. sand		Sacks cement	Sand cu. ft.	Pebbles cu. ft.
8	6 x 6	288	Four $\frac{7}{8}$-inch round rods	5.6	.7	2.1	7.8	.5	1.0	1.6
8	7 x 7	392		4.1	.95	2.9	5.7	.7	1.4	2.1
8½	7 x 7	416		3.9	1.0	3.1	5.4	.8	1.5	2.2
8	8 x 8	512	Four $\frac{7}{8}$-inch round rods	3.1	1.3	3.8	4.4	.9	1.8	2.8
8½	8 x 8	544		3.0	1.35	4.0	4.1	1.0	2.0	2.9
9	8 x 8	575		2.8	1.4	4.3	3.9	1.1	2.1	3.1
8	10 x 10	799	Four 1¼-inch round rods	2.0	2.0	5.9	2.8	1.4	2.9	4.3
8½	10 x 10	850		1.9	2.1	6.3	2.6	1.5	3.1	4.6
9	10 x 10	899		1.8	2.2	6.7	2.5	1.6	3.2	4.9
10	5 x 5	250	Four ¾-inch round rods	6.4	.6	1.9	9.0	.4	.9	1.4
12	5 x 5	300		5.4	.7	2.2	7.5	.5	1.1	1.6

smoother finished post can be produced and the molds, if duly cared for, will be much more durable than homemade ones. Larger posts, such as corner posts and gateposts, are generally cast in place, because of their weight, and wooden forms are erected in the excavation made for the purpose. The accompanying tables give sizes of posts and suitable reinforcing, as well as quantities of materials required.

Round rods are preferable to other forms of reinforcing for concrete posts, principally because this type of steel is most easy to obtain. Barbed wire and other scrap material used as reinforcement will not give such satisfactory results.

Sometimes a 1:2:4 mixture is used for concrete posts. This is all right, if the materials are very well graded and the concrete is mixed to proper consistency and very carefully placed; but a 1:2:3 concrete is better, because it will help to compensate for improperly graded materials and, being richer, will be more certain to bond securely with the reinforcement, if the concrete is placed right in the molds. Square or rectangular posts are in more common use than those of other shapes.

For fence posts, a little more water may be used in the mixture than is required to produce a quaky consistency. When placing concrete in the forms, it should be stirred by running a stick or rod along the edge of the form and by tapping or jarring the mold so as to cause the concrete to settle thoroughly to all corners and around the reinforcement, releasing air bubbles that may be in the mixture.

For ordinary line posts, coarse aggregate should not exceed half an inch in greatest dimension. Larger aggregate than this will make the concrete hard to place with respect to thoroughly surrounding the reinforcing rods. Unless these are thoroughly surrounded

by good dense concrete, moisture will get at them and, if rusting starts, bursting of the concrete will follow. Good, rich concrete will prevent this, because it has been proved that steel can be indefinitely prevented from rusting by imbedding it in good concrete. If mixtures used for concrete posts are of the right consistency, it is usually possible to remove the form from the post or the post from the form in 24 hours after placing concrete. No attempt should be made, however, to move the post until it is several days old. Cracks once opening up while the concrete is green will always remain and cause weakness.

After posts are a week old and have hardened under a covering of wet hay or straw, they may be set outdoors to finish hardening. They should not be used until they are at least 30 days old. What has just been said about fence posts applies literally to clothesposts. Gateposts and corner posts can be made of a 1:2:3 concrete in which the aggregate may be as large as three-fourths inch in greatest dimension.

Fittings for hanging gates should be cast in the post when the concrete is placed, pro-

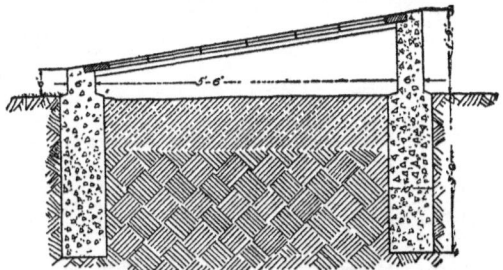

FIG. 430. A concrete hotbed is permanent, insect- and decay-proof, attractive and much more able to keep the heat in and the cold out than any temporary wooden structure.

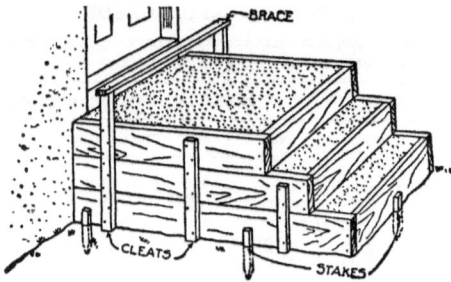

FIG. 431. A concrete walk and back steps keep muddy feet and many severe colds out of the house; they are both actually easy to make.

vision being made in the form to hold such fixtures in proper position. Gateposts and line posts should not be used until they, also, are at least 30 days old.

Hotbeds. Concrete makes permanent hotbeds. Any farmer who has had to replace the old wooden hotbed or cold-frame every year or two knows what this means. Excavation for hotbed walls should extend below possible frost penetration. Below ground the wall may be from 8 to 10 inches thick, but above it need not be more than 5 or 6 inches. No footing is necessary for hotbed walls.

If any one stretch of wall extends no more than 25 feet in length, reinforcement in the wall itself is not necessary; but reinforcing rods, 4 feet long and not less than one-fourth inch in diameter bent to the form of an "L" with each leg equal in length, should be embedded at the corners, to prevent cracking from changes in temperature as the result of expansion and contraction of the concrete. If hotbed walls are more than 25 feet long,

they should be reinforced throughout, or there should be provided at every 25 feet a joint in the concrete that will permit of contraction and expansion. This joint may be filled with some kind of tarred felt.

A 1:2½:4 concrete mixture is good for hotbed walls. If well spaded when placed in the forms, no surface treatment or finish will be necessary beyond that given by contact of the concrete with the forms.

Concrete steps. Concrete steps for the cellar or for the side or front porch are much preferable to wooden ones. To build steps and side walls for the cellar entranceway at one operation requires complicated forms; it is therefore, customary to build steps and side walls, or ramps, as they are called, separately. Either can be built first. A riser must be laid out just as in the building of wooden stairs, so that the number of steps necessary may be determined. Excavating must then be done until this riser sets at the right slope to bring the treads of steps level. The soil must be well compacted and must be excavated far enough back of the forms to permit at least 4 or 5 inches of concrete where the riser and tread join. It is often best to lay reinforcing rods in this backing, to help prevent the cracking. One-quarter-inch round rods, 8 to 10 inches apart, will be sufficient. A 1:2½:4 concrete may be used both for side walls and steps. If any difficulty is encountered in smoothing the treads or steps, this can be overcome by having ready a little 1:2½ sand-cement mortar to apply under a wooden float while finishing the treads. After the forms have been removed, if there are any imperfections in the work, these can be easily pointed up with sand-cement mortar.

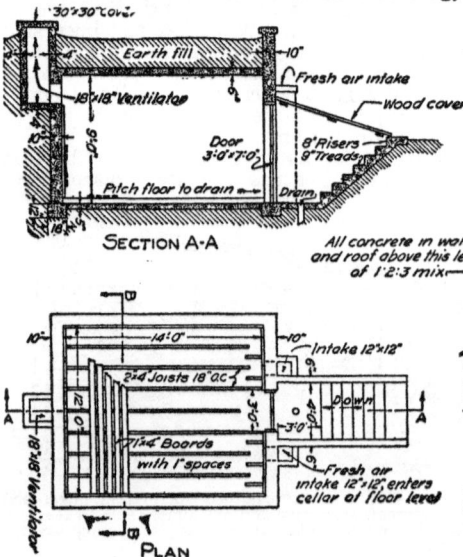

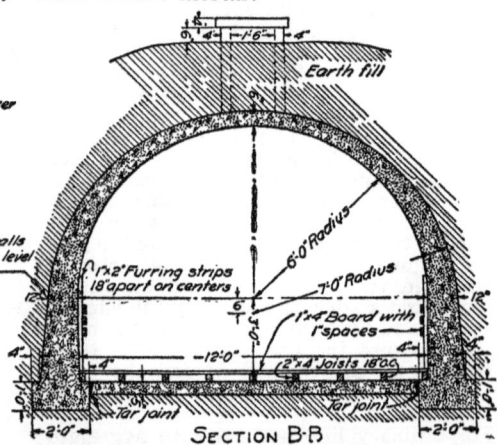

FIG. 432. A root and fruit cellar that is frost-tight, economical of space and materials, and a means of bringing down the cost of living. It requires 45 barrels of cement, 14½ cubic yards, sand and 22½ cubic yards of pebbles.

CHAPTER 26

General Repair Work on the Farm

By PROFESSOR R. P. CLARKSON (*see Chapter 14*). *One of the advantages of farm life is its independence. But this very independence makes necessary a certain self-reliance, an ability to look after one's self and one's belongings in emergencies. The making of repairs and the doing of odd jobs about the farm are illustrations of the application of such ability. This chapter mentions but a few of the possible tasks, listing the tools necessary for their accomplishment; it should, however, suggest some of the many directions in which a farmer may develop his efficiency.—*EDITOR.

EVERY modern farmer should have a good kit of tools and know how to use them. By so doing he may save many times over the value of a great deal of delay and frequent repair bills during a single season. Not only is it advisable to have the tools from the standpoint of making repairs, but, also, the possession of good tools often stimulates a man to use his spare time in working up various handy devices and arrangements about the house and barn.

Tools required. It is always difficult to pick out a list of tools required for any work so varied as that on the farm. There are certain essential ones, but the owner of a good kit acquires them gradually as the need is anticipated, and learns to know each separate tool, its value and its limitations.

In general, hand tools are sufficient for most of the work, provided a good blacksmith's forge is included. The materials worked on are almost always either wood, iron, or steel, or concrete, wood unquestionably predominating. Means must be provided to cut and shape these materials, to smooth them for easy handling and nice appearance, to repair them when fractured or bent, and to cover them with a protective coating when needed. Means should also be provided to connect together several pieces of the material, as with bolts, screws, nails, etc. To accomplish these things, the following tools should be acquired:

For Working in Wood

A cross-cut saw for cutting across the grain of the wood.
A keyhole saw for sawing interior holes in boards.
A rip saw for cutting with the grain.
A back saw for fine sawing as in cabinet and joinery work.
Several chisels for making joints, mortising, etc.
Several gouges for slotting, rounding corners, fluting, etc.
A wooden mallet for use with the chisels and gouges as well as for striking wood without marring.
A claw hammer for driving nails and pulling them out as well as for use in striking metal, etc.
A T-square for laying out work carefully.
A carpenter's level for plumbing and leveling.
A plumb line for erecting perpendiculars.
Carpenter's dividers for laying out circles.
A brace with several sizes of bits for boring holes; also a screw driver fitting the brace, for handling very large screws or screws in hard wood.
Small and large screw drivers.
A hand plane for planing up boards.
A jack plane for smoothing up the end grain.
A marking or scratch gauge for laying out parallel edges and lines in marking out work.
An awl for starting nails and screws, etc.
A rasp for smoothing up rough work.
A counter set for sinking screw heads.
An adze for shaping up rough work.

A draw shave for spokes and round work.
An ax.
A hatchet.
A putty knife.
Glue and brush.
Sand paper for finishing.
Assorted wire nails.
Assorted screws. (Beware of brass screws in hard wood, they twist off easily.)
A 2-foot folding rule.
A knife.
Several paint brushes.
Lumber crayons.
A bottle of shellac.
Carpenter's pencil.
Oilstone for whetting tools to a keen edge.
A grindstone for roughing tools to an edge.
An emery wheel for heavy grinding on tools and for bringing other material to shape.

The cost of all of the above tools and materials of good substantial quality will total about $35. It will pay to get the best steel in all cutting tools and to keep the tools always sharp, well-oiled and carefully placed in a tool chest or closet. A well-oiled tool will not rust, so it is a good plan frequently to wipe each tool with an oiled rag. Another sure way to prevent steel from rusting while laid away, is to dip it while warm in a solution of washing soda and then allow it to dry without handling.

For Working With Metals

The main purpose of any metal-working tool outfit is to enable the farmer to make necessary changes and repairs in his machinery. Tools desirable are:

A bench vise for holding material.
Copper vise jaws to hold pieces without scratching them or cutting their surface.
A hack saw for cutting metal, rods, pipe, etc.
Two dozen assorted blades for hack saw.
Two or three flat files, fine and heavy cut.
A rat-tail file.
Two or three triangular files for notching, sharpening saws, etc.
A ratchet drill for drilling any metal in place.
Assorted drills—size for tapping holes.
Centre punch for marking drill starts.
Ball peen hammer, machinist's style.
Riveting hammer, very light.
Set of taps with holders, for threading holes.
Set of dies for threading rods and pipe.
Soldering outfit with instructions.
A small bicycle wrench.
Two or three cold chisels.
An 8-inch and a 16-inch monkey wrench for nuts.
A 12-inch Stillson wrench for holding pipe.
A pair of wiremen's heavy pliers.
A pair of gas pliers.
A pair of heavy shears.
A gasoline torch for soldering and for thawing pipes, melting lead for joints, etc.
A small hand forge for welding metals, bending metal, and simple blacksmith work.
Blacksmith tongs.
Cutting chisel for metal.
A pair of iron or steel bench clamps.
Shears for cutting sheet metal.
Heavy blacksmith hammers.
Assorted bolts and nuts.
Several dozen assorted copper and soft-iron rivets.
Assorted lag screws.
Emery cloth.
Cotter pins.
A few odd pieces of pipe and metal rod.
Washers.
Some sheet tin, brass and copper.
6-inch calipers for measuring diameters.
A 12-inch steel scale.
A scratch awl for marking metal.
Some white chalk for coating metal so that scratch marks may easily be seen.

The above materials should cost about $60 for good-quality tools and materials, provided only small sets of taps, dies, drills, etc., are purchased.

A liberal allowance for a complete wood and metal-working outfit as listed above, including a tool closet or chest, would be $100. Such an outfit would enable a man to make or repair almost anything of metal or wood provided no machine work was required. In most cases, especially repair jobs, nothing more than skilful hand labor is needed.

The Use of Tools

There is a wide difference between the use and the abuse of tools. No tool should ever be used for a purpose for which it is not intended if such use tends to have an injurious effect on its quality or ability to do its work. Tools are expensive and each one is entitled to receive proper care, and careful use, to be regularly oiled and sharpened when in service, and to be carefully stored when not in use. Every cutting tool should be sharpened after continued use so that it will be in proper shape when next needed.

A Few Don'ts

Don't use a wrench as a hammer.

Don't use a Stillson wrench on nuts.

Don't use a cross-cut saw to cut with the grain.

Don't use nails for permanent work under strain or twist—use screws.

Don't expect a tool to stay sharp forever without grinding as well as whetting.

Don't take anything apart until you thoroughly understand how it works.

Don't try to bend and work metal when it is cold. Better heat it, even in a bonfire or kitchen stove.

Don't be afraid to tackle any reasonable task—common sense and good tools will go a long way and save you money. Besides that, you are learning something by doing the work.

Don't be afraid to write to manufacturers, your farm paper or the agricultural colleges.

You can get good advice and suggestions gladly given without cost.

Metal Working

Repairing cast iron. The main purpose of the blacksmith forge is to prepare metals: (a) for bending steel and iron to shape; (b) entirely changing their form by hammering; (c) cutting heavy metal; (d) welding broken parts together. Cast iron does not lend itself to any forge treatment. It is brittle, does not bend without breaking, and does not weld at all under forge temperatures. It may, however, be welded by a blow torch with the oxyhydrogen or oxyacetylene process or by electrical methods. The cost of thus getting broken cast-iron frames, flywheels, and similar parts mended is less than the cost of similar new parts. With proper skill in mending, the strength of the weld should be even greater than the previous strength of the piece. More important now, however, is the fact that the delay in waiting for a new part is avoided.

There are welding shops in almost every town and city of any size, usually in connection with machine shops or automobile garages and repair plants. A small welding outfit for home use will cost $50 and upward, depending upon how extensive a set of tools is selected. Of course, with such an outfit in hand, it is possible to pick up considerable repair work on surrounding farms and thus make the outfit pay for itself. The work is simple and quite easy for any man, if care is taken and directions carefully followed.

The process employed is usually as follows: The parts are carefully fitted together and arrangements made to hold them in place. They are then marked to guide in refitting them. The edges are beveled to make a sort of reservoir, and the torch applied to them together with the welding strips of iron metal. This metal melts, the fractured faces of the pieces become molten, and the whole mass fuses and welds together at the melting-point temperature. This process purifies the cast iron more or less at the point of welding, with the result that it is better and stronger metal, less brittle than the cast iron itself.

The use of the forge. One of the first essentials in forge work is a proper fire. It must be bright, with a good forced draft that can be applied at will, usually by means of foot or hand bellows. For the best work a good grade of coal is required, although almost any soft coal may be used if the gases are burned off before the metal is put in the coals.

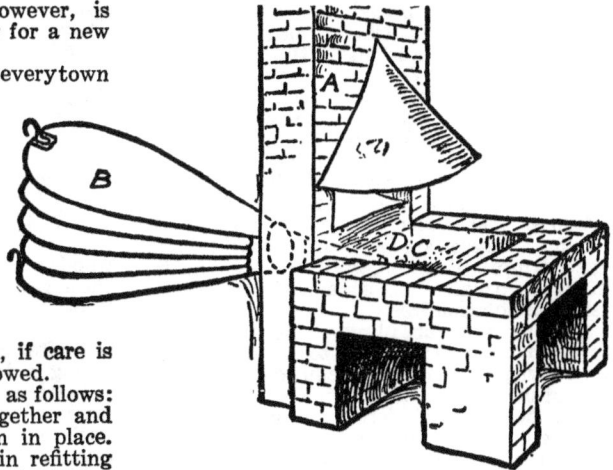

FIG. 434. Handy portable forges can now be bought very reasonably; but where this is impossible such an affair as shown here is efficient, cheap, and easy to make.

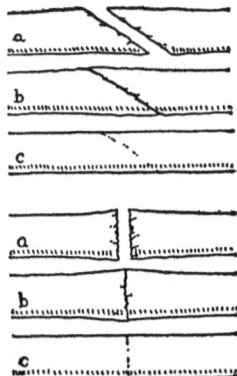

FIG. 433. Three stages in lap welding (*above*), and butt welding (*below*): *a* the ends roughly shaped then upset; *b* the fresh, enlarged weld; *c* the weld hammered down to the size of the rest of the bar.

Shavings are usually used for kindling. A bucket of water should always be handy to sprinkle around the fire and keep it confined to a very small area. A good forge fire for most work can easily be covered with a dinner plate.

For bending metal, heat it to redness and reheat whenever the color fades. In shaping-up metal it should be brought to a very bright red—almost a white heat, and should not be worked after the color has come down to dark red, or a poor job with cracked edges is likely to result. For welding, a white heat is required. The metal should be "spitting" fire, but great care should be taken not to burn the iron—that is, don't heat it so far that it is consumed. Where at all possible, the weld should be made (as it almost always can be) in one heat.

For welding the metal must be clean and at a welding heat; and it must be struck sharply by the hammer. The first blow of the hammer tells the story. That blow should do most of the welding and should be struck where the metal is thinnest because that is where it cools quickest. Some practise should be had before attempting work on any important piece. Make the practise weld, allow the metal to cool and then bend it with the hammer right at the weld. If the weld separates it is usually because the metal was either not clean or not hot enough. To be sure of the part being clean it is best to jar it on the anvil before striking it with the hammer.

In a lap weld, the two pieces are tapered so that, when joined, they will be slightly larger than the main part of the piece. Then after welding they can be reduced by hammering. Round rods are usually joined by butt welding; that is, the ends to be welded are "upset" or made bulky by previous heating and hammering on the end. They are then brought to welding heat and driven together endways by a sharp hammer blow.

Drill work. A good ratchet drill will drill any thickness of metal and any size of hole up to perhaps 3 inches in diameter or more, by hand. It costs less than $10 and weighs only 2 or 3 pounds, being very simple in construction. It can be used anywhere to drill holes without having to take the machine apart, if the ratchet can be braced. It can also be used as a bench drill by means of clamps with which to fasten the work.

In drilling metal use oil freely. It lubricates and keeps the drill cool, and prevents the loss of its temper and hardness particularly on the cutting edge. Lard oil is good for wrought-iron drilling. For cast-iron drilling, the graphite in the iron furnishes lubrication so that water can be used for cooling. Unless care is taken, however, this will result in rusting the iron. Machinists, therefore, usually mix soda with the water, using the solution freely.

Tapping and threading. The same rules apply to cutting threads as to drilling. Both are cutting processes and require care for the sake of the tool as well as of the result to be accomplished. Don't force the cutting tool. Cut light chips. Be careful not to bend or break the cutting instrument. It is always brittle because of its temper and can easily be broken by carelessness.

If a tap is broken in the hole, heat the piece and frequently the tap can be withdrawn. If not, drill into the tap with a small drill, drive a square shaped piece of steel lightly in the drill hole and turn with a wrench. Sometimes the tap will have to be drilled out to the size of the original hole and the shell of the tap left pried out with a cold chisel.

Sheet-metal work. The working of sheet metal is an art in itself but by taking time almost anything desired can be made by means of shears, a wooden mallet, a straightedge, and a soldering outfit. It is best for an inexperienced man first to make a paper pattern cut to the shape of the part desired and then to fold, bend, or roll this pattern with edges overlapping to the finished shape to be sure it meets the requirements. Then lay the pattern flat and mark the sheet metal to exactly the same shape. Mark on the metal the bending lines of the pattern and in every way make it agree with the tried-out paper pattern. Then cut out the metal with shears. Bend straight lines by laying the straightedge on the bending line and hammer the metal with the mallet against the edge.

For rolling to shape use a piece of pipe or a log to bend the metal around. With odd or complicated shapes, cut out a piece of

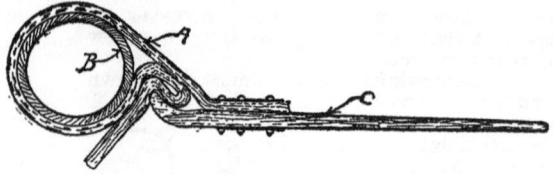

FIG. 436. When a Stillson wrench of the right size is lacking, this emergency affair will often make up or loosen refractory unions. B is the pipe; A is a heavy strip of canvas or belting: C is a tough piece of hardwood to which A is firmly bolted.

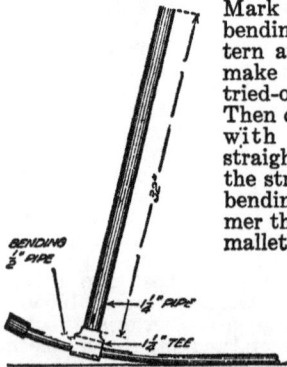

FIG. 435. Mechanical problems on the farm are varied and unusual and call for special tools. A pipe bender can be made of a strong tee with a pipe handle screwed into it.

wood to hammer the metal against. Overlap the edges and fasten temporarily with clamps, rivets, bolts or even by nailing through into a strip of wood on the inside. Then solder, and after cooling, remove the temporary fastening if desired.

Patterns for casting. Patterns may be made of wood, metal, concrete, plaster of Paris or any material which can be finished reasonably smooth. These patterns are taken to the foundry, sand is molded around them, the patterns are withdrawn and the molds filled with the molten metal. The resulting shape is then just like the pattern. Broken parts can be mended temporarily and used as patterns in getting new castings. Complicated parts to be cast require special knowledge of the foundry processes in order that the patterns may be made so that they can be withdrawn from the mold without breaking up the sand.

Frequently it is possible to make patterns by forming a mold with sheet metal or cardboard and pouring in wet plaster of Paris. This fills the mold and gives a solid pattern to the shape of the mold. This solid pattern can then be used in the foundry for a few castings, though it is not a good permanent pattern. Concrete or mortar can be used in the same way, of course, but takes much longer to harden.

Carpenter Work

In all work with wood, study the grain of the piece and work with the grain as far as possible. Otherwise no marked lines can be followed, the wood will split on the grain, and the piece will be spoiled. Cut lightly and often. Don't attempt to break off a piece until it is sawed or cut through completely. Don't use sandpaper except to give a light smoothness to the finished pieces, and do that with the sandpaper wrapped around a good block of wood. Neither sandpaper nor emery cloth are cutting tools for shaping; both are finishing tools.

In all work mark out the piece completely and keep outside the lines so that the completed piece will show the part of the lines along the edges. For fine work the lines should be cut lightly with a knife and pencil marks should not be used at all. For rough work use a carpenter's pencil or a lumber crayon.

The farmstead is the hub of the farm; it should both dominate the fields that surround it and receive its inspiration from them

The first purpose of a farmhouse is to shelter the farm family; it should, therefore, be an attractive home in all that the name implies

THE FARM IS BOTH A HOME AND A PLACE OF BUSINESS. THE DWELLING ESPECIALLY SHOULD MEET BOTH NEEDS

A $3500.00 MINNESOTA FARM HOUSE

FIRST PRIZE
HEWITT & BROWN
ARCHITECT — MINNEAPOLIS

FIRST FLOOR PLAN

SECOND FLOOR PLAN

BASEMENT PLAN

SIDE ELEVATION

REAR ELEVATION

SECTION

Reproduced by courtesy of the Society

DESIGN AND PLANS OF A LOW-PRICED FARMHOUSE THAT WON FIRST PRIZE IN A COMPETITION HELD UNDER THE AUSPICES OF THE MINNESOTA STATE ART SOCIETY

Paints and Paintings

The coming of spring should be a signal for painting everything that needs it, whether house, barn, fence, or machinery. Particularly should machinery be looked after and, emphasis cannot be too strongly laid on this point, for few things are so neglected as machinery on the ordinary farm. Not all paints are of equal value for these different jobs. What is good for iron is not good for concrete, and the paint which is so satisfactory on the house may be of little value for the wagons. A paint for woodwork consists of some dry material for coloring, a lead or a zinc base, a drier, and a vehicle or liquid. It is the vehicle which is often wrongly chosen, and in some ready-mixed paints, the vehicle is the part which is most likely to be adulterated. For outdoor work, except decorations, boiled oil is considered to be the best. For indoor work, linseed oil and turpentine are preferably used. A little drier, litharge for dark paints, and sugar of lead for light paints, should be added to each batch of paint mixed.

Undoubtedly linseed oil paints are more expensive than others, but they are well worth the difference in price. This oil enables the paint to spread well, dry hard and opaque, and leave a protecting skin over the wood surface. If adulteration is practised with resin oils, mineral oils, or fish oils, the paint will either remain sticky forever or will harden quickly only to soften again in a week or 10 days. Particularly should dark-colored paints be looked upon with suspicion unless purchased from a thoroughly reliable dealer, because such paints when cheap usually contain only unrefined resin oils which soften up within 2 weeks of the first drying. Thereafter they never harden again but continue to give constant trouble.

One of the best paints for roofs and machinery is a mixture of red lead and linseed oil. Another good metal paint is known as asphaltum varnish. It may be purchased ready for use, and when applied leaves a splendid-wearing shiny black surface which thoroughly protects the metal.

For painting jobs requiring the covering of a large surface, the paint may usually be sprayed on with much less labor than if the application is made with a brush. Almost any paint may be so applied if it is made thin enough. Use the ordinary spraying apparatus kept for orchard spraying and occasional disinfecting or whitewashing. It may be readily cleaned and will suffer no injury from such use. Probably whitewash is more often applied in this way than any other paint. Particularly in covering fences and out-buildings the spraying method means a great saving in time. Yet ordinary whitewash is not as economical as cement whitewash for it requires frequent renewals, while the cement wash often remains satisfactory even several years' wear. The combination is best made in the following proportions: Mix together one peck of white lime, a peck and a half of hydraulic cement, 6 pounds of umber, and 4 pounds of ochre. Slake the lime first, mix it with 2 ounces of lampblack moistened with vinegar, then add the other ingredients. Allow the paint to stand for 3 hours or longer, stirring occasionally. The addition of half a pound of Venetian red renders the appearance more pleasing and adds to the value of the paint. If ordinary whitewash is used at all, the addition of a little glue or a small amount of flour mixed with boiling water and poured in while hot will prevent it from rubbing off so readily.

For finished interior work, varnishes are best. They give an extremely hard surface, which protects the wood beneath, and they are easy to clean thoroughly. It is not advisable for any one but an expert to attempt to mix them at home, for many good ones are on the market, as well as many worthless mixtures called varnishes. True varnish is a solution of resins or gums in some suitable liquid such as alcohol or oil of turpentine mixed with linseed oil. Those in which alcohol acts as the solvent are spirit varnishes and are inferior in many ways to the oil varnishes, chiefly because the alcohol evaporates entirely, leaving the varnish so hard as to easily crack and chip. The oil varnishes, on the other hand, should never get brittle.

Repairing Floors and Foundations

Repairing old flooring. Often it is desired to cover old floors or, with the installation of bathroom fixtures in the house, to make a waterproof floor. For this latter purpose the various modifications of Sorel stone are highly recommended. Its strength and hardness exceed those of any artificial stone yet produced; at the same time it is one of the cheapest. For stable and stall floors it is also of considerable value, for it is sanitary, easy to clean, and wears well.

There are almost as many different varieties as there are users of the stone, for every one makes some little change in the details of mixing. The fundamental thing is to mix in with the various filling substances an "oxychloride binder" which is nothing more or less than a solution of magnesium oxides and chlorides. This is used to moisten the filling substances in the same way as water is used in concrete work. The fillers may be almost anything—shredded wood or cloth, sawdust, asbestos, sand, ashes, pebbles, etc. You may buy the material all ready for use from any large paint shop or hardware store and do the work yourself, or you can get any of the companies selling the substances to do the work for you. You may, if you like, mix the ingredients and make your own stone.

If you buy the material ready to use, you will get two packages. In one kind, the packages contain powders which must be mixed together after which water is added. In the other kinds of composition, one powder and one liquid are purchased and the two are mixed. In either case, the mixture is made somewhat stiffer or thicker than ordinary cement and is spread on the old floor, or on the flooring built to receive it, to a depth of half an inch. The surfacing must be well done and not left rough. It will "set" over night and will then be hard enough to walk on but the floor should not be really used for 3 or 4 days. Probably several months will elapse before the floor reaches an even color all over. From time to time it will be necessary to remove the white blotches caused by the chemical action going on in the floor material, by simply washing them. After a time the floor will be stone hard and, of course, fireproof and waterproof. This latter characteristic is a highly desirable one, for almost all other artificial stone, in common with brick and concrete, is very porous and open to the absorption of water. Any desired color may be added to the composition, the earth colors giving the best results.

Although this flooring has not long been receiving the attention of private builders, it is not a new thing. There are hundreds of patents for different mixtures, and one kind or another has been used on the floors of railroad cars, in public buildings, and similar places for at least 20 years. Recently some of the important patents expired, and since 1908 as many as 50 companies manufacturing composition have come into existence.

The ingredients and methods of mixing one of the best compositions are as follows:
50 parts (by weight) of calcined (burned) magnesite.
15 parts of dolomite (marble dust).
5 parts asbestos (shredded).
15 parts sawdust.
$2\frac{1}{2}$ parts silicate of magnesium.
11 parts of earth colors.

Mix the above powder very thoroughly and add a mixture of equal parts of water and chloride of magnesia until the proper consistency is obtained. Frequently, to make a better union of the elements, the above powder is added to $1\frac{1}{2}$ parts of muriate of ammonia.

In another composition flooring the materials are mixed at the shipping point, the receiver or user adding simply water and burned or calcined magnesite. In this the specific materials are:
85 parts magnesium chloride solution.
36 parts of any filler such as sawdust, ashes, etc.
25 parts of infusorial earth or fossil flour.
Add to the above:
100 parts of pulverized burnt magnesite.
$43\frac{1}{2}$ parts of water.
Desired coloring material as red oxide, ochre, etc.

All of these substances are cheap. The mixtures as retailed by manufacturers cost about 15 cents per square foot of floor surface for the substance and an equal amount for doing the work. Unless you are willing to take great pains with the laying and finishing of the floor, an expert should be allowed to do it.

By using the above mixture but substituting large pebbles or stones for part of the filling material suggested, a first-class concrete is obtained. The same mixture, also omitting the filling material and coloring matter, and adding the proper sand or sharp, small stones, may be used for the formation of grindstones, emery wheels, etc.

Stopping leaks in concrete. If old work is to be protected, surface coat-

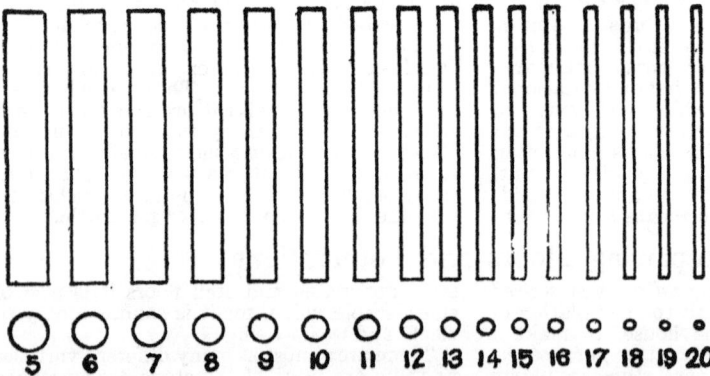

Fig. 437. Wire is a most important article in farm construction and repair work. These are the commonly used sizes shown lengthwise and in cross section, actual sizes. Of course the strength and stiffness of any size depends upon the material of which it is made.

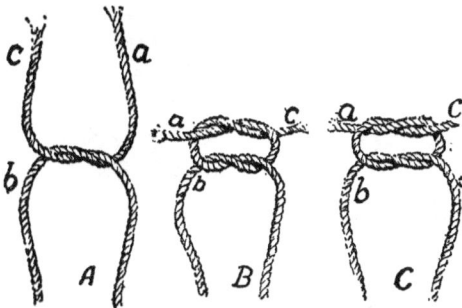

FIG. 438. A knowledge of how to use rope is a fundamental need in farm repair work. This shows how both the granny knot (B), which is insecure and unreliable, and the square knot (C), which is safe and efficient, start the same way (A).

ings only can be used, their object being to fill the pores already referred to. Four substances are commonly used for this, namely: neat cement, asphalt, paraffin, and an alum-soap compound. This last is used in what is known as the Sylvester treatment, and is one of the most effective. For surface coating a hot castile soap solution is made by dissolving three quarters of a pound of the soap in one gallon of hot water. An alum solution, of one half a pound of alum to 4 gallons of water, is then prepared. The substances are thoroughly dissolved and applied alternately to the wall which must be perfectly dry. The hot soap solution is first applied with a flat brush, care being taken to avoid leaving bubbles. After this coat dries for 24 hours, a coating of the alum water is put on and allowed to dry for a similar length of time. In this way, alternate coatings to the extent desired may be used, allowing a full day to elapse between each two coatings. A chemical process takes place between the sub-

stances used, the resulting compound plugging up the pores in the cement. The cost of this process, including two coatings of each material, will be from 35 to 40 cents per square yard.

Paraffin, although rather expensive, is often used for small jobs. It may be melted and applied while hot, the walls also being slightly warmed, or it may be dissolved in some solvent such as benzol, xylol, or even benzine of the common kind, these liquids quickly evaporating. Several coatings will be needed, and each coating will cost in the neighborhood of 50 cents per square yard.

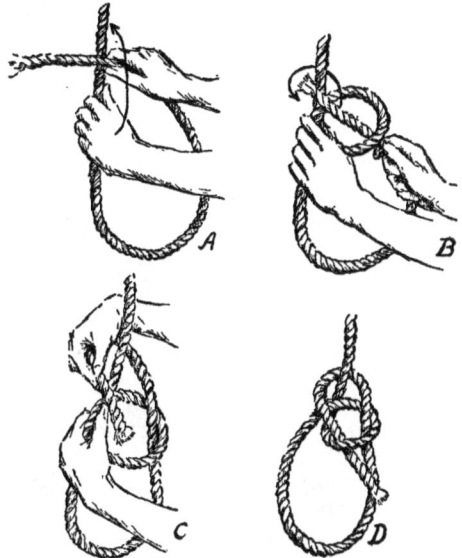

FIG. 439. The four steps in making a bowline or nonslipping loop—a most useful knot for both sailors and farmers.

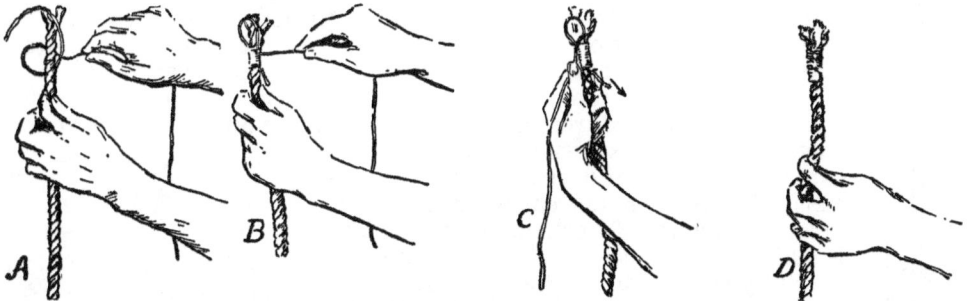

FIG. 440. The end of a rope, like that of any worthy task, should be well finished. This shows how to whip a rope's end so as to prevent raveling and at the same time avoid a bulky knot

CHAPTER 27

Dynamite and Its Uses on the Farm

By J. H. SQUIRES, *Agronomist and Editor of the E. I. duPont de Nemours Co., Wilmington, Del. Before he took up this particular line of work he had several years' experience with the U. S. Department of Agriculture. Previous to that he had been student and assistant in the Agricultural College of New York at Cornell University and Professor of Agronomy at the New Mexico Agricultural College, and has made a personal study of the agriculture of all parts of the United States. His training has therefore included the practical farm point of view as well as the study of the chemical, physical, and purely agricultural problems that confront the agricultural scientist. In common with many large commercial enterprises and organizations, the manufacturers of explosives have progressed far along the road of supplying accurate information and practical assistance to all who request them. It is with a knowledge of this condition in mind that we can heartily subscribe to Mr. Squires's further suggestion, as follows: "In many cases the uses of explosives are somewhat complex, either as regards the loading or the firing. Unfortunately, the amount of specific literature available on this subject is not as great as might be desired. When confronted with a difficult proposition, the safest procedure is for the amateur blaster to take up the matter by correspondence with some one of the leading manufacturers of agricultural explosives or a local professional blaster to whom he may be referred by the company consulted."—EDITOR.*

WHILE some black powder is used for different purposes on the farm, the chief agricultural explosive is dynamite. This is packed in round, paper-covered sticks or cartridges measuring about $1\frac{1}{4}$ inches in diameter and 8 inches in length. Each such cartridge weighs about half a pound. While there is a large number of kinds of dynamite and similar explosives made for general use, only a few are especially adapted for the farm. The commonest of these are: (1) the low-freezing extras and nitro-starch grades, for general work; and (2) straight nitroglycerin dynamite, for blasting ditches in wet soils without a blasting machine and for mudcapping very hard boulders. Dynamite is packed in wooden cases containing either 25 or 50 pounds net and is sold by the pound.

What dynamite is. The modern dynamites differ materially from those manufactured a few years ago. In appearance they resemble fine sawdust that has been slightly moistened with oil. Most dynamite is really fine wood pulp which has been allowed to absorb varying amounts of nitroglycerin or some other explosive compound. When exploded, dynamite is changed from a solid to a gas having a much larger volume than the original explosive. It is the pressure of this gas that does the work. If the gas is allowed to escape easily out of a bore hole, the amount of work done by the blast is materially reduced, so all holes should be carefully tamped to confine the gas.

Tamping. By tamping is meant the confining of the charge in the bore hole made for loading. This is done by first packing a few inches of moist soil lightly over the charge, and then filling the rest of

FIG. 441. A paper plug is sometimes used between charge and tamping.

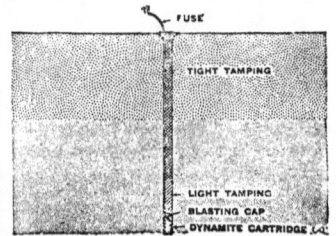

FIG. 442. A charge properly tamped, lightly at the bottom, tightly above

298

FIG. 443. One way in which dynamite can reduce the wear and tear on the machinery, the teams, and the physical and mental strength of the farmer. It has been estimated that every stump in a cropped field takes a toll of from 25 to 50 cents. At this rate the use of dynamite soon pays its cost.

the hole and packing it solidly. The few inches of light tamping act as a cushion to protect the charge from shock, so the rest can be forced in as hard as possible, using a wooden stick, such as an old hoe handle, in one hand. For mudcaps (p. 452) the mud placed over the charge acts as tamping and should be as heavy and thick as possible to better confine the explosive action. The force exerted by exploding dynamite is equal in all directions and works just as hard in an upward as in a downward direction.

Keeping explosives on the farm. While dynamite is a high explosive, it can, with reasonable care, be used on the farm with no more danger than attends a great variety of farm operations. It is sent from the manufacturer to the buyer by freight thus receiving rougher treatment in transit than is ordinarily given it in use. The law requires that it be removed from the freight station within 48 hours after its receipt. In hauling dynamite care should be exercised to guard against any unnecessary roughness, and against danger from fire. Explosives should not be hauled in the same vehicle with either blasting caps or electric blasting caps. The boxes should be protected with blankets or similar covering, to afford additional safety.

FIG. 444. Take with you at any time only as much dynamite as you will use. Carry caps and fuses in separate compartments

When stored on the farm, dynamite should be kept in a dry place, such as a little-used or isolated house, where it will be safe from the weather, and from meddlesome persons or animals. Thus stored it can be kept in perfectly good condition for a considerable length of time. When large amounts are kept constantly on hand, a special bullet-proof, weather-proof house or box should be provided.

Blasting caps require the same storage conditions, being quickly affected by moisture; but should never be stored with any kind of explosive. Fuse may be stored or hauled with either dynamite or caps, as it is not explosive; however, it must be kept dry. Neither caps nor dynamite should be stored in a blacksmith shop or in any place where there is danger from sparks.

Principles of Farm Blasting

Only the required amount of explosive should be taken to the scene of work at any

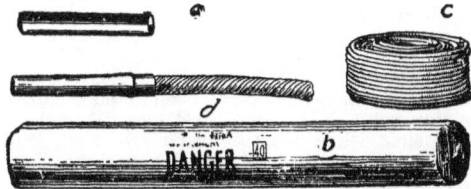

FIG. 445. What does the work. a, blasting cap; b, stick of dynamite; c, roll of fuse; d, cap crimped to fuse

one time. It should be handled as gently as possible, and by painstaking workmen.

For farm dynamiting in which caps and fuse are used, the tools needed are few and, usually, on hand. A crowbar or thick drill is useful for punching holes in nearly all kinds of blasting. A $1\frac{1}{2}$- or 2-inch auger with a long shank is good for both sampling soils and making holes under stumps and boulders, and for deepening drilled or punched holes. A cap crimper is needed for crimping the caps to the fuse. A tamping stick is easily made from a sapling or a broken tool handle.

In electric blasting, for the removal of large stumps and boulders, or for ditching in dry soil, a blasting machine is needed. This apparatus is made in different sizes to fire at one time different numbers of electric blasting caps. The smallest one recommended is that firing 10 charges; on the other hand, sizes larger than the 30-charge capacity machine are seldom needed except in extensive ditching work. These machines are small electric generators and are long-lived. Their cost is small as compared with their advantages in making it easier to do better or more difficult work. The electric current is conducted from the blasting machine to the blast by two 14 gauge copper wires enclosed within a single insulation. This double wire, if well cared for and kept straight, will also last for a long time. It should be not less than

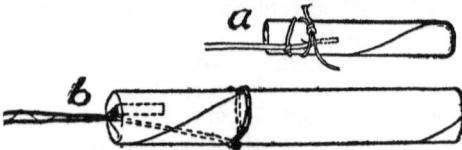

FIG. 446. Fastening the cap to the cartridge. *a*, blasting cap and ordinary fuse; *b*, electric cap and wires

250 feet long to enable the blaster to work at a safe distance from the blasts.

All curious or unnecessary onlookers who might cause trouble should be kept away from blasting operations; otherwise they may confuse the blaster or do something that might be dangerous for themselves or others. Farm blasting has now progressed to the point where many expert blasters are devoting a part or all of their time to it. By obtaining the advice and assistance of a competent blaster one can often effect a considerable saving in the amount of explosives used or improvement in the quality of the work done. One of the noticeable misuses of explosives by both beginners and experienced users is the loading of excessive charges for any given amount of work. By careful loading, the amount of dynamite used can often be reduced and with it the total cost of the work.

There are several distinct classes of dynamite, each of which is manufactured in different strengths, which are expressed as percentages. Thus they vary from 20 to 60

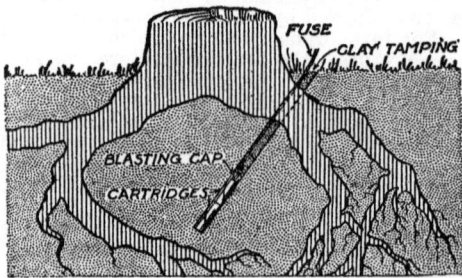

FIG. 447. Usual placing of charge for blasting a small stump

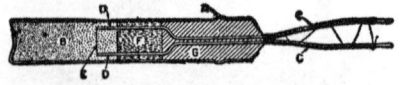

FIG. 448. Section of an electric blasting cap: *a*, copper shell; *b*, chamber containing explosive; *c*, copper wires (insulated); *d*, bare ends of wire in charge; *e*, "bridge" or fine wire joining ends of (*d*), which is heated by the electric current; *f*, plug to hold wires in place; *g*, filling material.

per cent. The lower strengths should be used whenever possible, as they do many types of work cheaper than the higher strengths. The substitution of the 20 per cent for 40 or 50 per cent grade in blasting stumps from heavy soils is proving economical. The low-freezing classes of dynamite are to be recommended for all farm work except blasting ditches without a blasting machine, and mud-capping hard boulders.

The use of dynamite on the farm, although it involved more than 25,000,000 pounds in 1917, is yet in its infancy. There is still much experimental work to be done before either the full benefits or the definite limitations of farm blasting are determined. Blasting is not recommended as a general "cure-all" for the farm, but as a specific treatment for a number of adverse conditions.

Practical Blasting Methods

Unlike black powder, dynamite is not fired or "detonated" by means of a spark, but by an intermediate agent, such as an explosive blasting cap or an electric blasting cap. The former is a little copper cylinder, closed at one end, and charged with a small amount of a powerful explosive. It is fired by means of safety fuse, which is a small chain of powder tightly enclosed in a strong covering. In use, the fuse is cut long enough to reach from the buried explosive to a few inches above the ground. It is not safe to use short pieces of fuse, as they do not give the blaster sufficient time to get a safe distance away after lighting the blast. The blasting cap is fastened securely to the fuse (Fig. 445*d*) by means of a special tool called a cap crimper. A hole is then punched in the side or end of a cartridge of dynamite, into which the cap with fuse attached is inserted and securely tied (Fig. 446.)

An electric blasting cap is more elaborately constructed and is fired, not by a fuse, but by an electric spark brought to it through small wires. It may be placed in the dynamite in exactly the same way as the blasting cap, or secured by a special loop (but *not* a half-hitch) of the wires (Fig. 446). A dynamite cartridge into which a blasting cap or electric blasting cap has been fitted is called a "primer." Only one blast can be fired at one time when blasting caps are used, but any number can be fired at the same time by means of electric blasting caps. For all classes of agricultural blasting one primer is sufficient for each charge, no matter how

FIG. 449. Making a hole for a charge under a stump with a soil auger

many cartridges are used in the hole; but with the single exception of ditch blasting by the propagated method (p. 302) every type requires this one primer in *every* charge or hole.

Stump blasting. For the removal of stumps, dynamite is used with cap and fuse as well as with electric blasting caps. For small stumps (Fig. 447), and for the large stumps of the Pacific Coast, a hole is punched well under the heaviest part of the stump and loaded with the required amount of dynamite, which may vary from one to many cartridges. Blasting caps and fuse are generally used in this case.

Large or hollow stumps, especially those with spreading roots, are best blasted by distributing charges, one under the stump and others under the main roots (Fig. 450). This method requires the use of electric blasting caps so that all the charges

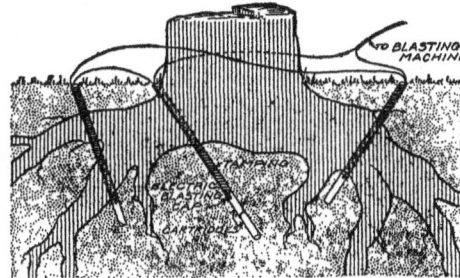

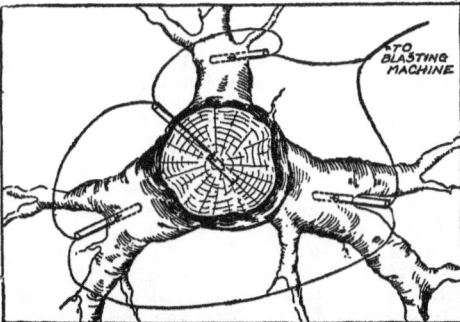

FIG. 450. Distribution and connection of charges for blasting a large stump with spreading roots; viewed from the side (*above*) and from the top (*below*).

may be fired at exactly the same time. Tap-rooted stumps present two possibilities: (a) A hole may be bored into the tap root and loaded as in the case of a small stump (Fig. 447); or (b) two holes may be punched on opposite sides of the tap root (Fig. 451), loaded, and fired at the same instant by means of an electric blasting machine.

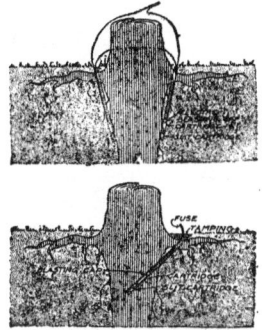

FIG. 451. Two ways of distributing charges under a tap-rooted stump.

Many variations are made in these methods of loading stumps to suit local soil conditions and the nature of the stump to be removed. Large amounts of dynamite are used in connection with stump-pulling machinery for splitting stumps before or after pulling. This combination method is especially satisfactory in heavy clearing. For splitting before pulling, a single hole is generally used and the charge materially reduced as compared with that required for blowing the stump entirely out. Trees may be blasted down by following the rules given for loading for stumps, but the amount of explosives must be considerably increased.

Boulder blasting. For breaking up and disposing of field boulders, dynamite is used in three ways: (a) mudcapping, which consists of placing a charge of dynamite on top of the boulder and covering it with a heavy cap of mud, (Fig. 452), requires the least labor, but the largest amount of explosives; for very hard boulders the use of the higher strengths of dynamite is recommended. (b) Snakeholing, which consists of punching a hole in the ground under the boulder (Fig. 452), requires a medium amount of both time and labor; the low strengths of dynamite prove best. (c) Blockholing, which is loading a hole drilled into the boulder (Fig. 452), requires the greatest amount of labor and the smallest amount

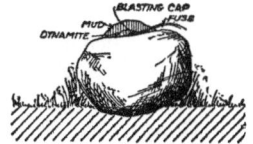

FIG. 452. Three ways to blast a boulder: from above, mud-capping snakeholing; and blockholing (see text).

FIG. 453. Large boulder, such as often interferes with tillage, ditching, or tree planting, before (*left*) and after (*right*) the explosion of a mud-capped charge of dynamite. It is now good foundation material in addition to being out of the way.

of explosives; the low strengths are more economical. Combinations of the mud-cap and snakehole methods are frequently used for large boulders. The method em-

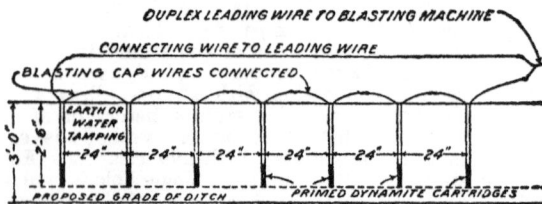

FIG. 454. Diagrammatic section showing how to place and connect up a series of charges in blasting a ditch

ployed should depend on the nature of the boulders encountered. Flat, easily broken boulders can be handled by the first method; large, difficult ones may require the last method.

Ditch blasting provides a new method of digging ditches and one that has proven successful under a great variety of conditions. There are two methods: (a) The propagated method, which can be used only in water-soaked soils, consists of punching holes to about the desired depth of the ditch and loading each with one or more cartridges of 50 or 60 per cent straight nitroglycerin dynamite. The soil must be saturated and the holes not more than 24 inches apart. A line of holes many feet long, when properly

loaded, is fired by a single primer, usually placed near the centre of the section. (b) The electrical method differs in that it can be used in either wet or dry soil. Lower strengths of low-freezing explosives are used, and an electric blasting cap in each hole. The holes can be spaced farther apart sometimes, in large ditches, as much as from 30 to 48 inches. For wide ditches two or three lines of holes are used. By this method ditches up to 6 feet deep and 15 feet wide have been successfully blasted. Blasts are frequently made in hard ground to loosen the ground to make digging by hand or scrapers easier. The application of ditch blasting to the straightening of streams has proven quite successful, as the work is quickly and economically done.

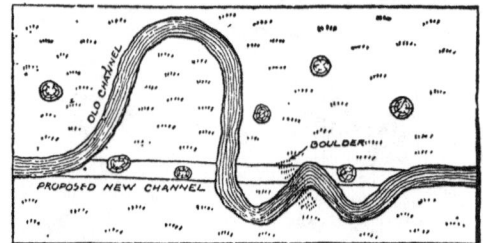

FIG. 456. A stream may be straightened and, incidentally, various obstacles removed from its new course, by means of dynamite.

Rafts and sandbars also may be blasted out of streams.

Loosening the soil before planting trees. The use of explosives in planting trees is recommended on tight and hardpan soils. Its purpose is to loosen the hard ground in such a manner as to permit: (1) better drainage and water storage; (2) better root growth; and (3) better soil aëration. The blasts are usually made with one fourth pound of a low strength, low-freezing dynamite placed at a depth of 30 to 40 inches or in the hardpan. A hole is made in the ground to the desired depth and loaded. The blast, in addition to cracking the soil, springs a pothole or cavity at the bottom (Fig 457), which must be filled with soil to overcome the danger of the tree settling after being planted.

FIG. 455. A surface outlet ditch dug with dynamite

This is usually done by emptying the hole and filling the cavity as shown in Fig. 458, which also illustrates the correct method of setting a tree, including the use of fertile top soil around the roots. Blasting in preparation for the planting of shade trees or flowering plants is done in exactly the same way as in the case of trees.

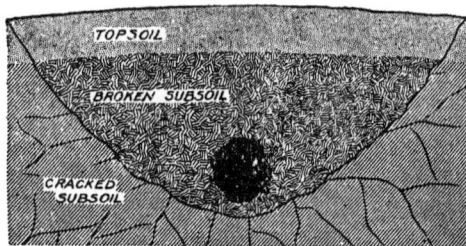

FIG. 457. In blasting before tree planting, the explosion loosens the soil, but also creates a pot hole at the bottom which, left unfilled, might injure the tree more than the other results could benefit it.

Summing the matter up, it may be said that the use of dynamite in the orchard accomplishes the following seven results:

1. It mellows the ground to a depth of 5 or 6 feet and throughout a circular area 10 to 20 feet in diameter, making it easy to dig the hole and plant the tree correctly.

2. It creates a porous, water-absorbing condition in the subsoil that makes the tree drought-proof, stopping the big, first-year loss.

3. It makes root growth easy and makes tons per acre of new plant food available, hence speeds up the growth of the tree and makes it fruit one to two years earlier.

4. It creates drainage and prevents stagnation of water on surface.

5. In old orchards that were planted by the old methods and have ceased to bear well, it is of great value in rejuvenating the old trees, causing them to yield heavily.

6. It destroys fungus, nematode, and other orchard soil diseases, hence makes it possible to plant new orchards where old ones have been removed without waiting several years to rest the land and get rid of the diseases.

7. At a cost little or no greater than that of old-style planting, it causes at least a year's earlier return on the investment in new orchards, and greatly increased returns thereafter as compared with spade-set orchards.

Subsoil blasting. This, like blasting for tree planting, is not recommended on loose or open soils, but only on tight clays or hardpan. Its purpose is to open and aërate the soil so as to afford a better moisture reservoir; to induce better soil sanitation; and to increase the possibilities of extensive root growth. The work is usually done in the fall when

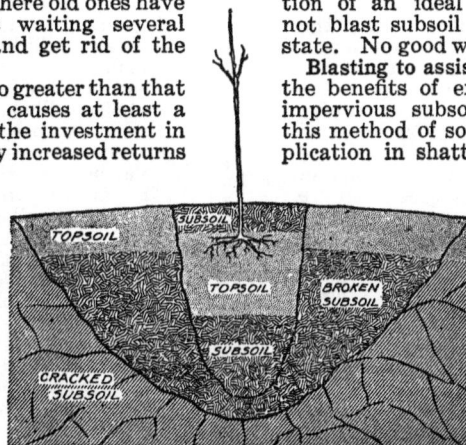

FIG. 458. To get the best results, therefore, dig out the centre of the loosened soil, fill in the pothole (enriching the subsoil with manure if possible), plant the tree, putting top soil around its roots, and firm all down securely. (Illustrations in this chapter, except Fig. 449, by courtesy the E. I. du Pont de Nemours Powder Co.)

the soil is dry. The standard charge is one fourth pound of a slow-acting, low-strength dynamite. The holes are usually made from 28 to 40 inches deep and about one rod apart. Such subsoil blasting has also been used to good advantage as a means of loosening hill soils to guard against erosion. Deep subsoil blasting is practised, especially in the bottoms of ponds and other wet places, to break through the underlying hardpan and permit the excess water to drain away through still deeper layers of gravel or sand.

The duration of the benefits thus obtained is not as yet clearly worked out, but it appears from tests that they should be effective for a considerable number of years. So far they have been more marked during the second and third years after blasting than during the first. If proper attention is given to the surface it is quite unlikely that they will return to their former state within the life of man. When heavy, deep-rooted crops such as alfalfa, for instance, are used to supply humus and increase the supply of nitrogen, the deep roots left to decay in the soil will guarantee the permanency of the benefits. Planting of a deep-rooting crop should always follow subsoiling.

On sour, wet land, where the clay is very sticky, it will be found an excellent practice to use considerable amounts of lime in order to sweeten the soil and to assist in the formation of an ideal crumb structure. Do not blast subsoil while it is in the sticky state. No good will come of it.

Blasting to assist drainage. Aside from the benefits of explosives for shattering impervious subsoils as above described, this method of soil tillage finds ready application in shattering subsoils to assist other drainage methods that are not giving satisfactory results. Tight clays and hardpan soils where tile drains are found to function poorly and establish drainage courses slowly, if at all, are found to be greatly improved by shattering the subsoils between the lines of tile. The same is found to be true in soils that are but little affected by open drains, where the water is held in too large amounts in the subsoil quite near the drains. Care, however, must be taken not to blast nearer than seven or eight feet to a tile drain.

The reason why deep tillage or subsoil plowing is desirable is that all the soil below the bottom of the ordinary plow cut, in other words, everything below six to eight inches, is still in its primeval condition. It has never been disturbed. Chemical analyses of soils down to a depth of twenty feet show that on the average acre there are tons of plant foods which become available only when roots can penetrate to them, or when ascending moisture brings them up to the roots that cannot get down.

Gully filling. This use of dynamite is for the purpose of blowing down the banks of the gullies so that the loosened soil can be easily dragged to the low places. Sometimes heavy blasts will throw much of the bank into the bottom of the old gully so that the earth will not require a second handling. The practice finds ready application especially in the southern hill lands. To correct for shallow washes, the deep loosening of the hard subsoil should extend out on each side for some distance.

If broken boulders, stumps, logs or any other loose material is placed in the bottom of the gully before it is filled the effect will be that of a deep, well-laid tile drain. Through this the excesses of water will be discharged without injury to the surface. In many cases, it will be well to straighten up the bottoms of the gullies and lay permanent subdrains before the filling is commenced.

For wide, shallow gullies, where the entire surface has been lost, but where the cutting has not been deep, the treatment is deep subsoiling with the spacing of the holes decreased to ten or twelve feet. In filling gullies large amounts of unaërated subsoil are exposed, and care should be taken to add humus either in the shape of rank-growing green-manure crops, vegetable litter or rough manure. Old corn stalks, forest leaves, or mildewed straw can be used to good advantage.

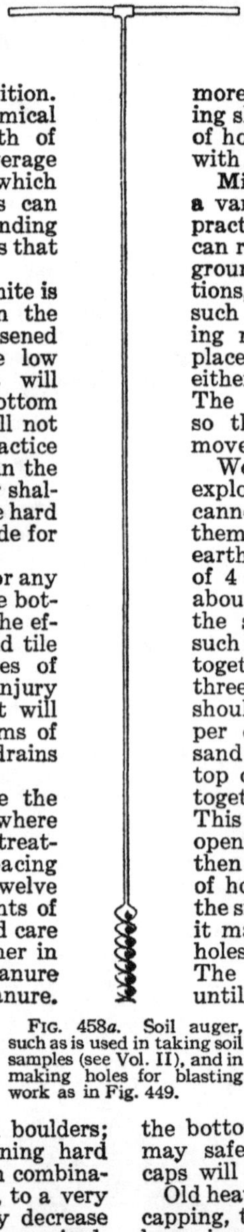

Road work. Explosives find a ready place in general road work for: (1) clearing away stumps and boulders; (2) opening ditches; and (3) loosening hard ground for grade improvements. In combination with road machinery, they can, to a very large extent, be made to materially decrease the amount and cost of hand labor required. This method finds application not only in the building of new roads, but in repairing and improving old roads. Blasting is a highly satisfactory method of getting many kinds

FIG. 458a. Soil auger, such as is used in taking soil samples (see Vol. II), and in making holes for blasting work as in Fig. 449.

of obstructions out of the way of the road drag. The methods of use are quite similar to those given for the same classes of field work.

In starting in a rock cut the ordinary practice is to drill the holes a few inches below the desired grade. These holes should be spaced back and apart a distance about equal to the depth of the cut, unless the holes are more than six feet deep, when the spacing should be about six feet. Each row of holes should be fired simultaneously with electric caps and a blasting machine.

Miscellaneous uses. Dynamite finds a variety of additional, special uses in practically all classes of farm work. It can readily be used for loosening up hard ground encountered in digging foundations, cellars, pit silos, and wells. For such work, modifications of subsoil blasting methods are used. The holes are placed much closer together and may either be deeper or more heavily charged. The object is simply to loosen the soil so that it can easily be shoveled or moved with a scraper.

Wells are often sunk with the aid of explosives through rock or ground which cannot be dug to advantage without them. When rock is reached and the earth above is properly supported, a circle of 4 or 5 drill holes should be started about half-way between the centre and the sides of the well and pointed at such an angle that they will come closer together near the centre when they are three or four feet deep. These holes should be loaded about half full of 40 per cent dynamite with damp clay or sand tamping packed firmly above to the top of the hole, and then exploded all together from the surface by electricity. This shot will blow out a funnel-shaped opening in the centre, and the well can then be made full size with another circle of holes drilled straight down as close to the sides as possible. If the well is large it may be necessary to drill a circle of holes between the inner and outer circle. The above process should be repeated until the well has passed through the rock or has been sunk to the necessary depth. Do not in any case enter a well until all the fumes of the last blast have come out. If in doubt, lower a lighted candle to the bottom; if it continues to burn the well may safely be entered. Electric blasting caps will give the best results.

Old heavy machinery can be broken by mud-capping, the same as boulders. Occasionally large iron vessels are broken up by being filled with water and shot by suspending a small charge of dynamite in the water in such a way that the explosive is not in contact with the metal.

PART V

Farm Buildings and Their Equipment

THE problem of locating, designing, constructing, and equipping his buildings is one of the most difficult that the farmer is called upon to solve. This, of course, implies that he is desirous of combining utility, moderate cost, and attractive appearance in the final result. In no respect is he more independent of outside influences and restrictions, than in his right to follow his own preferences as to style and arrangement. Presumably he has no near neighbors whose sensibilities he is either likely to offend or under any obligation to take into account. Farm houses are not built in blocks like city residences; their lines need not resemble one another nor even harmonize. The whole question resolves itself into satisfying himself and getting the most and the best for what he is able to spend.

However, there is another side to the matter. Whatever he may *think* he prefers, there remain those fundamental principles that have gradually been formulated after years of experiment and usage, as to what really is best, most efficient and most pleasing under average conditions. Only a genius is clever enough to disregard conclusions based upon the judgment of many generations; it is unlikely that any farmer is sufficiently a genius in architectural matters to be able safely to lay down new, unusual rulings as to how he shall plan and build. It is for this reason that farm buildings do, after all, resemble one another in the main, and often in most of their structural details. Because of this resemblance it is possible to set down, in relatively small space, a large amount of valuable information, suggestion and reference matter regarding them.

The chapters treating of this complex and very important department of farm engineering, are arranged in a definite, logical order: First, there is discussed the farmstead as a whole, its location and its arrangement; next, two chapters are devoted to the farm house, which after all is the most important building on the farm—probably no other worker spends as much of his time in the immediate neighborhood of the place where he sleeps and eats as does the farmer; the next two chapters deal with barns as a group, with reference, first, to their actual construction, and, second, to their equipment with labor saving devices; the remaining chapters treat the different classes of farm buildings as they may be grouped on a basis of purpose and usefulness.

It is no longer necessary, as it once was, for the farmer to hew his own timbers and pegs, split his shingles, saw out his lumber, and, with the help of his neighbors, build every barn and shed he requires from foundation to roof tree. Much and valuable assistance is now available in the form of manufacturers' coöperation and modern improved materials and methods. But it is still necessary for him to know what sort of buildings he shall erect; where and how they shall be erected; and finally, how to judge whether, when completed, they are what they should be. It is to help him to attain this knowledge that the following articles have been prepared and brought together.—EDITOR.

305

CHAPTER 28

Planning the Farmstead Layout

By H. H. NIEMANN, Manager of the Architectural Department of the Louden Machinery Co., whose work is very largely the helping of practical farmers in all parts of the country toward the solution of their building problems. Most of his younger days were spent on the farm. About 1893 he took up architecture, specializing, since 1900, in the designing and superintending of farm structures. During this time he has laid out a number of well-known farms about Chicago, taking charge of the locating and construction of roads, drainage systems, fences, etc., as well as all the buildings. He has been in his present position since 1912; for a number of years he has also been a regular contributor on farm building construction to the "American Carpenter and Builder"; and for 3 years he has been on the Committee on Farm Structures of the American Society of Agricultural Engineers, serving as Chairman in 1917.—EDITOR.

THE farm group, or farmstead, is a very important part of the farm, and deserves careful study as to its location on the farm, the buildings it requires, and the grouping of buildings, drives, and yards. The grouping of farm buildings cannot be standardized, because each farm has some special requirements which need individual study. A large number of rules may be suggested for such grouping; but, before any of them is applied, the following local factors must first be considered in connection with each farm: (1) climatic conditions; (2) topographic conditions; (3) soil conditions; (4) the type of farming; and (5) the kind of labor used.

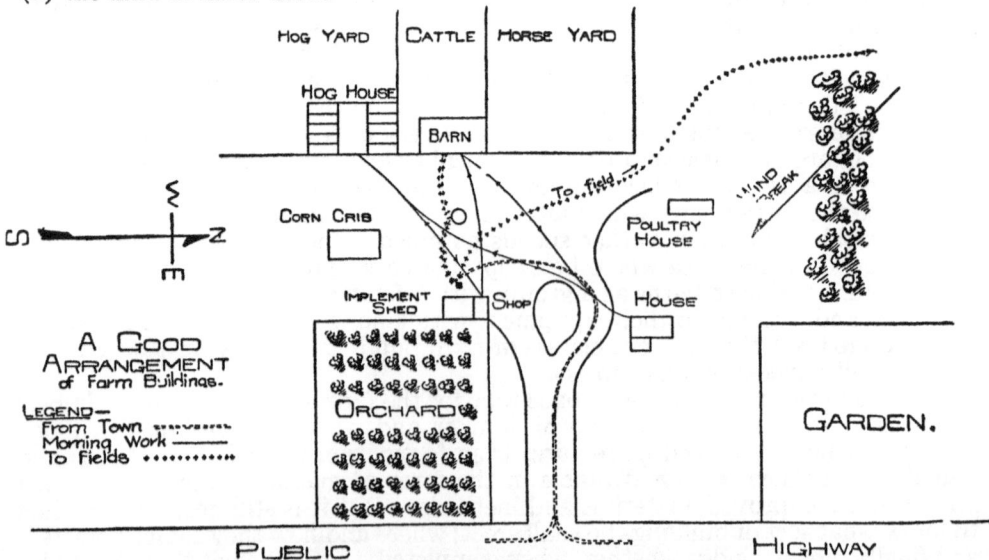

FIG. 459. A well-arranged farmstead showing the direct and comparatively few routes that have to be taken in doing the daily work. Compare with Fig. 460. (Both from Ia. Bulletin 126)

306

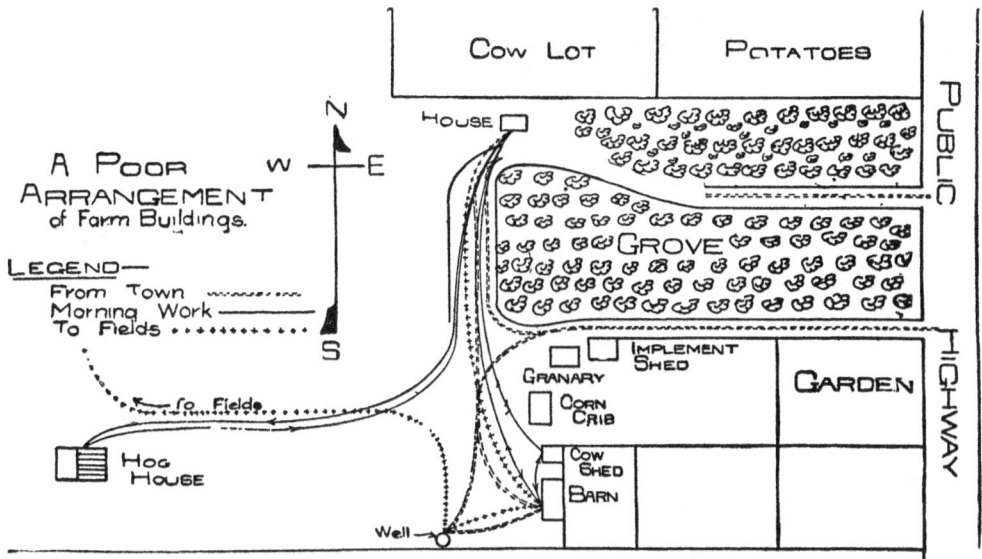

FIG. 460. Sometimes there are reasons for an apparently poor arrangement; more often they are the result of lack of thought. In either case they should be corrected as soon as possible. It is one thing to lose half an hour, unavoidably, once a month; it is another thing to lose half an hour every day the year round, especially when a little planning would prevent it.

Local Factors in the Grouping of Farm Buildings

Climatic conditions. In the northern part of the United States, and particularly in the northeastern section, it is necessary to provide warm quarters for all livestock. These quarters must consist of substantial buildings, constructed to resist storm and conserve animal heat and must have adequate light, drainage, and ventilation.

In locations subject to deep snowfall, ample provision should be made to protect from snow, by covered enclosures, the passages leading from one building to another; and the buildings should be grouped as close together as their fire-resisting qualities will permit. In the central states, fewer precautions may be taken against snow. The farming is more diversified, and calls for a larger number of buildings, which makes grouping more difficult. In the Rocky Mountain region, the farm group will consist chiefly of stock shelters and small storage bins for grain. In the West, where the air is dry, more attention should be given to fruit-drying sheds, packing houses, granaries, etc. In the South, the equipment of buildings need not be so extensive. Special attention should be given to grouping the buildings on a high place where drainage is good; manures and wastes should be located at a greater distance from the group; and more attention should be given to protection against flies and other insects.

Topographic conditions. It is not advisable to locate the farm group between hills, for several reasons. First, because of "trapped air"; the group must have a free air current. A valley generally lacks air drainage as well as water drainage. Cold air will settle between hills. The night temperature in a valley is always colder than at a location where there is more circulation. Further, in warm weather, valley air becomes very sultry.

Second, because of difficult land drainage. Valley land is always the last to dry off after a storm and the last to thaw out in the spring. It will catch the largest amount of snow, and hold it longest.

Third, because of short days. The morning sun's rays are later in reaching the valley, and the evening rays are cut off earlier, making the days shorter, colder in winter, damper in spring, and warmer in summer, from lack of air circulation. In this respect, hills on the north and south of the group are not so objectionable as hills to the east and west.

The extreme top of a hill is the most healthful location for man or beast, but it sometimes has objections which should be considered. Storms are very severe, requiring special bracing in the construction of buildings; and windbreaks around the yards and stockpens are necessary. All feed and other supplies must be drawn uphill to their destination. The first objection

Fig. 461. A windbreak both protects and beautifies the farmstead

may be overcome in the manner suggested, and the second is not serious, unless the hill is long and steep. On a prairie or slightly rolling country, the highest point on the farm is always the choice for the farmstead, provided its location is not too far from the centre of farm operations.

The advantages are long days of sunshine, good air, good drainage and, in most cases, good views of the farm.

Hillside with southern slope the ideal location. A hillside as the location for the farm group may be ideal or it may be objectionable, depending on the amount and direction of its slope. Just over the brow of a hill on the southern slope is an ideal location for the farm group, because it gives protection from cold storms, and is open to the sunshine.

A southern exposure for all buildings containing livestock makes an ideal condition, as far as climate is concerned, because here the ground dries out fast after a rain and the warm air currents are received.

Soil conditions. As the farm buildings, drives, and yards occupy a not inconsiderable space, which becomes unavailable for agricultural purposes, the question of soil should be considered. It pays to set the group a little farther away from the highway, if the land along the highway is more fertile than some other soil on the farm which has no objectionable features as to slope or location.

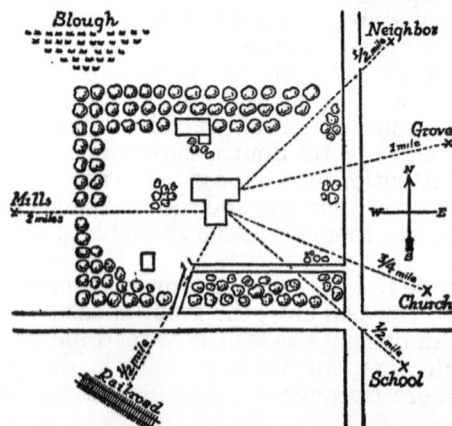

Fig. 462. The planting around the farmhouse should do at least three things: protect from severe prevailing winds; block out unsightly views; and frame the vistas that are beautiful or that tie the farm to its neighborhood and friends. (Ia. Exp. Station.)

Some of the barren spots on the farm, composed of soil which is not fertile, may also be objectionable for the yards, if it is of a dense nature that will not drain or dry out quickly; for example, clay or gumbo. These objections may be overcome by filling on top with sand, gravel, or cinders, after the surface has first been well tiled.

Type of farming. The type of farming that is to be done will affect the kind and number of buildings more extensively than their location, although, the more diversified the farming, the more centrally the buildings should be located.

A stock or grain farm may have the farm group at one corner, or at the centre; but where the farming system includes dairying, grain, meat stock, gardening, and small fruit, the location must be very carefully studied with a view to the elimination of unnecessary fences, drives, etc.

Kind of labor used. On the small farm, where most of the work is done by members of the family or by permanent help, housed in or near the residence, the buildings may be grouped closer to the house, with all walks, as far as possible, covered with shelter roofs. Where the work is done mostly by temporary labor housed in isolated tenement houses, all the barns may be located farther away from the residence; but the labor headquarters should be near enough to the barns for the chores to be conveniently performed.

The Site and the Buildings

After noting the foregoing requirements and applying them to your particular farmstead problem, carefully considering the slope of the land with relation to drainage and exposure, and the position of the site in regard to the highway and market, it will not be difficult to decide whether your farmstead shall face north, east, south, or west. Topographic and soil conditions should determine the proper location; the type of farming will then determine the number of buildings required; the size of the farm will determine the size of the buildings; and the value of the land and its permanence for farming purposes should determine the type of construction.

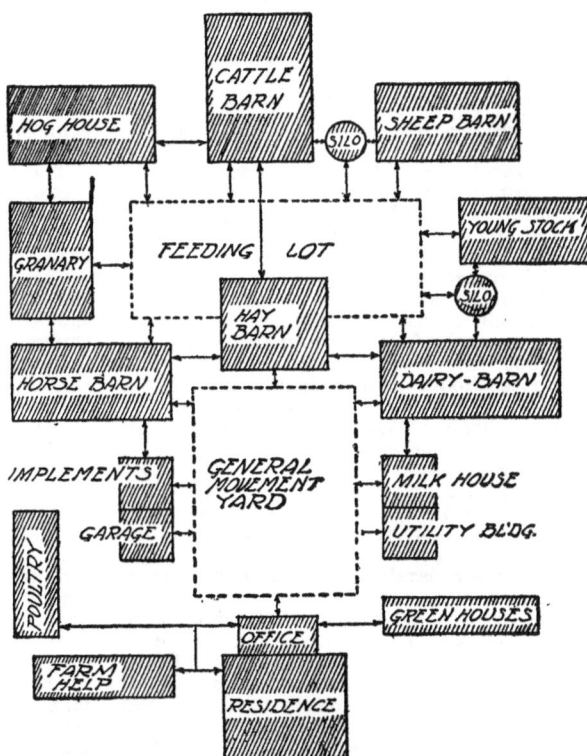

Fig. 463. Ideal arrangement of the farmstead on a large, general farm. A smaller enterprise would permit the combining of some of these buildings with a corresponding additional saving of time and steps.

A farmstead designed for a south front is not practical for a north front; neither is one designed for a west front practical for an east front. The compass points must be in the mind's eye from start to finish. They should be considered in grouping the buildings in proper rolation to one another, in the proper facing of each building, and in the planning ot its interior arrangement.

To place the buildings properly, the first consideration should be their relation to one another and to the farm in general. Farm routine and convenience to drives, fields, and highways should also be studied. Fig. 463 serves to illustrate, by means of the arrows, the relative connections of the buildings with one another. In determining their respective locations, the suggestions offered in the following paragraphs should prove useful. It is, of course, impossible to tell how any one farmstead should be arranged without carefully studying its problems. The aim should be to come as near as possible to the ideal arrangement outlined.

FIG. 464. This is not a very attractive view. If the farmhouse looks out only upon such scenes, either it should be moved and rearranged, or the farmyard and buildings should be made more pleasant to look at.

The residence. This should always remain the most important building of the group. Its location should be slightly higher than that of any of the other buildings and should afford a supervising view of the farm.

With regard to the highway the residence should be beyond the dust zone, which, in some localities, has been estimated to extend 650 feet north, 65 feet south, 250 feet east and 320 feet west of the highway. The residence should have a good view of the highway and of other points of interest. Its location should be such that prevailing winds will carry barn odors away from the house.

The general-purpose barn. The small farm, where only a few head of each kind of livestock are kept, does not justify separate buildings for the various purposes, but all stock (except, perhaps, the chickens and hogs, which have separate sheds), feed, and grain are placed in one barn.

The location of this barn should be very central, on well-drained land, and with convenient access to and from the house, fields, pasture, highway, and feeding yards.

It should be so placed with relation to the compass that all livestock may have the maximum of sunshine and the best of protection from winter storms.

The dairy barn. This barn, in which is produced the most nutritious of food for human beings—a food which is, however, the most susceptible to contamination and foreign odors—should be considered the most important barn on any farm where dairy products are produced. It need not always be the most expensive, nor the largest barn in the group, but it should have the choicest location with relation to drainage, air circulation, sunshine, and everything pertaining to its sanitation. It should have, also, convenient access to the hay and feed barn, exercising yards, and milkhouse. It is advisable to keep the dairy barn isolated from all other barns containing livestock.

The horse barn. Until the power tractor shall have entirely displaced the work horse, provision will have to be made for the latter among the buildings on the farm. The

horse should be housed in a barn which may also contain feed bins, hay storage, bedding storage, etc. The location of the horse barn should be selected with reference to convenience to the general yard, implement and vehicle shed, cultivated fields, highway, exercise yards, and granary.

The cattle barn. The barn for beef cattle, dry stock, and young stock should be located adjacent to a roomy feeding lot, so that the stock may freely enter or leave the barn as they choose, when the weather will permit. It should have large sliding doors on the south, and should form a good windbreak and shelter on the north of the feeding lot. A cattle barn is not complete without a silo, which must be convenient to the barn and feeding yard.

The construction of the barn and hay racks should make the feeding as nearly automatic as possible.

The foregoing remarks relating to the cattle barn apply also to the sheep barn.

The granary. The granary is a very important building on most farms, and its location should be studied with the view of having it convenient to the lanes leading to the grain fields. It should be convenient, also, to all barns containing livestock, and to the feeding lot. Its contents being very valuable, it should be far enough away from other buildings to prevent loss by fire. Sunshine and fresh-air circulation are very necessary.

The hoghouse. Where hogs are kept in connection with cattle, the hoghouse should be so located that the hogs may be turned into the cattle-feeding lot and allowed to pick up all the food which the cattle would waste. Also, the hoghouse should be quite near the corn-crib, for convenience in winter feeding, and should have sunshine and good air circulation. The location should also be such that prevailing winds will carry the odors of the hoghouse away from other buildings, especially the milkhouse, dairy barn, and residence.

The poultry house. The poultry buildings of most farms are sadly neglected, because they are built with the idea of making possible the existence of a few hens, which may, to a greater or less extent, aid the housewife with her daily menu, through the supply of a few eggs and an occasional fryer.

The very fact that the poultry plant of the farm is, in most cases, turned over to the women folks to look after, is sufficient reason why its location and convenience should receive careful attention.

The vital points to be considered are: (1) healthfulness of location; (2) convenience of situation with relation to the residence; and (3) care in design, with reference to convenience in feeding, egg gathering, etc.

(1) The location of the poultry house should be a healthful one for the fowls, dry, and with good air and plenty of sunshine. Provide ample glass or canvas-covered open-

A complete layout of farm barns. *At left*, horse stables; *at right*, cow and hay barns (connected) *in the middle foreground*, manure pit and shed; *in the right foreground*, hog house

Range of horse, storage, dairy and cow barns on a Vermont farm, all facing south, close to-gether and directly across the drive from the dwelling

FARM BUILDINGS MUST BE PLANNED AND PLACED WITH REFERENCE, FIRST, TO THEIR PURPOSE, AND, SECOND, TO THE NATURAL CONDITIONS OF THE LOCALITY

311

Rightly placed, between conveniently arranged buildings, a single track can carry both feed and litter conveyors

Concrete and metal are the safest materials for both stock and stockman, and the cleanest and most sightly as well

THE PURPOSE OF MODERN BARN EQUIPMENT IS TO MAKE CONDITIONS SAFE AND SANITARY, AND THE LABOR OF THE DAILY CHORES AS LIGHT AS POSSIBLE

ings on the south side, also fresh-air intakes and foul-air outlets operating independently of the doors and windows so that proper ventilation may be assured. If the ideal location is not high and dry enough, build the house high and fill in with gravel, cinders, or other porous material, to keep the place dry.

(2) The poultry house should be conveniently located with relation to the residence; and the walk between the two should be protected from rain, snow and mud. If a shelter roof is out of the question, a cement or plank walk should, at least, be provided,

and this should be sufficiently elevated to drain and to keep off drifting snow.

(3) Careful design in detail with regard to convenience in feeding, cleaning, disinfecting, the division of flocks, egg gathering, etc., is important. A hallway, from which feeding and cleaning may be done without entering the room containing the flock, equipped with tilting or sliding feed troughs and automatic feeders, is simple to build, and can be so arranged, with feed bins overhead, that the feeding can be accomplished by pulling a lever that will open the spout from a hopper-bottomed bin to the feed trough.

Buildings Under One Roof

The idea of grouping all buildings under one connected roof is a good one and has points in its favor, when the size of the farm will justify such arrangement, and when the farm is permanent and valuable enough to justify the construction of semifireproof buildings having effective fireproof walls between them.

Connected farm buildings should be built around an open court, which may be subdivided by walks or fences so that part may be used for livestock and part for a general yard with convenient gateways to the farm roads and the highway.

If the farm does not justify building the structures with masonry walls and fire-resisting roofs, the idea of connected buildings should not be considered. The walks between the several buildings may, however, be protected by shelters which could be quickly wrecked and cleared away in case of fire.

CHAPTER 29

The Farmhouse:
Its Location and Design

By RUBY WESTLAKE FREUDENBERGER, of Columbia, Missouri, who was born and reared on a farm in Boone County, that state, but whose Maryland and Virginia ancestors make her in name and fact a Daughter of the American Revolution. After graduating, with two degrees, from the University of Missouri, she taught mathematics there for two years. From 1911 to 1917, she lived in Carson City, Nevada, where her husband was Chief Engineer of the Public Service Commission, and where she was named by the Governor as member of the State Commission for the Promotion of Uniform Legislation in the United States. She has, however, at all times, been deeply interested in the social and home problems of country life and the conditions that give rise to them, carrying on personal investigations, working out principles and methods, and contributing valuable articles on different phases of the subject to rural life publications. In connection with the technical engineering problems of home making, she has had the coöperation of Mr. Freudenberger (see Chapter 18) with a lifetime of training and practical experience behind him.—EDITOR.

THE farmhouse is a distinct type of building. Its uses are as definite and characteristic as those of a shoe factory or a flour mill, and it should be planned and located with the same specific regard to the ends of its use as those buildings are. It should be highly specialized both in its plan and in its design.

The farmhouse is now being accorded the expert scientific thought which its importance has long warranted, and which promises, at least, to put it upon the same plane of efficient development with the purely utilitarian farm buildings.

The Farmhouse is Both a Home and a Business House

While the farmhouse is a dwelling, some of the controlling considerations in other types of homes have no bearing here: it has its own peculiar set of conditions, and its problems must be faced and solved for its own peculiar needs. Suggestions and help may be gathered from a study of other dwellings; but in this the characteristic demands and distinctive quality of the farmhouse must not be forgotten. A study of houses reveals so many beautiful ones, pleasing because of perfect adaptation to surroundings and uses, that it is difficult not to be carried away by their charm and appeal, and to remember that they are not farmhouses and that they do not meet the fundamental requirements of the farmhouse.

The farmhouse is unique in that its purposes are twofold: it must serve both as a home and as a business house. As neither purpose must be subordinated to the other, the farmhouse must combine in its arrangement the essential demands of both.

The requirements of the farmhouse as a home differ very little from those of any other house, whether in city, town, or village, because the ultimate aims and broad purposes of real homes are the same everywhere. A house in which a home is to be established must provide for the physical comfort and health of the inmates, and for social activity commensurate with the standing and mode of life of the family; and it must also meet the need and desire for beauty in surroundings and aid in satisfying the aesthetic and intellectual demands of human existence, besides providing for pleasures and cultural pursuits. A good house is necessary to a good home, and a good home is necessary to proper human development.

Wherein the farmhouse differs from other homes. It is, therefore, not in essentials that the farmhouse should differ from other homes, but in a more rigid elimination of nonessentials. Life is simpler and less complex in the country than in the city; hence, the farmhouse should be simplified in arrangement and free from all useless ornamentation and complicated construction. Thus it will be easier to get about in and will require less labor to keep clean.

It is in the arrangements for lightening the housework, and easing up the strain on the housewife, that the problem of the farmhouse differs most from that of city houses; for the farmhouse is practically servantless. Every detail of the house should be designed, as far as possible, with a view to conserve the time and strength of the housewife, and to make it possible for her to provide clean, cheerful, and inviting home conditions for her family. Since she must be the home-maker as well as the housekeeper, she must be spared some vitality and spirit to devote to the care and training of her children, and to enable her to maintain a home standard in keeping with the demands of the culture and progress of the times.

In its industrial aspect, the farmhouse is not only the workshop of the house-wife, but it is also the vital centre from which radiate all of the activities of the farm. It is a fundamental unit in the group whose aggregate is the farm plant. Its right placing and arrangement will mean much for the success of the farm as an industrial institution; for these will be factors in determining the whole plan of the farm operations, and the consequent labor program and efficiency.

Problems Connected With the Choice of a Site

The great diversity of natural conditions of soil and climate in this country has given rise to varied systems of agricultural pursuits, each section developing and following that system best adapted to its needs and possibilities. These, in turn, bear strongly on the requirements, arrangement, and location of the farm-house and on its position in the general layout of the farmstead.

Natural climatic conditions have a direct bearing, too, on the location of the farmstead; for it must be so placed as to furnish protection from cold and heat, snow and rain, dust, wind, insects, and unhealthful vapors.

The contour and the general surface conditions of the site must be such as to permit an economical and convenient arrangement of the group of buildings. Clearly, a bleak, rocky point or the side of a steep incline would not meet this demand.

Selection of a site. The time-worn adage "Circumstances alter cases" is nowhere truer than in choosing a site for the farmstead; there are so many factors to be taken into account, so many objectives to be sought, and so many conditions, natural and artificial, to be reckoned with. Some of these may be controlled or modified, while others, such as the natural properties of the site, are permanent and fixed, and

Fig. 465.   The farmhouse is necessarily isolated from its neighbors; it should therefore be sheltered from storm and wind—a real haven of refuge and comfort.

must be accepted. These may be taken advantage of and so incorporated into the plan as to serve useful ends and add to the general charm and efficiency of the whole.

The dual purpose of the farmhouse complicates the problem of the site, as well as that of the plan of the house. The relative importance of the different elements involved varies in the different sections, and even for each individual farm. Hence, any plan must involve compromises and concessions, the giving up of some advantages to secure greater ones, or to allow the incorporation of features appealing to individual tastes.

The size of the farm; the system of farming followed; general climatic conditions; surface contours, bearing upon air and water drainage; the nearness and relation of the farm to the public highway; the source of the water supply; the outlook and general surroundings; and, finally, personal preferences—these are some of the more important points to be considered in selecting a site for the farmhouse. Probably no arrangement can be worked out which will secure the ideal adjustment of all of these details; but systematic study of their relations to one another and of their influence upon farm life will obviate many of the inconveniences and discomforts existing on farms to-day.

In the northern and northeastern sections of the United States, where the winters are long and severe, compactness of arrangement of the farm buildings is necessary. Here a site should be chosen that will provide some natural protection against snowdrifts and the sweep of winter winds, and will permit such a grouping of the buildings that the feed and the stock may be housed close at hand and be reached without the attendants having to go out of doors. The outbuildings, such as laundries, woodsheds, and storerooms, might advantageously form a series leading from the house to the barns.

FIG. 466. The farm home should belong to its surroundings; to this end stone obtained on the farm makes a highly satisfactory building material.

In the South, where opposite climatic conditions obtain, an open arrangement of the buildings allowing free passage of the winds is desirable, and the site should be so chosen as to take advantage of shade and the prevailing breezes to relieve from the heat of the long summers. Care must be taken to guard against odors from the barns being carried to the house, and special provision must be made to screen against flies, mosquitoes, and other insect pests.

On the wind-swept, treeless plains of the West, the problem, after the water supply has been provided for, is to find such protection as may be had from the dust and burning sun of summer, and from the snowdrifts, blinding storms, and freezing winds of winter. Angles and ells of the house may be so placed as to break the sweep of winds in winter and provide shady spots in summer. Outbuildings, also, will serve as shields, if located in proper relation to the house and yards.

In the Central West, notable alike for diversity of crops and diversity of climate, provision must be made against extremes of both heat and cold, against rainy seasons and periods of drought, against dust, mud, snow, hail, and wind. The site must be chosen with these demands in mind; and the great complexity of the problem makes its satisfactory solution a correspondingly greater achievement. The short, intensive cropping season of the Corn Belt gives rise to a distinctive plan of operations; and this, in turn, calls for a particular and suitable farmstead arrangement.

In mountainous sections, sheltered sites, shielded against snowslides and exposed away from the usual direction of storms, are naturally chosen. The selection of the site is a matter deserving serious thought and most careful weighing of the many points to be considered. Our pioneer forefathers located their homes with a view to the sole requisite of nearness to a water source. They bothered not at all about drainage, malaria-laden fogs, open views, or step-saving plans. Consequently, we find many old homesteads in low places near springs, or on the hilltops just above them.

Once the site is chosen, it becomes about the most permanent thing on the place. It probably fixes the abiding place of generations of human beings, and the silent influence of its conditions and surroundings leaves its mark on the natures and constitutions of those who pass in procession there. Often, sentiment alone perpetuates it as a home, with all its satisfying or distressing qualities.

If the site offers such natural advantages as will make easy and economical the providing of modern conveniences and the many small adjuncts that mean so much for the comfort and well-being of the workers, the farm home may secure them; but, if a deal of labor and expense, and planning and changing be necessary to have them, the family most likely drifts along without them through years of accumulated discomfort and weariness of body and soul. If the sunshine and cool breezes and beautiful vistas, and gorgeous sunsets just naturally go with the site, the farm family have the benefit and inspiration of daily existence under such influences; but if they must be secured by some rearrangement of buildings or trees, or by some extra expenditure of thought and effort, they are most often done without. The far-reaching effect of site conditions makes the choice of a location for the farmhouse the prime consideration of the home builder.

Desirable features of the farmhouse site. The building site should be airy, sunny, and well-drained. These ends are best obtained by selecting a high, open space, where the sun and air can reach the house. This is as important for the lasting quality of the house as for the health of the family. There should be some large trees for both shade and beauty; but they should be so situated as not to shut out the breezes, and the shade should fall on the yard, rather than on the house itself. Direct sunshine on a house is more tolerable in summer than reflected heat from the ground, and is much to be desired in winter.

One seeming variation from this rule should be a tree to provide afternoon shade for a west porch or door. This tree need not be so near the house, however, as to violate the general rule laid down.

A densely wooded yard is dark, damp, and moldy, and is often as badly in need of ventilation as a shut-up house. Trees close up or touching a house hold the dampness, and cause the roof or walls to rot.

Another important reason for choosing a high location is that this makes possible good drainage, and secures against damp yards, cellars, and foundations. So far as climatic conditions are concerned, the best location for a set of buildings is just over the

FIG. 467. Sunlight is as necessary in the home as in the barns. These trees are too thick and too close for either health or comfort.

FIG. 468. A bare, unadorned building is never a real home. It should have a background, a setting as well as a pleasing outlook.

brow of a hill on the south slope. This allows more sunshine for the yards and buildings in winter—a great advantage in securing warmth and dryness. The high location will get the cool summer breezes, and the air will be purer as well as cooler. Damp, foggy, cold air settles in the valleys and low spots in still weather, and these places have fewer hours of sunshine. Destructive frosts, too, often form in low places, when the uplands escape entirely.

It may be objected that the high points are also more exposed to the sweep of winter winds; but if the slope be to the south or if there be a slightly higher rise to the north or northwest, this objection does not hold. Besides, artificial windbreaks may be provided by planting trees or by judicious placing of outbuildings.

There are many more artificial means of protection against cold than against heat, hence we must take advantage of all possible natural elements for keeping buildings cool in summer, and warm in winter. The summer is the time of stress for farmer folk, the period of greatest demands on their energy and endurance; therefore, the greatest care should be taken to provide them with every possible comfort and help for this season. In the winter, more of their time and energy is available for the promotion of personal ease and comfort.

A high location necessitates, of course, a climb every time the farmer or his helpers come to the house from the fields or from market; and if the barn is on the same elevation as the house, all the crops to be stored in it, as well as all building materials and general supplies, must be pulled up the grade. This is considered so serious a drawback by some farmers—it is not regarded so seriously by the farm wife, usually—that they allow it to outweigh all of the advantages of the high site.

However, on big farms—and it is only such that produce large crops to be stored or moved—extensive stock feeding is usually carried on. The feeding should be done over various fields for the economical distribution of the manure and for soil building, aside from other considerations. Extra storage barns might be erected on a spot having the general level of the fields or road, so that a comparatively small amount of hauling need be done to the hill barn.

The high location commands a view of the surrounding country and makes use of the subtle influences of beautiful scenery and picturesque surroundings. Often the outlook alone would justify the choice of the most elevated spot. The road up to it may be rough and steep, but no height is ever attained without its climb. The aesthetic phase of the building site does not often receive the attention it deserves. Splendid views and inspiring landscapes are permanent features of the estate; their influence is a constant force; and they are ever-present sources of pleasure and aids to uplift.

The homestead on an eminence gathers something of impressiveness and dignity from its elevation and commanding position.

House should be well back from highway. The farmhouse should be well back from the main highway, alike for artistic reasons, in order to be out of the dust zone, and to secure comfortable privacy and seclusion.

Just how far from the road it will be necessary to place the house, to escape the dust, will depend upon the direction of the house from the road; for the direction of the prevailing winds determine the width of the dust area. For instance, a house north of the road needs to be farther back than one on the south side, as in most sections of this country the prevailing winds are from the south.

especially in summer—the dusty season. It has been estimated that, in order to be safely away from the dust, a house north of a well-traveled highway should be set back 40 rods; south of a well-traveled highway, 4 rods; east of the highway, 15 rods; and west of the highway, 20 rods. These distances, however, except possibly the first, are hardly great enough to meet the demands of the best landscape effects.

The advent of the automobile and its extensive use as a means of country travel have widened the dust zones and, at the same time, have removed some of the social reasons for placing the house near the road. The automobile, rural delivery, and the telephone have done away with the need of having the home so placed as to enable the farmer's wife to look up from her work and view the passing along the road as a social diversion and as a means of keeping informed regarding the movements of the neighbors.

Ample grounds and a private driveway desirable. No farmhouse, however beautifully designed, can give its best effect or show to its full advantage unless surrounded by ample grounds and given a setting and background consistent with its style and feeling. A farmstead cramped for yard space, and pushed up so close to the highway that it seems to be straining the division barrier, is fatally lacking in beauty, comfort, and sanitation. Nearness to the street or road may be advisable or necessary in towns; but it is out of place in the country, and out of keeping with the character and spirit of country life. The essential quality of the country is openness and spacious repose.

A most attractive arrangement is to have a private road from the highway leading to

FIG. 469. The New England colonial type of country house; severe, simple, well-proportioned and roomy

the house through a wooded pasture. Such a pasture, with the road winding its way among the trees and up the slope to the house beyond, enables the place to present to the world a stately and prosperous front, and assures a pleasing first impression upon visitors and passers-by. This approach to the house should be so provided that there will be no gates to be opened and closed until the one main entrance to the house is passed and the barnyard enclosure reached. "The one main entrance to the house" is here emphasized, because there should be but one public entrance, and the approach to it should be the driveway. There should not be a front entrance and a side entrance, because in the farmhouse the front entrance will not be used. Nothing superfluous belongs with the farmhouse, not even a front door. The house should be so planned as to throw the main entrance in the logical relation to the service rooms demanded by the conditions of farm life.

Location of the House with Relation to the Outbuildings

The house should be near enough to the highway to be plainly visible from it, and there should be open spaces in the woods allowing a clear view of the house from the road and of the road from the house.

Locating the house with reference to the source of the water supply is not so important now as formerly, except in the arid or semiarid regions. Deep wells, windmills, gasoline engines and modern mechanical systems for storing and distribution, make it possible to supply adequate water for almost any site. Bored wells are sometimes a too expensive source of water supply, especially in sections where it is necessary to go very deep; but, failing other sources, it is possible, with proper care and attention, to have anywhere in the area of rainfall clean, pure cistern water.

A rise of ground slightly higher than the house site will supply a spot for the storage of water and permit the use of the economical

gravity system of distribution. This rise will serve, also, as a windbreak, if in the right direction from the house. If, however, this elevation is so great as to shut in the house site, and cut off a good view, it will be more of a drawback than a benefit.

The building site should not be too near a boundary line of the farm; else, conditions beyond control are brought almost to the doorstep, and corn fields or stockpens may appear in unpleasant proximity. Even an intervening road does not always remove this restriction.

Position of farmstead with relation to the fields. On a small farm, the position of the farmstead with relation to the fields is not especially important, but on a large farm, this item moves far up in the scale. The many trips to the fields in the production and harvesting of the crops, and the inconvenient handling of stock, on a poorly planned farm often aggregate in the course of a year miles

Fig. 470. The Dutch colonial type is common in Pennsylvania and a few other localities, but it is hardly a typical farm style.

of unnecessary steps and days of wasted time. From the sole consideration of the desirability of being within the easiest reach of all the fields, the farmhouse should be located near the centre of the farm. This is not advisable, however, unless the public highway pass through the farm; for such a location would probably result in a demoralizing isolation, instead of in the comfortable privacy which should characterize the country home. The farmhouse being off the line of general travel, visitors would be few, and strangers rarely seen.

If the relation of the farm to the public road be such that the farmstead may be placed near the middle of one side, this will be a most satisfactory arrangement, especially if the farm be oblong and a longer side takes the home site. It is better to have the house in a corner of the farm, even though this be a long way from the farthest field, than to have it too far from the public road. This road is the channel of communication with society and the outside world, and this way lie church, school, neighbors, and market. Many more interests call from this direction than toward the fields, and the aggregate of trips out this way far exceeds those in any other direction.

Orchard and garden. The orchard and garden should be within easy reach of the house, and their location should be part of the original planning of the farmstead. The presence of suitable soil and exposure for these should be taken into account in the choice of the site. Further, an orchard behind or at the side of the house supplies an artistic landscape feature so effective and pleasing that it will cover a multitude of small blemishes; also, it provides the ideal poultry range.

The house should be so situated with relation to the barns as to be out of reach of odors. As a fire precaution and for protection from odors and flies, it should be on the windward side of the barn, and at least 250 feet away from it. To have the house and barn near each other saves time and steps for the farmer, and adds to convenience in the care of stock; and this is advisable so far as

the demands of sanitation and safety from fire permit.

The prevailing summer breezes are from the south and southwest; and it seems the wiser to relate the buildings with regard to summer conditions rather than to those of winter, because the house is open more to the breezes and, hence, is affected more by outside conditions at this time. From this consideration, the barn should be placed north or northeast of the house. It is true that the yards are drier and cleaner in the summer, and there are fewer troublesome odors than in the winter, so that a position south of the house is not so objectionable, if other considerations make it desirable. The *distance* from the house, more than the direction, will control the odor and fly menaces.

Barnyards and outbuildings. A more important point for consideration here is that the outlook from the house shall not be over the barnyards. As most houses are planned with the living rooms on the south and east, the location of the barns north or west of the house would be best for this particular object. However, modifying conditions may be brought in, whatever the relation of house and barn. The barnyards may be put north of the barns, if it is desired to put the barn north of the house; and the north fences of the yards may be so constructed as to serve as shields from cold winds, though this is not the most desirable arrangement for the comfort of the stock. Proper planting of judiciously selected shrubbery will effectively screen the barn from the house and still not interfere with the enjoyment of the distant view in that direction. A combination of a water tower and garage may also accomplish this end.

The dwelling should dominate the group of farm buildings. It should stand in the foreground, with the outbuildings grouped back of it in such a way as to give it support and prominence, at the same time conveying the impression of the proper relation of each building to the others and emphasizing the general unity. The outbuildings should be grouped, not scattered about promiscuously giving in the latter case the impression of a little town without order or system in its layout. The offense, to the sensitive eye, given by numerous buildings dotted in disorder around a house, is similar to that produced by shrubs and flower beds peppered over a lawn. No building except the house should stand out with distinct emphasis from the group effect. The same materials should be used, and the same general architectural design and system of roof lines should be followed in all the buildings, to secure the best artistic results.

The outbuildings should be grouped away from the windward side of the house, to reduce the fire hazard, if this can be accomplished without sacrificing too many other important requirements. All of these buildings, except the barns and poultry houses, should be near the dwelling; in fact, they should be incorporated in its general plan, and should connect with it by sheltered and paved or floored passages. The housewife should be able to reach all of her supplies and carry on all of her domestic activities without the need of exposure to the weather in either summer or winter. To make this possible, the laundry, the smokehouse, the fuel storage room, the general storeroom, and the well should be connected with the house.

This arrangement, of course, increases the fire hazard somewhat, though not greatly, as the barn is not in the group, and the passages leading to the outbuildings may be of flimsy construction, quickly removable in case of fire.

Have trees about the house. No country building site should be devoid of trees. If possible, a site for the farmhouse should be chosen having some large native trees; but, if this cannot be done, a few of a quick-growing variety should be planted and carefully guarded. Comfortable, shady spots in the yard will invite to the outdoor performance of many tasks usually done inside, and every means of encouraging outdoor living should be utilized.

A tree so placed as to shield the dining porch from the direct rays of the morning sun will add to the comfort of the breakfast hour. Also a tree to the west of the house, to protect a porch or door from the glaring after-

FIG. 471. Either dressed or field stone set in mortar combines well with wood trim and gives a warm, attractive, durable structure.

noon sun, is an invaluable adjunct. The general location of trees should be away from the house, and there should be abundant openings between them, to allow full benefit of the distant views. A spreading, stately tree in intimate sheltering relation to the house provides an adequate and pleasing decorative feature, enhancing the architecture, and adding to the home character of the group. Heavily wooded areas serve well as landscape features and as wind and storm screens; but they should be in distant relation to the house, else they will shut it in too much. They also harbor animals which prey upon the poultry and young stock of the farm and are obnoxious in many ways. Trees about the house should not be allowed to cut off a fine view nor make a break in a charming vista. They should serve merely as frames for the pictures of the landscape.

Design of the House and Site Intimately Related

There is such an intimate relation between the design of the farmhouse and the site upon which it is to be placed that each should be selected with the other in mind. If one's choice is fixed upon a particular type of house, he must look about for a suitable site; if the site is already selected, then it is the task of the designer to evolve a building in harmony with the location.

The latter is much the more usual process and the wiser one, for the design of the building is not nearly so important as a successful house. To be successful, it must be comfortable, convenient, and satisfying to the eye; it must meet the demands of the particular purpose of its construction; and it must fill its part of the requirement for a beautiful whole.

No ugly house is a success. Beauty is not expensive; it does not add to the cost of houses. It is a by-product of simplicity and efficiency in design and construction. It is often given away and ugliness bought. Beauty does not mean ornateness or pretension, nor is it a matter of elaboration. On the contrary, it is obtained for the farmhouse by simplicity of line and composition, structural honesty, harmony with its surroundings, and adaptability to its specific uses. "The line of beauty is the line of perfect efficiency."

The farmhouse is an all-year home, and it must be designed with definite regard to climatic conditions. Large porches, open passageways, and many provisions for outdoor living should go with houses for warm, mild climates. More compact designs, emphasizing arrangements for conserving warmth, such as put-

FARM KNOWLEDGE

FIG. 472. A farmhouse designed for southern conditions. The abundant windows, overhanging eaves and interior ventilating system afford valued protection against severe summer heat.

ting chimneys and fireplaces on interior walls, vestibuled entrances, and fewer porches, are suitable for cold, windy locations.

The design of the farmhouse must be made with a view to economical and simple construction. The cost must be kept low; and this may be best accomplished by eliminating useless decoration and such features as towers and swell ends, and jig-saw work for the cornices, porches, and railings. Often the building is done by the country carpenter, assisted by the farmer and unskilled farm labor. Fancy features, involving complicated carpentering, must be omitted for this reason as well as for economy and good looks. The house should be so designed as to have no angles and corners and cubbyholes wasting space and material and labor. Restricting the plan mostly to square spaces and rectangular effects will make this much easier.

Features which the design should include. The design of the farmhouse should include features especially favorable to outdoor living. It should be characterized by generously proportioned porches, suitable for use as a dining room or a living room, or the two combined. Sleeping porches also, will add much to the health and comfort of the family. The two-story porch of southern houses is a most pleasing feature for mild climates, as it provides delightful outdoor sitting rooms and bedrooms. Also, an open passageway leading to the "summer kitchen," when screened against flies, makes a pleasant summer dining room or living porch.

The farmhouse design should keep to simple roof areas, as the more valleys and hips there are, the more leaks there are liable to be. The steeper the roof slope, the better the drainage and, consequently, the more lasting the roof. Roof areas exposed to the north last longer than those to the south, hence for durability irregular roofs should have their long slopes and greater areas to the north.

A house should be typical of the life of the people who live in it; hence, the farmhouse should be unpretentious, and devoid of "fussy" features. The farmhouse is not the house of the country place, nor the suburban home of the city worker, nor the summer cottage for vacation time only; and it cannot be properly designed or planned in keeping with the requirements of any of these, though all are houses in the country.

In size and general outline, the farmhouse should be impressive enough to harmonize with the importance of the estate. It should be large enough to provide everyday comforts for all the members of the family, and should include reserve accommodations for times of extra demand. It should include a spare room, because the traditional hospitality of the farm home should not be lost, and the comfort of the family should not need to be disturbed every time there is an overnight visitor. The house should not be made large simply to look imposing, and it should not include rooms and features not demanded in an ordinary year's exigencies. Too large a house puts the investment out of proportion to the farm value, and ties up capital in useless buildings; but, on the other hand, there should be no suggestion of niggardliness about the farm home. Generous provisions for comfort, but nothing superfluous or wasteful, should mark the farmhouse plan.

Proportional cost of farmhouse. The proper proportion between the amount to be put into the farmhouse and the total farm investment, or the relation of the cost of the house to the farm income, is a very difficult problem.   It involves the personal feelings and tastes of each farmer regarding the quality and appearance of his home establishment.   Moreover, the actual monetary return from the investment in a good house and domestic conveniences, while a very real thing, will not be as high as the return from capital put into other equipment.   This is because the benefits and influences of a home cannot be measured in dollars and cents.   Where surveys have been made, it has been found, in a majority of cases, that the house represents the net farm income for 1 year.

Whatever the design chosen, the farmhouse should be simple and dignified. It should be set in neat, well-kept surroundings and guarded from neglect and misuse.

The design should not be too severe and inflexible; else, it will be out of keeping with the character of the country.   The farmhouse should have stateliness without austerity, dignity without rigidity.   It may well be of the rambling style, to secure light and air for all the rooms, as space is plentiful and the outlook open in all directions.   Indeed, the unrivaled advantage in locating and designing the farmhouse is ampleness of space and unblocked outlook.

General Principles Governing the Selection of the Style of House

Some broad principles may be laid down governing the style of houses suitable for certain sites; but this is a bit dangerous, because some striking peculiarity of a site may so dominate it as to reverse the usual laws, and any attempt to apply general rules without a thorough comprehension of modifying conditions may bring quite the opposite result from that intended.

Farmhouses should be two-storied.   In general, however, farmhouses should be 2-storied, or should have tall, steep gables, or long, sloping roofs or, in some way, present long, vertical lines.   Such lines blend with the effects of trees and hills.   A low, flat house in a rolling, hilly country appears dwarfed and squatty by contrast with the dimensions of the hills and the massive, towering effect of tall trees.   Clearly, a simple, flat bungalow is not suited to crown a stately hilltop.

On the plains, the same law of emphasizing the natural lines of the landscape holds; so that here architectural features should be handled so as to produce horizontal lines.   That is, roof areas should be long horizontally and not be broken by dormers; windows may be grouped; and even the manner of using the wall material, whether wood or masonry, may contribute to the horizontal effect.   A house may very well be 2-storied without conflicting with this requirement, and the rambling house may very easily be handled so as not to violate the need for vertical effects in the broken country.

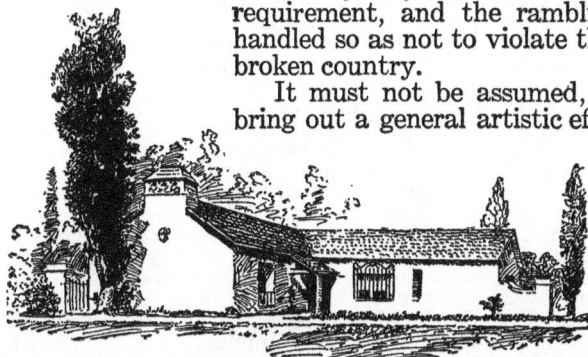

It must not be assumed, of course, that in the effort to bring out a general artistic effect, *all* of the outstanding lines of a group should be vertical or horizontal; for, quite to the contrary, there should always be a minor note of vertical lines to balance a mass of horizontal ones, or just the reverse. The flowing, sweeping lines of plain and prairie need the balance produced by the perpendiculars

FIG. 473.   The Spanish Mission type of architecture is well suited to, and has been largely adapted for, farm conditions in the arid sections

FIG. 474. The so-called half-timber construction on a basis of stone or cement is of English origin but is being widely used in this country.

of trees, towers, or windmills. The base lines of hills, in general parallel with the sky and horizon lines, give the needed balance to the up-and-down effect.

The horizon line of the landscape should serve as a guide to the sky line of the house, so that the general roof lines and contours simulate those of the surroundings. It is best, however, to attempt only the simple and obvious in roof lines unless the services of a competent architect are employed. The roof presents the most delicate problem for the artistic power of the architect, and its successful handling requires the exercise of his rarest skill and ingenuity. It is literally and figuratively the crowning achievement of the complete artistic whole.

Architectural Styles

The Colonial. There have been developed in the different sections of the country characteristic designs of dwellings, each having qualities of especial appropriateness to its native section. Each in some subtle way is an expression of the civilization and home spirit of its community.

The austere, inflexible New England Colonial style is a very difficult one to adapt to the needs of varying farm conditions throughout the country. The more graceful and gracious Southern Colonial, with its curving porticoes and stately columns, while a shade more adaptable than the New England Colonial, is impractical because of size and arrangement for any but a very few farm homes. The Spanish Mission style of southern California is distinctly local, though very charming where the climate and landscape make it suitable.

The Dutch Colonial style is a very successful modification of the New England Colonial. It is a cottage type of architecture for a 2-story house. The cottage effect is obtained by extending the roof down to the usual height for a 1-story building. The height necessary for the second-story rooms is obtained by using the gambrel, or double-pitch, roof, or by employing long dormers breaking through a single pitch from ridge to plate. The detached dormers only should be used with the gambrel roof.

This is a very beautiful domestic style, and, as it is more flexible in plan and design than the other types of Colonial, it is adaptable to a wider range of local conditions. The balancing of vertical and horizontal lines harmonizes this type with many varying landscapes.

Both the gambrel and the single-pitch types of Dutch Colonial are most excellent ones for farm use. Their roof systems are easily adaptable to all outbuildings; and both types have a brooding, homey air, coupled with the impression of complacent self-sufficiency, typifying some of the finest qualities of country life.

The half-timber style. The half-timber house is among the best styles of architecture for farm use. Perhaps no style better meets the demands for flexibility in farmhouses than this one. Half-timber work to-day only simulates the original method of construction given this name. In England, where the style originated many generations ago, the walls were constructed with timbers—some upright for supports, and some inclined for braces—and the spaces between these were filled with masonry. As now built, half-timber work consists, in part, of stucco walls paneled with boards. In a sense, this is a sham; but the varying line effects, which may be produced by skilful handling of the timbering, blend with so many varying landscape conditions that the style has a value by which it persists in spite of this drawback. When only vertical paneling is introduced, the type should not be used on treeless plains.

This style is wholly free from the restrictions as to symmetry and spacing which control the construction of the Colonial types. In these, the floor plans are made to conform to the windows and the gables and the entrance. The front door is exactly centred; all of the windows are of exactly the same size, and are exactly spaced with relation to the door and one another. In the half-timber house, the floor plan can be developed with sole regard to convenience and comfort; the chimneys can be placed where most needed; the windows can be high or low, large or small, grouped or scattered. Its well-braced stucco walls offer more than ordinary resistance to heat and cold, winds, and fire, and make it suitable for a wide range of climatic conditions. It is comparatively inexpensive, as the first cost is reasonable; and the upkeep cost is small, since weather conditions have but little effect upon its surface.

The bungalow. The bungalow is a widely used architectural style for farmhouses. It is a southern type especially suited, in this country, to the Southwest, where it has been developed and is very popular. It has long since broken over the boundary of the Pacific West and Southwest and is now the most popular of small houses in practically all of the states, regardless of climate or topography.

This wide popularity and ready acceptance are unmistakably due to innate merits. Chief among these is flexibility in plan, construction, and design, and its consequent adaptability to a wide range of conditions. It is essentially a country type of house. It is out of place in cramped, shut-in quarters. It belongs where it can open to the view and the breezes on all sides. It can be built of wood, brick, stone, or stucco or combinations of these; it can be constructed so as to provide the most rigid protection against severe winter conditions, or be made in keeping with the needs of milder climates; it can be of one story or 2-storied. It can undergo almost

FIG. 475. The bungalow is not the cheapest type of house to build; but it is easily cared for, and fits in admirably with the needs and conditions on many farms.

endless modifications without giving up its distinctive character or losing some of its most desirable features, such as the low sloping roof, the extended cornice, and the wide veranda. It offers better opportunity for individuality than the more pronounced styles of architecture. Its broad adaptability to many kinds of building sites eases the problem of the general artistic result. It can be made an inexpensive style, too, especially if designed with an upper story, which will lessen the cost of foundation and roof. The compactness of the floor plans, the omission of many partition walls, the plainness of finish, and the absence of ornamentation reduce to the minimum the cost for the interior. Its built-in fittings secure many conveniences within a small space; and these fittings cost less than a corresponding amount of furniture. The completed bungalow is almost furnished.

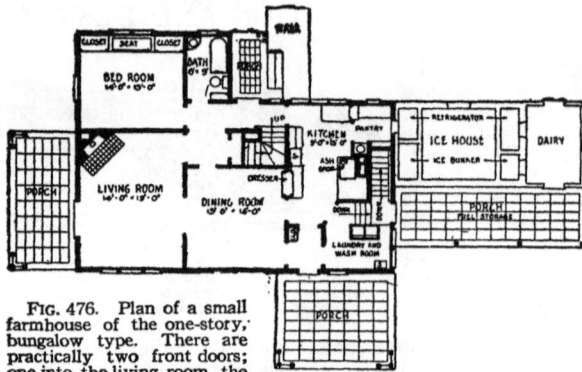

FIG. 476. Plan of a small farmhouse of the one-story, bungalow type. There are practically two front doors; one into the living room, the other into the hall, kitchen and wash room. Note the dairy and cold-storage house reached via the covered porch.

CHAPTER 30

The Farmhouse: Its Construction and Arrangement

By RUBY WESTLAKE FREUDENBERGER *(see Chapter 29). The construction and design of any house—and especially a farmhouse—are inseparably tied up with its location and arrangement and the materials of which it is made. Each of these details has a very important bearing upon the efficiency, as well as the comfort and happiness, of those who live in it. Fortunately, therefore, it was possible to have both aspects of this subject treated by one authority—an authority, moreover, who not only knows its principles, as a result of study and observation, but who also is familiar with their application, as a result of practical farm life and the varied experiences such life affords.* —EDITOR.

T HE chief aim in the construction of the farmhouse should be the production of a substantial, solid, weatherproof building. In every part, emphasis should be laid upon utilitarian features; and the money invested should go for the genuine essentials of stable construction, and not for showy, shoddy elements or attempted ornamentation. The farmhouse stands more alone in its buffetings by the elements than does the town house. Usually it has no near buildings or adjoining walls to shield it and help bear the strain of its weight. It is open on all sides to the straining, twisting force of the wind. It must provide its own resistance to cold and heat, dust and rain, sun and hail. Also, the danger from fire is greater in the country, because of the lack of fire-fighting facilities. Construction features designed to withstand these peculiar exposures should be introduced.

Construction Items Requiring Special Emphasis

Because economy is so important in building the farmhouse, only material and workmanship of good grade should be employed. The farmer can ill afford to pay the labor cost for flimsy construction and the putting in of poor material, soon to be repaired or replaced. A high first cost is cheaper in the end than repetitions of smaller outlays.

Emphasis should be laid upon such features as deep, solid foundations, double floors properly laid, fire stops, well-built chimneys, deep cornices, good plumbing, and ample guttering. All of these have a direct bearing upon the comfort and health of the family as well as upon the durability of the house.

326

Foundation walls and not piers should be used. The house perched upon piers is unsightly and unstable-looking as well as uncomfortable. The extra fuel and labor needed to warm and keep clean such a house would soon overbalance the saving by the omitted underpinning. Cold floors and cold feet are inevitable with such construction; for the cold winds sweep unchecked under the house, and, in cheaply built houses, a thin board floor, often with cracks in it, is the only structural protection provided.

FIG. 477. The money saved by using piers instead of a foundation will be spent in keeping the house warm. The best plan of all is to include a real cellar.

Necessary measures should also be taken to guard the foundation and basement wall from ground water and dampness.

Double floors afford protection against cold and heat, dust and wind. The subfloor should not be broken by partition walls, and it should be laid diagonally, so as to break joints with the top floor. The joists under the floor should be nailed together, where they lap, so as to make continuous ties across the building. A floor so constructed serves as an efficient wind brace for the building as well as a fire stop between the stories.

Fire precautions. It is especially important that some method be used to fire-stop every enclosed space in the framework of the farmhouse. Help for fire fighting is usually a long way off in the country, and every means of checking a fire in its early stages should be provided. Confining it within the space of its origin for even a few minutes, until help arrives, may mean the saving of the building. Fireproof guards should be placed about stoves, ranges, furnaces, and other containers of fire. Chimneys should be built from the ground up, not put on wooden shelves. They should be straight, not pulled over to extend through the peak of the roof. It is desirable, for the sake of the better draft, to have the chimneys built to the highest point of the roof; but this should not be secured at the risk involved in the inclined chimney.

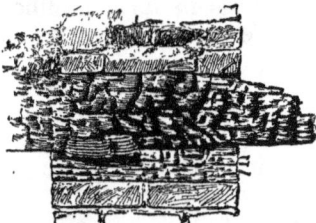

FIG. 478. Drawing (from a photograph) of poor, unsafe floor and chimney construction that is nothing less than criminal carelessness.

A low, wide cornice shades the windows and walls of the upper story, protects the side walls from rain and the wear of the elements, and carries the roof water away from the ground near the house. Ample guttering, properly put on and screened, in connection with the wide cornice takes care of the drainage from the roof and provides a supply of clean, fresh water for household use.

Good plumbing a necessity. Too much emphasis cannot be laid upon the necessity for good plumbing. The housewife who stands for hours at a time inhaling the poisonous gases from an untrapped drainpipe and sewer is more exposed to disease than if she had no kitchen sink and no system of plumbing. Safe plumbing for bathtubs and water-closets cannot be done inexpensively by present methods; and, in the effort to keep down the cost, farmhouse plumbing is often intrusted to unskilled workmen, with most disappointing results. A thorough understanding of the requirements of a sanitary equipment and a rigid adherence to these demands, are necessary for the installing of a safe plumbing system. Cheap plumbing is an expensive economy and not worth the risk.

FIG. 479. Unsafe roof and chimney construction is worse than interior carelessness simply because its consequences are harder to fight. Keep rafters away from the bricks and everything in good repair.

Exposed floors, ceilings, walls, and roof should be made so as to prevent transmission of heat both from within during winter and from without during summer. This is not necessarily expensive, and is desirable for both winter and summer. It is very probable that the saving in fuel would soon pay for the cost of insulating the walls of north, northeast, and northwest rooms, which, without such protection, would require from 10 to 20 per cent more heat than south rooms. Windows may be made extra tight by double-glazing. The amounts of heat that are transmitted through windows that are single and those that are double-glazed have been computed and found to be as 10 is to 6. At least, the north and west windows of much-used rooms might be economically double-glazed.

Good ventilation must be provided. Farmhouses are notorious for being poorly ventilated. Paradoxical as it may seem, it is regrettably true that fewer farm dwellings have a constant supply of fresh air in them than any other class of homes. Where there are open fires, and the walls, windows, and doors are especially favorable for air leakage, a reasonable supply of fresh air gains entrance in spite of utter disregard of provisions for ventilation. Some system for bringing in fresh air from the outside and allowing the escape of the foul air from the inside should be provided as an adjunct to the heating system of the house. The great number of cases of tuberculosis in farm families attests the fact that the outdoor occupations of country life do not obviate the need of good ventilation in the house.

Attic ventilation will make the house cooler. The circulation of air between the roof and the ceiling carries away some of the heat. Ventilators in the roof and in the soffit of the cornice should be provided for this. Also, there should be openings in the foundation to secure free circulation of air under the house, to prevent damp, moldy floors, and decay of the floor joists. Dampness, darkness, and stagnant air provide the most favorable conditions for dry-rot.

Choice of Material

The choice of material for the building of the farmhouse will be largely affected by local conditions. In building it is always best—from considerations of both beauty and economy—to use native materials.

Wood. Wood is by far the most generally used material for farmhouses. It is more widely distributed naturally than any other material; it is easier and cheaper to transport; it is more pliable and adaptable to various forms and uses; and, consequently, it is more easily handled, and its successful use does not require so much skill and training as other materials on the part of workmen. Wood lends itself to certain artistic effects which are difficult to obtain with other material, and the designing of a wooden building suitable for a dwelling is not quite so delicate a task as that of adapting masonry materials to this end. All of these things combine to make the wooden house the cheapest, for the same grade of workmanship, finish, and structural excellence.

On the other hand, wooden houses are less durable than those of brick or stone or concrete. Still the wooden house is not necessarily short-lived. Many may be found in this country to-day which have been standing for more than a century, and some for more than double that time. With proper-sized framing timbers, put together with due regard to strains and stresses, wooden houses of great strength and endurance may be constructed. Indeed, a well-constructed timber house will be more lasting and satisfactory than a poorly constructed one of masonry.

Because wood is the simplest of all building materials to handle and adjust to the needs of construction, skillful and careful woodworkers are more plentiful and less expensive than good masons. For cheap and medium-priced houses, wood is to be recommended.

The most serious objection to wood as a building material is danger of destruction by fire. In spite of every precaution and safeguard, this is always a menace. While it may be reduced, it cannot be removed—at least, not until fireproofing processes have become further perfected.

The necessity for painting, or in some way weatherproofing, is another disadvantage of wooden buildings. This is an expensive item and one that needs attention every few years. In some climates, wooden buildings so soon become shabby and worn in appearance, and require such constant attention to keep them looking well, that it is hardly desirable to use them.

The cost of a wooden house is subject to greater variation than that of one of any other material, because of the possibilities for greater variation in the quality of its construction. It may be either a flimsy, inferior affair or a very valuable, high-grade structure. There is a smaller range of quality in masonry than in wood construction.

Where local materials can be used to a large extent, and the rough, heavy work, such as hauling and excavating, can be done with farm labor at farm-labor prices, the cost may be much reduced. Framing lumber from country sawmills will cost only about half as much as the ordinary town lumber markets demand. It is possible, where the farmer superintends and manages the entire job himself, to build a frame house at a cost of about $300 a room, or from 8 to 10 cents a cubic foot of inclosed space. This gives a well-built, plain house with medium-sized rooms, having all of the accessories needed for comfortable living, but devoid of unusual or fancy features. It includes shingle roof, good cornice, sash weights, and window blinds. A slate or metal roof will add one-fourth of a cent per cubic foot to the cost.

Brick. Brick comes next to wood as a widely used material for country houses. At least, this was formerly the case; but stucco is now rapidly moving up in popularity. Brick is more costly than wood or stucco. It makes a durable, substantial building, warm in winter and cool in summer. In proper construction, it will meet most satisfactorily all of the requirements for a superior home. If well designed, it is suitable for almost any style of house, whether bungalow, Colonial or in combination with half-timber or stucco. A brick dwelling has an air of permanence and substantiality, putting it in especial harmony with country homestead conditions. It has greater distinction than a wooden house designed and built with the same care and skill. It is also suited to a wide range of climatic and landscape conditions. It is especially beautiful in sections where there is much foliage and many rolling grassy slopes, the constrast of brick with the richness of evergreens in the landscape being most pleasing.

The upkeep cost of a brick house is small. Such a house does not require painting and it is not easily susceptible to the wear of the elements. It does not require either weatherproofing or fireproofing. If given a slate or tile roof, it is practically fireproof from the outside.

In common with all ordinary masonry dwellings, the walls of the brick house should have a damp-proof course, above the ground line and below the floor joists, to intercept the rise of ground water, which, without this impervious layer, will rise to a height of several feet above the ground. There should also be one or more air spaces in the walls, to prevent the water absorbed by the masonry from penetrating to the walls and plastering. It is better if this space be near the outside of the wall, behind a veneer of masonry, as this will prevent the rain water from penetrating the greater depth of the wall.

The costliness of the brick house is the greatest obstacle to the use of brick as a building material for the farmhouse. The labor cost for construction is greater than for wood; and, if the brick must be transported, even a short distance, the freight charges soon amount to more than the cost of the brick. If local brick is available, however, this objection is overcome. Brick farm dwellings will cost from 2 to 3 cents more per cubic foot than wooden ones. The brick-

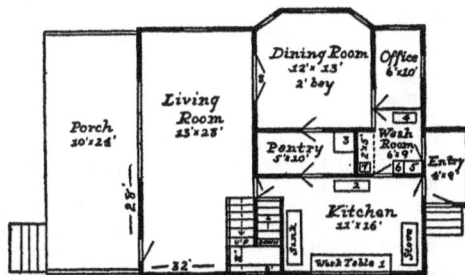

FIG. 480. First-floor plan of a set awarded first prize in a practical farmhouse contest conducted by the Iowa Agricultural College. 1, Built-in table cupboards and bins below; 2, wheel table; 3, iceless refrigerator; 4, wash basin; 5, chimney; 6, clothes chute, second floor to cellar; 7, toilet, screened. For second floor and cellar plans see figs. 481 and 486

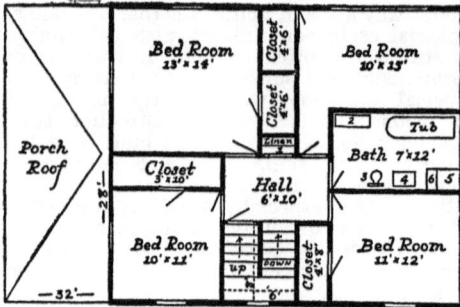

coated house has many of the merits of the all-brick one, and costs slightly less. However, the fire hazard is increased somewhat in this type of house.

Stucco. Stucco is a most satisfactory building material for dwelling houses and is fast becoming very popular. It is well suited to the farmhouse, because its cost is moderate; it is as nearly fireproof as brick or stone; it is warm in winter and cool in summer; and it is adaptable for use in all farm buildings, making a pleasing, harmonious farm group.

A stucco house is but little more expensive than a wooden one, especially if constructed with the outside plastering on metal lath attached directly to the studding, thus omitting the wooden and paper sheathing used in frame houses. In this method, before the interior lath is put on, the outside wall is back-plastered, which makes it about an inch and a half thick and completely incloses the metal lath. This gives a very rigid construction, and effectively braces the framework of the house. A more expensive construction is to put the usual sheathing on the studding and to use the stucco covering merely in place of the weatherboarding. Many authorities consider this an unnecessary expense, as the stucco wall is entirely efficient without it, and there is no practical reason for its use.

In some sections, the stucco house will cost no more than the wooden one. Where the metal lath is used for both inside and outside plastering, giving a permanent construction, it is found that the increased cost over the wooden house is only about 3 per cent. This addition would be more than consumed in an extra coat of paint for the wooden house. Sand and gravel for the concrete and stuccowork may often be obtained near at hand, and at no cost except for hauling. If the materials used in the stucco house are up to standard specifications, and the work is of good grade, the upkeep cost of the house is very small.

Stucco is a most responsive material and is readily adaptable to any architectural design. It is suitable for large houses or for small ones, for very simple ones or for those of a more massive, and pretentious character. It combines well with other building materials; it is susceptible of many different finishes, and it can be varied as to color and texture so as to obtain any desired result.

Stucco houses are suited to a wide range of landscape conditions. The possibilities for many color combinations in their development fit them for particularly beautiful results among the hazy pinks, purples, and grays of the western mountains and plains. Fittingly designed and properly constructed, they are attractive in any surroundings, and they have a dignity and reserve found in few other houses. They age gracefully, taking on with time a mellowness of tone which adds to their charm.

Stucco over hollow tile makes a satisfactory and durable construction, but is somewhat more expensive than plaster over wooden framing. These hollow blocks provide an air space in the walls, which helps to make the house less susceptible to outside changes in temperature. They are also of some benefit in preventing the passage of moisture through the walls; but they do not entirely obviate this difficulty, as they are popularly supposed to do. The method of manufacturing such blocks makes it necessary to use a smaller amount of water than is required for maximum density, and, as a result, the blocks are very porous.

Concrete and stucco are especially suitable for porch construction. Since porch floors, supporting walls, and steps are subjected to great exposure to weather, they should be of permanent construction. Materials particularly satisfactory in porch design and building have an added value for farmhouse use.

Stucco is also effectively used in remodeling farmhouses. This is done by what is called "overcoating," or simply applying a stucco finish over the old walls. This may be put on over the old weatherboarding, or the old covering may be removed and the metal lath applied to the sheathing or studding. This is an economical and satisfactory method of making a new house out of an old one. With the framework of the house in good condition, the result will be a building attractive in appearance and having all the essential qualities of an original stucco house. This method of remodeling eliminates the cost of painting and outside repairs.

Stone. The expense of construction with stone is so great as to be prohibitive of the use of that building material for farmhouses, if the cost of the house is to be kept within its proper proportion to the total farm investment. Even with the stone at hand, the expense of quarrying and dressing it, combined with the slower and more difficult construction required, will make the total building cost about 25 per cent above that for a wooden house. Field stones imbedded in cement make a substantial wall construction, cheaper than dressed stone. However, there are only occasional localities and conditions where this can be used. It is picturesque in effect, and demands special designing and combinations with other materials. It cannot be considered a standard or typical wall material.

Combinations of materials. Two or more of the various building materials may be advantageously employed in the same house. These combinations may be dictated by the architectural effect desired, by the outlay permitted, or by the available supply of the different sorts of material. Brick combines so well that it is nearly always found where more than one medium is used. Brick and wood form a very acceptable combination, where it is necessary to keep the cost below that of an entirely brick house. Brick and stone combine well, but make an expensive building. Brick and stucco give a good combination, and are hardly more expensive than brick and timber.

Recoating with brick for the lower story and with stucco or half-timber for the upper one is quite practical, and the results are good artistically. Brick and shingles, hollow tile and shingles, stucco and shingles are combinations often found. They are all satisfactory, and the determining factor of their choice should be local conditions affecting the supply of material or labor, or personal preference as to appearance. Combinations including shingles are especially good for Dutch Colonial designs, as in these the second-story walls are really roof surfaces. Different materials should be combined only in large or medium-sized houses, as the dividing lines break up the surface areas and decrease the apparent size.

Things to Remember in Planning the Farmhouse

The plan of the farmhouse must be based upon the daily program of the farm family. This differs in many essential points from that of town or city dwellers, and the differences should be kept constantly in mind in planning the farmhouse and its accessories. The demands made upon the time and energy of the members of the farm household are controlling factors in fixing their mode of living and in determining the material needs of their home arrangements. The chief business of farmer folk is work; and this usually injects itself, in some form or other, into all the hours of the day. The house is an all-day workshop. The country housewife cannot separate her hours of domestic duties from those given to social exactions or devoted to recreation and culture, as can the town woman. These all go along intermingled through the entire day. For this reason, the parts of her house devoted to these ends cannot be distinctly defined and set apart, as in the town house.

Consider the housewife. The plan of the farmhouse should look to such arrangements as will enable the housewife to meet the varied demands upon her with the least expenditure of time and energy. Very often the entire work of a farmhouse establishment falls upon one woman; if she have help at all, it is but occasional and meager. Only the most complete equipment and the most skillful arrangement of conveniences will make it possible for one woman to carry all of this and successfully meet all of the demands upon her vitality, strength, and resourcefulness. The plan of the farmhouse should be drawn to this particular end.

The numerous tasks and duties of the country housewife consume so much the larger part of her time and powers that, for her, the usual proportion between work and social pleasures for women is reversed, and so, also, is the relative importance of the living and service parts of her house. Her parlor is for occasional use only; her kitchen is the centre of her activities, and is usually occupied from dawn till dark. The house arrangement should be compact, with few halls and passageways, to save her steps and to enable her to carry on, or oversee, activities in different rooms at the same time. She may drop down in an inviting rocker in the living room and read the paper while waiting for the men to come from the fields for dinner, if the living room and kitchen adjoin, enabling her to keep an eye on dinner while waiting.

Use porches as much as possible. The farmhouse should be so planned that the family may live more on porches than in closed rooms. Every farmhouse should have a dining porch and provisions for outdoor sleeping. The dining porch, of course, must be screened against flies; and this will also shut out the dogs, cats, and chickens. This porch will, probably, very soon become the living porch, also. Many farmers, in planning their houses, omit porches because of the expense. It is true that the cost of a porch is about that of a room; but the service and benefit derived from a porch is not equaled by that of any room except the kitchen. The porch, fitted with hinged or removable sash to be used in cold weather, makes a comfortable and most attractive living room in all but very severe weather. With large inclosed porches, some rooms may be omitted or reduced in size, so that the porch need not really add much to cost of the house. The porch extending from the house and exposed on 3 sides gets more of the summer breezes than the built-in kind; but it costs more, and is less serviceable in protecting against rain and wind and sun.

Fig. 482. The farmhouse with plenty of porch space offers exceptional advantages, both in summer when the porches may be screened and in winter when they may be glassed in.

The recessed porch between two extensions of the house, or the inclosed one which is made an integral part of the house plan, affords all of the delights of outdoor living without the exposure to wind, sun, and insects accompanying the use of the open porch.

The farmhouse should be completely and effectively screened against flies. This is more important for the country house than for any other, if there can be comparative degrees in anything so imperative. The preparation and handling of food is going on so much of the time in the farmhouse, and occupies such a large and important place in the labor program, that flies are attracted in unusual numbers, so that screening is particularly necessary here. Indeed, the dining porch is utterly impractical unless effectively screened. If the only entrance to this porch be from the *inside*, the exclusion of flies will usually be more complete. Such an arrangement might work greater inconvenience in other ways, however.

Relation and Equipment of the Different Parts of the House

Because the social life of the farm family is so largely dependent upon its own members, and because home influences and conditions mean so much more for country children than for those who have the varied associations and opportunities of town life, all of the arrangements of the farmhouse should be conducive to easy sociability. They should provide constant conditions favorable to the natural injection of this spirit into whatever is doing about the home. The different parts of the house should be so related and equipped as to invite to pleasant companionship, and to offer, naturally, in the day's routine, the pleasures and comforts of a genuine home life.

The living room. There should be a large, airy living room with an open fire. This room should be centrally located and on the general lines of passage to the other parts of the house. A very desirable arrangement is to have the other downstairs rooms opening from this general room. The members of the farm family are usually tired at night, and they will not go out of their way to find sociability or diversion, but, if these are at hand, their benefits are reaped without extra exertion. This room, if sufficiently large, may well be a combination living and dining room. This arrangement, too, avoids the expense of a partition wall, saves steps for the housewife, and provides pleasant surroundings for the family meals. It is especially practical in connection with a dining porch, because the porch will be in use during the season when there may be many farm hands to be fed.

It may be objected that it will be difficult to keep this room clean on account of so much passing through it and because of its many uses. It certainly will entail no more work to get rid of dirt from 1 room than to clear away the same amount scattered over 2 or 3. Besides, the point of attack for indoor cleanliness is outside the house. If there are walks about the doors and yards, and provisions for cleaning muddy or dusty shoes and clothing outside, or for leaving them in an anteroom, no great amount of dirt need be brought into the house.

A very satisfactory arrangement is to have a small parlor, or annex, opening into the living room by double doors, so that the two may be thrown together, when necessary. This will supply the reserve accommodations, ready and near at hand, for use on more ceremonious occasions, or when more room is needed. This will also be a comfort to the housewife; for the room can easily be kept clean and in order and ready for the chance visitor. It will also make possible a desirable degree of privacy in the entertainment of visitors or among the members of the family. This latter is a much neglected provision in country homes; too often it seems impossible to have converse with the farmer, or with any of the household, except in the presence of the hired man and of every member of the family.

The general living room should not be a bedroom: the quality of both is lost by this arrangement. Inconvenience and discomfort result, and the real centre of the home-life influences is destroyed.

The living-room-dining-room combination is a very satisfactory one for the farmhome, but the dining-room-kitchen combination is one to be unreservedly condemned. The only recommendation that the latter ever had was that it saved the housewife some steps. If the dining room opens directly into the kitchen, there is really very little to this claim, especially as the kitchen must be larger, to be used this way. Besides, the strength exerted by the housewife outside of the kitchen, to the end of maintaining dining quarters for her family, cannot be better used. Economy of labor should be applied to some of the less important demands upon her.

All meals should be eaten in attractive, congenial, uplifting surroundings, for we seem to feed our souls as we do our bodies. The atmosphere in which we take bodily nourishment supplies at the same time in some subtle way, a food element for our spirits; and its quality goes to the essence of our natures. Kitchens at best are cluttered and unsightly after the preparation of a meal, and at no time does the equipment present an inspiring or restful appearance. In the summer time, the kitchen is the hottest and most uncomfortable place about the house, and the family should not be asked to take their meals there.

The breakfast alcove. When there is no help and the family is small, the breakfast alcove, flooded with the morning sun, is a pleasing and labor-saving addition. This is especially convenient for winter use, when

FIG. 483. A sunny breakfast porch is a boon for the family the year 'round, and an added benefit in harvest time when extra hands are boarded.

coziness and warmth are at a premium. It may become quite comfortably warm, from the kitchen range, before the fire is well started in the dining room.

The stairways. The stairway should go up from the living room or from a front hall. There should be a separate back stairway, or a short flight leading to a landing of the front stairs, so that access to the upper story may be had from the back part of the house without infringing upon the privacy of the front rooms. The need of this is evident when we consider that the farmhouse must often accommodate the family, visitors, and the hired help all at the same time. The cost of stairs is not great enough to weigh against the convenience of having them where needed. A very good flight of stairs may be had for $100, and simpler ones may run as low as $50. Box stairs for cellar or garret need not cost over $20 or $25.

The right construction of the stairway is a very important matter but one very often disregarded. The stairs are often made to fit into a certain space instead of an adequate space being provided for hygienic stairs. There is, so to say, but the fraction of an inch in difference between an easy stairway and a difficult one. Stair-climbing even at its easiest, is hard physical exercise and very wearing on women. The steep stairs of many small houses are responsible for countless cases of illness.

The height of the step on all proper stairs is about 7 inches; it may vary an eighth of an inch, but not much more. The tread, *not counting the nosing*, is exactly 10½ inches broad; that is, the notch on the horse is this size. This is the best possible stair proportion for the average person. A steeper one is too steep, and needs an unnaturally great muscular effort; one less steep is tediously slow to climb. One rule that has been developed to govern the proportion of stairs is that the sum of the rise and the tread shall lie between 17 and 17½ inches. Since the natural stride will not vary greatly, the one dimension should not be materially increased without decreasing the other by the same amount. Another good rule is that if rise and tread be multiplied together, the result should not be less than 70 nor more than 75 inches. This ideal proportion for domestic stairs does not, of course, hold for all stairs.

The bedrooms. There should be, at least, one downstairs bedroom, for the convenience of the family. The farmer is more subject to night calls than the town resident, as he must see to the comfort and well-being of his stock and poultry besides attending to the numerous small interests about the farmstead. Two downstairs bedrooms, opening into each other, or connected by the bathroom or a narrow hall, are really better than one. If there are small children or old people in the family, this arrangement is very desirable. In case of sickness, it saves many steps for the attendant, especially if she is also the housekeeper.

The bathroom. The placing of the bathroom is a very difficult problem in any house having but one. It is especially so where there are both upstairs and downstairs bedrooms. The bathroom should be so located as to be easily accessible from the majority of the bedrooms. If the house be a 2-story one and most of the bedrooms are upstairs, the bathroom should be on the second floor. In this case, the downstairs bedroom would be the spare room, used only for the occasional guest or in case of sickness, and consequently, it would be best away from the bathroom. On the other hand, if the family bedrooms are on the ground floor, and the less frequently used rooms are up-

stairs, the bathrooms should be downstairs. This latter arrangement seems to give the greatest convenience, all situations considered.

Cleanliness and neatness are nowhere more important than in the bathroom. To make these easily attainable, the walls and floor finish should be plain, washable, and impervious to water. Ceramic tiles are extensively used for bathroom floors in all but the cheaper class of houses. They are laid in cement upon a slab of concrete. As all of the work, except the actual laying of the tile, can be done by ordinary workmen, the cost of such a floor is not great enough to stand in the way of its use in so small a room as the bathroom.

Hired men's quarters. Many farmers prefer to provide quarters away from the farm home for the hired men, and this is the most desirable plan, when practical. On the vast majority of farms, however, there come times, when it is necessary to have some hired men housed with the family; and a room for this need should be provided when planning the house. This room should have an entrance apart from the family rooms. On the large ranches of the West, bunkhouses are always provided for the ranch hands, and there is no thought of including them in the family arrangements. This is a direct outgrowth of the farming and labor conditions obtaining there.

The kitchen. The location of the kitchen, its arrangement and surroundings, present one of the most difficult problems of the farmhouse plan. The kitchen is the "power plant" of the home, it is the workshop of the housewife, and should be her exclusive domain. It should not be a general passageway for the coming and going of all who are about; nor should it be the washroom for the family or for the men.

The kitchen should be well lighted, preferably from the north, as this gives an even light. Casement windows are best; and they should be high enough to permit the placing of working tables or sink under them while

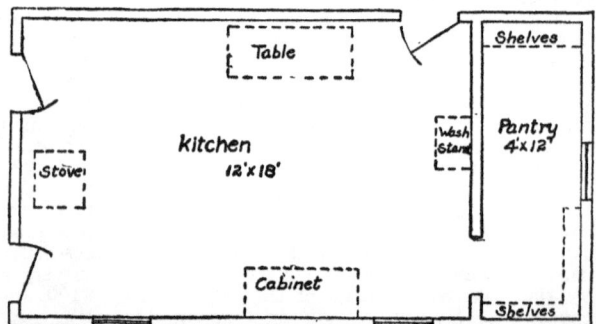

FIG. 484. A farm kitchen before its reconstruction and rearrangement. Its size and the scattering of its furniture mean miles of walking in doing the daily household tasks. Compare Fig. 485.

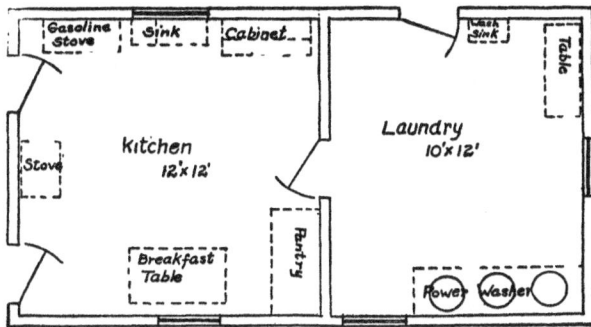

FIG. 485. The farm kitchen shown in Fig. 484, remodeled. Whether used as a dining room or not, the preparing and serving of meals is greatly simplified. The separate laundry prevents much crowding and confusion and provides space for canning and other emergency tasks.

escaping the whipping back and forth of the window curtains over the table top or in the face of the worker. There should be such a relation of its doors and windows as to provide cross drafts, to carry out the odors and smoke. If the windows are in the north, there should be a south door through which the breezes may come. There should also be a flue vent for carrying off the fumes. With these arrangements, the usual butler's pantry, interposed between the kitchen and dining room to prevent the odors penetrating to the rest of the house, may be omitted, thus saving steps for the worker. A built-in dresser or sideboard, with doors opening in both dining room and kitchen, permits prepared food to be passed to the dining room and the soiled dishes to be removed to the kitchen with a minimum of labor. This is also a convenient provision between the kitchen and the dining porch. The kitchen and the dining room opening directly into each other give the most convenient as well as the shortest route from kitchen range to dining table. An intervening pantry with two doors usually means two corners to turn, and this means an added expenditure of physical strength.

The kitchen entrance. The kitchen should be near the main entrance of the house and should be easily accessible from it. A good plan is to have a small entrance and stair hall, upon which open the kitchen, the living room, and the dining room. Then practically all of the passing in and out will be through this way, and the unused front entrance and the abused kitchen entrance will be done away with. The fact that nearly all approach to the farmhouse, by visitors and family alike, is to the side or back door is the logical consequence of prevailing conditions there. The kitchen is the centre of activity of the house workers and the place where they are the greater part of the time. It is as logical to go there to find them as to go to a professional man's office to see him during business hours. Also, the kitchen is the

room from which produce is loaded to be taken to market, and it is the unloading place for groceries and general supplies brought home from market; so that it is perfectly in keeping with local needs that the entrance near the kitchen should be the main approach to the house. This does not mean that the front entrance should be less attractive, but that the back one be more so; or rather, that they be combined into one both useful and inviting. However, the kitchen should have its own back door, opening into a small back yard, not approached by a driveway. This door should have a small work porch—a cool, secluded place, where many kitchen duties may be performed. There should also be a place to keep the garbage can and the swillpail, if the lack of more sanitary provisions render these necessary.

The popular kitchenette is not so well suited to the farmhouse as to other homes, unless it be helped out by a generous pantry, storeroom, and bin space. The country kitchen and pantry must provide storage space for large quantities of supplies for every-day use; otherwise, too frequent trips down and up the cellar steps will be required. With this provided for, the kitchen should not be large. For the greatest efficiency a kitchen should be only large enough to enable the workers to move about easily without interfering with each other. The sink, the table, and the stove should all be within immediate reach of the worker, or, at most, only a few steps apart. Much floor space, after all of the fixtures and equipment are placed, simply means unnecessary steps and wasted labor. The size of the kitchen must be determined, of course, by the number of workers it is expected to accommodate; but, in the average farmhouse, 12 by 14 feet will leave ample space, after all of the conveniences are installed and the built-in cupboards and dressers completed. The more compact the arrangement, and the more services one piece of equipment is made to fill, the less the work required.

The interior finish of the kitchen should be plain and simple. There should be no fancy moldings or ornate paneling, with grooves to catch dust and greasy vapors and thus add to the difficulty and work of keeping the room clean and sanitary. Where moldings are used, they should simply be plain 1 by 3 inch strips rounded at the edges. The walls should be covered with some washable material, either a paint that will stand soap and water or a commercial wall covering having a washable finish. Both wall and woodwork should be finished in a light color, preferably white. The floor should be covered with linoleum, or have a finish which is impervious to water and grease and dust, and which may be kept clean by light moppings. An unsurfaced wooden floor should never be used.

One level for all rooms on a floor. There should be but one floor level for all of the lower rooms of the house and the adjoining buildings. Similarly, there should be only one level for the upstairs rooms. Different floor levels, with short flights of steps between rooms, may make a picturesque architectural feature; but they do not belong in the farmhouse, where conservation of the housewife's strength is the controlling consideration. The refrigerator; the pantry or kitchen storeroom; the kitchen fuel supply; the well pump, if running water is not in the house—all of these should be on the kitchen-floor level. Going up and down steps requires much greater expenditure of energy than movement on a plane, and this needless expenditure of strength should not be imposed upon the farmer's wife.

The laundry. The laundry, whether connected with the kitchen or in a detached building, should have the kitchen-floor level. Many housewives prefer the detached laundry, and this certainly has advantages in fire protection for the house and in relief from the smell of suds through the dwelling. It is not so convenient when the laundry and cooking must be done by the same person, and it adds a few items to the construction cost. There is no good reason for putting the farm laundry in the basement. This practice is simply a town-house feature carried into country construction without thought as to its adaptability. It merely has the one advantage of being under the house roof and of saving so much expense thereby. But the proper degree of economy in the actual outlay for the farmhouse is only that which is compatible with the best use of the natural advantages of the country. The countrywoman should not be asked to go down and up the basement steps to do her laundry work, probably keeping an eye on the dinner cooking above at the same time, and then to carry the clothes, wet and heavy, up the steps to be hung out. The opporunity for drying the clothes in the open air and sunshine should not be foregone.

A paved court, under an extension of the house roof, serving among other uses as a drying yard in stormy weather, is a most comfortable and useful adjunct to the laundry, although not possible where the cost must be kept at the lowest.

The cellar. A necessity for the farmhouse is a large, dry, well-ventilated cellar, preferably under the house. It should be planned with the same forethought and study as that given to any other part of the structure, and every detail of its arrangement and construction should be carefully worked out in relation to the house plan. It should

contain a fireproof furnace room, large enough to contain a winter's supply of fuel. The space required for this will depend upon the size of the house and the kind of fuel used, but an average size is 14 by 16 feet. More storage room of every sort is needed about the country house than elsewhere. The vegetable and fruit room in the cellar should not be less than 16 by 16 feet for the average

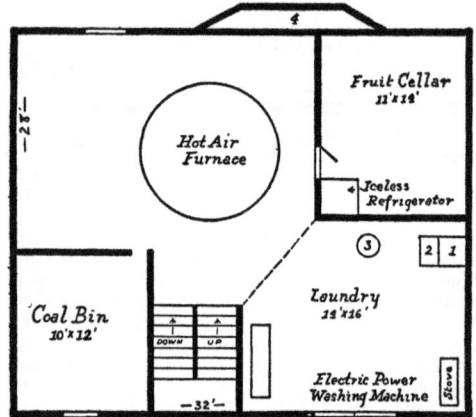

FIG. 486. Cellar plan of prize farmhouse (see Fig. 480). Note Mrs. Freudenberger's objections to the location of the laundry as suggested here.

farm. It should be fitted with hanging shelves and racks, and should be so protected from the furnace room as not to become heated from it. The cellar entrance should be under cover and easily accessible from both the inside and the outside of the house. The cellar stairs should have an easy slope.

Washroom for the men. A washroom for the men should be provided, both for their convenience and for that of the kitchen worker. This room should have an outside entrance, and passage from it to the dining room should be by way of a hall or porch, so as to avoid the necessity of the kitchen or living room being made a public passageway. Clothes hooks and hatracks should be placed here. The very dirty field clothes of the men may be exchanged here for more suitable ones for the table, thus saving the carrying of a great deal of dirt into the house.

The attic. A finished attic is unnecessary in the farmhouse. It is largely impractical as a storeroom, and the need of conserving ground space is not great enough to warrant the call to climb an extra flight of stairs. The attic space may be made effective in protecting the rooms below from heat, if proper ventilating devices are installed to induce air circulation through it. It is sometimes economical to place flooring over the attic joists for the heat it will save in cold weather.

Clothes closets. There are no more

valuable spaces in a house than the clothes closets, yet they are very often omitted from the farmhouse plan. Every bedroom should be provided with a fair sized, well-ventilated clothes closet. Proper ventilation will often prevent the dampness sometimes found in closets along outside walls, especially where the wall construction is poor. Closets need to be carefully planned, because it is the handling of the details of their construction with a view to the efficient use of all the space that keeps their cost down to a fair share of the whole. Ample closets and built-in dressers may often be provided by using spaces, which would otherwise be wasted, under stairways, low roofs, or in angles of irregular houses. Three-cornered closets, or those that project into the room, making extra corners, are not so desirable as those so built as to leave straight, entire wall spaces.

The fireplace. No one detail of the farmhouse will mean more for the cheer and comfort of the home than an open fire. All the weight of sentiment is in favor of the fireplace, for the hearthstone is the symbol of the home. Doubtless, sentiment has kept the fireplace in many houses where it was not needed, or was not adequate as a source of heat. Improvements in construction have eliminated many of the annoying defects found in the oldtime fireplace. A fireplace, properly constructed, is an efficient heating device and may take its place among the distinctly utilitarian features of the house. For greatest efficiency, chimneys for fireplaces should be placed against interior walls. This permits the heat absorbed by the chimney to be given out inside the house, instead of being wasted on the outside air. A fireplace designed for use rather than for looks need not be very large. For a small fireplace, an opening 24 by 30 inches is usual, and one 30 by 30 inches will heat a medium-sized room. Larger openings, however, look better, as well as serve better, for larger rooms. The flue opening should contract to about 5 inches in width, and a damper may be placed in this, to reduce it further for economical and efficient heating. There should be an offset, or shelf, just above the throat, to prevent down drafts from driving the smoke into the room.

Fireplaces are efficient ventilators, when in use, and will give valuable service without fires, if the dampers are open. As a decorative feature, the fireplace and mantel, susceptible of such a variety of treatment, have no rival.

How the House Should Stand

There are certain well-defined principles concerning dwelling houses and their position with relation to the points of the compass; and these are based upon considerations of sanitation and hygiene, as well as upon household routine and the fixed laws of the elements. It is a very simple matter to state these rules, but it is a much more complex problem to plan a house in which all of the desired objects can be attained. In general, the living rooms and porches should face the south and east. The dining room, if used also as the breakfast room, should have an east exposure; otherwise, it may very well be upon the west. The kitchen is best lighted from the north, but it should have a south breeze, if possible. It is most comfortable as a workroom, if placed on the west, so as to have shade in the forenoon. The less used parts of the house, such as halls, closets, storerooms, and stairways, should be on the north and west sides. In these positions, if against outside walls, they protect other parts of the house from cold winds and the hot afternoon sun, and allow the more pleasant exposures for the much-used rooms.

Building tradition and widely established custom to the contrary notwithstanding, a north porch has many merits. It makes a most delightful living room for many months of the year, receiving the early-morning and late-afternoon sun, but being shaded during the heat of the day. Not the least of its delights is the view that it affords of summer sunrises and sunsets. During the few midwinter months, when it is so exposed to the wind and snow as to be of little use as a living room, it protects the house from cold and saves heat and fuel.

Windows and outside doors. Windows should be generously provided for in the farmhouse plan, and they should be of medium size and fair proportions. They should be more numerous on the south, east, and west sides, so as to admit more sunshine and less cold wind. If placed too high above the floor, as was common in old-time houses, or made so narrow as to be little more than slits in the walls, as is sometimes done in cheap houses to-day, they are ineffective for the admission of sunshine into the house. If too large, they appear unwieldy and pretentious; for all farmhouse windows should be made to open, and this is impractical for very large ones. Their height should be so determined as to have the transom, or division between the sashes, come above the average eye-line, so as not to cut in two the landscape seen through the windows.

The grouping or non-grouping of the windows will be determined somewhat by the design of the house, and by the desired distribution of the light within the rooms. It is admitted, however, that grouped windows give more satisfactory lighting results than the same window area scattered. Casement sash are especially good for windows that are high above the floor. These need be only half as large as double-hung windows, to give the same amount of ventilating service. They are simple in construction, may be easily screened, and help to make a pleasing exterior.

Outside doors should be few in number and be placed on the warm sides of the house. They should open on to porches, or be protected by hoods or by small entrances, partly inclosed or not, but always covered. No outside door should ever be made without at least a small platform for it to open on and a roof extension to protect this and the steps. Doors opening on to inclosed porches are best protected.

The dooryard. The proper care of the farmhouse and its proper enjoyment are greatly dependent upon its immediate surroundings. The first necessity of these is a dooryard fenced against stock and poultry. This need not be very large, especially the back yard, but it should belong to the house alone. No yard or doorstep can be kept clean and attractive if overrun by chickens; and grass and shrubs and flowers soon give way to such invaders.

FIG. 487. The fireplace is a symbol of home and the family and as such should have a place in the farmhouse. But it is not an efficient heating apparatus

CHAPTER 31

Farmhouse Equipment

By K. J. T. EKBLAW, *Professor, and in charge of Agricultural Engineering, Kansas State Agricultural College. He was born on a farm in Champaign County, Illinois—the heart of the Corn Belt—and lived there for 21 years, attending country school, then high school and then teaching for two years. Later he studied engineering and architecture at the University of Illinois and at Yale. returning still later to teach agricultural engineering at the former institution. He has also lectured extensively before Farmers' Institutes and is author of "Farm Structures," "Farm Concrete," a number of station bulletins and many contributions to agricultural periodicals.—* EDITOR.

AFTER the site of the farmhouse has been selected, its design decided upon, the material of which it is to be built chosen, and its plan and general arrangements settled, there yet remains for consideration the general equipment of the dwelling.

The health, comfort, and convenience of the farm family and, indeed, the actual livableness of the farmhouse, are so largely dependent upon the methods or systems of heating, lighting, water supply, plumbing, and sewage disposal employed, that too much emphasis cannot be laid upon the importance of installing each of these in the best manner possible.

Heating

The modern farm home (except in the South) is incomplete without a central heating system. The fact remains, however, that a great number of homes are still heated by the more old-fashioned methods of fireplaces and stoves: and it is quite likely that both of these will continue in use for many years to come. The need for efficient heating systems has long been felt; and, with the rapid development of other modern conveniences for the farm home, there has been a coincident increase in the number of modern heating systems installed.

Fireplaces. Fireplaces, as a means of heating, are perhaps the most inefficient of all, though a century ago they were accepted as almost standard. The cost of fireplaces, as built in those days, was almost insignificant; for the entire fireplace could be constructed with common labor, which was extremely cheap, and, of course, the materials them-

339

selves were right at hand and could be obtained at almost no cost whatever. Fireplaces are also exceedingly wasteful of fuel; but, again, in pioneer days fuel could be obtained at no expense other than that of the labor involved in cutting the wood and transporting it to the home. The greatest disadvantage of the fireplace lies in its inefficiency; for, of the heat produced in the combustion of fuel, only a very small per cent is utilized in heating the room in which the fireplace is placed.

Fireplaces are still quite often met with, but they are designed mainly for the cheer which accompanies open-hearth fires. It is true that they may be made quite useful in the late days of fall and in the early days of spring, when the air within the house needs to be slightly tempered, in order to be really comfortable. Another advantage of the fireplace is the ventilation which it furnishes; most of the heated air goes up through the flue and, of course, fresh air must be drawn into the room to replace it. There are on the market special ventilating fireplaces in which a passage extends from the exterior of the house through the fireplace and opens into the interior above the fire; the heat of the fire naturally induces a circulation of air through this flue, and, as a result, warm fresh air is discharged into the room.

Stoves. The inconvenience of taking care of fireplaces, coupled with the great amount of dust and dirt attendant upon their use, led to the development of a heating device in which the fire could be inclosed within a receptacle and controlled to some degree. The first so-called stoves which were made to meet these requirements were re-

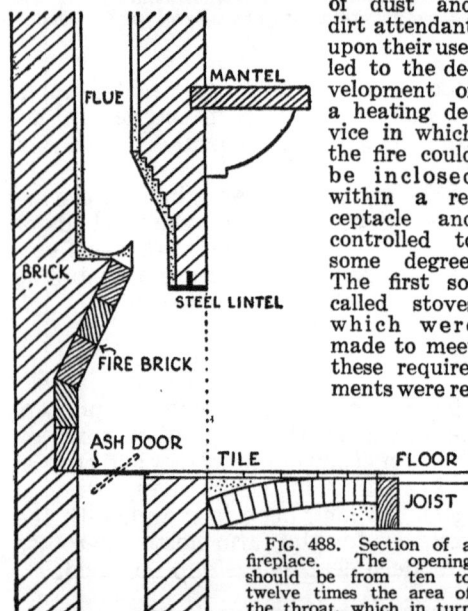

FIG. 488. Section of a fireplace. The opening should be from ten to twelve times the area of the throat, which in turn should not exceed that of the flue. Thus if the flue is twelve inches square (144 square inches) and the throat 36 inches wide, the latter should be but four inches deep and the fireplace opening approximately 42 x 44 inches.

ally enclosed fireplaces, and it required a little experimenting before a real stove was produced. The use of stoves is advantageous because the receptacle containing the fire is included almost entirely within the room to be heated, and the only communication with the outside air is through the smoke pipe; the loss of heat through the stovepipe is quite small when compared with the loss of heat from the fireplace. Aside from the heat lost through the smoke flue, all the heat radiation from the stove is utilized in warming the air of the room. A certain amount of ventilation will result, since, of course, air is required for the combustion of the fuel, and the stove must have a constant supply of a small quantity of fresh air, thus inducing a circulation.

The disadvantage of stoves is the inconvenience of struggling with the dust and dirt which always accompany them. Then, too, it is difficult to heat more than one room with a stove unless an extremely large one is used; and, in such cases, the distribution of the heat is certain to be irregular. The employment of a large number of stoves involves a corresponding increase in the labor of attention; and, where a large house is to be heated throughout, a considerable waste is likely to result.

Stoves are useful for certain purposes and, indeed, sometimes can hardly be replaced, as for water heaters, for laundry and laboratory purposes, and for heating small buildings in which the installation of a more elaborate equipment would be inadvisable from an economic standpoint.

Hot-air systems. The difficulty in caring for a large number of stoves, and the loss in economy resulting from their use, brought about the development of what might be termed the first modern heating system, which in reality was simply a large stove, but differing from stoves in that the heat distribution could be definitely controlled.

A hot-air system consists essentially of a heat generator surrounded by a tight sheet-metal case having an inlet for cold air and one or several warm air outlets by means of which the heat is sent through the various portions of the building to be heated.

The cold-air inlet may be within the house at some point near a door or where the degree of heat is of minor importance. It is sometimes connected with the exterior; and, in the best installations, the cold-air duct, as it is called, is connected with both the interior and the exterior of the house, and a damper is installed, so that the air supply may be obtained from either place. The location of the exterior opening of the cold-air duct should not be on the side of the house which is subject to the direct action of prevailing winds; for, in cold, windy weather, the force of the wind may be sufficient to drive more air through the system than is necessary, resulting in a waste of heat. Should this action occur, even when

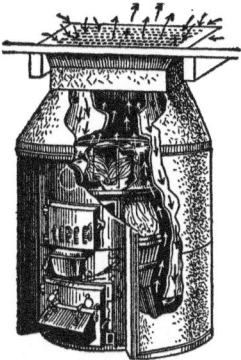

FIG. 489. Pipeless hot-air furnace in part section

the opening is on the protected side of the house, the d a m p e r should be turned so that part or all of the air will be taken from the interior.

The warm-air ducts leading to the various rooms are, in the best arrangement, t a k e n off the upper part of the casing from a section shaped like the base of a cone. This permits of a more direct flow of air, and eliminates the sharp turns which are necessary when the hot-air passages are taken off directly from the horizontal top of the casing.

Size of pipes is important. The size of the warm-air pipes is important. None should be less than 6 inches in diameter, and for an average room, 8 inches should be the minimum. Since a 10-inch pipe is about the maximum practical size, a large room should be supplied with two. The carrying capacity of circular pipes varies as the square of the diameters and is influenced also by friction, so that two 8-inch pipes are approximately equal to one 10-inch pipe in efficiency.

Warm-air pipes should be covered with insulating material, to prevent loss of heat. Connections between the vertical pipes, or stacks, and the rooms may be made by means of either floor or wall registers. The latter are better in that they are cleaner and interfere less with the placing of furniture; but, the installation of a wall register means another sharp turn in the pipe and a consequent reduction in the velocity of the air flow.

The total cross-section area of the cold-air duct should be about the same as the total cross-section area of all the warm pipes. The former, being one large pipe, has a lower frictional resistance than the combined warm-air pipes. It is usually brought down to the basement floor, and enters the furnace casing at the bottom.

For houses of the bungalow type, or those in which there are large openings between the rooms and open stairways between the floors, the pipeless type of hot-air furnace is easy to install and deservedly popular. Instead of several pipes it has a single duct leading from the top of the casing to a large floor register directly above. From this the warm air circulates through the house gradually cooling, sinking, and re-entering the furnace either through the same register and an outer jacket in the casing, or through a return air shaft built to open either into the cellar or on the main floor. In connection with this type one or more combination floor and ceiling

registers may be employed to carry surplus heat from one room to another above, and thereby improve the circulation.

A hot-air system is a desirable method of heating because it is easily cared for and permits full control of the heat distribution. Heat reaches the different rooms just as soon as the air starts circulating, which occurs within a few minutes after the fire is kindled. This system also removes any possible danger of frozen pipes or radiators whether caused by accident or carelessness.

The main objection to a hot-air furnace is that it is dusty and dirty in its operation, especially when the air supply is taken from inside the house. In the latter case, too, there is the possibility that the air, being used over and over, may become poor or "vitiated"; but this can, of course be easily prevented by ventilating. On the whole, hot-air systems are more efficient, cleanly and economical than a number of stoves of an equal heating capacity, but less efficient than the more elaborate steam or hot-water systems.

Steam-heating systems. The steam-heating system was the next stage in the development of modern heating systems. The essentials of a steam-heating system consist of a furnace, surrounded by a casing containing water, and a system of distributing pipes. In operation, the water must be brought to the boiling point before steam is produced. Since the system is inclosed, a still higher temperature will produce a small pressure, which will drive the steam through the distributing system. In the rooms

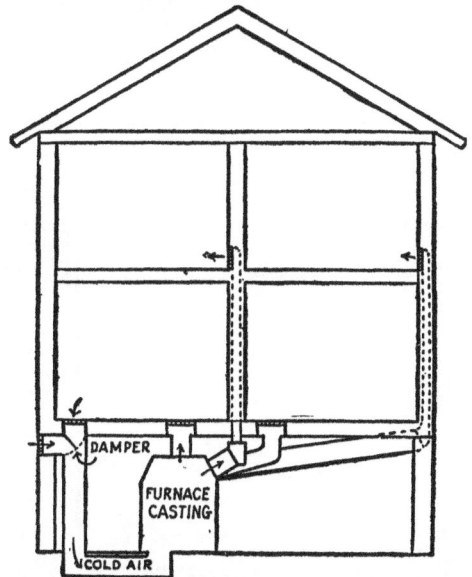

FIG. 490. Diagram of simple hot-air system. It is best to keep all vertical stacks in inside walls where a better circulation of air can be secured.

to be heated, the steam is circulated through radiators, where a part of it is condensed in giving off its heat; and this condensation is returned to the boiler and again vaporized.

Various types of steam boilers for heating purposes are manufactured, from small single-unit types to large horizontal tubular boilers of immense capacity. For residences, however, the most popular type is the sectional boiler, which is made of vertical or horizontal sections so combined as to form continuous heat and water passages. The number of sections can be adjusted to meet the heating requirements.

Different kinds of piping systems have been evolved to meet various requirements; the simplest system is known as the 1-pipe system. In this, a pipe is taken off the top of the boiler and carried to within a few inches of the cellar ceiling. From here it is given a pitch downward as it passes around the cellar until it reconnects with the furnace at the bottom. From this pipe are taken off branches or risers leading to the various radiators, being connected to the bottom of the radiator in each case. As many risers as needed may be taken off. The steam is forced up into the radiators; condensing there it returns as water through the same pipes that carried it up as steam.

For very large installations the more expensive and more complicated 2-pipe system may be used. This differs from the single-pipe system in that a separate pipe is provided for the return of the condensed steam. There may be a supply, or flow, pipe and a return pipe for each radiator stack, or a single supply pipe may carry steam to all the radiators, and a single return pipe,

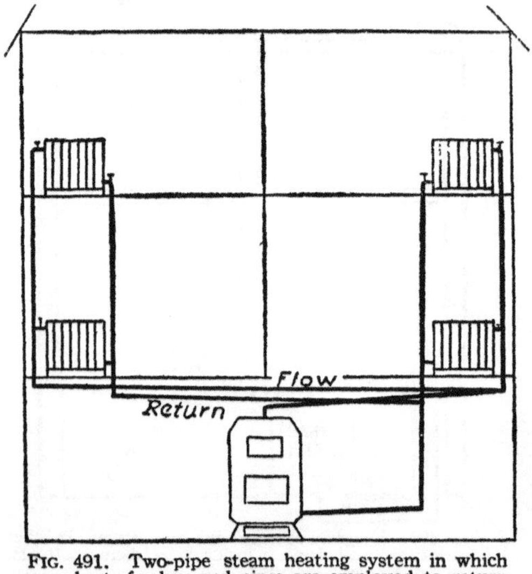

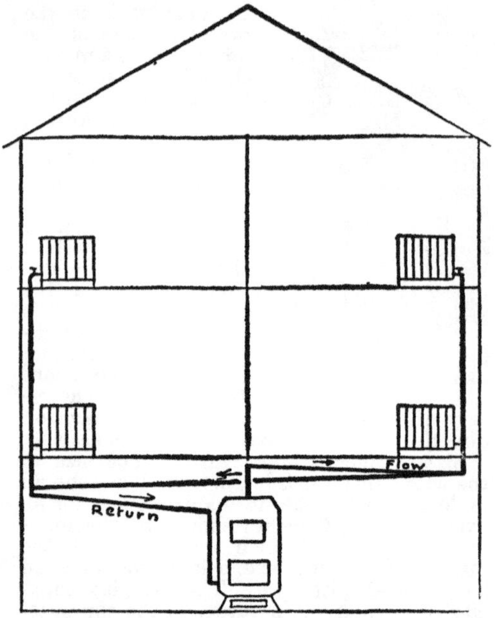

FIG. 492. Single-pipe steam heating system in which the condensed steam returns through the same pipes that first distribute it.

connected with all the radiators, may take care of the condensation.

Vacuum-vapor system. In a modification of the steam-heating system, known as the vacuum-vapor system, certain economy results from the production of steam under vacuum. In a system of this kind, a slight pressure is first developed in the boiler, which drives out the air within the piping through specially designed valves which allow the exit of air, but not of steam. When the air has all been expelled, the pressure is allowed to drop and the steam already in the pipe is condensed, forming a partial vacuum. With this reduced pressure, it is possible to generate steam at a temperature below 212 degrees, thus permitting the circulation of steam at a low temperature with a resultant saving in fuel. It is claimed that the economy effected by the vacuum-vapor system results in the saving of 15 per cent of the fuel used in an ordinary steam-heating system.

The steam-heating system is an exceedingly popular one. It is simple in its operation and, when once the water has been raised to the boiling point, the heating of the rooms is very quickly accomplished. On the other hand, as soon as the temperature of the water drops below the boiling point, the circulation of steam drops immediately and the rooms as rapidly cool. Also, unless the piping system is carefully designed and installed, there are likely to be annoying noises incident to the operation of the system. Even with great

FIG. 491. Two-pipe steam heating system in which a second set of valves and pipes are employed to return the condensed steam to the boiler.

care, knocking is likely to occur, especially where the boiler is overloaded, so that the supply of steam into distant radiators is insufficient; in an effort to furnish the heat required, the steam is condensed too rapidly, a slight vacuum is formed, and the incoming rush of steam to fill the vacuum causes knocking in the radiators and pipes.

Hot-water-heating system. In the hot-water-heating system, the boiler and all the distributing pipes are filled with water, and the whole body of water must be heated. Of course, the circulation of the water will begin as soon as heat is applied, because that portion of the water adjacent to the fire-box walls will become heated and will decrease in density and flow upward, and other water will flow down to take its place. In this respect, the hot-water system is superior to the steam system. It takes longer to get up heat with the hot-water system because a considerable period of time must elapse before the water is heated to a temperature high enough to insure adequate radiation.

The piping methods employed for hot-water systems are quite similar to those used for steam except that the 1-pipe system is usually not practicable except in very small installations, and then only when large pipes are used; for both the flow and the return must be carried on in the same pipe. Two pipes are therefore commonly used in hot-water heating, in order that there may be a complete and positive circulation; and connections are made with top and bottom of radiators to further facilitate circulation. One essential feature of the piping system for hot water is the inclusion of an expansion tank located above the highest radiator; for, since water increases in volume with increase in temperature, some means must be provided in which the expansion can be relieved. Sometimes the tank is closed, and provided with a safety valve regulated for a few pounds' pressure. The additional pressure simply results in increasing the temperature of the water, and this permits of the use of a smaller amount of radiating surface. As in the case of any safety valve, this part of the system should be inspected occasionally so as to be sure it is in working order.

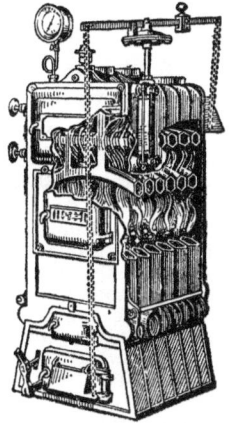

FIG. 494. Sectional type boiler for steam or hot water heating system

In operating a hot-water system, it is advantageous not to drain and refill it oftener than is really necessary. When water is heated a sort of sediment or "precipitate" is likely to form in it and this, accumulating upon the inside of pipes and radiators gradually reduces their efficiency. The oftener fresh water is added, the more of this sediment there is likely to occur.

Radiators. The radiators that are used in the steam and hot-water-heating systems are made either of cast iron or pressed steel, the former being much the more common. They are made in almost every conceivable size and shape, so as to fit every possible condition. They are made also in various designs, from perfectly plain ones to those highly ornate.

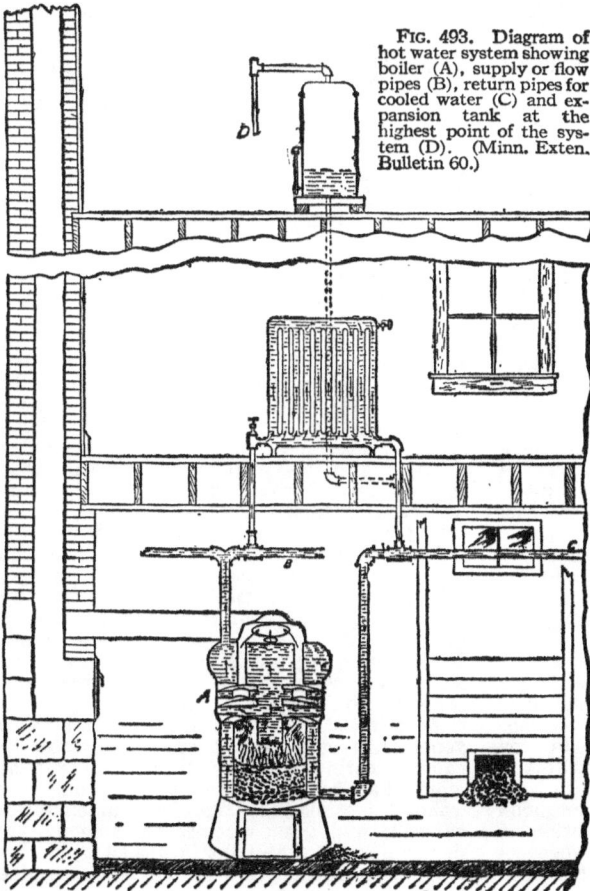

FIG. 493. Diagram of hot water system showing boiler (A), supply or flow pipes (B), return pipes for cooled water (C) and expansion tank at the highest point of the system (D). (Minn. Exten. Bulletin 60.)

Lighting

The development of modern lighting systems has been a rather slow evolution, though more has been accomplished within the last half-century than had been accomplished in all preceding time. Development has been chiefly along the line of perfecting a few ideas, rather than in the originating of a number of entirely different projects.

Candles. Candles are still used as a means of illumination. They were for many years the only means of lighting available for pioneers; and, while inefficient, they possess a certain charm of their own in that the light is necessarily produced in small units. To obtain any degree of illumination from them, it is requisite that the lighting units be rather widely distributed, thus preventing bright glare at any one point. Candles are made principally from tallow or from paraffin wax, the latter being a petroleum product. They are made in various sizes and are usually sold by the pound. Their chief use in modern times is, to a very small extent, as portable lights and, to a greater extent, for decorative lighting in residences.

Kerosene. The use of kerosene as an illuminating oil began about the middle of the nineteenth century, when the kerosene lamp, so familiar to all, was first invented. It was for perhaps 50 years an almost universal means of illumination in residences; and even yet in rural districts it is very widely used. The light furnished by a kerosene lamp is yellow in color and of a not particularly desirable quality. It is cheap, however, but the care of the lamp is an annoying and an inconvenient feature.

Kerosene may be burned in a lamp provided with an ordinary wick burner in which the oil is carried up by means of capillary action. A more efficient method utilizes the vaporization of the oil, combustion occurring within a mantle, thus forming an incandescent burner similar to that used with ordinary illuminating gas. The disadvantage of the latter method lies in the difficulty of vaporization of the comparatively heavy kerosene.

Gasoline. Many attempts have been made in recent years to utilize gasoline for lighting purposes, with more or less success. Since gasoline is a rather volatile oil, it is necessary in order to utilize it advantageously, to vaporize it and mix it with air. It may be used in individual lamps, the base of which is a reservoir for the gasoline and in the burner of which the vaporization and carburetion of the gasoline is accomplished. Such lamps are, however, not at all safe; for the flame is in rather close proximity to the oil itself, and explosions may result.

In addition to lamps, there are 3 methods commonly made use of in the utilization of gasoline as an illuminant; these are respectively known as the "cold" process, the

Fig. 495. Gasoline lighting plant. The gasoline stored outside is drawn into the drum-like mixer, vaporized, and distributed through the house, the apparatus being operated by the weight of the metal, rock-filled tub which is wound up like a clock weight at regular intervals.

"hollow-wire system," and the "central-generator" system. In the cold process, the air is passed over the gasoline and absorbs the gasoline vapor until it becomes saturated. It is then sent around through ordinary pipes much as any other gas is handled. In the hollow-wire system, the gasoline container is partially filled with compressed air and the gasoline is forced around to the various lamps through a hollow wire and then vaporized in the lamp. Such a system is not very desirable for residence use, since it requires several minutes to get the burner heated up sufficiently to accomplish a vaporization of the oil. The central-generator system is so called because the gasoline is vaporized by heat at a central point and there mixed with air before being sent around through pipes to the place where it is to be consumed. All that is necessary with a system of this kind is to turn on the gas and light it.

Acetylene. Acetylene is a gas which is generated through the absorption of water by calcium carbide. It is a form of illuminant which has been widely used, not only for isolated installations, but for municipalities as well; and great success has attended its development. The light furnished by acetylene when burned in a mantle is a very fine

A combination storage and dairy barn eminently suitable for northern sections

A combination hay and horse barn on a successful, practical, Iowa farm

FROM GOOD BUILDINGS COME PROFITS; AND THESE IN TURN MEAN BETTER BUILDINGS

A feed lot and general feeding barn for baby beef production

A feeding shed and shelter for hogs in the field

Single-story, concrete cow barn on a farm producing certified milk

Stallion and sales stables on a commercial horse breeding farm

THERE IS A BEST TYPE OF BUILDING FOR EVERY PURPOSE. THE FIRST TASK IS TO DETERMINE IT; THE SECOND, TO BUILD IT CAREFULLY AND WELL

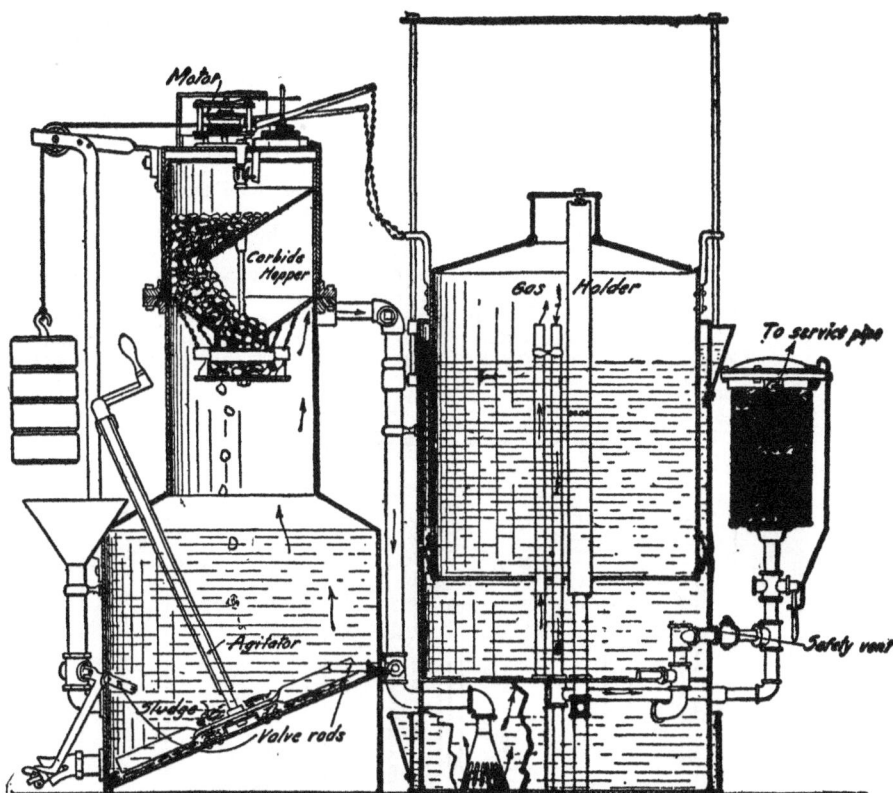

white light closely resembling sunlight. It may also be burned in the ordinary open burner, but the light then furnished is of a very inferior quality. The advantages of acetylene as an illuminant are that it is cheap, not only in first cost and in installation of the necessary equipment, but also in maintenance. Its disadvantages are that it is dangerous, unless handled with great care, and is not so flexible an illuminant as electricity.

There are two systems by which acetylene is used as an illuminant. One of these is known as the "carbide-to-water system," in which the carbide in rather small pieces is fed into a reservoir of water. It is advantageous in that there will always be an excess of water, consequently the carbide will be fully utilized. Experience has shown that it is the safest system, and practically all installations are of this kind. The other system is the reverse of this the water is fed to the carbide; it is used mainly in small lamps, such as bicycle lamps and the like.

As mentioned previously, acetylene is usually burned under a mantle. The piping should be carefully done, in order that there may be no waste of the gas. In the installation of an acetylene system, every precaution should be taken to eliminate any possibility of an accident. When, from any cause, an excess of carbide is fed in the generator, unless some sort of a safety valve is provided, the pressure may increase to such an extent that spontaneous ignition will occur, with a resulting explosion. Perhaps the safest installation is that made in a pit at some distance from the residence to which the gas is to be supplied. If such a pit be properly made and adequately protected, the danger in the use of acetylene will be practically eliminated.

Natural gas. Natural gas, though occurring in many scattered localities through the country, is not a universal illuminant. In regions which are so fortunate as to have natural gas, the fullest advantage may be taken of its use, since it is normally very cheap indeed and, when properly handled, is a very efficient form of light. The only precaution to be taken in the use of natural gas is to make the pipe installation with the greatest care, so that no leakage may occur.

Electricity. Electricity for illumination may be derived from two sources: (1) the services of a regular power company, and (2) an isolated system. When power-company service is utilized, the rate is extremely varied, depending not only upon the original cost of the production of current, but upon the distance the current must be transferred and the amount used. Such service, when supplied by a reliable company, is decidedly advantageous and has many points to commend it.

Unfortunately, however, everyone is not so situated as to be able to take advantage of service of this kind. Recourse must then be had to an isolated electric system, which is a very satisfactory method of illumination. This consists essentially of a generator, a storage battery, and switchboard, together with the accompanying wiring system by which the current is carried around to the various points where it is to be used. The size of the system will depend upon the voltage and upon the current required.

Both of these systems have been fully described and discussed in chapter 18 on "Electricity on the Farm."

One of the most recent developments in the way of isolated electric-lighting outfits consists of a combination unit with all the essentials of the isolated system included in such a way as to permit of an automatic operation. When the engine is started, it drives the dynamo as a generator and forces current into the storage battery. As soon as

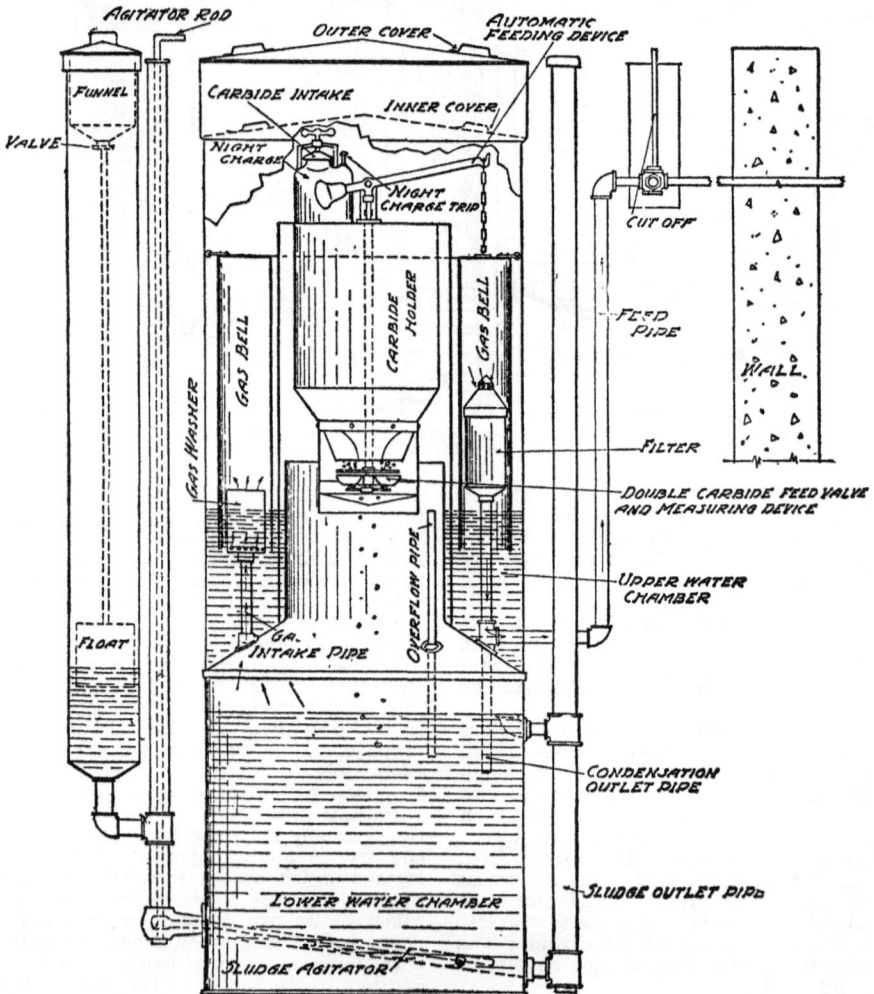

FIG. 497. Single-unit acetylene generator which should be buried in a concrete pit at least twenty yards from the house. (Minn. Extension Bulletin 58.)

the voltage of the storage battery shows a certain reduction, a circuit is closed, which directs a current into the dynamo, operating it as a motor, which, in turn, starts the engine. As soon as the engine is running under its own power, it, in turn, drives the dynamo as a generator and the recharging of the battery occurs.

Cost. Investigations conducted at the Kansas State Agricultural College show that, neglecting original cost and depreciation, the comparative cost of operating various systems of lighting for 80-candle-power illumination, is as follows: Kerosene, 0.4 to 4 cents; gasoline, 0.25 cent; acetylene, 1.25 cents. Electricity power-company service, 0.8 cent; in isolated plants, 1.33 cents; city gas, at $1.40 per thousand, 40 cents.

Water Supply

On the great majority of farms, the water used for various purposes is derived either from wells or from springs; only in comparatively rare instances is it obtained from municipal systems. The dangers attendant upon the use of cisterns, a pond or other reservoir, or an open stream, are generally recognized; and such sources of supply are usually to be carefully avoided.

Springs. Springs are generally satisfactory sources of water supply, since the water obtained from them seeps through many strata of soil and porous rock, which act as effective filters. They are of fairly wide distribution, though in plains regions they, naturally, are scarce, owing to the fact that their origin must be at a considerably higher level. The fact that a spring flows full and clear does not insure purity, however; for springs are subject to contamination, not only at their origin, but by coming in subterranean contact with other contaminated streams, and at their outlet. If there is any suspicion as to the quality of spring water, it should be analyzed, to ascertain whether pollution exists or not. The origin of springs should, as far as possible, be carefully examined.

The flowing spring itself requires careful protection, to insure sanitary operation. A good fence surrounding it will exclude stock. It should be dug out and walled up with stone or brick masonry, the water being led into the pit through a screened tile. Perhaps a better material than stone or brick is reinforced concrete. The pit should be, say, 3 feet square and 2 or 3 feet deep, and provided with a cover to exclude leaves and trash, which so often pollute a spring. The water may be conducted to the house through pipes

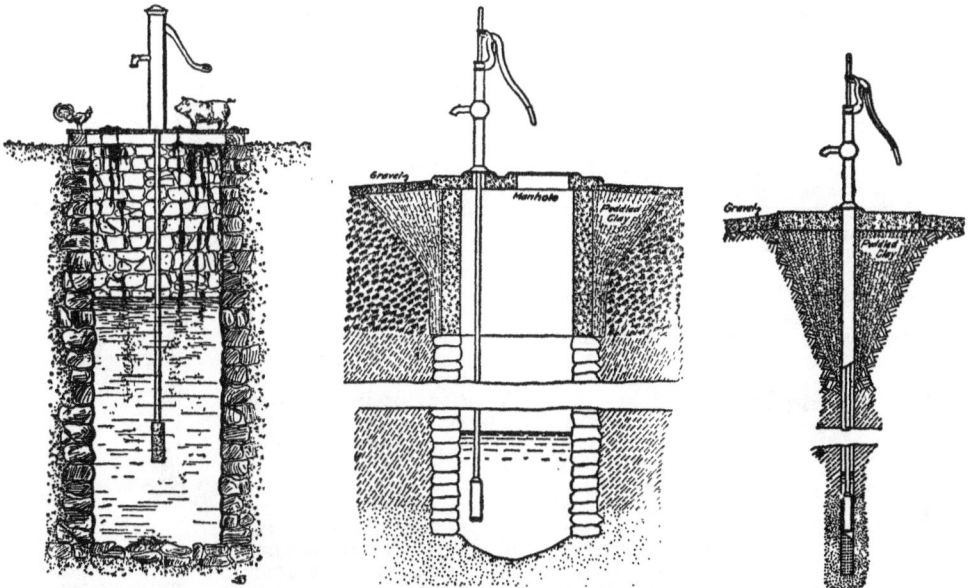

Fig. 498. A badly polluted well (*left*), an all too common sight on the farm; and suggestions as to how to protect both dug wells (*centre*) and driven wells (*right*) from similar trouble. Concrete is invaluable for this purpose

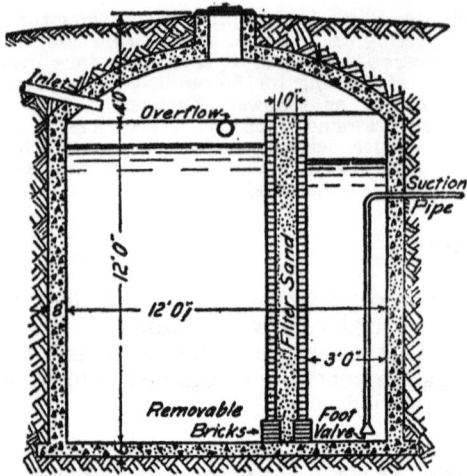

FIG. 499.   Section and plan of a cistern with partition filter made of two rows of brick with sand between

by gravity, if the spring be elevated high enough; otherwise, a hydraulic ram may be found advantageous (See Chapter 16).

Wells. Wells may be either deep or shallow, a depth of, perhaps, 30 feet being arbitrarily taken as the separating point between the two. Deep wells are usually bored or driven, while shallow wells are more often dug. The former are less likely to receive contamination than the latter, on account of the increased depth of soil through which the surface water must filter before striking the water-bearing stratum.

Fuller, in "Domestic Water Supplies for the Farm," gives the following concise description of wells:

"Dug wells are generally circular excavations 3 to 6 feet in diameter. They are adapted to localities where the water is near the surface, especially where it occurs in small seeps in clayey materials, and requires extensive storage space for its conservation. Bored wells are wells bored with various types of augers from 2 inches to 3 feet in diameter rotated or lifted by hand or horsepower. They are usually lined with cement or tile sections with cemented joints and often with iron tubing. They are adapted to localities where the water is at slight or medium depths and to materials similar to those in which open wells are sunk. Punched wells are small holes, usually less than 6 inches in diameter, sunk by hand or horse-power by dropping a steel cylinder slit at the side so as to haul and lift materials by its

spring. They are adapted to clayey material in which water occurs as seeps within 50 feet of the surface, but not at much greater depths. These wells should also be lined with tile, iron tubing, or sheet-iron casing. Driven wells are sunk by driving downward, by hand or horsepower apparatus, small iron tubes, usually $1\frac{1}{4}$ to 4 inches in diameter and provided with point and screen. They are adapted to soft and fine materials especially to sand and similar porous materials carrying considerable water at relatively slight depths, and are particularly desirable where the upper soil is likely to be polluted.

"Cemented rock or brick linings protect the well from pollution, except at the bottom, as long as the walls are not cracked. They also prevent the entrance of sediments and animals and do not impart a taste to the water. Iron casings are used in both rock and unconsolidated materials. They are usually used in deep wells. They may be either iron tubing 1 to 4 inches in diameter, or sheet-iron casings 4 to 16 inches in diameter, with snug joints. They are adapted to wells of all depths in which water is obtained from a stratum below the casing or from a stratum between cased sections or in case it is decided to procure water from a number of strata."

Wells should be given a further protection from pollution by surrounding them with watertight curbs, in addition to the casings or

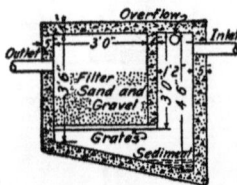

FIG. 500. A small concrete filter that can be used to purify the water supply of any farmhouse.

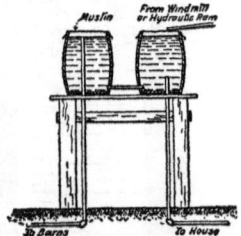

FIG. 501. A simple, home-made arrangement for obtaining clean water and some pressure which, of course, increases with the height at which the barrels are placed

linings mentioned above, and covering with a reinforced concrete cover, which should be 6 or 8 feet square, with an outward slope to carry the water away from the well. It is advisable to lay a circle of tile below the circumference of this cover, into which the surface water may seep, flowing away through a conductor tile to a connection with a regular drain. In this way all surface water may be effectually excluded.

Amount of water needed. In households where an adequate supply of water is available and convenient, and modern pumping systems are installed, the consumption of water is several times as much as where a pump is the only means of obtaining it. This in itself is a fair indication of the value of modern systems. Water consumption varies with the climate, the season of the year, the extent of various household operations, the household equipment, the number of animals on the farm, and the number of persons in the household. Fair estimates of the amount needed per day by each person and animal are as follows: person, 30 gallons; horse, 8 gallons; cow, 8 gallons; hog, 2 gallons; sheep, 2 gallons. From this it is evident that on an average farmstead, with 6 persons in the family, and with stock consisting of 30 head of cattle, 12 horses, 20 hogs, and 20 sheep, nearly 600 gallons of water are required per day; this should be increased, to provide for extra water used for lawn sprinkling, cleaning, and other contingencies, by at least 10 per cent.

Pumps. Two kinds of pumps are commonly used in farm wells—the lift pump and the force pump. In the former, the water is lifted by a piston or plunger attached to the lower end of the pump rod. Theoretically, the distance which a pump of this

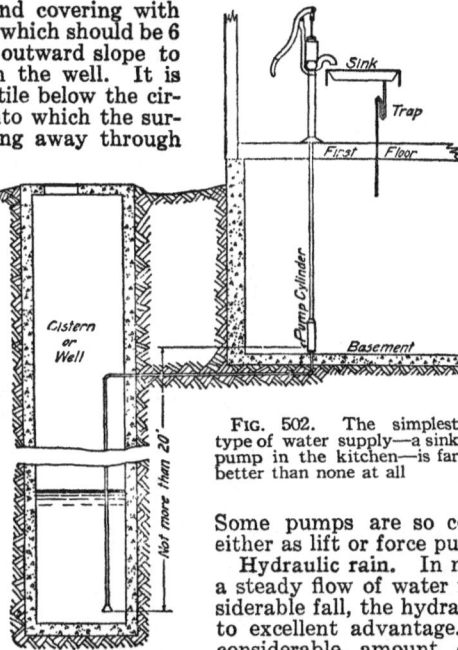

FIG. 502. The simplest type of water supply—a sink pump in the kitchen—is far better than none at all

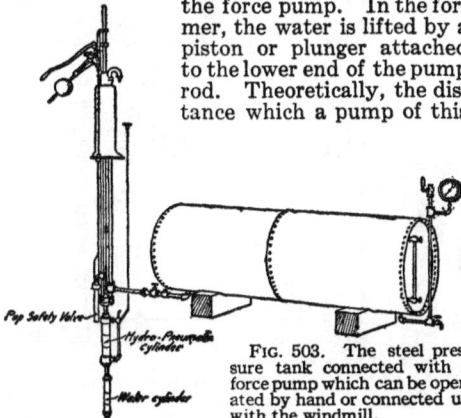

FIG. 503. The steel pressure tank connected with a force pump which can be operated by hand or connected up with the windmill.

kind should raise the water is a little more than 32 feet, a distance equal to a height of water-column giving a pressure equal to atmospheric pressure; but practical considerations reduce this to about 25 feet.

The force pump differs from the lift pump in that the raising of the water is done by exerting a pressure upon it with the plunger, the height to which the water can be raised depending upon the amount of pressure exerted. Some pumps are so constructed as to act either as lift or force pumps as desired.

Hydraulic ram. In many localities, where a steady flow of water is available with considerable fall, the hydraulic ram may be used to excellent advantage. The ram wastes a considerable amount of water, depending mainly upon the height at which it must be discharged, or, perhaps more correctly, upon the ratio of the depth of fall to the height of discharge. When the ram is operating, water flows through the drive pipe, increasing in velocity until the momentum is sufficient to close the outlet valve. As a result of the impact of the suddenly stopped flow, the discharge valve is raised and a portion of the water is forced into the discharge pipe, reducing the pressure thereby and consequently allowing the discharge valve to fall and the outlet valve to open, when the cycle is again repeated. An air dome is usually connected to the discharge pipe, which provides not only a cushion to relieve sudden mechanical strains, but equalizes the flow at the discharge end (see Chapter 16).

FIG. 504. When the sink is some distance from the stove, a faucet on the boiler provides a handy supply of hot water.

Elevated tanks. The elevated-tank system of water supply is a very common one, being used both in large and in small installations. It consists of a tank, located either on a tower, on an elevation of the ground, or in the top of a building, and is connected by properly arranged piping systems, with the pump from

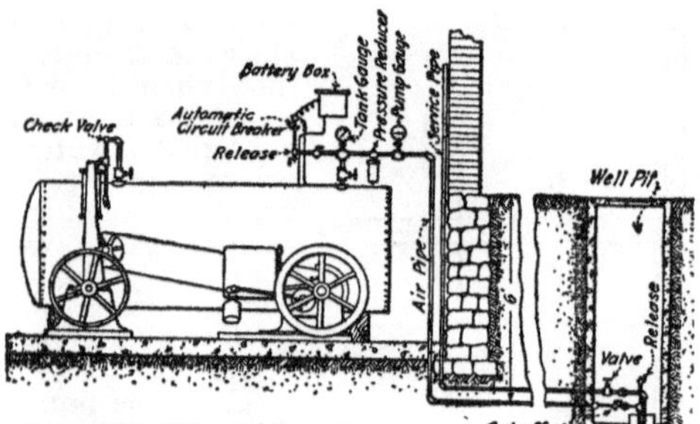

FIG. 505. The pneumatic pressure tank system run by gasoline engine power is a most desirable equipment for the farm.

which water is received, and with the buildings to which water is delivered. It is very simple and comparatively inexpensive, and delivers water at a pressure depending upon the difference in elevation between the tank itself and the point of delivery. It has certain disadvantages, however, in that the water stored in it is likely to freeze in winter and to become stale in summer; besides, a number of accidents have occurred from the failure of tank supports.

A force pump is used to elevate the water into the tank, which should be equipped with a gauge or overflow, to indicate when the tank is filled. The pump may be operated by hand; but when the tank is designed to store a supply sufficient for several days, some sort of a power-driven pump is preferable. The pipes conducting the water to and from the tanks should be of not less than 1-inch diameter. Three-quarter-inch pipe is sometimes used; but the friction loss is so great in the small pipe that it is much better to use the larger size, which costs only a little more originally, and is no more expensive to install. The tank itself may be of wood or metal, preferably the latter; and, where possible, as is the case when the tank is installed on a hill, it is best to make it of reinforced and waterproofed concrete.

Hydropneumatic system. In an attempt to produce a water supply system that would furnish an adequate supply of water under good pressure and yet eliminate the bad features of the elevated tank, the hydropneumatic system was evolved. Essentially, it consists of a powerful force pump, an airtight tank, and a distribution system. With all outlets closed, water is pumped into the tank, compressing the air contained therein until the desired pressure is obtained. Then, when any faucet on the distributing system is opened, the compressed air expands, and forces the water out through the opening. Of course,

all water pipes are taken off at the bottom of the tank.

Usually, the pump is of the duplex type, and can pump either air or water or both. This is desirable; for it sometimes happens that long use of the tank will deplete the air within the tank, and an additional supply must be furnished. Then, too, it is sometimes desirable to increase the pressure within the tank, in order to have available a large quantity of stored water at a higher pressure. The usual pressure carried in the tank is not to exceed 100 pounds gauge, 70 to 80 pounds being common. The pressure decreases as the water is drawn off, until, when the tank is empty, it is practically nothing. In the majority of installations, the pump is power-driven, using either gasoline or electricity with automatic shut-off; but it is practicable to have hand- or windmill-operated pumps, especially for less expensive outfits.

The tank is generally located in the basement of the residence, where there is no danger of freezing, and where the stored water is kept reasonably cool. From 75 to 80 per cent of the total capacity of the tank is used for water storage, the remaining space being utilized by the compressed air. A tank of from 1,200 to 1,500 gallons capacity will hold several days' supply of water for household use, exclusive of stock requirements.

Pneumatic system. The difference between the pneumatic system and the hydropneumatic system lies in the fact that the former does not provide for the tank storage

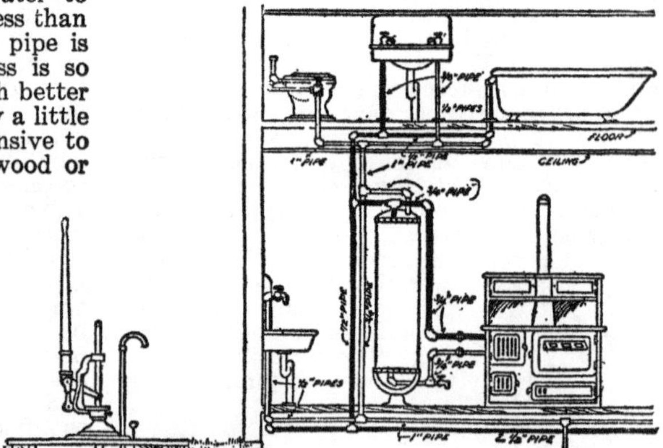

FIG. 506. Diagram of the arrangement and essential fittings of a simple but complete hot-and cold-water system.

FIG. 507. No wonder girls left the farm when housekeeping meant drudgery like this and worse—

FIG. 508. Especially when simple, inexpensive plumbing can make a kitchen as convenient and pleasant as this

of water, but for compressed air exclusively. The elements of the pneumatic system comprise an airpump, a tank for compressed-air storage, an automatic pump, and a distributing system. The air is compressd by the pump to a pressure generally slightly exceeding 100 pounds gauge and stored in the airtight tank. Thence a pipe leads to the automatic pump, which is a double cylinder contrivance with valves so arranged that when it is submerged in the well or cistern, one of the cylinders is filled with water by external atmospheric pressure. As soon as the filling is accomplished, the compressed air is automatically released, to exert pressure upon the water in this cylinder, driving it through the water mains, while cylinder Number 2 is filling with water. When Number 1 is empty

and Number 2 is filled, the action is reversed, the water in the latter being forced out by the compressed air.

This system is the most recently developed of all water-supply systems, and possesses many advantages. It may be used either in dug or in bored wells, successful operation being obtained with the pump located even as deep as 125 feet. The only places that wear occurs are in the air pump and in the automatic pump, but in both cases the worn parts are replaceable. No water is stored, to freeze or grow stale, and a constant supply is possible, the only attention that is required being that involved in maintaining the air pressure in the compressed-air tank; and, with an electrically operated air pump, even this may be automatically controlled.

FIG. 509. An abundant supply of good water makes for a better farm and healthier plants, animals and people. It is one of the most profitable of all investments

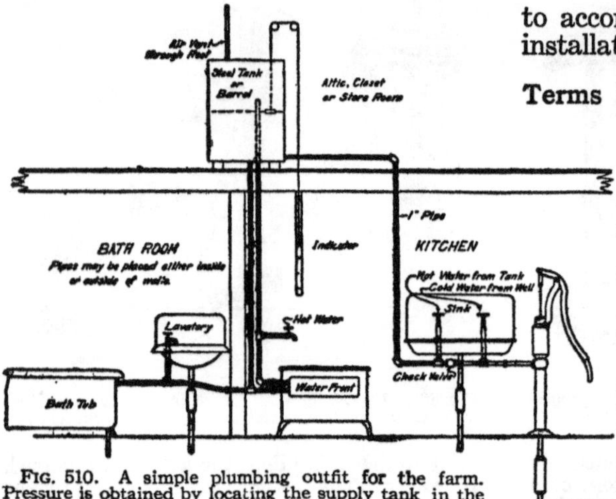

FIG. 510. A simple plumbing outfit for the farm. Pressure is obtained by locating the supply tank in the attic, and filling it by means of a hand force pump. Hot water pipes are shown in black.

Plumbing

The words "plumbing" and "plumber" have their origin in the Latin word for lead (*plumbum*) since early workers in pipe used lead extensively. Later development has, to a great extent, eliminated lead from plumbing work, though in certain places a satisfactory substitute for it has not been found. Modern plumbing fixtures quite generally consist of brass or steel pipe, usually nickel-plated; and it is only in occasional instances that lead pipe is used.

Though the journeyman plumber has an elaborate list of items in his equipment, it is possible for an amateur to accomplish simple plumbing successfully with only a few tools. Fortunately for the latter, manufacturers are now making fixtures which may be connected with the piping system by means of simple unions of various kinds, so that he does not need to know how to make a "wiped" joint, a knowledge of which constituted a standard test for the early plumber. A pipe wrench, a pipe vise, a cutter, a reamer, a rule, a diestock and dies, and some white lead should, when combined with ordinary skill, be sufficient

to accomplish ordinary work in simple installations.

Terms Used in Plumbing and Their Meanings

Soil fixtures. Fixtures which receive human discharges; they include water-closets, urinals, and sometimes slop sinks.

Soil pipe. The pipe connecting a soil fixture with the soil stack.

Soil stack. The vertical pipe into which soil fixtures, such as closets and urinals, discharge.

Trap. An arrangement providing a water seal separating the interior of the house from the discharge portion of the plumbing system. The trap is an essential part of every fixture, for it prevents sewer gas from entering the house.

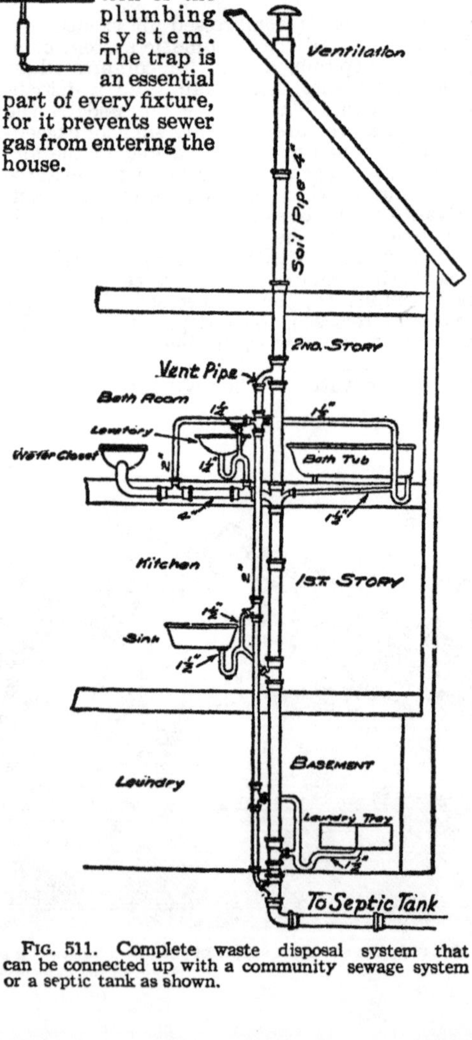

FIG. 511. Complete waste disposal system that can be connected up with a community sewage system or a septic tank as shown.

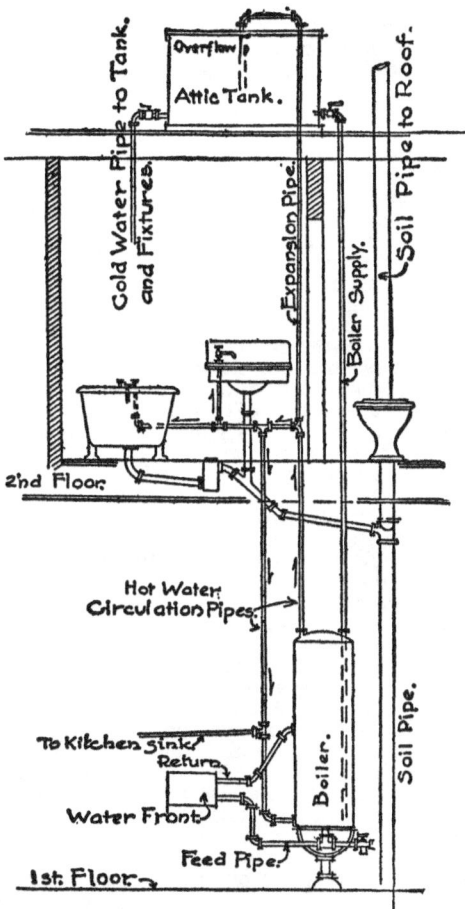

FIG. 512. Somewhat more elaborate plumbing system than that in Fig. 510; in this case the water tank is on the third floor and the bathroom, complete, on the second. Sinks, tubs, etc., are connected with a single soil pipe requiring a sewage disposal equipment.

Vent pipe. The connecting pipe between fixtures and the vent stack.

Vent stack. A vertical pipe, connecting with the outer air, into which vent pipes discharge.

Waste fixtures. Plumbing fixtures, such as lavatories, tubs, sinks, etc., which do not receive any human discharge.

Waste pipe. The pipe connecting a waste fixture with the waste stack.

Waste stack. The vertical pipe into which waste fixtures discharge.

Systems of plumbing. Two systems of plumbing are in common use—the single-pipe system and the 2-pipe system. In the former, 1 pipe is used both as a combined waste pipe and soil pipe and as a vent stack; it is almost universally used in residence installations. The 2-pipe system provides 1 pipe for discharging and a separate one for a vent stack. While theoretically more nearly ideal than the 1-pipe system, it is more expensive and not so practical. Separate waste, soil, and vent stacks are rarely used in residences.

When the single-pipe system is used, it is well to use drum traps instead of the ordinary bent pipes for traps, since the latter are subject to siphonage, resulting from the aspiratory effect of a large volume of water being discharged down the waste pipe, as when a closet is flushed. The seal in traps may fail from other causes, such as evaporation, capillary action when a string or piece of cloth catches on the discharge lip of the trap, or other accidental occurrences.

Fixtures. Plumbing fixtures vary in the material used in their manufacture. Bathtubs may be made of solid vitreous ware, which is exceedingly handsome and durable, but expensive. Just as serviceable ones may be made at a much lower cost of heavy, stamped sheet iron, porcelain-enameled on the inside and painted or enameled on the outside. The same is true of lavatories. Sinks are usually made of enameled iron, though more expensive installations sometimes include earthenware slop sinks. Laundry tubs are almost always of plain soapstone. Made of this material, they are more durable and not so likely to become damaged as those of enameled iron. Water-closets are generally manufactured of vitreous ware, their complex shape precluding the possibility of stamping or enameling; and, besides, there must be no danger of their becoming insanitary as a result of chipping, which would be likely in the case of enameled metal.

For sanitary reasons, plumbing fixtures should be as simple in design

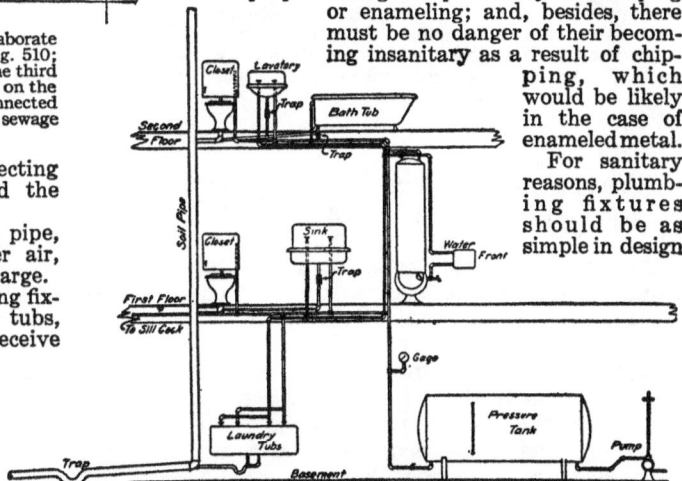

FIG. 513. A system quite similar to that shown above, but in which water pressure is obtained from a pneumatic tank filled by means of a hand-power force pump.

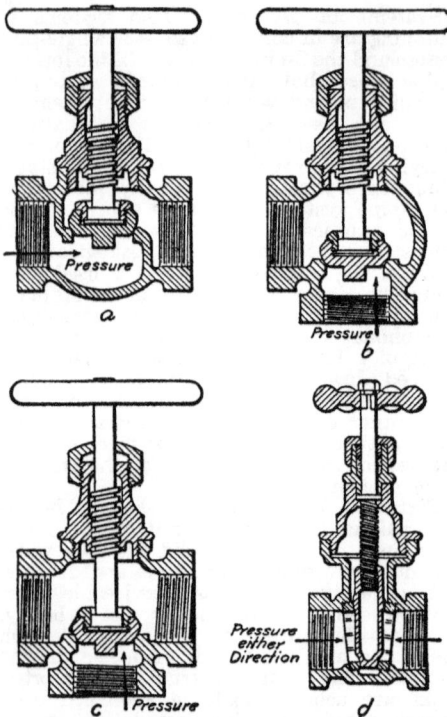

FIG. 514. Commonest types of valves used in heating and plumbing work: *a*, globe; *b*, angle; *c*, cross; *d*, straightway or gate.

as possible. The surfaces should be smooth, so as to admit of easy cleaning; and no acid or substances containing acids should be used on them. The water-closet, especially, should be simple and perfect in construction, should contain no movable parts, and should be furnished with a flange by which an absolutely perfect joint may be made with the soil pipe.

Two types of closets are to be recommended —the "wash-down" and the "siphon-jet." Of these the former is the cheaper; the latter is a newer development and has the advantage of a rapid and an almost noiseless flushing.

Pipe. Heavy cast iron pipe, not less than 4 inches in diameter, is used for soil stacks and for connection to the sewage-disposal installation. Lighter steel pipe, not less than $1\frac{1}{4}$ inches in diameter, may be used for fixtures other than the water-closet. Care must be taken to have the waste pipes at least as large as the outlet of the fixture.

In fitting pipe, measurements should be made carefully, consideration being given to the length of thread, or distance which a pipe will screw into the fitting. Standard dies are made with the thickness of the die equal to the proper length of thread. White lead is used to insure a tight joint, it being spread on the external, *not* on the internal thread, before parts are screwed together. The use of a wheel cutter in cutting pipe leaves a thin rim or burr extending into the pipe and reducing its cross section. This burr should be carefully removed with the reamer as soon as the pipe is cut.

Requirements. The requirements of a good plumbing installation may be summarized as follows:

The fixtures must be simple in construction and action, made of good materials, and sightly in appearance.

Tight joints must be made, to prevent leakage of water, waste, or sewer gas. Every fixture must be provided with an effective trap which can be easily cleaned. Pipes must have sufficient slope to insure a rapid, self-flushing flow. Provision must be made for cleaning every part of the system, especially soil and waste pipes that may become clogged.

Care in installation and an occasional inspection are necessary, if continuously satisfactory operation is to be secured.

Sewage Disposal

Sewage may be defined as waste matter, either excretory or fecal matter or slops, discharged through a system of pipes known as a sewage system. Farm sewage may be of human or of animal origin, or a product of cleansing operations in the household, or waste from the milkroom or dairy.

The proper disposal of sewage is of prime importance, for several reasons. In the first place, sanitation requires its removal; for sewage may contain elements exceedingly dangerous to the health of individuals or to the community at large. In the second place, sewage contains much that is given off during the process of putrefaction. Finally, common decency requires that no offensive materials be allowed to accumulate in the region of the household or farmstead.

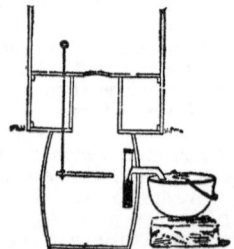

FIG. 515. One of the simplest possible arrangements for human sewage disposal, insuring proper protection of the health of family and neighbors.

The development of rural sewage-disposal systems has been remarkably rapid in recent years; and it is now entirely possible to dispose of sewage in a clean, satisfactory, and scientific manner with as much success as accompanies the disposal of city sewage. So far as the plumbing installation is concerned, it is the same in both the city and the country homes; but no street sewer is available in the country, as in the city, and other outlets for discharge must be found.

Many dwellings are not equipped with a water supply and plumbing system which permit of the use of a liquid carrier for wastes. It may be impracticable, on the score of economy or for other reasons, to make such an installation; and, in such a case, special methods must be adopted to get rid of wastes.

Household wastes. It is possible to dispose of waste from the kitchen and laundry in a satisfactory manner, if occasional care be given to the installation. The simplest way is, of course, to pour it out loose, porous soil, in different places successively; but this may result in supersaturation, or "waterlogging," of the soil, and in summer time may produce a feeding ground for flies. A plan to be more recommended is to employ subsurface irrigation, pouring the wastes out into a system of tile laid with loose joints, from which it seeps out and percolates through the soil to another system of similar tile laid 18 to 20 inches lower, from which it is carried away. Such a system may be made to furnish some moisture to a small garden, or portions of it, without harm. All liquid waste, from dishwashing, cooking, or laundry and cleaning operations, may be discharged into the tile.

Privies. Where a water-borne sewage system is not practical, some special means must be devised to dispose of human excremental material. The privy as it is too commonly built is to be abhorred: it is insanitary, offensive, ugly, and sometimes positively dangerous. It may contaminate water supplies, and it affords an ideal breeding ground for flies. With proper construction and supervision, however, it may be made an

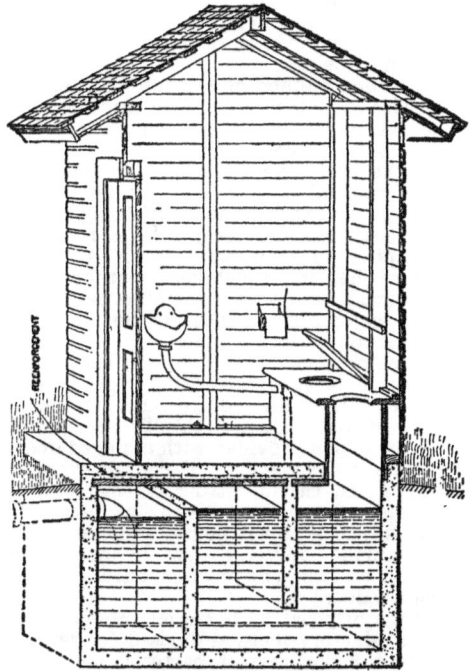

FIG. 517. Septic tank principles applied to a simple privy vault. There is no reason why any farm should be without at least this degree of convenience.

acceptable and reasonably satisfactory agent in human sewage disposal. It should be well built, with adequate ventilation and with all openings carefully screened to exclude flies.

The excrement may be taken care of in 2 ways—(1) by the dry method, which is the

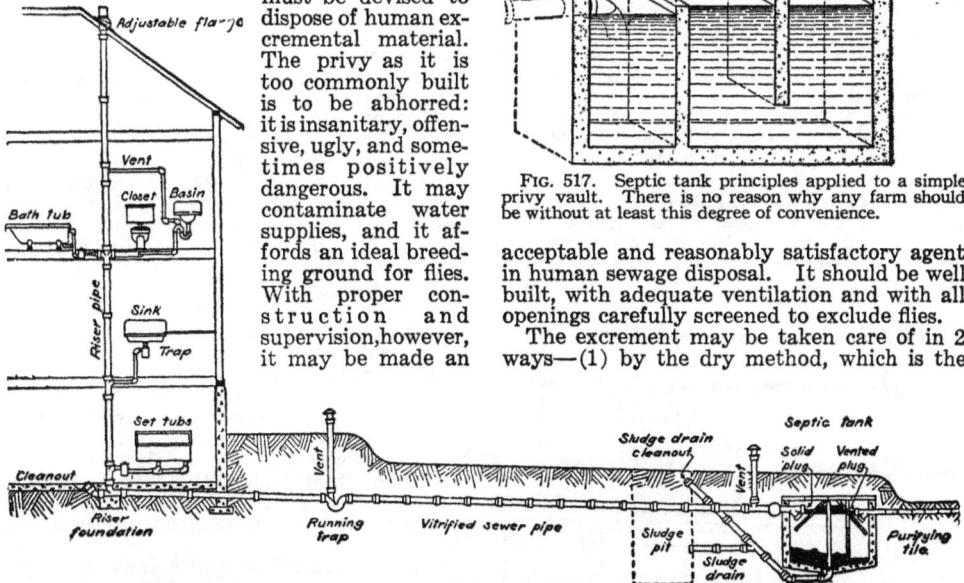

FIG. 516. Complete sewage system making use of the one-chamber type of septic tank which discharges into a line of purifying tile run under the garden or lawn

simpler, but requires constant attention, and (2) by the wet method which is more or less automatic and requires attention only at intervals. In the former a receptacle is provided which receives both the fecal matter and the urine; and each time the privy is used a small quantity of dry, fine earth or ashes or fresh-slaked lime is sprinkled over the deposit, to deodorize it. The receptacle must be watched and removed when necessary, the contents being buried in the soil at some place from which no harmful contamination can occur. Sometimes the receptacle is not entirely closed at the bottom, so that the urine may drain away, then to be carried to an underground system of loose tile, and absorbed into the soil.

The second method of equipping a sanitary privy with some means of liquid disposal may be carried out in one of several ways. The simplest is to place a rather large receptacle below the seat, the receptacle to be filled perhaps half full of water, which may or may not be supplied with some disinfecting agent; when necessary, the receptacle is removed and emptied in some safe place. A better but somewhat more elaborate method is to build a small water-tight chamber below the seat, in which septic action, to be described subsequently, may take place.

In construction, a sanitary privy should be, perhaps, 4 feet square, with tight walls, which may, if so desired, be built double to afford insulation from cold. The rear wall below the seat is provided with a hinged door, to facilitate the removal of containers; and the close and careful fitting of this door is important. Sometimes, such a privy is located adjacent to the house, and communication is made between the two by means of a vestibule, or, less to be recommended, directly through a door.

Sewage disposal by bacterial action. The most satisfactory method devised for sewage treatment makes use of bacterial action in accomplishing its reduction. Briefly, this action is as follows: Three kinds of bacteria operate in the disposal of sewage: aërobic bacteria which require the presence of oxygen for their existence; anaërobic bacteria which die in the presence of oxygen, and can work

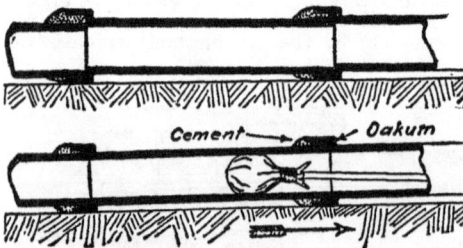

FIG. 518. Sewage tile laid incorrectly (*above*) and correctly (*below*). The interior course must be smooth and unbroken.

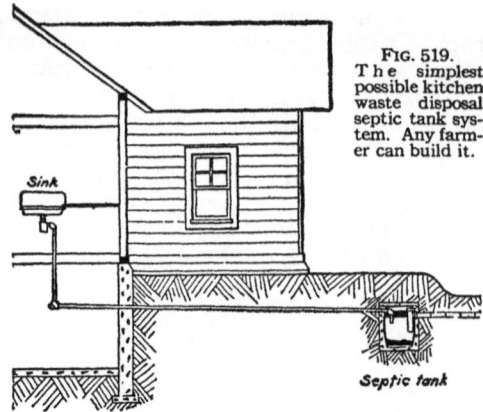

FIG. 519. The simplest possible kitchen waste disposal septic tank system. Any farmer can build it.

only where it is entirely excluded; and facultative bacteria, which live and operate under either of the above conditions.

When sewage is discharged into a retention chamber, it is first subjected to the action of the anaërobic bacteria; liquefaction to a certain degree occurs, the suspended organic matter breaking up into liquid and gaseous compounds with a slight precipitation of insoluble matter or sludge. The product of this decomposition is then carried out into the soil, which contains a certain amount of air and in which the aërobic bacteria can live and work. Here the sewage is purified by the oxidation of the decomposed compounds, rendering them usable as plant food. The facultative bacteria operate mainly in the first stage of the reduction.

Cesspools. The cesspool occupies a rather anomalous position in the field of sewage disposal. It usually consists of a simple pit loosely lined with brick or stone. Sewage is discharged into it directly, and it is evident that there will result an overlapping of reduction processes. Since the sewage will not be absorbed immediately, some anaërobic action is likely to occur, while that portion absorbed into the surrounding soil will be subjected to oxidation by aërobic bacteria. However, neither action will be even approximately complete.

The cesspool may operate with apparent success for an extended period; but sooner or later it will fail, owing to the soil surrounding it becoming water-logged and thus preventing aërobic action. Often, especially if the location of the cesspool be in very sandy, porous soil, it may operate well for years; but it is never safe, for the leachings from it are likely to be transmitted into strata from which the water supply is drawn. The cesspool is never to be recommended, except as a last resort.

Septic tanks. The septic tank is simply an elaborated cesspool in which provision is made for the proper accomplishment of the

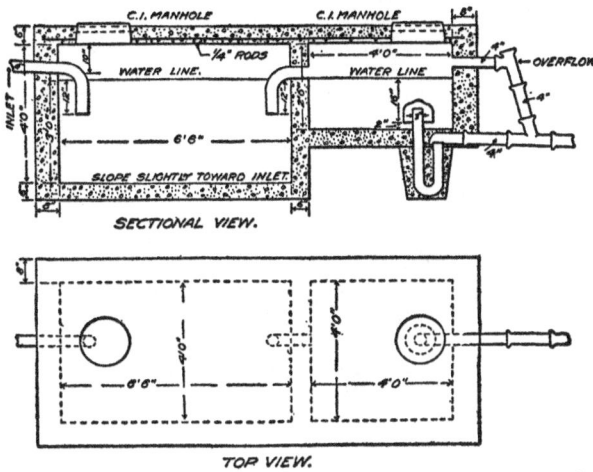

FIG. 520. Two-chamber septic tank with automatic siphon which periodically empties the smaller chamber. The liquid leaving such a tank is practically harmless.

bacterial reduction of sewage. In its most elementary form, it consists of a water-tight chamber of sufficient size to insure the retention of sewage for at least 48 hours, with a baffled inlet and outlet and a provision for the disposal of the effluent from this chamber. A better and more satisfactory form, and one which is more generally built, combines with the first chamber a second one in which is located an automatic siphon for the intermittent discharge of sewage, which prevents the soil from becoming water-logged. The decomposition of organic material is effected in the first chamber; the second stage, or oxidation, occurs in the subsurface distribution system into which the siphon discharges the liquefied sewage.

For an average family of 8 persons, the first chamber, or septic tank, should be 4 feet wide, 6½ feet long, and 4 feet deep. The second chamber should be 4 feet square, with a minimum depth of 2½ feet; and a 3-inch siphon should be used. The effective depth of sewage in the septic tank is only 3 feet, because of the inlet and outlet pipes; so that care must be taken that this is not materially diminished as a result of sludge accumulation.

Proper arrangement of the irrigation system to receive the effluent from the septic tank is necessary for successful operation. True, in some cases, the effluent is discharged upon the surface or in shallow trenches if the soil be porous; but this may ultimately prove undesirable. The subsurface method is much to be preferred. It consists of 3- or 4-inch ordinary agricultural tile, laid with open joints, at a depth of not exceeding 18 inches. The joints are protected by coverings consisting of pieces of broken tile, to prevent earth from falling in. In some especially

close soils, it may be necessary to surround the tile with loose gravel, for the purpose of more rapid and complete absorption. The proportion of tile to the quantity of effluent discharged should be about 1 foot to 3 to 5 gallons of effluent, depending upon the character of the soil. It might be supposed, from the shallowness of the soil above the tile, that the effluent would freeze; but experience indicates the contrary.

A septic tank will not begin to operate satisfactorily until the septic chamber is filled; and even then it may be necessary to inoculate it with anaërobic bacterial cultures from another tank or cesspool, to correct the "sick" condition. A thick, heavy scum on the surface usually indicates good operation, and this should be disturbed as little as possible; the inlet and outlet should be protected by baffles, so that any discharges into or from the tank will not affect the scum. Often, a septic tank will operate well for years without any attention; but occasionally it may be necessary to remove sludge accumulations or to clean the siphon and discharge pipe.

It must be remembered that the reduction of sewage by this method may not be absolutely complete, nor is it certain that all disease germs will be destroyed; consequently, reasonable care should be given to the tank and to the subsurface system, in order that no harmful results may occur.

Creamery sewage. The disposal of wastes from a cheese factory or any milk-handling or dairy manufacturing establishment is one

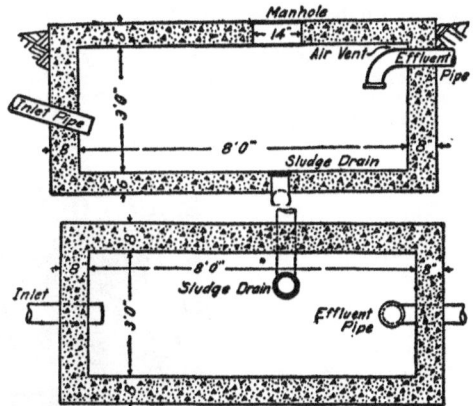

FIG. 521. Lengthwise section and plan of a single-chamber septic tank. If the discharge can be run out into a sandy or gravelly soil, it is entirely satisfactory, but in any case the accumulated sediment or sludge must occasionally be removed.

FIG. 522. Before (*left*) and after (*right*) the coming of modern plumbing and a sewage system to the average farm home. Inconvenience, endless daily chores and personal carelessness give way to greater comfort, less labor and better health and habits when such a change occurs

of the most vital factors in its installation. Quite often it is impossible to connect with a city sewer system, and no large stream of sufficient steady annual flow is available to afford adequate dilution; the ordinary household septic tank just described cannot adequately meet the demands for sewage disposal from a creamery or similar building.

Creamery sewage differs from ordinary sewage in that it contains curd, butter fat, oils, and considerable lactic acid, resulting from the cleansing of milk utensils and dairy product machines; and decomposition is somewhat slow. The tendency of lactic acid to retard or suppress bacterial action is so great that every effort should be made to reduce its quantity, and to exclude as much as possible from the septic tank.

It is recommended that creamery sewage be allowed to remain in the septic tank for a minimum of 3 days; a longer period being advisable, but necessitating a larger and consequently more expensive tank. Practice

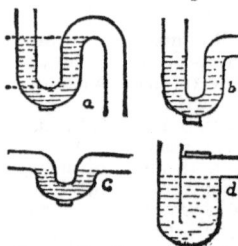

FIG. 412. Types of waste traps: *a* S-type; *b* half-S or P-type: *c* running trap: *d* drum or non-siphon type. In each case the object is to prevent the return of impure gases from the sewage system

indicates that from 1 to 1½ gallons of sewage results from the production of 1 pound of butter, so that it is evident that, even in an ordinary small creamery of 1,000 pounds butter output, a much larger tank is necessary than for simple household requirements.

A sewage-disposal system for a creamery is constructed on practically the same principle as for any other purpose, the main variations occurring in the size of the tank and in the method of caring for the effluent from the septic tank proper. The ordinary subsurface irrigation system is usually inadequate, and a special filter bed of porous material may be provided. For a 1,000-pound capacity butter manufacturing plant, a septic tank of approximately 600 cubic feet content will be necessary; allowing for a 2-foot depth of scum and sludge, a tank 7 feet deep, 6 feet wide, and 20 feet long, is required. The filter bed should have an area of approximately 2 square feet per gallon of effluent, or about 3,000 square feet—say, 30 feet wide and 100 feet long.

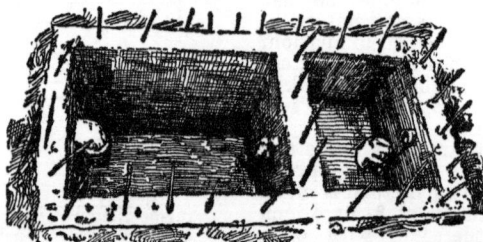

FIG. 514. Looking into a partly finished concrete, two-chamber septic tank. The iron rods will be bent over to help reinforce the concrete cover. Note in the right-hand chamber the siphon, which empties it as often as it becomes full, and thus gives the sewage the longest possible time to decay and purify under the action of bacteria

CHAPTER 32

Farm Buildings: Their Construction

By H. H. Niemann (see Chapter 29), whose training and experience for a number of years have equipped him to design farm buildings, plan their location, arrangement and construction details, and superintend their erection from the basement to the weather vane. In discussing the farm group, he naturally included some consideration of that essential feature—the farm house. In this chapter he restricts himself to the discussion of farm work buildings, or rather the principles of construction that are common to all of them. The particular features and requirements of special types of houses planned for special purposes are treated in the chapters immediately following.—EDITOR.

BEFORE one can decide what kind of materials are best for, or what particular type of construction to adopt in, the erection of a farm building, the requirements of the structure must first carefully be considered. Sometimes it is not practical to arrange and construct the contemplated building so as to fulfill all of these. In such cases the builder must be his own judge in making a selection.

Requirements of Good Farm Buildings

Keep contents of barn dry. The first requirement of any farm building is that it shall keep out rain and dampness. Animals must be kept dry to keep them warm; foodstuffs must be kept dry to prevent mold; and tools and implements must be kept dry to prevent rusting.

Keeping the animals warm and dry will save a considerable portion of their energy. Keeping mold, mildew, and fermentation out of grain, hay, feed, and other crops by storing them in a dry barn is very important. Rust and corrosion in implements, tools, and vehicles can be most economically prevented by keeping them in a dry building.

Pure air, proper temperature, and good light essential. Pure air is as essential as food for the well-being of animals. In fact, all animals require more pounds of air per day than they do of food and water combined.

Air once breathed is not only full of poison, but is saturated with moisture. Unless proper ventilation is provided, this moisture will soon increase the humidity of the air in a barn to such an extent that water will drop from the ceiling and run down the side walls.

In a warmly built barn, a smaller amount of the food consumed by the livestock is used to produce heat, and a greater amount for the production of meat or milk, tending to convert the boarder cow into a profitable producer.

In order to keep milk cool, or to preserve ice, the dairy or the icehouse must be built to resist changes in temperature.

The necessity of light is very much neglected, even in many of our so-called up-to-date barns. We all know that unrestricted sunlight is nature's greatest destroyer of bacterial life, that it is essential for the healthy development of all

361

362 FARM KNOWLEDGE

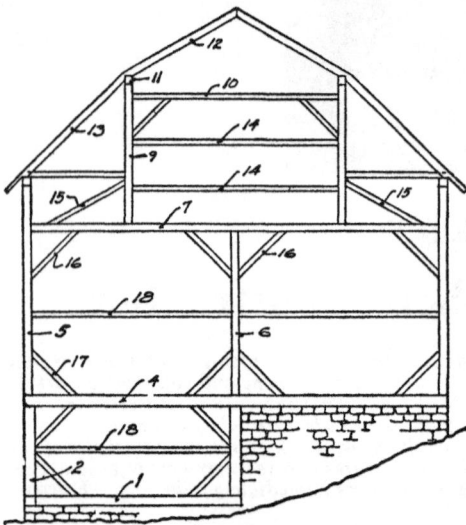

FIG. 525. Bank barn showing end framing of the heavy timber type. Though common in the early days this is now practically never employed because of the difficulty of getting heavy material. (See page 375 for names and sizes of parts designated here by numbers).

livestock, and that it is also the cheapest disinfectant obtainable. Knowing this is it not "penny wise and pound foolish" to bar out sunshine because glass costs a few cents more per square foot than lumber?

Cleanliness and comfort for the animals. The most effective aid in keeping the barn clean is to so build and equip it that the labor and time required for cleaning will be reduced to the minimum. In its construction, avoid all unnecessary angles, corners, cracks, and crevices. Use materials that will produce smooth surfaces and be as far as possible impervious to moisture. The equipment for cleaning and for the removal of litter should be sufficient and effective.

Cleanliness is the one paramount requirement, where articles of human food are concerned. There is no excuse for the unsanitary dairy or for the disease-laden and rat-infected buildings often used to shelter poultry and other livestock.

Prevent the waste of grain and the spread of disease by making the barn and all other farm buildings proof against rats, mice, and other vermin. Use masonry construction with plenty of cement below ground and as far above ground as its cost will permit. All hollow-masonry work, such as tile or cement block, should occasionally have one layer filled solid with cement mortar, or a layer of slate or shingle tile should be inserted in the courses, so that no air space in walls or floors will form a continuous passage for mice.

Provide all stalls with simple and sanitary ties or stanchions which will allow the animal to make freely any of its natural motions, protect itself against insects and other enemies, and to be comfortable either standing or lying down. Make provisions in the barn for supplying the animals with an abundance of fresh, clean water at all times.

Stalls should be so grouped that bedding may be conveniently supplied both for comfort and for the absorption of liquids. Stall gutters with ample slope should be provided, with sanitary trap and strainer of a type easily removable for cleaning, or for flushing tile which leads to the manure cistern, with which every barn should be provided.

Economy in feeding and watering. Economy in feeding and watering is of great importance. It includes not only saving of feed and water, but saving of the labor of feeding also. Arrange feed storage and mixing space convenient to the feeding alley. The greatest economy in feeding is secured by having the stock arranged in 2 rows of stalls, all facing the feed alley running lengthwise between the rows. A feed room with bins on the sides or above, with hopper bottoms and spouts, located at one end of the feed alley, large enough for sheller, fanning mill, grinder, and mixing bins, and having good outside light, dustproof doors to feeding alley, and entrance doors from the outside is most convenient. It is also advisable to have silo and hay chutes in the feed room rather than direct in the feeding alley, in order to avoid dust in the stable.

This combination of an open shed facing south, a tight fence around the feed lot and a wind-break beyond, insures comfort for the animals and rapid profitable gains

A feeding barn for fattening beef cattle, in which hay is stored in the centre; the feeding racks and bunks are located under cover, around it

COMFORTABLE, CONTENTED ANIMALS DO THE BEST WORK, GIVE THE MOST MILK, MAKE THE QUICKEST GAINS AND RETURN THE LARGEST PROFITS. THE BUILDINGS LARGELY DETER-MINE THE DEGREE OF THEIR COMFORT AND CONTENTMENT

Under some conditions butter making for market does not pay. This is one such condition

Inside the creamery on a practical Massachusetts dairy farm, which has enabled it to secure and hold a high-class, profitable city milk and cream trade

MANUFACTURING WORK ON THE FARM AS ELSEWHERE, IF IT IS TO BE PROFITABLE, MUST BE PLACED ON A BASIS OF SCIENTIFIC EFFICIENCY

Labor-saving devices and conveniently arranged farm buildings mean the elimination of much irksome farm work, and, consequently, the shortening of working hours.

Construct hay-storage rooms and implement rooms as free of posts and other obstructions as economical and durable construction will admit.

Economy of space often involves a saving in expense and in the convenience of using it. Concentrate all feeds so that their distribution will be central in relation to all stalls. Arrange all the tie stalls nearest the feed room and all pens and box stalls farthest away from the feed rooms so that the largest bulk of feed will be carried the shortest distance.

Arrange cleaning alleys with doors convenient to the manure pit. Arrange tool room so that a definite place may be assigned to each tool

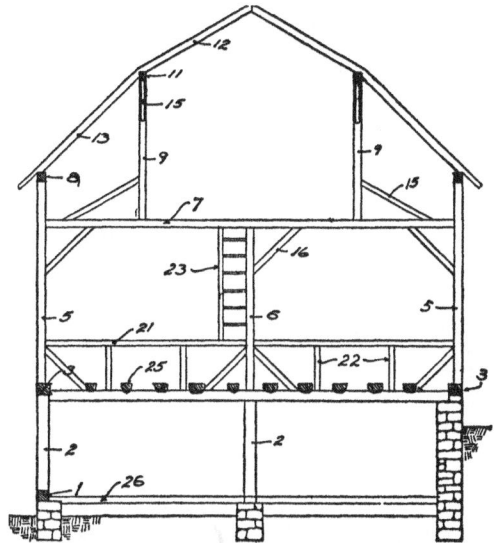

FIG. 526. An intermediate type of barn framing. The basement would be improved by including a bridged space between wall and bank as shown in Fig. 583b.

and implement, and that any tool may be taken out or replaced without disturbing other tools or implements.

Economy in construction. To secure economy in construction use local materials, so far as the requirements of the building will permit. If the farm contains a bank of good building sand, all foundation work and most of the lower-story walls can be built as cheaply of concrete as lumber purchased at retail prices. Should the farm contain field stone, it may be used with economy in the walls and foundation of any kind of farm building. Where no sand, gravel, or stone is convenient, brick or hollow tile may be most economical. Most of the local woods which grow in the woodlot may be used with economy and satisfaction for all farm buildings.

For average conditions, a plank frame superstructure is more economical than one built of heavy square timbers.

The roofing question is an important one, hinging, to a large extent, on local markets. It is advisable to procure quotations on several good roofing materials available in the locality, before a decision is made.

While economy in construction is desirable because it keeps down the original cost of a building, strength and durability are also desirable because they insure permanence and low expense for upkeep. Quite often it is possible to procure a suitable material which combines economy with strength and durability. Time is always well spent in investigating local market conditions before deciding definitely what material is the most efficient. To build for permanence may not add greatly to the original cost; and the saving in upkeep, insurance, and depreciation will soon make the permanent building the cheapest.

Protection against fire. Fire causes most of the building losses on farms, and has destroyed in the United States in a year buildings worth nearly half as much as the new ones erected in the same period. Fire insurance may repay the farmer the worth of the burned portions of a building; but it does not make up for the total loss incurred by the exposure of livestock to storms, nor for other inconveniences and expenses incurred during the rebuilding or restoration of the damaged premises.

Use fire-resisting materials, such as stone, brick, concrete, or tile as much as possible. If wood construction is used for the superstructure, place the buildings a safe distance apart—100 feet or more is advisable. When, for the sake of convenience, or from lack of space, or owing to climatic conditions, it is necessary to connect the buildings, build fire walls with fire-resisting doors between all buildings.

Design should be pleasing and logical. A pleasing appearance is by no means a minor consideration, and it can be obtained without additional expense. Beautiful landscapes defiled by unsightly structures should be a thing of the past. Substantial, sightly buildings give the farm an appearance of prosperity.

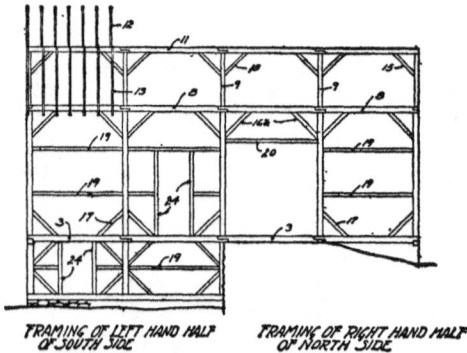

TRAMING OF LEFT HAND HALF OF SOUTH SIDE   TRAMING OF RIGHT HAND HALF OF NORTH SIDE

FIG. 527 Other details of the heavy timber type of barn framing

The tendency to follow the design of existing buildings in a neighborhood is no doubt the greatest hindrance to progress in barn construction. Defects are perpetuated and improvements ignored in copying the arrangement and construction of old barns, even if these have been found handy and durable. Farming methods are changing rapidly, and the old type of barn will soon be classed among relics.

Each building should be carefully planned for the service it is to render, regardless of previous practice. Time and effort expended in analyzing the needs of this service and in planning to meet these needs in the best possible manner will save labor and expense as long as the building is used. Compact arrangements pay dividends by saving labor.

Building Materials and Their Proper Use in Construction

It may be well said that in the construction of farm buildings there is a proper place for all building materials and that each material should be applied in its proper place. No building material meets all of the requirements mentioned in the preceding paragraphs, but the one best suited to its particular purpose should be selected from the many available. Below are given certain rules and standards with which the various materials must comply in order to meet the building requirements, together with suggestions as to how the materials should be applied and what materials may be used with the best results.

Excavations for buildings. The ground must be excavated before putting in foundations, either for basement or for trenches, to place the masonry below the frost line. The nature of the ground must be carefully noted, after excavations have been carried down to a depth where no frost will occur under the walls; the kind of subsoil will determine the width of the footing or lower course of stone or concrete upon which the weight of the building will rest. For building purposes, the various kinds of soil may be classified as follows: bed rock, clay, gravel, and sand.

Bed rock is generally safe to build on for any kind of structure, provided the foundation beds are kept level.

Gravel, even when mixed with small boulders, may also be considered perfectly reliable for any ordinary structure.

Sand will carry very heavy loads, provided it is confined; but great precautions must be taken to confine it properly, and also to keep running water from it, as the action of the latter will soon wash it away.

Clay, when compact and dry, will carry large loads, but water should be kept from it, both under and around the structure, because wet clay is pasty and semiliquid, causing the foundation to slip and settle unequally.

Clayey soils should be thoroughly tiled inside and outside of the foundation footing, and gravel or cinders used for filling in above the tile and against the foundation.

The silt, slush, and decayed vegetation of the marshy lands in the southern states require spread footings.

In all cases, the base of the foundations should be so spread out as to keep the pressure per square foot of footings within the safe limit. The accompanying table gives the safe bearing power of soils, and may be used by calculating the weight of the building and its contents and spreading the foundations to cover the required area for its safe load.

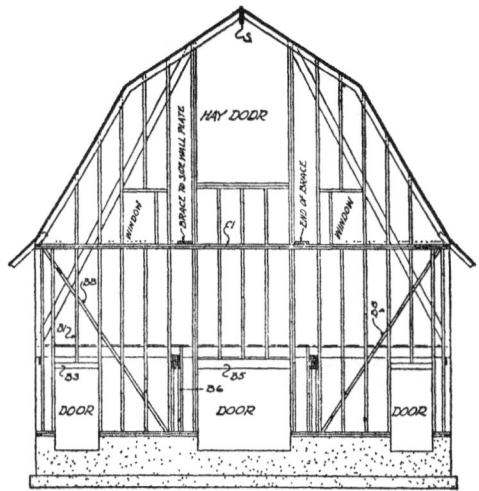

Fig. 528. End framing of a 36-foot wide barn built on the braced-rafter type. This design makes the use of a hay hoisting rig and track (S) practically essential.

BEARING POWER OF SOILS IN TONS PER SQUARE FOOT

KIND OF MATERIAL	MINI-MUM	MAXI-MUM
Rock, equal to stone masonry	25.25	40
Clay, dry, in thick beds	4	6
Clay, soft	1	2
Gravel or coarse sand, well cemented . .	8	10
Sand, compact and well cemented	4	6
Sand, clean and dry .	2	4
Quicksand, alluvial soils, etc.	0.5	1

Foundation footings and walls. By spreading the load of the structure over a larger area of bearing surface, the weight of the building is more evenly distributed. Timber is sometimes used for spread footings where a large bearing surface is necessary; this is perfectly sound construction, if the ground is continually wet, because timber will not decay when kept saturated with water.

The best method of using plank under a masonry wall is to use 4 x 12 inch plank, cut to short, uniform lengths and placed crosswise in the trench so that the grain of the wood runs at right angles with the length of the wall. The masonry wall should be placed in the centre so that the planking will project equally beyond both faces of the wall.

Good concrete is unquestionably the best material for footings, even if stone or brick is used for the foundation walls. It is continuous, without joints, and has the advantage of arching itself over any soft places in the soil that would allow other materials to settle, cracking the walls. Concrete footings are generally made by excavating the bottom of the wall trench to the exact width required for the concrete and with a true, level bottom, so that the concrete may be poured directly into the trench without the use of forms.

Foundation walls are mostly built out of stone, concrete, brick, or tile. Either of these may be used, provided it is of first-class material, for any kind of foundation; and the selection may be made by comparing their cost in the locality.

Materials Used in Masonry Construction

All walls, whether of brick, tile, or stone, should be laid in cement mortar, or lime-cement mortar mixed in the proportions hereafter given.

Lime. Lime is the product of the burning of limestone. Both cement and lime should be kept in a dry place, as exposure to the air causes lime to air-slake and cement to "take a set" and become useless.

Cement. Portland cements are composed of pure clay and pure lime combined in certain definite proportions, thoroughly mixed, and calcined, then ground to a fine powder, and packed in sacks. These cements will stand 4 parts of sand to 1 of cement, for cement mortar.

Sand. The sand used in making mortar should have no mixture of clay or loam in it, but should be clean and sharp particles of quartz or other disintegrated rock. These particles enter readily into the irregularities of the surfaces of stone or brick, thus forming a perfect bond.

Lime mortar. For lime mortar, the lime should be slaked by pouring water on it. The water is very quickly absorbed by the lime in the process of slaking; care should be taken to keep all the pieces wet by adding water as fast as the lime will take it up.

Lime mortar is prepared in much the same

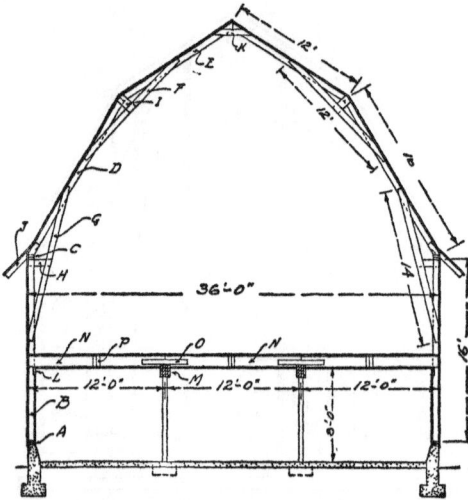

Fig. 529. Section of the barn shown in Fig. 528, showing the abundant mow space and absence of interfering supports, rafters, etc.

way as pure cement mortar. For mixing lime mortar, a bed of sand is first made in the mortar box, and the lime is distributed as evenly as possible over it, both the lime and the sand being first measured, in order that the proper proportions may be obtained. The proportion of sand and lime usually specified is 3 of sand to 1 of lime.

It is considered better to make lime in large quantities; then to leave it in piles for use as it may be needed, after stirring and tempering.

Cement-and-lime mortar. For this mortar the lime should be first slaked and then allowed to stand until it is thoroughly cool. Mix the cement and sand together dry, using 1 measure of cement to 4 measures of sand; to this add the slaked lime, mix thoroughly, and reduce with water to the proper consistency. Cement-and-lime mortar should be used as soon as mixed, before the cement sets.

Cement mortar. This should be mixed in the proportion of from 3 to 4 parts sand to 1 of cement. It is advisable that these parts be actually measured, so that there will be no question about having the proportions just right. After the sand and cement are thrown on the platform, they must be thoroughly mixed by shoveling the two materials together, at least twice, so that the cement may be thoroughly incorporated with the sand. Sufficient water should now be added to make a stiff paste, and the mortar must be immediately conveyed to the work and used, as the cement sets very rapidly and, after it is once hard, the mortar cannot be used again.

Brick. Brick may be called an artificial stone, manufactured in small pieces for convenience in laying. The principal ingredients in brick are clay and protoxide of iron. In the eastern states, most of the common brick are made $7\frac{3}{4}$ x $3\frac{3}{4}$ x $2\frac{1}{2}$ inches; in the western states the common brick are made $8\frac{1}{2}$ x $4\frac{1}{4}$ x $2\frac{1}{2}$ inches.

Good building brick should be well burned and quite hard. It should have a ringing sound when two bricks are struck together. A good brick should not absorb more than one tenth its weight in water, nor should it break under a crushing load of less than 4,000 pounds per square inch.

Brick masonry. No brick walls, even for the smallest buildings, should be built less than 8 inches in thickness. Eight-inch walls should never be more than 10 feet, nor more than one story in height. All walls that are subject to lateral pressure from movable materials piled against them, from heavy wind pressure or from the outward thrust of roof rafters, should never be less than 12 inches in thickness. No 12-inch walls should be more than 22 feet in height or more than 2 stories high.

Walls for 1-story buildings having many windows, such as dairy barns, should have either 12-inch walls or 8-inch walls reinforced at intervals with pilasters. These pilasters may be built hollow and, in the case of livestock barns, may be used for ventilation flues. Where pilasters are to support concentrated loads from the ends of girders, roof-trusses, etc., they should be built solid and with cement mortar.

Brick laying. All brickwork must have a good bed of mortar between all bricks. Brickwork, therefore, consists of both brick and mortar. The strength of any piece of brickwork depends on the quality of the brick, the strength and quality of the mortar, the way in which the bricks are laid and bonded, and whether or not the bricks are laid wet or dry.

Brick should be laid in a bed of mortar at least three sixteenths and not more than three eighths of an inch in thickness. Every vertical joint between ends of brick and all spaces between the outer and inner courses of brick must be filled in solid with mortar, so that no air space will exist within the walls.

Stone masonry. If rubble stone can be procured on the farm or near by, it will no doubt make the most economical wall; and it may be used from the footing up to the window-sill level in barns where a large number of windows are necessary. In many cases, too, it may be carried up to the hay-mow floor level or to the roof.

It does not pay to carry the stonework up between the windows where they are spaced close together, because the amount of labor required in dressing the stone to square corners will cost more than other and more appropriate materials.

The most satisfactory manner of measuring stonework is by the cubic foot or cubic yard.

The cord (128 cubic feet) is sometimes used. The perch is a measure that is often misleading, because its value ranges all the way from 16 to 24 cubic feet, according to local custom.

It requires about one third cubic yard of mortar to lay up 1 cubic yard of rubble-stone work. This amount of cement mortar will require about two thirds of a barrel of cement and three tenths of a cubic yard of sand.

The safe bearing load on rubble-stone masonry is calculated at 80 pounds per square inch, if lime mortar is used, and 150 pounds if cement mortar is used.

A wall can be built very economically of field stone or boulders by proceeding as in laying concrete. Set in the plank forms to the thickness of the walls, fill in with cement mortar about 3 inches deep, then place the stones in this mortar, packing them as close together as possible, and placing the smaller stones between the larger ones. After a layer of stone has been placed in the form, put in more mortar, covering the stone over with mortar several inches deep, and then proceed with another layer of stone as before. This can be done by unskilled labor and the work will progress very rapidly. One side of the form is built up the full height of the wall to start with; and the side from which the stone is laid is built up one plank at a time as the wall is laid, so that there will be no unnecessary lifting of materials.

Concrete masonry. In estimating the amount of material required for barn or basement walls, the accompanying table compiled by a well-known cement company gives information for 3 heights of concrete walls. The proportions of the ingredients are 1 part of cement to 2½ of sand and 5 of stone or gravel.

DIMENSIONS OF WALL			MATERIAL PER 10 RUNNING FT.		
Height Feet	THICKNESS		Cement Sacks	Sand Cu. Ft.	Gravel Cu. Ft.
	Bottom Inches	Top Inches			
6	6	6	6	14½	29
8	10	8	12	29	58
10	15	10	24	57	114

The materials required for 1 cubic yard of concrete of various mixtures may be found by the following table, published by another cement company:

MIXTURES			QUANTITIES OF MATERIALS		
Cement	Sand	Gravel	Sacks of Cement	Cubic Yards of Sand	Cubic Yards of Gravel
1	2	3	7.0	.52	.78
1	2	4	6.0	.44	.89
1	2.5	4	5.6	.52	.83
1	3	5	4.6	.51	.85

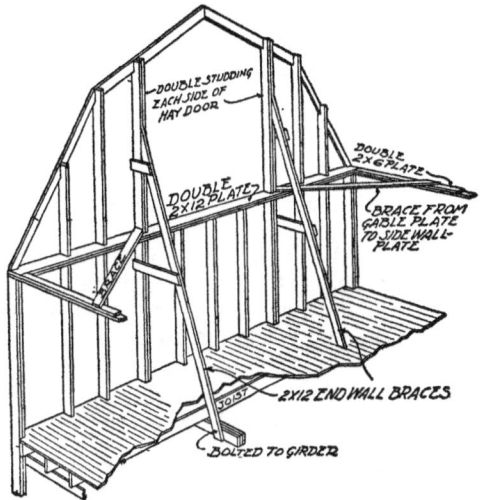

Fig. 530. Detail of end of barn built on the braced rafter plan, showing how to brace the gable end walls to resist the hay pressure (see p. 377).

Concrete, when properly handled, is recognized by the farmer as satisfactory material for all forms of building work. The advantages of good concrete are: protection against fire and decay; absence of repairs; saving of insurance; practical indestructibility; cleanliness; and vermin proofness—all resulting in the greatest economy. Good concrete is water-tight; walls built of it are light in color, reflecting light to all parts; and it leaves no small corners, crevices, or pockets in which dirt may collect.

The use of concrete by farmers has increased enormously, but it has been pretty largely restricted to the building of floors, walks, foundations, and retaining walls. The reason for this is that it has been necessary to erect wooden forms, making heavy, cumbersome work at great expense. To-day, adjustable steel forms, that can be used for any type of solid or double wall, may be purchased at such a low price that concrete walls can, in most places, be built for about the price they would cost if built of wood.

The double-wall construction, with a continuous air space for insulation against temperature and dampness, meets all the requirements for any type of farm building.

Small 1-story farm buildings may be built strictly fireproof by making the roof of concrete on stiffened, expanded metal. The latter can be purchased in lengths up to 12 feet. The expanded metal is supported by the walls and on steel or reinforced concrete beams between the walls at intervals of 6 to 10 feet. Concrete is spread on top of this expanded metal to a thickness of about 2 inches and, when hard, the underside is also plastered with concrete, so that the metal is entirely imbedded in the slab. When thoroughly set, this slab becomes very strong; and the top surface is then coated with asphalt or some other elastic roofing paint to make it thoroughly waterproof. This method of roof construction is very economical, is stormproof, waterproof, and indestructible.

Carpentry

The rules which govern the grading of the various kinds of lumber sold on the market are determined by the various lumber associations. These associations publish small pamphlets containing the grading rules, which may be obtained free upon request. For additional information regarding the properties of lumber, application should be made to the United States Department of Agriculture, Forest Service, Washington, D. C.

Standard sizes of dressed lumber. Finishing lumber is dressed to the following: 1 inch S. 1 S. or 2 S. to $\frac{13}{16}$; $1\frac{1}{4}$ inches S. 1 S. or 2 S. to $1\frac{1}{16}$; $1\frac{1}{2}$ inches S. 1 S. or 2 S. to $1\frac{5}{16}$; 2 inches S. 1 S. or 2 S. to $1\frac{3}{4}$ inches. 1 x 4 S. 4 S. is $3\frac{1}{2}$ inches wide finished; 1 x 5 S. 4 S. is $4\frac{1}{2}$ inches wide; 1 x 6 is $5\frac{1}{2}$ inches; 1 x 7 is $6\frac{1}{2}$ inches; 1 x 8 is $7\frac{1}{2}$ inches; 1 x 9 is $8\frac{1}{2}$ inches; 1 x 10 is $9\frac{1}{2}$ inches; 1 x 11 is $10\frac{1}{2}$ inches, and 1 x 12 is $11\frac{1}{4}$ inches.

Flooring. The standard of 1 x 3, 1 x 4, and 1 x 6 inches D. and better is worked to $\frac{13}{16}$ x $2\frac{1}{4}$, $3\frac{1}{4}$, and $5\frac{1}{4}$ inches; $1\frac{1}{4}$-inch flooring is $1\frac{3}{32}$ inches thick; $1\frac{1}{2}$-inch flooring is $1\frac{11}{32}$ inches thick, the same width and matching as 1-inch stock.

Drop siding. D. and M. is worked to $\frac{3}{4}$ x $3\frac{1}{4}$ and $5\frac{1}{4}$ inches face, $4\frac{1}{2}$ and $5\frac{1}{2}$ over all. Worked ship-lap, $\frac{3}{4}$ x 5 inches face, $5\frac{1}{2}$ over all.

Ceiling. Ceiling is worked to the following: $\frac{3}{8}$-inch ceiling, $\frac{5}{16}$ inch; $\frac{1}{2}$-inch ceiling, $\frac{7}{16}$ inch; $\frac{5}{8}$-inch ceiling, $\frac{9}{16}$ inch; $\frac{3}{4}$-inch ceiling, $\frac{11}{16}$ inch. Same width as flooring. Working of ceiling is beaded centre and edge with slight bevel on groove edge. The bead on all ceiling and partition is depressed $\frac{1}{32}$ inch below surface line of piece.

Bevel siding. This is made from stock S. 4 S. worked to $\frac{13}{16}$ x $3\frac{1}{2}$ and $5\frac{1}{2}$, and resawed on a bevel.

Ship-lap and barn siding No. 1 common. 8, 10, and 12 inches is worked to $\frac{13}{16}$ x $7\frac{1}{8}$, $9\frac{1}{8}$, and $11\frac{1}{8}$ inches.

D. and M. No. 1 Common. 8, 10, and 12 inches is worked to $\frac{13}{16}$ x $7\frac{1}{8}$, $9\frac{1}{8}$, and $11\frac{1}{8}$ inches.

Grooved roofing. 10 and 12 inches S. 1. S. and 2 E. is worked to $\frac{13}{16}$ x $9\frac{1}{2}$ and $11\frac{1}{4}$.

Dimensions are worked to the actual sizes shown in the accompanying table:

ACTUAL SIZES OF LUMBER DIMENSIONS, IN INCHES

(SOUTHERN YELLOW PINE MANUFACTURERS' ASSOCIATION)

FOR SURFACED ONE SIDE AND ONE END

BREADTH DEPTH	2 INCHES	4 INCHES	6 INCHES	8 INCHES	10 INCHES
4 inches	$1\frac{5}{8}$ x $3\frac{5}{8}$	$3\frac{5}{8}$ x $3\frac{5}{8}$			
6 inches	$1\frac{5}{8}$ x $5\frac{5}{8}$	$3\frac{5}{8}$ x $5\frac{5}{8}$	$5\frac{5}{8}$ x $5\frac{5}{8}$		
8 inches	$1\frac{5}{8}$ x $7\frac{1}{2}$	$3\frac{5}{8}$ x $7\frac{5}{8}$	$5\frac{3}{4}$ x $7\frac{3}{4}$	$7\frac{3}{4}$ x $7\frac{3}{4}$	
10 inches	$1\frac{5}{8}$ x $9\frac{1}{2}$	$3\frac{5}{8}$ x $9\frac{5}{8}$	$5\frac{3}{4}$ x $9\frac{3}{4}$	$7\frac{3}{4}$ x $9\frac{3}{4}$	$9\frac{3}{4}$ x $9\frac{3}{4}$
12 inches	$1\frac{5}{8}$ x $11\frac{1}{2}$	$3\frac{5}{8}$ x $11\frac{1}{4}$	$5\frac{3}{4}$ x $11\frac{1}{4}$	$7\frac{3}{4}$ x $11\frac{1}{4}$	$9\frac{3}{4}$ x $11\frac{1}{4}$

FOR SURFACED FOUR SIDES

BREADTH DEPTH	2 INCHES	4 INCHES	6 INCHES	8 INCHES	10 INCHES
4 inches	$1\frac{1}{2}$ x $3\frac{1}{2}$	$3\frac{1}{2}$ x $3\frac{1}{2}$			
6 inches	$1\frac{1}{2}$ x $5\frac{1}{2}$	$3\frac{1}{2}$ x $5\frac{1}{2}$	$5\frac{1}{2}$ x $5\frac{1}{2}$		
8 inches	$1\frac{1}{2}$ x $7\frac{3}{4}$	$3\frac{1}{2}$ x $7\frac{1}{2}$	$5\frac{1}{2}$ x $7\frac{1}{2}$	$7\frac{1}{2}$ x $7\frac{1}{2}$	
10 inches	$1\frac{1}{2}$ x $9\frac{3}{4}$	$3\frac{1}{2}$ x $9\frac{1}{2}$	$5\frac{1}{2}$ x $9\frac{1}{2}$	$7\frac{1}{2}$ x $9\frac{1}{2}$	$9\frac{1}{2}$ x $9\frac{1}{2}$
12 inches	$1\frac{1}{2}$ x $11\frac{1}{4}$	$3\frac{1}{2}$ x $11\frac{1}{2}$	$5\frac{1}{2}$ x $11\frac{1}{2}$	$7\frac{1}{2}$ x $11\frac{1}{2}$	$9\frac{1}{2}$ x $11\frac{1}{2}$

BOARD FEET OF LUMBER IN TIMBERS OF VARIOUS SIZES

SIZE IN INCHES	LENGTH IN FEET								
	10	12	14	16	18	20	22	24	26
2 x 4	$6\frac{2}{3}$	8	$9\frac{1}{3}$	$10\frac{2}{3}$	12	$13\frac{1}{3}$	$14\frac{2}{3}$	16	$17\frac{1}{3}$
2 x 6	10	12	14	16	18	20	22	24	26
2 x 8	$13\frac{1}{3}$	16	$18\frac{2}{3}$	$21\frac{1}{3}$	24	$26\frac{2}{3}$	$29\frac{1}{3}$	32	$34\frac{2}{3}$
2 x 10	$16\frac{2}{3}$	20	$23\frac{1}{3}$	$26\frac{2}{3}$	30	$33\frac{1}{3}$	$36\frac{2}{3}$	40	$43\frac{1}{3}$
2 x 12	20	24	28	32	36	40	44	48	52
2 x 14	$23\frac{1}{3}$	28	$32\frac{2}{3}$	$37\frac{1}{3}$	42	$46\frac{2}{3}$	$51\frac{1}{3}$	56	$60\frac{2}{3}$
4 x 4	$13\frac{1}{3}$	16	$18\frac{2}{3}$	$21\frac{1}{3}$	24	$26\frac{2}{3}$	$29\frac{1}{3}$	32	$34\frac{2}{3}$
4 x 6	20	24	28	32	36	40	44	48	52
4 x 8	$26\frac{2}{3}$	32	$37\frac{1}{3}$	$42\frac{2}{3}$	48	$53\frac{1}{3}$	$58\frac{2}{3}$	64	$69\frac{1}{3}$
4 x 10	$33\frac{1}{3}$	40	$46\frac{2}{3}$	$53\frac{1}{3}$	60	$66\frac{2}{3}$	$73\frac{1}{3}$	80	$86\frac{2}{3}$
6 x 6	30	36	42	48	54	60	66	72	78
6 x 8	40	48	56	64	72	80	88	96	104
6 x 10	50	60	70	80	90	100	110	120	130
6 x 12	60	72	84	96	108	120	132	144	156
6 x 16	80	96	112	128	144	160	176	192	208
8 x 8	$53\frac{1}{3}$	64	$74\frac{2}{3}$	$85\frac{1}{3}$	96	$106\frac{2}{3}$	$117\frac{1}{3}$	128	$138\frac{2}{3}$
8 x 10	$66\frac{2}{3}$	80	$93\frac{1}{3}$	$106\frac{2}{3}$	120	$133\frac{1}{3}$	$146\frac{2}{3}$	160	$173\frac{1}{3}$
8 x 12	80	96	112	128	144	160	176	192	208
10 x 10	$83\frac{1}{3}$	100	$116\frac{2}{3}$	$133\frac{1}{3}$	150	$166\frac{2}{3}$	$183\frac{1}{3}$	200	$216\frac{2}{3}$
10 x 12	100	120	140	160	180	200	220	240	260
12 x 12	120	144	168	192	216	240	264	288	312
12 x 14	140	168	196	224	252	280	308	336	364

Heavy flooring. For 2 and $2\frac{1}{2}$ inches, matching, the thickness should be $\frac{3}{8}$ inch less than the rough material. The tongue should be $\frac{3}{8}$ inch thick and $\frac{3}{8}$ inch long. For 3-inch and thicker matching, the tongue should be $\frac{1}{2}$ inch thick and $\frac{3}{8}$ inch long, and the thickness of the stock should be $\frac{3}{8}$ inch less than the rough material.

Heavy ship-lap. This is worked to the same thickness as heavy flooring. The lap is $\frac{1}{2}$ inch long, occupying one-half the finished thickness of the piece.

Timbers. These are worked to the following: 4 x 4 and larger S. 1. S. or S. & E., $\frac{3}{8}$ inch off each face surfaced; S. 3 S. or S. 4 S., $\frac{1}{2}$ inch off each face surfaced.

Plastering lath. No. 1 measure 2 inches in thickness to every 5 lath, green; the minimum thickness of any one lath is not less than $\frac{1}{16}$ inch green, and is not less than $1\frac{7}{16}$ inches in width, green, length 4 feet, $1\frac{5}{8}$ inches thickness to every 5 lath, dry, and do not measure less than $1\frac{7}{16}$ inches in width, dry. They are not more than $\frac{1}{2}$ inch scant in length when dry.

To determine the number of board feet in timber, it should be remembered that a board foot is the measure of a piece of timber 12 inches long, 12 inches wide, and 1 inch thick. Thus a block of timber 12 inches square on the end, and 1 foot long, would contain 12 board feet.

Shrinkage of timber. Timber shrinks but very little lengthwise of the grain when drying, but crosswise the shrinkage may be quite considerable. The soft woods, such as pine, spruce, or cypress, shrink evenly, with but little cracking; but the hard woods, such as oak and hickory, are often subject to injury in shrinking.

APPROXIMATE SHRINKAGE OF TIMBER PER FOOT OF WIDTH IN DRYING

KIND OF WOOD	SHRINKAGE	KIND OF WOOD	SHRINKAGE
Ash	.60	Horse chestnut	.72
Basswood	.72	Locust	.72
Beech	.60	Maple	.60
Birch	.72	Oak	1.20
Box elder	.48	Pine, hard	.48
Cedar	.36	Pine, soft	.36
Cherry	.60	Poplar	.60
Chestnut	.72	Spruce	.36
Cypress	.36	Sycamore	.60
Elm	.60	Tamarack	.48
Hickory	1.20	Walnut	.60
Honey Locust	.48		

Ordinary lumber waste. In the use of ordinary lumber on walls, floors and ceilings, the following percentages should be added to the actual measurement of the surface to be covered, in order to allow for lapping, matching, etc.:

Battens, 1 x 4, placed 6 inches on centres, only $\frac{3}{4}$ of surface measure is needed.

Battens, 1 x 6, placed 8 inches on centres, only $\frac{3}{4}$ of surface measure is needed.

Ceiling will be same as flooring (see below).

	PER CENT
Flooring, 3-inch matched	50
Flooring, 4-inch	33
Flooring, 6-inch	20

Papers and felts are usually listed sufficiently below the actual contents of the roll to allow for lapping.

	PER CENT
Sheathing, common, laid horizontally on walls without openings	10
Sheathing, common, laid horizontally on roofs without openings	10
Sheathing, common, laid diagonally on buildings with usual openings . . .	17
Sheathing, tight, 6 inches, laid horizontally	20
Sheathing, tight, 8 inches, laid horizontally	15
Sheathing, tight, 10 inches laid horizontally	12
Sheathing, tight, 6 inches, laid diagonally	25
Sheathing, tight, 8 inches, laid diagonally	17
Sheathing, tight, 10 inches, laid diagonally	12
Siding, drop.	20
Siding, lap, 4 inches to weather . . .	50
Siding, lap, 4½ inches to weather. . .	33

To estimate the quantity of sheathing or of ship-lap, approximately, calculate the exact surface to be covered, deducting openings; then add the following percentages:

	SHEATHING PER CENT	SHIP-LAP PER CENT
For floors. . .	1/7 or 15	1/6 or 17
For side walls .	1/6 or 17	1/5 or 20
For roofs . . .	1/5 or 20	1/4 or 25

WEIGHT OF WIRE NAILS NEEDED PER 1,000 FEET OF LUMBER

KIND OF MATERIAL	DISTANCE APART OF JOIST, OR STUDDING NAILING SPACE, IN INCHES	NUMBER OF NAILS TO EACH BOARD, EACH NAILING SPACE	SIZE OF NAIL	POUNDS OF NAILS
Framing 1 x 4	16	2	10d. common	65
" 1 x 4	24	2	10d. "	45
" 1 x 6	16	2	10d. "	43
" 1 x 6	24	2	10d. "	30
" 1 x 8	16	2	10d. "	32
" 1 x 8	24	2	10d. "	23
" 1 x 10	16	3	10d. "	39
" 1 x 10	24	3	10d. "	27
" 1 x 12	16	3	10d. "	32
" 1 x 12	24	3	10d. "	23
" 2 x 6	24	2	30d. "	53
" 2 x 8	24	2	30d. "	40
" 2 x 10	24	3	30d. "	48
" 2 x 12	24	3	30d. "	40
Ship-lap 1 x 8	16	2	10d. "	36
" 1 x 8	24	2	10d. "	26
" 1 x 10	16	3	10d. "	43
" 1 x 10	24	3	10d. "	30
" 1 x 12	16	3	10d. "	35
" 1 x 12	24	3	10d. "	25
Flooring 1 x 4	16	1	8d. "	27
" 1 x 6	24	1	8d. "	12
" 1 x 6	16	2	8d. "	36
" 1 x 6	24	2	8d. "	24
" 1 x 8	16	2	10d. "	36
" 1 x 8	24	2	10d. "	26
Ceiling 3/8 x 4	24	1	6d. finishing	6
" 3/8 x 6	24	1	6d. "	4
" 7/8 x 6	24	1	8d. common	12
Siding 1/2 x 4	16	1	8d. finishing	15
" 1/2 x 6	16	1	8d. "	10

Estimating quantities of nails. The accompanying table gives the number of wire nails in pounds for various kinds of lumber, per thousand feet board measure, allowance being made for loss of covering surface due to lap or matching of material. The sizes given are as rated on the market.

Shingles, per 1,000, require 3½ pounds of 3d., or 5 pounds of 4d. nails. Lath, ordinary, per 1,000, studding spaced 16-inch centres, 8 pounds of 3d. common wire nails. Bridging, per set for 2 x 10 joists spaced 16-inch centres will require 26 pounds of 8d. common wire nails per 1,000 linear feet of bridging.

Framing studding will require 15 pounds of 10d., and 5 pounds of 20d. nails, per 1,000 feet of studding.

Framing joists will require approximately the following amounts of 20d. nails per 1,000 feet:

Frame buildings, 16-inch centres	15 lbs.
Brick buildings, 16-inch centres	10 lbs.
Brick buildings, 12-inch centres	12 lbs.

Finish seven eighths inch, will require about 20 pounds of 8d. finishing nails per 1,000 feet, while 1¼ inches will require 30 pounds of 10d. finishing nails per 1,000 feet.

Loads on Structures

Sometimes it becomes necessary to calculate the loads on a structure, in order to determine the exact size of timber required. The loads on a structure may be divided as follows: The *dead load* is the weight of the material of which the structure is composed. The *live load* is the weight of the various articles in the building that are not a part of the stucture itself. The *snow* and *wind loads* are, as the names imply, loads that are caused respectively by snow and wind.

In calculating the loads on a structure for the purpose of determining the size of timber or thickness of walls that may be required, with the view of building safely but without waste, the live load must be added to the dead load in figuring the total load on a floor beam. The snow and wind loads must be added to the dead load for figuring the total load on a rafter. All loads combined must be figured in calculating the size of foundation footings.

The accompanying tables (on this and the next page) will be found useful in calculating the live and dead loads on structures:

WEIGHT OF BUILDING MATERIALS PER SQUARE FOOT

NAME OF MATERIAL	POUNDS PER FOOT
Corrugated (2½ inch)	
No. 26 galvanized iron . .	.99
No. 28 galvanized iron . .	.86
Felt and pitch roofing without	
sheathing	3
Glass ⅛ inch thick . . .	1¼
Lead ¼ inch thick	6 to 8
Lath and plaster (ordinary) .	6 to 8
Sheathing 1 inch thick	
Hemlock	2
Spruce	2
Yellow pine	3
Chestnut	4
Maple	4
Ash or Oak	5
Sheet iron 1-16 inch thick .	3
Shingles 5 inches to weather .	2
Steel roofing	1
Tile, Spanish roofing . . .	8½
Tile, plain	18
4-ply gravel roofing . . .	5½
3-ply asphalt roofing . . .	6 to 10

WEIGHT OF BUILDING MATERIALS PER CUBIC FOOT

MATERIAL	POUNDS PER FOOT
Brick work	125
Concrete	140
Earth, dry	90 to 100
Iron, cast	450
Limestone	146 to 168
Masonry, limestone . . .	150
Quicklime	53
Sand, dry	90 to 100
Steel, structural	489.6
Tile	110 to 112

To find the correct size of beam required to support farm produce, first find the total weight of such live load per square foot of floor area upon which it is to rest. Then find the number of square feet of floor area supported by one beam, and multiply this by the live load per square foot just referred to. This gives the total live load to be supported by the beam and by referring to the second table on page 374, the right size of beam can be located at once.

WEIGHT OF FARM PRODUCTS PER CUBIC FOOT, IN POUNDS

PRODUCT	WEIGHT	PRODUCT	WEIGHT
Barley	38	Hay, Alfalfa, in bales	12½ to 14
Beans	48	Hay, Clover, in bales	14
Beets	44	Hay, Clover, loose	4.6
Bran	16	Land plaster	80
Buckwheat	39	Potatoes, white	48
Butter	59	Potatoes, sweet	41
Cabbage	40	Rye	45
Carrots	40	Meal	37
Cheese	30	Middlings	32 to 38
Clover seed	48	Milk	65
Corn on cob, husked	56	Oats	26
Corn on cob, unhusked	58	Peanuts	18
Corn, shelled	46	Silage (at top)	19
Corn meal	38	Silage (at 36 feet)	61
Cottonseed	25	Straw	19
Eggs	68	Wheat	48

SAFE LOADS, IN POUNDS, UNIFORMLY DISTRIBUTED FOR YELLOW-PINE BEAMS

(Supported at Both Ends)

SPAN IN FEET	2 x 6 1⅝ x 5⅝	2 x 8 1⅝ x 7½	2 x 10 1⅝ x 9½	2 x 12 1⅝ x 11½	2 x 14 1⅝ x 13½	2 x 16 1⅝ x 15½
6	1,714	3,047	4,488	7,163	9,872	14,020
8	1,285	2,285	3,666	5,372	7,404	10,515
10	1,028	1,829	2,933	4,298	5,923	8,412
12	857	1,523	2,444	3,582	4,936	7,010
14	734	1,306	2,095	3,070	4,231	6,008
16	642	1,142	1,833	2,686	3,702	5,256
18	....	1,016	1,629	2,388	3,291	4,505
20	....	914	1,466	2,149	2,961	4,206
22	....	....	1,333	1,954	2,692	3,823
24	....	....	1,222	1,791	2,469	3,505
26	....	....	....	1,653	2,278	3,235
28	....	....	....	1,535	2,115	3,804
30	....	....	....	....	1,974	2,804
32	....	....	....	....	1,851	2,628

Loads above heavy horizontal lines are calculated for both strength and stiffness. Loads below heavy horizontal lines calculated for strength only, and will deflect more than one thirtieth of an inch per foot of span, and should not be used with plastered ceilings.

Barn framing. In case the framing timbers for the barn are to be cut out of the woodlot with a portable saw, it may be best to use the old style of heavy-timber construction in some cases, because this takes less pieces and thereby reduces the amount of sawing, although the actual number of board feet of lumber is more than required for a plank-frame barn.

Figs. 525 and 527 show the method of building the heavy-timber barn; and below is a list of the parts, giving their names and average sizes used. The numbers at the left of each item in this list coincide with the numbers in the illustration.

PARTS OF A HEAVY-TIMBER BARN

	NAME OF PART	SIZE		NAME OF PART	SIZE
1.	Basement sill	10 x 12 inches	15.	Purlin braces	3 x 4 inches
2.	Basement posts	12 x 12 "	16.	3-ft.-run brace	3 x 4 "
3.	Main sill	10 x 10 "	16½.	2½-ft.-run brace	3 x 4 "
4.	Cross sill	10 x 10 "	17.	3½-ft.-run brace	3 x 4 "
5.	Main post	8 x 8 "	18.	End girts	4 x 6 "
6.	Centre post	8 x 8 "	19.	Side girts	4 x 6 "
7.	Main beams	8 x 10 "	20.	Door girts	4 x 6 "
8.	Main plate	8 x 8 "	21.	Breast girt	6 x 8 "
9.	Purlin posts	6 x 6 "	22.	Breast-girt studs	3 x 4 "
10.	Purlin beams	6 x 6 "	23.	Ladder post	3 x 4 "
11.	Purlin plate	6 x 6 "	24.	Door posts	4 x 4 "
12.	Upper rafters	2 x 6 "	25.	Overlays, top and ends	
13.	Lower rafters	2 x 6 "		flatted to	6 "
14.	Purlin girts	4 x 6 "	26.	Sleepers	6 x 6 "

Plank, Braced-rafter, and Plank-truss Types of Construction

Plank construction. The steady increase in the price of lumber and building materials has necessitated a closer calculation of their strength. Economy prescribes that each piece shall be only as large as needed to withstand safely the strains to which it will be subjected, and that it shall be so placed that it will be strongest in its allotted position. In the largest and best barns built to-day, you will seldom see timber thicker than 2 inches. This is partly due to the fact that small dealers carry a limited assortment of sizes, and, to a greater extent for the present-day calculations of architects.

Most modern barns are built with self-supporting roofs, as this type of construction eliminates heavy beams and posts and reduces cost. This type of roof resembles the hull of a boat

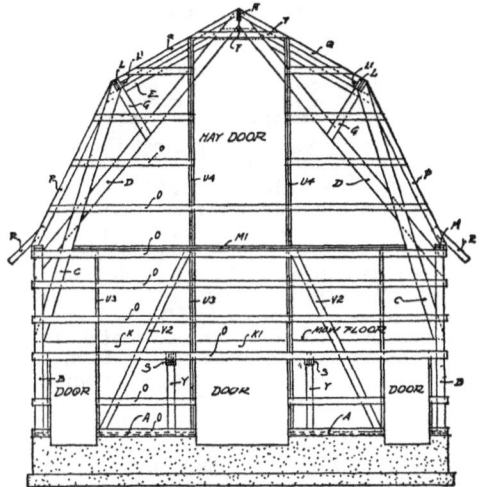

FIG. 531. End framing of a barn 36 feet wide with 16-foot side walls, built according to the plank-truss type of construction.

turned upside down, and consists of built-up plank arches reinforced with splice braces at angles, spanning from one side wall to the other. This roof, usually, has 4 surfaces, the lower 2 being steep, and the upper ones about quarter-pitch. Many make the mistake of calling this type a "hip roof." Its proper name, however, is "gambrel roof" and it is known also as "curb roof" and "mill roof."

Fig. 531 illustrates the construction of a favorite type of modern barn, which consists of a frame structure, the frame of which is built entirely out of planking not over 2 inches thick and on a concrete foundation which extends far enough above floor and outside ground level to prevent moisture from coming into contact with the wooden sill and frame.

The sill should be well bolted on top of the concrete foundation; and the studs are 2 x 6-inch in size for barns of ordinary dimensions, and spaced 16 to 24 inches on centres, the 24-inch spacing being preferred because any stock-length boards can be nailed thereto without waste. The studding is generally of 14- or 16-foot lengths, and has a doubled 2-inch plate spiked on top, which ties the studs together, keeps them in a straight line, and forms a sill for the rafters.

The floor joists of the hay-mow floor are made of 2 x 10- or 2 x 12-inch joists, as the weight of hay may require, and are spaced the same as the studding, so that the end of

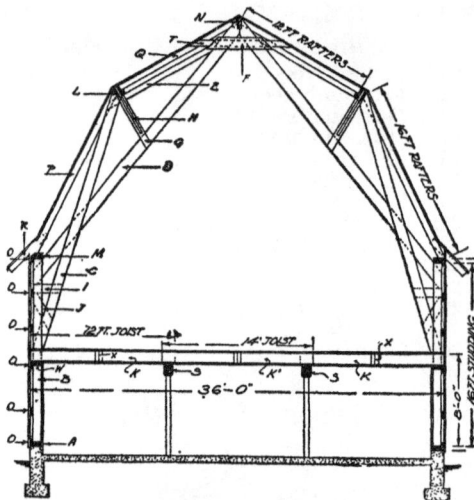

FIG. 532. Section of barn illustrating plank-truss construction, which, at the cost of a little storage space, makes use of lighter and therefore more easily obtained materials, than do other types of barn framing.

Stay bracing. As this article is written more particularly for the inexperienced builder, it may be well to mention that as soon as the studding are set in place, they should be well braced against wind, and as soon as the joists are in place, more braces should be added. These braces should remain until the siding is in place and the roof has been completed, when they may be taken out.

Framing the roof. In framing the roof one set of rafters is carefully laid out on the hay-mow floor or other convenient level platform; and, after the exact length of each piece is computed, these are used as patterns, and the required number of pieces are cut from this one set of patterns. When all rafters, braces, ties, and collar beams have been cut, each set of rafters, braces, ties, etc., is spiked together, so as to form a complete arch rib which will reach from the plate of one side wall to that of the other.

The best method of procedure is to build all of these arches laid flat, one on top of the other, on the building. The ends of one arch (the heels of the lower rafters) are raised to the wall plates and then the point of the

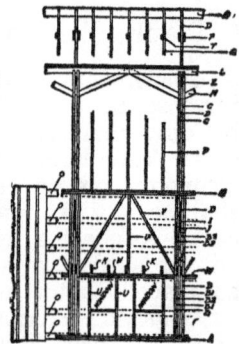

FIG. 533. Side elevation of plank-truss barn construction showing details.

arch is hoisted to a vertical line, plumbed and spiked into place, and braced.

Each arch is nailed to several sheathing boards, which are used as guides and ties to secure the arches as soon as they are raised; and each arch is braced to the studding as soon as set in place. These arches can be raised and set in place by 3 or 4 men, while with the old method of heavy purlin-and-post construction, 10 or 15 would be necessary to help hoist the heavy frame.

This type of frame has the advantage of requiring less material and labor than the heavy timber frame; it is just as strong, and it forms a mow without any obstruction.

each joist may be spiked against the side of the studding and at the same time rest on a 2 x 6-ledger or "ribbon," which is notched 1 inch into the studding and continues the full length of both side walls with as few joints as possible. Three lengths are generally required to reach from one side of the barn to the other. The ends of the middle tier of joists are spiked and lapped against the ends of the 2 outer tiers, so that each set of joists forms a continuous tie from one side wall to the other, to take up the outward thrust of the roof; and the joists are supported under the lapped ends on a set of girders, built up out of 4 thicknesses of 2 x 10- or 2 x 12-inch joists, built up continuously from one end of the barn to the other with as few lengths of plank as possible and all end joints broken, so that there will be not more than one end joint at any one place along the length of the barn. These floor beams are supported by posts or, preferably, iron columns, which are so placed that they will intersect with the line of stanchions and the partitions between the stalls, and rest on concrete piers below the concrete floor.

Braced-rafter construction for basement barns. If a superstructure of the braced-rafter type is built on top of a basement of wood or masonry walls, the joist of the floor which also forms the basement ceiling are set in place and the flooring is laid to within about 12 inches of the outside edge. This floor can then be used as a working platform upon which to build the braced-rafter barn.

Each set of studding, lower rafters, and upper rafters are all completely nailed, together with the braces and tied at the corners, so as to form a complete arch; and these arches are raised into place one at a time, starting at one gable end which is raised and braced first.

With this method of raising, no scaffold is required, but each arch should be well stay-lathed as soon as it is raised and plumbed. It saves labor in framing, as all the heavy nailing is done on the floor instead of in the air.

In this type of barn, the double plate at the top of the studding is omitted, and this joint is tie-braced just as the joint between the lower and the upper rafters is framed.

The narrow matched siding running horizontally is preferred, as it ties the arches together at the studding, just as the sheathing ties them together at the rafters. If vertical siding is desired, place continuous 2 x 4-inch nailing girts on the face of the studding, and break joints so that the girts will form a continuous hoop, or tie, around the building.

Gable-end bracing. The gable ends of the braced-rafter barn should be braced by using double 2 x 12-inch plates on top of the lower studding; and these gable plates should be braced across the corners to the plates of the side walls with 2 x 10 plank or with two 2 x 6's nailed together, so that one sets on edge and the other lies flat, thus being stiffened against hay pressure from both directions. The gable above the plate is braced by running 2 x 12-inch braces from the studding to the girders supporting the mow floor. These braces may be braced back to the gable, to counteract the hay pressure (Fig. 530). Fig. 528 illustrates the method of framing the ends.

Ventilation flues for braced-rafter barns. Wherever foul-air flues are to be built into the side walls, they may be easily formed by substituting 2 x 12-inch studding in place of the regular 2 x 6-inch size, to form the sides of the flue; also, 2 x 12-inch, should be used for the lower rafters and braces at the plate at all places in the roof containing these flues.

After the siding and sheathing have been placed on the outside face of the studding and rafters respectively, cover this surface (inside the flue) with building papers and over this paper surface place another thickness of matched boards, running vertically, to form the outside wall of the flue.

Construct the wall of the flues exposed to the interior of the barn, built in sections, by nailing together two thicknesses of dressed and matched boards, with building paper between, the boards of one thickness running at right angles to those of the other thickness; and, after well nailing, place these sections against the interior edges of the 2 x 12 studding and nail in place, so that the boards on the inside of the flue run vertically and those on the outside run horizontally.

The bottom of the foul-air flues should have an air intake opening equal in area to that of the flue; and there should be another opening of equal area at the ceiling of the room which is to be ventilated, provided with an air-tight wooden cover, hinged at bottom, to swing out, and provided with a ratchet for regulating and closing this opening.

The flues should extend, in an air-tight construction, to the underside of the roof at such places as ventilators or cupolas are placed on the roof. The flues should be kept in as nearly a vertical and straight line as possible.

The air from the mow should never be taken out at the same cupola or ventilator that is expected to suck the air out of the flue. A separate ventilator for each flue will always give better results than two flues run into one ventilator.

Plank-truss construction. In certain parts of the country, it is advisable, owing to the long rainy seasons, to have barns with large floors which may be used for threshing or for unloading the hay wagons within the barn during wet weather.

This barn floor, or interior driveway, is most convenient if placed on a level with the mow floor, running crosswise of the barn. This requires very large doors in the side of the superstructure, which makes it necessary to have the side walls above the mow-floor level from 14 to 16 feet high.

These large openings also make it necessary to concentrate the support of the roof at certain points (between the wide doors) on the side walls. These conditions have developed the plank-truss type of barn construction (Figs. 531 and 532).

Framing the plank-truss barn. The entire dead load, wind load, and live load imposed on the superstructure of this barn is carried by the plank trusses, which are spaced from 12 to 16 feet apart. The total weight is consequently very considerable and the trusses must, therefore, be built carefully and strong and out of good, sound material.

Also the long members require extra strength against lateral-deflection.

The various members of each truss are bolted together, and upon each of these bolts rests a large responsibility. As a chain is no stronger than its weakest link, so this truss is no stronger than its weakest joint. The bolts should be carefully placed, and provided with large, thick washers at the head and nut ends, so that the tightening of the bolt will not crush the wood fibers about the bolt into worthless pulp.

The trusses, after carefully assembling, are raised into position and well braced; after this the studding is framed and set up (Fig. 533); then the purlin and its braces are placed, from a scaffold built on the mow floor; then the rafters are set in place, and the barn is ready for the inclosing lumber.

The ends are framed as shown in Fig. 531. Since the entire weight of this barn is carried by the trusses, it is not necessary to have so many wall studding between the trusses; and, because the studs are set farther apart, it is economy to use vertical siding on this type of barn.

This type of construction has been so successful in the sections of the country requiring a barn floor, that, through wide publicity, it has come into very extensive use in other parts of the country where the barn floor was not called for and has, consequently, been omitted. Under such conditions, the plank-truss barn will cost about 15 per cent more than the braced-rafter construction.

The One-story Type of Barns

A great movement for better barns is now sweeping over the country. Better barns, however, not always mean large, expensively built, pretentious barns.

The farmer who desires to start with a small capital and, at the same time, build well and along scientific lines whatever he does build, may be interested in the 1-story type of barn construction, which will cost very much less to build per head of livestock capacity or per ton of hay capacity than either of the 3 double-story types previously described.

This 1-story type of construction consists of a 1-story shed for livestock, detached from or attached to a 1-story shed for hay storage. Some of the advantages of this type of barn are the following:

Both sheds need not be built at the same time. If capital is limited, the shed for the livestock may be built first and, if necessary, the hay may be stacked in the yard under a tarpaulin for the first season. This reduces the original cost to a very small amount.

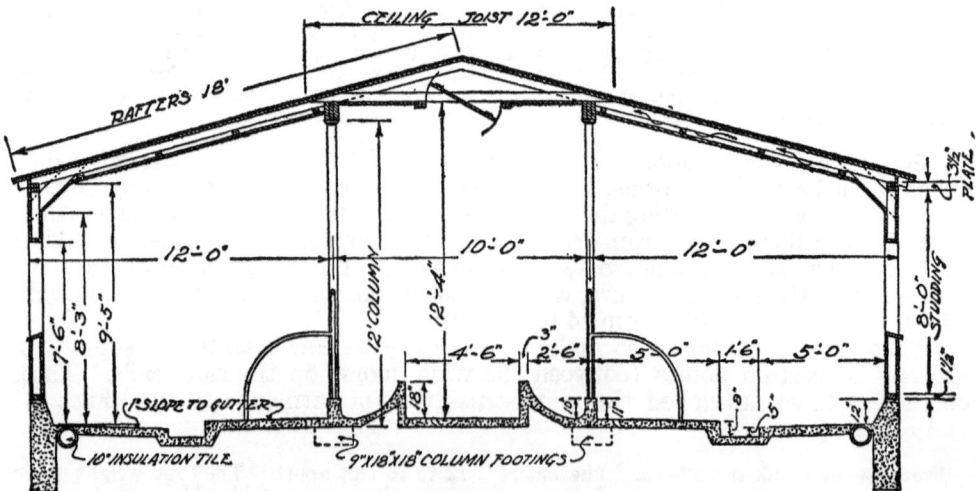

FIG. 534. Section of a single-story dairy cattle barn with concrete floor, metal interior equipment, double walls providing ventilation flues. It is arranged so that the cows face inward toward a central feeding alley. (See Chapter 34 for advantages and disadvantages of this plan.)

The hayshed, built to the north of the stockshed and at right angles to it, forming an L, gives protection from north winds to the stockshed and the exercise lot.

The storage of the hay on the same ground level with the stock saves labor in filling the mow, because the hay track is about 11 feet nearer the ground—a saving of about 60 per cent in hoisting the hay to the track.

Where first-class milk is produced in the 2-story barn, requiring hay to be thrown down from the mow above to a dustproof room at one end of barn, the 1-story barn saves labor in feeding hay, because it is not necessary to climb up a ladder to get to the mow and then to carry the hay to the chute.

It is more sanitary, because more dust is excluded from the cow barn by eliminating the hay chute, and fewer odors from the stock will reach the hay. The ventilating system is better, because the ceiling is arched and air can be taken out at the centre, when so desired. The greater height of the ceiling gives a greater air capacity per animal.

Risk by fire is lessened, because flames and heat will not reach the stock quickly, giving more time for driving out the stock; and, when the wind is in a favorable direction, there is a chance of saving the stockbarn when the hayshed burns.

Risk of storm is less, because the stock is in a low building, and the haybarn does not extend so high in the air.

Smaller timber may be used, because a light roof is all that the low side walls have to support; this, together with the fact that no heavy wooden hay-mow floor is needed, reduces the cost very much.

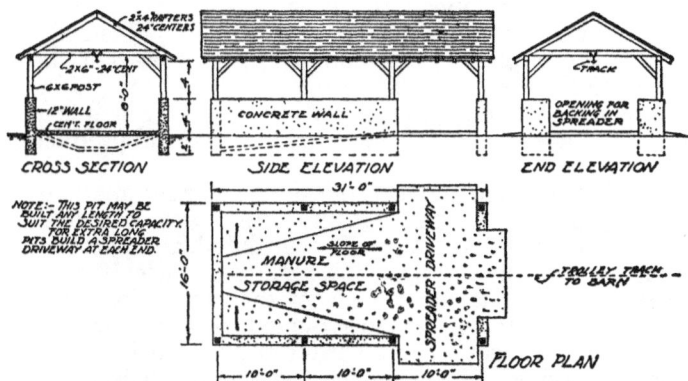

FIG. 535. Plans and dimensions of a concrete, covered manure pit, which in connection with well-built livestock barns and a manure spreader, enables the farmer to get the maximum benefit from his manure supply.

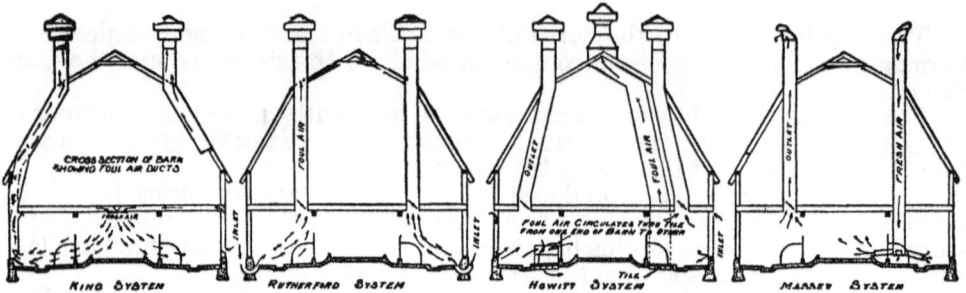

FIG. 536. Diagrammatic sections of combination dairy and storage barns illustrating the principles and method of operation of the four main types of ventilation systems

CHAPTER 33

The Equipment of Farm Buildings

By M. A. R. KELLEY, Agricultural Engineer, in charge of the Service Department of the Louden Machinery Co. He was born and reared in an Iowa farming community and, with the exception of one or two years, has been always in touch with farming and farm problems. Graduating from the Iowa State College, he received degrees in both Mechanical and Agricultural Engineering. He then was for four years in charge of the agricultural engineering work at the University of Missouri, after which he entered upon his present activities. He is a member of the American Society of Agricultural Engineers and, in 1916 and 1917, was chairman of its Farm Building Equipment Committee.—EDITOR.

WHY is it that so many farmers who take much pride in the exterior appearance of their farm buildings manifest so little concern in their interior arrangements? On the latter depend, to a very large extent, the health of the stock and the output and value of dairy and other products. When one considers that the barn or stable is occupied by the animals for at least half the year, it is to be regretted that, until recently, so little attention has been paid to such matters as barn ventilation, water supply, and lighting, each of which is a vital factor in the prevention of disease among livestock and in successful production and profitable farming.

While much improvement along these lines has been made during the past few years, a journey through the farms in any of the states will reveal to the visitor the existence of numbers of dark, damp, ill-ventilated, and poorly equipped barns, which are obviously behind the age with regard to agricultural progress. The aim of the present chapter is the remedy of such conditions by offering to the farmer (1) some useful hints on what to avoid in farm-building equipment and (2) some practical suggestions, set forth as simply as possible, in regard to cleanliness, sanitation, and the saving of labor, which together contribute so largely to increased production and to success in farming generally.

Ventilation

The importance of proper ventilation is now recognized by all livestock farmers. It has been fully demonstrated that good ventilation in a barn means less food consumption, greater production, and healthier animals. There is no single factor of greater importance in the control of tuberculosis than fresh air; and for this reason, if for no other, the question of ventilation is of vital interest to all dairymen.

380

There are 4 good reasons for ventilating a barn: (1) to maintain proper temperature; (2) to remove foul air and the moisture of respiration; (3) to prevent spontaneous combustion in the mow; and (4) to provide pure, fresh air for the animals.

A dairy cow requires, for proper ventilation, approximately 3,600 cubic feet of air per hour, which is only a little more than that which the natural heat from her body will raise from 0 to 50 degrees F. Thus, if flues of proper size are used, it will be easy to maintain a temperature of 45 to 50 degrees F. in the stable. The temperature in a dairy barn should not be allowed to drop below freezing.

If the air in a stable or dwelling is not changed with sufficient frequency it becomes so damp as to interfere with the respiratory organs. Dampness in a barn is an indication of poor ventilation. Each animal respires 10 pounds of water every 24 hours, or, in other words, 20 cows will give off a little less than a barrel of water a day.

Many fires on the farm, usually resulting in a total loss of building and contents, are caused by the storage of green leguminous hay in the barn. These fires are caused, by confining the heat and the gases which are given off during the curing of the hay. With proper ventilation, these gases are carried off as formed and do not reach the high temperatures which cause them to ignite.

Oxygen is just as important a part of the nourishment of cows and horses as the food they eat. It is essential to have plenty of pure air for the lungs of all animals and to prevent the moisture of respiration from collecting on the walls.

Four forces which produce ventilation are: (1) wind pressure; (2) wind suction; (3) aspiration, or wind across the top of flue; and (4) difference in temperatures. These 4 forces must be considered in the design of any system of ventilation.

Ventilation systems. The most common systems of ventilation are the Massey, the Howitt, the King, and the Rutherford. These are sometimes supplemented by the use of ventilating windows or canvas curtains. In all of these systems, inlet flues are used to bring in fresh air, and outlet flues are used to remove the foul air and odors (Fig. 536).

The Sheringham valve, or ventilating window, is used primarily to supplement the King system during mild weather. It consists of a window tilting in at the top and provided with a shield on either side, forcing the air to go up over the top, thus preventing direct draft on animals.

Ventilation through muslin is due to current movement through the meshes which, in a good grade of muslin, are very small. If cheesecloth is used, it is short-lived. In spite of the fact that cloth is commonly used, it cannot be considered an efficient ventilator.

The Massey system has an inlet pipe running from the top of the barn and opening at the bottom of the stable. A cowl is placed on this pipe through which the wind blows downward, giving a circulation of air. When the wind is blowing, air is distributed in small pipes around the stable. This system is not in common use, as it depends principally upon the wind for its action.

In the Howitt system of ventilation, the air enters the barn through louvers or screened openings extending the full length of the barn and located near the ceiling. It passes down between the studding and out through a continuous slot just above the sill with two separate sets of foul-air flues, one set of outlets at the floor and one with openings at the ceiling. This system can be regulated to give good results in any kind of weather.

The King system of ventilation, which is the predominant system in the United States, has for its principal idea a number of small intakes and one or more large outlets. The fresh air enters above the sill, rises between the studding, and enters the stable at the ceiling. The outlet flues start near the floor, pass up inside of the barn, through the mow, and through the ventilator on the roof. The area of the outlets is generally made two thirds of or equal to the area of the inlets, depending on climatic conditions.

In the Rutherford system, which is quite commonly used in Canada, the positions of the air inlets and outlets are just

Fig. 537. Ventilating window with side shields or cheeks which prevent drafts. Opening inward, this style can be permanently screened on the outside.

the reverse of those in the King system. The fresh air enters the inlet flues at ground level, turns downward, passes inside, and rises into the stable. The outlet flues start at the ceiling, pass up through the mow, and out at the ventilator. The total area of outlet in this system is made twice that of the inlet.

Importance of proper flues. In order to secure good ventilation, it is important that the flues be made of proper size. The accompanying table gives the size of flue recommended for use with the Rutherford and King systems respectively.

| ANIMAL | SIZES OF OUTLAY FLUES | |
	RUTHERFORD SYSTEM	KING SYSTEM
Horse. .	20—24 sq. in.	36—40 sq. in.
Cow . .	16—20 " "	30—34 " "
Swine. .	8—10 " "	12—14 " "
Sheep. .	6— 8 " "	8— 9 " "

The constructon of the flues in both systems is quite similar, since they must be air-tight, but they differ in arrangement. In the King system, it is desirable to have intakes on all sides, spaced about 10 or 12 feet apart, and made as straight as possible with corners rounded. The inlet flue should be provided with a screen on the outside, to keep out birds and trash, and with a door, or register, on the inner end, to regulate ventilation. It should start far enough above the ground to prevent snow from closing the opening, and the outer end of the flue should be at least 3 feet below the inner end.

The foul-air flues should be spaced not over 40 or 50 feet apart, and should start about 18 inches above the floor. Each flue should be made air-tight and provided with a sliding door, or damper, for regulating the size of the opening. It should be made as straight as practicable and should rise above the ridge of the roof. High outlet flues are desirable, as the heat effect and suctional effect of the wind increase with the height.

Although the King system of ventilation has been used to some extent for years, there is, apparently, considerable ignorance concerning its principles. It is highly important that the size of flues be properly proportioned to suit the contents of the stable and climatic conditions. None of the systems is automatic but all of them require intelligent supervision in order to secure the best results; and these can only be obtained by a knowledge of the basic principles involved.

In the King system, the assumption is that respired air, laden with moisture and carbon dioxide, is heavier than air and falls to the floor, where it is drawn out by the action of the wind over the top of ventilator, assisted by the movement of air caused by the difference in temperature of the air inside the flue and outside of the barn.

The foul-air flue is provided with a valve, or door, which opens into the flue near the ceiling and permits warm air to go directly out during mild weather. This valve is an important item and should not be omitted, as it enables the system to be adapted to the weather conditions.

The question is not, Do I need ventilation? but, How can I secure best results? A careful study of the foregoing paragraphs should help in answering this question.

Water Systems for the Barn

There is no factor of greater importance in farming and, perhaps none more often neglected, than that of providing suitable drinking water for farm animals. Our ideas of sanitation are advancing, however; and, in the future, greater care in the production of all foodstuffs will be demanded. Especially will this be the case in regard to a pure water supply in the dairy, since it bears such an intimate relation to the general health of mankind and particularly to that of the young child.

Water storage. There are many sources from which a good supply of water may be obtained, and when this has been secured the question of storage comes. Two systems of water storage are especially connected with the barn; (1) the elevated tank on the silo structure and (2) the tank within the barn.

(1) The masonry silo, of tile or concrete, provides an excellent means of supporting an elevated water tank. In this way, the additional cost of elevation is very little and the tank does not interfere with the use of the silo. This form of water storage provides the farmer with the luxury of a city waterworks and good fire protection on his own grounds.

The water tank should not be made larger than necessary and its construction should be left to those who are familiar with this kind of work. The accompanying cut and diagram (Figs. 538 and 539) show how such a tank has been utilized and the construction of the same. This system is as yet but little known; but, with the increased use of the masonry silo, we may expect to see a greater number of elevated tanks in the future.

There is, however, one drawback in using

this system in northern states: it is difficult to keep the pipes from freezing during cold weather. This difficulty may be partially overcome by insulating the pipes with straw or sawdust. If fresh water from the well is pumped up each day, the warm water from the well will tend to lessen the danger from freezing.

The arrangement partly illustrated in Fig. 539 tends to keep pipes from freezing: A three-fourth inch pipe is placed within a 2-inch water pipe, the upper end running above the water level, and the bottom end running to the base of the silo. The lower end is provided with suitable fitting and a shield opening into the pipe. Below this shield a lantern may be hung on cold nights the warm air from which, rising through the pipes will diminish the chances of freezing and in case of freezing will also help to thaw them out.

(2) A tank placed inside of the barn on the mow floor may be used for water storage. In this position it can be packed with hay, which lessens the danger from freezing. Eave troughs placed on the barn can easily be run into this tank. Such water is usually cleaner than that obtained from the roof of a house, as the barn roof is free from soot. If a filter be provided, the water may be used for human consumption.

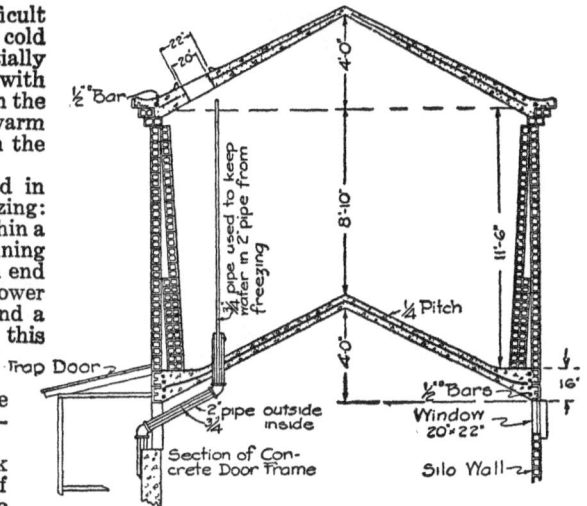

FIG. 539. Section of a hollow tile water tank built on top of a tile or concrete silo, the roof of which is shown

FIG. 538. Water tank elevated on masonry base which can be used as a silo or storage building.

Watering stock. The methods of watering stock on a farm may be divided into 2 classes; (1) the open-tank method in the feed lot, and (2) the various methods inside the barn.

(1) The water tank in the feed lot is widely used. It is convenient, and low in first cost; but when water is stored in it for any length of time, it becomes stale, and affords an excellent breeding place, for disease germs. Diseased animals slobber in the tank and infect other animals. Water is known to be a carrier of tuberculosis, glanders, and contagious abortion, all of which are greatly dreaded by herdsmen. It has been estimated by some authorities that 65 per cent of the tuberculosis found among cattle in the United States has been transmitted through water.

A green, mossy formation gathers on the side of an open tank, but this may be prevented by cleaning and scrubbing the tank once a month. Another method which does not involve so much labor, but which should not entirely replace the periodic cleaning, is the use of copper sulphate in the water. Place a small amount of the crystals in a cloth bag and drag this back and forth through the water a few times. This will prevent the mossy formation.

Outside open water tanks are objectionable in the winter, as it is difficult to keep ice from forming in them. A cow should never be required to drink icy water, as she will drink only enough to satisfy her thirst and never enough to keep up her milk production. This is a direct economic loss.

Covered tanks help to keep the water from freezing. There are many forms of tank heaters on the market, but they are more or less troublesome. Most of them need to be fired from the top, and it is unhandy to take out the ashes. Water heaters are sometimes built into concrete tanks; but these also, are more or less unsatisfactory, since the heat causes the concrete to expand and contract, and this soon results in a leak. Where steam is used in a dairy, a pipe may be run to the tank and live steam turned on. This will warm a large tank of

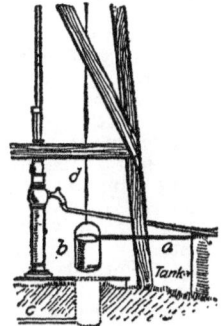

FIG. 540. Handy automatic device for controlling a windmill. When the tank overflows through a into the bucket (b), it sinks and stops the mill by pulling the shut-off wire (d). But a small hole in its bottom lets the water gradually run off through the drain (c). When b is empty, the vane action lifts it and holds it up until the tank again overflows.

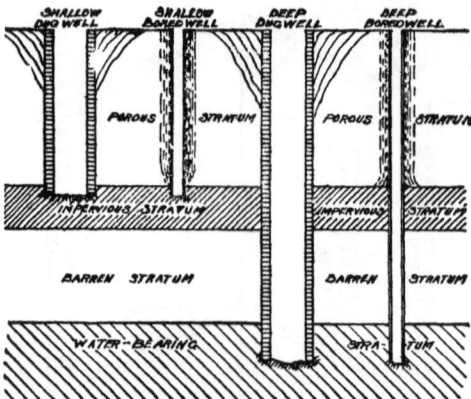

FIG. 541.　Well sections show comparative chances of contamination in four types.　The deepest water supplies are usually the safest.　(Farmers' Bulletin 549.)

water in 5 minutes and, if covered, it will keep warm for some time.

It is not advisable to store large quantities of water in an open tank inside a stable, as the water will quickly absorb the stable odors and be unfit for use.　If open tanks are used in a stable, small tanks should be used and means provided for renewing the supply frequently.　The writer recently visited two barns which were fortunate enough to have running spring water under pressure in the barn.　This is an excellent method for watering, but very few farms are so fortunately situated.

A dairy cow requires 8 gallons of water in the production of 10 gallons of milk, besides the water she needs to keep up her body. It does not pay to use high-priced feed to increase milk production and neglect the opportunity of obtaining an increase cheaply by providing an abundance of pure water. It will pay any dairyman to see that his cows are not required to go outside in stormy weather to drink.　It has been found that water of a temperature of 55 to 60 degrees F., or the temperature of well water, gives best results.　Watering in a concrete manger is better than driving cows outside during bad weather, but this method is far from sanitary.　Water is usually run into the manger twice a day.　All the cows drink from the same supply and often do not secure all they want and at the time they want it.　The water washes down dirt, salivated feed, slobber, etc., to the cows at the lower end of the manger.

Gravity water bowls.　By the gravity system of water bowls, one bowl usually serves two cows.　This system requires only one main water pipe.　The bowls are clamped to the stall post and at a uniform height.　One end of the water main is attached to a governing tank containing a float valve, which controls the height of the water in the bowls.　The

water flows freely and intermingles between the bowls.　There is no outlet except through the cows.　The water bowls usually contain less than a thirsty cow will drink; hence whenever a cow quenches her thirst, water is drawn from the governing tank and adjacent bowls.　Lids on the bowls help to keep out dirt.

This system is very convenient, but dangerous.　Animals with a contagious or infectious disease drink and deposit germs in the bowls, where they mingle with the saliva and feed and surge back and forth in the pipes. These germs soon propagate, causing the bowls and pipes to become unsanitary, and the pipes are very difficult to clean.

It is said that, when cows have free access to water at all times they drink at least 10 per cent more at night after eating than they do in the daytime.　Under these conditions, it is not hard to see why those farmers who

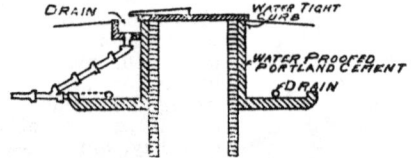

FIG. 542.　A simple but effective method of protecting a dug well, which also keeps the ground around it dry

have installed automatic water bowls in their barns have been able to increase their milk production 5 to 15 per cent.

Automatic water bowls.　Automatic water bowls (Fig. 543) are of two types.　One has outlet pipes or drains, the other has no drains.　Both are actuated by some form of trigger work, which is operated by the cow.　The one without the drain provides for an individual drinking cup; but this is little better than the gravity bowl, as the feed and saliva collect in the bowl and are stirred up each time a cow drinks.　In order to be strictly sanitary, water must be admitted at each drinking, and all excess water drained away when the cow has finished drinking. The inlet pipe should be above the highest level in the bowl, and in the bottom of the bowl a drain should be placed.

Water pipes should be placed in trenches to prevent them from freezing in cold climates, and they should be provided with frost-proof hydrants.　All

FIG. 543.　Individual drinking bowls are more costly than a common tank for all the stock, but one valuable cow saved from disease by means of them would more than balance the account.

FIG. 544. A concrete watering trough is permanent and can easily be kept clean. It should be conveniently placed and well sheltered.

pipes should be of galvanized wrought iron and of such sizes as to allow a free flow of water. Not more than one outlet should be used on a half-inch pipe; and all valve and drain cocks

should be set in gravel, so that they will drain easily.

The tuberculin test is of value in detecting a diseased animal, but it does not prevent her from infecting another. We have seen how contagious abortion and tuberculosis may be transmitted through methods of watering and feeding, hence we cannot be too careful in choosing our system of watering.

In estimating the amount of water necessary for barn use, the following data will be of value: A horse requires 8 to 10 gallons per day; a cow, 10 to 15 gallons; a hog, $2\frac{1}{2}$ to 3 gallons; and a sheep, about 2 gallons. A dairy cow requires water in proportion to the amount of milk given, the ratio being about $2\frac{1}{2}$ pounds of water for each pound of milk.

Labor-saving Appliances

The modern agriculturist no longer builds in a hit-and-miss fashion. He carefully plans his buildings and equipment so as to secure the greatest possible returns on his investment. Breeders of high-grade livestock find that it pays to use first-class equipment, complete in every detail, as this makes a favorable impression on prospective buyers. Producers of certified milk, too, find that they are able to secure more customers and better prices by using sanitary equipment. But the wide use of modern barn equipment on the average farm is largely due to the saving of feed and labor and the additional comfort for the animals resulting from its employment.

Good barn equipment is no longer considered a luxury, but a necessity which pays daily interest on investment by saving material, time, and labor. The various labor-saving devices for the barn may be divided into 3 classes: (1) conveniences for handling feed and litter; (2) conveniences for handling stock; and (3) miscellaneous conveniences, which help to increase the efficiency of the barn.

Hay tools. The hay carrier was, perhaps, the first important labor-saving device invented for use in the barn. It made possible the storage of large quantities of hay in the barn and has introduced more economical types of construction for hay storage. Indeed, no large modern barn is complete without one. The scarcity of labor during the haying season and the increased value of hay make it imperative that we use every means available to facilitate the handling and the saving of the hay crop.

The original hay carrier was operated on a wooden track,

FIG. 545. An excellent example of the use of labor saving appliances. Four men and an extra team could not unload and stow hay as fast or as well as this one man can by using the hay carrier rigging and hoisting engine.

and this is still used in some localities. The steel track, of which there are many forms on the market, makes the carrier much easier to operate, and is greatly superior to the wooden one. Simplicity, durability, and strength are 3 important factors to consider in the selection of a hay carrier. It should be simple, so as to be easily understood and operated. It must be durable and strong enough to support the loads. Breakage during the haying season is expensive because of loss of time in repairing and this delay may very easily cause a partial or a complete loss of the hay crop, in case of a sudden storm.

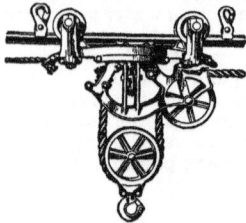

FIG. 546. The hay carrier is one of the most important items in the storage barn equipment.

Hay forks. The different hay forks which are suitable for barn use may be divided into 5 types: (1) single harpoon; (2) double harpoon; (3) triple harpoon; (4) grapple fork; and (5) sling.

The single harpoon (Fig. 547a) is the original hay fork. It will do good work in long, heavy timothy hay and where the hay is carefully loaded and handled in small bunches. It is not so successfully used in clover or alfalfa, especially when the hay is dry. The double harpoon (Fig. 547e) has been on the market for years, and will do good work in timothy hay and under average farm conditions. The triple harpoon (Fig. 547c) is a later invention. This fork is much stronger than the double harpoon, and will lift larger loads and bring them closer to the track. It is durable and a light fork to handle, and is good for general conditions.

The grapple fork (Fig. 548) is the most widely used type of all, and works similar to a pair of ice tongs. It handles all kinds of forage under practically all conditions. It can be used for alfalfa, clover, straw, or grain in bundles. The 8-tine fork may be used for manure. This fork increases in popularity each year.

The sling (Fig. 549) is used for handling hay in large loads. It handles the hay quickly and cleans the rack perfectly. It consists primarily of a set of parallel ropes, held apart by wooden spreader bars, and is made in a great variety of styles. It is built into the load at the time of loading. It requires large door openings and good clearance over beams.

The various types of forks require different sizes of hay doors, in order to get good clearance. The single harpoon may be used in a door as small as 5 x 7 feet. The 6 x 8 size is better, as it causes less binding. A door 8 x 10 may be used for the double harpoon, while the triple harpoon requires a door 9 to 10 feet wide and 10 feet high. A door 9 to 10 feet wide and 10 to 11 feet high should be used with the grapple fork; and, when slings are used, the door should be 10 x 12 feet.

Power hoist. The power hoist (Fig. 545) is rapidly coming into use where large quantities of hay are handled. It is conveniently and easily operated, and saves much time and labor during a busy season. It displaces a man and a team which are necessary when hoisting with horses. The horsepower required depends upon the size of the load and

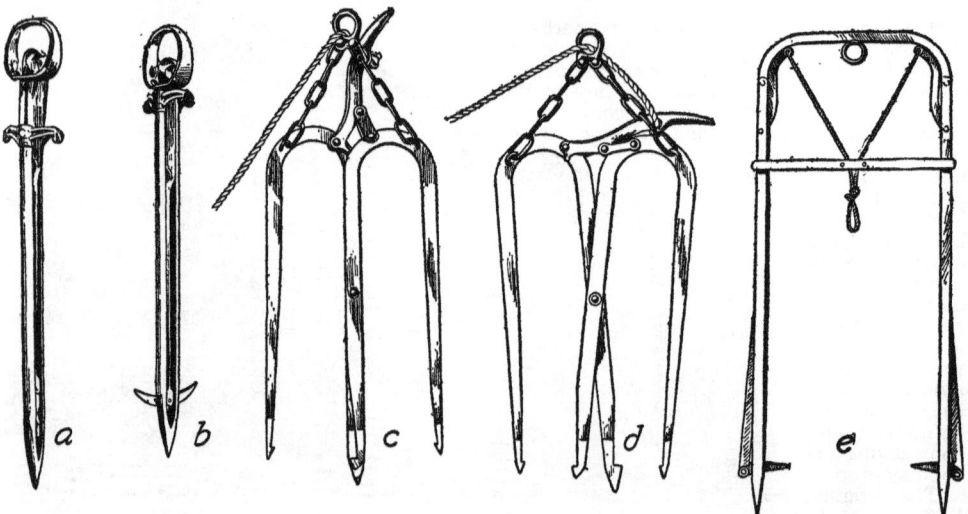

FIG. 547.   Types of hay forks: a and b single harpoon, open and closed; c and d, triple harpoon open and closed; e double harpoon closed. This type is opened by pulling the rope loop in the centre

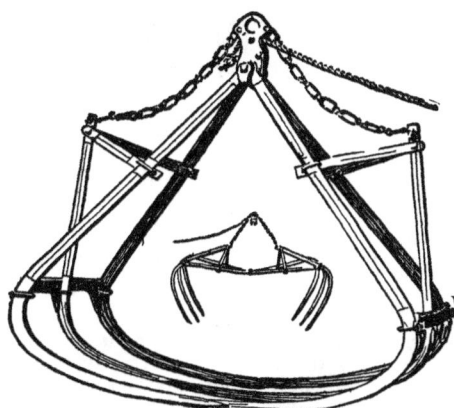

FIG. 548. The grapple type of hay fork is probably the best of all for mixed and miscellaneous hays

the speed of the engine. It is usually from 3 to 5.

The value of the power hoist is greatly increased by the fact that after haying season it can be mounted with a portable engine and used for general farm purposes, such as digging wells, elevating grain in sacks or boxes, constructing silos, hoisting silage from pit silos, etc.

Feed and litter carriers. Anything that adds to milk production or makes it more profitable, is of importance to the dairyman. Litter and feed carriers are great conveniences which aid in cutting cost by reducing the time and labor required for the barn chores. The hay carrier cuts labor and time during the haying season, and is now recognized as a necessity; but feed and litter carriers are of even greater importance, since they save time and labor in the daily tasks.

By using a litter carrier, manure may be dumped directly into a manure spreader and taken to the field with only one handling. It enables a boy to do a man's work, and thus decreases the hired help needed. It has often been proved that it cuts the labor of cleaning a barn in half. The old wheelbarrow is still used in many barns, but it is a disagreeable method of handling manure. By this method the manure is usually dumped in piles near the barn, requiring a second handling before it is put in the field. With the modern litter carrier, manure may easily be deposited 50 feet or more from the barn. The sanitary laws of many states require the observance of this distance, which takes the filth away from the barn and helps to decrease the number of flies.

Carrier tracks. In order that both feed and litter carriers may last longer and be operated more easily under heavy loads, the trolleys used should have roller bearings and should run on a solid steel track. There are two general types of steel tracks used for litter carriers; namely, the rigid steel track, of which there are many forms on the market, and the steel-rod track. The steel-rod track may be used in the small dairy barn, and is cheaper; but it is not advisable where curves and switches are necessary. The rigid-steel track may be used anywhere, and it provides a good strong support for the heavy loads. Moreover, it is much easier to load a carrier on a rigid track.

There are now on the market several combination track systems which combine the advantages of the rigid track inside the barn with the advantage of the rod track outside the barn. By this system enough momentum can be given the carrier to take the load away from the barn and return the carrier by gravity without the necessity of the operator leaving the barn. This system is suitable for small installations; but the rigid-track system is best for large installations, where numerous switches and curves are required.

Feed carriers may be run on the same track as litter carriers and may be either of two types. Of these types one runs on tracks like litter carriers; the other has a feed box placed on a truck which runs on the floor. Both types are widely used. The floor trucks may be conveniently used when the feed is stored in the barn and the floors are of the same level. Outside the barn, it is much easier to use the track carrier, which is more generally employed than the floor feed truck. These carriers are made in various sizes and forms, to fit the conditions found on different farms. A carrier will hold enough silage to feed 25 or 30 cows; and by this means one man can feed 100 cows as quickly as another man can feed 50 or less by the old basket method.

Carriers may also be obtained for handling milk cans and other forms of merchandise. These help to cut down the time and labor of transporting milk from the barn to the dairy.

Mangers. Sanitary dairy barns have been made possible by the use of concrete mangers and steel stalls. Concrete is rapidly replacing wooden mangers and other wooden and, therefore, more or less unsanitary parts in the dairy barn. Wood forms an excellent harbor for disease germs and other bacterial life, which penetrate the cracks and crevices and lie dormant until conditions favor their propagation.

A properly constructed manger is a very important factor in maintaining the health and comfort of the animal. Wide, flat-bottom mangers are not advisable, as it is very hard for the cow to reach her food in them. In attempting to do so she often slips on the floor and injures her knees, which is the cause of big knees. She may also easily strain herself and cause abortion.

The shape and size of the mangers determine whether a cow can eat in comfort. The bottom should slope and be rounded, so that the feed will roll down within easy reach.

The surface of the concrete should be made as smooth as possible and all corners be rounded off, so as not to hold dust and filth. Three different mangers are widely used throughout the United States and have become standard among dairymen. One of these is a manger which is adopted under crowded conditions, and it is as small a manger as can be conveniently used. This manger has the disadvantage of a raised or high feed alley, and the food is pushed out into the alley with the same results as previously mentioned. The cows attempt to reach it, and in so doing very often injure themselves. This type of manger is preferred by some farmers, however, because it is easy to sweep the food back into the manger. This is bad practice, as dirt and other filth are thus drawn into the manger and mixed with the food. Another manger is of standard size for the average cow and for average conditions, while a third is used for large cows and where a higher manger is desired. The last-mentioned type is easily kept clean; the feed is always within reach of the animal, and the curve of the manger is the approximate curve described by the cow's nose in raising and lowering her head. The first and third types are shown in Vol. I, Fig. 513.

Stanchions. Comfort for the dairy cow pays big dividends. A cow is a nervous animal, and, when ill treated, will not produce well. When held in a stiff and rigid stanchion, she will not give her full quantity of milk; and, when held in this way for any length of time, she will soon become stiff and lame. The old-style wooden stanchions were not only inhuman and uncomfortable, but they were unsanitary also.

The best-paying piece of barn equipment is the modern tubular steel stanchion. The

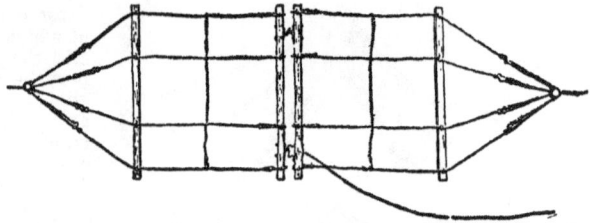

FIG. 549. Hay slings do more and cleaner work than any of the forks, and the trouble of arranging them on the load in the field is not worth mentioning.

chain-hung flexible steel stanchion is now recognized as an ideal cow tie, as it embraces the features of safety, comfort, cleanliness, and convenience. The round-sloping-end stanchions are best, as they are much safer than the flat-bottom stanchion.

Steel stalls. The greatest protection against disease is sanitation. Sanitation in the dairy barn has been made possible by the advent of the modern steel stall and the use of concrete mangers and floors, which make it easy to keep the barn clean. Steel stalls are strong and durable and do not obstruct the light or ventilation. The stalls should be simple, with all fittings of the dustproof type, so that they may be easily wiped off and kept clean.

The partition is an important part of a cow stall and is necessary to prevent a cow from stepping on her neighbor or crushing her teats when she is lying down. It is also useful in protecting the milker and in preventing the cow from turning sidewise and soiling adjacent stalls. The single-bend stall partition is the best as it is stronger than the triple-bend type, and with it there is less liability of injury to the cow.

It is important that the length of the stall be properly proportioned to the cow, in order

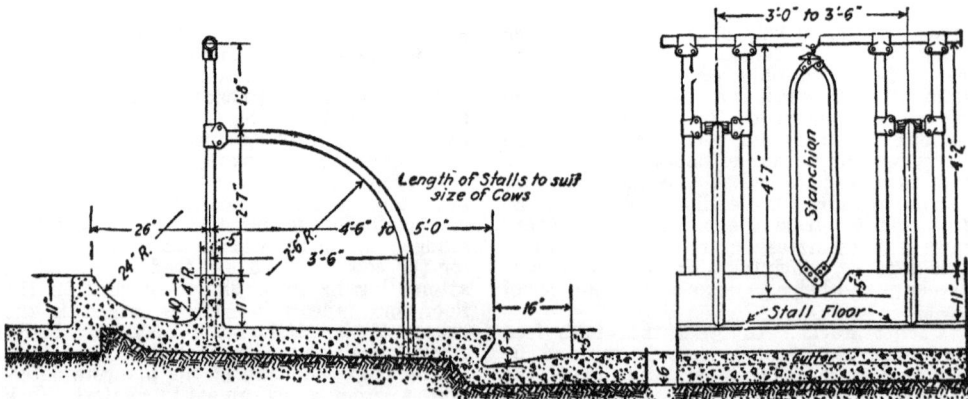

FIG. 550. Side and end views of a modern concrete floored cow stall and metal stanchion giving average dimensions or those that are uniform for all classes of stock

that she may be easily kept clean. The stall should be of such length that the cow may stand easily on the platform and all droppings may fall into the gutter. There are many alignment devices on the market for securing this result. These are an added expense, however, and are not necessary if the stall on one end of the row is made to fit the smallest cow and that on the opposite end is made to fit the largest cow, all intervening stalls being made accordingly. The cows may then be assigned to the stalls most convenient for them. Alignment devices are, however, very convenient where it becomes necessary to change the length of stall after the concrete work is finished.

The table shown below gives the sizes of stalls recommended for cows of various breeds, and for heifers.

Steel pens also aid materially in maintaining sanitation in the barn, as they admit the maximum sunlight and do not obstruct ventilation. Calves cannot do well in dark and damp pens, but they grow rapidly when kept in clean and well-lighted pens. This is a safe, sanitary, and humane way of caring for calves. Steel maternity pens are easily kept clean and easily disinfected. This is a big factor in reducing navel trouble in young calves.

Milking machines. The modern milking machine is one of the greatest labor-saving devices ever invented for dairymen. It solves the difficulty of securing efficient help and replaces cheap labor with skilled. It is now considered a part of the necessary equipment of all large dairies, and it has been an important factor in the development of the dairying industry. It has been developed to a degree that makes its use entirely practicable in dairies having more than 10 cows. There is, however, still room for improvement. It has no harmful effect on the physical condition of the cow; and, as regards quality and quantity of milk, its use competes successfully with hand milking. It is of great importance that the rubber parts and all parts which come in contact with the milk be thoroughly cleansed after using.

All milking machines are somewhat complicated and require the exercise of mechanical ability on the part of the operator. Like

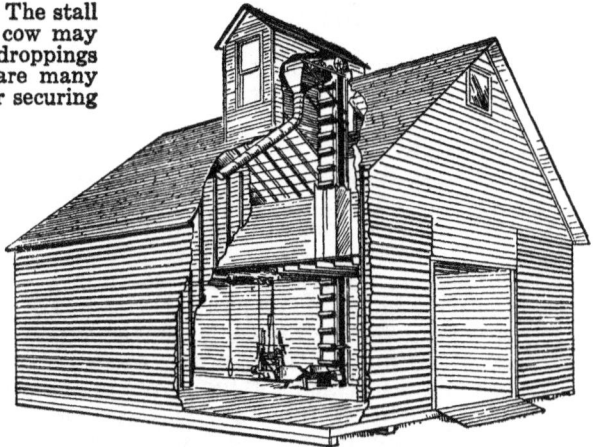

FIG. 551. Granary partly cut away to show grain elevator in place. The movable chute may be directed into different bins. There is usually a trap door and dumping arrangement by which wagons can be unloaded into a bin from which the buckets elevate the grain.

other mechanical devices, they require intelligent supervision in order to secure the best results. The milking machine must be operated and adjusted to meet the needs of the individual cow, and its success or failure depends upon the ability of the operator.

Grain elevators. The grain elevator (Fig. 551) is another machine which has effected a great saving of labor, especially during harvest. It is meeting with much favor in the Middle West, where large quantities of grain are stored in barns and granaries. It may be used for all kinds of small grain and for ear corn. Two general types are employed: the portable elevator on the outside of the barn, and the stationary elevator inside, with some form of a dumping device. Both types are widely used, and selection is determined by local conditions. However, when erecting a new set of buildings, provision is usually made for the stationary type. The elevator may be run either by horsepower or by motor. The most popular type of farm elevator is composed of a series of buckets or cups fastened to an endless chain running over a set of parallel shafts and pulleys.

Among the numerous other labor-saving devices which materially help to increase the efficiency of the barn, but which need only be mentioned are: Sliding doors with birdproof tracks and roller-bearing hangers, making door operation easy and obviating noisy and broken hinges; hay-carrier returns and pulley-changing devices; overhead feed bins with hopper bottoms; hay chutes; platform scales and spring balances; elevators for silo chutes, etc.; all of which assist in reducing chore labor on the farm.

BREEDS, ETC.	SIZE OF COW STALLS	
	Width	Length
Heifers	2 ft.–9 ins. to 3 ft.–0 ins.	3 ft.–6 ins. to 4 ft.– 0 ins.
Jerseys	3 ft.–2 ins. to 3 ft.–4 ins.	4 ft.–6 ins. to 4 ft.– 8 ins.
Guernseys	3 ft.–4 ins. to 3 ft.–6 ins.	4 ft.–8 ins. to 4 ft.–10 ins.
Shorthorns	3 ft.–6 ins. to 4 ft.–0 ins.	5 ft.–0 ins. to 5 ft.– 2 ins.
Holsteins	3 ft.–6 ins. to 4 ft.–0 ins.	5 ft.–0 ins. to 5 ft.– 6 ins.

Manure Waste and Conservation

Former Assistant Secretary Vrooman, of the U. S. Department of Agriculture, is authority for the statement that "our billion-dollar manure waste is the world's greatest economic leak." This assertion is based upon reliable statistics and is a careful and conservative estimate. Here, then, is a place where every farmer can assist in cutting down a huge economic loss. Some of this loss is unavoidable, but the greater part can be saved. Notwithstanding the great value of manure, there is probably no material on the farm in which there is so great and needless a waste.

Manure waste. The principal losses to farm manure occur in the barnyard and are caused by weathering, leaching, heating and rotting. The rains wash out a large part of the fertilizing elements; the sun burns out more; and bacterial action during the rotting process drives off the valuable ammonia gases. Fully 50 per cent of the fertilizing value of manure may be lost in this way.

The accompanying table, giving data compiled by the Indiana Agricultural Experiment Station (Circular 49) shows briefly why manure is of such great value as a fertilizer. The chemical analyses were made from many samples from various experiment stations and may be said to be truly representative.

QUANTITY, COMPOSITION, AND VALUE OF MANURE FROM DIFFERENT
CLASSES OF ANIMALS

	HORSE	DAIRY COWS	STEER	SHEEP	SWINE
Pounds of manure produced per day per 1,000 pounds live weight	35—45	70—80	40—50	30—40	40—50
Pounds per ton nitrogen	11.8	9.7	13.8	27.5	15.2
Phosphoric acid	5.6	5.4	5.6	9.9	9.5
Potash	14.6	9.4	10.5	22.7	14.6
Value per ton on basis of analysis* . . .	$ 2.84	$ 2.21	$ 2.90	$ 5.83	$ 3.49
Tons of manure produced per year per 1,000 pounds live weight	7.0	12.7	7.5	5.5	7.3
Value of manure produced per year per 1,000 pounds live weight*	$19.88	$28.07	$21.75	$32.06	$25.48

*Computed on the basis of the following prices: Nitrogen, 15 cents per pound; Phosphoric acid, 3½ cents per pound; potassium, 6 cents per pound.

Manure represents fertility which has been drawn from the soil by crops. Nearly 80 per cent of the fertilizing value of crops can be returned to the soil in the manure. The manure which is returned to the soil not only serves to cut down the drain on it, but also helps it in many other ways. It increases the supply of humus, adds plant food, and makes other plant food in the soil available, besides aiding in the development of soil bacteria. It also helps the soil to warm up earlier in the spring, decreases soil washing, improves the drainage, and enables the soil to receive and retain more moisture for the growing crops. It is reported by the experiment station at Rothamsted, England, that the residual effects of manure can be noticed for 40 years after application upon land that has been continuously cropped.

Conservation of manure. Manure begins to deteriorate the very hour it is dropped. The sooner it reaches the field, the better. Tight floors are valuable agents in the saving of manure, and a good concrete one in the barn will soon pay for itself in the saving of fertilizer elements.

A manure spreader is the best and most economical method of applying manure to the soil. As a farm implement, it is second in importance to the self-binder only. It is always ready for its part. It distributes the manure more easily and evenly and renders possible the spreading of specific quantities.

While the direct-to-the-field method is preferable, it often happens that this is not convenient or possible. Bad weather and busy seasons sometimes interfere. A manure pit should, therefore, be provided, to hold the excess manure formed during such periods.

If we allow manure to stand in a heap, we shall find it divided into 3 layers: the fresh manure on the top, the rotted manure in the middle, and at the bottom decomposed manure with a very offensive smell. The latter has lost a large part of its fertilizing elements, and is not in a condition to give best results.

The bacterial actions which go on in a heap of manure are caused by two kinds of bacteria—aerobic, which live in the presence of air, and anaërobic, which can live without the presence of air. The manure pit should be deep enough to permit both kinds of bacterial action, and not too deep to prevent them. For this reason, a manure pit should not exceed 3 or 4 feet in depth.

Manure in a pit should be spread uniformly, kept moist, and well packed. The wetting down can be easily accomplished by using a pump, preferably one of the diaphragm type. The bottom of the pit should drain to one place, so that the liquid which has separated out may be pumped back over the manure.

Since manure is very low in phosphorus, acid phosphate is sometimes mixed with the manure in the pit, at the rate of 1 pound per day per head. This helps to increase the fertilizing value of the manure, and prevents the breeding of flies.

How to construct a manure pit. In constructing a manure pit, the following requirements should be kept in mind: (1) it should be permanent and water-tight (for this purpose, concrete is an admirable material); (2) all corners should be rounded, so that the manure will pack well; (3) it must not be built too deep to prevent bacterial action; (4) surface water must be kept out of the pit; and (5) the driveways in and out of the pit must not be too steep.

The size of the manure pit will depend upon the length of time the manure is to be kept in storage. Manure should not be stored longer than 8 or 10 weeks, as in that time bacterial action is carried too far. It is preferable to empty the pit every 2 or 3 weeks.

A pit 12 by 12 by 4 feet deep will provide

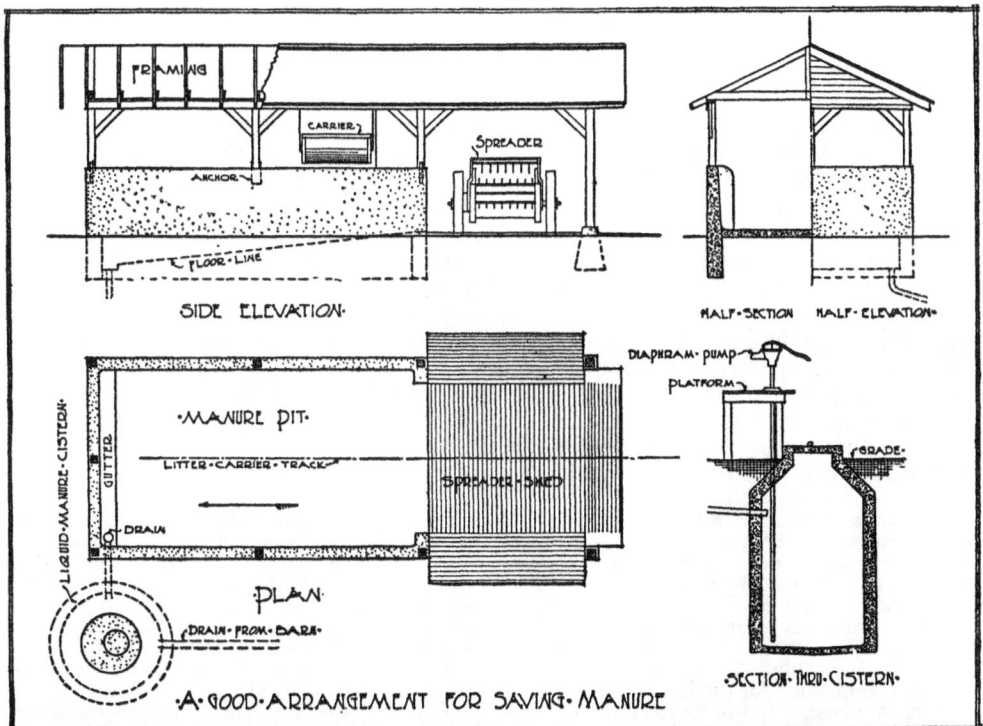

FIG. 552. Plans of a manure pit and liquid manure cistern by means of which all the plant food may be saved and the manure kept in the best possible condition

FIG. 553. A manure carrier is a great time- and labor-saver. This is a simpler structure than is shown in Fig. 552, but wouldn't you prefer using it to—

storage space for manure from 100 head of cows for 2 weeks.

Every particle of liquid manure should be saved, as with all farm animals 43 per cent of the nitrogen and 60 per cent of the potassium passes off in the urine. The liquid-manure loss constitutes by far the greatest waste of manurial elements on the farm. The annual potassium loss in the United States in liquid manure is alone more than a million dollars.

A tank or cistern to hold the liquid manure should be provided, and this should be below the level where the animals stand, so that the urine may drain into it. In warm weather, this tank should be emptied every 3 or 4 weeks, and kept tightly covered, to prevent the escape of ammonia gases and consequent loss of valuable manurial elements. A cow will pass daily 15 to 20 pounds of urine; a horse, 8 to 12 pounds. This data may be used in determining the proper size of cistern or tank to build. A round tank, necking at the top like a cistern, is the best, as it has a smaller surface for evaporation.

FIG. 554.—Doing this every day?

Lighting

During the winter season most of the barn chores are done with the aid of some form of artificial light; and this is particularly the case on the dairy farm, where such light is required most of the year. A good lighting system in the barn is a paying investment: it means that the chores can be done much easier and better, and with the least expenditure of time and labor.

The old-style kerosene lantern is still quite commonly used, regardless of the fact that the carrying of a lighted oil lantern in a barn among highly inflammable material is a very dangerous practice. It will be remembered that Mrs. O'Leary's cow kicked over a lighted lantern in a barn and caused the great Chicago fire; and similar accidents may happen in any barn where the old-style lantern is used. An enormous fire loss on farms is caused every year by the careless use of lanterns. When a lantern is overturned, the oil it contains adds fuel to the conflagration; and a fire once started in a barn usually burns itself out.

If it is necessary to use a lantern of any kind, some form of electric lantern should be procured. The electric lantern is more convenient to use, does not blow out in the wind, and is entirely safe. It works on the same principle as the electric flash light. It consists of a small electric bulb, with a suitable reflector, and derives electricity from dry cells. Although there is still room for improvement in them, electric lanterns have reached such a state of development as to make their use entirely practicable on the farm.

Acetylene gas and electric light. Two other sources of light are available for the barn, and these far surpass any portable light. They are acetylene gas and electricity.

Acetylene lights can now be obtained with the flame properly protected and with electrical igniting devices which make them safe for use in the stable. They are not, however, to be recommended for use in the hay mow.

Just as electric lights have replaced the kerosene lamp in the city home, they will eventually replace the dangerous kerosene lantern in the barn. The electric light is the one which strongly appeals to the farmer.

A statement was recently made by a well-known cattle breeder that electric light in his barn had paid for itself in one night. By having a good light he was able to save a valuable animal which was sick, whereas, if it had been necessary to depend on a lantern, he would have lost it.

Usually, electric lights in the barn are easily provided for, since a gas engine is commonly needed to pump water or run a feed grinder. It is very convenient to have electric light, as a wire may be run to any place where a light is needed. When lights are placed in feed and litter alleys, they materially cut down the time and labor required to do the chores. They are also convenient and safe to use in the hay mow; for they may be placed up out of the way, and they will furnish a good light in the mow.

When a building is wired for electricity, only such material and fittings should be permitted as will meet the required electrical standards. Switches should be provided at convenient places, so that the lights may be easily turned on and off. This is one of the advantages of electric lights because a three-way switch may be used on the line between the house and the barn, which will enable the lights to be turned on at the barn and off at the house. This is a great convenience in case it is necessary to go out in the night to attend a sick animal or to investigate disturbances in the barnyard. The electric light is always in its place, and it is not necessary to stumble around in the dark to hunt for a lantern.

EQUIPMENT USED IN THE CARE OF LIVE STOCK

By C. F. GOBBLE, Assistant Professor of Animal Husbandry in the Purdue University School of Agriculture, where his several years of teaching have supplemented a wide, practical farm experience.—EDITOR.

The equipment used in the care of livestock includes a number of the smaller implements which, by the employment of power in their operation become valuable labor-saving appliances. Among these are, for example, the machine sheep shears and machine horse clippers.

While in large sections of the country, these appliances have not yet come into wide use, it is only a question of time when their employment may be expected to become general.

Sheep-shearing machines. The necessity of shearing sheep at least once a year has caused the manufacture of several machines for this purpose. Formerly, the work was done with the common hand shears, which are quite like a pair of scissors, except that the power is applied between the fulcrum and the blades, and the blades spring open when the pressure is removed.

These shears are cheap, do the work in a satisfactory manner in the hands of a skilled workman, and are indispensable on any farm where sheep are raised, for trimming and tagging the flock. For shearing, however, they are being gradually replaced in most sections of the United States by machine shears. With these, less skill is required, the work is done more quickly and easily, fewer cuts are made in the skin, and more wool is obtained, because it is cut more closely. The machine shears are, of course, more expensive, but this is soon offset by the saving of time and wool.

The power for these machines is supplied by hand, belt, shaft, or motor. The hand machine used by the small sheep-owner is light, compact, durable, and easily operated and adjusted. One boy or man is required to furnish the power in addition to the one handling the shears. The cutter head works on the principle of the sickle on a mowing machine, the important parts being the cutter and the comb. The comb, made up of several long teeth with sharpened edges, passes through the wool at the surface of the skin, dividing it into small bunches for the cutter, which is ordinarily made up of 3 teeth with sharpened edges. The cutter, oscillating at a high rate of speed over the surface of the comb, shears the wool almost as fast as the cutter head can be passed over the body of the sheep. As long as the comb is held flat on the surface of the skin, the wool will be sheared close to the body without danger of cutting the skin, as the teeth in the comb are much longer than those of the cutter and act as a guard. The power is transmitted through a flexible shaft.

Formerly, the chief objection to this machine was the difficulty in getting the cutters and combs sharpened. This has been largely overcome by the manufacture of a grinder, which consists of a revolving disc run by crank and gears.

With the other power machines, the cutter head and flexible shaft are the same as in the hand machine, the point of difference being in the source of the power. These machines are of necessity more expensive, and, therefore, are practicable only where large numbers of sheep are to be shorn, as on the Western ranges. The power may be derived from a steam or gasoline engine by belt or shaft or from an electric motor.

Horse-clipping machines. Clippers are used to remove the horse's winter coat after the weather turns warm in the spring; to trim the hair on the udder, flank, and thighs of dairy cows in milk; and to shorten the hair on the head, tail, and legs of some beef animals in preparation for show.

A clipper, similar in type to that used by

barbers, but larger, is used by some stockmen. The large hand clipper which is operated by both hands is, however, employed more extensively than the lighter type. These are satisfactory for small jobs, but where the whole animal is to be clipped, or several are to be trimmed, the time necessary for such an operation renders such hand tools impracticable. To meet this condition, machines are being manufactured that are very similar to the sheep-shearing machines, except that the teeth in the cutter head are much shorter and more numerous, and that the power is transmitted through a chain which turns in a flexible tube. The chain in the tube gives to the machine more flexibility than the flexible shaft, and works satisfactorily, since less power is required to cut hair than to cut wool.

The power for these machines may be applied in exactly the same manner as for sheep-shearing machines. In fact, the horse-clipping attachments may be used on the sheep-shearing machine, or the sheep-shearing attachment on the horse-clipping machine, thus making it possible for the farmer to add either one or the other at a small additional cost.

Power dental float. This is an attachment to the above clipping machine. The disc file, rotating at high speed, quickly files the teeth to the desired length. Such an instrument is desirable where enough horses are kept to warrant the purchase of a clipping machine.

Grooming machines. The motor brush is a cylindrical brush attached to a flexible shaft, which, when rotating at a moderate rate of speed, makes it possible for one man to groom several horses in a shorter time and with less work than with the common currycomb and brush.

With the vacuum machine a vacuum is maintained in pipes, by which the dirt is drawn out of the hair and conveyed to a receptacle whence it can be removed. Motor brushes and vacuum cleaners have not as yet been very generally adopted, even in large stables.

Sprayers. Many farmers fail to recognize the value of disinfectants in keeping lots and barns sanitary. Spraying apparatus is as useful in the care of livestock as of fruit, and the same equipment may be used for both. The tank, or tub, sprayer is probably most useful for the average livestock farm, as it can be operated by one man, and can be easily wheeled in and out of stalls. It consists of a tank, or half-barrel, set on wheels, a pump and agitator, a hose and nozzle.

Poultrymen and small stock farmers might prefer the bucket pump, which can be used with a common bucket and is less expensive.

Hog oilers. Patent hog oilers are manufactured on the principles that if a hog has lice he will scratch, and that, if crude oil can be applied to the part that itches, the lice will be eliminated.

There are two general types of oilers—the revolving-drum and the tank-valve. With the former, a solid iron drum is so placed that the under side is immersed in the oil at all times. When the hog rubs against it the drum turns, bringing up a coating of oil which is applied to the hog where most needed. With the latter type, a tank is supported by one or more standards or legs. When the hog rubs against the standard, a valve is opened, which allows small quantities of oil to run down to the point of itching.

There are a number of different makes of oilers on the market that give more or less satisfaction. The experience of users of them seems to indicate that, in general, they help to keep the lice in the herd under control; but they could hardly be expected to clean up a badly infested herd, as it seems some hogs refuse to use the oiler. Trouble, too, has been experienced with some oilers from the valves clogging with mud; and some will be effective with pigs of a certain size, but cannot be used successfully by small pigs. Even with these objections, however, there is no doubt but that a patent oiler would be a profitable addition to the hog equipment on many farms, especially where no dipping vat is available.

Earmarkers. There are in use on farms three general methods of marking the ears of purebred livestock: (1) the ear label, or tag; (2) the notch; and (3) the tattoo mark. All are used to a greater or less extent. Sheepmen ordinarily use the metal-ribbon label; hogmen use the ribbon, the button, the tag, and the notch; the notch being used almost universally in marking litters. The tattoo is used more by cattlemen than by other breeders, although the metal-ribbon label is also extensively used by them.

The metal ribbon requires the use of a punch, similar to a leather punch, which cuts an oval hole through the ear, to allow the

FIG. 555. Feeding is a task that must be done at least twice daily, the year 'round; every possible method of lightening it is worth trying. This wheeled, slop barrel makes hog feeding considerably easier than it would otherwise be.

FIG. 556.   By making stock take care of itself we save both time and money.   Oilers keep hogs free from parasites and thus contribute to their comfort, health and increased growth.

insertion of the label.   The hole for the button is made in the same way, with the exception that it is round and smaller.   The tags are fastened to the ear with a common hog ring.   The tattoo requires a special marker, which consists of a pair of pincers with one jaw made to accommodate a set of removable letters and numbers.   The letters and numbers are made of steel points which, under pressure of the pincers, penetrate the skin of the ear and allow the deposit of sufficient tattoo oil to mark the ear permanently.

Teeth nippers.   In order to eliminate sore mouths and sore teats, caused by the long, sharp teeth of young pigs, two types of nippers are in use.   One is simply a small pair of pliers that break off the sharp points; the other is supplied with sharp edges which cut the teeth off at the desired length.

Dehorning instruments.   There are two general methods for dehorning mature cattle; namely, by the use of the saw and by the use of the knife.   The most improved knife has two blades ground with concave-angle cutting surfaces which form two opposing shears and thus give equal pressure from 4 directions. The power is furnished by the 2 handles having corresponding eccentric gears that work in the teeth of a tapering rack attached to the movable blade.   When the handles are brought together, the power of their combined leverage is transmitted through the rack to the sliding knife, which, working against the stationary knife, shears off the horn.   This dehorner has the advantage of doing the work with more speed; but it is more expensive than the saw, which, for the small stockman, will give satisfactory results.

Sheep-docking irons.   Lambs are docked when 1 or 2 weeks old, in order that they may be kept clean around the hind quarters and that the ewes may be more easily bred. A sharp knife or chisel and a mallet will do the work, but the accompanying loss of blood has caused many shepherds to adopt one or other of the types of docking irons.   These irons, being hot, sear the cut surface of the tail and eliminate the loss of blood, thus lessening the danger of losing some of the lambs.

There are two general types of docking irons —one very similar to the common chisel, and the other constructed on the plan of hoof parers used by blacksmiths.   Both types have heavy jaws and blunt cutting edges.   Neither is expensive.

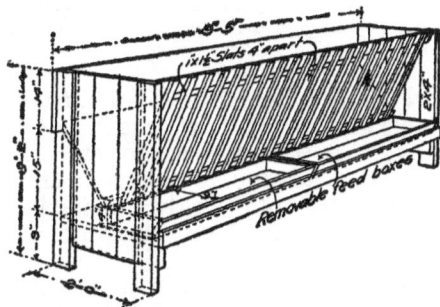

FIG. 557.   A convenient feeding rack for sheep to be used either in the barn, the feed lot or on pasture. The grain boxes being removable are more easily kept clean.

CHAPTER 34

Special-Purpose Barns

By JOHN M. EVVARD, *Associate Professor and Assistant Chief in Animal Husbandry, and Chief in Swine Production, Iowa State College, who was born on a farm in Livingston County, Illinois, and received his agricultural education there and at the universities of Illinois and Missouri. For three years he was Assistant to the Dean of the College of Agriculture of the University of Missouri, having charge of the animal husbandry problems in nutrition. In 1910, after a year of travel through the Middle West studying farm and livestock conditions, he joined the staff of the Iowa College, where he has since carried on his investigational and instructional work in livestock production. Both his training and his work have therefore made him especially able to discuss the principles underlying the construction of the farm buildings that have as their special task, the housing of farm animals.*—EDITOR.

IN THE planning and building of special-purpose barns, one should keep clearly in mind two main considerations. The first is the particular kind of livestock to be housed, for each should be sheltered so that it may do its work to the best advantage. This means that the barn must be planned from the inside out, and not from the outside in. The second is the convenience and preference of the livestock man himself, important factors being the saving of labor, low relative cost, pleasing appearance, etc. Of course the local conditions and requirements must be met in a practical manner. This means that there must be many averages struck, until the final result meets all needs as nearly as possible.

We must bear clearly in mind the use or function of the kind of livestock housed. Dairy cows, for instance, produce milk for human food, the finished product being drawn from the animal machine right in the barn. In the case of a pig or a steer it is different; we build up a machine and then we eat it, and the product is not gathered right in the barn but must be slaughtered and prepared either on the farm or in a packing house. The brood sow, or cow or mare performs still another function—that of producing young for future usefulness. With sheep, the wool is the product and must be kept clean; moreover labor is required in putting it in shape for use. Finally, the horse may be kept solely for work. All these differences mean the building of vastly different structures; but there are, nevertheless, some general considerations which are essential for all classes of live stock. Though in some instances these have already been mentioned in earlier chapters, they may well be reviewed here.

Ice harvesting offers an unexcelled opportunity for successful and profitable community co-operation

But each individual farm should have its ice house, and, if possible, a cold storage building adjoining it

THE FARMER'S FUEL IS IN HIS WOODLOT; A SUPPLY OF ICE IS HIS FOR THE TAKING. WHAT OTHER PROFESSION OFFERS SUCH PERQUISITES IN ADDITION TO ITS NORMAL RETURNS?

A portable farm brooder house that can be changed into a colony house as the birds outgrow the need of artificial heat

Inside a well-built, well-arranged poultry house of the long or commercial type. Note roosts, droppings platform, nest boxes and elevated feed hopper and drinking vessels

CAREFUL HOUSING OF POULTRY PAYS. BECAUSE A FLOCK SURVIVES NEGLECT IS NO REASON WHY IT SHOULD NOT BE GIVEN CONDITIONS UNDER WHICH IT CAN THRIVE

The Main Considerations

1. **Protection from severe cold and heavy winds.** Animals can stand still cold much better than cold winds, but buildings reinforced by windbreaks furnish sufficient protection from both. This, of course, includes protection from exceptionally heavy storms, such as tornadoes, or driving hail which ofttimes plays havoc with live stock.

2. **Correct temperature.** This does not mean ordinary room temperature for human beings around 68 to 70 degrees. Heavily wooled sheep, for instance, are perfectly well off in freezing weather, even with the temperature around zero Fahrenheit. They can largely regulate their own temperature if given correct feeds, and they do much better so than if housed in a warm barn. The fattening steer with heavy covering of fat, hide and coat, does splendidly in snappy, bracing weather around 20 to 30 degrees F., better indeed, than in the very hot summer months.

The hog on the other hand needs protection and must be warmly housed. An average temperature of 68 to 70 degrees, though not always possible, is not far wrong for a growing pig. Fattening hogs stand quite cold weather but they, too, should have a tight house and a nice warm bed. The horse driven continuously throughout the winter should have a fairly warm place, or else be carefully blanketed after every drive. "Stocker" colts and horses running in the field can stand much cold; however, on real cold days all horses should be amply protected.

But there is one class of all kinds of farm animals that needs considerable warmth, namely the new-born ones. In the case of lambs, colts and calves, this is true for only a short time; little pigs, on the other hand, need much protection for many weeks after farrowing.

Warmth is mainly secured by tight walls and roofs; plenty of windows to admit sunshine; low ceilings (especially for pigs and sheep); and artificial heating.

3. **Dryness.** The amount of wet that our domestic animals can stand, particularly beef cattle and horses, is remarkable yet they gallantly respond to a good dry building. A dry, solid, slightly sloping floor which does not attract moisture is in order, to make possible a dry bed.

To make a building dry, it is essential that the roof and walls be tight, the doors well hung and properly framed; the windows correctly jointed with whole panes of glass and frames snug. The roof is all the better if sheathing is placed underneath. Sliding doors are usually more convenient than swinging doors, but they have the drawback of not being easily made tight against rain, drafts and drifted snow.

The condensation of water which sometimes occurs on the under roof of buildings is undesirable for live stock, keeping the air damp and the bedding wet, and sometimes dripping on the animals' backs. A metal roof quickly causes condensation because, when the moisture-laden air rising from the animals hits the cold roof, it contracts in cooling and the moisture is quickly deposited. Inasmuch as the animals constantly give off moisture through their lungs, the surface of their bodies, and their excreta, proper ventilating devices are essential to prevent its accumulation.

4. **Ample drainage.** Good drainage is essential to health; and the slope should always be away from the building. Many open sheds shelter stagnant pools or mud wallows where disease germs multiply, and which should be prevented by drainage. Both surface and underdrainage should be looked to. Briefly, therefore, the drainage should be from the inside out, and then from the outside away to some distant point. Of course the outlets must first of all be planned for.

5. **Abundance of light.** The saying "Darkness breeds disaster" has much truth in it. Light paves the way to an inspiration to clean

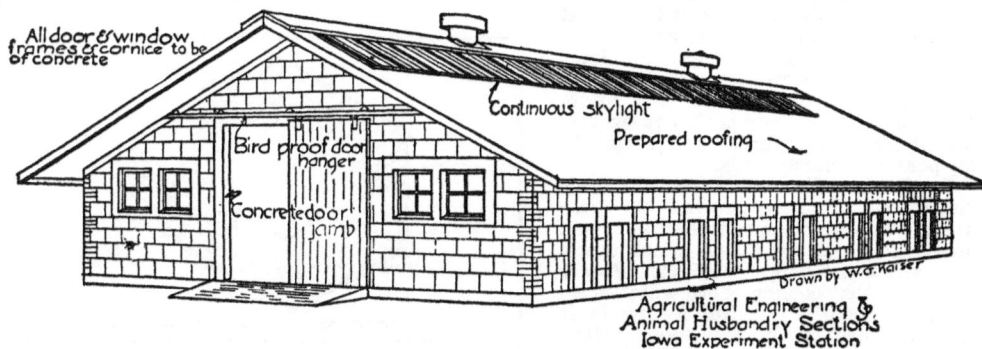

FIG. 558. Buildings for livestock should protect from the weather, and provide drainage, ventilation, dryness and sunlight. The Iowa Community Sunlight hog house is a strikingly successful structure in all these respects.

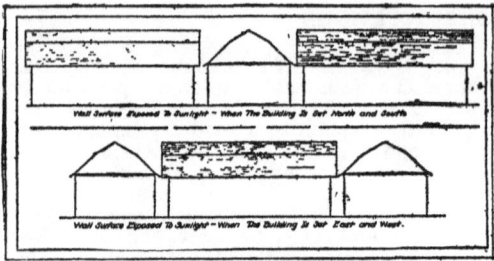

FIG. 559. Showing how a building placed north and south (*above*) receives more light along its sides where the most windows are, than one set east and west (*below*.)

up; darkness offers no incentive because it does not disclose what is needed. Dark stables are dangerous to man as well as beast, for horses are more liable to kick in the dark, and mules also are more unruly.

Light, even diffused light, is a great germ destroyer, and our modern barns are like our modern factories, being exceptionally well lighted with the windows properly placed so as to admit maximum daylight. White walls are becoming more and more popular because with them we can almost double the amount of light; solid partitions are being discarded because they darken the stables and barns and are unsanitary; other schemes are being used to advantage such as splayed windows, skylights, light-colored (concrete) floors, etc.

6. **Direct sunlight** is coming into its own as a universal germ and disease destroyer, costing practically nothing and always available if the right opportunity is presented. It promotes dryness, warmth, ventilation, diffuse lighting, sanitation, and drainage. It instills general vigor, muscular strength, high vitality, and good color in the animals. Young animals respond favorably to light as do plants. Perhaps not so vividly so far as color is concerned, but certainly so as regards growth and development. The forenoon sunlight is particularly important with livestock; after a relatively long, dark and perhaps cool and damp night, it is of great value as a morning bracer or reviver.

In securing direct sunlight, buildings built east and west are not nearly so favored as those built north and south. The southern exposure is always eagerly sought for cattle yards, sheep corrals, pig pens, and horse yards.

7. **Shade.** Protection from the sun's direct rays is necessary all the year 'round. Too much sunshine is not best, nor is too much shade; we must, therefore, strike the happy medium. Proper shade has that cooling effect which is essential. There is a restful feeling in it as contrasted with the open, bright rays of the sun. Considerable load is taken from the nervous system when it

is provided. Furthermore it keeps the flies partially controlled.

8. **Abundant ventilation.** All animals demand an abundance of fresh, pure air. They are continuously giving off moisture and carbon dioxide, a poisonous gas. This means that the barn air is steadily becoming laden with products that are unhealthful. The essential thing, therefore, is to supply a circulation of fresh air without drafts in such a way as to remove the foul air and not interfere with the comfort of the animals.

The air change required by animals is variously estimated, but it is generally conceded that the air should change at least 3 and upwards to 6 times or more an hour for best optimum results. The average amount of air inspired and exhaled per hour is estimated for mature animals approximately as follows:

Horse . . . 100 to 200 cubic feet, an average of 150
Cow . . . 80 to 160 cubic feet, an average of 120
Swine . . . 40 to 100 cubic feet, an average of 70
Sheep . . . 20 to 50 cubic feet, an average of 25

This does not mean that this much air has to be provided every hour, but it does mean that this much air passes into and out of the lungs in that period.

Estimates as to the amount of fresh air which should be allowed the different animals per hour and the cubic feet of barn air space they should be allowed, are as follows (although accurate experimental evidence to give them weight is lacking):

MATURE ANIMAL	CU. FT. FRESH AIR PER HOUR	CU. FT. SPACE IN BARN
Horse . . .	4,000	900
Cow . . .	3,000	700
Hog . . .	1,500	250
Sheep . . .	800	150

Ventilation can not be secured to great advantage without system, hence there have been developed two great ventilating systems, the King and Rutherford, both exceptionally

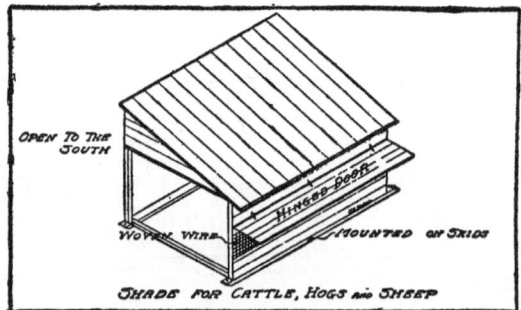

FIG. 560. The low cost of a movable shelter like this is repaid many times over in the added comfort of the stock

good. The former is particularly popular in the Middle West, and the latter in the northern, colder climates where it is said to work to good advantage. (See Chapter 33.)

Some schemes for securing ventilation—schemes which do not interfere with the safety and comfort of the animals—are based on (a) the use of cupolas which have upwardly slanting slats which tend to pull the current of air upward, deflecting it from the barn proper (a screen over it tends to keep birds out of the hay mow and barn proper); (b) the use of windows hinged at the bottom and swinging inward from the top are useful as are also the ordinary windows; (c) the use of double doors, the upper and lower half of which can be closed separately are used widely; (d) the King and Rutherford ventilating systems which provide means for the air to enter as well as to escape from the building, the general principle under which they work being that warm air is lighter than cold air.

Livestock, whether work animals such as the horse or the ox, fattening animals such as the hog and steer, milk-producing animals such as the dairy cow, and goat, or breeding females, cannot do their best without an abundance of fresh air—air that is continuously and gently circulating over their bodies.

9. **Sanitation.** We are living in a sanitary age because we now appreciare what sanitation means; this is nothing more nor less than keeping things healthy through the elimination of disease germs when they gain access, but primarily the prevention of their getting a foothold or even gaining an entrance.

Cleanliness is highly important, and a building, to be kept clean, must be free and accessible to the broom, the spray, and the pitchfork. To this end walls, roofs, and floors should be tight and smooth.

Dust, which carries countless millions of bacteria including many disease-producing ones, should be so far as possible prevented. In the dairy barn, for instance, dust from the hay mow is undesirable; to prevent it tongued and grooved lumber should be used between the floors, and the hay should be brought down through hay chutes, carefully screened off from the barn proper.

The prompt removal of litter is highly important. The easier this can be done the more likely it is to be done properly, completely and quickly. The litter carrier (p. 392) smooth floors, and a handy stall arrangement are of considerable help. Let us forget that some animals, such as the pig, live and therefore breathe, eat, drink and sleep close to the ground. For this reason if for no other, the floors, and the yards, stalls, and pens in which livestock are kept should be carefully kept clean of objectionable material.

10. **Safety.** The modern slogan "Safety First," is of particular importance in connection with correct livestock housing. It is not uncommon nowadays for a man to own a

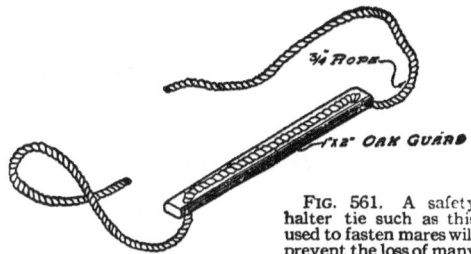

FIG. 561. A safety halter tie such as this used to fasten mares will prevent the loss of many a foal that might otherwise get the rope around its neck and be strangled to death.

dairy cow worth $1,000 or a beef bull, worth $2,000 or herd boar worth $1,200 to $1,500. Such high capitalization makes it necessary that we provide for the safety of the animals, by protecting them from mechanical injury and disease. Fenders that save the little pigs, safety doors that cannot strangle or injure animals; proper placing of manger and halter ties; large box stalls, and countless other details help bring about this result.

Here are some features which need special consideration: Avoid *low doorways*, and narrow ones, if too low horses may bump their heads and develop poll-evil; if they are narrow, horses and cattle may knock down their hips. Avoid *high sills*, particularly in brood sow pens, because the mere dragging of their bodies over these sills while the sows are heavily pregnant may result in dead pigs at farrowing time. Avoid *"slick," slippery floors*, particularly in dairy barns, or we may expect animals to be injured by seriously spreading their legs. Avoid *unstable construction* that results in barns being blown down when heavy winds come, ventilators, windows, and doors that easily jar loose, etc. Avoid weak or damaged and unrepaired partitions; stallions have been ruined by getting their feet through such partitions and breaking their legs; hogs and sheep have strangled themselves in dangerous partly boarded places of this sort. Avoid *"chuck" holes* in the floor which are a source of constant danger to sucking pigs. Avoid *steep, treacherous inclines*, particularly those unequipped with side rails or walls, and proper cleats; many a horse has been made unsound through carelessness in this connection. Avoid low mangers into which horses and cattle can jump; make such structures of just the right height for the animals that are to use them. Avoid sheep barns that can not be fenced and closed tightly against dogs or other enemies. There are hundreds of such little details that must be given attention if the greatest possible safety is to be assured.

11. **Comfort.** The ease and contentment of livestock counts for much. The dairy cow is highly sensitive to her surroundings, and must be comfortable, if the milk pail is to be filled. This involves such points as the size

Fig. 562. The comfort of these cattle is increased by the height of the feeding bunk which is built especially to fit them. They would be even better off if in place of the knee-deep mud there was provided a dry, straw-covered concrete pavement.

of the gutter and its distance from the manger; and the height of the manger with respect to the animal it is provided for. Handy stop boards can be placed in the pig pens so as to keep the bedding in a certain corner and thus assist in maintaining a warm, soft bed. Plenty of space should always be provided because crowded animals are not at ease and never crowd naturally except to keep warm. Everything possible should be done to induce livestock, especially dairy cows, fattening animals and hard-worked horses, to lie down and rest. Much more energy is liberated when animals are standing.

12. **Convenience.** What is one man's convenience is another man's inconvenience, and each should plan to suit himself; yet there are some details that are effective aids in all cases. For example, a modern complete watering system saves steps, saves pumping, and the need of constant attention. Handy, well-located feed bins, root cellars, silos, mows and hay chutes simplify work and save time. Feed alleys should be of ample width so placed that the feeding can be done with the least expenditure of time and labor. Work rooms where milk is separated, where tools are handled and repaired, or where harness is kept, should receive attention as to location and arrangement. Feeding boxes and hay racks, if carefully placed, have an important effect in saving exertion. Well-placed doors and windows are easier to open and close, and insure satisfactory ventilation. Litter carriers save time and do their work cleanly with the least hand *labor*, and the same is true of hay and silo carriers. When vehicles are stored in the horse barn, handy exit doors should be provided so that the horse can be attached inside the barn. Offices and general headquarters for the farmstead located in one of the barns, are often a great convenience not only for the livestock husbandman but for the women of the farm as well.

The convenience of the animal should also be considered, particularly if it is self-fed. How much more likely it is to eat generously if the feed is handy and accessible. And how much more likely it is to drink plenty of water if the trough is under cover and handy to the place where it sleeps.

13. **Serviceability.** That barn which is most useful every day in the year is worth a great deal more than the special-purpose barn which can be utilized only at certain seasons or for special functions. The more continuously it is used and the more results it can accomplish, the less costly it is per day of usage. The barns can be so built as to shelter different kinds of animals, or, with slight changes, to provide space for various operations none of which may be important enough or last long enough to justify a building of its own.

14. **Sufficient size.** Overcrowding is poor policy. The barn must, therefore, be large enough to house all the animals, and each stall or pen must be of sufficient size to meet the demands made upon it. Old animals, of course, require more room than young ones, stallions more than mares, and bulls more than cows. In arranging for fat cattle, we must make allowance for the hogs which follow and which are usually an essential adjunct.

As an approximate estimate of the number of square feet of floor surface which should be available for different animals standing and lying, the following figures may be useful:

KIND AND SIZE OF ANIMAL	SQ. FT. REQUIRED PER HEAD
SWINE	
50–100 pounds.	3–4
100–200 pounds	3–7
200–300 pounds	6–9
300–400 pounds	8–11
400–500 pounds	10–13
One sow with litter	40–60
Three or more sows with litters, each	20–60
Boars, mature, single	30–50
BEEF CATTLE	
Calves and steers:	
75–150 pounds.	7–11
150–300 pounds	10–13
300–500 pounds	12–16
500–800 pounds	14–19
800–1,100 pounds	17–24
1,100–1,400 pounds	23–35
Breeding cows:	
Single box stall	80–120
Cows with calves	25–50
Bulls, breeding, box stalls	100–160
DAIRY CATTLE	
Calves, heifers, and steers	Same
Breeding cows	Same
Bulls, breeding.	Same
Milking cows alone, stanchions	15–30
SHEEP	
50–100 pounds.	4–7
100–150 pounds	6–8
150–200 pounds	7–10
Over 200 pounds	9–14
Mature bucks	9–14
Ewes with lambs	10–14
HORSES	
In stalls.	20–50
Brood mares in box stalls with or without colt	100–160
Stallions in box stalls	120–200
Young, growing horse stock in open sheds:	
500–800 pounds	16–25
800–1,000 pounds	20–28
1,000–1,200 pounds	24–38

15. Durability. That building which will successfully withstand the effects of rain, snow, and other elements and the wear and tear of actual usage so as to stand up in the best fashion for the greatest number of years, is the most durable. Heavy construction helps toward this end, as do fireproof materials, and paint both increases the length of life and makes a structure more solid.

16. Storage facilities. In practically all of our farm buildings some storage space is essential. The hay mow above, to one side, or at the end is ever a pressing consideration; bedding to best serve its purpose must be kept dry and intact; grain bins properly arranged and handily located are necessary; and working equipment nicely covered in some handy place lasts better and is always available.

With the increased use of gasoline and electric power, there is a noticeable tendency to store grain overhead in general livestock barns, although there is a marked tendency to erect single-story buildings for swine and dairy animals. Where power is not available, hay is often stacked up in the central portion of the barn, or else to one side, the main point being that it comes clear to the ground. The general idea is to provide adequate storage conveniently within the barn so long as it does not interfere with the general efficiency from the standpoint of the animals to be housed and of the caretaker who does the work.

17. Reasonably low first cost. This should be in line with the kind and amount of service rendered. Extra service may require extra expense, but the value received should always be carefully noted. By keeping down the height of the walls, and by increasing the floor space under a certain roof, we save materials and keep down costs.

18. Minimum maintenance cost. The building that costs the least to erect is often the most expensive to maintain, and vice versa. The livestock must pay dividends, and, therefore, must be housed in such a manner as to make this possible.

19. Pleasing appearance of the farmstead has already been discussed; the same general principles apply to each building as well as to all of them together.

20. The ability to meet local conditions. Many of the points already discussed in this and other chapters have a bearing here, but the importance of the requirement as a whole should be emphasized.

21. Economy of time has reference to the owner as well as the livestock; the energy, health and well being of all concerned must be considered. Thus there should be kept in mind: nearness to pasture and shade, convenient arrangement of feed, water, bedding, etc., and such schedules or other plans as will lead to the avoidance of unnecessary steps in doing any work around the building.

The Housing of Swine

Pigs need protection from severe cold and heavy winds more than most animals: they should be kept in dry quarters because they are very susceptible to rheumatism and kindred ailments; they should have abundance of sunlight, both because it is especially invigorating to young growing pigs and because it cleanses the quarters. In summer they need shade from the very sunlight that is so important in winter. Their houses should be sanitary, for of all animals that carry a heavy risk, swine are the leaders. There should be abundant space, because pigs are inclined to overcrowd, particularly in cold weather. The first cost should be quite low because the average hog is worth much less than the average horse or cow. And because, as noted above, the hog lives close to the ground, it is always subject to infection unless its bed be kept clean, dry and sanitary.

There are two types of hoghouses: the large, centralized, community or stationary house, and the small, individual or movable house.

The community house contains a large number of pens which may or may not have movable partitions, and is a comparatively large, elaborate and durable structure. The individual house has but one pen, is small, simply constructed and comparatively durable. As

FIG. 563. The Iowa Gable Roof movable hog house with sides and roof sections raised to give maximum ventilation and provide maximum shade for summer conditions.

representatives of these types we may consider the Iowa Community Sunlit House (Fig. 558) and the Iowa Movable House (Fig. 563), both produced and used by the Iowa Experiment Station.

The following general considerations must be borne in mind in settling upon a correct site for the hoghouse, whether it be of the community or of the movable type: (1) Economy of labor and time; (2) sufficient drainage; (3) sunny exposure; (4) southern slope; (5) protective windbreaks; (6) nearness to pasture and summer shade; (7) suitable elevation; (8) prevention of odors reaching dwelling; (9) lessened risk from disease and infection.

The Community vs. The Movable House

Advantages of the community house as compared with the movable type are:

1. Time and labor required may be less, since facilities for feeding and general management may be more conveniently arranged under one roof. The sunning of the quarters can be done without extra work such as lifting the side doors; ventilation is more readily controlled; the stock can be more easily shown to prospective buyers; a horse and wagon are not needed in distributing feed, etc.; there is no moving of houses; and repairs are more simply made.

2. Durability is greater as would be natural with a permanent house.

3. Lighting by direct and diffuse sunlight is more practically arranged because of the greater height and generally larger dimensions.

4. Ventilation may be better because its principles can generally be carried out to better advantage in a large building.

5. The general equipment required is usually less and more compact.

6. Closer attention to the herd is possible.

7. The herdsman suffers less exposure, being always under shelter.

8. Feed storage, water supply, etc., can be more handily arranged.

9. Sanitation is better provided for since concrete and tile construction can be used,

FIG. 564. One of the vital essentials in a hog house is abundant sunlight. The double skylight in the gable roof community house provides it to a highly satisfactory degree.

and litter removed more easily. Moreover wide bands of sunlight can sweep across the entire interior.

10. Vermin may be eliminated because of

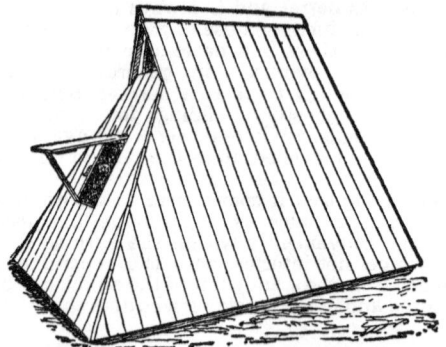

FIG. 565. The tepee type of individual house has the advantages of height in the centre and inclining sides which afford protection to little pigs, but in general it is less popular than either the gable or A-shaped type.

the solid floors. There is often much trouble with rats under movable houses.

11. Selection of a site is simpler, because only one is needed.

12. May prove more serviceable because a large house can house a greater variety of animals than a small one.

13. Artificial heating is easier because a single stove is needed as compared with a number of heaters for scattered movable houses.

14. One common feeding floor and a common wallow can be used. It is generally impracticable to have a number of wallows for different movable houses.

15. Danger may be less than when there are several houses in one yard, because huddling and piling up will not be so likely to occur in a large, warm, permanent house; pen divisions can easily be arranged.

16. Provides quarters for keeping track of the operations, such as an office.

17. Fire and other risk may be lessened by using masonry construction.

18. Grouped swine become better acquainted than those widely separated.

19. Makes the saving and best use of liquid manure easier.

20. Advertising value is greater because of its size, permanence and general impressiveness.

Disadvantages of the community, as compared with the movable, house are:

1. Location not easily changed.
2. Isolation of sick animals practically impossible.
3. Sanitation may be made difficult on such ground as: (a) New surroundings can not be chosen at will; (b) rotation of pastures and paddocks not easily accomplished unless the swine are made to walk a considerable distance to the various fields; (c) floors are sometimes damp; (d) dust is likely to be more abundant because so many more animals are present on a small area, etc.
4. Construction is more complicated.
5. Not so practical and economical for beginners or owners of small herds.
6. Much higher first cost.
7. Artificial heat in individual pens not so easily supplied. It is a case of heating the whole house or going to the extra trouble of partitioning off certain portions.
8. More fencing is required to provide similar range and pasture conditions.
9. Greater probability of using it simply as a special-function house particularly for farrowing; this lessens its serviceability.
10. Risk may be greater because the permanent house is usually located near to other buildings where the chance for fire is greater and where lightning is more liable to strike. Hail may damage such a house as the Iowa community with its open windows.

In general, the selection of the type of house should resolve itself into a happy combination of both types, the advantages of one over-

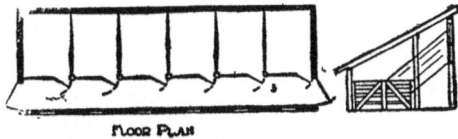

FIG. 566. Plan and section of shed roof community hog house with feeding alley, single row of pens and windows in front.

coming the disadvantages of the other, and both working together for a more harmonious housing system.

Types of Community House

The shed roof type (Fig. 566) which is a very simple structure.

The half monitor (Fig. 569) has some advantages as regards sunlight. This type faces the south, the light being secured in both series of pens by a double set of windows.

The round type (Fig. 567) is very little used in practice because it is difficultly arranged and the sunshine is hard to secure, neither does it usually fit in well with sur-

rounding buildings, pastures, and paddocks.

The modified community gable-roof type (Fig. 569) which is fairly popular, receives its light from two rows of windows on the same roof side.

The regular Iowa community sunlight type (Fig. 558) devised by Professor J.

FIG. 567. Partial plan and section of round community type. Unless used for storage the central space is apt to be wasted.

B. Davidson in conjunction with the author, which has a number of advantages including the following: (1) One can build it with a triple row of pens if need be. (2) More sunlight and its better distribution is possible because it comes directly through the roof into the pens and bathes the house thoroughly. Practically every portion is in the direct sunlight during some time of the day (Fig. 568). (3) Low walls are made possible, which is a very important consideration. (4) Hollow-tile construction means extra air spaces within the walls to keep the interior warm. (5) Ventilation is insured by means of the over-head windows which are adjustable, as well as through the ventilators which appear in the roof. (6) The floor may be built of tile, thus providing an air space between the layers of concrete. In the central feeding alley, ordinary solid cement can be used because the hogs do not ordinarily

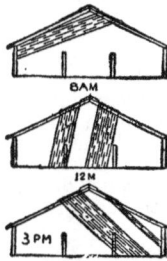

FIG. 568. Section of Iowa Sunlight type house at three hours of a March day, showing how the skylights permit bands of sunshine to sweep clear across the building.

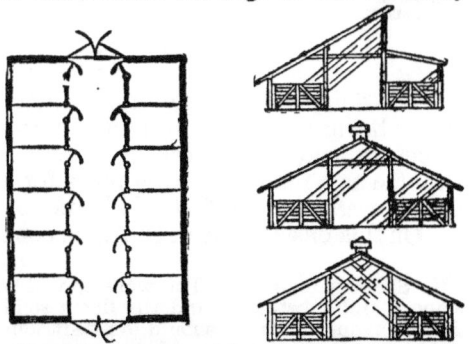

FIG. 569. Sections and floor plan (which is the same for all) of the Half Monitor (*top*), Modified Gable (*centre*) and Iowa Sunlight types of community house, showing comparative amounts of light received.

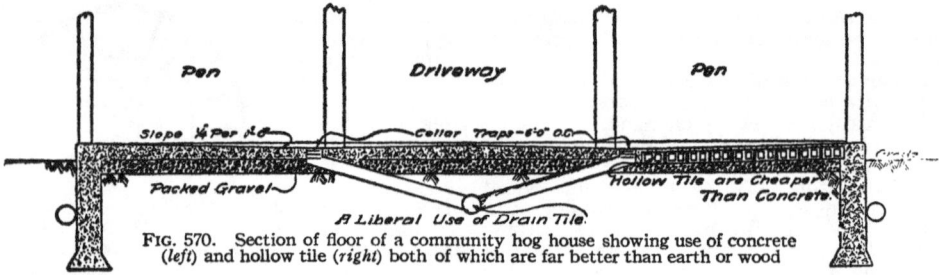

FIG. 570. Section of floor of a community hog house showing use of concrete (*left*) and hollow tile (*right*) both of which are far better than earth or wood

have to lie there. In the building of a combination concrete and tile floor, a simple, somewhat heavy layer of concrete is put upon a firm foundation, after which the rectangular clay tile are put in and covered with a second layer of concrete. Round tile may be used but not so advantageously.

The Movable House

A good type of movable house is the Iowa gable-roof (Fig. 571). As compared with the A-type (Fig. 572), it (a) provides more open space with the same floor area; (b) gives more vertical space for self-feeders and other equipment; (c) makes shade doors possible; (d) provides for easy ventilation both summer and winter by means of exits underneath the peak ends of the roof, as well as the windows and doors; and (e) is easier to move because a rope can be hitched around the upright side

FIG. 571. The Iowa gable roof movable house in winter. Glass sash in the roof keeps the interior dry, bright and warm.

wall more easily than around the slanting sides of the A house.

The greatest virtues of the A-type are: (a) It provides its own fenders in its slanting

FIG. 572. The A-type is the simplest of all hog houses. The hinged side makes cleaning easy but reduces the protection afforded in wet, stormy weather.

side walls (But in providing these fenders much room is lost). (b) It is cheap to build although not much more so than the gable-roof house, everything considered. (c) It is very easy to build, and easy to care for, especially if made with a door in the side wall.

The swine farm that raises sufficient hogs to justify the use of both community and movable hog houses, is particularly fortunate.

The Housing of Dairy Cattle

The dairy cow is essentially a machine, the raw materials it uses entering, the refuse being discarded, and the finished product being produced within, the very structure that houses the machine. We must therefore provide a barn that is suitable as a manufacturing place for a perishable and easily contaminated product, as well as a comfortable home for the animals.

On this basis, some of the main considerations in building a dairy cow barn are:

1. **Sanitary features.** The walls should be smooth and substantial and the floors solid. In the mangers there should be partitions between each two cows, and gutters should be provided to drain away the liquid manure and to catch the solid excrement. The floors should be kept clean and dry, and made of materials that contribute to this end. Odors should be kept down by providing adequate drainage. Whitewash may be used liberally both to make things sanitary and to improve the lighting.

2. An abundance of light and as much direct sunlight as is reasonable. One square foot of flat glass surface to 20 to 25 feet of floor surface is one allowance; another is 4 square feet to the cow. The whole interior should be as open as possible.

3. Safety and comfort. This means that stalls of the proper dimensions should be provided as well as careful methods of fastening the animals within. Slippery floors must be avoided and the cows should not be made to lie on damp, wet concrete but preferably on creosoted wood or cork platform blocks.

4. Ventilation demands special emphasis. Two general schemes of ventilation are used, one admitting the air at the middle of the barn over the alley, and taking it out near the floor at the side walls. The other admits the fresh air at the floor level of the outside walls, passes the impure air out overhead in the central portions of the barn. It is a good principle to have the air enter so that it will pass over the cows' heads towards the rear rather than from their backs forward. Where the cows face inward, the first of the two systems is excellent; where the cows face outward, the second may be preferable, although either might be used to advantage in either case depending upon conditions.

5. Miscellaneous considerations, which may be grouped under the following heads; (a) Provide rooms for office, washing, storage of clothes and weighing (and perhaps the separating) of the milk. The latter should not be done in the barn itself. (b) Running water is desirable, particularly in the milk and washrooms. (c) Power should be available, particularly if a milking machine is used. It is well to allow space for a milking machine and the power to run it in building, even if the apparatus is not to be installed immediately. (d) A silo and storage room for concentrates should be near at hand. (e) Hay storage is most convenient when easily accessible, whether above or at the end or side of the barn. (f) Not more than 2 rows of cow stalls should run lengthwise in the barn; a larger number makes the lighting poor and ventilation difficult. (g) Most good barns are built north and south so as to secure maximum sunshine.

Arrangement of Cows

Many arguments are advanced in favor of the cows facing each other, that is toward a central alley, some of these being:

(1) Sunlight in the rear of the cows keeps the floor where they walk drier and safer.

(2) The light is brightest at the end of the cow where the milking is done, making the work easier for the milker.

(3) Feeding can be done along one central alley.

(4) The alley can be at the highest level making the drainage toward the rear of the cows—a desirable sanitary condition.

FIG. 573. Section of barn shown below

(5) The centre of the barn can be kept clean and more attractive to the eye of the casual visitor.

(6) The cows do not have to face directly into the light.

(7) It is easier to turn the cows out and handle them generally when the tying and untying are done along the central alley.

(8) Feeding can be done to greater advantage. If the cows face outward, either there is no feeding alley or it is likely to be narrow, because two are required.

On the other hand, some authorities advise

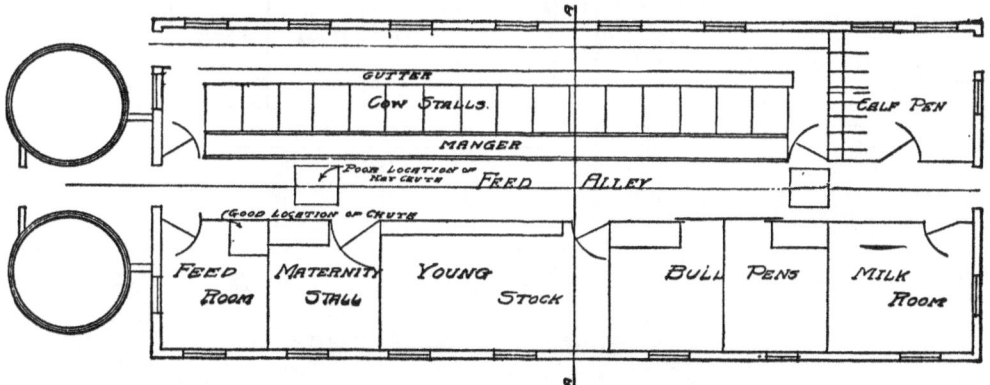

FIG. 574. General dairy barn embodying the features noted on the next page

that the heads be placed outward toward the wall for these reasons:

(1) Manure removal is centralized, all being taken from one alley.

(2) Milking is done more easily, particularly with milking machines, since less shafting is required and also less shifting of the machine from one side of the barn to the other.

(3) Cows come in together through the central alley, staying together until they reach their respective stalls, instead of entering on two sides and perhaps meeting.

(4) Walls are not spattered with manure.

(5) Brings the udders of cows in line where they can be compared and the general appearance of the herd studied. However, the light is not so good for such an examination as if they were along the wall.

Both the schemes are used in practice, and, depending upon local conditions, both are satisfactory and good. It is largely a matter of individual preference, and of adapting the most efficient scheme to the particular location.

Types of Cow Barn

There are two types of cow barns. One is a general type which takes care of the entire herd and also stores the hay. The other type is developed primarily for milking purposes and consists simply of two rows of cow stalls within a well-lighted, abundantly ventilated structure (Fig. 534). In this case the milk room may or may not be a part of the barn proper. The former scheme is best adapted to the average farm, but for specialized dairies, the latter is worthy of a thorough trial. Fig. 574 shows a dairy barn in which these particular points may be emphasized:

1. It is a general-purpose barn, housing all the animals as well as the milk and feed room. An office may be included also.

2. There is a central feeding alley with the cow stalls and calf pens to the left and the box stalls for young stock, maternity animals and bulls, and also the milk room, to the right.   3. The hay chute is well located, coming down in the feed room so as to avoid dust in the cow stalls. A poor location for the hay chute is also marked.

4. The silos are placed at the north (the shady) end of the building.

5. The calf pen and milk room are placed on the south (the sunny) side.

6. The calf pen is placed directly across from the milk room, making it easier to feed either whole or skimmilk. This has some disadvantages in so far as odors being carried to the milkroom is concerned, but it is excellent from the standpoint of convenience.

7. The hay storage space is conveniently located above.

8. The feed room is handy to the silos so that silage can be mixed with the grain to good advantage.

On the dairy farm, the sires are sometimes housed in a general-purpose barn, but if a number of bulls are to be kept, a special set of bull pens is highly advantageous. Fig. 575 shows the south elevation and floor plan of such a structure suitable for 5 bulls each kept alone. Note that they have a little open run on the south, a part of which is paved so as to give them solid footing just as they step out of their pen. The feed alley is conveniently arranged along the front. Hay and bedding may be stored above. Sliding doors are used to the pens because they are strong and safe. The feed room is on the north side in a central portion, making it handy for feeding as well as for the unloading of feed from the outside.

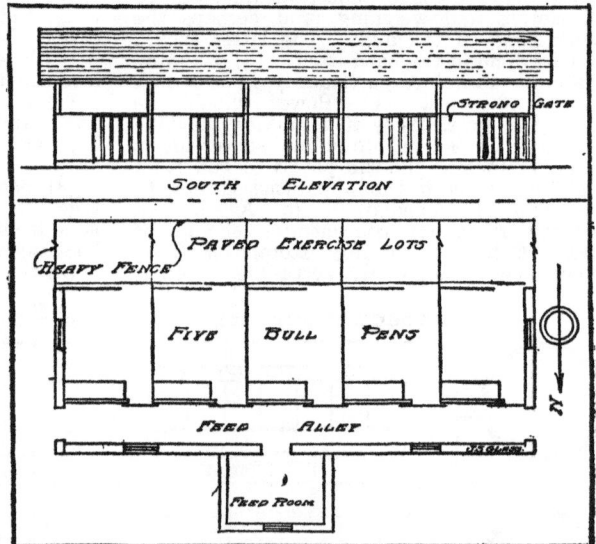

FIG. 575.   Elevation and plan of bull pens especially adapted to meet the needs of a pure bred breeding establishment.

The Housing of Beef Cattle

Beef cattle are housed much more simply than dairy cattle, four general types of barn being used, the first for the breeding herd, the second for the fattening cattle, the third for a combination of the above, and the fourth for temporary shelter in the pasture. Fig. 576 shows a good general-purpose barn for beef cattle. It might be even better if the cows were changed to the west, and the calves, etc., to the east side. The features to be noted in this barn are:

1. A central feed and service alley wide enough to take care of the cattle as well as the feeding operations.

2. Stalls for breeding cows (preferably on the west side) provided with open partitions high enough to keep the cows nicely separated.

3. Abundant windows on both east and west sides, the barn itself running north and south.

4. Calf pens and bull pens (preferably on the east side) where the animals may run out for sunshine. These are connected with outside runs which are not provided for the breeding cows. The latter can be taken in groups through the service alleys.

5. Silos are placed on the north to furnish protection.

6. The feed room is separated from the silos, being placed next to the office for convenience sake. Ordinarily the feed room would be placed next to the silos, but on many beef-cattle farms silage is fed alone. This arrangement could easily be changed.

7. The office is placed on the south where it is nice and warm.

8. Abundant hay storage is provided for above—this is quite essential.

Fattening shed. A feeding barn for fattening beef cattle is shown in Fig. 577. This barn is particularly well adapted for the storage of a large amount of hay. It also gives complete protection to the cattle. Note that the feed bunks for silage and grain feeds are arranged along the outer wall. A feed carrier track could easily be erected between the silos and these bunks. Fattening cattle such as steers which are bought on the open market can be advantageously housed in a very simple open shed facing the south, preferably arranged with hay storage above for convenience sake.

Fattening steers running in groups should be provided with the following conditions: (1) Plenty of open air and sunshine. (2) sh lter from cold winter rains and snows and also hot sunshine. (3) A dry bed to sleep upon and solid footing on arising. Yet it is surprising how well steers will do even if they have to go *knee deep in mud* out to the feed bunks. (4) A place for the hogs following the steers to consume the voided and otherwise wasted grain. It makes considerable difference in the housing of a bunch of steers whether or not an extra hog or two is going to sleep with them. These hogs must be provided for either in a separate structure or in a protected corner of the same building; (5) Handy hay and feed. Hay is best provided and also fed inside, where the cattle can eat and be contented. (6) Open water should always be provided. It is best al-

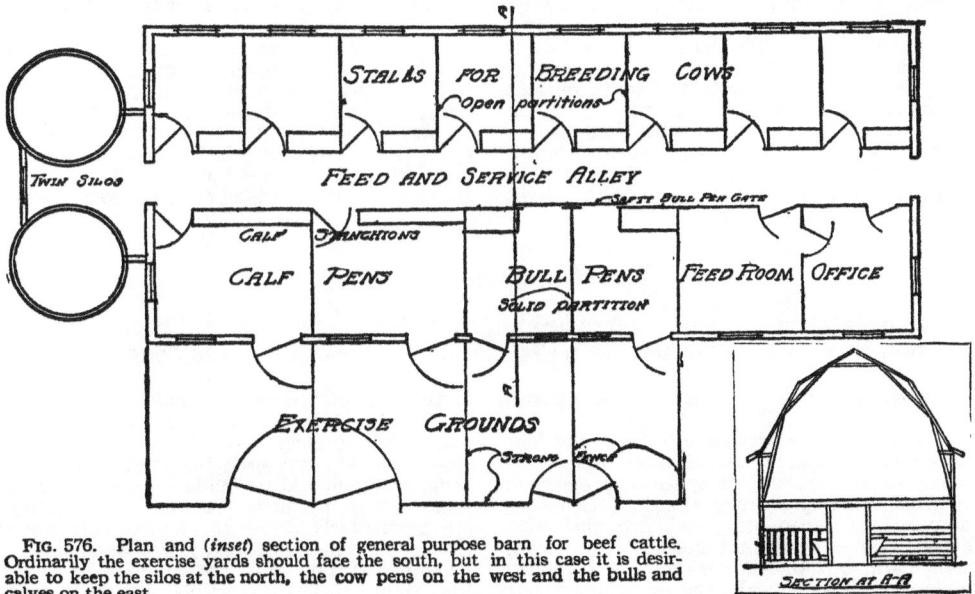

FIG. 576. Plan and (*inset*) section of general purpose barn for beef cattle. Ordinarily the exercise yards should face the south, but in this case it is desirable to keep the silos at the north, the cow pens on the west and the bulls and calves on the east.

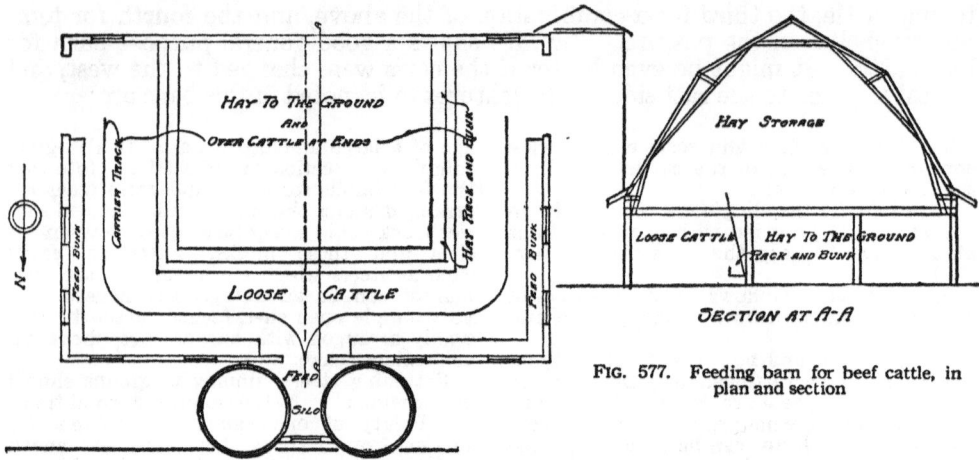

FIG. 577.   Feeding barn for beef cattle, in plan and section

lowed inside but a short distance outside is all right, provided the animal has protection from the winds while going to the watering trough. (7) The steer place should be quiet and well away from unusual noises. Steers coming from the range country are not accustomed to the congestion found on the average farm. They often are afraid of men on foot, and shy at other animals simply because they are unusual. A wild steer is made *often* more wild by his environment, whereas for best results his surroundings should be quiet, and in harmony with his nature.

Points of a good feeding shed. A shed which provides these essentials is shown above. Note these points in connection with it: It faces the south. Hay and bedding are stored above. The feed rack for hay is handily placed at the north side of the pens. The watering trough is to the south outside of the shed and near by. The paved run extends out a few feet from the shed proper to provide solid footing; here the feed bunks can be placed. Windows are placed in the front or south elevation so as to provide light

in the hay mow; if the hay mow is dispensed with they provide for the shed interior. Note further that this shed is wide and open with very little to disturb or interfere with the steers as they lie in it. The straw chutes are placed in a favorable position for throwing the bedding directly to the point where it is needed.

A handy cattle shelter built in the woods or in the field or in the pasture is of much importance on many farms. Such a building that is in use on one large farm and which has given excellent satisfaction at a comparatively low first cost embodies the following noteworthy features: (1) It is made of tile blocks, which provide a warm structure. (2) It faces the south. (3) The glass windows provide an abundance of sunlight within. (4) The open entrance door, large and spacious, is placed in the southeast corner. This permits the early morning sun to strike directly into the barn and also insures that southwest winds (which prevail in the Corn Belt) will not blow into it. Storage for hay may be provided on the second floor.

The Housing of Sheep

In general two types of barns are built for sheep; one a general barn, housing the breeding flock and other sheep in general, and the other is a fattening shed.

Some important features of a general sheep barn are as follows:

(1) Keep things particularly dry under foot because sheep can not stand much mud. (2) Sheep require more floor space in proportion to overhead hay storage than do cattle or horses. (3) Sheep being well provided with a warm covering do not need a very warm, snug barn, but the barn should nevertheless be built with warm walls and so arranged that

it can be kept open on the south and east by the handy placement of easily opened large doors and windows. (4) Sheep need protection from rain and snow particularly, and arrangements should be made for getting them under cover whenever storms come. They are particularly prone to stay outside when they should be inside. (5) The walls should be smooth inside so as not to catch and tear the wool. (6) Convertible panels are a handy asset in the sheep barn, particularly when they are interchangeable and can be used to make pens of different sizes. (7) Ordinary movable feeding bunks and hay racks can be largely used although some permanent racks are in order. (8) A root cellar or silo depending upon which kind of roughage is fed, should be attached. (9) Stove-heating arrangements should be provided for lambing time. Warm lambing pens are also in order. These should be tightly closed off, making provision, of course, for ventilation.

Points of a good general barn. A general L-shaped sheep barn with exercise pens to the south and east is shown below. These points should be noted in connection with it:

1. It is built so as to give north and west protection, which is desirable.
2. The silo is placed on the northwest corner handy to the continuous feed alley running around in front of all sheep.
3. The feed room is in the corner handy to the silo and also handy to both ends of the L.
4. A shepherd's room and office is provided next to the feed room on the east where he is close to the ewe and lamb pens.
5. Provision is made in the eastern portion of the L for lambing pens, movable partitions or pens being supplied; these may be also used for rams.

6. The ewe lambs are run in one bunch and the breeding flock in another. If certain members of the breeding flock need special attention they can be taken to the ewe and lamb pens.
7. Exercise pens are provided on the south and west where there is an abundance of sunshine and warmth; this is much better than if the exercise pens were on the north and west where it is cold usually and oftentimes bleak.
8. The exit doors are best if of the sliding sort, but those next to the ewe and lamb pens should be snugly fitted so as to make these pens warm. The ewe lamb, and breeding flock exit doors can be kept open most of the time.

A fattening shed for sheep, whether bought or raised on the farm and prepared for fattening, may be a very simple structure. Hay storage can be provided above the shelter proper although some prefer to have a big feed and passage-way in the front where this is stored. A scale should be placed out of the way, on the north where it can connect with the central feed and passageway. Such a shed is designed simply to furnish protection for the fattening sheep which do best if kept largely in the open, but which need a warm bed underneath a waterproof, snowproof, windproof shelter.

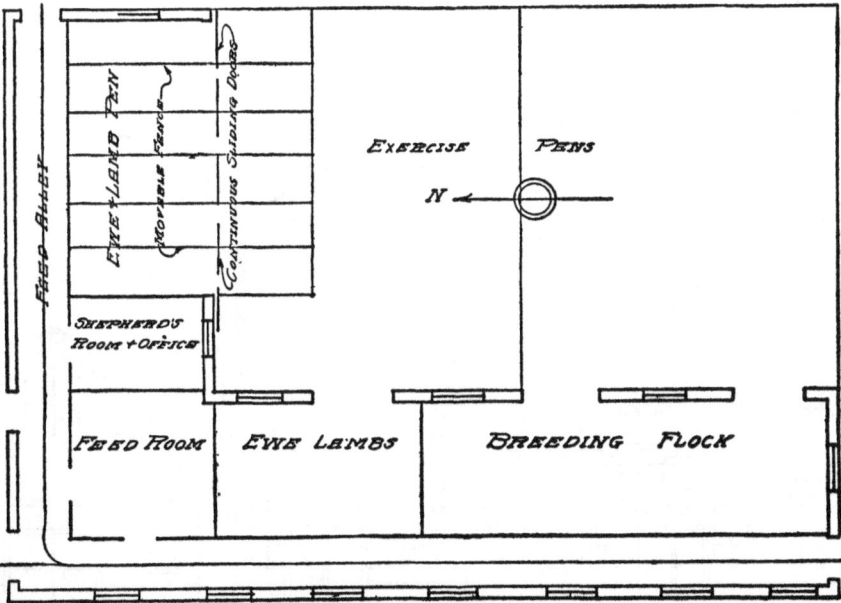

FIG. 578. Sheep barn plan. Sheep need but little protection from cold, except at lambing time, but they must be kept dry and given yard room where they can obtain plenty of exercise

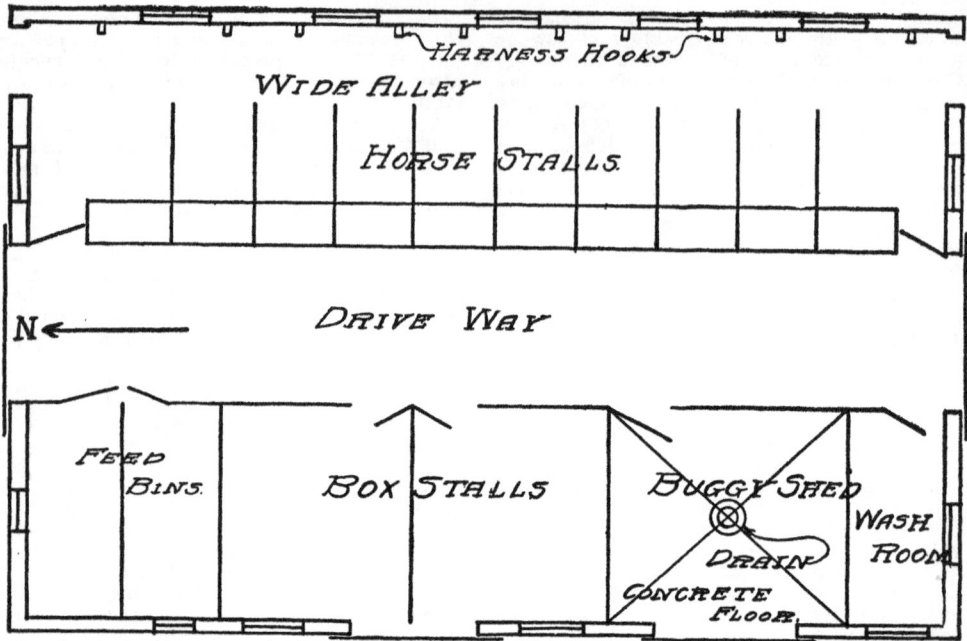

FIG. 579.   Plan of a general work horse barn.   A harness storage and repair room would be an additional advantage, but for daily use harness is best kept behind the stalls

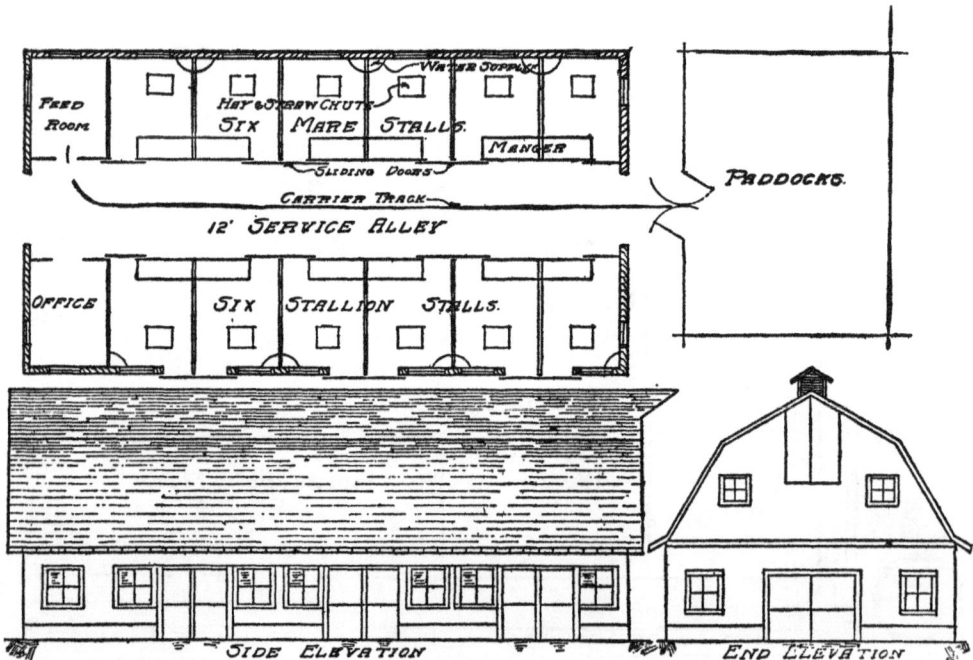

FIG. 580.   Plan and two elevations of a stallion and mare barn.   The professional breeder must build attractive structures in which to receive buyers as well as efficient ones in which to care for his stock

The Housing of Horses

Horses as a class of farm livestock differ from the meat or milk producing animals in that they are kept in large measure for working purposes. This includes of course, their driving and riding, for, though types of recreation for man, from the horse's standpoint these are forms of work. In building a barn for horses, the following points should be given as careful attention as possible:

1. The barn should be handy, and particularly safe and comfortable. Horses being strong, stalwart animals require safe construction or they are likely to injure themselves. A handy driveway and feeding passage, particularly one in the centre with the horses facing it is in order. This is a much safer arrangement than where the horses face out and have to be fed from the "heel end." A place for harness should be provided, preferably to the rear of the horses although handy harness rooms conveniently placed may be used for special harnesses. It is much easier to take the harness from a peg directly behind the horse than to run half way across the barn with it. We must remember that the horse is, or at least may be, a kicking animal, hence the partitions should be tight and solid. Solid, strong floors are also essential. A centre dividing pole where horses stand by pairs in open stalls, are often of advantage. The mangers should be fairly deep yet easy to clean. The floor should be slightly roughened to prevent slipping and safety doors are to be commended.

2. Protect the horse from drafts, particularly after it has come in from work. Ventilation should be natural and easy, rather than draughty and violent.

3. Provide an abundance of light. Sometimes it is well to be able to darken a portion of the stable so as to keep the flies down, but special arrangements can be made for this. The windows should be simply protected by means of iron bars or heavy screen wire to prevent breakage.

4. The floors should be solid but not hard and unyielding. Horses that are compelled to stand on concrete floors often puff up considerably in their feet. Wood creosote blocks are good and heavy wood slatted flooring is excellent.

5. Smooth walls prevent injury and offer no places against which the horses can rub their tails and scrape the hair off.

6. Exercise paddocks should be provided so that they can be turned out easily, preferably on grass. Hard-worked, heavily fed horses should be given their exercise on off days so as not to develop kidney trouble, particularly azoturia.

7. Spacious stalls are highly desirable. Single stalls should be at least 4 feet wide and preferably 5 feet by 8 to 10 feet deep. Double stalls should be at least 8 feet wide, preferably 9 or 10 with the same depth as singles.

General horse barn. The ground floor of a general horse barn is shown in Fig. 579. Some features which may be emphasized here are: (1) the central driveway; (2) the horse stalls all on one side; (3) work stalls on the east or early morning sun side; (4) handy entrances to these stalls from the alley way; (5) convenient harness hooks at the rear of the horses; (6) feeding done from the centre; (7) feed bins at the north end of the barn; (8) box stalls on the west side having outside communication to paddocks; (9) buggy shed, washing floor and wash room on the west side to the south; this washing floor can be used for washing horses as well as for wagons. Note also that the buggy storage room opens to the outside. A watering trough may be placed in the wash room if water is available. The central driveway is big enough to drive through; this is quite important particularly in filling the feed bins. The hay chute is best arranged to come down right in the middle of the driveway, preferably close up to the horse stalls on the east.

A good stallion and mare barn is shown in Fig. 580 in which these features should be noted: (1) A spacious service and feed alley. (2) Roomy mare stalls on the west, each provided with a hay and straw chute close to the wall. (3) The mare stalls open to the service alley, whence the mares are taken to paddocks on the north side for exercise. (4) Stallion stalls on the east with separate exits; these may be used for mare stalls. (5) Mangers are placed in the stalls next to the service alley. (6) Sliding doors are arranged so that one door covers one stall openings and is stopped in such a manner that only one of these exits can be kept open at a time; this is advantageous in preventing the possibility of two stallions or two mares getting out at the same time. (7) The abundant lighting through windows placed on the east, west and south sides. (8) The office is placed on the south side as is the feed room; the latter may however, be placed on the north side if deemed advisable. (9) A water supply controlled by a float system is present in each stall, a common float controls all of the concrete troughs, one of which supplies each two stalls.

This is a very handy barn and has been found especially adapted to those farms where a considerable number of mares and stallions are kept.

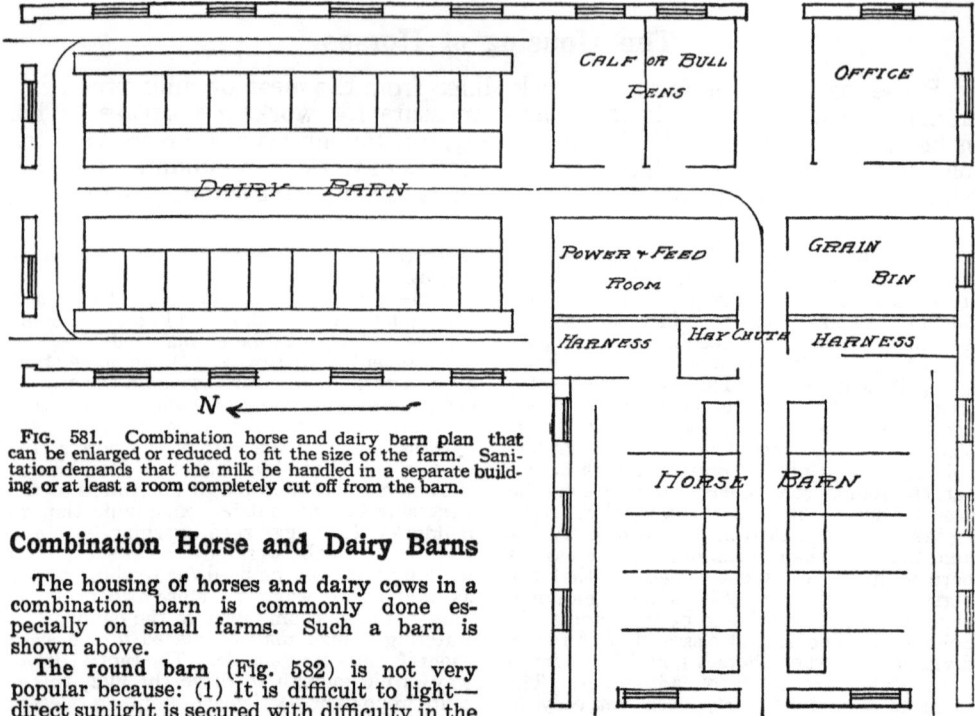

FIG. 581. Combination horse and dairy barn plan that can be enlarged or reduced to fit the size of the farm. Sanitation demands that the milk be handled in a separate building, or at least a room completely cut off from the barn.

Combination Horse and Dairy Barns

The housing of horses and dairy cows in a combination barn is commonly done especially on small farms. Such a barn is shown above.

The round barn (Fig. 582) is not very popular because: (1) It is difficult to light—direct sunlight is secured with difficulty in the interior portions. The north half of the barn is inclined to be somewhat cold at times because it is out of reach of the sunshine. (2) It is hard to arrange satisfactorily. (3) It is not in keeping with the average farmer's scheme of things; he is in the habit of working on a basis of squares and oblongs and a circular arrangement seems odd and uninviting. (4) The stalls are of an odd and somewhat inconvenient shape. (5) It is somewhat difficult to install litter carriers, milking machines and so on because of the circular tracks needed; of course it can be done, but less easily than in a rectangular barn. (6) The centre is usually utilized as a silo; this makes the silage handy, but it has the disadvantage of supplying a double set of walls for the silo which is an expensive procedure. A silo can stand very well by itself without having the protection of an expensive barn wall. (7) It is a difficult shape with which to combine additions and other buildings.

On the other hand a round barn is com-mendable because it furnishes a maximum of inside area per unit of wallspace; it is also easily built by those who have had experience.

The bank barn is less popular than it was in years gone by, before the great advantages of light, sunlight, sanitation, and ventilation were so thoroughly understood and appreciated. Has the disadvantage of being difficult to light, particularly on the side next to the earth bank (usually the north side) and there it is liable to be dark, damp and unhealthy. On the other hand such barns are very warm and comfortable in severe cold weather, and quite cool in the summer time.

If a bank barn is to be built, Fig. 583 pictorially shows the right and the wrong methods of constructing it. Assuredly a bridge should be provided from the barn across to the other bank, and windows placed in the side of the barn next to the bank; this will permit the sunshine going through the bridge to pass into the barn.

This farmstead is a success because it is efficient as well as attractive

The Iowa sunlight community hog house—a practical building for practical farms

An inexpensive, convenient commercial house for a specialized poultry farm

IT IS NEITHER THE COSTLINESS NOR THE CHEAPNESS OF A BUILDING THAT MAKES IT A SUCCESS
THE POINT IS: DOES IT GET RESULTS?

Twin silos and a single-story cow barn on a farm where architectural effect was given careful consideration

A hollow-tile silo and beef cattle feeding barn built first of all for practical results.   Inset shows a handy feed wagon for carrying mixed silage and grains

THE SILO, ASIDE FROM BEING AN INVALUABLE ASSET ON THE STOCK FARM, IS OFTEN A STRIKING FEATURE OF THE FARM GROUP.   THERE IS A TYPE FOR EVERY PLACE AND EVERY PURSE

Although the types of barns designed to shelter the different classes of farm animals have throughout this chapter been discussed as separate units, there will often be need to combine the essential features of two or more of them in one structure. This may result from the kind of farming followed, from the condition and means of the farmer—whether owner or tenant, from the difficulties of the location with regard to the obtaining of materials, or from the demands of economy. The small farmer will not, of course, require as costly or as extensive buildings as the one who operates

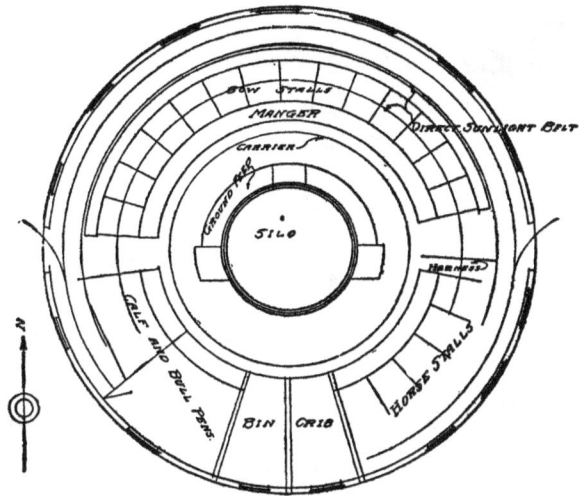

FIG. 582.   Plan of a round barn which under certain conditions is an economical and efficient type

a large place and maintains valuable herds and flocks of high grade animals. But on the other hand, the diversity of his limited interests will call for a good deal of careful planning so that all his stock may share one building and still enjoy the conditions that are most beneficial to each of them.

There are so many different functions that barns may perform, and so many different kinds of structures that are admirably adapted to such performance that it is difficult to describe any particular types that are entirely suitable for more than a single set of conditions. The aim of the writer has been, therefore, to give general principles and outline certain practices so that the livestock man may have a somewhat definite basis upon which to work in building a barn most acceptable to himself and to his animals under his particular, and sometimes peculiar local conditions.

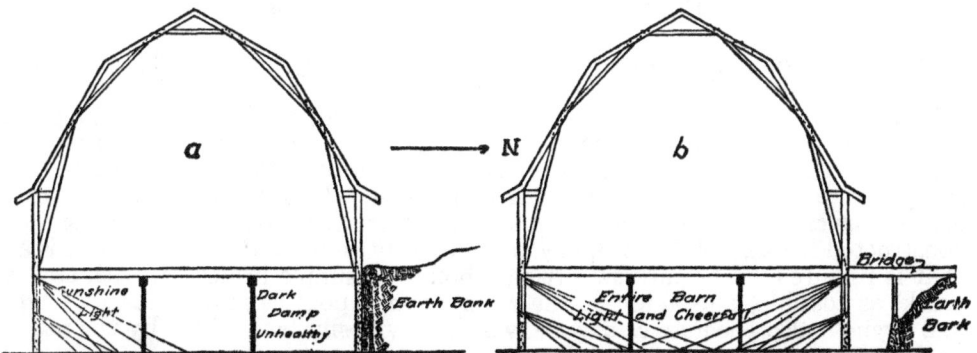

FIG. 583.   The bank barn is a common and valuable type in northern sections. The hill into which it is set provides much-needed protection, but light and ventilation must also be provided for. a shows the usual method of building such a barn; b shows a much better method which should always be followed if possible.

CHAPTER 35

Farm Poultry Buildings

By PROFESSOR H. L. KEMPSTER *of the College of Agriculture of the University of Missouri, who in 1911 organized its Poultry Department, of which he is now in charge. Previously, with the rank of Assistant Professor, he had been in charge of the Poultry Department of the Michigan Agricultural College, from which institution he graduated in 1909. Before entering college he lived and worked on the southern Michigan farm on which he was born. He has prepared a number of experiment station publications, all of which reflect the farmers' view point, without which, as Professor Kempster says, "one couldn't fit in with the atmosphere of the Middle West." The modern cry is for every family to produce as much of its food as possible. A flock of hens is an invaluable help in doing so; but to serve its purpose it must be well cared for and housed. Details of its care are treated in Volume I; what the farmer need know about farm poultry buildings is given here.*—EDITOR.

POULTRY housing plays an important part in successful poultry keeping. If, during inclement weather, the hens are not kept comfortable they will cease to be productive. Young stock will not thrive, unless given the proper protection. Success in handling fowls depends upon 3 important factors: (1) the kind of stock; (2) the poultryman himself; and (3) satisfactory environmental conditions, such as feeding and housing. If the stock is poor, or the poultry keeper is careless or unsympathetic, the results will be unsatisfactory. The poultryman should remember that good housing of poultry is only one of the factors contributing to success, and that, unless it is accompanied by intelligent feeding and personal responsibility, discouragement will result. On the other hand, even though a good ration be fed, there will be times when proper housing is necessary for economical production.

Essentials of a Poultry House

Comfort is one of the essential features of a poultry house. This is equally true in summer and winter. Other points which should be considered are simplicity of construction, economy of building material, efficiency of lighting and ventilation, and the convenience of the attendant. Due regard should also be given to the location and dryness of the house. Money is often spent unnecessarily in providing expensive building equipment. Unduly artificial conditions are neither essential nor desirable in successful poultry raising. A plain, simply constructed house, well-lighted, dry, and properly ventilated without drafts is all that is required. The interior fittings should be simple in design, with as few cracks as possible, thus aiding in the suppression of poultry mites. Lack of light and dryness also encourages disease, which may easily be avoided by the use of poultry houses properly designed.

Location. Convenience of location and arrangement is essential to economy of time in the care and management of poultry. Many chicken troubles may be traced to the selection of a site or soil not adapted to efficient sanitation. Perfect dryness is essential. This is quite largely controlled by the type of soil and the contour of the ground. A light soil, such as a sandy loam, not so sandy, however, but that it will produce an abundance of green food for forage, is most desirable. The lighter, more friable soils can be cultivated more easily than clay and at any time of the year. They also drain quickly, and warm up early in the spring. Heavy clay soils are objectionable, because it is more difficult to keep them sweet and sanitary, as they dry out slowly. The possibilities of cultivation and reseeding are also restricted when extreme dryness and baking occur. The fact that such soils remain muddy longer renders conditions disagreeable for attendants, and a greater number of dirty eggs may result from muddy feet.

A low spot is unsuitable for a poultry house, because surface water is apt to accumulate and damp air always settles in such places. Land which is naturally wet, either because of the nature of the soil or because of springy conditions, should be properly drained. Muddy quarters cause fowls to consume large quantities of filth. Dampness also results in unhealthy flocks. A windbreak should be provided, as it affords protection from the winds and the sun. If possible, the house should be located on a south or an east slope, where the ground will dry and warm up quickly in the spring.

In selecting a location for a poultry house, the farmer usually chooses the one nearest to his home, so that the housewife may conveniently care for the flock. This accounts for the usual location of the poultry house, halfway between the house and the barn, where it is easy for the hens to overrun not only the farm buildings, but the kitchen porch as well. The indiscriminate throwing of feed also encourages the birds to inhabit other buildings than their home. If the farm poultry house is located so as to make it natural for the hens not to overrun the farm buildings, there will be little trouble, provided they get enough to eat at home. Frequently, poultry can be encouraged to run in an adjoining field or orchard by a simple arrangement of the fences.

Yarding. As perfect sanitation is one of the requisites to success in poultry raising, the larger the yards the more easy it will be to maintain healthful conditions. While there is no set rule as to the amount of yard space required, if wholesome conditions are to be maintained, 150 square feet of yard space should be allowed each bird. Farmers neglect this important essential. They allow only room for the house, never realizing that the unduly artificial conditions afforded by grassless, hard, filthy yards are not conducive either to health or to economical production. From the standpoint of the farmer, yardage may be desirable at times. In the early spring, the poultry may be confined while the crops are getting a start, after which they may be turned loose. Occasionally the poultry is yarded in order to protect the near-by crops. Then, too, during the brooding period, old stock should not be permitted to run with the young, as the latter will not thrive, if compelled to pick their living with mature fowls. Grain crops may often be grown upon the same ground on which the poultry flock is running, with very little injury to the crop. Corn is especially adapted to such a practice, and it furnishes an abundance of shade, when needed. By plowing the yards occasionally, the soil is exposed to the sunlight. thus tending to destroy disease germs and, especially, intestinal parasites, which latter always frequent yards in which poultry is closely confined. In addition to making the yards more healthful, the growing of crops utilizes the droppings. This sweetens the ground and reduces the feed cost. Permanent sod runs cannot be maintained, if more than 400 hens are kept on an acre. Sod runs, however, do not furnish as much green food as do forage crops. Wheat or rye can be sown in the fall; oats or barley, in the early spring, followed by oats and rape, succeeded by rape and buckwheat. In this way, an abundance of succulent feed can be provided.

FIG. 584. General view of the "Missouri" house, designed to combine the desirable features of all special types in one building in which the farm flock can be housed safely and economically under average conditions and in any section of the country.

Fig. 585. The "open-front" house provides plenty of ventilation without draughts so long as the other three sides are tight. This may require double wall construction in cold sections where cloth sash may well be used in front in addition to the wire screening.

While the ideal of the farmer should be to fence the poultry out instead of in, and in this way to exclude them from the garden and the dooryard, a little forethought will enable him to follow out some of the above suggestions with little expense without sacrificing convenience.

Height. The height of a poultry house should be such that the attendant may work conveniently. This will afford conditions satisfactory for the birds. It is also best to have as little air space in the house as possible. If a roof of sufficient pitch be used, the back, or north, side of the house need not be more than 4 feet 6 inches in height. It is usually desirable not to have the height of the south side less than 6 feet nor more than 8, except in houses of special design. Where light is admitted from the south side only, the front should be a little less than half as high as the house is deep. This affords a satisfactory arrangement of lighting.

Width. Generally speaking, narrow houses are more expensive to construct for a given amount of floor space than are wide ones. It is also possible to reduce the amount of floor space allowed each bird with the wider houses. With modern types of ventilation, it has become more important to arrange the roosts at a considerable distance from the front. No house should be less than 12 feet deep, unless constructed for special purposes. The nearer square a house is, the cheaper it is to construct. A house 16 feet deep probably satisfies the requirements as well as any, although many stations are recommending houses as large as 20 to 25 feet square.

Size. The size of the house will depend upon the number of hens that are to be kept. The common rule in poultryhouse construction is to allow 4 square feet to a bird. A house 20 feet square is sufficient for 100 hens. Each hen will have plenty of room. In fact, under mild climatic conditions, 125 could be housed temporarily with no ill effects. On the other hand, it would not be advisable to crowd the house where weather conditions demand close confinement over a long period. A house containing only 150 square feet of floor space, while rated to house 35 birds, will be crowded if more than 25 birds are housed in it. It is thus seen that the larger house is the more useful.

Light. Properly lighted houses are necessary for satisfactory production. Sunlight is the cheapest disinfectant possible. Dark quarters are conducive to filth, dampness, and disease. A dark house is also unpleasant, and, in addition, shortens the period that a hen can feed. This may mean all the difference between poor and good egg production. Experimentally, it has been shown that, by artificially lengthening the day during the winter, the egg production more nearly approaches that which is expected in the spring. Where the open-front, or muslin, type of ventilation is used, a safe rule is to allow 1 square foot of glass to every 12 or 15 square feet of floor space.

Too much window space is not advisable. A house with such an excess is subject to extremes in temperature, due to warming up on sunshiny days and to the radiating of a corresponding amount of heat at night. In winter, the high temperature causes the combs of the chickens to become tender and more liable to freeze than in houses of more uniform temperature.

A common mistake is to place the windows too low. Direct sunlight is more effective in the middle of the pen than near the front. In general, the height of the tops of the windows should be a little less than one half the width of the house, that is, in a house 16 feet deep, the tops of the windows should be 7 feet high. The windows should be so distributed that there will be no dark corners. In fact, there are no objections to placing windows on the east, west, and north sides. With such an arrangement, there are no dark corners; and in this way the laying of eggs on the floor is discouraged. Also, when light comes from one direction, the hen always faces in that direction when she scratches, and, in consequence, there is a gradual movement of the litter toward the back side. With light evenly distributed, hens face in all directions. There is no piling up of the litter in dark corners, because there are none. It is interesting to note that to an increasing extent windows are being installed on all sides of poultry houses. Windows may be covered with wire screens, so as to keep the poultry in and the sparrows out.

Ventilation. An efficient system of ventilation is the most essential feature of a poultry house. Birds cannot be comfortable if drafts are present in the house; and yet there is no class of livestock that demands so much fresh air as do chickens. This is because the liquid excreta of fowls is passed off almost entirely through their lungs, for they have no sweat glands and their feces are comparatively dry. Unless a poultry house is well ventilated, it is sure to be damp; and in a damp house there are sure to be frozen combs in winter. A damp, cold atmosphere is much more harmful than a dry, continued cold of lower temperature. Other indications of poor ventilation are frost on the windows, walls, and roof, and a "chicken smell" or closeness in the house, due to the foul odors and tough, leathery litter. Such conditions will result in an abundance of colds, a lack of thrift, weakened vitality, and poor egg production. When they occur, better ventilation should be secured.

Probably the most popular type of ventilation is that employing cloth curtains or frames. A cotton frame allows the air to work through the cloth and yet keeps the chickens out of a draft. The cloth also admits an abundance of light. The cotton-front house is efficient as long as the cloth remains dry. If it becomes wet, it retards the movement of air. Even when cloth frames are used, the frames should be raised a greater portion of the day, so as to air out the house. The frames may be hinged at the top or side, or made to slide up and down like any window. A moderate-sized frame is more easily handled and permits more accurate control than a larger one. Where cloth is used in houses 16 feet in depth, the amount of cloth is the same as of glass, that is, 1 square foot of cloth to every 12 to 15 square feet of floor space. If the house is 10 feet deep, use 1 square foot of cloth to every 20 square feet of floor space; and, if the house is 20 feet deep, use 1 square foot of glass to every 10 square feet of floor space. The cloth curtains should be located on the south side of the house, farthest from the roosts.

Open-front ventilation. Another type of ventilation is that known as the open-front. This type consists of an open space on the south side, covered with wire screen only. The popular size of house for the open-front is 20 feet square. It is not adapted for houses of less depth. On the south side of such a house is an opening 3 feet wide one foot from the floor, running the entire length. In the original open-front types, the south side was only high enough to permit the opening. The roof was of unequal spans, the front span being twice as long as the back one, the ridge of the roof running east and west. The windows were put in the west end, and a door in the east. During winter, the success of ventilation of this type depends upon having the sides, back, and roof of the house entirely air-tight, so that there may be no drafts. The wind will drive in a short distance, but never to the roosts, which are on the north side and located above the top of the open space. There is a gradual outward movement of the air within, thus insuring an abundance of ventilation. The open-front type has an advantage over all other types, in that it requires no adjusting, never plugs up, and always works. This type adapts itself to temperature changes without constant attention, and in this way reduces to a minimum

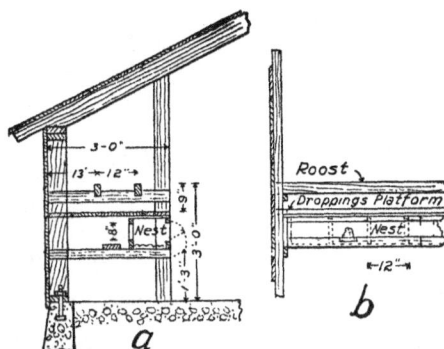

FIG. 586. Desirable arrangement of roosts, droppings board and nests, with dimensions, shown in section as seen from the end (a) and front (b).

the labor of caring for the house. It probably meets the requirements of a simple, effective farm poultry house more nearly than any other type that has been previously designed.

An item often overlooked in the construction of poultry houses is the summer ventilation. This is particularly the case in the South. It should not be overlooked in any country where the temperature gets uncomfortably warm. There should be openings on the north side, in order to insure a free circulation of air during warm weather. One of the advantages of having windows on more than one side is that by removing the windows the openings are made.

Tight walls necessary. The walls of a poultry house should be tight. This is necessary, so as to avoid cross drafts. Nothing seems to affect the health of the birds more quickly than does a cold wind. Colds are sure to develop, and these may pave the way for more serious infections. While wood is more frequently used, hollow tile or hollow concrete blocks may be employed, although they are much more expensive. Unless there is a hollow space in the cement construction, the walls are apt to sweat and the house become damp. The same is true of stone construction.

When wood is employed, it makes little difference what form of the material is used, so long as the wall is such that the air cannot sift through in winter. Matched flooring, sided up and down, makes a tight joint; and, in many sections, car siding can be obtained at more reasonable prices. With matched material, a tighter joint may be obtained by painting the joints when the wall is being laid. There will also be fewer cracks if well-seasoned lumber is used.

It is equally necessary that the roof be tight also. Sheeting the house with an extra ceiling on the inside is unnecessary. For roofing, probably the most common material is some brand of roofing felt, which is easily laid and which forms a tight roof. It should be laid on a smooth surface. If shingles are used, a cheap grade of roofing paper should be put down first, otherwise the wind will sift through the shingles. This necessitates laying a fairly tight roof before laying the shingles. If a metal roof is used, the roof should be prepared as for shingles, that is, the roof boards should be covered with a cheap grade of roofing paper. If this is not done, the house will be extremely hot in summer, and in winter there will be moisture condensation on the roof, resulting in a damp house.

Interior arrangement. Economy of floor space, simplicity of construction, the fewest new-fangled ideas possible, and convenience in handling the fowls, are all points to be observed in the arrangement of the interior of a poultry house. All fixtures should be portable and easily removable, so as to make thorough cleaning possible and easy; also, they

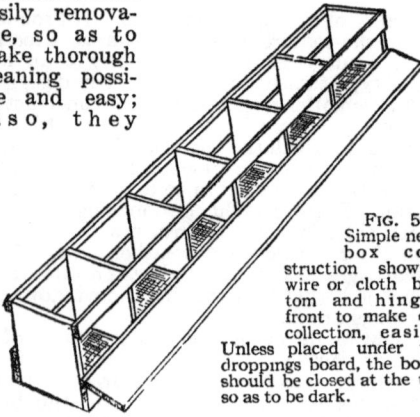

FIG. 587. Simple nest-box construction showing wire or cloth bottom and hinged front to make egg collection, easier. Unless placed under the droppings board, the boxes should be closed at the top so as to be dark.

should be up off the floor, so that the entire floor space may be used by the hens for scratching purposes. There should be as few cracks and crevices as possible, for in these filth accumulates and parasites thrive. Feed boxes, or hoppers, should have sloping tops, so that birds cannot roost upon them. In fact, there should be no place in the house for the hen to perch except the roost.

Floors. One of the main things that contribute to the securing of a dry house is a satisfactory floor. Of the 3 types of floors—earth, board, and concrete—the earth floor is perhaps the most common, and certainly the cheapest to construct. The most satisfactory floor, however, is of concrete. It should be kept covered with litter to prevent the fowls from developing sore feet, but other than this, its only objection is the original expense. Earth floors have the following disadvantages: the presence of dust in the house, which is irritating and disagreeable to the chickens as well as to the caretaker; the possibility of rat invasions; the necessity of removing a portion of the foul earth each year and replacing it with fresh; and the

more frequent cleaning necessary, because an earth floor soils the litter and becomes sour and filthy quicker. Obviously, the earth floor involves more labor and expense in care and upkeep than does a concrete floor.

In constructing a floor of either type, care should be taken to have the floor level at least 6 inches higher than the surrounding ground, so that the floor will not become wet from surface water outside. Some provision should also be made to prevent the rise of water to the surface of the floor by capillary attraction, which takes place like the rise of oil in a wick. Some coarse material, such as cinders or coarse stone (cinders preferred), will serve the purpose. Tamp well a 4-inch layer of cinders and on top of this put 4 inches of concrete and a finishing coat or, to make a really satisfactory earth floor, on top of the cinders put a 2-inch layer of damp clay. Pack hard, smooth well, and let dry. After it has dried, it should be covered with an inch of loose earth, which tends to discourage hens from digging holes in it. If the foregoing suggestions are carried out, there is no reason why the floor of the poultry house should not be dry.

Roosts. Roosts should be located in the warmest part of the house. With the open-front and cloth-front types of ventilation, they should be located on the north side. If placed on a level, there will be no tendency for the hens to crowd to the top roost, as is the case when they are placed ladder fashion. Roosts should be smooth, free from cracks, firmly placed, and not less than 2 inches in diameter. Poles or 2 x 4s placed on edge, with the upper corners planed off make sensible and satisfactory perches. They should be placed not less than 15 inches from the wall, at least 20 inches from the roof, and about a foot apart. From 8 to 10 inches of roosting space should be allowed for each bird.

Droppings platform. When droppings accumulate on the floor, the house becomes unclean. The birds are continually running over the droppings and, in this way, the feed is contaminated. There is also less space for the hens to use for scratching purposes. A platform to receive the droppings increases the size of the house and renders conditions more healthful for the fowls. This platform should be 6 inches below the roosts, so that the droppings may be removed without removing the roosts. If the platform is built of smooth material, the droppings can be easily removed.

Nests. Darkened nests are desirable, for they furnish seclusion to the hen and lessen the amount of egg eating. A too dark nest, however, encourages broodiness. It is also important that the nests be well ventilated during the summer. Portable nests facilitate cleaning. A common method of installing nests is to place them under the droppings platform at the front edge. Such an arrangement may place the nests so low that the floor space underneath will be little used for scratching purposes. This should be avoided. Where light comes from the south side only, the space at the back of the nests may be dark, and the hens will thus be encouraged to lay on the floor. Nests under the droppings platform are the easiest to construct. Twelve-inch boards are cut into 14-inch lengths and set on edge 13 inches apart; 1 x 4-inch strips are nailed along the lower edges, and 1 x 2-inch strips along the upper edges. Between these strips on the front, an 8-inch door is placed. The bottom of the nest should be covered with hardware cloth, 5 meshes to the inch. This allows all dirt and dust to drop to the floor, and makes frequent cleaning unnecessary. The nests rest on cleats at each end. Along the back is nailed a board upon which the hen may jump when entering. One nest should be provided for every 5 or 6 birds.

Partitions. Where all the poultry is quartered in one house, a partition is necessary to separate the pullets from the old hens. Later on, a separate pen is convenient to use for special breeding pens and in many cases is used for sitting hens during the spring. In order to prevent drafts, long houses need solid partitions either of wood or screen covered with cloth. These should be placed every 50 feet. In some houses, the solid partition extends only to the edge of the droppings platform. All partitions should be solid at least 2 feet high, to prevent the male birds from fighting through the partition.

Water stands. It is advisable to put watering utensils upon a stand at least 18 inches high. They can be built attached to the wall, or as separate stands. This prevents straw and refuse from the floor from collecting in the water, making it more sanitary. Solid tops collect more filth than do those made of slats or of wire.

Roosting closets. At one time, it was recommended that a curtain frame be arranged so as to drop down and inclose the birds in a small space at night. This has since been found unnecessary and the practice has fallen into disuse.

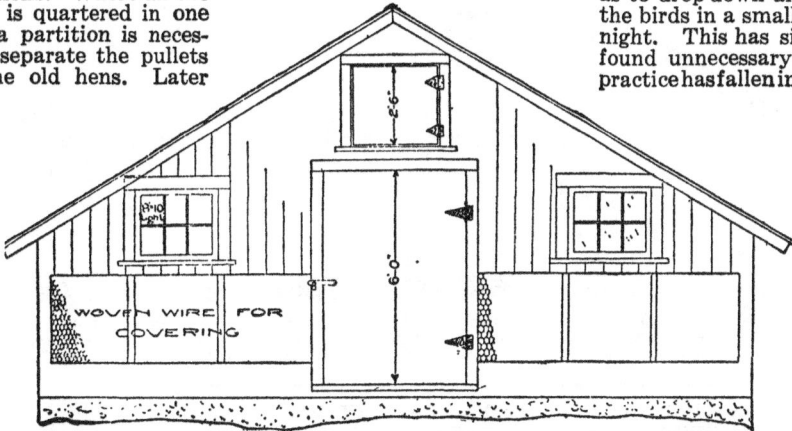

FIG. 588. Diagrammatic front view of the Missouri house showing glass windows, openings to be covered with netting and when desired, equipped with cloth screens, and the window through which straw is packed into the upper story.

Types of Poultry Houses

The "Missouri" house. The "Missouri" house, so called by the designer, the writer, attempts to combine the desirable features of all poultry houses in a house adaptable for the average farm flock. While designed for Missouri conditions, it has been successfully used in New Hampshire, as indicated in the second annual report of the New Hampshire Poultry Growers' Association, which says: "During the winter the birds not only kept in the best condition, but gave good egg production." This shows that the house is suitable for cold climates, and experience at the University of Missouri justifies its being recommended even for warmer climates.

Since the average farm flock numbers from 100 to 150 birds, this house is 20 feet square. The ridge of the roof runs north and south, the roof being of the gable type, that is, of equal spans. In such a house, there is an advantage in having the end face the south, for it permits the windows on the south to be placed higher than would be possible in the side walls. The walls are 5 feet high, the roof being 11 feet high at the peak. The south end contains a door in the centre and windows each 2 by 3 feet on each side. These windows are placed above a 30-inch-wide opening which is 1 foot above the floor. This opening extends the entire length of the south side on each side of the door. It is covered with wire netting, which keeps the hens in and the sparrows out. In stormy weather, this can be closed by means of a cloth curtain, if deemed advisable.

Light. On the east and west sides are 2 windows each 2 by 3 feet. There is another in the back. Those on the sides are placed as high as possible, but the one on the north is situated next to the floor underneath the droppings platform. The advantage of even lighting has already been mentioned.

Ventilation is by the open-front method, supplemented by the straw loft which has been a common feature in poultry houses for years. To form the loft, the joists or collar beams which tie the roof together are placed just high enough to afford sufficient head room. These can be covered with 1 x 4s placed about an inch apart, or poles may be used. The loft is filled with two or more feet of straw which is an absorbent, removing damp air from the house. It also protects the fowls during the hot weather by acting as insulation. Such a house is at least 4 degrees cooler during the hottest part of the day than other types of houses. Of the straw

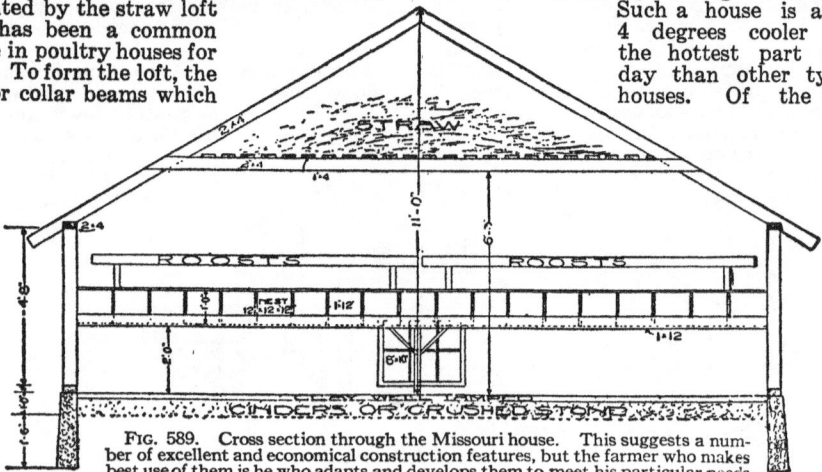

FIG. 589. Cross section through the Missouri house. This suggests a number of excellent and economical construction features, but the farmer who makes best use of them is he who adapts and develops them to meet his particular needs

BILL OF MATERIAL FOR MISSOURI POULTRY HOUSE

USE	PIECES	SIZE	GRADE	BOARD FEET	COST	TOTAL
Framing	19	2 x 4 – 16	1	203	$2.50	$ 5.08
Rafters	22	2 x 4 – 12	1	176	2.50	4.40
Plates and sills	8	2 x 4 – 20	1	107	2.50	2.63
Studding and framing	1	2 x 4 – 10	1	7	2.50	.18
Roosts	5	2 x 4 – 20	1	74	2.50	1.85
Finishing	6	1 x 4 – 10	2	20	2.00	.40
Finishing	1	1 x 4 – 16	2	6	2.00	.12
Finishing	4	1 x 4 – 10	2	14	2.00	.28
Finishing	8	1 x 4 – 12	2	32	2.00	.64
Floor for loft	52	1 x 4 – 14	2	260	2.00	5.20
Sides, car siding	....	1 x 6 – 10	....	600	2.00	12.00
Roof and droppings platform ship-lap	....	1 x 8 – 12	....	720	2.25	16.20
Shingles	....	....	....	5½M	3.00	16.50
Sashes	7	6-light 8 – 10		....	.60	4.20
Front and over windows, wire netting	....	3 – 32	....	....	.04	1.28
Hinges	1 pair	....	....	....	.10	.10

Materials, excluding nails and foundation		$ 71.11
Labor		28.28
Foundation, 3 cubic yards at $6		18.00
TOTAL		$117.39

loft, Professor Herner says: "It is quite dry in the coldest weather and is easily ventilated." As compared with the "shed-roof" type, he says: "The shed-roof house is always damper and colder in winter and, while the gable roof will cost more, the difference in efficiency is well worth the additional cost."

Materials. The walls are constructed of car siding, which is similar to flooring, except that it is beaded. The roof is made of ship-lap covered with shingles. It is necessary to have a tight roof. The roosts are placed on the north side, as suggested on a previous page.

The cloth-front type. The cloth-front poultry house (Fig. 590) is representative of that type of ventilation. This particular house is 14 by 24 feet. It will house comfortably 70 hens. For a flock of 100 hens, it would be built 14 by 28 or 16 by 24. This house is constructed upon a concrete foundation 6 inches wide, 8 inches above ground, and extending under ground to below the frost line. A concrete wall is advisable

FIG. 590. A simple but well-constructed colony house of the cloth-front type. This will house a large enough flock to supply the farm needs and a generous surplus besides.

for any permanent structure. It adds durability to the house, and eliminates the trouble from sagging and warping which may accompany houses with poor foundations. The south side is 6 feet and 8 inches. The studs are put 2 feet apart. The roof known as the "combination type" is best adapted to wide houses, the front span being one half the length of the back span. If the side walls were 6 inches higher, a shed or "shanty" type of roof could just as well be used.

The front consists of four windows, each having for its upper sash a cloth frame which slides up and down while the lower sash is a 6-light 10 x 12-inch glass window. This allows 1 square foot of glass and cloth to every 16.5 square feet of floor space. The cloth frames are of a size convenient for the control of ventilation, it being possible to open as many as necessary to supply a sufficient amount of fresh air, varying the number of open frames with the nature of the weather. When open, there are no drafts on the birds, because they are on the floor. In extreme cold or stormy weather, the frames are closed at night. This arrangement of cloth frames has one serious objection which is that when the curtains are dropped down they cover the glass. This interferes with the direct sunlight. Hinging at the side or end and opening in will remedy this.

The roosts located at the back side are made of 2 x 4-inch material placed on edge, the upper corners being rounded off. They are set in notched boards at the ends, and run the entire length of the north side. Droppings platform and nests are placed beneath. The rest of the equipment is a feed box 4 feet by 16 inches with sloping top and a water stand. The feed box has two compartments, one for ground feed and the other for whole grain.

USE	PIECES	SIZE	BOARD FEET	COST	TOTAL
BILL OF MATERIAL FOR A HOUSE OF THE SHED-ROOF TYPE 14 BY 28, 8 FEET HIGH IN FRONT AND 6 FEET HIGH AT BACK					
Rafters	15	2 x 4 – 12	160		
Plates and sills	10	2 x 4 – 14	93⅓		
Studding, Front.	7	2 x 4 – 16	74⅔		
Studding, Back	7	2 x 4 – 12	56		
Studding, Ends	2	2 x 4 – 14	18⅔		
Centre supports.	2	2 x 6 – 14	28		
Centre posts.	1	2 x 4 – 16	10⅔		
			441⅓	$2.50	$11.02
Siding	..	....	600	2.50	15.00
Sheeting for roof	..	....	506	2.50	12.65
Door framing	1	1 x 6 – 16	8		
Door casing	1	1 x 4 – 16	5⅓		
Door sill	1	1 x 8 – 4	5⅓		
Window frames	4	1 x 6 – 16	32		
Window casing	4	1 x 4 – 16	21		
Window sills.	1	2 x 8 – 12	16		
Finishing	4	1 x 4 – 16	22		
Corner boards	4	1 x 4 – 14	18⅔		
Roost-pole supports.	3	2 x 4 – 14	26		
Roosts	6	2 x 4 – 14	56		
Droppings platform.	..	....	112		
Nests	..	....	60		
			400	2.50	10.00
Rough total					
Sashes	4	6-light 10 x 12	....	....	3.00
Roofing paper	..	....	6½ squares	1.75	11.38
Hardware	..	....	....	....	4.00
					$67.05

Cost of **poultry houses.** The cost of a poultry house will vary with the material used and the economic conditions of the country and particularly of the community. According to figures given on a previous page, the cost of materials was about 70 cents a bird. Lewis, in "Productive Poultry Husbandry," gives bills of material for several types of houses and states that the material cost is from 88 cents to $1.12 a bird. The accompanying bill of material is for a house of the shed-roof type to hold 100 hens.

Buildings for Special Purposes

Incubator cellar. An incubator cellar should not be subject to temperature changes and yet should have provision for adequate ventilation. The humidity should be relatively high and no direct sunlight should strike the incubators. Under farm conditions, the basement of the house or a cave cellar is frequently used. A basement window covered with cloth or a cupola ventilator in the cave cellar will furnish adequate ventilation for 1 or 2 machines.

An incubator cellar should be placed 4 or 5 feet in the ground. The ceiling should be comparatively high, and the walls, if not double, should be constructed in such a way that the room is not influenced by outside temperature. Concrete or hollow tile is suitable in this connection. The superstructure may be used for any purpose, such as a feed house, laying house, etc. The extra space will add to the efficiency in utilizing the building.

The windows in an incubator cellar will necessarily be placed high and should be double, the outer sash being hinged at the top and the inner sash at the bottom. If neither sash is opened more than 45 degrees, the air will circulate through the cellar, but will not strike the incubators.

Brooder houses. In planning a brooder house, the type of heating determines the kind of house to build. There are 3 popular methods of heating brooders: (1) lamp brooders, which use kerosene oil; (2) coal-burning brooders; and (3) hot-water brooding systems.

Lamp brooders. For lamp brooders, a small portable colony house is generally used. A house of this kind may be used in the spring for brooding, as a colony house for housing chickens on the range in summer, and as a special pen for breeding hens in winter. A good size of house for this purpose would be about 7½ feet by 10 feet. It is best not to have it too large, as it will be clumsy and difficult to move. It should be placed on runners for that purpose. There will be sufficient headroom if it is made 4½ feet high at back and 6½ feet in front. A door, 2½ feet wide, should be placed on the south side. There should also be 2 windows, preferably one on each side

FIG. 591. A good type of brooder house for the farm. The runs are partly covered so that some shade may be provided. The soil in them should be kept healthy by frequent spading and planting to a quick-growing crop.

of the door. These should be placed 18 inches above the floor. Two 6-light 8 x 10-inch glass windows will furnish sufficient light. The window frames should be made a foot longer than necessary, and this space covered with a cloth frame for ventilation purposes. A board floor is essential in a portable house, as it keeps the chicks dry and makes the house ratproof. By means of the portable house the chicks can be raised near the house, and close attention given them. Later, they can be moved to the range, such as an orchard, cornfield, etc., where an abundance of shade, green food, and insects will furnish conditions ideal for economical and rapid growth.

Coal-burning brooders. The brooder-stove method of brooding involves the use of a larger house than the portable house described above, although the latter house may be used. Any room 12 feet square is of sufficient size for a brooder stove. The room should be so arranged that the chicks may have outside runs. In fact, it is practically impossible to raise chicks, unless they have access to the earth. A plan often employed is to partition off a part of the laying house and use this for brooding. The objection which might be raised to such a plan is that the ground surrounding a permanent poultry house is sure to become contaminated, and conditions will not be favorable for satisfactory growth. There is no reason why a yard for chick raising should not be attached to the permanent house; and, if due precautions are taken to keep the mature stock off, it will serve for brooding purposes for a long time. Yard cultivation will lessen the danger of trouble from intestinal parasites. With the coal-burning brooder stove, a series of pens may be attached to each other, in which case the construction would be the same as for a house for laying purposes.

Hot-water brooders. The third type of brooder arrangement is that employing hot water. Such a house will vary in length and width, depending upon its capacity and whether hovers and pens form a single or a double row. The method of heating is by means of a series of hot-water pipes running underneath the floor of the pens. These pipes are boxed in, so that the chamber is completely isolated. There is a partition in the box corresponding to the partition in the pen. This makes each pen a unit by itself. From this heated chamber the heat is conveyed to the hover above by means of pipes about 6 inches in diameter. The hover is covered with a circular disc, 2 feet in diameter, covered with felt folds, which retain the heat. There is thus a continual flow of fresh air into the hover. With any type of brooding, it is objected that a warmed floor causes weakness of legs and loss of vitality. In modern hot-water types, the heated compartment is separated from the hover floor by an air space, thus avoiding the heated floor. The warm air enters the hover at the top, so that, following nature, the heat comes from above.

A hot-water brooding house must contain an alley for convenience, a row of hovers next

FIG. 592. Details of a modern hot-water brooding system. One hover is cut away to show how the warm air from the pipe box is liberated above the chicks' backs. Note the air space above the pipe box to prevent the floor from getting too hot.

to the alley and, corresponding to the hovers, a series of pens. This alley should be 3 or 4 feet wide, the hover chamber 3 feet wide, and the pen as long as may be deemed necessary, usually about 6 feet. A single row of hovers would necessitate a house about 14 feet wide; a double row, a house at least 22 feet in width. The house with a single row of hovers would run east and west, thus permitting runs on the south. A house with 2 rows of hovers would run north and south, having runs on the east and west sides. A gable-roof house is best adapted for a long brooder house. There should be liberal lighting. The alleyway should be about 20 inches lower than the hover floor, to facilitate cleaning and to afford room for the hot-water pipes. The water is kept at practically constant temperature by means of automatic regulation of drafts. In addition, the amount of heat allowed any compartment can be controlled by dampers in the pipes.

The colony house. Farm poultry keeping is quite largely a colony-house method. Usually, the entire flock is kept together, and one house is used. Commercially, the colony-house system has been employed in the raising of young stock by means of portable houses. For housing mature fowls, the system is being extensively used in Rhode Island, where it is called the "Little Compton" system of poultry farming. Briefly speaking, the plan employed is this: The colony houses are comparatively small and portable. They are scattered over the fields, which may be cultivated. The poultry flock enters into the system of crop rotation, usually following the hay crop. In this way, birds run on the same field only a year at a time and at intervals of 3 or 4 years. The advantages of this system are: (1) conditions favorable for chickens; (2) economical feed cost; (3) improvement of the land through added fertility; (4) less danger of soil contamination; and (5) outbreaks of disease can be more easily controlled. Colony-housed stock does not require so many precautionary measures as that kept in long houses; for the more closely chickens are confined, the greater amount of work necessary to keep conditions sanitary. The disadvantages of such a system are: (1) added labor in bad weather, it being disagreeable for the caretaker to go from house to house; (2) it frequently requires more fencing, if colony houses are stationary; (3) the poultryman will not make as close a study of individual birds, as it is impracticable to trapnest his stock; and (4), where flocks are permitted to mingle, he does not have complete control over matings.

For stationary houses, the size of the unit may be as large as suits the individual poultry keeper. The colony houses described on a previous page are for 100 hens. The Yesterlaid Farms at Pacific, Missouri, are using a colony house 30 by 60 with 2 floors, housing 1,000 birds to a house.

Long houses. Long houses for poultry keeping are representative of what is known as the intensive system. This type necessitates restricted yards and extensive fencing. It is primarily adapted to the housing of laying stock. The advantages of such a system are favorable conditions during inclement weather for birds and attendant, convenience for the caretaker in the daily routine, and economy of land. The disadvantages are yard contamination, increased cost of fencing, and extra labor in caring for the yards. The long house is a commercial proposition.

A good type of house of this kind is the "Commercial" poultry house which is 18 x 180 feet, with a feed house at the end. It is divided into pens 18 feet square and will house 700 hens. The front is 7 feet 8 inches high, and the back 4 feet 8 inches. The roof is of unequal spans, the front span being one half the length of the back span. The roof is comparatively steep, having 1 foot rise to 2 feet horizontal run. The pens are connected by a series of doors, thus compelling the poultryman to mingle with the birds where he is able to study their needs more carefully.

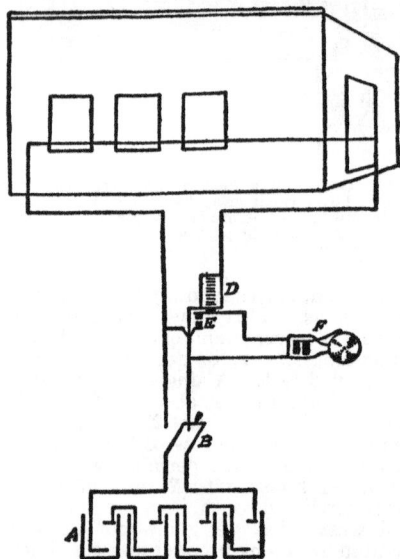

FIG. 593. An electric burglar alarm on the poultry house is a good investment. In this plan the switch (B) is closed at night thus completing the circuit through the wiring and batteries (A) but not through the bell (F). If the circuit is broken by the opening of a door or window in the hen house, the magnet (D) releases the armature (E) which springs over, closes the bell circuit and alarms the household.

The cloth-front type of ventilation is used in this house. In the centre of the south side of each pen is a glass door made by hinging two 9-light 9 x 12-inch windows, thus affording a door which can be opened for cleaning purposes. On both sides of the door are cloth frames 3 x 5 feet. These are placed 4 feet from the floor and, when open do not allow drafts on the birds. At night, there are no drafts on the birds, because the frames are closed. No matter how cold the weather, at least one of the frames is opened every day, unless there are storms from the south. The arrangement of droppings platform, perches, and nests follows very closely the instructions given above under "Interior arrangement." The roosts are set on 2 x 4-inch pieces in the form of a frame, which is hinged at the back and may be raised, thus rendering the droppings platform easily accessible for cleaning. Often, it is desirable to raise the roosts, so as to force lazy hens off them to the floor.

The cost per hen of materials for such a house will be practically the same as for the colony house.

FIG. 594. A well-built corn crib of the common slat type. This is better than those found on many farms, but unless protected against rats and kept in repair, it may permit considerable waste.

CHAPTER 36

Storage and Work Buildings

By PROFESSOR K. J. T. EKBLAW (see Chapter 31). Like all manufacturers, the farmer has to provide for the storage of his raw materials, his implements and tools, and, to some extent, his finished products. He is at a disadvantage, however, in that these vary greatly in bulk, weight, nature and condition, and consequently require several types of structure. Futhermore, they are sometimes of comparatively small value, in proportion to their extent; wherefore their shelter cannot be of too expensive a type. Professor Ekblaw discusses here some of the principles to be kept in mind in the construction of adequate but practical and economical storage buildings. The second group of structures referred to in the chapter title consists of those in which the farmer works over his raw materials and products, with the exception of those in which his "machines", that is, his livestock, are kept and fed. These have been covered in Chapters 34 and 35.—EDITOR.

I T IS undoubtedly the fact that too often the construction of the storage and work buildings on a farm does not receive the careful consideration which its importance claims.

The questions of location, size, material, and interior arrangement should all be thoroughly thought out, so that needless expense may be avoided, and the greatest amount of usefulness secured from the several buildings. These buildings usually include some or all of the following: hay barn, granary, implement shed, milkhouse, creamery, and cheese factory.

Hay Barns

The purpose of a hay barn is to provide shelter for hay and facilities for curing it under the most favorable conditions. One of the first requirements is ample ventilation. With the ordinary barn, the freedom of the circulation of the air is so great that no especial provision for ventilation need be made.

Primarily a hay barn is simply an inclosed shelter; and, since the load of the contents does not come upon the structure itself to any extent, the building should be designed mainly to resist wind pressure, both external and internal. The interior of the structure is of necessity rather open, to admit of the unhampered handling of the hay. Open framing precludes the use of crossties braces, the which, in many other structures, greatly strengthen the framing; to overcome lack of strengthening frame members, great care must be exercised in the design.

429

FIG. 595. A common type of hay barn in regions where comparatively little wet, stormy weather is encountered

Location. The hay barn may be located either as one of the buildings of the farmstead or as a separate building in the hay field itself. It may be advantageous to have the hay stored in close proximity to the feed lots; but, in many cases where hay is the main crop on the farm, it is not fed, but is marketed; and then, of course, it is not necessary to store it in any place other than in the field.

Size. The size of a hay barn is governed by the amount of hay to be stored and by certain structural limitations. Ordinary loose hay will occupy approximately 500 cubic feet per ton; when packed, this will be reduced by 50, 75, perhaps even 100 cubic feet. With timber framing, it is possible to make barns as wide as 48 or 50 feet; anything above this is not practicable. With the plank frame, it is possible to make the trusses more than 40 feet wide; but it is not advisable, on account of the difficulty of securing adequate bracing when the framing is made sufficiently strong. Such circumstances require the use of an unduly large quantity of material which renders the cost excessive.

Construction. Since the framing of the barn is built up from the ground, it is unnecessary to have a continuous foundation. The frame consists, usually, of built-up units, sometimes called "bents," which are from 10 to 14 feet apart and are supported at the sides upon masonry piers. These should have footings large enough to carry the weight of a half of a bent and the portion of the walls and roof which is supported by it. On ordinary soil, a load of approximately one ton may be allowed per square foot of footing.

Two types of framing are in more or less general use. One, known as the "timber frame," is built up of pieces of timber varying in size from 4 by 4 inches to, perhaps, 6 by 8 inches. The members of the frame are joined by means of mortise-and-tenon joints. In the other type, no lumber thicker than 2 inches is used, but the pieces are combined in such a way as to form strong and rigid trusses. In general, the latter type is much the more desirable, since it is economical of material, admits of careful designing for strength, and leaves the interior open.

Walls. It is common practice to cover only a portion of the sides of the building. The lower half of the walls may well be left open; and, in fact, many barns are built with no wall covering. However, this is hardly to be recommended, because an undue amount of the hay is exposed to the weather. Plain 1-inch siding, 10 or 12 inches wide, nailed to 2 by 6-inch girts at 5-foot intervals makes an entirely satisfactory wall. The covering of the cracks with ogee battens is optional.

Roof. Wooden hay barns, as well as other types of barns, are ordinarily made with a gambrel roof, sometimes erroneously called a "hip" roof. The main advantage of the gambrel roof over the ordinary straight-pitch roof is that it permits of the arrangement of the truss members to produce maximum strength; besides this, it affords much more mow space (Fig. 597 inset.)

The covering of the roof may be either of wooden shingles or of shingles made of prepared roofing. Prepared roofing itself is not a desirable material to use on a gambrel roof, because the lower section of the roof is

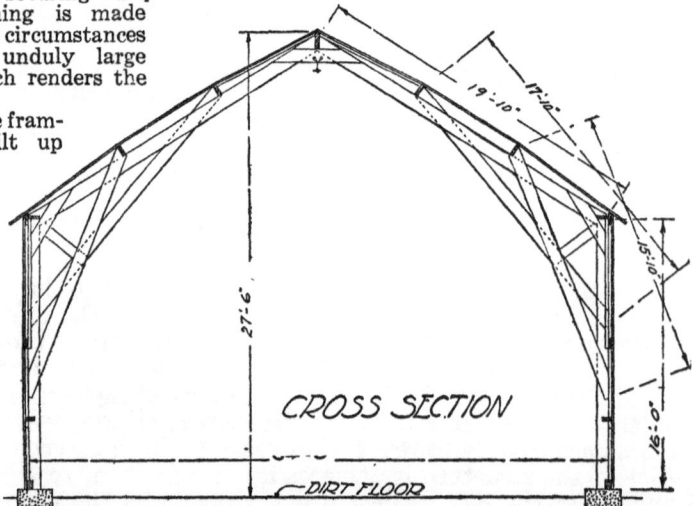

CROSS SECTION

DIRT FLOOR

FIG. 596. Section of a barn built on the straight-pitch, braced-truss principle solely for hay storage. In northern sections building material is saved and greater protection given to the stock by storing hay in the upper part of the horse and cattle barns.

very steep-pitched, and such roofing has a tendency to sag, when used under such circumstances.

Equipment. It is possible to obtain hay barn equipment in such complete and careful designs that practically no manual labor is involved in the handling of the hay. If a hay barn is built in the field, the hay can be brought in from the windrow directly to the barn, there hoisted, and transported to almost any portion of the barn. The equipment within the barn usually consists of a central longitudinal track suspended from the roof at the ridge. A carrier, to which is attached the swing or the fork, runs upon this track; and, when the swing or fork is loaded, the hay is moved either by horsepower or by means of an engine-operated hoist.

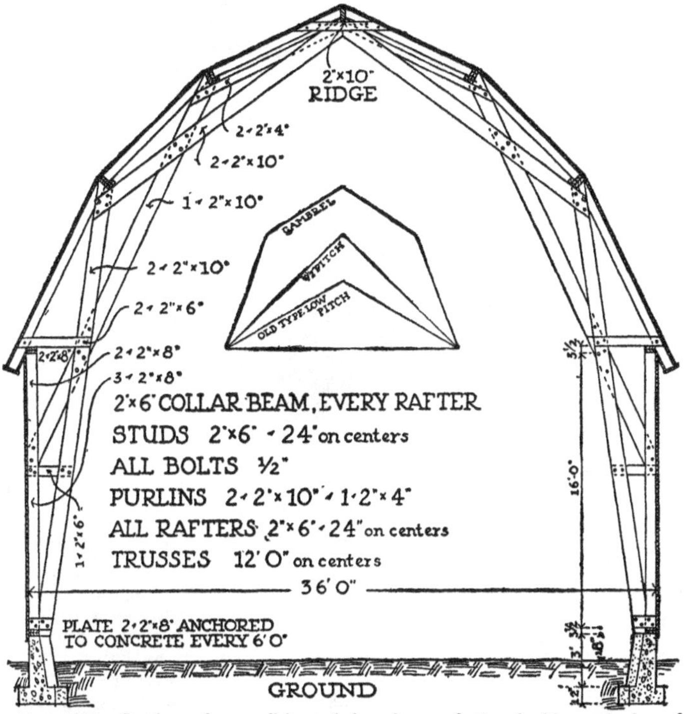

The hay barn is quite often an isolated building, and for this reason should be provided with adequate lightning protection. It is claimed

FIG. 597. Section of a well-braced hay barn of the double-gambrel roof type, built on a secure concrete foundation. The inset shows the space gained for storage by using a gambrel roof rather than the single high- or low-pitch type.

that the vapor arising from hay that is going through one of the stages of curing, known as the "sweat," acts as a conductor of electricity. Such claims have never been fully substantiated. Nevertheless, the lightning-rod equipment should be carefully designed, and every sharp ridge or corner should be protected by conductors. The aërial terminals should be of some non-corrosive material and should be placed at intervals not exceeding 25 feet. The lower end of the conductors should be well grounded by means of a soldered connection with a galvanized iron pipe extending far enough into the soil to reach the level of permanent moisture. A better but more expensive ground is made by connecting the conductor with a metal plate, buried 5 feet in the earth, and encircled by a bed of charcoal which is retentive of moisture.

Granaries

The term "granaries" should include all structures which are used for the storage of grain. Common usage sometimes limits the application of the term to such buildings as are tightly walled, and in which are stored only the so-called small grains. Corn is commonly stored in a building called a "crib," whose walls are more or less open, in order that some ventilation may be afforded.

The granary is a farm building whose requirements are somewhat peculiar. The building must be designed to resist not only vertical loading, but lateral pressure as well; for small grain acts, to a certain extent, as a fluid, and its lateral pressure is sometimes very great. A number of investigators have conducted experiments to determine the bottom and side pressure of grain bins of various kinds. Their general conclusions are that the lateral pressure is from 30 per cent to 60 per cent of the vertical pressure and that neither of them increases to an appreciable extent, if the depth of the bin becomes greater than $2\frac{1}{2}$ or 3 times the diameter or width. Both the lateral and the side pressure depend, to a certain extent, upon the weight and the character of the grain. The failure of many granaries is due to the fact that proper consideration is not given to the lateral pressure, which at the bottom of the bin may be very great; and, as a result, the walls are burst out at this point, leaving the superstructure of the building practically unsupported, while the whole structure becomes racked and twisted.

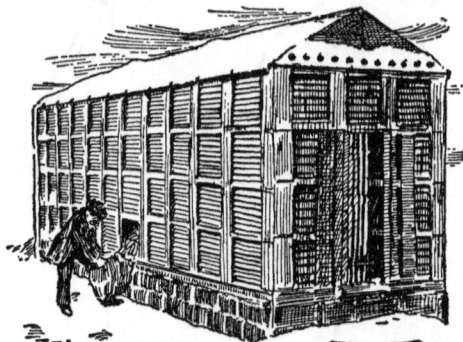

FIG. 598. A hollow-tile corn crib that is weather-, rat-, and fire-proof, and in the long run no less expensive than a more temporary type.

Location. In the majority of cases, the granary is located at the farmstead, since much of the grain is fed, and that portion of it which is marketed can be taken just as easily from such a location as from one in the field, where it would probably be occupying valuable ground. Granaries containing feed for livestock should, of course, be built near the feed lots. It is always well to provide ample space around the building, to allow of the free handling of teams.

Size. The width of a granary for small grains depends entirely upon the design of the building. Small grains are likely to heat, especially when first stored, and for this reason cannot usually be stored for any great length of time without some resulting deterioration in the quality of the grain. Common sense would indicate that 12 or 14 feet is a reasonable width for structures of this kind. For ear corn, the width of the crib is dependent upon the climate. In the heart of the corn belt, the common practice is to make the cribs 9 feet wide. Farther east, it is necessary to reduce this to 8 feet or even to 6, while in the western corn-raising states it is found practicable to increase the width to 10 feet without any damage resulting. The reason for thus adjusting the width to the climate is that the corn, when first husked, is somewhat damp, and, unless provision be made for rather rapid drying, it will rot. The drier air of the western states permits the use of wider cribs.

The height of the granary is variable. If the grain is to be unloaded into the building by hand, about 12 feet is the maximum, but, if elevating machinery is to be utilized in unloading the grain, the height may be any desired. On farms where great quantities of grain are to be stored, it is quite common to find cribs that are 16, 20, or even 24 feet high at the eaves. With the ordinary interior portable elevator, perhaps 16 feet should be the limit; but, with interior elevators installed as permanent equipment, great depth of bins is not only practicable, but desirable, since the additional capacity is obtained simply by putting on additional wall material, the same roof and floor serving for a high building as well as for a low one.

It is well to make the length of the granary such as to admit of the use of standard lengths of lumber, though this requirement does not hold where other types of building material are used. When lumber is the material employed, a practical unit length is 16 feet, and the crib may be made up of as many unit lengths as desired.

Shape. On small farms, a simple rectangular building is, perhaps, the best. The roof may be of the shed type or of the ordinary gable type, the former being the simpler. Double granaries are made by building 2 single granaries a few feet apart and extending the roofs so as to cover the open space between, which forms a sort of passageway. This scheme has been elaborated until what has become almost standard construction on large grain farms is a building consisting of 2 corncribs with a driveway, 10 to 14 feet wide, between them. At a height of about 10 feet the driveway is floored over with tight flooring, and the upper portion of the inclosure is used as storage for small grain.

Construction. The foundation of the granary must support not only the weight of the superstructure itself, but a large part of the entire load of the inclosed grain. The vertical load upon the floor in deep bins is very great, while the weight of the superstructure is perhaps not excessive. The load due to the friction of the grain on the walls is transmitted to the foundation, and this additional load is, perhaps, greater than the weight of the superstructure itself. This can readily be understood when it is explained that grain in a bin forms a sort of an arch in settling, and that the portion of the grain above a height equal to $2\frac{1}{2}$ or 3 times the diameter is supported mainly by the friction upon the side walls.

The foundation should be a continuous well extending below the frost line and should have a footing not less than 18 inches in width. In high buildings the footing should be 24 inches wide. The floor may be of ordinary sill-and-joist construction, but a concrete floor is generally to be recommended. It should be not less than 6 inches thick, and, when properly made of a rich mixture of clean materials, it is practically waterproof. The floor should slope to the rear of the bin, so that drainage may be provided for any excessive storm moisture which may accidentally gain access to the interior.

Framing. It is essential that the framing of the crib be very strong, if the building is to be at all durable. Vertical studs not less than 2 by 6 inches should be located at intervals varying from 18 to 30 inches, depending upon the depth of the bin and the amount of lateral pressure produced. These studs may be toe-nailed to a plank sill which, in turn, is anchored to the foundation; but a better plan is to use studding sockets, imbedded in the concrete, which are made to receive the ends of the studs and afford a firm and secure fastening. The studding sockets avoid the inconvenience and difficulty caused when it is necessary to replace the plank sill. In fact, if the end of the stud be treated with some wood preservative, no replacing is likely to be necessary before the whole structure shall have fallen into decay.

The framing over the driveway must be exceptionally strong, especially if the driveway is a wide one. It may be necessary to double the studs up to the small-grain-bin floor, and the joists extending across should be carefully designed, so that there may be no danger of bending and collapse. It must be borne in mind that grains such as wheat and rye are quite heavy and that a load of many tons is put into the bin.

The wall covering for structures containing small grains must, of course, be tight. The material that is ordinarily used is either shiplap or drop siding, the latter being preferable. For covering the sides of corncribs, ordinary 1 x 6 boards are entirely satisfactory and should be applied with a 1-inch space between adjacent boards. A special siding is made in which the edges are beveled; this affords just as great an amount of opening in the walls, but the overhanging bevel assists in preventing storm water from being driven into the crib. Slightly more material is necessary when beveled siding is used than when plain boards are employed.

Roof. In combination cribs such as have been just described, steep-pitched roofs of not less than half pitch are the rule. It can readily be seen that the steeper the roof, the greater will be the storage space underneath and the greater the space afforded for the installation of conveying equipment, which is an essential part of the modern granary. The roof covering may be either of wood or asphalt shingles.

Ratproofing. The question of ratproofing

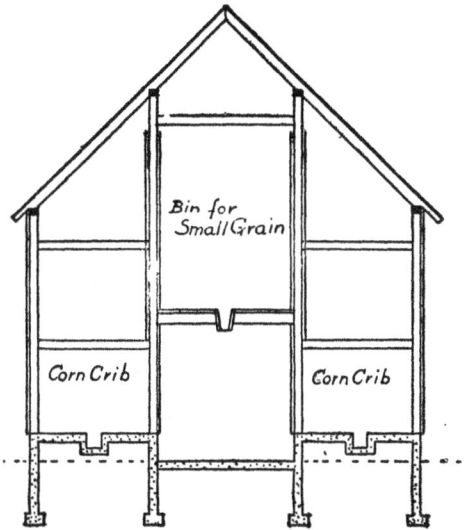

FIG. 599. Section of a combined corn crib and granary that could advantageously be equipped with a grain elevator (Fig. 551)

a crib is an important one, since an immense amount of grain is damaged or destroyed annually by rodents. A very efficacious method of ratproofing, described by the United States Department of Agriculture, is as follows: A heavy galvanized wire screen of quarter-inch mesh and about 28 to 30 inches wide is nailed on the outer edge of the studs before the siding is applied. This is brought down close to each corner, and effectually prevents the admission of rodents under the space which is covered. To prevent the animals from climbing above this space, a strip of galvanized iron is tacked on around the exterior of the building, on the outside of the siding, at the same height as the top edge of the galvanized wire screen. This strip is about 8 inches wide, and positively prevents any rats from crawling up the side of the crib, since they can obtain no foothold upon the metal. The strip should extend around the entire outer boundary of the crib, covering walls and doors, as any unprotected spot entirely destroys its value. Of course, no objects such as a board, neck-yoke, or anything upon which a rat could climb, should be left leaning against the crib to provide a pathway for the animal.

Equipment. The modern development of grain-handling equipment has brought into use some exceedingly ingenious and convenient apparatus, and the old, laborious method of shoveling grain has been rendered entirely obsolete. Grain brought to the granary is dumped into elevators, by which it is carried to the top of the crib, and thence, by means of horizontal conveyors or flexible tubes, is delivered into any desired portion of the structure. The exterior portable elevator has an inclined conveyor, which may be supported upon the roof of the building with its upper end extending over an opening provided in the roof. This opening is usually covered with a cupola or trapdoor. Where all the grain produced on a farm is stored in one building, a permanent vertical elevator with accompanying horizontal conveyors may be installed. In the case of the elevated bin, provision for emptying it may be made by simple valves in the bottom. In corncribs, a trough extending the length of the crib is constructed in the floor. When the bin is filled, this trough is covered with boards; and, when it is desired to empty the bin, a conveyor is slipped into the trough from one end of the bin, so that the grain may run by gravity into the conveyor and by it be carried away.

Implement Sheds

Many farmers do not realize that much of the deterioration of their machinery and implements is due to insufficient protection from the weather and to lack of proper care. It has been found in the case of somewhat expensive machines, as the grain binder, for instance, that the average life is but 5 years, whereas with proper care it should be at least 15 years. It has been estimated that the cost of proper implement sheds would be saved in 4 years by this prolongation of the life of machines. The importance of these buildings to the farmer will thus be readily seen.

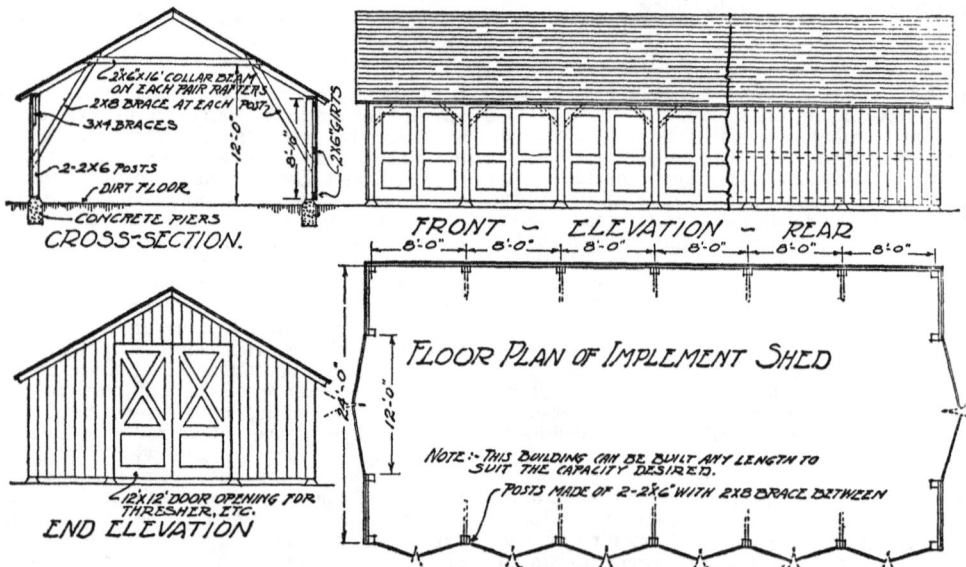

Fig. 600. Plans and dimensions of an implement shed that can save many times its cost in the increased life it gives to the machines kept in it

Location. The location of the implement shed is the first point that the prospective builder should consider, since, in the natural course of farm work, the horses are taken from the barn, hitched to the implements, and then driven to the fields. It follows that the implements themselves should be kept in some place which is intermediate between the barn and the fields. The direction of the prevailing wind must also be taken into consideration, in order that the doors may, as far as possible, be located on the leeward side. The garage is sometimes connected with the implement shed; consequently, it is usually desirable to have the location not too far from the dwelling. If a repair shop or a power plant be incorporated in the same building, the same recommendations would apply.

Requirements. Since the building is designed primarily as a simple shelter, and no part of it, except the floor, is subjected to any great load, it is necessary in designing the superstructure to make it only strong enough to resist snow and wind loads. The interior

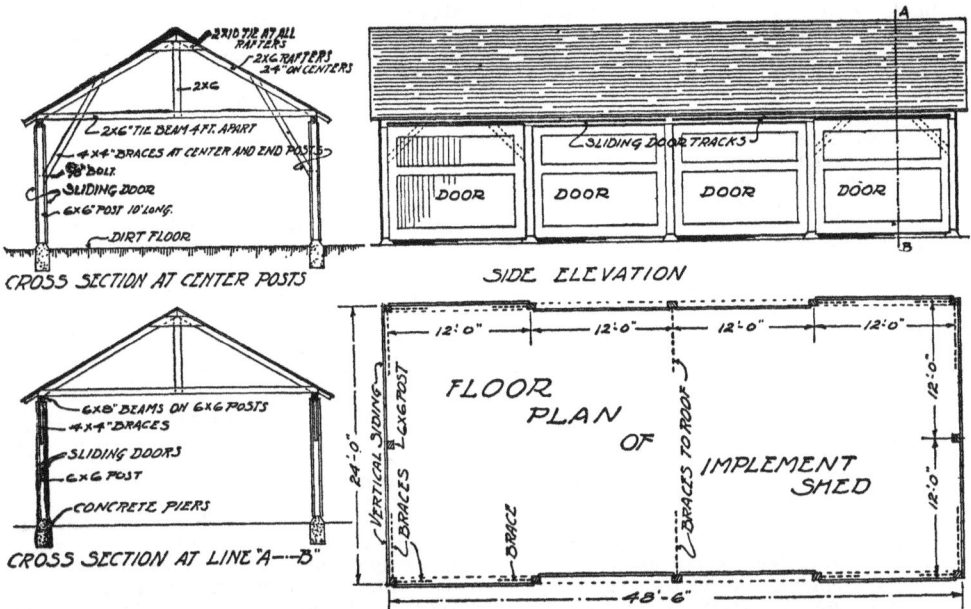

CROSS SECTION AT CENTER POSTS

SIDE ELEVATION

CROSS SECTION AT LINE "A--B"

FLOOR PLAN OF IMPLEMENT SHED

FIG. 601. This shed has greater depth than that shown in Fig. 600, and sliding doors which often work more freely than hinged ones. Light implements and parts can be stored on the rafters, whether floored over or not

should, if possible, be kept free from supporting posts, in order that there may be the minimum interference with the arrangement of implements. The building should not be too narrow and its length not so great as to produce an awkward appearance in the structure itself.

Materials. Probably the best material for the construction of an implement shed is wood, although, where a large number of machines are to be sheltered and where a heavier investment is warranted, it may be found advantageous to use some of the more permanent building materials such as brick, hollow clay blocks, or concrete. Since simplicity and economy are the fundamentals of construction, it is doubtful if the average implement shed justifies the use of the more expensive permanent materials. The floor of

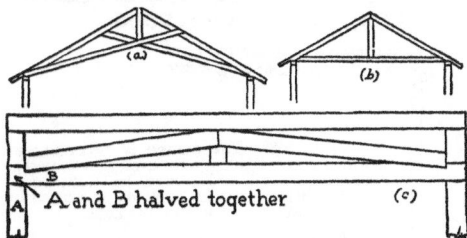

FIG. 602. Simple roof framing for machine sheds (*a* and *b*) the latter providing extra storage space; and (*c*) one style of trussed lintel for a wide shed door.

the shed may be made of concrete for the sake of durability; but it is expensive, and sometimes dulls the sharp edges of certain cutting implements which may be run over the floor. A wooden floor has neither of these disadvantages; but, on the other hand, it is subject to decay, particularly so since it is usually preferable to construct the floor rather close to the ground.

Size. Experience has shown that, on the average farm, sheds 18 feet and 25 feet in width are most advantageous in the placing of machines; and both of these widths are such as to admit of the economical use of standard lengths of lumber. Of course, any variation from these figures is permissible, if a better arrangement can be secured thereby. The length can be adjusted to meet the required floor space, although the general proportions of the building must be kept in mind for appearance sake.

It is usually advisable to build an implement shed of one story only. The use of a 2-story structure is ordinarily not advantageous, since either a hoist or a ramp must be utilized in getting an implement into the second story, and both devices are likely to be unsatisfactory and difficult to operate. Under some circumstances, however, as, for instance, when the shed is built on the slope of a rather steep hill, it may be possible to gain access to the second floor from the slope above the building, while the first floor may be entered from the slope below, as is very often the case in bank barns. If a 2-story building is erected, care must be taken to design the framework so as to withstand the extra loading which is placed upon it. The usual arrangement in a building of this kind results in the placing of the lighter and more easily handled machines in the upper story.

Interior arrangement. The interior arrangement of the machine shed will depend, to a great extent, upon the particular use to which the building is to be put. If it is to be used simply as an implement shed, a rectangular building with a free interior will be the most desirable. If a shop or power plant be added, it may be placed in an extension of the building at one end, the same arrangement holding good if the building is to include a garage. However, too many extensions in one direction are likely to make the building unduly long; consequently, it may be well to make some of the additions in the form of wings or ells. Where local requirements are important, the position of the garage or the repair shop may properly govern the arrangement.

In planning a machine shed, the natural procedure is to make a list of the various implements to be housed, to ascertain their space requirements in width, length, and height, and then to determine the total floor space required. Considerable economy of space may be effected by using care in the arrangement of the machines. For instance, spike-tooth harrows may readily be disposed of by hanging them upon the wall. Small hand tools, walking plows, drills, and so forth may be slipped into the corners between other machines. In this way, the total floor space required may be reduced as much as 10 to 15 per cent.

When making a tentative arrangement of the machines within the designated area, the natural sequence in which they will enter into the farm operations, as well as the number of times per year they will be used, should be kept in mind. For instance, a binder that is used but once a year may well occupy a corner farthest from the door, while a mower, being used more or less throughout the entire season, should be placed next to the door, where it will be readily accessible. Implements like wagons and spreaders are used to such an extent that it is advisable to keep them in a place where no time will be lost by having to take them out of and put them into a shed. When so disposed of, they may be drawn by the horses directly under the shelter and left there when the horses are unhitched. Similarly, no time will be lost when the implements are used again.

A repair shop or a power plant built in connection with an implement shed usually requires a little better construction than in the shed itself. The repair shop is likely to be used during wintertime and, consequently, must be heated. It follows that the walls must be built tight and should be insulated, in order that the waste of heat be not too great. Ample light must also be provided, which necessitates the use of several windows. Such a room must also be ceiled. A plan that may be followed to good advantage, if a seed storeroom be needed, is to make a 2-story structure, with the workshop below and seed storeroom above. While the seed room should have a tight floor, any loss of heat through the workshop will only add warmth and dryness to the storeroom above. A garage in connection with the implement shed should be so located that there will be a door between the workshop and the garage, thus making possible the utilization of the same heating unit in both rooms.

An addition to the main part of the shed

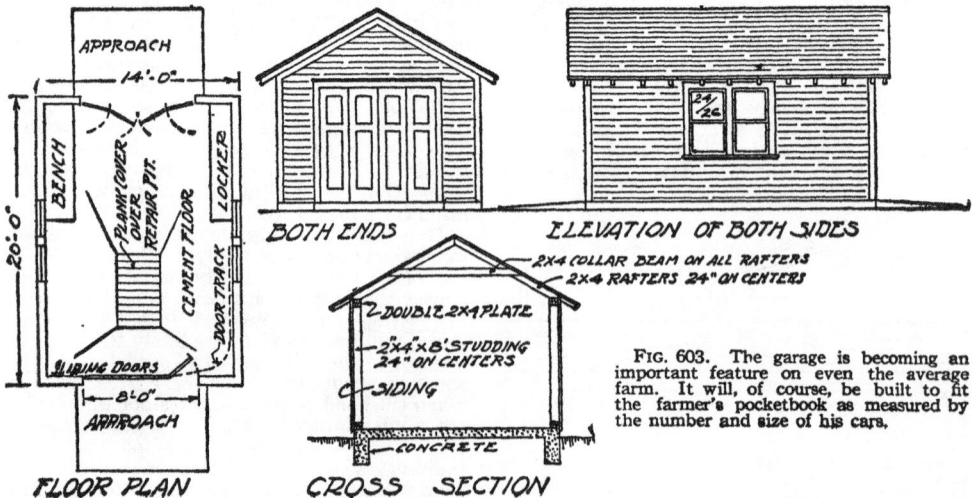

BOTH ENDS

ELEVATION OF BOTH SIDES

2X4 COLLAR BEAM ON ALL RAFTERS
2X4 RAFTERS 24" ON CENTERS

APPROACH
14'-0"
BENCH
PLANK COVER OVER REPAIR PIT
LOCKER
CEMENT FLOOR
DOOR TRACK
20'-0"
SLIDING DOORS
8'-0"
APPROACH

FLOOR PLAN

2 DOUBLE 2X4 PLATE
2"X4"X8' STUDDING 24" ON CENTERS
SIDING
CONCRETE

CROSS SECTION

FIG. 603. The garage is becoming an important feature on even the average farm. It will, of course, be built to fit the farmer's pocketbook as measured by the number and size of his cars.

is often provided for such implements as wagons and spreaders, consisting simply of a roof supported on posts 9 feet apart. The wagon or spreader is driven under this structure and the horses are then unhitched, leaving the implement where it stands. Quite often, the implement is kept under separate cover near the barn, where the litter carriers may be emptied directly into it.

Framing. Two methods of framing may be followed, depending upon the type of wall covering that is to be used. In one method, 6 x 6 posts, set upon concrete piers with 4 x 6 plates on top, form the main part of the frame. On the outside of the posts, at the top, bottom, and middle, 2 x 6 girts are cut into the posts, flush with the outside surface. These provide a nailing for the wall covering which, in this case, consists of 1-inch boards, usually 10 or 12 inches wide, placed vertically, the cracks between the boards being covered by ogee battens.

If it is desirable to have horizontal siding, it will be necessary to use, instead of the posts, 2 x 6 studs 2½ or 3 feet on centre, the 4 x 6 plate being placed on top of these. In regions where the wind is never severe, it may be possible to use 2 x 4 studs, but such a procedure is likely to weaken the structure greatly.

Some objection to the use of horizontal siding has been made, on the ground that the horizontal joints between the boards are likely to retain moisture and thus cause early decay. The joints between vertical boards are self-drained.

If the heavy post framing be employed, the only foundation required will be the supporting piers; but, if the alternate method of framing be used, it will be desirable to have a continuous foundation wall, which need be only 6 inches wide and, perhaps, 2 feet deep; for no great weight comes upon it.

Roof. The rafters supporting the roof must be strongly self-braced, since the presence of interior supporting posts makes the handling of machines difficult in a shed less than 18 feet wide. A simple crosstie, with a kingpost at the centre, is usually sufficient; but in wide buildings it may be necessary to utilize some sort of a simple truss, to prevent sagging of the roof.

The roof covering may be either shingles or prepared roofing. In the first case, the roof sheathing consists of 1 x 4 boards spaced 2 inches. In the second case, it will be necessary to use close sheathing. Shingles should not be used on roofs of less than one fourth pitch, for on roofs of low pitch they are subject to rapid decay, due to slow drying.

Doors. In a building of this kind, there should be as few doors as possible. In a small machine shed, one door is usually sufficient; but this will have to be rather wide, in order to accommodate such machines as rakes and wide drills. In some arrangements of sheds, it may be found desirable to have one entire side equipped with doors; but extra doors always increase expense, not only in first cost, but in maintenance as well. Swinging doors wider than 4 feet should not be used, on account of the difficulty of securing adequate supports for the hinges. Even with good supports, the weight of the door will often cause it to sag. Since it is usually not practicable to have the main door less than 10 or 12 feet wide, it follows that the plate above the door must be extra strong. It may consist of three 2 x 10's nailed together on edge, or of a simple triangular truss whose rise is not more than one foot in the entire span.

Milkhouses

The primary purpose of a milkhouse is to provide a place in which the milk may be properly taken care of immediately after milking. Experience has shown that quick cooling after milking prevents the growth of bacteria, for warm milk is an almost perfect medium for bacterial growth. Also, the flavor of milk is improved by quick cooling. The milkhouse should be kept separate from the barn, in order that the milk may not absorb odors from the silage and manure. Since it is necessary that the utensils used in milking be kept clean, an excellent place in which this may be accomplished is the milkhouse.

Location. The location of the milkhouse may be either near the barn, or, as is commonly the case with small dairies, nearer the house than the barn. Much will depend upon who has charge of the care of the milk. In certain localities, where springs or running artesian wells are available, the location of the milkhouse may depend upon the accessibility of the water supply, because a supply of running water is such a decidedly good feature that the fullest advantage should be taken of it. Since the building is usually rather small and inconspicuous, in comparison with the rest of the farm buildings surrounding it, it usually does not obtrude itself upon the view; consequently, no matter where it is placed, it is not likely to prove an undesirable factor in the general appearance of the farmstead.

Size. The size of the milkhouse will depend mainly upon 2 things, namely, the size

FIG. 604. This kind of milk house brings neither satisfaction nor profits—

of the herd and the method of disposal of the product. On the ordinary farm, where only a few cows are kept, a building of 100 to 150 square feet of floor space will be amply large. The size of the building may be increased as much as desired to accommodate a greater production of milk. If the product of the cows is sold as whole milk, a smaller building will be needed than when the product is sold as cream or butter; for, in the latter case, additional provision must be made for accommodating additional equipment, such as a cream separator and a churn. The building is almost always of one story only, sometimes partially below ground, an arrangement which renders excellent drainage imperative.

Arrangement. A typical arrangement for a milkhouse of a rectangular shape is to have it divided into 3 sections, one being partitioned off to provide a washroom for the milkers and one as a room for the cleansing of the milk utensils. The second section accommodates the boiler and provides space for the storage of fuel. The third section is the milkroom proper, and is usually located between the other 2 sections. Provision must, of course, be made for ample light and adequate ventilation, since both of these are valuable sanitary influences.

Equipment. The method of disposing of the dairy product will naturally control the amount and variety of equipment installed in the structure. If the owner sells the milk, it will be necessary to have a cooler, which usually consists of a vat, or tank, provided with running water of a temperature not exceeding 50 degrees or which may be kept at that temperature by adding ice to the water. Very often springs furnish water which is of itself sufficiently cold; but, if this cannot be had, and ice must be provided, it will be necessary to include in the milkhouse some arrangement for ice storage. If the milk is sold in bottles, some additional equipment in the way of a sterilizer, a small bottler, and a refrigerator for the storage and rapid cooling of the bottled milk is necessary.

FIG. 605.—This kind of milk house brings both

When cream is sold instead of the milk, the cream separator is an essential item of equipment, and the wise dairyman will include a Babcock tester. When the cream is churned into butter before being sold, a

churn is, of course, necessary; and, if the amount of butter produced is at all extensive, some sort of power for operating the churn and some of the other machinery will be required.

The washroom for the milkers should be provided with sanitary lockers for their clothes, and a lavatory and a shower bath would be decidedly advantageous. In that portion of the room given over to the cleansing of utensils, the equipment should include a large sink so set that not only may a supply of water be available for it, but steam, also, for sterilization of the utensils. There should also be installed a slotted rack of adequate size upon which the utensils may be placed for draining after having been thoroughly cleansed. In many milkhouses, a closet is built between the washroom and the milkroom, into which the utensils, after cleansing,

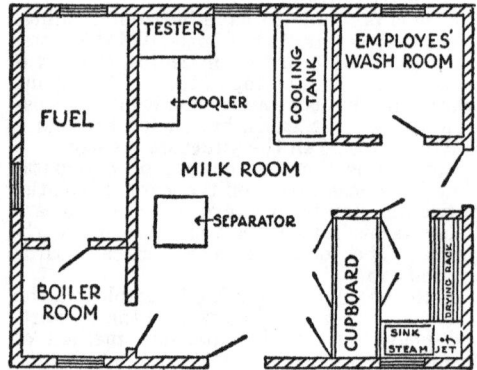

FIG. 606. Floor plan of a small milk house adapted to the needs of a general farm carrying on a moderate dairy business.

are put from the washroom side, being taken out as needed through doors on the milkroom side.

The boiler room need not be very large, since the only thing to be placed in it is the vertical boiler of 2- to 5-horsepower, which will furnish the steam and hot water necessary for cleansing purposes. The fuel room should be large enough to contain a suitable amount of fuel. In most regions, the fuel will consist of coal, the use of which is accompanied by dust and dirt; but certain sections of the country are especially favored with natural gas or with crude oil, either of which provides a very cleanly fuel.

Construction. In a small building of this kind, a very simple foundation, consisting of a continuous wall extending below the frost line and with a footing not exceeding 1 foot in width, is all that is necessary. It will be advantageous to have this foundation of concrete and to make it practically continuous with the floor which should also be of con-

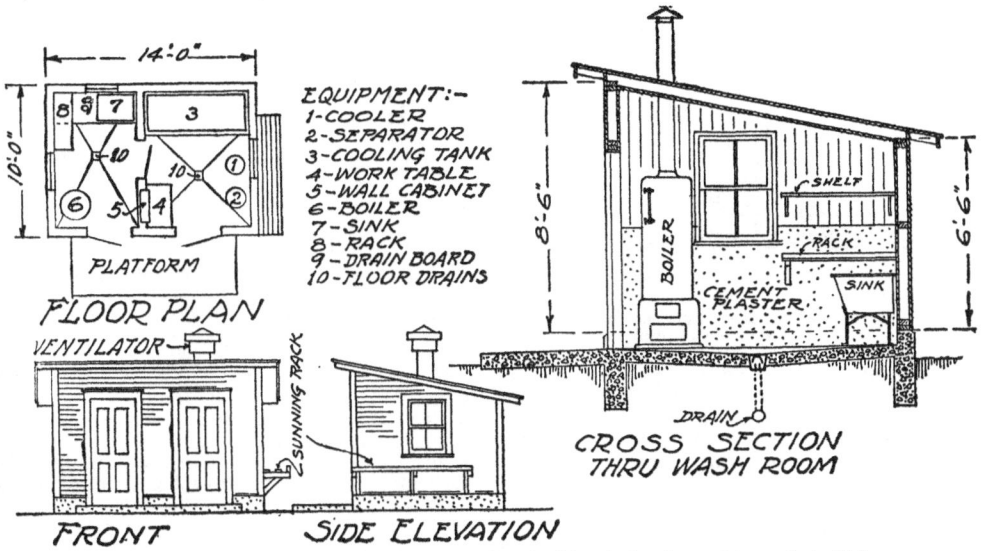

EQUIPMENT:-
1-COOLER
2-SEPARATOR
3-COOLING TANK
4-WORK TABLE
5-WALL CABINET
6-BOILER
7-SINK
8-RACK
9-DRAIN BOARD
10-FLOOR DRAINS

FLOOR PLAN

FRONT   SIDE ELEVATION

CROSS SECTION
THRU WASH ROOM

FIG. 607. Plans of a two-room milkhouse equipped with a boiler for washing and sterilizing

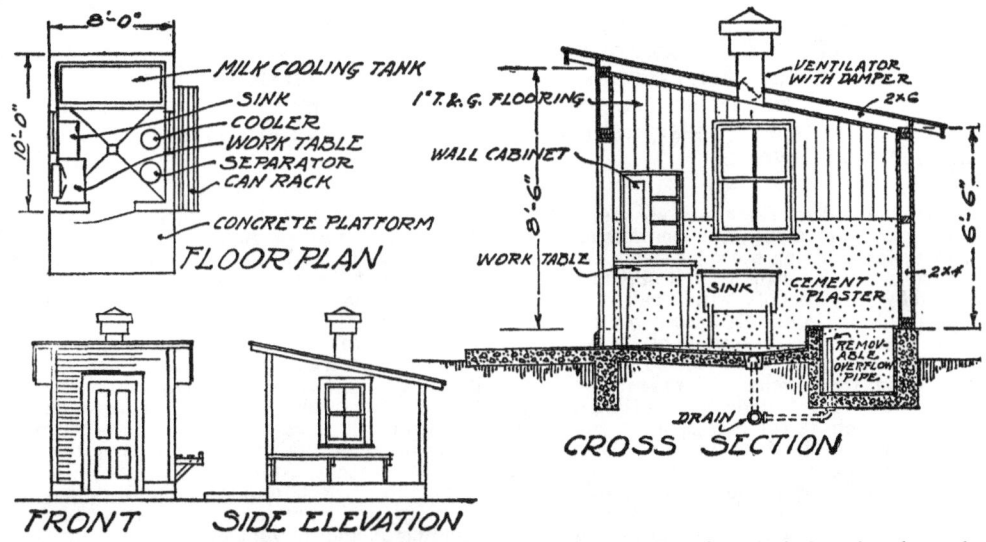

MILK COOLING TANK
SINK
COOLER
WORK TABLE
SEPARATOR
CAN RACK
CONCRETE PLATFORM

FLOOR PLAN

FRONT   SIDE ELEVATION

CROSS SECTION

FIG. 608. Plans of a small one-room milkhouse in which milk and cream for a limited local trade can be handled promptly and in a sanitary manner

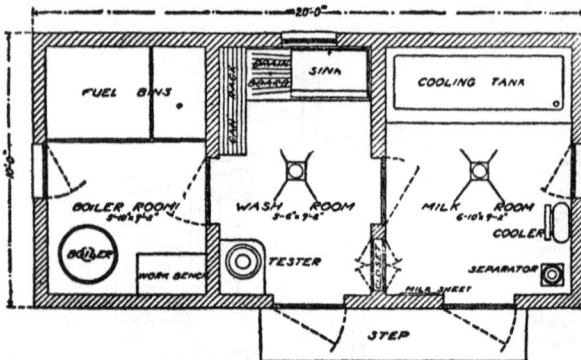

FIG. 609. Floor plan of a larger, more complete dairy house including a boiler room which can be used as a shop

crete. There may be some objections to this material because it is so cold and so hard to walk upon; but any such objections are more than counterbalanced by the sanitary value possessed by concrete, since it may be readily kept clean. The floor should not be made perfectly level, but should be provided with a drain toward which the entire floor should slope.

The superstructure may be of wood, in which case the studs are attached to the foundation by the use of stud sockets. Normally, wood will be the cheapest material; but, for the sake of durability and sanitation, it may be found desirable to build practically the entire structure of concrete. There are on the market various forms of metal wall reinforcing which may be combined with concrete so as to form a wall sufficiently strong and substantial for a building of this character. The inside walls, whether of wood or cement construction, should, of course, be kept as smooth as possible.

It must be borne in mind that sanitation is the keynote in the construction of any kind of dairy buildings; for recent careful investigations indicate that contamination of milk occurs to a greater extent in the handling of the milk after it comes into the milkhouse than in the barn itself.

Creameries

These buildings are found almost entirely in regions which are primarily fitted for dairying, although occasionally a creamery may be located in regions of general farming where the transportation is good. Indeed, the ideal situation for a creamery is in the centre of a good dairy production community, with good roads and good facilities in the way of railway transportation. It must be remembered that, usually, there is an investment of several thousand dollars in buildings of this kind and, since they are generally owned by a number of individuals, or stockholders, it is necessary to have some assurance of profits in order to satisfy the numerous owners.

Location and requirements. As far as local conditions controlling the location are concerned, the building must be away from dust and dirt, which means that it should be protected by some sort of a windbreak, if possible, and should be some distance away from the road. An adequate supply of water is essential; this must be obtained either from springs or from wells. There must also be opportunity for· the disposal of waste and sewage, which, in the case of dairy buildings, is especially difficult to handle. If an adequate supply of running water is not available, it will be necessary to store considerable quantities of ice; and, for this reason, the presence in the vicinity of a body of good water, from which ice may be harvested in winter, is a decidedly advantageous feature. Of course, the cooler the natural water supply, the less the ice that will be needed.

Arrangement. It is difficult to enumerate any definite principles which should control the arrangement of a creamery. In general, however, it may be said that a refrigerator is a necessity and that its natural location should be at the north end, since this is likely to be the coolest part of the building. Storage rooms will, of course, be needed in close proximity to the refrigerator and, consequently, the whole north end of the building may be given over to refrigerative operations. The intake for the purchased dairy products will, normally, be at the other end of the building and, usually, will occupy a considerable space. An exterior unloading platform, about 4 feet from the ground, should be provided, and a portion of the floor within the building should be built up to the same level; by this arrangement it may be possible to utilize gravity, to some extent, for conveying the raw material to the other parts of the plant. When cream or milk is purchased, it, of course, must be tested for butter-fat content. This requires a testing room, which should be

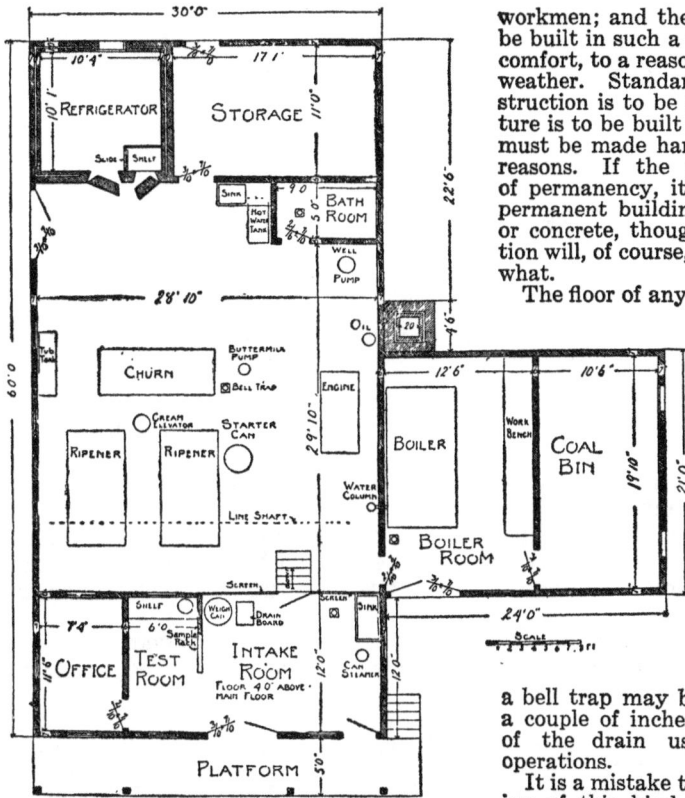

FIG. 610. Floor plan of a creamery such as a community of dairy farmers could coöperate to build, support, and profit by.

workmen; and the structure must, therefore, be built in such a way as to provide for their comfort, to a reasonable degree, in all kinds of weather. Standard double-wall building construction is to be recommended if the structure is to be built of wood. The inside walls must be made hard and smooth for sanitary reasons. If the project shows any degree of permanency, it may be well to use more permanent building materials, such as brick or concrete, though the expense of construction will, of course, be thereby increased somewhat.

The floor of any dairy building should be of concrete. If heavy machinery is to be placed upon it, it should be constructed very carefully, with a proper sub-base covered with not less than 6 inches of a good, rich mixture of concrete. The corners of the floor should be rounded, in order that no recess may be formed for the accumulation of dirt. A slope of, perhaps, 1 inch in 8 or 10 feet should be made toward some point at which a good drain provided with a bell trap may be located. A depression of a couple of inches in the immediate vicinity of the drain usually facilitates cleaning operations.

It is a mistake to make the walls of a building of this kind low. Ventilation is absolutely essential, because great quantities of steam are usually produced and must be carried off. In the summer time, the air is likely to become oppressive if the walls are low. A height of at least 14 to 16 feet is to be recommended, and as many windows as possible should be put in. Since it is ordinarily necessary to attach line shafts to the wall and ceilings for the operation of the various machines, care must be taken that adequate strength is provided to support rather heavy loads.

In the wintertime, some form of heat must be provided. This may be furnished by the exhaust steam from the boilers; if this is not sufficient, the boiler should be designed large enough to furnish an additional supply of live steam for heating purposes.

The equipment for houses of this kind will depend upon the size of the plant; but the various items will be practically the same for either a large or a small plant, varying only as to size. The equipment for a creamery in a community where the milk from, perhaps, 1,000 cows is handled, is given in the accompanying list which has been taken from Bulletin 224 of the Wisconsin Agricultural Experiment Station.

near the intake and should also be in close proximity to the office where the records are kept. The intake, testing room, and office may well occupy the south end of the building.

This leaves the main workroom to be located at the centre of the building, which is its logical location, in order that it may fit in with the most efficient scheme of sequence in operations.

Provision must also be made for a boiler room; and there must not only be room for the boiler itself, but there must be some storage room for fuel. It is always a good idea to have a small repair shop, for quite often the creamery is a more or less isolated building, and to get repairs done quickly may prove rather difficult unless they can be done at the plant. Some sort of a wash-room should also be provided for the workers in the plant, as well as locker accommodations for their clothes.

Construction. A dairy building of this kind calls for a good type of construction. It operates the year 'round with a force of

10-horsepower steam engine	can steamer	water heater	moisture-test scale
15-horsepower boiler	butter printer	cream elevator pump	moisture-test apparatus
small truck	starter can	cream-testing scale and	salt-test apparatus
platform scale	print scale	equipment	acidity-test apparatus
weigher for buttermilk and	tub paraffiner	water bath	office equipment
skimmilk	tub tank	24-bottle turbine tester	cleaning utensils
churn	water tank	conductor head and spouts	shafting and pulleys
two ripeners	buttermilk tank	60-gal. weigh can	repair equipment

SOME CHEESE FACTORY DETAILS

By PROFESSOR J. L. SAMMIS (*see Volume I, Chapter 44.*) *Although the cheese factory, like the creamery, is not a farm structure, it is one about which the farmer, especially the dairy farmer, should have some knowledge. In many cases he may have a coöperative interest in its management; in others he may depend upon its success for a large percentage of his returns. In any case he should be interested in its efficiency and sanitation and know how they can be maintained and improved.—*EDITOR.

It is a little more than 60 years since the first cheese factory in America was established, in Herkimer County, New York. The early factories were constructed in nearly every case of wood, and were somewhat makeshift in character; but the rapid development of the factory system was accompanied by a corresponding improvement in the buildings in which the industry was carried on. To-day, the thousands of cheese factories in the United States include a large number of buildings which are of a substantial character and are constructed on sound sanitary principles.

The most careful attention should be given to the internal arrangements of the cheese factory, so as to insure perfect cleanliness, without which, indeed, no cheese factory has a justifiable right to exist.

Location and dimensions. The use of a lot, large enough for the factory and affording also a vegetable garden for the maker, may often be secured free of charge from a farm owner who wishes the factory located near by. The building should be planned large enough to admit of using an additional vat or two, when the natural increase in the milk supply makes this necessary. The curing room at an American cheese factory may be built one half to two thirds as large as the makeroom, since cheese are commonly shipped every week, as soon as the surface is well dried. The rooms above the factory may be used partly for the storage of cheese boxes and partly as living rooms for the maker's family, the latter being a great convenience for the maker in going to and from meals, etc.

The intake, whey tank, coal bin, and cheese-shipping door should be located conveniently with reference to the driveway.

Curing room, whey tank, and boiler room. The curing room should be located on the north side of the makeroom, as this affords the best protection to the cheese against temperature changes. The intake, if located on the east side, will be then best lighted in the early morning and protected from north and west winds in winter. The whey tank should be located within sight of the intake, but far enough along the driveway to afford room for several teams to stand. The boiler room should be close to the intake or within easy access from it, so that the maker can conveniently get up steam and take in milk during the early morning.

The makeroom. The dimensions of the makeroom should be arranged to fit the vats and other machinery. In most cases, and especially where more than 2 vats are used, it is best to place each vat with one end toward the intake, so that a single long conductor pipe may be used for all vats.

A cement floor, extending for a foot or more up the wall and provided with sufficient pitch and with bell drain traps, is far better than a wooden floor. The walls and ceiling of the makeroom should be finished and smooth, so as to be kept clean without difficulty. High ceilings, plenty of light, and screened windows and doors are necessary. All pipes or utensils used for holding milk or other food for man should be built and located so that all parts can be readily seen and cleaned daily. In Wisconsin, where factories and makers, after passing satisfactory inspection, are licensed by the Dairy and Food Commissioner, the sanitary requirements as to the construction and operation of a factory are described in printed rules and suggestions.

The whey tank and its care. A stinking whey tank which can be smelled by passersby is both intolerable and unnecessary. The whey tank should be emptied and scrubbed regularly; and an accumulation of spilled whey on the ground below the outlet should be prevented, as this may be the principal source of bad odor. To this end, a cement block, perhaps 8 feet square, with a good pitch toward the middle, should be located below the whey-tank outlet. A drain tile, with an iron grating opening at the lowest point in the centre, will allow all spilled whey to be quickly scrubbed down and carried away in the drain, thus avoiding a foul-smelling mudhole at this point. With a properly built septic tank, many factories in Canada and elsewhere dispose of their scrub water, etc., without creating a nuisance.

The accompanying suggested equipment for a cheese factory, also abstracted from Bulletin 224 of the Wisconsin Agricultural Experiment Station, may prove helpful.

SUGGESTED   CHEESE-FACTORY   EQUIPMENT

10–20-horsepower boiler	curd forks
4–8-horsepower engine	curd pails
force pump and jack	curd strainers
600-pound platform scales	dippers
60-gal. weigh can	strainer dippers
conductor head and spout	curd mill
strainer rack	cheese knife
Babcock tester	cheese trier
2 sample jars for each patron	churn
acidmeter	cream ripener
Wisconsin curd test	butter printer
Marschall rennet test	cream-testing scale
sampling dipper	whey separator
hoisting crane	counter scale
600-pound cheese vat	paraffin tank
curd agitators	wash sink
double presses for cheese	starter can (40-gal.)
50 cheese hoops	tin pails
horizontal curd knife	galvanized iron pails
vertical curd knife	cleaning equipment
hand rakes	office equipment
	shafting and pulleys
repair equipment	

Fig. 611. Harvesting the ice crop. Every farm that can do so, should make this one of its regular operations

CHAPTER 37

Icehouses and Cold-Storage Houses

By PROFESSOR R. P. CLARKSON (*see Chapter 14*). *Aside from the advantages it offers as a means for keeping foods of all kinds in good condition in hot weather, there are two reasons why every farmer that can should cut and save his own supply of ice. In the first place it can be done at no cost whatever save that of his labor and hauling expense; secondly, the work can be done only during winter when no other jobs demand immediate attention. Moreover there is always a market for any surplus, either in the nearest village or among his neighbors. The use of canned goods and the tendency to buy meats of the butcher rather than to raise, kill, and dress them, have to some extent dimmed the importance of cold storage facilities. But to balance this, the increasing popularity and use of concrete and hollow tile offer the inducement that an efficient icehouse can be more easily and more cheaply built and a supply of ice more easily gathered and more economically kept than ever before. Practically the whole story of how to do it is told in the next nine pages. Every farmer within reach of a pond and freezing weather should learn it, act upon it, and profit by it.—*
EDITOR.

FOR more than 100 years traffic in ice has been carried on until it has so outgrown natural supplies that artificial-ice plants have sprung up all over the country and artificial ice can be sold in most sections cheaper than natural ice. Since about 1830, when the export trade reached its height, practically all tropical supplies have been produced artificially on the spot, whereas before that time thousands of tons of ice were shipped around the world. Notwithstanding this situation, the individual farm icehouse, storing up the natural ice from some small nearby pond, is a very desirable thing for every economical farmer. The lack of a natural pond or of a freezing climate does not, by any means, shut him off from the advantages of cold storage. Artificial-ice plants, refrigerating methods without ice, and other methods of preservation, are all open to him at comparatively small cost.

Icehouses for the farm. The advantage and convenience of having a good supply of ice is far beyond the small cost of gathering it, not only to the general farmer, but also to the dairyman, the country merchant, and the rural dweller. Often a vacant shed, a corner of the barn, an unused cellar, an empty silo, a vegetable storehouse, a dry well, or even an old cistern in the ground, when properly cleansed and fitted up in accordance with the principles here given, will serve as a satisfactory icehouse for many years. The successful icehouse is not necessarily the most expensive one. In southern Virginia a hole dug in the ground entirely above the water level and lined with native clay held ice satisfactorily

444

FIG. 612. Concrete blocks, being provided with interior dead-air spaces, make an excellent icehouse.

through the fall. Its only covering was of leaves and pine boughs. This was the type used by the Romans in the early ages to preserve snow, and it is now quite common in many parts of this country.

As another extreme of simple construction, a farmer in New York state built four walls of a single thickness of board supported by upright green poles freshly cut from the woods. He filled it with ice surrounded by a foot of sawdust, using a layer of sawdust for a floor and another layer for covering. It had no roof, doors, nor windows, and the ice kept all summer without much waste. This type is now being suggested to the natural-ice dealers.

Secret of keeping ice. It is obvious from these two examples that building material, whether wood, earth, stone, brick, or concrete, may not be the deciding factor in the keeping of ice. The secret is in the strict observance of four principles all of which finally reduce to one, namely, good insulation. First, there must be good under-drainage to carry off the melted ice, otherwise it would form a conductor of heat to the remainder of the ice stored, and would gradually melt it from underneath. Water melts ice much faster than air, for the latter merely affects the surface while the former penetrates throughout. Second, there must be perfect ventilation at the top of the ice in order that the covering of sawdust, straw, hay, moss, or leaves may be kept as dry as possible so that it will not form a conductor for the heat from the air and melt the ice on top. Third, the ice must be packed so as to prevent the circulation of air through the mass, for there is certain to be some heated air enter into the house when the doors, windows, ventilators, or top are opened. These currents of air rapidly warm up, while dead air does not readily become heated because of the fact that air is a very poor conductor of heat. Fourth, good insulation at the sides and bottom must be carefully provided.

Size and capacity of house. The size of the house needed may be determined from the fact that a ton of stored ice occupies approximately 42 cubic feet of space. The average size of house for a small farm is about 10 feet high from the ground to eaves, with an inside area 12 by 14 feet. After allowing for the space occupied by the sawdust around and under the ice, this will give room for the storage of from 25 to 28 tons of ice. A cubic foot of solid ice weighs close to $57\frac{1}{4}$ pounds, so that 35 cubic feet would weigh a ton. From this we can estimate the amount possible to cut from a pond. The thickness of the cakes usually cut ranges from 6 inches in the central states to 16 and even 20 inches in the north; probably 12 to 14 inches is the average. The cakes are cut in various sizes, also; perhaps 12 by 16 and 16 by 16 are common sizes, but this is not important. Assuming cakes 12 inches thick and 12 by 16 inches, there will be 26 of them to the ton, each one weighing $76\frac{1}{3}$ pounds. In the field, allowing for breakage and waste, a surface of 50 square feet will harvest at least 45, and possibly 50, tons of 12-inch ice.

FIG. 613. By building the icehouse wholly or partly underground both its first cost and the loss of ice by melting are lessened.

The Ice Crop

Care of the ice field. The essential thing is to provide for clear, clean ice of sufficient thickness. Before the water freezes it must be purified as far as possible. Sources of possible pollution must be removed, all rubbish taken away from the field and branches or floating logs cleared out. These factors carefully looked after will insure clean ice. Motion of the water during freezing not only expels the air but also promotes the growth in thickness. By expelling the air the ice is made clear, and sometimes it is desirable to induce a gentle current all through the field during ice-making weather for this purpose as well as for the additional thickness gained. The method of doing this in the case of land-locked ponds is to provide an outlet which must be readily controlled, as too rapid or violent motion will retard the growth of ice.

FIG. 614. Removing snow with a scraper before cutting ice.

After freezing has taken place the watchfulness of the farmer is properly directed toward increasing the thickness of the ice and keeping it clean. The handling of snow is important. Except in warm weather, snow should be removed as soon as possible, as it prevents the escape of heat from the water and thus retards growth of the ice. In soft weather, however, the snow is desirable to act as a blanket in shielding the water and ice from the heat of the sun. In case of rain, also, the heat of the rain is largely used in melting the snow and thus does not affect the ice so much.

Clearing off snow. After a thaw the snow and water on top of the ice freeze and form a porous cloudy layer. If not too thick and it is near the cutting time, this layer is not altogether a detriment. It detracts from the quality of the ice, to be sure, but it makes the handling of the ice easier. It prevents breakage of cakes in packing either for storage or in shipment.

For removal of snow when desired, there are two main methods. If the ice is strong enough, horse-drawn scrapers are used. If the ice is thin, at every 8 or

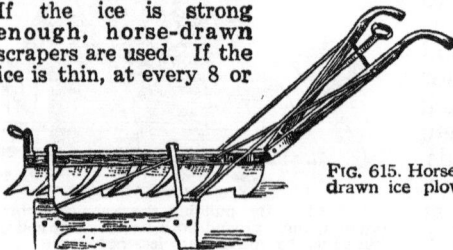

FIG. 615. Horse-drawn ice plow

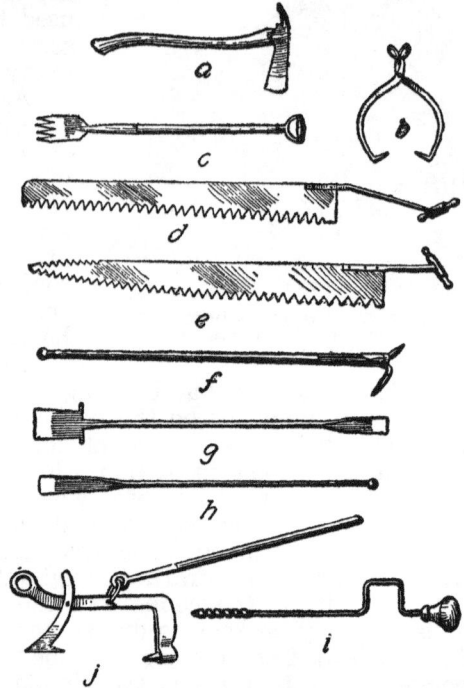

FIG. 616. Hand tools for ice harvesting: *a*, tapping axe; *b*, tongs; *c*, trimmer bar; *d*, saw; *e*, saw for "breaking out;" *f*, ice hook; *g*, bar for "breaking out"; *h*, splitting chisel; *i*, ice auger; *j*, grapple iron.

10 feet holes are cut in the ice through which water rises and floods off the snow by melting. Then, as the ice thickens, the snow ice which may be formed is planed off if necessary.

Tapping the ice. Continued soft weather or a thaw during the growth of the ice is one of the serious times with an ice field. If water washes on to the ice, it must be removed at once, and usually the only way possible is by tapping holes in a number of places so that the water will flow off and the ice, being lighter, will be raised. It is not worth while doing this if only an inch or two of water stands on top, as this may be allowed to freeze and may then be planed off.

Harvesting the ice. The first move is to inspect the field thoroughly and mark all shallow or dangerous places. The field is then laid out with a marker, which is really a hand plow made to cut a light groove along the line. The horse plow is set in the groove and run back and forth until the ice is cut more than half through. This is done with all lines in one direction and then with all lines in the other, the pond being thus cut into squares.

The next move is to open a channel by deeply plowing the groove on either side and sawing through completely with an ice saw. This channel section is then split into cakes which are usually pushed under the remaining ice in large ponds to open up the field quickly. In small ponds, however, they may be floated to shore and stored.

With an open channel to shore, sections of perhaps 100 squares are sawed off and floated, the sections being split into cakes just before being pulled out of the water. These floats, however, must not be left too long, as the grooves will flood and freeze up.

Tools required. A field planer, ice auger, tapping axe, snow scraper, marker, horse plow, ice saw, breaking bar, splitting chisel

ice hooks, trimmer bar, ice adze, several grapples, loading tongs, and packing chisel constitute a reasonably full and quite inexpensive equipment for the proper handling of any considerable quantity of ice. For a very small field it is possible to do without most of these tools. Hoes and scoops or shovels may be substituted for the planer and scraper. A hand marker and an ice saw will cut the field up. A pair of tongs will lift each cake out and carry it to the drag or the incline leading to the house and an ordinary axe will trim the cakes. Use of the tools mentioned above save time and labor, especially the former, which is of real importance in harvesting during uncertain weather.

Building the Ice House

Material of building. Having determined upon the size of house and the outlay of money that can be afforded, it remains to determine the material to be used and the plan to be followed. Beyond any reasonable doubt wood is better in many ways than stone, brick, or concrete for icehouse construction, although any of these may be used with satisfaction if the ice is packed far from the walls and well insulated from them by 10 or 12 inches of sawdust. The only objection to wood which any one can have is its tendency to rot under the continued influence of moisture inside and dryness outside. For this reason cypress is to be highly recommended as a serviceable wood, although pine will last for some years and is quite generally used.

Underdrainage and foundation. For a foundation concrete is best, all things considered. Let it go into the ground below the frost line and extend a foot above ground, to keep the sills dry. Unless the soil is well drained, there should be a main ditch with side branches cut in the floor, covering the whole space below the ice, the main ditch leading out on the lower side. Fill the ditches with broken stone, crockery, brick, or clinkers, and spread a thin layer over the whole floor. On top of the stone place a layer of straw covered with a thickness of coal ashes. On top of the ashes may be placed floor boards with cracks between them to allow free drainage of the water from the melted ice. More often, however, the boards are

dispensed with and an 8- or 10-inch layer of sawdust put directly on the ashes, the ice being packed on that.

Side walls and insulation. The walls may be either single or double, and should be built with matched boards or papered with tarred roofing paper. I would recommend both. The paper is cheap, costing $4.50 to $5 for a 500-foot roll, so that it does not add much to the cost of the work, while it certainly gives a much better house. If the single walls are papered the papering should be on the outside, of course, while if the building is made with double walls the papering should be on the sides within the air space. Double walls are much better for insulation, and may easily be made by nailing the boards on both sides of the 2 by 4 joists used as uprights. This leaves a 4-inch dead air space between the walls, which should not be filled with sawdust nor with anything else. The best insulator is dead air, and the purpose of sawdust, felt, wool, shavings, and such substances is merely to keep the air dead—that is, these substances prevent circulation of air by catching small quantities in the spaces between the particles. The use of these substances is not to be recommended, either in icehouses between walls or in the walls of cold-storage boxes. In either case the filling would become damp and remain so, thus rotting the construction from the inside. In

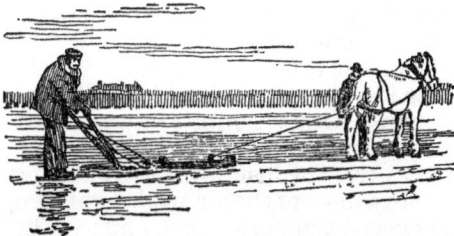

Fig. 617. The ice is cut about half way through into blocks. These are split apart, floated to shore and loaded upon sleds or stored in an icehouse if near by.

cold-storage boxes it will also absorb and retain odors, making the box unfit for holding eatable produce. Furthermore, when damp, such fillings are reasonably good heat conductors.

"Pocketing" the air. In the air space between the boards, in the icehouse construction, every 3 or 4 feet up there should be a strip of tarred paper tacked, to form a horizontal partition, thus preventing any up and down circulation of the air. The result of this construction is that the ice is surrounded by walls consisting of a large number of boxes containing dead air. These boxes will be from 3 to 4 feet square and 4 inches thick—the thickness of the air space.

House sills. The sills of the house should be laid directly on the concrete foundation and in close union with the concrete, to prevent entrance of the air between them. In my experience it has been found desirable to lay a coating of tar or asphalt on the foundation walls and on this put the sills, thus making an air-tight job. There must be no entrance of air underneath the ice. It is true that a small amount will enter through the drain, if the latter is not trapped, but this is not sufficient to do any harm. In

a commercial icehouse of large size, however, the drain should be of tile and trapped as it comes from under the icehouse. Preferably, too, there is a drain around the foundation on the outside, both of the drains being brought together and led away to a lower level.

Roof and ventilation. The roof for a small building may be of almost any material to shed the rain, keep off the sun, and provide good ventilation. The latter feature is the most important one in connection with building the house. The ventilators should be closable and kept closed on foggy days and nights. For this reason trap-doors on the sides and roof are preferable. The roof should be a V-shaped or hipped roof, with trap-doors at each end and at the ridge. Near the top of each end wall arrange a small door. Each fine, dry day open one of these doors and the opposite trap, so that the air may circulate freely and keep the top dressing or covering of sawdust dry. This top dressing should not be too thick, the practice being to have it from 8 to 12 inches. The dressing must be looked after and kept dry at any cost. It will be found helpful, although troublesome, to divide the top layer of sawdust by a thick layer of newspaper.

Packing and Keeping the Ice

Packing the ice. In packing, the first layer is commonly placed on edge rather than laid flat. There is no less wasting that way, for, although each cake wastes less, there are more cakes on the floor. Sometimes this plan is followed

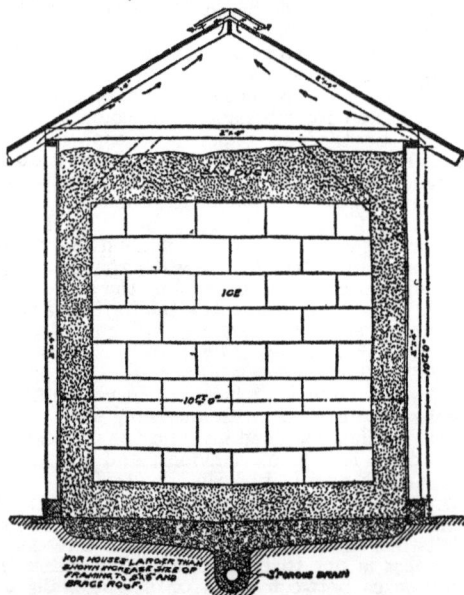

FIG. 618. Section through a small, wooden farm icehouse, showing sawdust insulation, drain for carrying off moisture, and ventilator openings for letting out the warm air.

throughout, the advantage being that in breaking the ice out there is less adhering surface between the cakes. It is harder to pack this way, however, and the liability to undue side-wall pressure is greater. At least every third layer, no matter how packed, should be laid so as to break the joints of the previous layer that there may be no circulation through the mass. The packing can be done up to within 6 inches of the side walls if a double wall is used, and up to within 8 or 10 inches with a single wooden side. As already stated, if concrete, stone, or brick be used, there should be from 10 to 12 inches left around the sides. In every case the space left should be filled with sawdust lightly tamped into place, but not rammed tightly. Hard tamping forces the sawdust down so solidly as to remove most of the air, while light tamping keeps the mass porous but yet held together tightly enough to retain the air and prevent its escape or circulation.

Finally, it should be said that the cakes must be cut as true as possible, and no small pieces or broken cakes should be allowed to enter the house. The ice should be packed in freezing weather, so that the cakes will be dry and not freeze together in the house. Each cake should be kept an inch or an inch and a half from its neighbor on every side. Some ice dealers and others make a practice of filling this space between cakes with snow or broken ice so as to further prevent circulation of air through the ice. This has many advantages. It makes really a huge solid cake of ice in the icehouse but by reason of the weakness of the joints, it makes "breaking out" of cakes comparatively easy.

Care of crop when stored. Whenever the icehouse is entered, warm air is necessarily admitted as the doors are open. The ventilators should then be open as the warm air will cause vapor to collect above the ice, and this should be permitted to escape at once. The dressing on the top will need occasional attention, as it must be kept dry. Drains should be inspected, to see that they are not clogged.

"Breaking out" the ice. As the ice is removed, the top dressing should be kept in place. It is usually a good plan to take cakes from a number of layers at one time, the courses or layers being kept in a sort of a steplike series of tiers, the top tier being worked back the farthest, the next layer not so far, and the third still less. This allows greater ease of operation in every way. The cakes are usually pried out by a bent bar and separated from adjacent cakes of the same tier by a long-handled chisel, both tools being specially made for such purposes. Occasionally it is necessary to use a saw to separate the cakes at the sides. A special pointed saw is designed for this purpose.

Cold-Storage Houses for the Farm

Advantage of cold storage. Most progressive farmers have learned the value of the individual icehouse on the farm, yet many have not realized that the most economical way of using the ice cannot be developed without a properly constructed cold-storage chamber. Creamery and coöperative cold-storage chambers are now quite common, and their importance is realized. As the farmer observes them in use for commercial purposes he will undoubtedly come to appreciate the value to him of a similar house at home built on a smaller scale.

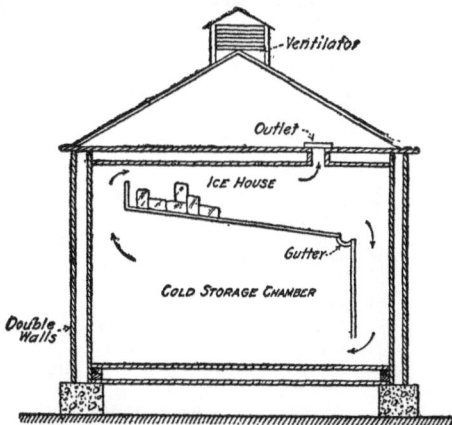

FIG. 619. Section of a combination ice- and cold-storage house for the farm. The slope of the floor of the ice chamber is exaggerated here; it should be merely enough to carry the water into the gutter. Note the double walls, floor and ceiling and the provisions for ventilation and the escape of warm air.

The great advantage of cold storage lies in the fact that produce need not be shipped to market immediately, but can wait a favorable time and a favorable market. Especially is this true of fruit crops, where the market is apt to be glutted during the usual delivery season, and a hold-back for a few weeks will help to equalize the supply. What is true of market shipments is equally true of meat and produce for home consumption. Purchases may be made at a favorable time and in considerable quantity. The produce raised on the farm for home use can be held over a longer period and waste more nearly eliminated.

Insulation. The details of construction may, as in the case of the icehouse, be widely varied to suit particular needs. There

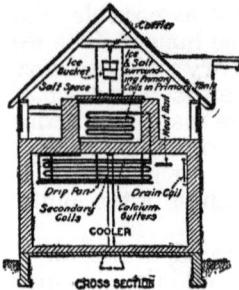

FIG. 620.   Cross-section through a small cold-storage house.

best form of protection. As the air within the space must be *dead* air, the walls *must* be airtight to give satisfaction.   There are many other ways of insulating, as by filling the space between walls with some so-called "non-conducting" substances, such as the following, which are named in the order of their desirability: hair felt, slag wool, wood ashes, chopped straw, charcoal, cork, and others. The insulating properties of these substances are largely owing to the fact that they enclose in the tiny spaces between their individual particles small amounts of dead air which cannot escape. That air is the insulator. For this reason the substances cannot be packed solid, and they should be tamped lightly into place rather than rammed hard.   For cold-storage work it should be borne in mind that there is more to be considered than merely keeping the products. They must be kept properly and untainted. Something must be chosen for insulation which does not readily absorb moisture and odors.   There is no one substance which does not do this to some extent. If the building can be built with matched boards and the dead-air space lined with tarred paper, the space need not be filled with anything. In fact, a filling would be a decided detriment. Absolutely dead air is essential, however.

are certain fundamental principles which can be laid down for guidance, however, close adherence to which will mean success in construction. Satisfactory insulation can be obtained through the use of double walls for the cold-storage chamber, in this way providing a dead air space between the walls, as that is the

Detrimental effects of dampness. Moisture has the property of absorbing many gases and impurities from the stores, so that it is very desirable that the air in the food chamber be kept as dry as possible and that the moisture which it does take up be removed.   In this way the air may be purified.   The way in which this is accomplished is by providing proper circulation of the air in the storage chamber, thus cooling the stores by circulation of the cold air in contact with them rather than by radiation. Unless cooling is done in this way, the moisture which the air contains will be deposited on the stores and not on the ice.   This, of course, will cause some of the packed material to become tainted.

Circulation of air.   To secure a good circulation in the storage chamber it is only

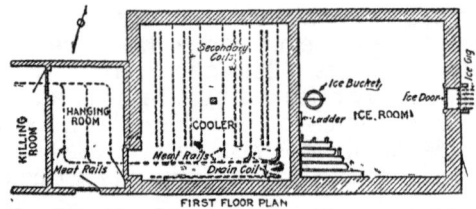

FIG. 622.   First floor plan of the cold storage building shown in the two preceding figures

necessary to appreciate the fact that cold air falls and warm air rises.   All that needs to be looked out for is to have the icebox above the level of the storage-space floors and to introduce the cold air at the bottom of the storage space and provide an outlet and return at the top of the chamber, to allow the heated air to go back to be cooled and deprived of its moisture.   For a small chamber it will be satisfactory if the cold air is allowed to enter all along the lower edge and the warm air taken out at the upper and diagonally opposite edge.   This will make it necessary for the air to cross and circulate all through the storage space before reaching the outlet.   In a larger chamber, the cold air could be introduced at the centre of the floor and taken out at each of the upper side edges. In a still larger room, the cold air may be introduced along two side edges at the bottom and allowed to pass out through two side edges at the top.   Shields or deflectors, which may be made of wood painted with enamel, should be placed so as to prevent the cold air, as it warms up, going from the inlet directly to the outlet without circulating through the room.   These deflectors should slope from the bottom upward and be placed just over the cold-air inlets, so that as the cold air warms it will

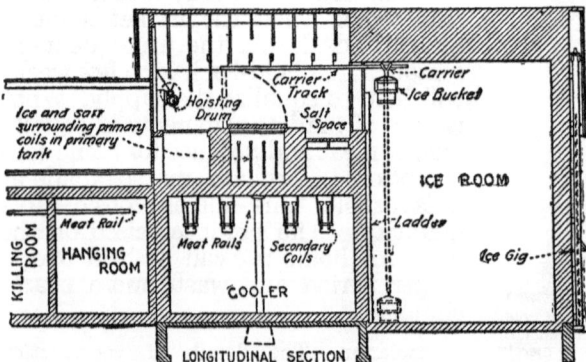

FIG. 621.   Lengthwise section through the same house which is cooled by brine and constructed as part of a small slaughter house.

rise along the deflectors toward the outlet. Care must be taken not to place the deflectors so as to pocket any warm air—that is, they should not be made so that any body of warm air will be caught in an upper corner and have to go downward to escape. Deflectors are necessary only where the outlet is nearly over the inlet and a path from one to the other does not lead through or near the centre of the storage space.

Ventilation. Ventilation is essential, but, except in very large rooms, it is satisfactorily taken care of by the opening and closing of the entrance door.

Types of cold-storage houses. The usual cold-storage house is a 2-story affair with the icebox on the second floor and the storage chamber below. Flues are provided to admit the descending cold air at the bottom of the storage chamber, and to take out the warm ascending air from the top. The water from the melting ice is carried away in drains which must be carefully trapped and sealed as in the case of the icehouse.

The cold-storage house can be built like the icehouse, described above, as to foundations, walls, and roof. A solid foundation is essential with strong floor beams and posts. The walls and floor should be double with an air space to provide insulation. The top ceiling should also be double. There should be a ventilator in the roof and a controlled opening through the ceiling of the icebox.

Temperature maintained. Usually the temperature best adapted for storage purposes is one of from 4 to 10 degrees above freezing. The lower temperatures, 34 to 36 degrees, are generally used, but some fruits do better at a little higher temperature. For freezing meats and poultry, temperatures much below 32 degrees must, of course, be maintained. This is done by means of a salt refrigerating tank.

Refrigerating tank. In place of the icebox, a water-tight sheet-metal tank is used. The tank must be fairly large, in order to have a large radiating surface. There should be about 4 square feet of radiating surface to 25 cubic feet of cold storage space. A drain is arranged for the tank, the bottom being sloped toward the drain, to allow the water to run off. As moisture will condense on the surface of the tank, provision must be made to drain the floor below. Usually a tray or pan is placed below. The tank is usually placed right in the cold-storage room, raised well above the floor. It is regularly supplied with ice and salt and, properly cared for, will maintain a zero temperature.

Artificial refrigeration. There are numerous machines on the market for maintaining a low temperature without the use of ice. The method used is scientific. Ammonia, which is naturally a gas, is very highly compressed, its temperature lowered, and the ammonia liquefied. The liquid is allowed to escape through a tiny valve into expansion coils and, being relieved from pressure, it vaporizes again into a gas. This requires heat, just as when water is boiled. The heat is supplied to the ammonia from the surrounding objects and they become colder and colder as the process continues. This action is like that of any vaporizing liquid. If alcohol be dropped on the hand, the hand becomes cold as the alcohol vaporizes, because the alcohol abstracts the heat from the hand causing vaporization to take place.

Now, in the case of the ammonia, if the expansion coils be placed in brine, the heat will be abstracted from the brine and pans of pure water placed in the brine will be frozen. If the coil be placed in a cold-storage chamber, the temperature of the chamber will be lowered and kept lowered as long as the process continues. The ammonia gas is withdrawn from the expansion coils, compressed, and released again and, with a slight leakage, is used over and over again.

The process described is not always followed out in every detail, but is typical of artificial methods.

Storage without ice. Root and fruit cellars are quite common in all parts of the country. A cave in the side of a hill is commonly used in New England for a potato cellar and, in southern Europe, similar caves in the mountain-sides have been used for the storage of oranges. The essentials of such cellars are dryness and proper ventilation. Dampness always promotes decay, and lack of ventilation will frequently result in sweating. The sweating of fruit and vegetables upon sudden removal from such a cellar or from cold storage, always takes place, and the moisture formed must be removed before packing the produce for sale. To prevent sweating, bring the

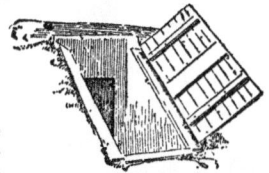

Fig. 623. The essentials of an iceless cold storage cellar are dryness and sufficient ventilation.

temperature of the produce gradually to the temperature of the outside air.

For indoor storage of root crops, a sand covering is very successful. A dry floor is essential. A layer of sand is spread and then the roots are spread in layers, each layer being covered with sand. The storage space must be kept ventilated. In place of sand, sawdust, hay, excelsior and sometimes lime is employed. The whole object is to keep air away from the fruits or vegetables so stored. Keep the air away, keep the produce dry, have good ventilation to prevent sweating, and a considerable degree of success is certain.

Packing stores. The packing of stores in cold storage is a science in itself and can be learned only by experience. One general rule, however, is of value and will take care of most difficulties, namely: pack the stores fairly close together and leave a space between them and the walls so as to allow a path for the circulating air. Never pack up close to the walls.

Duration of cold storage. It is difficult to give more than a general idea of the length of time for which any specific produce may be kept economically and profitably. This depends greatly on local conditions. Meats and fowl can be kept for 15 or 20 days and, when frozen, are, of course, frequently kept longer. Butter and eggs, cabbage, turnips, potatoes, etc., may be kept for months. Most fruits will keep a month; grapes, pears, watermelon and citrus will keep very much longer; and apples will keep six months.

FIG. 624. Not a bomb-proof shelter, but an earth-covered potato cellar. Note the open ventilators; there is a similar set on the other side which can be opened when the wind is from the side opposite to them.

CHAPTER 38

Silos: How to Build and Use Them

By J. KELLY WRIGHT, who for 8 years has been connected with the Missouri State Board of Agriculture in the capacity of Lecturer. During this time he has made a special study of silo types and uses, especially with reference to the needs of the average, practical farmer. He was born and reared on a Missouri farm, graduated from the State College of Agriculture, and served for 4 years as School Commissioner of Boone County. Among the many developments in agricultural education, which his farm training has enabled him to bring about, was the organization of the first boys' corn show held in his state.—EDITOR.

THE silo is a container in which to store or "can" certain field crops in order that they may retain their feeding value and succulence (juiciness) and be relished by livestock. Feeders have long recognized the value of succulent feeds such as green grass. Feed stored and preserved in the silo is called ensilage, or "silage."

Silage is not a new feed, but in the United States it has been known less than 50 years. According to ancient writers, it was a common practice of the Greeks and Romans to preserve green feed and even grain in underground pits. This method has also been in use for hundreds of years in northern Europe where the uncertainty of the weather and the low temperature make it difficult to cure hay. However, it attracted little attention until in 1877, when a French farmer, Goffart, published a book giving the results of 25 years' experience in preserving green feed in this manner.

The first silo in the United States is said to have been built in Michigan, in 1875, by Manly Miles. Not long after this, agricultural experiment stations in America began to make feeding experiments with silage and to urge the building of silos. From this beginning the number of silos has increased until it is estimated that by 1917 there were 400,000 in the United States.

How and why it pays to use a silo. All classes of livestock feed better under summer conditions when feed is plentiful, succulent, and palatable, and surroundings are comfortable. By the use of a silo, summer conditions can more nearly be maintained during the winter months. Silage is the best cheap roughage. It can be fed with profit to all kinds of young, growing stock, to mature cattle of all classes, and, under certain conditions to other grown animals. When fed as a part ration, it is good for fattening lambs and for breeding ewes. It is also a good feed for colts and horses not doing heavy work, and has been fed successfully to fattening mules. As a tonic to keep hogs in good condition it may well be fed to them once or twice a

FIG. 625. The sketch at the top of this page shows a hard, wasteful, unprofitable way to feed roughage. How much better for both man and animals to feed silage —with or without grain—in bunks, in a sheltered yard, like this!

453

A SILO WITH EVERY BARN

1-WILL KEEP MORE STOCK

**2-INSURES SUCCULENT FEED
        WINTER AND SUMMER**

3-SAVES THE WHOLE CROP

4-PREVENTS WASTE IN FEEDING

5-HELPS UTILIZE OTHER FEEDS

6-SAVES STORAGE SPACE

7-IS FILLED AT SLACK SEASON

FIG. 626.   What a silo does (International Harvester Co.)

week. In short, all stock like it and no better feed can be found on the general farm.

Silage is equally invaluable as a summer feed when, during periods of drought, pastures are short. Its use at such times often prevents a decrease in the flow of milk from the dairy cow, and a loss of weight in the beef animal. Experienced dairymen who know that when once a marked drop in milk production takes place, it is almost impossible, even with the best feeding, to bring a cow back to her previous high production, and feeders of beef cattle who have found it expensive and difficult to overcome a "slump" in weight caused by scarcity of feed, will appreciate what this means.

In the future the silo is sure to prove a yet more important factor in beef production. The great ranges of the West are being cut up as land advances in price; instead of the big, coarse, rough steer produced on cheap land and fed on cheap grain the ranges are supplying a smaller steer of better quality. These conditions, together with a great shortage of the world's meat supply, demand the most economical handling of feeder, stocker and breeding cattle. The silo will make possible such a system.

With a silo the number of head of stock that could otherwise be carried per acre can be doubled. More roughage can be stored in less space and fed out with less labor. Crops can be harvested and stored at a season when farm labor is not busy, and when the weather is good. The silage can often be fed indoors and exposure to weather be avoided; and even though it is fed in the open, the silo is a far better place to have kept the feed than a wet, muddy, snow-covered field could be. On many a farm the silo has lessened the feed and labor bill from 5 to 20 per cent and at the same time has made life more pleasant for the farmer, his boys and his hired men.

It is difficult to point out any single factor that alone can determine whether or not a man should have a silo. Hence it is difficult to say just how small a herd of dairy or beef cattle would justify one. Few good dairymen can afford to be without a silo. Every 160-acre farm near a market for dairy products could carry, among other classes of livestock, 10 or 12 good dairy cows and on every such farm a small silo would pay.

The general farmer with 20 or 25 head of cattle and horses to feed can certainly make good use of a silo; the general farmer who feeds a car-load or two of cattle each year will surely find a silo profitable over a series of years. Owners of some of the larger farms of the Corn Belt, and of some of the western ranges, find 25 or more silos none too many.

The larger a man's acreage in proportion to his business ability, the less use he can make of a silo. For the

PROFIT PER COW PER MO.

| GRAIN RATION | $2.46 |
| SILAGE RATION | 5.86 |

COST OF 100 LBS. MILK

| GRAIN RATION | $1.06 |
| SILAGE RATION | 69 |

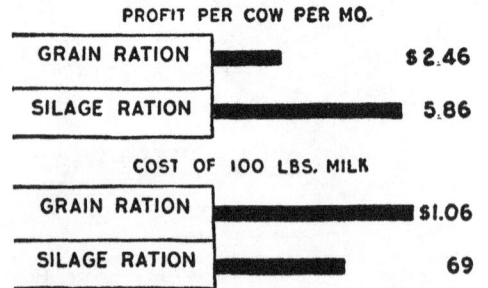

FIG. 627.   The money value of a silage ration as compared with a straight grain ration (International Harvester Co.)

larger his acreage the greater the amount of roughage at hand; and the greater the amount of roughage, the more of it a man can waste without seeing the need for a silo.

FIG. 628. Why do this of a winter morning when by filling a silo in summer you can feed better roughage, under cover, when snow flies?

General Principles of Silo Construction

Size and capacity. The height of a silo should never be less than twice its diameter. The taller the silo is in proportion to its diameter, the more silage it will hold for the reason that a greater weight can be placed on any given area of surface, and more silage can be packed into it. To illustrate: the heights of two silos of equal diameters, one 12 feet high, the other 36 feet high, are as 1 to 3; but the amounts of silage that they will hold are about as 1 to 5.

The diameter of a silo should be determined by the number of head of stock to be fed, and the height by the number of days in the feeding period.

On warm summer days the silage on top, when exposed for more than a day, will begin to spoil; it therefore is necessary at such times to feed off from the top from 1 to 2 inches a day in order to keep the silage fresh and sweet. It will readily be seen that if the diameter of a silo is great and the number of head of stock to feed is small, the removal of enough silage each day to prevent spoiling might result in a waste of feed. It is much better to have two small silos than one very

CAPACITIES OF SILOS OF VARYING SIZES IN TONS

DEPTH OF SILAGE (FEET)	INSIDE DIAMETER IN FEET				
	10	12	14	16	18
25	36	52	68	96	122
28	40	61	81	108	137
30	44	68	90	115	150
32	50	72	95	126	162
34	53	77	108	142	171
36	57	82	114	158	194

large one, especially when the number of stock is small and the feeding period long. Again, if two small silos are built in preference to one large one, one silo can be left undisturbed until needed, perhaps for summer use when pastures are short.

The number of pounds of silage that an animal will eat in a day depends largely upon its size; for mature cattle it varies from 25 to 40 pounds. Since a mature beef animal will eat about the same amount of silage in a day as a dairy cow of the same size, the accompanying tables (from Bulletin 103 of the Missouri Experiment Station) are of value generally.

After knowing the capacities of silos of various sizes and the length of time the silage in each will last with a given number of animals to feed, the next question is the number of acres of corn required to fill a silo of given dimensions. The following figures show the average amounts of silage that an acre of corn will make:

Corn making	30 bushels per acre	6 tons
" "	40 " " "	8 "
" "	50 " " "	10 "
" "	60 " " "	12 "
" "	80 " " "	16 "
" "	100 " " "	20 "

Thus a silo 14 x 30 feet will hold about 100 tons of silage which will feed 25 head of cattle about 180 days; to fill it will require about 10 acres of corn averaging 50 bushels to the acre.

Shape. Experience has shown that round silos or those nearly round are rather to be recommended than "square" or even 6-sided ones. Silage can settle and pack more closely around the wall of a round silo with the inside wall smooth and straight and such a silo can more easily be reinforced to resist the pressure of the silage. A few farmers, however, have filled square-cornered, strongly-built bins or granaries with green feed, particularly corn, and in some instances have produced a fair silage.

FIG. 629. No farm that supports a silo need ever be guilty of burning corn stalks as, unfortunately, some farms are.

RELATION OF SIZE OF SILO TO LENGTH OF FEEDING PERIOD AND SIZE OF HERD

NUMBER OF COWS IN HERD	180-DAY FEEDING PERIOD			240-DAY FEEDING PERIOD		
	ESTIMATED SILAGE CONSUMED (Tons)	SIZE OF SILO (Feet)		ESTIMATED SILAGE CONSUMED (Tons)	SIZE OF SILO (Feet)	
		DIAMETER	HEIGBT		DIAMETER	HEIGHT
10	36	10	25	48	10	31
12	43	10	28	57	10	35
15	54	11	29	72	11	36
20	72	12	32	96	12	39
25	90	13	33	120	13	40
30	108	14	34	144	15	37
35	126	15	34	168	16	38
40	144	16	35	192	17	39
45	162	16	37	216	18	39
50]	180	17	37	240	19	39

Construction details. There are just two things that cause silage to spoil: (1) if the air gets to it, it will rot; and (2) if it becomes too dry it will mold. For these two reasons the wall of the silo must be tight enough to practically keep out the air and prevent the escape of moisture. In other words, the silo must be "almost air-tight." Control the two factors, air content and water content, and it makes little difference as to what the silo is made of, so far as its ability to keep silage is concerned. It is not enough, however, for the silo to keep air from and retain moisture in the silage. It must be strong enough to resist the pressure of the silage, not for one year, but for several; it must be solid and rigid enough to maintain its shape and efficiency for a long time; and it must have weight and durability enough to withstand the action of wind and weather, especially when empty.

Kinds of Silos and Their Construction

Good silos have been built in a variety of ways and from many different ma terials. The first silos in the United States were made of stone or brick, with thick, substantial walls. These were expensive, and wooden silos were successfully tried. Silos can be grouped in two classes: (1) *temporary*, and (2) *permanent*. Whenever the cost of a temporary silo equals or approaches that of a permanent silo, the farmer should buy the permanent one.

Wooden Silos

The wooden-stave silo. Probably the silo most commonly used is of the wooden-stave type. This is a good silo and will keep silage with as little loss from spoiling as any on the market; but not better than some others.

The stave silo, as it comes from the company, is ready to be put together. However, the purchaser must have prepared a foundation on which to set it. This foundation can be made of stone, brick, or preferably, concrete. The wall of the foundation should be from 8 to 12 inches in thickness and should extend from 2 to 3 feet into the ground, and from 1 to 1½ feet above it. A 32-foot continuous stave with a 4-foot foundation makes a good combination for a silo 36 feet in height. It is a good practice to make the foundation wall a foot thick at the bottom letting the out-

side slant in to 8 inches at the top, so that the inner wall is kept perpendicular. After the foundation wall is completed, a concrete floor should be laid. It should be concave, that is, several inches lower in the centre than around the wall.

A wooden chute should be built around the doors to prevent scattering of the silage when it is thrown down from the top.

Stave silos vary in cost according to the size and to the quality of the lumber. There are also differences in doors and in many minor parts. A continuous stave silo of good quality 16 x 32 feet, together with the foundation, will, under normal conditions, cost about $375.

The wooden stave silo demands more attention during summer when it is usually empty and apt to be neglected than at any other time. At this season when the weather is hot and dry, the staves shrink and the hoops get loose—sometimes so loose that the silo will

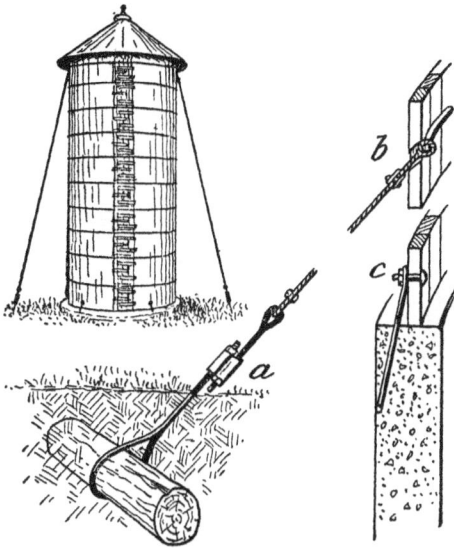

FIG. 630. Stave silo with guy wires; also details of *a*, turnbuckle and dead man; *b*, method of fastening wire to top of silo; *c*, additional brace support at base.

fall down—"fall to staves." When the hoops and staves of a silo become loose, a slight wind will sometimes blow it down, even if it be anchored. On the other hand, if the summer has been hot and dry and the hoops of the silo have been tightened several times, it will be necessary to loosen them when the silo is filled in the fall. If this is not done the moisture from the silage will swell the staves and may cause them to "buckle" in places, and sometimes let in the air or break the hoops.

Considerable difficulty is experienced with wooden stave silos in semi-arid sections where the weather is very dry and strong winds prevail. In other sections where material for building silos is scarce or not to be had unless it is bought, and where experience in building silos is limited, the wooden stave silo is a good silo to buy. When buying one, follow the instructions supplied by the maker or dealer regarding its erection and care.

Wooden-hoop silos. These are simply stave silos in which the hoops are of wood instead of metal. There are several varieties, but practically all are made of a good quality of pine flooring. The hoops are made four ply, of $\frac{1}{2}$ x 4-inch white oak lumber, which will bend to conform to the circle required.

There are 2 methods of making the hoops. Perhaps the better and most common method makes use of a floor space or platform on which can be marked out a circle with a diameter 2 inches larger than the diameter of the proposed silo. After this has been done, wooden blocks 2 x 4 x 10 inches are nailed to the floor or platform 2 feet apart around the circle, inside and just flush with the line.

Opposite each of these blocks and $4\frac{1}{2}$ inches from it another block 2 x 4 x 8 inches is spiked down with the spike through the middle of the block. One corner of each of the outside blocks is sawed off at an angle of 45 degrees, the beveled or sloping side toward the centre of the circle. These outside blocks turning on the spikes serve as clamps to hold the hoop material firmly against the ends of the inner circle of blocks. In other words, the 4 layers of hoop material are placed between the 2 circles of blocks and are held in place by the outside blocks until they can be nailed together into one hoop. In this manner the required number of hoops can be made. They can be distributed 30 inches apart from the top of the silo one third of the way down, but from this point to the foundation they should not be more than 24 inches apart.

The foundation for this type of silo is the same as for any other. Perhaps the ordinary concrete foundation is best.

When the foundation and hoops are made and everything is in readiness for raising the silo, all of the hoops are placed on the foundation. Inside them and against the foundation wall 5 scantlings or poles are set up, equal distances apart around the wall, and anchored by means of cross braces and guy wires to hold the silo in shape until it is completed.

Beginning with the top hoop, each hoop is then raised to its place and temporarily held by crosspieces until the staves can be nailed on. From the point where one side of the door is to be, the staves are nailed on just as flooring is laid. In the event that some staves are short and joints must be made, the joints should alternate. The staves should be nailed on until only a 2-foot space is left for the door.

The doors are made of the stave material, or flooring fitted in crosswise of the continuous opening. An extra stave is nailed on the inside 2 inches

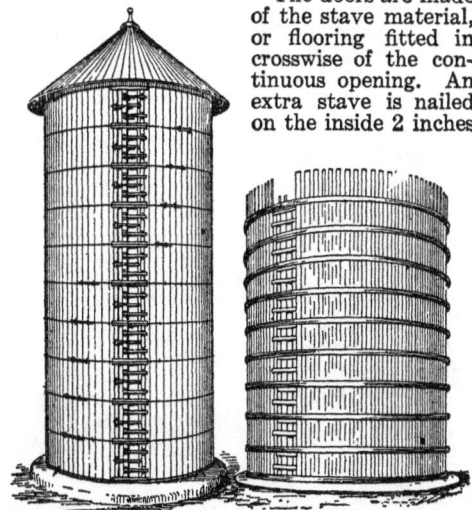

FIG. 631. Stave silos. One with roof and iron hoops completed, the other with wooden hoops, unfinished

back from the opening on each side to form the "door jambs." In addition, 2 staves are set in the opening one on each side, and fastened temporarily to the inner side of the hoops to form door cleats when the cross pieces are nailed to them. The cross pieces are fitted in from the bottom to the top of the opening and nailed to the staves mentioned. After the continuous opening is completely closed, the spaces between the hoops are numbered from the ground up. Then the cross-pieces are loosened from the top downward, and at each hoop the upright staves holding the doors are sawed through midway between the 2 hoops. This makes a separate door for the space between each pair of hoops, and each door will fit if kept in its place. The numbers previously placed upon them permit this to be done.

In sections where snow and sleet make feeding operations disagreeable, the silo should have a roof. Since the wooden-hoop silo is a homemade affair, any style of roof that the builder's ingenuity and taste may suggest, will suffice. The chute also can be made of any light material available.

The wooden-hoop silo can, under normal conditions, be built for about one dollar per ton capacity. However, there are modifications of the wooden-hoop silo which generally add to its convenience but also to its cost.

Instead of making wooden hoops, many farmers buy nuts and have the local blacksmith make and thread hoops of $\frac{1}{2}$ to $\frac{5}{8}$-inch iron rods to fit them. Such hoops can be adjusted from time to time. Many farmers have gone one step farther by making the walls of the silo in sections; a silo of this kind with iron hoops can be taken down or put up in a very short time and can be kept indoors when empty.

On account of the low cost, its efficiency in keeping silage, and the ease with which one handy with tools can make it, the wooden-hoop silo is popular in many sections. While it should be remembered that such a silo is, after all, only a temporary affair, it should

Fig. 632. A wooden, eight-sided silo made of short strips which are purchased ready milled, then laid horizontally and locked together. The corners are finally covered with protector strips.

also be remembered that it is worth what it costs and more, and that it is within the reach of farmers of moderate means. The use of such a silo in most cases make possible the purchase of a more permanent one.

Among other types of good wooden silos which, on account of lack of space cannot be described in detail, are the "Tung-Lok," a manufactured silo; the crib silo, a homemade, 8-sided affair made of 2 x 4-inch lumber laid flatways, each layer being nailed securely to the one beneath it; and the Buff Jersey, another homemade sort made of 2 x 4-inch lumber, not tongued and grooved but set up as staves and supported by iron hoops. The last mentioned silo has not given general satisfaction.

Concrete Silos

The solid or monolithic. This is one of the permanent silos. When built properly, it will last more than a lifetime; it will not blow over; it needs no guy wires; it will neither dry out and fall down nor burn down.

If proper care is taken in making it, the solid concrete silo will keep silage perfectly; the statement that it "will not keep silage," is now seldom heard. It is true that some concrete silos have cracked, and that in some silage has spoiled. But if properly reinforced, the concrete will not crack. And if the correct proportion of cement, sand, gravel or chats and water is used to make the mixture airtight, this type of silo will preserve silage perfectly. One common mistake has been the use of too little cement; another has been the failure to get the right amount of water into the mixture. If the mixture is too dry, there will be porous places that will admit air; if too wet, the cement and sand will "run," leaving the aggregate without enough cement to prevent cracking and the entrance of air. However, any man who has had experience in making concrete walks, watering troughs, etc., can build a concrete silo; if he has had no experience at all in making things of concrete, his safest plan is to secure the services of some one who has. (See Chapter 25 on Concrete on the Farm).

Regardless of who builds the silo, he must remember: (1) to use enough cement; (2) to use enough reinforcing material. The mixture of cement, sand, and gravel (or crushed rock or chats) generally used is 1 of cement, 2 of sand, and 4 of gravel (a 1:2:4 mixture).

For reinforcing, woven wire has proven very successful, though half-inch iron rods distributed both up and down and around the silo are better. A woven wire fencing (38-inch, No. 9 wire, with a 5- or 6-inch mesh) answers the purpose very well.

The following estimate of cement, sand, gravel and woven wire is for a solid 6-inch concrete silo, 16 x 32 feet, made of the 1:2:4 mixture just mentioned.

Portland cement . .	220 sacks
Sand	15 cubic yards
Gravel	30 cubic yards
Woven wire (38-inch fencing, 40 rods) .	2,090 square feet

The following table shows the amounts of cement, sand, and gravel for silos of different sizes:

MATERIALS	DIMENSIONS OF SILO IN FEET		
	12 x 28	14 x 30	16 x 32
Cement, barrels .	37	45	55
Sand, cubic yards	11	13	15
Gravel or stone, cubic yards. .	21	26	30

The forms for building concrete silos can be homemade or bought. If made at home they will cost about $50. The steel forms on the market are serviceable, easily handled, and can be rented out for enough to cover the first cost. A good plan is for several men in a community to share equally in making or buying the forms.

The cost of a solid 6-inch concrete silo, 16 x 32 feet, under normal conditions, will vary from $350 to $450, depending upon the price of labor and cement and the distance that the materials must be hauled.

The expense of maintaining the solid concrete silo is practically nothing. During the summer when the silo is empty its walls become very dry. For this reason they should be wet thoroughly before new silage is put in (this precaution should be taken with stave silos as well). Wetting the walls prevents absorption of moisture from the silage. Just as the inside of the stave silo should have a treatment of creosote once in 2 years, so the concrete silo should have a thin coat of cement and water of the consistency of whitewash, every 2 years. It will stop up all pores and keep the wall smooth.

The concrete-block silo. It is not claimed that the concrete-block silo will keep silage any better than the solid-wall type; however, the concrete-block silo has one advantage over the monolithic type, namely, the blocks can be made at times when other work on the farm is not pressing. These are of dimensions to suit the builder, and are hollow and made with a groove in one side through which passes

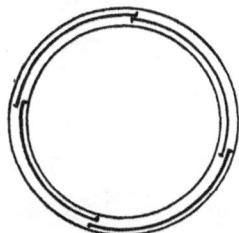

FIG. 634. The solid concrete silo, carefully made, is on of the strongest, handsomest, and most durable of all

an iron reinforcing rod. This type of silo must be well reinforced to prevent cracking. Many silos of this type are in use. Their average cost is about the same as for a monolithic, or good stave silo.

Hollow-Tile Silos

The hollow, glazed-tile silo is one of the best. On account of the material from which it is made and its resistance to fire and storms, it is classed as a permanent silo. Not only is it efficient, but also, in common language, it is a "good looking" silo in keeping with other first-class farm buildings.

This type should be considered by the man who wants a permanent silo and who has not on his farm or within easy reach the material for making a concrete silo. Especially should the hollow tile be considered by such a man if his experience in making things of concrete is limited, and if a professional concrete constructor is not available. Any good mason can build a hollow-tile silo.

The hollow-tile silo is made of hollow,

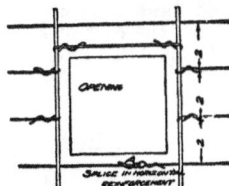

FIG. 633. How to place reinforcing rods in a solid concrete silo.

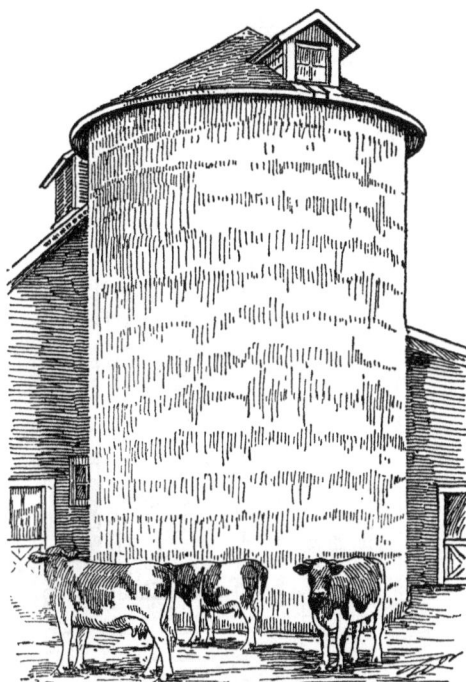

FIG. 635. Details of silo reinforcing around door

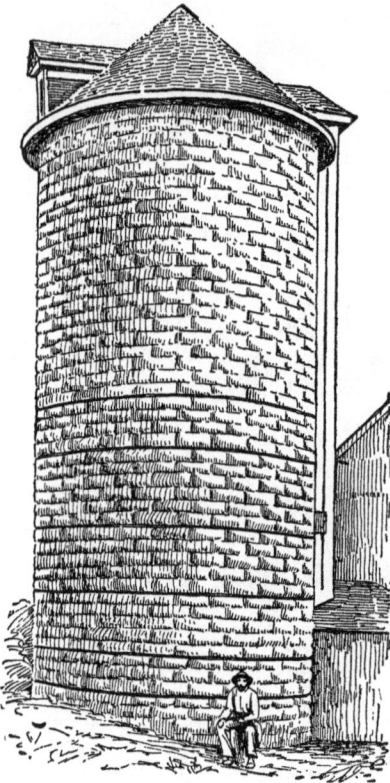

FIG. 636. The concrete-block silo is well suited to the needs and constructive ability of the average farmer. It must be well reinforced.

vitrified tile blocks reinforced usually by iron bands which fit into the mortar between the blocks or in grooves made in them. The curved blocks are laid on an ordinary concrete foundation such as would be made for any other type of silo, and when laid properly make a perfectly smooth wall. When the blocks are properly glazed, they are impervious to air and moisture. The hollow spaces in the blocks serve, it is claimed, as a protection against changes in temperature.

There are several makes of hollow-tile silos, both homemade and patented, among them the Iowa and the Dickey. Regardless of name, the best are those that are best glazed and best reinforced. Under normal conditions, the cost of a first-class glazed hollow-tile silo will be very little more than that of a first-class one-piece stave or monolithic concrete silo.

Other Types of Silos

The Gurler. For the small farmer of limited means, and particularly for the man who is struggling to pay off a mortgage and cannot buy a high-priced, permanent silo, the Gurler is highly satisfactory. This silo gets its name from Mr. H. B. Gurler, of Illinois, the first to build one, but it is sometimes called the "plastered" silo. It is homemade and the cost is low, since native lumber can be used in its construction. A silo of this kind with a capacity of 100 tons can, under normal conditions, be built for $100 to $150.

The foundation is made of concrete extending from $1\frac{1}{2}$ feet to 2 feet into the ground, and the same distance above. Before the foundation hardens, a sill is laid in the top of the concrete. To this sill 2 x 4-inch scantlings or studdings set 18 inches apart are nailed. To the inside of the studding, extending round and round, is nailed half-inch sheeting of native lumber; either elm, sycamore, cottonwood, or oak will do. Inside of this and running with the sheeting are nailed laths, also homemade if desired. If to be bought, steel laths are better. To the laths is applied a half or three-quarter-inch layer of cement plaster. There should be a cement floor, lower in the centre than around the wall. As soon as the plaster lining hardens, the silo can be used, but in order to protect the inner wall, boxing should be put on the studding outside, and painted. Vents or holes should be made in the boxing near the bottom and in the inner wall near the top to allow a free passage of air between the walls. This will prevent wood mold from forming and destroying the sheeting. A roof is not absolutely necessary, for enough water to injure the silage is not likely to fall into the silo, and snow does not injure the silage but merely makes it unpleasant to handle.

This type of silo, although temporary, when well built, will keep silage perfectly and will last 10 or 15 years, according to the material

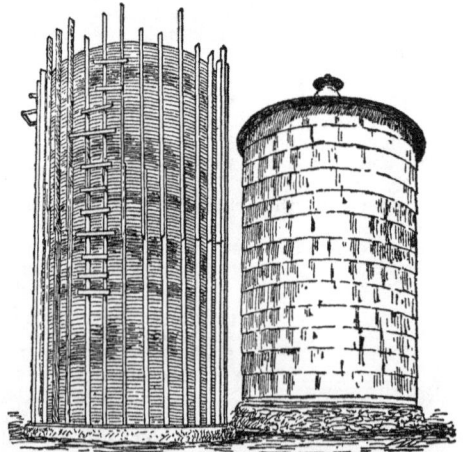

FIG. 637. The Gurler silo, though simple and low priced, is highly satisfactory. If sheathed with metal (*right*) and roofed over, its life is greatly lengthened.

used and the care given it. It will not dry out and collapse, and there are no hoops to keep tightened. Where all the material for the Gurler silo must be bought, a silo 16 x 32 feet can be built in normal times for about $225. If native lumber sawed from timber on the farm can be used, the expense will be less, perhaps not more than $125 to $150.

Another type of silo very similar to the Gurler is in common use. In it instead of the laths and cement plaster, a layer of tar paper is used, and inside this another thickness of half-inch sheeting. A silo of this kind is even cheaper than the Gurler. It will keep silage well and last from 10 to 12 years, perhaps longer, depending, of course, on the material used.

The Minneapolis or panel silo. The framework or reinforcing material of this patented silo is made of steel studs, which are set up perpendicularly and held in place by short wooden staves fitted in horizontally between them, and by iron hoops which hold the parts of the silo together. The staves are short and tongued and grooved. It is claimed by the manufacturers that this manner of using the stave permits little shrinkage of the wood and little difficulty in keeping the hoops tight. Whenever a stave appears defective, it can be cut out and replaced by driving all of the staves above it downward and inserting a new stave at the top.

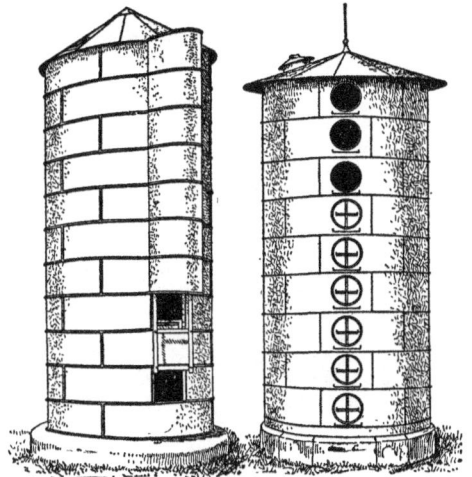

FIG. 638. Gurler silo construction showing: (A), plan; (B), section and details of inside finish; and (C,D,E,F) different methods of finishing the outside. (Wis. Bulletin 214.)

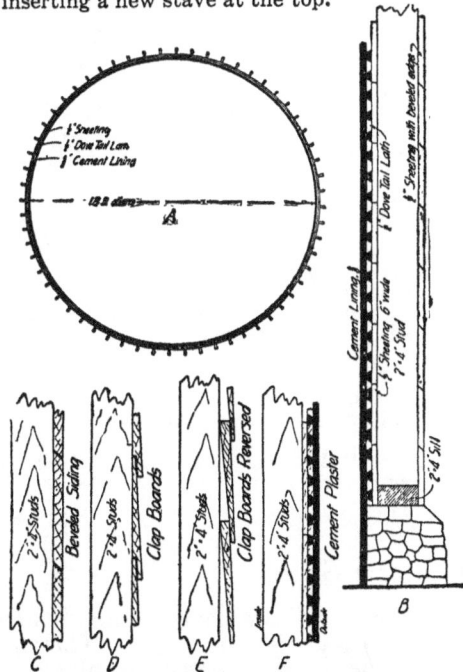

FIG. 639. Metal silos illustrating two types of door construction, that on the left being continuous with an enclosing chute.

The steel silo. There seems to be considerable demand for this type of silo especially in sections where dry seasons and strong winds seriously affect wooden silos. The steel silo in large sizes is more expensive than some other manufactured silos. However, it is a good silo, and if painted on the inside with asphaltum every year, it will keep silage perfectly. On account of the thin wall, silage will perhaps freeze in a steel silo a little more quickly than in some others.

The plastered steel silo. This is made of steel frames on which is fastened heavy woven wire. On the outside and inside of the woven wire, 2 inches of cement plaster are applied.

The Pit Silo

This type, as the name implies, is built mainly below the surface of the ground. However, when a considerable portion extends above ground, it is referred to as a *semi-pit* silo; when such a silo is placed in a bank and the retaining walls serve as a chute, it is called a *bank* silo. Pit silos can be built and used successfully in sections where the water-table is well below the surface. It is not advisable to build a pit silo in water-bearing soils.

The pit silo has gained its popularity in semi-arid sections, although, under proper conditions, it can be successfully used in humid regions. It is becoming more popular in parts of the north and northwest, where the winter weather is very cold and where silage freezes considerably in other types of silos.

The manner of constructing the pit silo is simple. The top of the silo is built first; the ordinary foundation for any other type of silo would serve well for the top or "collar" of a pit silo. First a circle having for its diameter the diameter of the desired silo is laid off, on the ground. Around the edge of this circle every 2 or 3 feet, are driven 2 rows of stakes $1\frac{1}{2}$ feet apart, until about 1 foot of each stake remains above ground. Then half-inch sheeting, or any kind of board material that will bend, is bent around and nailed to the stakes, one strip being nailed on the outer side of the inner row of stakes and the other on the inner side of the outer row. The dirt is then dug from between the two rows of stakes to a depth of 2 or 3 feet and the trench filled to the top of the sheeting with concrete which is allowed to set. This makes a good concrete collar or a good foundation if the silo is to extend some distance above the ground. When the collar is made, the earth inside is dug out to a depth of 5 or 6 feet, and a coat of cement plaster from 1 to $1\frac{1}{2}$ inches thick applied to the earthen wall. Then another 5 or 6 feet of earth are dug out, and another band of plaster is applied. In this way, 5 or 6 feet at a time, the pit silo is built. If the pit silo is to extend above ground, the extension can be of staves, concrete, concrete blocks, or hollow tile reinforced as in the ordinary overhead silo.

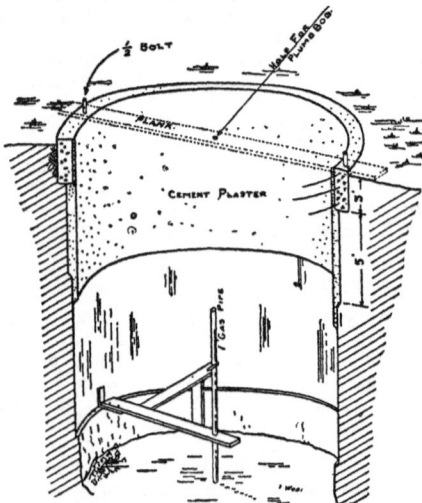

FIG. 640. Section of a pit silo showing collar, lining, anu simple apparatus for keeping the size and shape accurate. (Neb. Bulletin 39.)

The most common method of removing the earth in digging pit silos is by means of an ordinary hay carrier track and tubs made from barrels. Horse power can be used for hoisting. If the hay carrier arrangement can not be had, an ordinary block and tackle or windlass can be used.

The greatest disadvantage of the pit silo is encountered in getting the silage out. However, this inconvenience is offset by the low cost of the silo and by the fact that it is long-lived, easy to fill, nearly air-tight, below the frost line, warm in winter, and cool in summer.

Of the various methods used for hoisting the silage out of the pit silo, the windlass is perhaps the most general.

At the time of filling the silo, and until the cells or tissues in the silage are dead, a poisonous gas (carbon dioxide) is formed. This gas is heavier than air, and for this reason is held in a pit silo unless it is open wide at the top so that currents of air from the outside can carry the gas out. A week or two after the silo is filled, the formation of gas is so slow that there is no danger from this source. The danger comes at the time of filling, especially if the silo is partly filled one day and the work finished a day or so later. At this time no one should get into the silo until a lighted candle or lantern has been lowered into it. If the light goes out there is danger. The danger is equally great in any overhead silo at or near filling time if the doors are in place for several feet above the level of the silage.

The cost of a pit silo 16 x 30 feet, full pit with roof and hoisting apparatus, is given in Nebraska Bulletin 39 (February, 1917) as $1.41 per ton capacity; the itemized cost is as follows:

Labor	$ 62.85
Board	18.00
40 sacks cement	20.00
8 loads sand	4.00
100 feet half-inch wire cable	2.50
150 feet three-quarter-inch rope	4.00
Rope	4.00
Carrier for cable	4.00
Rope and post to erect cable	1.50
One-half-inch by 12-inch bolts, top wall	1.00
Boxes to lift dirt	5.00
Lumber for roof	20.00
Total	$146.85

The Care and Use of a Silo

Care in building. Great care should be taken that the walls be as nearly air-tight as possible and, if concrete, reinforced so they will not crack. The silo must be plumb, and its walls smooth and true. If it leans, the silage will settle away from the one wall and the air will get in. If the walls are left uneven and rough the silage will not settle properly.

Care in filling. In order to exclude the air, the silage must be packed carefully around the walls when the silo is filled. There is a tendency for the heavier pieces to fall in one place while the lighter ones fall a greater distance from the distributor. In order that the silage be uniform throughout the silo and settle uniformly, the weight must be kept uniformly distributed.

Each farmer should own his own blower and distributor. However, since it is more economical to use a large cutter than a small one, it is well for several neighbors to buy and use a large cutter jointly.

FIG. 641. Hoisting rig for pit silo digging. The plank is swung to one side when the wheelbarrow is lowered.

It is not advisable to buy a cutter with a knife less than 14 inches in length and a 17-inch knife is better. In order to keep all hands going and everything humming, the cutter should have a capacity of not less than 10 tons per hour, while 15 tons is better. It often happens that a cutter too small, choking down, getting out of order, and keeping expensive labor idle, is the cause of the high cost of filling a silo. The cutter should be set to cut the corn into half-inch lengths, and when so set should run right along without choking down. If it will hold up when set at this, it is not likely to fail when set to cut in longer pieces.

An engine, if not owned on the farm, might be hired or owned jointly. There is nearly always a traction engine in the neighborhood that can usually be hired at from $6 to $10 per day. A gasoline engine of not less than 8-horse power will do, unless the cutter is unusually large. It is always economy in the end to have an engine that can pull all that is required and a little more.

The number of wagons needed will depend upon the capacity of the cutter and the distance that the corn must be hauled. A driver will be needed for each wagon, and 2 men will be needed in the field to help load. If a corn binder is used, a man and teams to run it will be required in the field.

One man will be needed to run the engine, but he can help some at the cutter. Another man will be required all the time at the cutter. Two men should be in the silo and they should, above all things, be conscientious, active men, for on them depends the keeping of the silage.

If by waiting for the grain to mature, the stalk, blade, and shuck are allowed to begin to dry up, it may be necessary to add water to the silage as the silo is filled, or to run a small stream of water into the blower as the silage is elevated into the silo. There will be little or no danger of adding too much water; the danger will be, in all probability, in not getting enough.

When the silo is full, the top should be leveled off, tramped down firmly over the whole surface, and wet down thoroughly. The silage should be tramped down once every day for several days after filling.

If the silage has been tramped down properly around the walls of a good silo, there will be little or no spoiled silage around the wall. But the silage on top, if not protected by some sort of covering, will spoil down to a depth of from a few inches to perhaps a foot or more. To prevent this loss from spoiling, some men run a lot of wet straw through the cutter and blower, covering the silage to a depth of about a foot. Another practice is to sow oats on top of the silage, thick enough so that when they sprout they form a very thick mass of roots which provides a good tight cover. A few men have been known to "salt down" the silage when through filling. This, they claim, is a good way to prevent decay of the silage at the top. A barrel of salt is sufficient for a silo 16 feet in diameter. Other men have spread several inches of sawdust on top of the silage, and in this way kept the air out. Still another very good practice consists of running a half-inch layer of pitch over the silage just as soon as the filling is completed. The pitch does not adhere to the wall of the silo enough to prevent its settling with the silage; it keeps the silage as well as any cover in use; and it is not expensive. When the silo is opened, the layer of pitch can easily be broken up, thrown into a barrel or box and used again.

If it is not convenient to follow any of the practices mentioned, a few loads of corn, from which the grain or ears have been removed, may be run into the silo at the last, and wetted down. This is a very common practice; if adopted, the loss will not be great even if the silage does spoil down to a depth of a foot. The cost of filling silos in the Corn Belt is from 75 cents to $1.00 per ton.

Care in emptying. When the silo is opened, all spoiled silage should be thrown off where stock cannot get it. The middle should be kept full, particularly in hot weather. If a "feather edge" of silage is allowed to remain around the wall in hot weather it will spoil and make unpalatable all the good silage with which it becomes mixed. The silage should be removed from the silo at a uniform depth each day, the exact depth depending on the number of animals and their appetites (unless on limited rations).

Every one knows that spoiled fruit, whether canned or not, is not fit for food, and should not be eaten. Yet in spite of this men have fed spoiled silage to livestock. Often nothing serious comes of it, yet frequently we hear of some man who fed spoiled or rotten silage and, as a result, lost some of his stock. The feeding of moldy silage is always a bad practice, as it may cause the death of the animals that get it. Freezing does not seriously hurt silage, but it should not be fed while frozen. However, because frozen silage is hard to get

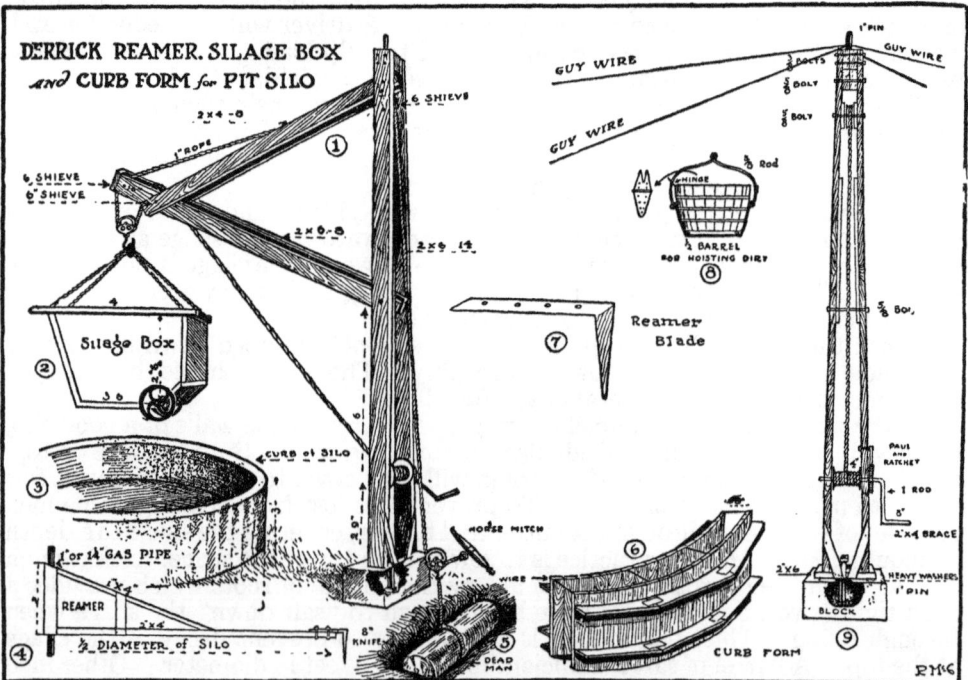

FIG. 642. Dimensions and arrangement of apparatus required in building and using a pit silo. (International Harvester Co.)

out of the silo and hard to handle, we never like to have it freeze at all if it can be prevented.

It should be remembered that one should never take stock away from good silage and put them on dry feed. In other words, a man should figure just how long his silage must last to carry his stock to grass again, and then feed with that date in mind. When fed silage, stock should have all they will clean up but not more. Thus in order to make it hold out until grass comes, silage feeding should not begin too early in the season. Of course, the man with both a winter and summer silo need not worry about this.

Best Crops for the Silo

Corn. One of the greatest advantages of using a silo is perhaps the ability it affords of saving the corn crop. When land was cheap and there was an abundance of coarse roughage which had little economic value it was not so serious when a big per cent of the crop was wasted. However, when land and labor are high in price, it is imperative that we save all of the corn crop and follow a system of farming which requires less labor per unit of production. Considerably less labor is required in harvesting and feeding corn as silage than in any other way.

From 30 to 40 per cent of the feeding value of the corn plant is in the stalk, blade, and shuck. Practically all of this is lost when the corn is harvested from the stalk or from the shock. The expense of putting corn into the shock and shucking it out is double what the stover is worth. This loss of feed and money can be prevented by use of the silo.

The grain from an acre of corn yielding 40 bushels, when valued at 60 cents a bushel, is worth $24. On this basis, the stalk, blade, and shuck (representing 40 per cent, or two fifths of the feeding value) are worth $16. But ordinarily "stalk fields" sell for about $1.00 an acre. Here, then, is a loss of about $15 per acre, that can be prevented by the use of a silo.

It has been stated that "upon the basis of total food value 2½ tons of silage are equal in feeding value to one 1 of timothy hay." One acre of 50-bushel corn will make 10 tons of silage, the equivalent of 4 tons of timothy hay. On the basis referred to, with corn selling at $1.00 a bushel, a ton of silage is worth $6.70. Calculated in this way, an acre of corn yielding 50 bushels per acre when put into the silo is worth $67, while the grain at $1 a bushel is worth only $50. However, if 10 tons of silage is equivalent in feeding value to 4 tons of timothy hay and takes the place of the hay in the feed lot, then it is worth what the 4 tons of hay would sell for on the market. Silage takes the place of high-priced hay.

In some parts of the country, silo owners grow a special silage corn, which produces much foliage and little grain. Farmers in other localities, hearing about this corn, sometimes ask if a special corn must be grown for the silo. The answer is "no." The corn that grows best in a locality is the corn for the silo in that locality. Nothing is gained by filling the silo with corn that produces an inferior ear or none at all. The more grain the better and richer the silage, other things being equal. While an acre of so-called "silage corn" may yield heavily as silage, the feeding value of an acre of such corn would not necessarily be greater than that of ordinary field corn carrying more grain.

Some people when they see silage for the first time are surprised to find that the color is not as they had expected, green. This idea of having a green-colored feed, together with lack of experience, has resulted in the filling of many silos with green, immature corn, carrying a high per cent of water and a low per cent, comparatively, of nutrients. Experience has taught us that if we want the best quality of silage, we would better wait for the dent to form in the grain, even though the stalk and blades dry up a little.

There are a number of cases where silos were not completed at the time the corn was at the right stage for silage. The corn was cut and put into shock, where it completely dried out. Then it was run through the cutter into the silo with plenty of water, and made good feed. However, this practice is not the best. It is mentioned to bring out the point that cutting the corn too green is not advisable, and that, by the addition of plenty'of water to the silage, corn can be allowed to mature before it is siloed, thus enabling us to secure the greatest amount of nutrients from an acre. Dry stover may require an equal weight of water to make good silage. As to the right stage at which corn should be cut for the silo, Prof. C. H. Eckles of the University of Minn-

Fig. 643. Nebraska pit silo showing hay loader and track used for removing silage

Fig. 644.   Twin pit silos and a combined hoisting device for both.   (This and Fig. 643, Neb. Bulletin 39)

esota (see Volume I, Chapter 41) says in Missouri Bulletin No. 103:

"The proper stage to cut corn is when it shows the first sign of ripening. In a year of normal rainfall, this is when the husks first begin to turn yellow at the end of the ear, while the leaves of the plant are still green. At this time the kernels are entirely past the milk stage and are glazed and dented. Silage made from such corn does not develop so much acid as when cut in a less mature stage, although it still develops a sufficient amount to preserve it."

Sorghum. After corn, sorghum ranks next as a crop for the silo. It yields heavily; sometimes, on good land, it will make from 12 to 15 tons of silage to the acre, although the yield is more often less than this. However, it yields about the same number of tons per acre as corn, soil, cultivation, season, and everything else being equal. There is considerably more water and more sugar in the sorghum plant than in the corn plant, and for this reason sorghum silage is apt to be sour.

Where sorghum is grown for the silo, it is a common practice, and a good one, to grow some corn to put into the silo with it. Best results with sorghum as silage have been obtained by making a half-and-half mixture in which a load of sorghum is run through the cutter, then a load of corn, etc. This practice is particularly advisable if the corn is a little dry. The drier corn will have a tendency to offset the sourness due to the sorghum. On the other hand, some farmers, upon finding their corn for the silo too dry, have added a little sorghum on account of its having more "juice" in it, thus making sure that their silage would not be so dry as to mold. Quite a number of extensive cattle feeders have made the statement that for making beef they would just as soon have silage made from corn and sorghum, half-and-half, as silage from corn alone. Sorghum, like corn, should be mature; the seed should ripen before the plant is put into the silo.

Kafir. There are sections where the drought-resisting kafir grows as well as, or better than, ordinary field corn, and there it is recommended for the silo. It was found at the Kansas Station that kafir silage ranks second to corn silage as a milk producer. The plant seeds heavily and has abundant foliage, yielding from one third to one half more tonnage per acre than corn and about the same as sorghum. Kafir, like sorghum, should be mature before being put into the silo. Feterita, another one of the sorghums, also makes good silage.

Cowpeas. All of our common field legumes, such as clover, alfalfa, soybeans, and cowpeas have been tried out in the silo and can all be used, but the resulting silage is of an inferior quality. This is particularly true of clover and alfalfa, but cowpeas make fairly good silage especially if mixed with corn. All the legumes, when siloed singly, undergo a change that gives them the appearance of having rotted; when siloed with some other crop like corn or sorghum they do not show this "rotted" condition. Moreover by mixing the nitrogenous cowpeas with corn which carries a high per cent of carbohydrates, there is produced a richer, more growth-making feed. The proportion should be 1 load of cowpeas to 2 of corn.

However, every man who feeds silage should, if possible, feed some legume hay with it, in which case it is advisable to refrain from putting cowpeas into the silo. It is not an uncommon practice to feed silage straight without any hay, but stock, when fed silage, nearly always do better if they have access to some leguminous hay.

How to Determine the Weight of Silage in the Silo

Sometimes we like to know how many pounds or tons of silage remain in the silo after we have begun feeding. Feeders have been heard to say: "If I had known that my silage would run out before grass was good enough for pasture, I should have fed a little lighter." Sometimes, too, when the silage is partly used out of a silo, we wish to sell the remainder. The accompanying table (from Wisconsin Bulletin 59) shows the computed weight of well-matured corn silage at different distances below the surface, and the total weight to those distances, 2 days after filling.

To illustrate: Suppose John Blank has a silo 16 x 32 feet inside dimensions. This silo, after the silage settled, contained 24 feet of silage. He fed out 14 feet of silage from the top, leaving 10 feet remaining in the silo which he wished to sell to a neighbor. How many tons had he to sell?

From the table it will be seen that one square foot of silage to a depth of 24 feet weighs 862 pounds. One square foot of silage to a depth of 14 feet is 407 pounds. Since this 14 feet has been fed out, subtract 407 pounds from 862 pounds which leaves 455 pounds, the weight of 1 square foot of the silage from a depth of 14 feet to the bottom of the silo, a distance of 10 feet. Then, if we multiply this 455 by the number of square feet in the surface area of the silage, the product will be the number of pounds of silage remaining in the silo. In order to find the number of square feet in the top surface of a silo, find the diameter, take half of it, multiply it by itself, and multiply this product by 3.1416. The diameter of Mr. Blank's silo is 16 feet; half of this is 8 feet; eight times 8 equals 64; 64 times 3.1416 equals 201.06 square feet, the area of the top surface of silage in the silo. Multiplying 455 by 201.06, we get 91,482.3 pounds, the weight of silage remaining in the silo. Dividing this by 2,000 (the number of pounds in a ton) we have 45.74, the number of tons of silage Mr. Blank has for sale.

WEIGHTS OF SILAGE AT DIFFERENT DEPTHS

DEPTH OF SILAGE	WEIGHT PER CUBIC FOOT AT DIFFERENT DEPTHS	TOTAL WEIGHT TO DEPTH GIVEN OF 1 SQ. FT.	DEPTH OF SILAGE	WEIGHT PER CUBIC FOOT AT DIFFERENT DEPTHS	TOTAL WEIGHT TO DEPTH GIVEN OF 1 SQ. FT.
Feet	Lbs.	Lbs.	Feet	Lbs.	Lbs.
1	18.7	18.7	19	45.0	619.7
2	20.4	39.1	20	46.2	665.9
3	22.1	61.2	21	47.4	713.3
4	23.7	84.9	22	48.5	761.8
5	25.4	110.3	23	49.6	811.4
6	27.0	137.3	24	50.6	862.0
7	28.5	165.8	25	51.7	813.7
8	30.1	195.9	26	52.7	966.4
9	31.6	227.5	27	53.6	1,020.0
10	33.1	260.6	28	54.6	1,074.6
11	34.5	295.1	29	55.5	1,130.1
12	35.9	331.0	30	56.4	1,186.5
13	37.3	368.3	31	57.2	1,243.7
14	38.7	407.0	32	58.0	1,301.7
15	40.0	447.0	33	58.8	1,360.5
16	41.3	488.3	34	59.6	1,420.1
17	42.6	530.9	35	60.3	1,480.4
18	43.8	574.3	36	61.0	1,541.4

INDEX OF SUBJECTS

INDEX OF TEXT ILLUSTRATIONS

FIG. 643. A well-made loose stone wall or fence. See page 492

APPENDIX

Practical Farm Fence Construction

By PROFESSOR H. C. RAMSOWER *of the Department of Agricultural Engineering of the College of Agriculture of Ohio State University, who was born and raised on a farm in Licking County, that state, and whose close contact with practical conditions is indicated by the fact that since 1909 he has managed his own farm. After graduating from the college in which he now teaches, he took special courses in Mechanical and Civil Engineering, having his present work in mind even at that time. In addition to developing his department he has traveled extensively through farming communities; was for 5 years Professor of Agronomy; has written a text book on Farm Equipment; and has made a special study of farmstead improvements.—*EDITOR.

THE prices of all kinds of fence materials have been steadily advancing during the last two decades and conditions created by the late war brought them to a level scarcely dreamed of a few years before. From this level they will probably not greatly recede. The problem of fencing the farm, therefore, has become more serious than ever and hence more deserving of careful study.

Amount of fence should be reduced. The total number of rods of fence on the average farm could and should be materially reduced. It is frequently an unnecessary expense to fence fields used in a regular crop rotation. If pastured at some time during the rotation, fences are seemingly necessary but it might be more economical to provide a permanent pasture if by so doing the amount of fence to be built and maintained could be reduced.

Fence repair. The average farmer will readily recognize that the old adage, "A stitch in time saves nine" is nowhere more applicable than in connection with farm fence. It is a splendid custom to have handy a tool box with nails of different sizes, a hatchet, a pair of pliers and a small saw so that it can be set in the wagon or auto when trips are made over the farm. A loose board on a gate may need tightening, a panel in a fence may need straightening, or a wire may need to be spliced, each of which tasks done in time is of small moment, though it may save hours and dollars later. Frequent whitewashing or painting of gates, especially around the farmstead, is much to be desired and not only adds to the life of the structures but gives a touch of neatness and finish that brings a certain return in dollars and cents.

Essentials of a good fence. A good fence is one that serves long and well the purpose for which it was built. Beauty rather than utility may have been the primary requisite as in the construction of a lawn fence. Or, utility may have been the only consideration as in the construction of a brush fence to turn sheep. Or, again—and this is most frequently the case—the good fence must both

489

Farm Fences in the United States

PERCENTAGE OF VARIOUS TYPES USED IN CERTAIN SECTIONS

AREA	WOVEN WIRE		BARBED AND SMOOTH WIRE	HEDGE	WOOD	STONE
	HIGH (OVER 42 IN.)	LOW WITH BARBED WIRES				
Western Dak., Kan. and Neb. and northern Minn.	5.5	10.2	84.0	0.03	0.3	0.0
Eastern Dak., Kan. and Neb. and southern Minn.	8.8	20.0	63.0	6.4	0.6	0.6
Iowa	8.0	45.5	43.5	2.1	0.9	0.0
Missouri	13.8	49.4	27.2	5.6	3.8	0.04
Wisconsin	13.5	33.4	49.8	0.04	2.3	0.8
Illinois	11.4	41.7	29.0	12.4	5.5	0.0
Michigan	55.9	11.8	11.9	0.6	19.7	0.0
Indiana	53.3	18.0	12.9	1.6	14.1	0.05
Ohio	59.8	3.8	7.0	1.2	27.9	0.05

From Bul. 321, U. S. Department of Agriculture, 1916.

serve its utilitarian purpose well and also harmonize with, and serve to enhance the beauty of, the farm and the farmstead. A good fence, therefore, is one that is built of strong, durable material and that is possessed of a measure of beauty.

Types, Materials, and Construction

Brush fence. The brush fence was used in early days when the land was uncleared and when the space occupied by the fence was not important. It served especially well in confining sheep. Except in the more sparsely settled regions where timber abounds it is no longer to be considered. It harbors insects and vermin and is a menace from fire.

Root fence. The root fence really followed the brush fence and represented a further stage in the clearing of land. The stumps and roots had to be piled in preparation for burning and it was generally desirable to leave them in piles for a year or so until they became thoroughly dry. It was quite natural, therefore, that they should be piled into fence rows to serve a real purpose against the time when they should be burned and replaced by a better fence.

Rail fence. The worm-rail fence was the first real fence built by early settlers in timbered regions. The first work to be done was that of clearing the land for crops and generally the best of the timber was split into rails. It is not uncommon even now to find on farms black walnut rails large, straight and in a perfect state

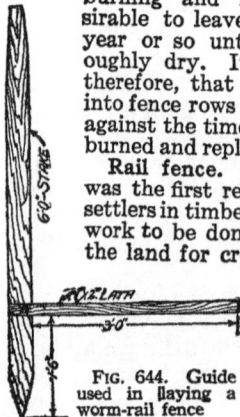

Fig. 644. Guide used in laying a worm-rail fence

of preservation though in continuous outdoor service for three quarters of a century or more. Rails were generally cut in 12-foot lengths with the diameter approximately 4 to 6 inches. A rail is nearly always larger at one end than at the other so that in laying up a fence the small end should always be laid forward to keep the corners of even height.

The laying of a worm-rail fence must proceed in a systematic way if the line is to be straight and the worm regular. Stakes should be set along the middle line throughout its length. A guide (Fig. 644) is then constructed by nailing a lath a little more than 3 feet in length at right angles to and about 18 inches from the bottom of a 6-foot stake. With the stake set on the middle line and the lath at right angles to it, the end of the lath marks the point where a corner should fall. With the rails lapping about one foot a 12-foot rail will thus lay a panel about 8 feet long. A narrower worm could be used which would increase the panel length but the fence would not stand so erect and would be more easily blown over.

Ground-chunks are generally laid beneath the corners. They serve to keep the rails from coming in contact with the ground and to increase the height of the fence. These chunks are about 2 feet long and from 8 to 12 inches in diameter being usually knotty portions of the tree or small logs, that are hard

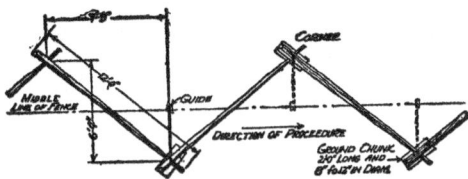

FIG. 645.   Diagram of a section of rail fence to show distances, construction, use of guide, etc.

to split. If the fence is used to confine hogs either the ground chunks must be omitted or pieces of rails must be laid under the fence.

Eight rails are usually required to lay the fence to the proper height. It can be made more secure by laying short rails on top of the fence and across the worm and locking them in place by still shorter pieces. Or, just the corners may be locked by setting pieces the height of the fence in the corners. Sometimes stakes and riders are utilized to make the fence still higher. Cross stakes are set at the middle of each panel then rails laid from one set of stakes to the next. Sometimes where locks are not used, the top 3 or 4 rails are wired together to prevent stock pushing them off.

While this type of fence served its purpose well when land was cheap it occupies so much space as to be objectionable on high priced land. Also it is easily blown over especially following a rain when the rails are wet and slippery. Then, too, the fence row tends to grow up to weeds, briers and underbrush, and harbors insects and rodents.

The post-and-rail fence (Fig. 646b) followed the worm fence in development as it made more economical use of the rails. The posts were mortised and the rails flattened at the ends and slipped into the mortises as the posts were set. A five-rail fence of this kind is sufficiently high, occupies less ground than the worm-rail fence, will not blow over and will last for the life of the posts. The tedious work of mortising the posts may be done away with by wiring the rails to the posts (Fig. 646a). A No. 10 or 11 galvanized wire is passed through a staple in the top of the post, carried to a point near the ground and stapled to the post. A rail is then placed between the post and the wire which is then drawn up tightly and stapled above the rail. The second rail is then placed and the wire stapled, and so on to the last rail. Two men are needed in erecting this fence, but it is found very satisfactory.

Hedge fence. A wide difference of opinion exists as to the desirability of the hedge fence for farm use. It unquestionably has its disadvantages. It is rather difficult to get started and several years are required for its growth; it requires prompt and careful trimming at least twice each year; it robs the soil of both fertility and moisture for several feet on either side; it becomes unsightly if

not properly cared for; and once beyond control it is a long hard task to cut it down and grub out the stumps. On the other hand, the good hedge fence properly set, correctly trained, and carefully trimmed is an everlasting fence which does not decay, does not blow over, is not crowded over by stock (which it will not injure) and adds beauty and freshness to the landscape. However, the good hedge fence is the exception rather than the rule.

The osage orange is universally used for hedge fence. It is a fairly rapid growing plant, hardy, not very susceptible to insect injury, produces heavy foliage and is sufficiently thorny to repel the advance of stock. Sheep will sometimes though not often, eat the foliage. Plants may be obtained from nurseries.

The first step in setting out a hedge fence is to clear the old fence away, plow a strip at least 4 feet wide, and thoroughly fit the soil. Plants will generally do better if set in the spring of the year although fall setting will give fair results if a moist time is chosen. A line should be stretched accurately along the fence line and a shallow furrow drawn with a hoe. The plants are then set 6 inches apart, water being used the same as in garden planting. A temporary fence must be placed next to the hedge as the plants must be carefully protected for three years at least. Meanwhile they should be cultivated and missing plants should be replaced the second year.

An old method of training the plants so as to secure a tight bottom to turn hogs and sheep consists of binding the plants down at an angle of approximately 45 degrees from the vertical and fastening them in this position during the third or fourth season. The branches should then grow upright making a fairly tight bottom for the hedge, However, this is a slow, difficult task and not always a successful one.

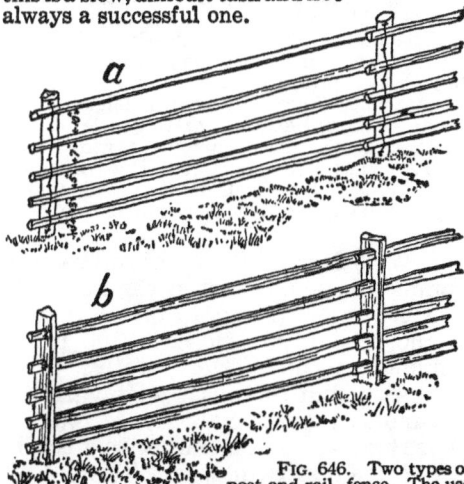

FIG. 646.   Two types of post-and-rail fence. The use of wire in a does away with the mortising of the posts in b

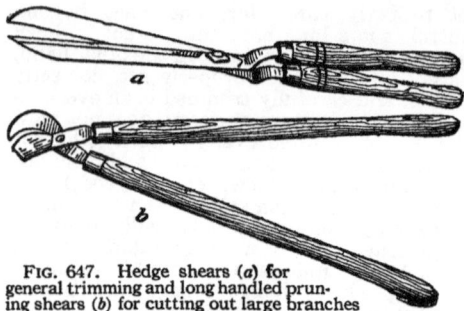

FIG. 647. Hedge shears (*a*) for general trimming and long handled pruning shears (*b*) for cutting out large branches

A better though somewhat more expensive plan is to erect a 30-inch woven wire fence close to the hedge during its third year, with posts set every 30 feet along the hedge line. To insure that the plants will stand erect, two or three smooth wires are stretched along the side opposite to the fence and fastened both to the posts and to the fence too if necessary. It is especially desirable to have such a wire near the top of the fence as it helps to hold the hedge plants more securely when they are being trimmed. This fence insures absolutely hog-tight construction.

A hedge will generally need its first pruning the fourth season and this should not be neglected as it will cause the plants to branch more freely. For trimming use a large pair of hedge shears, a pair of long handled pruning shears, and a hedge knife or corn cutter, the latter being the better if the hedge is trimmed frequently enough to keep the branches small.

A hedge should not be allowed to grow higher than 4 feet. Even the most unruly stock will seldom attempt to jump a hedge of this height and a higher one is more difficult to trim. There is a tendency to leave a hedge a little higher at each successive trimming, so usually it is necessary every 3 or 4 years to take heavy pruning shears and cut the top of each plant back to a proper height. Side branches should also be carefully trimmed so as to keep the fence narrow—a width of 8 or 10 inches being enough. One man can trim 25 to 40 rods of such a hedge in a day. A hedge should be trimmed regularly in June and September in the Central states.

Stone fences. In regions where there is an abundance of stone, more particularly limestone and sandstone, this material may be utilized in fence building. A neat and most effective wall can be laid with flat stone if made some 2 feet thick at the bottom tapering to 10 or 12 inches at the top. Large stones reaching through from one side to the other should be laid at frequent intervals especially near the top to bind the wall into one solid mass. Smaller stones can be used as fillers on the inside. (See Fig. 643.)

The height of such a wall is seldom more than 4, and frequently not more than 3 or 3½ feet. Sometimes the fence is finished by setting a row of large stones on edge along the top. This gives a more jagged and forbidding appearance, and the stones thus set are not so likely to be dislodged as when laid in a flat position. This kind of fence is common in the New England States (where, however, field boulder walls are also common) and in some regions of the Middle West. The amount of labor required for its construction is out of all proportion to its value, unless the stones are in nearby fields and must be removed for agricultural reasons. Like the worm fence it takes up valuable space and becomes a weed-grown menace unless looked after. The regular masonry fence of cut and matched stone set in cement mortar is another type that is restricted to country estates and suburban homes. In beauty and permanence it is of course a desirable boundary along a highway, but its cost cannot be justified on purely practical grounds.

Concrete rails. These, in combination with heavy concrete posts, are used in only a limited way in fence construction. Where some general scheme of beautification is being followed they are sometimes advisable. The proper reënforcement of such rails so increases their cost as to make it almost prohibitive except on elaborate country estates, which are about the only places on which they are found. Where unusually strong paddock fences are desired heavy timber planks are to be preferred. In an occasional case of this sort heavy iron piping, not less than 3 inches in diameter, is used with large posts of either wood or concrete.

Wire fence. The table on page 490 shows that *woven wire* is by far the most common type of fence in the Middle West. If there were statistics to show the proportion of this to all other kinds of fence erected during the last 5 years it would probably be close to 100 per cent. The use of barbed wire is especially common in the Western states.

Although there is a wide variation in the size of wire used in woven fences, the modern tendency is to use No. 9 throughout. This makes a more expensive fence and one that is harder to erect, but its increased length of life more than offsets these disadvantages. The use of stay wires as small as No. 12 or No. 14 is to be discouraged.

The Rusting of Wire Fencing

The rapid rusting of modern fencing wire has been observed by all users of woven wire fence. Scientists have long been searching for the cause, but as yet no satisfactory explanation has been offered, though there seems to be some evidence which tends to point to the modern method of making the steel from which the wire is manufactured. Years ago nearly all the steel used in making fencing wire was manufactured by the puddled iron process, in which the metal was carefully and thoroughly worked by hand. However, it was only after the introduction of the Bessemer and open hearth processes of steel manufacture that woven-wire fencing came into prominent use, largely because these processes made possible cheaper steel.

After steel is taken from the Bessemer converter or from the open-hearth furnace, it is poured into large molds and forms ingots of solid steel. It is then carried to the rolling mills and reduced by constant rolling to a size approaching that of a lead pencil; finally it is drawn cold through appropriate dies and gradually reduced to wire of the desired size.

To *galvanize* steel fence wire is to coat it with a layer of zinc. The wire to be galvanized is first drawn through an oven to properly anneal it. Thence it is placed into a tank of weak acid to clean it properly, and to provide a soldering solution. Next it is drawn through a tank of molten zinc, a portion of which clings to it. Finally the wire is drawn through asbestos wipers in order to produce a uniform coating. Undoubtedly the thickness of the zinc coating is a factor in the lasting quality of the wire.

In spite of this coating, however, the wire rusts, and investigations carried on by the United States Department of Agriculture in 1905 and reported in Bulletin No. 239 seem to indicate that the element manganese, which is added to the steel, whether made by the Bessemer or the open hearth process, just before it is poured from the furnace, is at the bottom of the trouble. When a wire rusts it pits, and an explanation frequently given, though not yet regarded as an established fact, is that the pitting is due to electrolytic action set up because of the unequal distribution of manganese in the wire. The fact remains that modern fence wire does rust out rapidly, though it is believed that manufacturers are doing everything in their power to improve its quality. The farmer's chief protection lies in buying a fence made of nothing smaller than No. 9 wire; even larger sizes may be advisable.

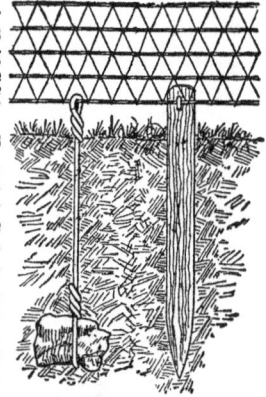

FIG. 648. Two methods of anchoring a wire fence between posts and in low spots

The height of woven wire fencing varies considerably. What is known to the trade as a 10-47 fence, that is, one having 10 wires and measuring 47 inches in height, is most common in Ohio, Indiana, and Michigan. Such a fence, if intended for horses, is not complete without a barbed wire on top. The low fence, with 2 or more barbed wires, is most common in Iowa, Wisconsin, Missouri, and Illinois. It makes a good cattle fence, but where it is used to fence in horses it is somewhat more dangerous than the 47-inch fence with but one barbed wire on top.

Barbed wire should be used with the utmost care. The practice of placing it on old fences, and especially near the ground, when horses are to be pastured or inclosed, cannot be too severely condemned. Horses maimed and ruined for sale if not for service are to be found in every community where barbed wire is used. It has, to be sure, a real place on top of a well-constructed fence 42 or 47 inches high, but placed lower than this it is a serious danger.

When erecting a wire fence, one

FIG. 649. Individual wire stretcher in use and (*a*) in detail to show how it grips and tightens a wire for stapling

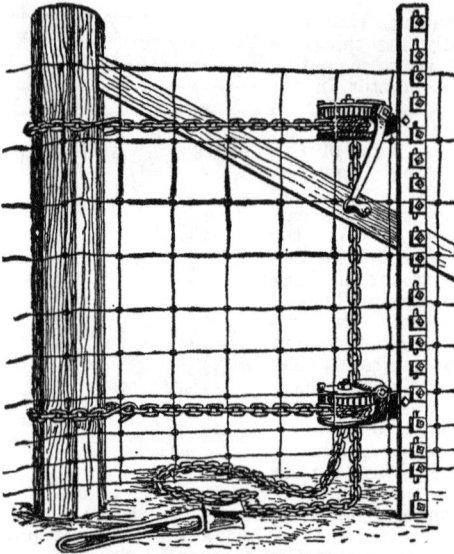

FIG. 650.   Good type of wire fence stretcher with chains and windlass at both top and bottom of clamp-bar

should remember that it is to be in place for a number of years, and that the degree of tightness will determine in no small measure the length of its life.   Every effort should, therefore, be made to secure it properly at both ends after stretching it as tightly as possible.   Fig. 650 shows one of the best types of stretcher.   The chains are fastened to the top and the bottom of the clamp-bar, each chain being controlled by a separate windlass.   This makes it possible to keep the fence plumb, which is not possible when there is but one windlass.   The clamp-bar should be so placed that it will stand about three feet from the end post when the fence is finally stretched. The fence should then be cut off and the wires tightened by means of an individual wire-stretcher (Fig. 649) as each is stapled to the post.   The wires should be bent around the post and wrapped around themselves, as staples are usually not sufficient to hold them; if steel or concrete posts are used, the wires must, of course, be wrapped.   This is a rather difficult task, but it may be accomplished in such a way that there is but little slack in the fence after the stretcher is removed.

Home made wire fences were in much more common use several years ago than they are now, the extensively manufactured woven wire having virtually replaced them.   In one home-made type a fence of 10 smooth wires was first erected.   Then by means of special tools either wire or wood vertical stay-rods were woven into the fence at intervals of from 1 to 4 feet.   This type, as well as others of similar design, was quite satisfactory, but labor is now so scarce and so high that the individual can scarcely compete with the large factory, especially if any considerable amount of fence is needed.

Wire fence is, of course, stapled to the line posts if of wood.   To concrete or steel posts it is fastened in a number of ways (Fig. 651).   In *A*, which is the most common method, the fastening wire is passed around the post.   In *B* it is looped over the line wire, passed through a hole in the post then bent around either side of the post and fastened to the line wire.   The sketch at *D* shows a fence locked in place by a No. 9 locking wire run through staples put in place as the post is made. In any case only about half the wires are fastened to each post.   In the case of one patent concrete post (*c*), the fence is fastened with

staples driven into a strip of composition cement cast in the face of the post when molded, and which takes and holds a staple well.   Wood rails for a picket fence or boards are sometimes nailed to posts of this sort.

Fig. 656 shows two common methods of fastening wire to steel posts.   In *a* the fastening wire is wrapped around the post; *b* shows a round, galvanized iron

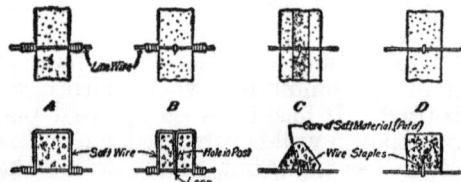

FIG. 651.   Four methods of fastening wire fencing to concrete posts (see text).   Upper figures show front view; lower figures show sections across posts

post in which small ears are partially cut out of the face of the post and bent around over the wires. Sometimes a staple is inserted through a hole in the post and clinched; and in another case in which the post is triangular in shape horizontal grooves are cut part way through the post along the apex of the triangle. Each line wire is then slipped into one of these grooves and a nail is dropped in between it and the back of the post, that is, on the inside of the triangle.

Wooden Fences

Board fence. The comparatively small amount of regular field fence now made of boards finds its chief use about barn yards where wire fence can scarcely be used with safety. The panels are usually 14 feet long with posts 7 feet apart. The boards are generally 1 by 5 inches and may be sawed from local timber, either hard or soft wood serving the purpose.

The regular field fence is 5 boards high spaced from bottom to top as follows, 3, 5, 7 and 10 inches. The bottom board is placed 3 inches from the ground if hogs are to be confined. If the fence is to turn only grown horses or cattle the bottom board may be omitted. Joints should be broken, that is, some of the boards should lap from one panel to the next, to give a more rigid construction.

Picket fence. This, like board fence, is no longer generally used in the field. It has, however, a real place around gardens and farmsteads. The patterns after which it is built vary greatly and can be made to suit the individual taste of the owner.

A **picket-and-wire fence** is occasionally used for field purposes. In one type 3 double strands of No. 12 wire are stretched and loosely stapled on posts one rod apart. The upper pair is placed so as to fall 6 inches below the top of the pickets, the lower, 6 inches above the bottom, and the middle pair half way between. The pickets are then woven into these wires. A special wrench is provided to twist the wires between the pickets which are spaced 2 to 2½ inches apart. A string may be stretched along the top line of the

FIG. 653. Section of well-built board fence showing broken joints and good spacing

pickets so as to secure an even line. If the pickets are of equal length a board may be placed on the ground beneath each panel and the pickets placed upon it to give an even line at both top and bottom.

Pickets for this type of fence are usually sawed ½ by 2 inches and 4 feet long. Local hard wood such as beech, maple, chestnut, oak, etc., may be used.

The *wood-rail* picket fence is still more common. The rails are generally 2 by 4 inches by 14 feet, two being used for each panel; the posts are set 7 feet apart. The pickets

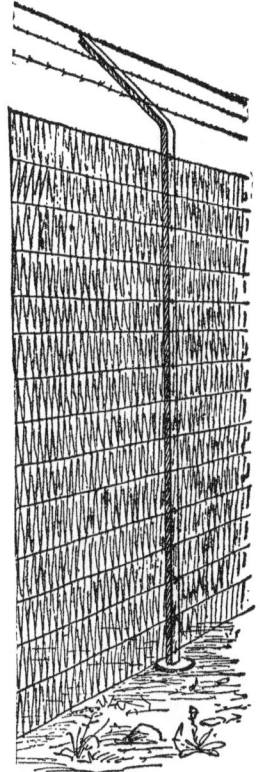

FIG. 652. A fence of woven and barbed wire and steel posts, useful where extra security is desired

are generally ⅞ inch by 3 or 4 inches by 4 feet, but for lawn purposes they may be narrower and shorter.

Portable fences. It is sometimes convenient or necessary or both to have a portable fence. It may be used in a barn lot or for confining stock, especially sheep or hogs, on temporary pasture. One good type in common use is shown in Vol. I, Fig. 89. Hurdles of chestnut or other hard wood, about 7 feet long, the posts of which are driven into the ground and held together with long oak pins, are frequently seen on Eastern farms and estates. They are highly satisfactory for confining horses and for enclosing exercising rings, race tracks, etc. They are not as easily moved as the sheep hurdle just referred to, and they are of course unnecessarily heavy and high for confining hogs, sheep or other small animals.

Where the custom of hogging down corn is followed and it is found desirable to fence off a part of a field, a 24-inch woven wire fence may be unrolled, set up along a corn row and tied to the standing corn. If a supply of old fence rails is at hand they may be used to good advantage in building temporary fences.

Fence Posts

Since from 85 to 95 per cent of modern fence is built on posts the question of post material is of vital importance. Fence posts may be made from either of three materials: wood, steel, or concrete, but the type to choose is not easily determined.

Wood posts. Wood has been the most common fence post material in past years and no doubt will continue in the lead for several years to come. The kind of wood used in a given locality will be determined in part at least by the kind of timber common there. Yet some woods formerly much used for posts have become so valuable for other purposes that as post timber they have been largely replaced by other kinds even though the latter must be shipped in from a distance.

The relative durability of the common post timbers has been investigated only in a limited way. The accompanying table is the summary of a study made in the central part of the United States by the Department of Agriculture. For purposes of classification the region was divided into

Fig. 654. How a heavy cross piece is used to anchor a fence post

AVERAGE LIFE AND COST OF DIFFERENT KINDS OF FENCE POSTS USED IN THE MIDDLE WESTERN STATES

| KIND OF POST | AVERAGE LIFE | | AVERAGE COST IN ALL AREAS | | AVERAGE COST IN EACH AREA | | | | | | | |
| | | | | | NO. 1 | | NO. 2 | | NO. 3 | | NO. 4 | |
	NUMBER ESTIMATED	YEARS	NUMBER ESTIMATED	CENTS	NUMBER ESTIMATED	CENTS	NUMBER ESTIMATED	CENTS	NUMBER ESTIMATED	CENTS	NUMBER ESTIMATED	CENTS
Osage orange . .	789	29.9	774	22	105	25	326	24	320	17	23	18
Locust	464	23.8	465	24	501	26	21	22	29	18	14	18
Red cedar. . .	557	20.5	574	29	346	29	97	31	104	27	27	21
Mulberry . . .	88	17.4	82	19	45	20	25	17	12	15		
Catalpa . . .	48	15.5	45	17	15	17	17	17	13	18	10	18
Burr oak . . .	97	15.3	90	15	10	10	54	15	26	15		
Chestnut . . .	94	14.8	91	15	91	15						
White cedar . .	1,749	14.3	1,709	18	642	18	459	18	374	19	274	16
Walnut . . .	60	11.5	56	13	6	5	11	13	39	12		
White oak . .	1,242	11.4	1,218	12	333	14	389	11	421	12	75	13
Pine	41	11.2	37	18	12	23	7	22	3	11	15	12
Tamarack. . .	67	10.5	64	9	6	16	26	8	7	9	25	9
Cherry . . .	9	10.3	9	8	7	8	2	8				
Hemlock . . .	10	9.1	9	12	3	20	6	8				
Sassafras . . .	19	8.9	17	14	11	15	6	10				
Elm	15	8.8	15	12	6	10	5	9	4	15		
Ash	69	8.6	58	10	17	11	2	10	15	10	24	10
Red oak . . .	22	7.0	24	7	6	7	10	8	8	4		
Willow . . .	41	6.2	33	7	1	12	2	7	25	7	5	9
Concrete (e s t i-mated) . . .	42	48.0	121	30	53	30	48	29	19	31	1	35
Stone	11	36.3	15	35					4	38	11	35
Steel (estimated).	131	29.9	219	30	82	30	71	29	54	30	3	30

From Bulletin No. 321, United States Department of Agriculture, 1916.

areas as follows: No. 1, Ohio, Indiana, and Michigan; No. 2, Wisconsin and Illinois; No. 3, Iowa, Missouri, southern Minnesota, eastern Kansas, and eastern Nebraska; No. 4, northern Minnesota, the Dakotas, western Kansas, and western Nebraska. The results are based only on common practice and are not claimed to be scientifically accurate.

From the standpoint of durability osage orange is really in a class by itself with locust and red cedar following, there being no great difference in the lasting qualities of these two. It is generally considered that a post timber which shows an average life of less than 10 years should not be given serious consideration.

The Ohio Agricultural Experiment Station published in Bulletin 219 (1910) the results of personal examinations of over 30,000 fence posts located in 292 different kinds of fences. This investigation rates post timbers in the order of their durability as follows: osage orange, locust, red cedar, mulberry, white cedar, catalpa, chestnut, oak, and black ash, which is very similar to the arrangement based on the Department of Agriculture's investigation.

The Ohio Bulletin gives the following general conclusions based on the results of the investigation:

1. There is no difference which end of the post is put in the ground except that the larger end should have preference.
2. From data collected so far, seasoning does not seem to have any marked effect on durability. It is hoped that the matter will be investigaged further.
3. Timber that grows rapidly in the open is not so good as that of the same variety grown in the woods.
4. There is some evidence that it is not good to cut posts just as a tree begins to grow in early spring.
5. The wood at the centre of the tree is not so good as that just inside the sap wood.

Preservative Treatment of Wood Posts

There is no question but that the life of posts can be prolonged by proper treatment and some of the methods used are perhaps worthy of trial on many farms. Rotting is caused by the growth of fungi, or low forms of plant life, which feed upon the tissues of the wood. They require heat, light, moisture, and food for their development, all of which conditions are supplied near the surface of the ground; wherefore posts rot off at this point. A post rots but little beneath the surface of the ground because the lack of air prevents the growth of fungi there, while above ground the lack of moisture prevents their growth. It is not possible to control the moisture, the heat, nor the air supply, but if the food supply can be cut off the fungi cannot develop. The most feasible method of destroying the food supply is to fill the tissues of the wood with creosote, a coal tar product.

This method is best carried out by heating the creosote in a large kettle, setting well-seasoned posts into it, leaving them there for an hour or so, and then removing them to a kettle of cold creosote where they are left for several hours. During this time the liquid soaks into the posts, the depth depending upon the nature of the wood. The hot bath is of itself fairly effective, but the cold bath following is essential to a thorough treatment. Painting the lower portion of the posts with hot creosote will probably prolong their life long enough to pay for the treatment, the estimated cost of which is about 15 cents per post. When the cheaper timbers are used it may reasonably be expected that it will double or even treble the life of the post. It probably will not pay to treat the better classes of hardwood posts, since the denseness of the wood prevents the absorption of the creosote in beneficial quantities.

Charring the post is a very effective method of preventing decay. To do it place that portion of the post which is to go into the ground in a bed of hot coals and leave it until thoroughly charred. This process also destroys the food supply of the fungi and prevents their getting a start.

Painting the lower portion of the post with an oil paint is effective so long as the paint lasts, but it soon scales off. The beneficial effects of this treatment are probably not worth the time and trouble it takes.

Zinc chloride has sometimes been re-

FIG. 655.   Home-made outfit for treating wood posts with hot creosote

commended as a preservative but it is little used. It is usually applied by boring holes in the posts and filling the holes with the solution, which is slowly absorbed by the wood. Being easily soluble, the solution next the surface of the post soon loses its effectiveness.

The table at the right, from Circular No. 117, United States Department of Agriculture, illustrates the value of the preservative treatment of lodgepole pine posts in Idaho. It should be remembered that the posts treated were of pine which, being a soft wood, absorbed a relatively larger amount of the preservative than would a hard post. The value of the treatment would, therefore, be greater for soft than for hard woods.

PER POST	UNTREATED	TREATED
Initial cost . . .	$0.06	$0.06
Cost of treatment .	.00	.15
Cost of setting . .	.12	.12
Total cost set . .	.18	.33
Estimated length of service (years) .	4	20
Annual cost (allowing 6 per cent interest on investment) about	0.05	0.03

Steel posts. Steel posts have been in more or less common use for 15 or 20 years. During this time both their material and their form have been widely changed so that there are now a great many different designs on the market. It is impossible to say which type will prove most satisfactory for all conditions.

The V-shaped post made of soft steel about ⅛ inch thick is likely to rust at the top of the ground in a few years. Furthermore, it is rather easily bent and, once bent and straightened, it seldom possesses its former strength.

The round galvanized steel post seems to last fairly well but bends quite easily. Other forms of the galvanized post have been designed to remedy this weakness but with indifferent success.

The high carbon, stiff steel post seems to be giving a large measure of satisfaction. It will spring back to shape if bent and in many designs at least is supplied with an enlarged bearing surface in the form of a wing beneath the surface of the ground which tends to hold it erect even though a strain comes upon it when the ground is wet.

Nearly all steel posts are easily driven by means of a sledge hammer, which makes it possible to set a long line of them in a very few hours—a decided advantage. The best way to set steel posts is to load them in a wagon, proceed along the line of the fence and drive the posts from the rear end of the wagon. A special driving cap is provided for driving most metal posts so that the top will not be crushed.

Concrete posts. Because of its durability, concrete is a splendid material for fence posts for which purpose it will, without doubt, find a wide use in the future. The greatest care must be exercised, however, both as to choice of materials and as to their handling if success is to be assured. The materials required are cement, sand, gravel or crushed rock, and reënforcement. The cement must be one of the standard brands of Portland cement; natural cement should never be used. For directions for making concrete posts see Chapter 25.

The home-made cement post is not cheap in first cost. With cement at 40 cents per bag the cost for this item alone will be from 8 to 10 cents. The reinforcing material will cost about 10 cents per post at pre-war prices for steel. If sand, gravel and labor are taken into consideration the cost will run to a high figure. As a matter of fact it is a real question whether or not the farmer can afford to

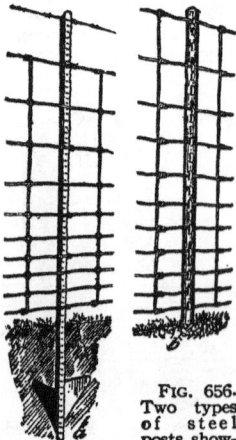

FIG. 656. Two types of steel posts, showing also two methods of fastening fence (see text)

make his own concrete posts when factories are so well equipped to do the work. It may be real economy to purchase the posts especially if a factory is near so that heavy freight rates are avoided.

In purchasing commercial posts one should, if possible, be assured by the experience of others, that the posts are giving satisfaction. The kind and method of reënforcement should be investigated as well as the materials used.

(FIG. 657. A concrete fence corner unit which if well built should last 100 years

Setting Fence Posts

The life of any fence requiring posts depends in large measure upon the kind of post used and the way in which it is set. Especially is this true of wire fence and inasmuch as 85 per cent. of fence now erected is of wire, chief consideration will be given to setting posts for this type.

Post holes should be dug in the spring season when the ground is wet if the most rapid progress is to be made. There is a variety of post hole diggers on the market. One of the best types is the auger with which a hole can be quickly dug if the ground is fairly moist and free from stones. A 16-inch ditching spade is a valuable tool for starting post holes.

Posts should not be set when the soil is filled with water, but they can be set much more easily if the soil is moist, *not wet*. If absolutely necessary to set them when the ground is really wet, do not stretch wire fence on them until the ground is more solid and they are more firm.

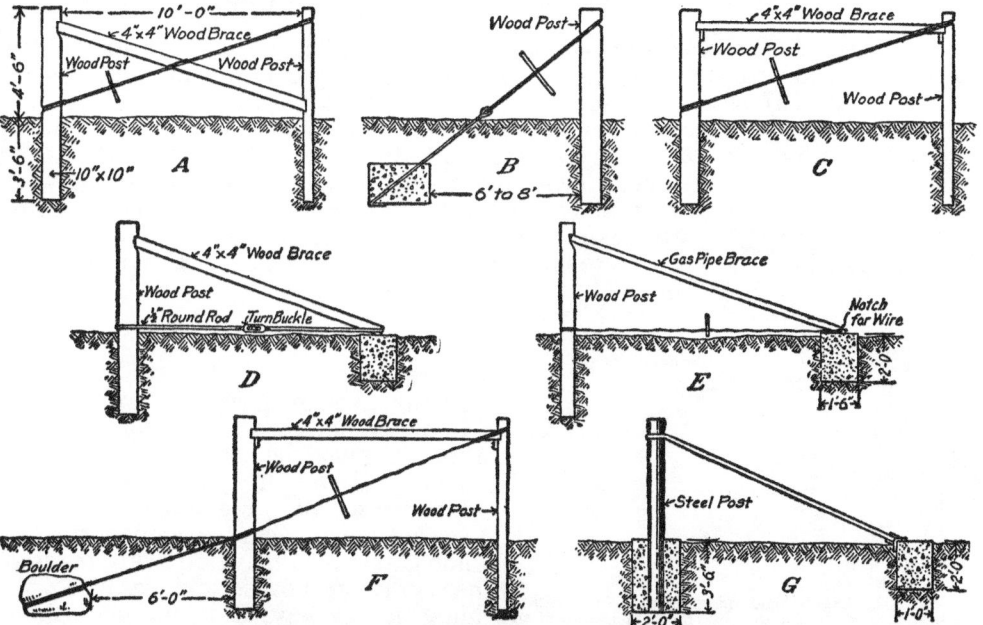

FIG. 658. Different methods of bracing end posts. The problem of making corner posts secure is solved by carrying out such methods of bracing as are shown here, but in two directions

Wood end posts should be 8 feet in length and should never be set less than $3\frac{1}{2}$ feet deep. Line posts should be 7 feet long and should be set $2\frac{1}{2}$ feet deep and never more than 20 feet apart, while one rod is better.

The end post in a wire fence is most important; too frequently it is poorly set and inadequately braced. Fig. 658 shows at A and C the most common methods of bracing end posts. The diagonal position of the brace at A will tend to lift the post out of the ground: hence, the construction shown at C is better. No. 9 fencing wire is wrapped around the post as shown, at least two and, better, three strands being used. The brace wires are then twisted to the proper tension.

The method shown at F is a good one if the brace wire going beyond the end of the fence, will not be in the way. In this case the wire itself is wrapped around a stone buried in the ground; sometimes a concrete "dead man" with a $\frac{1}{2}$-inch iron rod embedded in it is used in place of the boulder. The objection to either of these methods is that the wire soon rusts off at the surface of the ground; however, a pipe may be slipped over the wire to protect it and prevent this for a time.

The braces shown at D and E are excellent, that at D being especially good in that it provides a way to tighten the brace when it becomes loose as it is sure to do. It is more expensive and more trouble to put in, hence is seldom used. Braces of this kind can be purchased from makers of steel posts. The wire in E soon becomes rusted so that it breaks easily if an attempt is made to tighten it.

Steel posts make excellent end posts; a method of setting a patented steel post is shown at G. The brace is of steel and fits into an angle-iron socket which rests on the corner of the concrete block. The upper end is attached to a ring clamp which slips over the post. The brace is tightened by pounding the ring clamp down on the post and fastening it by means of a bolt provided for that purpose.

Concrete end and corner posts are highly desirable. They may be molded separately and afterwards set in place and braced by any one of the methods shown. Fig. 657 shows a reinforced concrete corner post 8 by 8 inches, with concrete and wire braces to two other posts. Fig. 659 shows details of form construction for a braced concrete post made in place. The form can be easily taken down and used elsewhere. The brace may or may not be molded with the post; it is probably better to make it separate, in which case the post as well as the concrete footing should be recessed to receive the ends of the brace.

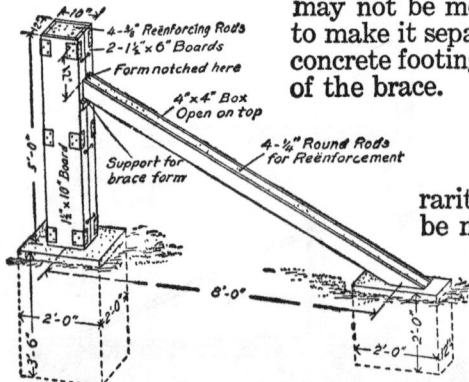

FIG. 659. Details and measurements for making a braced concrete end post. The form is hinged so that it may be set up and taken down with the least possible trouble and delay

Gates

A neat, substantial gate well hung is a rarity on the average farm. A gate should be made of strong material but at the same time it must be reasonably light in weight. Several designs for home made gates are shown in Fig. 661.

At A is shown a common type of slide gate which may be quickly made of heavy, rough boards and satisfactorily hung in this way; it is, however, more difficult to open than a hinged gate.

The designs shown at *B* and *C* are excellent, well braced and substantial yet light in weight. Bolts, rather than nails, are used throughout. The hinges are large and heavy; it is always a mistake to use small hinges.

FIG. 660. A patent wire and metal pipe-frame gate hung on substantial, well-braced posts

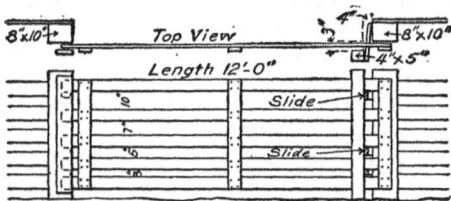

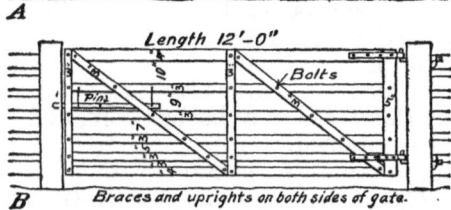

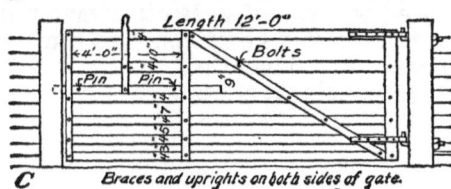

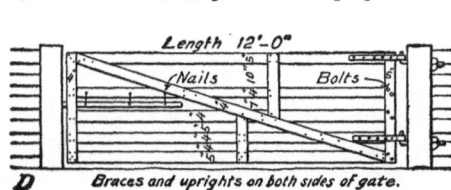

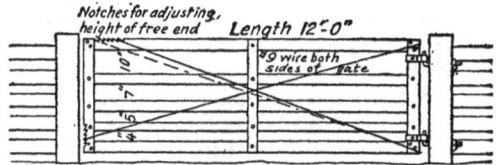

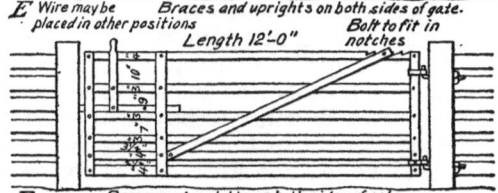

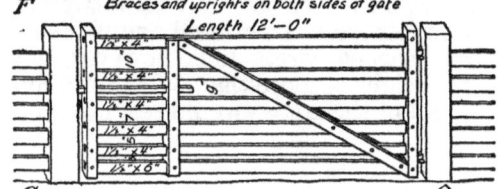

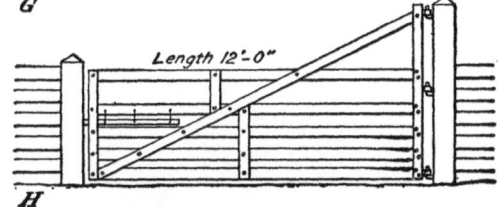

At *E* is shown an adjustable gate; that is, if the outer end sags it can be raised and held in position by adjusting the wires as shown. At *F* is shown another method of accomplishing the same result. At *G* is shown a mortised gate made of extra heavy materials. At *H* is shown a gate in which the boards are mortised into huge posts.

There are several types of patent gate on the market which have considerable merit. Those made of boards rather than of wire are in general to be desired. The wire gates usually have a steel pipe frame which if once bent out of shape, is difficult to straighten. A wooden gate with a board or two broken is much more easily and quickly repaired than a damaged steel gate.

There are also various types of so-called self-opening gates and appliances which make it possible for one to open and close a gate without getting out of a vehicle. These are quite common in certain sections, both in home-made and ready made, patent styles. However, they are rather expensive and tend to get out of repair more readily than simpler gates.

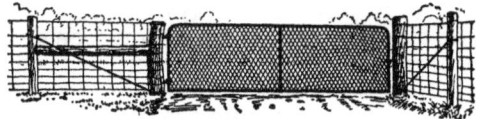

FIG. 661. Various types of farm gate that can be made at home. The constant usage to which a gate is subjected makes a strong, simple, and relatively light construction practically essential

A gate should always be hung to a large, well-set post.   In boring holes for
the hinges a vertical line should be drawn in the middle of the post by means of
a plumb line.   It is sometimes a good policy to set the bottom hinge ½ inch off
of this vertical line on the side toward which the gate will open.   This arrangement will tend to raise the end of the gate from the ground as it is being opened,
a desirable thing to do, especially in winter when there is snow on the ground.
Also, difficulties arising from the sagging of the gate are thus prevented.

Legal Aspects of Fencing

Unfortunately but few states have comprehensive, well-defined laws relating
to the construction and maintenance of farm fence.   Aside from mere legal statements, custom and good practice (which should, of course, be the basis of equitable laws) dictate that the responsibility of line-fence construction and maintenance is a joint one between abutting property owners.   In early range history
it was necessary for freeholders desiring enclosed homesteads to fence out stock
pasturing on the range.   In the early history of states which had passed the free-
range period it became necessary—as, indeed, it is the custom in many places
even to-day—for owners of stock to fence their animals in.   In the absence of
such fences, owners were liable for damages done by their stock to property of
adjacent freeholders.

Since livestock of some kind is now kept on virtually all well-managed farms
the duty of constructing line fences rests upon adjacent owners alike.   An
extract from the laws of Ohio on this point reads as follows: "That the owners
of adjoining lands shall build, keep up and maintain in good repair all partition
fences between them in equal shares, unless otherwise agreed upon between them,
which agreement must be in writing and witnessed by two persons."

The Ohio statutes further say that if any party refuses to comply with the
above law the township trustees shall view the land and after proper procedure
shall let the contract to build such fence, the cost being assessed against the offend-
ing party and, if not paid within 30 days, assessed against his property as taxes.

And again "No person or corporation shall be permitted to have any willow
fence or any other line fence, except that known as the osage or blackthorn hedge,
or construct or cause to be constructed a partition fence from barbed wire, unless
written consent of the adjoining owner be first obtained."

Under the same statutes it is not necessary to obtain permission of the adjoining owner to use not more than two barbed wires if the lower one is not less than
48 inches from the ground and if they are placed on fence of some other material
than barbed wire.   It is, however, unlawful to permit a hedge fence to remain
at a height of more than 6 feet for a longer period than 6 months.

All of these laws would seem to represent reasonable practice.